HISTORICAL ATLASES

HISTORICAL ATLASES

THE FIRST THREE HUNDRED YEARS, 1570–1870

WALTER GOFFART

THE UNIVERSITY OF CHICAGO PRESS
CHICAGO AND LONDON

Walter Goffart was professor of medieval history at the University of Toronto until his retirement in 1999. He now lectures in history at Yale University. He is the author of *Barbarians and Romans, A.D. 418–584: The Techniques of Accommodation* and the prize-winning *Narrators of Barbarian History, A.D. 550–800,* among other books.

The University of Chicago Press, Chicago 60637
The University of Chicago Press, Ltd., London
© 2003 by The University of Chicago
All rights reserved. Published 2003
Printed in the United States of America
12 11 10 09 08 07 06 05 04 03   5 4 3 2 1

ISBN (cloth): 0-226-30071-4

Library of Congress Cataloging-in-Publication Data

Goffart, Walter A.
   Historical atlases : the first three hundred years, 1570–1870.
      p.   cm.
   Includes bibliographical references and index.
   ISBN 0-226-30071-4 (cloth : alk. paper)
   1. Historical geography—Maps.   I. Title: First three hundred years, 1570–1870.   II. Title.
   G1030.H3   2003
   911—dc21

                                                                    2002031039

♾ The paper used in this publication meets the minimum requirements of the American National Standard for Information Sciences—Permanence of Paper for Printed Library Materials, ANSI Z39.48-1992.

To the Centre for Medieval Studies,
University of Toronto

CONTENTS

List of Illustrations   xi

Preface   xiii

System of Citation and Abbreviation   xix

    Catalogues and Reference Works   *xix*

    Libraries and Other Map Collections   *xxi*

INTRODUCTION   1

1 SIXTEENTH- AND SEVENTEENTH-CENTURY ATLASES
   RELEVANT TO HISTORY   13

    Claudius Ptolemy and "Ancient Geography"   *13*

    The Comparison of Old and New Geography   *20*

    World Atlases: Vessels of Diffuse History   *22*

    "One Size Fits All"   *24*

    Nouvelle metode de geografie historique [*sic*]   *27*

    "Ancient Geography, Sacred and Secular"   *30*

    History Everywhere and Nowhere   *38*

    NOTES   *39*

2 THE MIDDLE AGES IN SIXTEENTH- AND SEVENTEENTH-
   CENTURY MAPS   51

    Bygone Kingdoms   *52*

        The Anglo-Saxon Heptarchy   *52*

        Where Is Austrasia?   *57*

        Austrasia Foreshadows Habsburg Austria   *61*

        The Rest of France   *66*

        The Empire of Charlemagne   *69*

        Mementos of a Shifting Province   *74*

        Southern Netherlanders   *76*

Poster Maps: English Battles and Far-Flung French Royalty *80*
Outside Northern Europe *86*
Frisia: Tidal Flats and Historical Cartography *92*
The Foreshadowing of Historical Atlases to 1700 *104*

NOTES *112*

3   FROM 1700, NEW DEPARTURES   *131*

Atlases Called "Historical" *132*
Moving on from Ancient Geography *135*
Innovative but Unrealized Atlases *140*
Atlases Reflecting the Course of Written History *147*
Universal History in Sixty-six Maps *152*
A Disconnected Fragment *155*
The First Fully Published Sequential Atlas *156*
The History of France in Sequential Maps *161*
The Lost Atlas of Johann Christoph Gatterer *168*

NOTES *174*

4   EIGHTEENTH-CENTURY MAPS OF THE MIDDLE AGES   *187*

Single Maps and Small Sets *187*
  Isolated Scenes *187*
    Barbarian Invasions *187*
    Carolingian Sidelights *194*
    Europe as Seen By . . . *196*
    Poster or Catalogue Maps *201*
  Divisions and Limits of Medieval Lands *205*
  "Outside Europe" *222*
    Byzantines, Crusaders *222*
    Asiatic Peoples *226*
    European Travelers to Asia *229*
  Getting It Wrong *231*
  Beginnings without Continuations: The History of France *239*
  Other Countries at Medieval Moments *250*
Historical Atlases *254*
  Delisle's Unrealized Outline of History *254*
  The Succession of Great Empires *256*
  World Chronicles in Maps *262*

The History of France Graphically Unfolded   268
A Geography of the Migration Age   272
Isolated Initiatives: Beaurain, Tomka Szászky, Andrews   274
Innovative Atlases and Perplexing Maps   280
NOTES   282

5   HISTORICAL ATLASES COME OF AGE   303

The Midas Touch of Émmanuel de Las Cases   303
Kruse's Europe at 100-Year Intervals   314
The Historical Atlas Still Unknown and Disavowed   321
Scholars in a Time of Ferment: C. G. Koch and C. Malte-Brun   323
Minor Works on the Sequential Plan   327
Lithography and Other Innovations of the 1820s   329
Ambitious History from the Weimar Geographical Institute   330
The Burgeoning of National Atlases   337
A Creative Burst: 1829–30   339
Edward Quin: End of an Era in Maps for History   343
Conservatism and Change in the 1830s   347
A New Format for Comprehensive Historical Atlases   350
Karl Spruner's *Historisch-geographischer Hand-Atlas*
(1837–46)   353
Packing Much in Little   356
Maps for History Attain Maturity   359
NOTES   361

6   NINETEENTH-CENTURY MAPS OF THE MIDDLE AGES   379

An Estrangement Not Yet Healed   379
A Canon of Maps for Medieval Europe   383
Medievalia in the *Atlas Lesage*   391
Tracks for Crusaders   394
Medieval and Modern in the Last Sequential Atlases of Universal
History   401
The Middle Ages in Maps of National History   405
Large Atlases Warmly Welcoming the Middle Ages   413
Spruner's *Hand-Atlas* and the Problem of Geographic Order   418
To Condense so as to Teach   422
Honorably Mentioned   427
NOTES   430

CONCLUSION   *443*

    Map Types and Written Glosses   *443*
    Recapitulation   *451*
    NOTES   *460*

Catalogue of Maps and Atlases   *463*

    Historical Atlases and Maps; Atlases with Historical Additions   *464*
        H, M, G—Sixteenth and Seventeenth Centuries   *464*
        H, M, G—Eighteenth Century   *472*
        H, M, G—Nineteenth Century   *500*
    Ancient and Ecclesiastical History   *540*

        A—Ancient, Sacred, and Comparative Atlases (Sixteenth to
            Nineteenth Centuries)   *540*
        E—Ecclesiastical and Sacred Collections (Exclusively)   *550*
    Other Noteworthy Works with and without Maps   *554*
        O—Other Relevant Atlases and Maps   *554*
        B—Books, Usually without Maps   *562*
Index of Maps and Atlases   *569*
Index of Secondary Literature Cited in the Notes   *577*
Subject Index   *589*

1. Abraham Ortelius, "Aevi veteris typus geographicus" (1590) *18*
2. William Lambarde, Map of the Anglo-Saxon Heptarchy (1568) 53
3. Wolfgang Lazius, Cover Page to the *Typi chorographici provinciarum Austriae* (1561) 63
4. Petrus Bertius, "Imperium Caroli Magni" (1623) 72
5. John Speed, "The Invasions of England and Ireland" (1601, 1627) 82
6. Pierre Duval, "La carte de l'empire des Sarrazins" (1665) 89
7. Johannes Mejer, "Frisia Borealis" ("Nortfrieslande") in 1651 and 1240 (1652) 96
8. Menso Alting, "Free Frisia before the Floods" (1701) 102
9. Philippe de La Ruë, "Sourie ou Terre sainte moderne" (1651) 108
10. Guillaume Delisle, the Byzantine Empire in the Eighth Century (1711) 136
11. Christoph Cellarius, "Germania Medii Aevi" (before 1707; published in 1776) 142
12. Guillaume Delisle, "Theatrum historicum ad annum Christi quadringentesimum, pars occidentalis" (1705) 144
13. Johann Matthias Hase, "Imperium Romanum sub Iustiniano I" (1743) 150
14. Universal History in Sixty-six Maps (ca. 1747) 154
15. A Sequential Atlas of World History in Published Form (1763) 158
16. Giovanni Antonio Rizzi-Zannoni, *Atlas historique de la France* (1764, 1765) 164
17. Surviving Sheet of a Twenty-four-Map Teaching Atlas (1775) 170

18.   Reconstructions of Medieval Features and Districts (1732)   198

19.   Third and Fourth Plans of Paris in Nicolas de Lamare, *Traité de la police* (1705)   206

20.   Reconstructions of Medieval Features and Districts (1768)   220

21.   Henri Liébaux, Gaul at the Time of Its Conquest by Clovis (1728)   244

22.   Peter Georgisch, "Regnum et imperium Merovingo-Carolingicum" (1732)   248

23.   A. Lesage (Émmanuel de Las Cases), *Genealogical, Chronological, Historical, and Geographical Atlas* (1801)   305

24.   A. Lesage (Émmanuel de Las Cases), Section on Germany from *Atlas historique* (Paris edition)   310

25.   Christian Kruse, Europe in A.D. 1000 (1802–10)   318

26.   Christophe Guillaume Koch, Europe in A.D. 1074 (1814)   324

27.   Friedrich Wilhelm Benicken, "Map for General History from the Downfall of the Western Roman Empire to Charlemagne (476–758)" (1820)   332

28.   Edward Quin, *Historical Atlas in a Series of Maps of the World As Known at Different Periods* (1830)   344

29.   E. Andriveau-Goujon, "L'Europe au temps des croisades" (1840)   398

30.   Adrien Hubert Brué, *Atlas géographique, historique, politique et administratif de la France* (1820–28)   407

31.   Karl von Spruner, the British Isles under the Anglo-Saxons, to 1066 (1837–46)   421

32.   Medieval Spain in Rudolph von Wedell, *Historisch-geographischer Hand-Atlas* (1849)   424

PREFACE

It occurred to me in 1986 to examine who originally devised the well-known map usually called "The Barbarian Invasions (A.D. 376–568)." Not only was this inquiry relevant to my work in early medieval history, but also I owed an article to a testimonial volume for a scholar whose research made a study of this map an appropriate tribute.

While engaged in this task, I learned how demanding studies of cartographic material can be. Historical atlases and maps proved to be largely uncultivated. It was indicative of their neglect that the most comprehensive existing repertory of historical atlases was limited to the libraries of post–World War II Poland. The Polish repertory, though excellent and useful, left one none the wiser about the other map collections of Europe and America. To plug the gap, I had to resort to labor-intensive, pragmatic methods that, though sometimes taxing, were also very instructive. In London, New York, and Washington, I stalked my prototype "barbarian invasions" map from one atlas to the next, with as much resourcefulness as possible and with eventual success.

By the time the investigation of that one map was finished, I had learned enough about methods of research in this area (formerly unknown to me) that it seemed a shame to close the file and return at once to the early Middle Ages. Historical atlases seemed inviting; what was said about their development was thin and lacked finality. They were an aspect of the history of historical writing, about whose early-medieval facet I had written a book. Besides, maps had fascinated me since childhood; I even have a good sense of direction. So encouraged, I decided to undertake the project whose results are presented here.

Contrary to the dictum according to which *c'est le premier pas qui coûte*, my first steps into map research were congenial and

XIII

often delightful. I very much enjoyed traveling to public collections and relished the certainty that each one would offer me something new. Even the last libraries I visited—in Leiden and Budapest—yielded, among other things, small, rare, early-nineteenth-century historical atlases that no one appeared to have heard of before. Just as I thought I had had enough, two not very distant American collections that I had not visited were brought to my attention. They will wait. Map hunting is exhilarating but could go on without limit unless curtailed. All sorts of obstructions have delayed this book, not least its extent and the difficulty of shaping its material. It is a relief that my most time-consuming venture has reached a fortunate end.

As I just mentioned, this project is off my beaten track. In its earliest days, when I hesitated, two very dear and learned friends, Elizabeth A. R. Brown (Professor of History, emerita, City University of New York) and Susan Reynolds, F.B.A. (London, United Kingdom), beat down my qualms and encouraged me to continue down the road of historical atlases. They were preaching to the almost converted, and I am very thankful to them for that.

Too many persons deserve acknowledgment for me to be able to commemorate them all. Some are mentioned in the notes and not here; some in both. Those whose names follow, alphabetically, contributed a brick, and sometimes several bricks, to the building of this book: Michael Idomir Allen (Assistant Professor of Classics, University of Chicago); Jeremy Black (Professor of History, University of Exeter, United Kingdom); William Blissett (Professor of English, emeritus, University of Toronto), Tim Brook (Professor of History, University of Toronto); Penny Cole (Castleton, Ontario); Ed Dahl (National Archives of Canada, retired); Jürgen Furjyik (Map Library, University of Göttingen); Milton M. Gatch (Union Theological Seminary, New York, emeritus); Dominique (Mme Rémi) Gounelle (Lausanne, Switzerland); P. D. A. Harvey (Professor of History, emeritus, University of Durham); Brian McGuire (Chair of Medieval History, University of Roskilde, Denmark); James Muldoon (Professor of History, Rutgers University, Camden); Thomas Noble (Director, Medieval Institute, University of Notre Dame); Dan O'Donnell (Professor of English, University of Lethbridge, Saskatchewan); Matthew Ponesse (Centre for Medieval Studies, Toronto); Rodney W. Shirley (Buckingham, United Kingdom); William Stoneman (Director, Houghton Library, Harvard University); Yolanda Theunissen (Osler Map Library, University of Southern Maine); Fred Unwala (Editor, Pontifical Institute of Mediaeval Studies, Toronto); Robert L. Volz (Chapin Library,

Williams College); Dirk de Vries (Curator, Bodel Nyenhuis Collection, Leiden University Library); and, not least, Armin Wolf (Max-Planck-Institut für Rechtsgeschichte, Frankfurt-am-Main), an outstanding student of maps dealing with history.

The frequenting of libraries and map collections gave joy to this project, along with wonderful memories, such as navigating the unfamiliar streets of Gotheborg before sunrise so as to catch the first train to Stockholm, spend a good day at the Royal Library, and twice cross the breadth of Sweden; or coming upon the beauty and calm of the Società geografica italiana in its villa on the Celian hill. Most of the librarians who helped me and made my visits fruitful did so anonymously; I hope that my appreciation will not be wholly hidden from them. I single out for special mention the map room staffs of the Bayerische Staatsbibliothek, Munich; the Biblioteca nazionale Vittorio Emmanuele, Rome; and the Biblioteca centrale de la Marina, Rome (whose maps are not segregated). Shoshana Klein, with her assistant, made me very welcome at the Eran Laor Cartographic Collection, Jewish National and University Library, Jerusalem. So, at the Royal Geographical Society, London, did Herbert Francis, whose fame as a map librarian is deservedly great, not least for his kindness. Early in this project I was granted a month's fellowship by the Hermon Dunlap Smith Center for the History of Cartography, Newberry Library, Chicago. Among those making my stay there memorable and instructive, I single out for gratitude Ruth Hamilton, Robert Karrow, and James Akerman. My credentials at the Bibliothèque nationale go so far into the past that I cannot reconcile myself to the new, humble suffix "de France." This atlas project led me for the first time to the BN department of Cartes et plans, a haven of tranquility, where the attendants who deliver atlases and maps are notable for courtesy and friendliness. The senior staff whom I gladly thank are Mireille Pastoureau (now at the Bibliothèque de l'Institut) and Catherine Hofmann. No map collection saw so much of me as that of the British Library. Not only are its resources vast, but, unlike most of the other collections I visited, it buzzes with service to the public. Scholars, a minority of that public, are well treated; I am very grateful to Tony Campbell (Map Librarian), Peter Barber, and Jeff Armitage. Others, whom I am unable to name, have also earned my thanks.

The East Coast of the United States has superb map collections, but geography tends to be cultivated elsewhere, in the Midwest and beyond. It was my good fortune to work on this project at the University of Toronto; it has both a Department of Geography and a well-cared-for Map Room, which makes up in current purchases for what it lacks in treasures. As I scribbled notes, I could look across the room at my colleague Professor Paul R. Magocsi (Chair of Ukrainian Studies, Toronto), compiling his eminent *Historical Atlas of East Central Europe* (Toronto and Seattle, 1993); he was actually doing what I was studying. Among my Toronto colleagues in the Department of Geography, I was much helped in getting started by William G. Dean (now emeritus), a driving force in the magnificent *Historical Atlas of Canada* (1987–93). Jock Galloway generously smoothed my access to the Royal Geographical Society. No one in Toronto contributed more to my work than Joan Winearls, the Map Librarian (now retired). Besides answering many questions of mine for many years, she organized the 1993 Conference on Editorial Problems in Toronto, entitled "Editing Early and Historical Atlases," and published its proceedings. The talk, then article, that I supplied on that occasion was the trial balloon for central ideas in my book. I have much to thank Joan for, not least the invitation to her conference.

The Social Science and Humanities Research Council of Canada awarded me a standard research grant, including one nonsabbatical term off (six months were deemed enough for writing the book, a view that probably expressed the limits of generosity rather than a literary assessment). The University of Toronto has often given me small, welcome grants for research expenses. The hospitality of the Newberry Library has already been mentioned. The Center for the Humanities at the University of Wisconsin—Madison opened its doors to me for a term. All these institutions have my sincere thanks, as do the many universities in Australia, Europe, and North America that welcomed my lectures on early maps for history.

Some helpers stand out for the scale of their services. For several years, starting with my visit to Vienna for map study, Johannes Dörflinger (Professor of History, University of Vienna) has fostered the development of my manuscript. His help and support have sustained me to the very eve of publication. In northern Europe and especially Germany, I could unfailingly rely on Beatrice La Farge (Institut für Skandinavistik, J. W. Goethe-Universität, Frankfurt-am-Main) to be my intermediary with Deutsches Bundesbahn. Long ago, I made friends with Danuta Gorecki (retired, University of Illinois Library) on the basis of a shared interest in late Roman taxation. When asked to help with maps, she disapproved of the diversion of my interests but did not stint her graciousness and openhandedness. No one has contributed more information, energy, and sympathy to this project than Mary Sponberg Pedley (Clements Library, Ann Arbor, Michigan), whom I have relied on very much for (inter alia) geographic and cartographic advice. It was by her invitation that, in 1991, I gave at the Clements Library the first of the sixteen lectures that this project has generated. I also very gladly acknowledge the invaluable contribution of Scott Westrem (Professor of English, Graduate Center, City University of New York). Far exceeding the obligations of a publisher's reader, he painstakingly scrutinized every line of my manuscript, not once but twice, noting obscurities, rooting out ineptitudes, suggesting changes. The final form of my manuscript results largely from his suggestions. I am enormously grateful to him. A final word of thanks goes to my wife. She has always been patient with my preoccupations and supportive of them, even one so prolonged as *Historical Atlases* has proved to be.

Four sets of abbreviations are used: those for (1) catalogues and reference works, and (2) libraries, are listed below. Abbreviations for (3) individual maps and atlases, and (4) secondary sources, are as follows. Throughout the volume, cartographic materials are referred to by shortened forms, consisting normally of a proper name (or, occasionally a title) followed by a year. The "Name, Year" abbreviations may be followed by a slash and the letter A, E, O, or B, referring to subsidiary sections in the Catalogue of Maps and Atlases (for an explanation, see pages 463–64). This short form of reference determines the alphabetical and chronological order of the full entries in the Catalogue of Maps and Atlases, and is the basis for the Index of Maps and Atlases. Full bibliographic details for secondary works are provided in the first note in which they are cited; thereafter, shortened forms are used. The Index of Secondary Literature Cited in the Notes lists the shortened forms and guides readers to the notes containing the first citation.

CATALOGUES AND REFERENCE WORKS

ADBiog.	*Allgemeine deutsche Biographie.* 56 vols. Leipzig, 1875–1912.
AGEphem.	*Allgemeine geographische Ephemeriden* (Weimar). 51 vols. 1798–1816. 2d ser. (prefixed *Neue*), 31 vols. 1817–31.
Arch. biog. franç.	*Archives biographiques françaises.* Microform. London, ca. 1988.
Biog. univ.	*Biographie universelle, ancienne et moderne.* 45 vols. Repr., Graz, 1966–70.
BL Maps Catal.	British Museum. *Catalogue of Printed Maps, Charts and Plans (to 1960).* 15 vols. London, 1967.
BN Impr.	Bibliothèque nationale. *Catalogue général des livres imprimés.* 231 vols. Paris, 1897–1981.

Catalogus mapparum	International Committee of Historical Sciences, Commission for Historical Geography. *Catalogus mapparum geographicarum ad historiam pertinentium.* Warsaw, 1933.
DBiogArchiv	*Deutsches biographisches Archiv.* Microform. Munich, ca. 1982.
Dict. biog. franç.	*Dictionnaire de biographie française.* 19 vols. to date. Paris, 1933–.
DNB	*Dictionary of National Biography.* 63 vols. London, 1885–1900.
Encyc. Brit.	*Encyclopaedia Britannica.* Unless otherwise specified, the 11th ed. (29 vols. Cambridge, 1910–11) is cited.
Engelmann	Wilhelm Engelmann. *Bibliotheca Geographica: Verzeichniß der seit der Mitte des vorigen Jahrhunderts bis zu Ende des Jahres 1856 in Deutschland erschienenen Werke über Geographie und Reisen.* Leipzig, 1858.
Geog.: Biobib.	*Geographers: Biobibliographical Studies.* Ed. T. W. Freeman, Marguerita Oughton, Philippe Pinchemel, et al. 21 vols. to date. London, 1977–.
GV	*Gesamtverzeichnis des deutschsprachigen Schrifttums, 1700–1910.* Ed. Peter Geils et al. 160 vols. Munich, 1980.
Harms, *Künstler*	Hans Harms. *Künstler des Kartenbildes: Biographien und Porträts.* Oldenburg, 1962.
Harms, *Themen*	Hans Harms. *Themen alter Karten.* Oldenburg, 1979.
Inst. néerlandais	*Répertoire de cartes publié par l'Institut royal des ingénieurs néerlandais.* 2d ed. 2 vols. The Hague, 1856–67.
Julien, *Nouveau catalogue*	Roch-Joseph Julien. *Nouveau catalogue de cartes géographiques et topographiques.* Paris, 1763.
Koeman	C. Koeman, ed. *Atlantes Neerlandici: Bibliography of Terrestrial, Maritime and Celestial Atlases and Pilot Books Published in the Netherlands up to 1881.* 6 vols. Amsterdam, 1967–85.
Laor	Eran Laor. *Maps of the Holy Land: Cartobibliography of Printed Maps, 1475–1900.* New York and Amsterdam, 1986.
LC Bibliog. of Cartog.	Library of Congress, Geography and Map Division. *The Bibliography of Cartography.* 5 vols., and Supplement, 2 vols. Boston, 1973–80.
LC List (no.)	Philip Lee Phillips. *A List of Geographical Atlases in the Library of Congress.* 4 vols. Washington, 1919–20. *Supplement.* Washington, 1955.
Le Long	Jacques Le Long. *Bibliothèque historique de la France.* New ed. by Charles Boullemier. 5 vols. Paris, 1768–78.

LGKartog.	*Lexikon zur Geschichte der Kartographie.* Ed. I. Kretschmer, J. Dörflinger, and F. Wawrik. 2 vols. Vol. C of *Die Kartographie und ihre Randgebiete: Enzyklopädie,* ed. E. Arnberger. Vienna, 1975.
Łodyński	Marian Łodyński. *Centralnykatalog Zbiorów Kartograficznych w Polsce.* 4 vols. Warsaw, 1961–68.
Mees	G. Mees. *Historische Atlas van Noord-Nederland.* 15 fasc. 1851–61. Rotterdam, 1865.
Moreland and Bannister	Carl Moreland and David Bannister. *Antique Maps.* 2d ed. Oxford, 1986.
Nebenzahl	Kenneth Nebenzahl. *Maps of the Holy Land: Images of* Terra sancta *through Two Millennia.* New York, 1986.
NUC	*National Union Catalog Pre-1956 Imprints.* 754 vols. London, 1968–81.
Pastoureau, *Atlas français*	Mireille Pastoureau. *Les atlas français, XVIe–XVIIe siècles: Répertoire bibliographique et étude.* Paris, 1984.
Pedley, *Vaugondy*	Mary Sponberg Pedley. Bel et utile: *The Work of the Robert de Vaugondy Family of Mapmakers.* Tring, Herts., 1992.
Quérard, *France littéraire*	J. M. Quérard. *La France littéraire.* 12 vols. Paris, 1827–64.
Quérard, *Litt. franç. contemp.*	J. M. Quérard. *La littérature française contemporaine, 1827–1849.* 6 vols. Paris, 1842–57.
Shirley, *Maps of Britain to 1650*	Rodney W. Shirley. *Early Printed Maps of the British Isles, 1477–1650.* Rev. ed. East Grinstead, 1991.
Shirley, *Maps of Britain to 1750*	Rodney W. Shirley. *Printed Maps of the British Isles, 1650–1750.* Tring, Herts., 1988.
Sommervogel	Carlos Sommervogel. *Bibliothèque de la Companie de Jésus.* New ed. 11 vols. Brussels and Paris, 1890–1932.
Woltersdorf	E. G. Woltersdorf. *Repertorium der Land- und Seekarten.* Pt. 1. Vienna, 1813.

Libraries and Other Map Collections

I list only collections that I visited. Libraries that supplied me with information by correspondence are acknowledged at the point where the information is used. When a visited library has several relevant divisions, the one I used is specified either by naming the division or by writing "-maps," "-main," or "-rare" after the library name. When an addition of this kind is made, it does *not* exclude use of other parts of the library, notably the main collection and map room. In keeping with the practice of map historians, call numbers are cited when available. My research took place before the reunion of the Deutsche Staatsbibliothek in Berlin, and before the

Bibliothèque nationale added "de France" to its name and before its printed books were moved to the Bibliothèque François Mitterand. Except for a minor adjustment in the Berlin listing, I have kept my nomenclature as it was when I visited.

AmstRijksprent	Amsterdam, Rijksprentenkabinet
AmstUL	Amsterdam, Universiteitsbibliotheek
ArchNat	Paris, Archives nationales
BAV	Vatican, Biblioteca apostolica Vaticana
BerlinDSB	(East) Berlin, Deutsche Staatsbibliothek, now called Staatsbibliothek zu Berlin, Haus Unter den Linden
BerlinPK	Berlin, Staatsbibliothek preussischer Kulturbesitz, now called Staatsbibliothek zu Berlin, Haus Potsdamer Straße
BL	London, British Library (maps, North)
BN	Paris, Bibliothèque nationale (Cartes et plans, rare)
BPL	Boston Public Library (rare)
BrusselsBR	Brussels, Bibliothèque royale Albert Ier
BSB	Munich, Bayerische Staatsbibliothek
Budapest	Budapest, Széchényi National Library
CalStLib	Sacramento, California State Library
CalStLibSutro	San Francisco, California State Library, Sutro Branch
CambUL	Cambridge University Library (rare)
CathU	Washington, D.C., Catholic University of America, Rare Book Room
Columbia	New York City, Columbia University, Butler Library (rare)
Cornell	Ithaca, N.Y., Cornell University Library (rare)
DumbOaks	Washington, D.C., Dumbarton Oaks Research Library (rare)
Göttingen	Göttingen, Niedersächsische Staats- und Universitätsbibliothek
GroningenUL	Groningen, Rijksuniversiteit, Bibliotheek
HABW	Wolfenbüttel, Herzog-August Bibliothek
Harvard	Cambridge, Mass., Harvard University Library (Houghton)
Jerusalem	Jerusalem, Jewish National and University Library (Eran Laor Cartographic Collection)
LC	Washington, D.C., Library of Congress (maps, rare)
LeidenUL	Leiden, Bodel Nyenhuis Collection, Rijksuniversiteitsbibliotheek
MilwaukeeAGS	Milwaukee, University of Wisconsin, Golda Meir Library, American Geographical Society Map Collection

MilwaukeePL	Milwaukee Public Library
NaplesBN	Naples, Biblioteca nazionale
Newberry	Chicago, Newberry Library
Northwestern	Evanston, Ill., Northwestern University Library
NYPL	New York Public Library (maps, rare)
Oxford	Oxford, Bodleian Library (maps, rare)
PhiladAmPhil	Philadelphia, American Philosophical Society
PhiladFree	Philadelphia, Free Library
PhiladLibCo	Philadelphia, Library Company
PhiladUL	Philadelphia, University of Pennsylvania Library (rare)
Princeton	Princeton University, Firestone Library (rare)
RGS	London, Royal Geographical Society
RLScot	Edinburgh, Royal Library of Scotland (maps only)
RomeAngel	Rome, Biblioteca angelica
RomeBNVEmm	Rome, Biblioteca nazionale Vittorio Emmanuele
RomeCasan	Rome, Biblioteca Casanatense
RomeMar	Rome, Biblioteca centrale della Marina
RomeSGI	Rome, Società geografica italiana
StockholmRL	Stockholm, Kungliga Biblioteket (maps only)
Toronto	Toronto, University of Toronto, libraries as specified
UCB	Berkeley, University of California Library (Bancroft)
UChi	University of Chicago Library (rare)
UIll	Urbana, University of Illinois Library (rare)
UKans	Lawrence, University of Kansas, Kenneth Spencer Research Library (rare)
UMich	Ann Arbor, University of Michigan Library (rare)
UMich-Clements	Ann Arbor, University of Michigan, Clements Library
UMinn	Minneapolis, University of Minnesota, Twin Cities Library
UTexas	Austin, University of Texas Library
UWis-Madison	Madison, University of Wisconsin (rare)
ViennaÖNB	Vienna, Österreichische Nationalbibliothek (rare)
ViennaTheres	Vienna, Theresianum
ViennaUL	Vienna, Universitätsbibliothek
Yale	New Haven, Conn., Yale University Library (maps)
YaleB	New Haven, Conn., Yale University, Beinecke Library
YaleBAC	New Haven, Conn., Yale University, British Art Collection Library

INTRODUCTION

Geography and history are not strangers to each other. Their association is of such long standing that the origin of historical atlases might be deemed a nonsubject. The *Theatrum orbis terrarum* of Abraham Ortelius (1570), often called the first modern atlas of geography, was in its entirety "dedicated to the understanding of history." Ortelius makes the point in his opening remarks: "All [lovers of histories] will readily affirm with us how necessary is the knowledge of regions and provinces, of the seas, the location of mountains, valleys, cities, the course of rivers, etc., for attaining [a full] understanding of histories. This is what the Greeks called by the proper name 'geography,' and certain learned persons (rightly) call *the eye of history*."[1] What holds for Ortelius holds, too, for the great Dutch atlases of the next century and for many more. Geography aided historical study; readers of literature, notably history, were those expected to profit most from collections of maps. Early world atlases cannot be denied the modifier "historical"; some of them even appropriated it in their titles. Nevertheless, they differ very much from what historical atlases have become since the eighteenth century.

Ortelius's *Theatrum* and its successors contain all-purpose maps potentially helpful to readers of history. A very different plan is encountered in the historical atlases contemporary to us. These works are specialized collections of "maps for history" designed specifically to illustrate moments of time—short or long, but in any case circumscribed and limited. Physical features, such as mountains, are dispensed with if doing so helps historical instruction. Visual aids for history matter, whereas topography is a subordinate concern. These specialized atlases give preference to "modern history"—the centuries since the fall of Rome—the period now subdivided into medieval and modern.

The historical atlases circulating today are diverse. Some, usually devoted to a single region or country, are lavish in thematic detail and in exacting scholarly preparation. Others perpetuate a long-standing fascination with the geography of the classical world and the Bible. Still others address a public attentive to military and naval affairs. There are no fixed rules for what is and what is not a historical atlas, any more than there were in the past. Authors and publishers try to surprise and enchant the potential clientele. But the historical atlases most representative of the genre today are devoted to general, or world, history, and their featured maps, in a Eurocentric context, illustrate medieval and modern European history. These representative atlases are the ones whose origins I propose to trace. I approach them through the examples now widely available. Some of them are at bookstores or listed at Web sites selling books; these are typical of the historical atlases in current use. Much more material relevant to my inquiry is in map libraries, where the atlases and loose maps of the past can be examined and appraised for historical content.

The agenda of collections of specialized historical maps in past centuries is most closely reflected by today's atlases of general, or world, history.[2] For North American readers, a serviceable example is the *Anchor Atlas of World History* by Hermann Kinder and Werner Hilgemann, published in two paperbacks between 1974 and 1978. The *Anchor Atlas,* derived from one issued in Germany in 1964, is not tradition-bound. It illustrates history with an array of thematic visual aids; geography is not privileged. Outline maps are supplied, colored but without physical relief. They and other images accompany a capsule history, or chronicle—a quite lengthy text—whose course guides the choice of illustrations. Chronology is not rigidly handled; time is skewed when the order of presentation is aided by doing so. Topographical maps, as one type of chart among others, carry only as much detail as seems called for by the topic they illustrate.[3]

The *Anchor Atlas,* with its array of visual and textual aids, is also attached to a narrower conception of maps for history. France, Scandinavia, India, and other lands are shown at turning points of their past, in chronological sequence. Other important occurrences are also mapped. The selection of subjects—generous, varied, and richly tinted—necessarily includes many items, such as the Empire of Charlemagne, the Crusades, and the great voyages of discovery, that have long been obligatory in similar works. Although Asia, Africa, and America are not neglected, European civilization in its established periods is treated as the core of world history. The French Revolution divides the two volumes. Volume 1 surveys the more distant past, including antiquity, the Middle Ages, and early modern times; volume 2 deals with contemporary history in a more leisurely way and greater detail. The *Anchor Atlas* is not isolated in its design. Recent competitors, skimpier in explanatory text, such as the Harper/Collins and the *Times* atlases of world history, rely more exclusively on maps, yet largely share its strategy.[4]

Historical atlases with a general scope—costly productions with exacting technical demands—occasion much cooperation among European and American publishers. The authors of the *Anchor Atlas* are German; their work was first published in Munich, also as a paperback, ten years before the American version. The translation made in England appeared there as *The Penguin Atlas of World History* and simultaneously as the *Anchor Atlas* in America, where Doubleday had acquired the rights to the Penguin version.[5] Such publication histories are not recent innovations. F. W. Putzger's *Historischer Schul-Atlas,* a celebrated and widely diffused collection, had its first edition in 1877 and, after more than a score of revisions, continues to be in print. The English translation that appeared in 1903 was soon superseded by a more ambitious offshoot: William R. Shepherd received permission to use Putzger's maps as the foundation for his *Historical Atlas* (1911), the standard setter for atlases in American classrooms for more than half a century. It was still being reedited as late as 1973.[6]

Some atlases of the kind in question here try to reach general readers, perhaps as coffee-table books or as narratives-in-pictures. Most are concerned primarily with the teaching of history.[7] Detailed chronicles, as provided by the *Anchor Atlas,* are not mandatory. Instructors are assumed to prescribe textbooks of their choice; the atlas complements the assigned text as an aid to learning the course of events. Because geography is an "auxiliary science of history," map-books have the same vital role for the comprehension of space as handbooks of chronology have for mastering time.[8] Encouraged by the expanding use of visual aids in classrooms, history teaching today is on good terms with maps. For better or worse, sizable parts of Shepherd's atlas, and of an even older, English collection, are available on the World Wide Web for consultation and for projection.

Not too rigid a meaning should be attached to the term "atlas." The great cartographer Gerhard Mercator (†1594) is considered responsible for giving "atlas" the sense familiar to us, but geography has never won exclusive use of the word and has certainly not been stripped of a former monopoly by "new developments." Even before the eighteenth century, "atlas" could refer to an outsize book with a variety of contents. Library catalogues bear out that, among atlas subjects, geography is outweighed by medicine, surgery, and anatomy. "Atlas" is even attested in 1761 meaning a single map—a synonym for *carte*.[9]

Atlases (in a context of geography) are collections of maps in book form, often including much written matter as well. Though no single format is mandatory, atlases tend to be associated with imposing dimensions, similar to those of loose maps in one sheet. There are even "monster atlases," mainly Dutch in origin, whose bulk stems, not from the size of the included maps, but from the exuberant combination of maps with illuminated engravings and other treasures.[10] Nevertheless, the subjects shown in even oversize atlases have a reduced format by comparison with the multiple-sheet maps that cartographers routinely produce. Early publishers, responding to the demands of consumers eager for atlas ownership but unable to afford un- abridged models, engaged in severe compression. Pocket atlases, ancestors in size of the *Anchor Atlas,* began to be produced toward the middle of the six- teenth century. In the golden age of massive, lavish atlases, the compact article sold very well.[11] Historical atlases, especially when destined for individual ownership, have inclined to ordinary, rather than oversize, book formats.

Historical atlases like Kinder and Hilgemann's, or Putzger's for that mat- ter, are a comparatively recent species. An anonymous expert writing in 1933 expressed the view that the seventeenth and eighteenth centuries had actively developed such collections: "Attempts were made to publish large, scholarly historical atlases and school atlases. The clearest characteristic of this period is that historical maps were surrounded by a substantial apparatus of syn- chronic or genealogical charts. The large atlas of Le Sage (Paris, 1807) may be regarded as crowning this series of works."[12] This appraisal, blessed by the In- ternational Committee of Historical Sciences, goes strangely awry. Early modern atlas-makers were preoccupied with "ancient geography," and the pages they produced, with rare exceptions, were not bordered by informative tables. Works anticipating our *Anchor Atlas* and its like were the outcome of isolated experiments, not of a collective effort. As for the Lesage atlas, it was first published in London, 1801 (not Paris, 1807); and far from capping past developments, it diverged sharply from the then prevailing design of special- ized historical collections.

Rightly seen, "scholarly historical atlases and school atlases" of the kind foreshadowing Kinder and Hilgemann's belong to the nineteenth century, and more to the middle of the century than to its beginning. What came before was an unplanned preparation, but not idle or lacking interest. Collections of maps with claims or aspirations to being historical are ubiquitous. The type that anticipates the collected maps for history on today's bookstore shelves was a mutant in a wider tradition that, from its beginnings, tended to be much more closely linked to history, or to what was thought historical, than geography has now become.

Existing accounts of the development of historical atlases, though more accurate than that of the anonymous expert of 1933, are few in number and usually brief. They presuppose that, in the golden age of atlas-making, "ancient geography" was historical while "modern geography" was not. They envisage the development of historical atlases as a straightforward expansion of sixteenth-century "ancient geography." They usually cite a small and repetitious sampling of evidence. Improvement is possible.[13]

On the assumption that the general historical atlases of today, mainly concerned with medieval and modern Europe, are normative for the genre, I set out to learn how specialized works marshaling geography solely to illustrate moments of the past came into being, and how they diverged from the earlier atlases helpful in a general way to readers of history. My initial (and continuing) task has been to acquire an adequate basis for determining which atlases are and are not historical and to assemble as exhaustive an array of relevant material as possible, something not done before. The public collections, large and small, that I have visited extend from Sacramento to Jerusalem, Stockholm to Naples, and many points in between. I could not identify and examine all maps for history between 1570 and 1870. More examples than I should like are regretfully marked "not seen," and other instances have undoubtedly escaped me altogether. The Catalogue of Maps and Atlases at the end of this book lists more than 700 items, gleaned from about sixty collections.[14]

Atlases alone do not permit comprehensive study of my subject: I have paid much attention to single maps. Some of the most prolific makers of maps for history, such as Guillaume Delisle (†1726) or Robert de Vaugondy, father and son (†1766, †1786), never consigned this part of their output to special atlases. Petrus Bertius (†1629) compiled two atlases of "ancient geography," but his map of the Empire of Charlemagne (1623), the first of its kind, remained an isolated part of his output. An ambitious eighteenth-century *Universal History* (1730–36) includes, in its Dutch and German translations, a notable series of unsigned map illustrations, most of them found nowhere else.[15] I thought it important to examine as many maps for history as possible regardless of whether they were gathered into books or not. Isolated cases have special value at times when specialized historical atlases are rare. Unavoidably, many single maps, especially in nineteenth-century works, have eluded me. There is much left to occupy lovers of old maps with historical subjects.

Maps with scenes from medieval history are central to my project. I pay comparatively little attention to maps of classical antiquity and the Bible. Ancient maps, classical and biblical, were the dominant historical type from the sixteenth to the eighteenth century and are still vigorous niche products in atlas-making. Despite their former prominence, they are a subordinate concern here. The priority I give to medieval subjects is not the result of my being a medievalist or a student of medievalism; it follows from the distribution of map subjects: the atlases whose origins I probe are predominantly of post-classical history. For a very long time maps illustrating the many centuries after the fall of Rome hardly ever showed subjects dating from after 1500. This omission was rapidly filled in after 1800. Until that change, the main focus for maps of "modern" or "new" history belonged to the medieval period. I concentrate on specialized maps of "modern" history, most of which (until 1800) concern the age now styled medieval.

I have not kept maps wholly separate from their makers, many of whom are appealing figures. Maps for history, as a collective enterprise, need the faces of those who designed and collected them. Not all, unfortunately, have left traces in biographical archives. The mid-nineteenth-century Frenchman responsible for the first 100-map historical atlas, a work widely available in map collections, is virtually faceless. The lives of many others, however, can be recaptured. They form a colorful and complex gallery.

Historical atlases are alive and well today. Some recent ones embody a high order of scholarship; others stand out for their artwork; a few have had enviable commercial success. The enterprise as a whole, however worthy its products, has steered clear of self-examination, either in the past or at present. Solitary experimentation has been preferred to incremental, group-based improvement. There is no how-to literature for potential atlas-makers. "Historical geography," a currently flourishing branch of geography with several learned journals and many interests, does not specialize in the study (or the making) of historical atlases and the potential amelioration of the genre.[16]

An examination of origins will, I hope, furnish a basis for reflection on what the geographical portrayal of the past involves and requires. Atlas maps are often variations of a single model designed long ago. New atlases pay little attention to distant prototypes, which are mostly unknown. The normal procedure seems to be simply to refine or embellish the versions of a recent predecessor. Determining where historical scenes come from should encourage critical assessment of what is being done and the development of new patterns. Maps for history come in a very limited number of designs. Most are "passive," that is, portrayals of momentary constellations (such as "Europe at the Treaty of Westphalia, 1648"); others are "comparative," a form once considered almost synonymous with history itself; still others are "synchronic," or dynamic, packing together events of a century or more (such as "The Expeditions of the Crusaders, 1096–1270"). There are additional, minor types, as well as a multiplicity of thematic maps, dating chiefly after 1870. The issue that has stimulated discussion is whether or not maps can depict historical change or developments; "the representation of changes in time is always a challenge for cartographers."[17] This is an important and necessary question. There should probably be more.

The branch of geography serving historical teaching and study has no unequivocal name. The *Lexikon der Kartographie,* the major current repertory of geographical terms, differentiates *historische Atlanten* (historical atlases) from *Geschichtsatlanten* (possibly translatable as "atlases for history").[18] *Historische Atlanten* are composed of maps having intrinsic documentary value—maps originating in the past and embodying cartography of that time, such as early city plans. A large-scale map of the Netherlands compiled by a government agency between 1850 and 1950 has recently been reprinted province by province in a twelve-volume atlas under the title *Historische Atlas [province name].* This series, whose maps have undeniable documentary value for 1850–1950, fits neatly into the *Lexikon*'s definition of "historische Atlanten."[19] But many researchers accustomed to normal word usage may imagine, until they open a volume of this series and verify its contents, that it provides a multivolume history-in-maps of the Dutch provinces. *Geschichtsatlanten* are our specialized historical atlases, composed of maps usually drawn long after the moment depicted in order to illustrate the geographical context of past events or conditions. Their value is not documentary but mainly illustrative and pedagogic. "Historical" and "map" cannot be lightheartedly combined. Anyone employing this compound needs to clarify whether the map in question has documentary value, as a historical source, or is a reconstruction, illustrating some part of the past.

Other terms or expressions have been proposed in this connection, notably "historical cartography," which its users equate to maps drawn specifically for historical studies. Such maps, one advocate claims, were initiated by Ortelius. "Historical cartography" lives as a convenience for today's scholars. No one has shown that it was a pursuit engaged in under that name by early cartographers. Defined as it is, "historical cartography" serves mainly as a conceptual bridge between the "ancient geography" of early modern mapmakers and the *Geschichtsatlanten* of today. "Historical cartography" cuts sharply across the terms proposed by the *Lexikon der Kartographie*. It also embodies ambiguities of its own and links pursuits that should probably be carefully differentiated.[20]

Problems of terminology do not end there. "Geography, historical— Maps" is the Library of Congress subject heading most useful for tracing historical atlases in library catalogues. Yet the current discipline of "historical geography" rarely if ever includes historical atlases among its many pursuits. As I pick my way through this quagmire, I use the adjective "historical" with as much restraint as possible. For single items I speak of "maps for history," an English equivalent of *Geschichtskarte*. Even if that equivalence is not attained, "maps for history" captures the idea of a cartography applied solely to the study and teaching of history. "Historical atlas," too, does not have a single meaning. It sporadically refers to collections that, by today's criteria, have nothing historical about them. Nevertheless, it is more secure as the equivalent of *Geschichtsatlas* than any other.

The material of my book is divided into two categories and three periods. The categories are (1) maps and map collections that are milestones in the origins of specialized historical atlases; and (2) individual maps illustrating moments and scenes from later-than-classical history. The first category is concerned with collections and the changes affecting them, whereas the second traces the accumulation of scenes and depictions of "modern" (medieval) history. Collected and individual maps both contributed to the parentage of a specialized cartography for history. The three time periods are the sixteenth– seventeenth, eighteenth, and nineteenth centuries. For each of them, I assign one chapter to the first category of material ("milestones") and a second chapter to the second ("illustrations"). The odd-numbered chapters, therefore, are concerned with the development of the genre, whereas the even- numbered ones record and comment on what, to simplify, I call maps of the Middle Ages. Some overlap between the two chapter types is unavoidable. I have tried hard to avoid repetition and apologize for lapses.

I follow the development of specialized maps and atlases for history to the point where they were an established, abundant, and recognizable commodity. This point had not been reached in 1800 but definitely was by 1870. That is where I stop. The decision to end where I do, omitting almost a hundred and fifty very productive years, has been helped by the appearance in 1997 of Jeremy Black's *Maps and History: Constructing Images of the Past*. Professor Black emphasizes the atlases and maps for history published in recent times. Most of his book discusses material from after the mid–nineteenth century. "Origins" are dealt with in about eighty pages, steered in part by a 1995 article of mine.[21] Professor Black is expert in the current decades of maps for history, not least in mapmaking techniques and the preoccupations of publishers. He underscores that mapmakers have been Eurocentric and have disregarded women and other disadvantaged groups. It would be hard to discourse about recent atlases without duplicating his work. With atlases until 1870, the same problem does not arise.

NOTES

1. Catherine Hofmann, "La genèse de l'atlas historique en France (1630–1800): Pouvoirs et limites de la carte comme 'oeil de l'histoire,'" *Bibliothèque de l'École des chartes* 158 (2000): 97–128, at 99, "la géographie tout entière, non seulement celle qui s'occupe des temps anciens, est ainsi mise au service de l'histoire." This holds true, Hofmann adds, from Ortelius (†1598) to Bourguignon d'Anville (†1782). This quotation from Hofmann is applied by her to early maps of modern geography rather than to Ortelius in particular. The second quotation in the text is translated from Hofmann, "Genèse de l'atlas historique," n. 6, and is from a 1572 French-language edition of Ortelius's *Theatrum,* p. 1.

2. For a valuable discussion and listing of historical atlases since World War II, see Armin Wolf, "Das Bild der europäische Geschichte in Geschichtsatlanten verschiedener Länder," *Internationales Jahrbuch für Geschichts- und Geographieunterricht* 13 (1970–71): 1–38.

3. Hermann Kinder and Werner Hilgemann, *The Anchor Atlas of World History,* vol. 1, *From the Stone Age to the Eve of the French Revolution;* vol. 2, *From the French Revolution to the American Bicentennial,* maps by Harald Bukor and Ruth Bukor (New York, 1974–78). For more on its publishing history, see below. Norbert Ohler, "Historische Atlanten— Tendenzen und Neuerscheinungen: Eine Auswahlbibliographie," *Militärgeschichtliche Mitteilungen* 22, no. 2 (1977): 148, attributes the great success of the *Anchor Atlas* to its close linkage of words and maps and its unusually universal outlook.

4. For the Harper/Collins and *Times* atlases, see the next note.

5. Original edition: Hermann Kinder and Werner Hilgemann, *DTV Atlas zur Weltgeschichte* (Munich, 1964–66). The English (Penguin) version was translated by Ernest H. Menze (Harmondsworth, 1974). The Kinder-Hilgemann atlas has enjoyed more than routine success; Italian and French translations preceded the English one, and there are many versions besides these.

Vidal de la Blache et al., eds., *Le grand livre de l'histoire mondiale* (Paris, 1986), lies behind the *Collins Atlas of World History* (London, 1987), equivalent to the *Harper Atlas of World History* (New York, 1987). The latest Rand-McNally *Atlas of World History* is an American version of the English R. I. Moore, ed., *Hamlyn Historical Atlas,* (London, 1981). Geoffrey Barraclough, ed., *The Times Atlas of World History* (London, 1978; 2d ed., London, 1984), has been widely translated and circulated. Frank Debenham, ed., *The Reader's Digest Great World Atlas* (London, 1962), is a basically English production (with little attention to history) from which a shorter U.S. edition was drawn (Pleasantville, N.Y., 1963); it includes a limited historical supplement.

6. For Putzger, 1877, see Armin Wolf, "100 Jahre Putzger—100 Jahre Geschichtsbild in Deutschland (1877–1977)," *Geschichte in Wissenschaft und Unterricht* 11 (1978): 702–18. For

Shepherd, 1929/O, see L. J. Cappon, "The Historical Map in American Atlases," *Annals of the Association of American Geographers* 69 (1979): 630–31. Its ninth edition appeared in 1973 in the United States, 1974 in the United Kingdom. Earlier still, Spruner, 1837 (a very celebrated atlas), had an anonymous English translation (Spruner, 1861), then was issued in English by the original publisher (Spruner, 1872). Koeppen, 1854, acknowledges use of six of G. W. Green's translations of Spruner's medieval maps. Dietrich Reimer of Berlin, publisher of the much praised Kiepert, 1859/A, supplied the maps for French, Italian, and U.S. editions.

7. Cappon ("The Historical Map in American Atlases," 623) states, "Historical maps do not begin to appear until the nineteenth century, in time to serve history as a new historical discipline." Cappon relates the "newness" of the subject to the founding of the American Historical Association in 1884 (626). His comment may apply to the United States, but even there, maps for history had been produced half a century before 1884, notably by Willard, 1828/O.

8. J. Keuning, "The History of an Atlas: Mercator-Hondius," *Imago Mundi* 4 (1947): 37. For Mercator's early and very wide ranging plans for what would become his atlas, see below, 41n. 17; see also Svetlana Alpers, *The Art of Describing: Dutch Art in the Seventeenth Century* (Chicago, 1983), 134. Genealogy ranked with chronology and geography as a third indispensable auxiliary to historical study; Herbert Butterfield, *Man on His Past: The Study of the History of Historical Scholarship* (1955; repr., Boston, 1960), 42. Handbooks of chronology seem to be more numerous than historical atlases. (Other auxiliaries were numismatics and antiquities, i.e., archaeology.) Levasseur, 1844, typically has the alternative title *Atlas du Dictionnaire des dates.* See also Hanno Beck, *Geographie: Europäische Entwicklung* (Freiburg, 1973), 136.

9. On Mercator and the term "atlas," see James R. Akerman, "From Books with Maps to Books as Maps: The Editor in the Creation of the Atlas Idea," in *Editing Early and Historical Atlases,* ed. Joan Winearls (Toronto, 1995), 3–48, here 19–20. The wide meaning of the term at an early date is recorded by the *Oxford English Dictionary,* s.v. "atlas, 3, 4." For the belief that the term originally belonged exclusively to geography, see Peter H. Meurer, *Atlantes Colonienses: Die Kölner Schule der Atlaskartographie, 1570–1610* (Bad Neustadt a. d. Saale, 1988), 17: "neuere Entwicklungen der Schriftengattung . . . [have carried the term 'atlas' over] auch auf Werke nichtgeographischen Inhalts." See also R. A. Skelton, ed., *A. Ortelius, "Theatrum orbis terrarum," Antwerp, 1570* (Amsterdam, 1964), v. For "atlas" as a single, cut-down map of Bordeaux, see *Journal des sçavans,* July 1761, 377.

10. On monster atlases (notably the van der Hem, now Prince Eugene, Atlas in Vienna), see Marijke De Vrij, *The World on Paper: A Descriptive Catalogue of Cartographical Material Published in Amsterdam during the Seventeenth Century* (Amsterdam, 1967), 12, 80–83; Cornelis Koeman, *Collections of Maps and Atlases in the Netherlands: Their History and Present State* (Leiden, 1961), 35–39, 64–69.

11. In 1548, Gastaldi published the first pocket-sized Ptolemy *Geography* (with its atlas): Nebenzahl, 79 (cf. Moreland and Bannister, 289). *Epitome theatri Orteliani* (Antwerp, 1601). Hondius, 1607, is a step away from the Mercator collection and points toward the future Hondius-Janssonius atlas.

12. *Catalogus mapparum,* x.

13. Ibid., ix–x; William G. Dean, "Sic enim est traditum," *Journal of the Historical Atlas of Canada* 1 (1980): 6–14; Johannes Dörflinger, "Geschichtskarte," in *LGKartog.* 1:266; Johannes Dörflinger, "Geschichtsatlanten vom 16. bis zum Beginn des 20. Jahrhunderts," in *Vierhundert Jahre Mercator, Vierhundert Jahre Atlas: Die ganze Welt zwischen zwei Buchdeck-*

eln, ein Geschichte der Atlanten, ed. Hans Wolff (Weissenhorn, 1995), 179–98; Jeremy Black, "Historical Atlases," *Historical Journal* 37 (1994): 643–48.

14. I have not taken account of three kinds of material: medieval *mappaemundi;* maps of battles, sieges, and campaigns ("military maps"); and the historical vignettes of the Vatican Map Gallery. Some *mappaemundi* have historical content, but the distinct cartographic tradition they belong to distances them from my subject. Military maps, though historical, have almost always been classified and collected apart from historical atlases. Like military maps, world maps decorated with the routes of navigators have historical content yet live in their particular context. As for the Vatican Map Gallery (1579–81), I have discussed its vignettes in detail in "Christian Pessimism on the Walls of the Vatican Galleria delle carte geografiche," *Renaissance Quarterly* 51 (1990): 788–827. The gallery maps are accessible in Lucio Gambi and Antonio Pinelli, *La Galleria delle carte geografiche in Vaticano,* 3 vols. (Modena, 1994); and Roberto Almagià, *Le pitture geographiche murali della terza loggia e di altre sale Vaticane,* Monumenta cartographica Vaticana 4 (Vatican, 1955).

15. Individual items at UCB-Bancroft and PhiladFree led me, by marginal markings indicating their provenance, to volumes with additional maps; see *Welt-Historie,* 1744; *Hist. universelle,* 1779.

Dr. Armin Wolf (Max-Planck-Institut für Rechtsgeschichte, Frankfurt), a noted map scholar and collector, sent me a copy of a loose map he had acquired relating to the Carolingian period. By good fortune, I came across an article guiding me to its source — the proceedings of an eighteenth-century academy containing seven more items of the same provenance; see Lamey, 1766.

16. The detachment of historical geography from maps and atlases for history seems implied by current journals, whose very varied contributions have little in common with the subject matter of my book. See *Journal of Historical Geography* (London, 1975–); *Historical Geography* (Northbridge, Calif., 1978–); *Geographica historica* (Bonn, 1970s–); *Historical Geography* (Japan, 1980–). Alan R. H. Baker, "On the History and Geography of Historical Geography," *Historical Geography,* no. 179 (1996): 1–24.

17. Hofmann, "Genèse de l'atlas historique," 102.

18. Werner Witt, *Lexikon der Kartographie,* vol. B of *Die Kartographie und ihre Randgebiete, Enzyklopädie,* ed. Erik Arnberger (Vienna, 1979), 201. Witt's distinction between *Historie* and *Geschichte* lends itself to single maps as well as atlases. Its shortcoming is linguistic; the contrasting terms for "history" do not translate into most other European languages.

19. *Historische Atlas. [province name] chromotopografischer kaart des rijks I:25.000,* ed. G. L. Wieberdink, 12 vols. (Landsmeer, 1989–90). Going by only the title, one would assume that each of these volumes was an atlas of the history of the province in question.

20. Peter H. Meurer, "Ortelius, or the Father of Historical Cartography," in *Abraham Ortelius and the First Atlas: Essays Commemorating the Quadricentennial of His Death, 1598–1998,* ed. Marcel P. R. van den Broecke, Peter van den Krogt, and Peter Meurer (Westrenen, Netherlands, 1998), 133, distinguishes the "history of cartography" (a self-explanatory term) from "historical cartography." The latter he equates to maps for history that are initially encountered in an Ortelian context.

21. Jeremy Black, *Maps and History: Constructing Images of the Past* (New Haven, 1997). I saw Black's book when mine was virtually finished. Black's chapters 1–3 (out of eleven) concern the periods I examine. The article of mine in question is "Breaking the Ortelian Pattern: Historical Atlases with a New Program, 1747–1830," in *Editing Early and Historical Atlases,* ed. Joan Winearls (Toronto 1995), 49–81, cited by Black at ch. 1, nn. 60–62, 64.

CHAPTER 1

Sixteenth- and Seventeenth-Century Atlases Relevant to History

CLAUDIUS PTOLEMY AND "ANCIENT GEOGRAPHY"

Though valued by, and considered helpful to, readers of history, sixteenth-century maps contained all-purpose geography and were not deliberately created for students of history. The sort of map most prized then—given a place of honor until well into the nineteenth century—was of "ancient geography," a product useful to readers of history, no doubt, but one also of importance to professional geographers. "Several maps of ancient geography" (*veteris geographiae aliquot tabulae*) are what Abraham Ortelius offered from 1579 onward as a supplement to his atlas of modern maps, the *Theatrum orbis terrarum* (1570). As epigraph for the collection of these special maps, Ortelius used the motto *historiae oculus geographia,* "geography (is) the eye of history," meaning that geographical material allowed history to be visualized. He had recited the same motto in reference to all geography in the foreword to the *Theatrum.* The history in question was not limited to antiquity.[1] An "ancient" map was not historical just because its subject was classical antiquity, nor were "modern" maps nonhistorical just because their subject was the designer's time.

The origin of historical atlases, as now conceived, presupposes that a "geography" called "ancient" or "old" signified ancient history. It follows from this reasoning that atlases that embody "ancient geography" may be freely said to contain "historical" maps and to form "historical" collections. Ortelius's supplement of *vetus geographia* is described in the twentieth century as a "series of maps illustrating ancient history, sacred and secular."[2] This substitution of "history" for "geography" is hasty and uncalled for: old geography was no less geographic than the "new" variety. The discipline of ancient geography fascinated geographers and mapmakers for a long time. "Old" maps differed from "new" in respect to their time period and to the sources on which each was based: "ancient" geography, comprising the ages before the fall of Rome, was grounded in ancient sources.

The Latin translation in 1406–7 of the *Cosmographia,* or *Geography,* by Claudius Ptolemy of Alexandria (†after A.D. 151) set in motion a new era in mapmaking. If ever known in the Roman west, it was for a millennium among the works whose authors were familiar to scholars as wise men but whose teachings were largely forgotten. In its earliest known manuscripts and printings, Ptolemy's work was accompanied by at least twenty-seven maps; with or without the accompanying text, they proved enormously influential.[3] Today's historians of cartography no longer scorn the maps made in the Middle Ages.[4] Even so, Ptolemy's second-century treatise was a crucial acquisition for fifteenth-century Europe. It gave access to unknown methods and ideals, among them, the use of a grid of latitude and longitude, an extensive set of coordinates, and the goal of uniformity of scale, projection, and coverage. There was much that Ptolemy did not know; practical seafarers rejected his authority for a long time; but, in the scholarly circles of the Renaissance, he made cartography continental. Typically, Ptolemy's outline of the Caspian Sea, though egregiously wrong, was authoritative until the 1720s, when Peter the Great obtained modern measurements. The twenty-seven maps that exemplified the master's teaching—one of the world and twenty-six of specific regions—were the prototype of world atlases. They dominated cartographic publishing for a century after their initial publication.[5] Ptolemy took learned Europe by storm.

The comparatively rapid obsolescence of Ptolemy's geography is almost as conspicuous as its enthusiastic reception. The proliferation of Ptolemy in manuscript and in print coincided with the time when European exploration and mapping of the world broke spectacularly out of the limits of the Ptolemaic coordinates. The *Cosmographia* did not long remain a source of immutable information. An early copyist of its maps added improved outlines of the northern lands, now an integral part of Christendom (1482); other copyists incorporated a map of the Holy Land, probably as a historical subject rather than as current geography. From the 1450s, supplements of several kinds enlarged the Ptolemaic atlas. The renowned Strasbourg edition of 1513 could include as many as twenty "new maps" (*tabulae novae*), almost doubling the original work.[6] Ptolemy might seem to be superseded by revisers, but he retained momentum. His *Cosmographia* continued for many decades to be a core worthy of being reprinted, improved, and enlarged.

The acquisition of Ptolemy by Latin scholars revived a science of geography, and it capped the ancient geographical literature known to Europeans. Barely one century after 1406, however, geography divided into ancient and modern branches without conceding primacy to the latter.

Ancient geography was cheerfully cultivated on the basis of the Alexandrian master and all other ancient authorities, such as the descriptive geographer Strabo (ca. 63 B.C.–after A.D. 21). Strabo was translated into Latin even later than Ptolemy and, because eminently discursive rather than mathematical, was much prized. Authorities like Pliny the Elder, Solinus, and many others were already known to scholars. Led by Ptolemy, the *antiqui* retained esteem even after the "great discoveries" had shown that the world contained much more land than had been known to them.[7]

Humanistic scholarship had a part in this reception of Greek and Latin authors, but reverence for the past was not the only cause. Practicing geographers relied on the information of these sources to complement more recent observations, whose superiority to those of an earlier age could not be taken for granted. Ancient geography continued to be studied with respect and care well into the eighteenth century; it was deemed to be the senior discipline. According to a committee of the Institut de France in 1808, "Ancient geography is not only, as one might imagine, an auxiliary or appendix for history; it is an essential and integral part of geographical science properly speaking." A well-wisher of the French mapmaker Pierre Lapie (1779–1851) regretted that Lapie's very busy life denied him the opportunity to master the learned languages well enough to be able to "raise himself above modern geography." The physical features, political boundaries, and place-names of the Greek and Roman world were taken hardly less seriously than modern ones; and the comparison of ancient and modern lands was (as will shortly be seen) a vital occupation of teachers and students of geography.[8] Ortelius contributed to this passion for comparison by an appendix to the *Theatrum* called "Synonymia locorum sive populorum . . . appellationes et nomina," a major effort to establish equivalences between ancient and modern place-names.[9]

Alongside the discipline of ancient geography, and in deferential proximity to it, a modern geography (on Ptolemaic foundations) took form out of

the mounting accumulation of new knowledge, including an entire hemisphere now opened to exploration. The "New World" was not the only hint
that information was extending beyond the bounds of ancient geography; the
Strasbourg Ptolemy contentedly affirms, on an added map: "This part
of Africa remained unknown to the ancients."[10] Although the expansion
of learning beyond its Ptolemaic limits was not universally welcomed—
Leonardo and Erasmus did not mention America in their extant writings—
it could hardly be disputed. With outstanding historical sensitivity, Ortelius
superimposed a Ptolemaic world map onto a bare grid of the world now revealed and dramatized in this way that the modest bounds known to Ptolemy
and everyone else down to 1492 encompassed a mere quarter of the globe's extent (fig. 1). (This superb *Geschichtskarte* will concern us again.) Committed
humanist though Ortelius was, he recognized that knowledge had grown, and
he devised a vivid image to convey its expansion. The notion of progress had
no part in Ortelius's accompanying comment; as became a Renaissance
moralist, he quoted Cicero about the entire world being now open to "our"
greed and rapacity.[11]

The conviction that geography came in two varieties, each resting on its
corpus of evidence, affected the scope of "old" maps. R. A. Skelton suggests
that the Strasbourg Ptolemy of 1513 was the first collection in which "the arrangement of an edition of Ptolemy plainly illustrated the distinction between ancient and modern geography; the Ptolemaic work was treated as a
historical atlas." He contends that, after 1548, the ancient maps became a "traditional adjunct, . . . a kind of historical atlas."[12] The outcome of such reasoning is that "ancient geography" is identified with history, as though it had
been stripped of its legitimacy as geography; and collections of maps for ancient geography are routinely dubbed "historical."

Even if the Strasbourg edition illustrates the divergence of one type of geography from the other, as old from new, many decades passed before the
lesson of this division was absorbed and a divorce took place in the design of
geography books. The two branches continued to travel together in sixteenth-
century Ptolemy editions: new was never far from old and vice versa. The ancient master was primary, whereas modern maps were included, some say, so
that each collection might be more attractive to customers.[13] The first narrowly modern atlases resulted from the suspicion that Ptolemy might have
been outgrown. Facts rather than technique were at issue: the world uncovered by recent explorers was so different from the one known to Ptolemy that
it no longer made sense to fit ancient and modern maps into the same volume. The Italian made-to-order collections generically attributed to Lafréri

(a French dealer in Rome, 1544–77) were for new maps.[14] Depictions of the ancient world needed a home of their own.[15] Ortelius's modern *Theatrum,* in its successive editions, is the outstanding sign of this parting of the ways. When maps of the ancient world entered Ortelius's collection in 1579, they were not integrated in the manner of the traditional "Ancient and Modern Geographies"; they formed an addition, the *Parergon.* This supplement, in its most developed form, contained twenty-six maps of regions of the "old" world—the staple of "ancient geography." Interspersed among them was a medley of seven scenes, some from literature, others adapted from recent publications, but mainly historical in substance. No map bore a date for what it depicted. Although the scenic part of the *Parergon* forswore Ptolemaic austerity, Ortelius's supplement emphasized, in a ratio of four to one, a body of information that was unmistakably geographic, information that no one yet, or for long after, considered to be of exclusively historical interest.[16]

Ortelius segregated modern geography from ancient, and Gerard Mercator was as eager as he to do the same. Ancient geography, in his view, needed to have its authenticity actively restored and safeguarded. When producing a Ptolemaic world map in 1578, Mercator carefully stripped it of the accretions by which editors had improved its geographic accuracy; the idea now was that its historicity deserved scrupulous respect and should, where necessary, be restored.[17] Another eminent scholar, Philip Clüver (†1622), also spoke out as an advocate of a purified Ptolemy, stripped of well-intentioned improvements and restored to the author's words.[18] Both modern and ancient geography were expected to benefit from these efforts to efface modern improvements from the work of the ancient master.

The gulf separating "ancient geography" from "history" is perhaps best illustrated by two specialized atlases compiled by Petrus Bertius, a professor at the University of Leiden. Bertius's two-volume *Theatrum geographiae veteris* (1618) combined Mercator's edition of Ptolemy with other major texts, including the regional maps of the ancient world from Ortelius's *Parergon.* Bertius's collection is a source book of "old" geography, destined initially for geographers. Bertius followed this up in 1628 with *Ancient Geography from Ancient and Creditable Writers* (*Geographia vetus ex antiquis et melioris notae scriptoribus*), an atlas in which the portrayal of each part of the world is derived from a classical authority, as, for example, "16. Arabia Eudaemon Out of Pliny Bk. 6, Chap. 28," or "1. The World as Conceived by Pomponius Mela."[19] This collection, though not necessarily useless for historians, was not designed to serve their needs. Bertius's atlases presuppose an ancient geography, a living subject to whose fanciers they would primarily appeal.

AEVI VETERIS, TYPVS GI

EVROPA.

Zona frigida,

SEPTENTRIO.

Circulus Arcticus,

OCEANVS HYPER
BOREVS.

Zona temperata,

EVROPA

Tropicus Cancri,

HISPA
NIA

GAL

MAVRITANIAE

Zona torrida, et

AFRI

Aequinoctialis, sive

LIBIA INTE

RIOR CA

MARE

OCCIDENS

fervorem a veteribus

OCEANVS

OCEANVS AE
THIOPICVS.

INDICVS

Tropicus Capricorni,

Zona temperata,

Circulus Antarcticus,

Zona frigida,

MERIDIES

AMERI

CA

EN SPECTATOR, PILAE TOTIVS TERRAE ICHNOGRAPHIAM. AT
TIS NONAGESIMVM SECVNDVM SVPRA MILLE QVADRINGENT. C

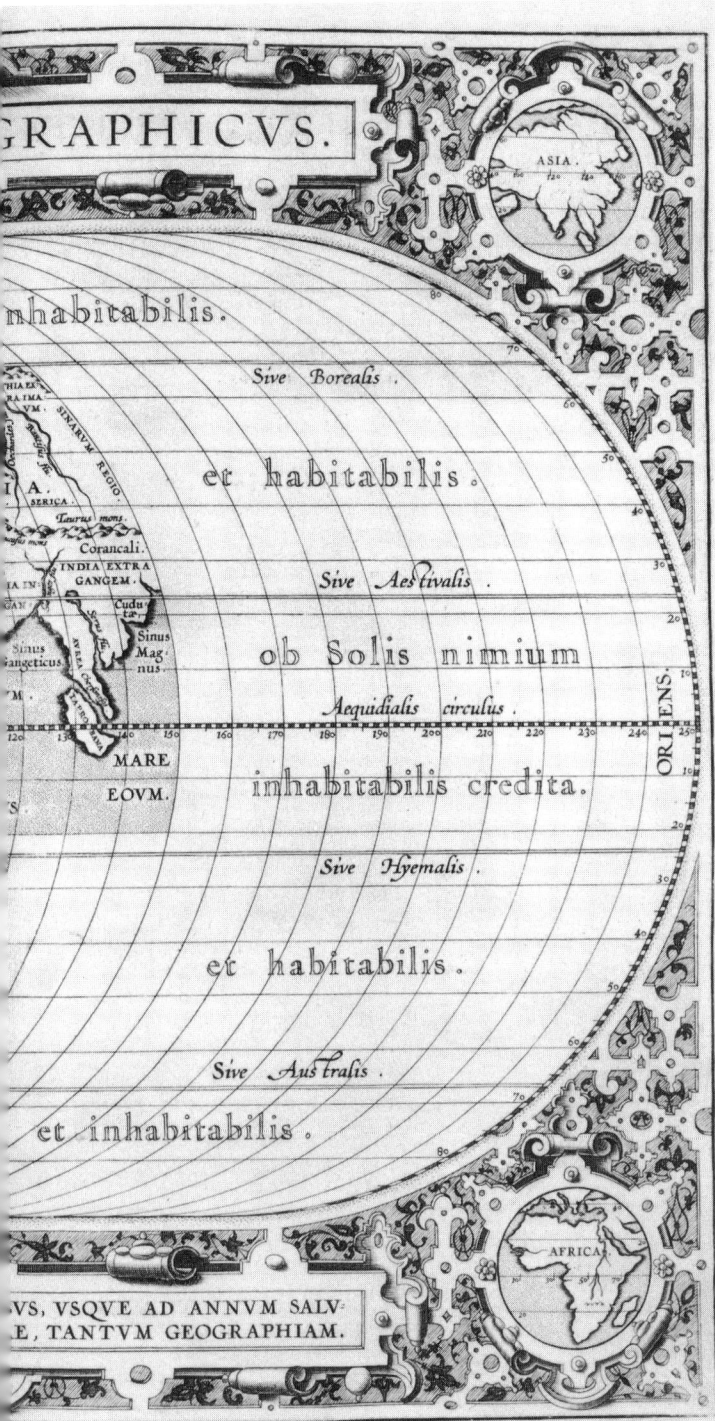

GRAPHICVS.

ASIA.

inhabitabilis.

Sive Borealis.

et habitabilis.

Sive Aestivalis.

ob Solis nimium

Aequidialis circulus.

MARE
EOVM.

inhabitabilis credita.

Sive Hyemalis.

et habitabilis.

Sive Australis.

et inhabitabilis.

ORIENS.

AFRICA.

VS, VSQVE AD ANNVM SALV:
E, TANTVM GEOGRAPHIAM.

Figure 1 Abraham Ortelius, "Aevi veteris typus geographicus" (1590). Ortelius added this map to the 1590 edition of the *Parergon*. Its too-brief title is amplified in the caption that runs across the lower edge. This is a map of the "old" world but is set on a grid representing the world as it had now become known. The map portrays the expansion of geographic knowledge since 1492. The contrast of the grid to the filled-in Ptolemaic world dramatizes the immensity of change. (Courtesy of the Beinecke Rare Book and Manuscript Library, Yale University)

The Comparison of Old and New Geography

Ancient geography was preeminent for centuries. Mercator and Ortelius gave it of their best, as to an object of the most serious concern; so did Nicolas Sanson (1600–67), the patriarch of French cartography. An even brighter star of the French school, Bourguignon d'Anville (1697–1782), won his most enduring fame in mapping ancient lands. He and a notable predecessor, Guillaume Delisle (1675–1726), prized modern geography for the light that it shed on ancient.[20] D'Anville's sole pupil, Barbié du Bocage, supplied a historical novel with scrupulous reconstructions of the topography of ancient Greece. The dialogue between the ancient and modern worlds preoccupied sixteenth-century geographers and their successors. Animated by the great importance that "humanistic culture attached . . . to the relationship between classical and modern geography," Ortelius took the unprecedented step of annotating his modern maps with references to classical authors. His initiative met with enthusiastic approval.[21] From the sixteenth century to the twentieth, atlases specifically designed to compare ancient and modern geography were in steady demand. "Comparative" even occurs sporadically as a direct synonym for "historical."[22] Editions of Ptolemy, such as that of Venice, 1562, positioned each modern map immediately after its ancient counterpart. The practice was already old. Francesco Berlinghieri, verse translator of Ptolemy's *Geography* and supplier of four *tabulae modernae,* published them so that they faced their "ancient" counterparts (1482).[23] Such books belong to the time when comparison was obligatory in geographical portrayals of the world. The resourceful Parisian Nicolas de Fer (1646–1720) welcomed juxtapositions not only of the ancient Holy Land to the new but also of the ancient desert monks of the Egyptian Thebaïd to their modern descendants in the new Thebaïd of La Trappe in Normandy.[24] De Fer notwithstanding, exercises in comparison remained faithful to their dominant sources and centered mainly on profane classical antiquity.

Comparative maps for classroom use were first compiled and published by the French Jesuit Philippe Briet. His *Parallels of Ancient and Modern Geography* (*Parallela geographiae veteris et novae*) in three quarto volumes (often bound as one) is primarily a geography textbook, illustrated by many small-size maps, ancient before modern. A fixed order of presentation is followed, starting with the seas (shown by a world map) and continuing with Europe and individual countries from west to east. Briet's *Parallela* has been called the best-selling historical atlas in seventeenth-century France. Its commercial success is more certain than the appropriateness of calling it historical. It is not a historical atlas to us, nor was it to its many buyers. The prized commodity it supplied was comparative geography. Briet stayed popular well into the Enlightenment.[25] Geographical comparison was deemed indispensable. Lenglet Dufresnoy, prominent in the eighteenth century for writing aids to history, censured the widely praised *Notitia orbis antiqui* of Christoph Cellarius for failing to engage in comparison.[26] School curricula rooted in the Greco-Roman classics unhesitatingly treated geography as an exercise in comparison of ancient and modern. The prolific geography teacher Edme Mentelle produced several collections of this sort, including an amply conceived *Géographie comparée, ou Analyze de la géographie ancienne et moderne des peuples de tous les pays et de tous les âges*, in one atlas plus seven volumes of text. Philippe de Prétot, claiming to respond to the recent enlargement of history teaching, interpreted "comparison" more broadly than usual, in a leisurely atlas.[27]

England was little less concerned with comparison than France. To cite only the nineteenth century, the noted map publisher A. Arrowsmith issued *A Comparative Atlas of Ancient and Modern Geography . . . for the Use of Eton School*. D. Blair prided himself, mistakenly, on being the first to present ancient and modern lands on the same map; this economical method had in fact been intermittently practiced since the sixteenth century. In the *Comparative Atlas of Ancient and Modern Geography* by Alexander G. Findlay, old and new came face-to-face in a sober form suited to a scientific age.[28] Comparative atlases, committed to showing ancient regions rather than historical moments, remained closely bound to the Ptolemaic prototype, the ancient geography par excellence. In keeping with the master's map of the world and twenty-six regional maps, comparative atlases rarely strayed from austere outlines of provinces and regions. They were a precocious tradition in atlas-making and, for centuries, appealed to an appreciable share of the clientele. The juxtaposition of ancient and modern undeniably conveyed a sense of history.

Nevertheless, "comparative" in the nineteenth century also had a less traditional sense. According to William Hughes (1817–76), the "alliance [of Geography and History] forms the basis of the study known as 'Comparative Geography,' a main object of which is the exhibition of the successive changes in the distribution of states, with their attendant alterations of frontier, which are presented by a particular region of the globe, viewed at succeeding periods of time." The branch of study and atlas-making that Hughes called "comparative" was generally known by this time as "historical." Hughes bore this out by adapting his maps from Spruner's standard-setting *Historisch-geographischer Hand-Atlas*.[29] Hughes fell short of gaining favor for this new terminology. Before fading altogether, it survived to the eve of World War I in reissues of atlases descending from his own: Philips's *Elementary Atlas of*

Comparative Geography and its companion "school" atlas of the previous year.[30]

WORLD ATLASES: VESSELS OF DIFFUSE HISTORY

Much of geography, as understood by learned circles during the sixteenth and seventeenth centuries, was ancient. Its main contribution to history took place (as long supposed) when it was compared with modern geography. Students of history could benefit from maps even without comparisons with antiquity. Today as then, readers of history or other literature obtain answers to many geographic questions they encounter—such as the location of a river or a country—by consulting an ordinary geographical atlas. Ortelius had eloquently spelled out the needed information readers of history would obtain from his (modern) maps. Early maps "were primarily general . . . containing not much more than coastlines, rivers, cities, and occasionally crude indications of mountains . . . [;] the mapping of any geographic information beyond what we today call base data was unknown."[31] The limited contents of maps usually included political boundaries. Thanks to hand coloring, territories and borders were particularly prominent. Place-names received so much attention that their encyclopedic accumulation sometimes impeded legibility.[32] The reverse side of each map was earmarked for printed commentaries, as by Ortelius: "[They] contain a medley of historical or legendary lore, speculative discussion of origins and place-names, notes on the topography and principal sights of the plan, information on government, commerce and the practice and distribution of trades."[33] Atlas maps (now as then) are almost always accompanied by letterpress. Ortelius's association of modern maps with the comments of classical authors has already been noted.

Early maps, sometimes called encyclopedic, are pervaded by unfocused history. Atlas maps are supposed to supply reliable geography without time limits; chronologically exact history is not their business. Three fascinating exceptions to this rule occur as insets in Ortelius's *Theatrum*, two of them referring to territories at moments in the thirteenth century.[34] An early delineation of Russia provided a rudimentary account of Russian government. Viewers of Ortelius's map of Germany—whose unity was more reminiscent of Ptolemy's image of "Magna Germania" than the political divisions of the sixteenth century—could imagine themselves looking into the past, in which older and newer place- and district names, many of them evocative of the past, were layered.[35] Historical atlases could not be urgently needed when geographical collections had appreciable historical content.

That these collections had a historical character was widely understood: atlas titles prove the point. A set of maps by Nicolas Sanson (†1667), posthumously published at Utrecht in 1683, declared itself to be a *Geographic and Historical Description of the Four Known Parts of the World* (*Geographische en Historische Beschryvingh der vier bekende werelds-deelen*). This naming practice endured. A similar title graces Daniel Lizars's *Edinburgh Geographic and Historical Atlas* (1831), composed exclusively of modern maps. Titles like these, explicitly invoking history, were not imperative to make the owners and readers of comprehensive collections of maps consider them historical: such content was implied.[36] The equivalence of "general" or "universal" atlases with "historical" ones remained standard into the nineteenth century, so much so that "historical atlas" in the new sense it was acquiring—a collection of specialized maps for history—sometimes clashed with general atlases and their familiar, diffuse history.[37] Tradition held its own. Alongside the already mentioned Lizars atlas, "history" is paraded in the title of an unremarkable collection—the *Géographie universelle, Atlas-Migeon historique, scientifique, industrielle et commerciale* (1866), associated with the prominent cartographer Alexandre Vuillemin. Of its thirty-eight maps, four are of ancient geography. The *Atlas-Migeon* could have omitted this explicitly "ancient" vestige and still have claimed to be *historique* on the strength of its maps of the current world. This happens in A. K. Johnston's *National Atlas of Historical, Commercial and Political Geography* (1851). The contents are modern only, and thematic maps of physical geography occupy the pages at the close that other atlases sometimes devote to "old" geography. Nevertheless, "historical" is stolidly in the title.[38] The tradition that "geography was the eye of history," so that any general atlas might claim to assist historical study, lived on in Victorian England.

The classing of comprehensive atlases as historical is perhaps best illustrated by the ambitious *Atlas historique* that the firms of Honoré and Châtelain published in Amsterdam from 1705 to 1720. This seven-volume work seems to be the first in whose title the word "historical" is directly paired with "atlas." Despite the new name, a contemporary reviewer rightly judged the atlas to descend from the major Dutch collections of the previous century and to share their geographical program. The title did not announce a new product; it simply said more compactly what the Utrecht publisher in 1683 had called a "Geographic and Historical Description."[39] The *Atlas historique's* complement of "ancient" maps is somewhat unusual, not in quantity, but in its selection of subjects; and the profuse commentaries on the realms and lands surveyed are not confined to boxes or to the reverse side of maps.[40] These departures from the norm were not meant to give the work a distinctive or intensified historical bent. Regardless of the title, the atlas's proportion of "history"—if understood as the affairs of the past rather than those of the present—is little more than an intermittent, unsystematic, and cursory background to present-day lands and peoples. Modern maps predominate. In keeping with long-standing practice, the Châtelain *Atlas historique* earns its name as all such atlases did from incorporating historical information in its maps, tables, and explanatory letterpress, and not from illustrating moments or periods of the past.

"One Size Fits All"

Early maps were either ancient or modern in content; greater chronological precision was rare. In each compartment, time was homogenized. Someone leafing through the *Theatrum* of Ortelius not only gazed upon the contemporary world but also took in all of modern history, everything since the fall of Rome, including the medieval centuries. A collection of "ancient geography," such as Mercator's edition of Ptolemy, was no more or less historical a product than Ortelius's "new" geography.[41]

There was another way for current maps to be "historical." In 1681 the French cartographer Guillaume Sanson, son of the famous Nicolas Sanson, published (with reprints reaching far into the future) *Introduction to Geography*.[42] The extended title spells out that geography had three parts: astronomical (spheres and projections), natural (mountains, boundaries between land and water, and other natural features), and historical. The plan of this elementary introduction has a narrow, geometrical quality not shared by Nicolas Sanson's large and varied output of maps. The *Introduction* indicates that, natural features aside, geography is "historical":

> Historical geography divides the surface of the earth in relation to civil history or sacred history. That is why we divide it into (1) civil or political geography (2) and sacred geography, (3) to which we add the geography of languages.
>
> Political geography divides the surface of the earth for us into empires, kingdoms, republics, peoples, and other sovereign states.
>
> Sacred geography divides [the world] . . . into the extent of the principal religions.
>
> And the geography of languages lets us know the extent of the most widespread languages on [the earth].[43]

A map entitled "La géographie des differentes espèces ou races d'hommes" was added by the Moulart-Sanson edition in 1719.[44] The absence of "ancient geography" is noteworthy in view of Nicolas Sanson's large number of maps of this kind. Moreover, Guillaume Sanson's censure of bad geographers implies that "political" geography entered into more details than just "sovereign states." Incompetents gave false boundaries, reared up nonexistent kingdoms, and jumbled the circumscriptions of justice, finance, and the armed forces, as well as ecclesiastical and temporal jurisdictions. This list of flaws anticipates, in its subjects, a type of Sanson map that will be looked at more closely in a moment.

The Sanson *Introduction* expounds geography, not cartography. Its laudable concern with the themes of religions and languages was rarely translated into maps in the seventeenth century. Its program for "historical geography" is unobjectionable, except for having more to do with "political institutions" or "administration" than with "history." The missing ingredient is chronology. It seems as though, in the Sanson vocabulary, something was historical because it was the outcome of history.

Chronologically specific scenes embodying geography were created before Ortelius. The unsurpassed theme is the Exodus, which appears in several superbly imaginative printed examples. But the characteristic historico-geographic maps designed in this period were much more limited exercises than those portraying the Exodus: maps, ostensibly timeless, became historical by the addition of inscriptions or legends. "The Travels of Saint Paul," included in the *Parergon*, is found earlier as a Bible illustration. Its designer took a Ptolemaic map of the eastern Mediterranean, then wrote the reference to Saint Paul in its upper margin.[45] No more was needed to turn geography into history. The same procedure, with appropriate adjustment of the geographical field, produced "The Expedition of Alexander the Great." The inscriptions made the difference between an ordinary "old" map and one historically specific. "New" maps lent themselves to the same treatment.[46] Eighteenth-century cataloguers of maps for history—Eberhardt David Hauber, Jacques Le Long (or his reviser), Lenglet Dufresnoy (or a reviser)—had an indulgent view of the available products they recommended to historians. Le Long's admirable list of scenes useful for medieval history includes a series of maps of kingdoms that once occupied the territory of modern France. Lenglet Dufresnoy cites maps specifically mentioning "Austrasia," while Le Long draws attention to Gerard Mercator's "Royaume d'Arles" (1580), Melchior Tavernier's "Ancien royaume d'Austrasie" (1642), and Pierre Duval's France in four sheets containing not only the "Royaume de Bourgogne et d'Arles" but also maps of the Kingdoms of Aquitaine, Neustria, and Austrasia (1671). Hauber, like Le Long, refers to Duval's four sheets as a portrayal of medieval France.[47] Overlooked by these cataloguers is the map headed "Neustria" in Ortelius's

Theatrum (1594): Brittany and Normandy are shown and named in the cartouche, but the collective term "Neustria," evoking a bygone Frankish kingdom, caps them both.[48] Unlike Le Long and Lenglet Dufresnoy, Hauber voiced critical opinions. In connection with the map of Austrasia he was preparing (1724), he refers to a forerunner: "[Tavernier] already published a map of the Kingdom of Austrasia, and the old Duchy of Lorraine, but he showed the old boundaries only after today's provinces, without the least *accuratesse*."[49] Tavernier, an engraver rather than a cartographer, used the same methods as the much more respected Mercator and Duval; for all of them, as for the anonymous creators of maps for Saint Paul and Alexander the Great, one map for each main epoch was enough. Provided the chosen geographical field was appropriately ancient or modern, a narrower definition was not needed: a mere inscription turned (for example) a corner of France as it was in the sixteenth century into seventh-century Neustria. These allegedly historical maps, "old" or "new," bring to mind the commercial slogan "One Size Fits All."

The conjunction of timeless (modern) topography with historical themes is most strenuously carried out in the *Typi chorographici provinciarum Austriae* by Wolfgang Lazius (1561). Senior to Ortelius's *Parergon* by nearly two decades, this eleven-map circuit of the Habsburg dominions has been called "the first ever historical atlas of Austria" and "one of the first attempts at producing a historical atlas."[50] The *Typi* should not be classed among maps for history. Lazius's great work belongs to the humanistic celebrations of one's country, or chorographies, pioneered by Flavio Biondo's *Italia illustrata* (1453); no attempt is made to reconstruct the past. William Camden's *Britannia* (1586) is another widely known example of chorography.[51] Lazius, a man of multiple achievements, was appointed as historiographer to the Habsburg emperor Ferdinand I in 1546. Lazius called the *Typi* a "work . . . about all the provinces of Austria and the beginnings of the Austriadic [i.e., Habsburg] family." Unlike Biondo and Camden, Lazius makes a dynasty and its possessions, rather than a country, the focus of his chorography. He dedicated the Austrian collection to his Habsburg patrons; the maps are in oval frames topped by the double-headed eagle.[52]

The *Typi* develops the one-size-fits-all combination of maps and history that we have learned to know. The lands Lazius exhibited in his maps were of the present day. Portrayals of contemporary inhabitants usually decorate the leaves that divide sections, as in the legends "Carinthian countryman," "Bavarian boatman," "Tirolian miner." Five Habsburg provinces are identified by modern names only, and three more by decoratively "old" names—"kingdom of the Boii," "kingdom of the ancient Suevi," "alpine

Rhaetia" (*Boiorum regnum, regnum veterum Suevorum,* and *Rhetia alpestris*)"—irrelevant to the contents of the (modern) maps. Lazius, like many successors, took current maps (which he personally drew and engraved) and decked them out with emphatic appeals to history, especially by means of inscriptions and other written additions. To cite only one instance, the foundation of the Ostmark by the dynasty of the Saxon Ottos (in the tenth century) is evoked alongside a map in which much attention is paid to Vienna and its sixteenth-century road net.[53]

Lazius's *Typi,* no mean accomplishment, will be looked at again. To call the work "historical" detracts from what it is, namely, a talented panegyric celebrating the sixteenth-century Habsburgs and their dominions. The geographic form helped Lazius to annex to the dynasty he served a glorious past from which it was excluded by time but not by place: the Merovingians (Austrasia), Charlemagne, the Ottos, the Babenbergs (Saint Liutpold). These are gently blended with the Habsburgs of the future, to whom they were connected only by topography. This is not historian's history (or geographer's, for that matter), but it is a gifted exaltation of Lazius's patrons and benefactors.[54] He showed that geography could be associated with history to serve current needs and not just to help scholars.

NOUVELLE METODE DE GEOGRAFIE HISTORIQUE [SIC]

Unmodified maps of ancient or modern lands could be and were made explicitly historical by the addition of a suitable title. As late as the eighteenth century (to judge from bibliographers like Le Long), this summary procedure was thought to be compatible with reputable cartography. More ambitious models were also undertaken, especially in connection with administrative boundaries. These ventures, though little more sensitive to chronology than lowlier efforts were, did aspire to cut time into smaller pieces.

Nicolas Sanson is the dominant figure in seventeenth-century French cartography. Because the large stock of maps he left to his heirs was often reprinted, he was a force in the map trade for more than a century after his death.[55] In addition to contributing to biblical and classical "ancient geography," Sanson produced an atlas with historical associations called, in its second edition (1651), *France, Spain, Italy, Germany, and the British Isles . . . Described in Several Maps and Various Geographical and Historical Treatises in Keeping with the Finest and Principal Distinctions That May Be Observed in All the Ancient and Modern Authors.*[56] The five kingdoms are dealt with on a uniform plan: each is assigned an ancient and a modern map, and each map is modified five times, for a total of ten per country. Parallelism from country to country was sought but necessarily adjusted in the light of local circumstances. Sanson's legend for "France" is an example of what he does:

> The maps of France are five of Ancient [geography] under the name "Gaul," and five of Modern. The ancient are, 1. Gaul in general and divided in two parts; 2. Gaul in four regions; 3. Gaul in seventeen provinces following Roman [organization]; 4. Gaul in several peoples following Ptolemy; 5. Gaul by the Roman Itinerary of Antoninus and the Itinerary [Peutinger] Table.—The modern are: 1. France in general; 2. France in twelve general governments; 3. France in dioceses of archbishops and bishops; 4. France by *parlements* [judicial districts]; 5. France by *généralités* [financial districts].

An abridged list of the titles for Germany suggests the range of variation:

> 1. Ancient Germany in general; 2. in four major peoples; 3. in medium-sized peoples; 4. in its least peoples following Ptolemy; 5. the part bordering Germany and south of the Danube described following Ptolemy and the Roman itinerary. Modern Germany: 1. in general; 2. by archbishoprics, bishoprics; 3. in three [nine] main parts; 4. by [administrative] Circles; 5. by its states, principalities, etc.[57]

Pastoureau comments, "This atlas is a convenient digest of historical and administrative geography."[58] It is a limited foreshadowing of historical atlases, recording divisions of various times but steering clear of dating anything. Its merit resides in moving beyond the single all-purpose map for each of the ancient and modern epochs and providing both with at least a set of noteworthy subdivisions. Multiplying the maps by five did not, in itself, form them into a historical ensemble, but it took a needed step beyond "one size fits all" in intimating the complexity of the past.

Collections like this, mainly focused on France but lacking dates, had some claim, until the mid–eighteenth century, to furnishing that country with a historical atlas. In addition to "The Five Kingdoms," Sanson compiled a specialized atlas of France in three parts: physical, administrative, and ancient. The administrative third had maps of provinces, general governments, *généralités*, financial districts, and *parlements*. Because each layer of administrative division had its own origins, Sanson's six maps in this collection were historical in a more complex way than the five maps of ancient Gaul, but the time element remained hazy. A generation later, Hubert Jaillot, much involved in capitalizing on the hoard of Sanson maps, issued a single sheet showing France "divided into Parlements, in which are distinguished the 'Cours des Aides' and Exchequers, with the Duchies, Duchy-Peerages, and Counties, both those that subsist and those that have died out." [59] In England, Christopher Browne met great commercial success in 1700 with a wall map entitled "England Showing Its Antient And Present Government being Divided as in the Saxon-Heptarchy also into Dioceses, Judges-Circuits And Countyes." Incorporating almost every circumscription that could be plotted, its crowding made it hard to decipher, but it was no less popular for that. [60]

The welcome given to Browne's map suggests that presentations of administrative divisions might have more practical than historical appeal. Nevertheless, the historical dimension was deliberate and avowed. In 1697, a member of the newly established Académie française, the abbé Louis Courcillon de Dangeau, published part of an atlas under the title *New Method of Historical Geography for Learning Easily and Long Remembering Modern and Ancient Geography, Modern and Ancient History, the Government of States, the Interests of Princes, Their Genealogy, Etc.*[61] Dangeau was a concerned teacher, intent on supplying aids for students. The *Nouvelle metode* has bright coloring and outline maps; also featured is a long section of administrative maps for civil and ecclesiastical districts in France. Many decades later, when French history obtained its first atlas emphasizing chronological divisions, the publisher, Louis Desnos, advertised it as "Part 1" immediately complemented by "Part 2"—a sixty-map *Tableau analytique* of current France.[62] The *Tableau analytique* was somewhat more varied than the collections of Sanson and Dangeau; its core, nevertheless, was a systematic rundown of French administrative, military, ecclesiastical, and academic boundaries. Resourceful in marketing his productions, as map sellers had to be, Desnos offered the freshly minted historical atlas in any form that seemed promising. One idea that seemed worth trying was to associate the chronological design of the new product with the older tradition of districts-as-history.

These maps of administrative boundaries were more promising than any other early initiative foreshadowing historical atlases. They anticipated the thematic maps that became plentiful in historical atlases in the 1920s. A work of the 1850s pointed the way. Louis Dussieux's *Atlas général* is probably the most massive French production of its time, visibly inspired by the equally massive (and much more famous) atlas of Karl Spruner. Dussieux's coverage of France, out of proportion with that of other countries, opens with thirty-eight thematic maps, five of them of *ancien régime* administrative boundaries, like those in Sanson, Jaillot, Dengeau, and Desnos. The other thirty-three range widely among mines, livestock, and other statistics of contemporary interest. At least in his handling of France, Dussieux's *Atlas général* was a worthy successor of Desnos's two-volume historico-administrative set.

"Ancient Geography, Sacred and Secular"

A decade after starting to publish the "modern" *Theatrum orbis terrarum*, Ortelius responded to the "pleas of [his] friends and *studiosi* of ancient history, sacred as well as profane," and compiled a set of maps of biblical and classical subjects, almost all of them designed by himself. He called the work "secondary to the *Theatrum*" (*Parergon theatri*).[63] For this achievement, Ortelius has long been called the creator of the historical atlas. A recent authority, Peter Meurer, sets the matter in narrower limits: "As an atlas of the ancient world, the *Parergon* part of the *Theatrum* is the absolute beginning of modern scientific mapping for history." He adds, Ortelius's "greatest significance . . . in the history of cartography resides in the elaboration of topographic maps with historical contents." Meurer is not alone in this assessment.[64] The aspect of Ortelius's work that he emphasizes is "scientific" mapping. Mapmakers before Ortelius had represented the Exodus and other biblical scenes; what matters to Meurer is that Ortelius carefully reconstructed classical antiquity on the basis of the ancient and modern sources available to him. This was serious, erudite mapping of the past rather than, for example, piously imaginative maps of Bible scenes.

These opinions, though cautious and true as far as they go, substitute the term "history" where Ortelius himself said he was providing maps of "ancient geography." Ortelius should be taken at his word. The welcome that his maps were expected to receive from lovers of history, his original audience, and the utility of his maps for reading history do not turn Ortelius into a historian or imply that the *geographia vetus* of his *Parergon* was any less geographical than the modern maps of his *Theatrum*. The importance attached to ancient geography for several centuries, and not just for historical reasons, is certain. Atlas after atlas, well into the nineteenth century, persists in including this commodity, if only as a supplement. Ancient geography was an authentic scholarly pursuit—not an old-fashioned or euphemistic way of saying "history." It was the science of the world as learned from classical authors, such as Strabo, Pliny, Ptolemy, and Solinus.[65]

Ortelius was a devoted antiquarian humanist. He owned a notable collection of coins, traded in antiquities on a modest scale, and won praise from learned men for the list of ancient place-names that he added to the *Theatrum* and later published separately.[66] He took personal responsibility for all the maps of his "ancient" appendix, starting with three in the first edition and swelling to thirty-three in the last with which he was involved. His pains were rewarded: the *Parergon* stayed in demand for more than a decade after the main *Theatrum* was overtaken by competitors.[67]

Ortelius's *Parergon* initiated several long-lasting developments. In its own right, it pioneered the repertory of maps that—to borrow a title from the Philadelphia map publisher Samuel Augustus Mitchell (1792–1868)—might be generically called the "Ancient Atlas, Classical and Sacred." Even Mitchell did not call the contents history. The ensemble launched by Ortelius was the common starting point for what, today, are our (separate) classical and biblical atlases.[68] Besides, Ortelius's decision to provide the *Theatrum* with a supplement of ancient geography invited imitation. Collections of "maps of all parts of the [current] world" came to be normally augmented by an additional section in which a more modest, variable number of maps of "ancient geography" were grouped. Occasionally, the addition was set at the beginning, as a prologue suggestive of chronological order, ancient before modern. Whether as appendix or as prologue, these supplements did not supply a different commodity from the main atlas: both were geography, one modern, the other ancient.

The selection incorporated into these appendices is sometimes arbitrary or skimpy. Wilhelm Janszoon Blaeu's *Theatrum orbis terrarum* (1639) ends its German section with "Ancient Germany" ("Germania vetus") and closes its coverage of France, more generously, with "The Empire of Charles the Great," "Ancient Gaul according to Strabo," and a second "Ancient Gaul." The fully grown atlas of Janssonius gathered ancient and biblical maps into a separate volume; its selection, heavily weighted toward maps of regions, differed little from Ortelius's. This, too, was an appendix, but destined for sale apart from the atlas as a whole as well as with it. A later example of this practice is the multivolume mid-eighteenth-century *Atlas* of Jean Beaurain, whose ending is dedicated to maps that inch toward looking historical.[69]

The highly selective appendix lived on. G. M. Cassini's three-volume *Nuovo atlante geografico universale,* whose publication straddled two centuries, ends with sixteen austere delineations of ancient geography. Nine items limited to Greece and Rome complement the thirty-nine current maps in L. Vivien de Saint-Martin's *Atlas universel, pour servir à l'étude de la géographie et de l'histoire anciennes et modernes* (1825).[70] Other ancient supplements, sometimes at the beginning rather than the end, are lengthier and more comprehensive but are always overshadowed by the modern maps. The annexed quasi-*Parergon* is encountered in world atlases as a fossilized or quaint feature of their varied designs.

The maps of the *Parergon* and its progeny are "ancient geography, sacred and secular," as the subtitle of the Janssonius atlas of 1652 puts it. They represent the learned branch of geography. Their commitment to the geography of the ages before Rome's fall stood in the way of an extension of their contents into the history of later times. They rarely overstepped their established chronological limits.

Ortelius's own "maps of ancient geography" are of three kinds: ancient lands, ancient scenes, and the Bible. The classes increased to four in the *Tabulae geographicae* published in 1699 on behalf of the episcopal seminary in Padua—a derivative, visually unimpressive collection but exemplary in organization.[71] The four Paduan classes are geographical maps of the classical world; maps "to elucidate ancient historians and poets" (*ad veteres tum historicos, tum poetas illustrandos*); maps based on the Bible; and, to close, an ecclesiastical geography consisting of patriarchates, bishoprics, and councils (the class absent from Ortelius). Two of these divisions are profane, two sacred; each pair has a main constituent and a complement. The four Paduan parts precisely match what early atlases knew as ancient geography.[72]

Offspring of ancient geography rather than specifically of the *Parergon*, classical and biblical map collections are still vigorously cultivated today. The appearance in 2000 of the vast, scientific, and cooperative *Barrington Atlas of the Greek and Roman World*, edited by Richard J. A. Talbert, attests to the vigor and seriousness that continue to sustain the mapping of classical antiquity. The Ptolemaic twenty-seven were the seedbed of these activities. Ortelius carried them to a new plateau in the *Parergon* by replacing Ptolemy's maps with depictions of the ancient world based on modern maps and refined by reference to ancient authors.[73] Even if Ortelius did not create the historical atlas, he definitely rejuvenated ancient geography and gave it the scientific impulse that has animated it ever since. Such was the discipline whose sources Petrus Bertius assembled in the *Theatrum geographiae veteris*.

It cannot be overemphasized that ancient geography primarily addressed lands rather than historical moments. Lands account for twenty-six out of thirty-nine items in the *Parergon* (1612), nineteen out of twenty-nine in an English *Geographica classica* of 1721. Karl Spruner's *Atlas antiquus*, an honorable nineteenth-century descendant of such collections (1850), was not indifferent to the established repertory. Spruner's stately array of twenty-seven plates assigns twice as many to regional geography as to historical moments.[74]

The predominance of topographical maps in collections of ancient geography affects their historical dimension. Contrary to Meurer's claim that such maps are "the core of a general historical atlas," it is only in atlases of ancient history, as a vestige of ancient geography, that maps of territories are consistently central.[75] In atlases of medieval and modern times, such coverage is comparatively unimportant: territories belong properly to geographical atlases. When Ortelius offered his maps to "*studiosi* of ancient history," he spoke in the spirit of the dictum *historiae oculus geographia*: his topographical maps—carefully reconstructed ancient geography—offered students of classical antiquity the geography they needed and served as geographic companions to learned reading. The *Parergon* supplied the same commodity for the "old" world as Ortelius presented for the modern one in the *Theatrum*. The maps in both the *Theatrum* and the *Parergon* were geography.

Whereas comparative atlases limited themselves to an austere diet of territories, the *Parergon* and its progeny almost always complemented classical lands with maps "to elucidate ancient historians and poets." These portrayals and their biblical companions include some of the earliest maps for history—deliberate reconstructions of scenes out of the past. They are better called "literary" geography, since they accord the same respect to the poetic wanderings of Ulysses and Aeneas as to the historical campaigns of Alexander, Pyrrhus, and Hannibal. They increased in number with the passage of time but had very modest beginnings: four classical ones in the *Parergon* are augmented by two out of the Bible.[76]

The few literary maps of Ortelius were restrained and prudent. He depicted the lands and seas the Argonauts traversed, for example, but refrained from supplying a track of the mythological adventurers' course. He did not even mark the routes of authentic generals leading their armies over documented terrain. Only the patriarch Abraham's journey from Mesopotamia to the Holy Land acquires an explicit line of movement and does so only in an inset. The Hondius-Mercator *Atlas minor* supplied a track for Saint Paul, as well as for Aeneas. Simple though the device was, it annexed the map to history much more intrinsically than an inscription would have.[77] Halfway into the seventeenth century, restraints were somewhat loosed. Pierre Duval (1618–83), Nicolas Sanson's nephew, was a major exponent of literary maps. He produced a small three-part atlas consisting of "maps . . . for well understanding the historians," followed by a time chart and a selection of "maps for [accounts of] modern itineraries and voyages." This plain collection, designed for the school trade, was a must in Louis XIV's France as a text for the few minutes that the curriculum allotted to geography. A richer selection of such maps was included in another of Duval's atlases.[78] In their wake, the maps of the 1699 Paduan collection trace the routes of several famous travelers: Xenophon's 10,000, Alexander the Great, Aeneas, and Ulysses. Other travelers continued to be shown only in geographic contexts.[79]

With tracks or without, maps like these anticipate the historical cartography familiar to us. Their efforts to portray moments of the past make up for the frequent defects of their execution. Ptolemaic and most *Parergon* maps depict a fragment of ancient territory as it presumably was during the whole span of antiquity; literary geography exemplified datable events, only occasionally diluted by imaginative literature. By the time of Spruner's *Atlas antiquus,* the repertory had been cleansed of Aeneas and his companions in fiction. Strictly territorial maps held their own, but maps of the dated, "literary" type and maps associated with defined historical periods had gained a prominent place.[80]

Much the finest maps in the *Parergon* are extensions of the geographical repertory: the Roman Empire; the ancient world compared to the new (both by Ortelius himself); and, only in 1624, a two-sheet religious "light of history" (by Haraeus, another Antwerper). All these contain much pertinent information; those by Haraeus might even be considered overloaded. As for Ortelius's "Map of Antiquity" ("Aevi veteris typus geographicus"), it is a masterpiece, saying much with very little, vividly impressing on the viewer how much the known world had widened since 1492 (fig. 1). These creations are "ancient geography," but they would not be out of place (artistic details aside) in future historical atlases.

The *Parergon* maps with the longest past have biblical subjects. They are the primary means by which early Christian and medieval maps were linked to modern ones. One of the earliest *tabulae modernae* added to the Ptolemy atlas had the Holy Land as its subject. Ortelius and others were guided by sixteenth-century Bible illustrators, who themselves were heirs of a rich pictorial tradition. The medieval copy of a late Roman road map called the Peutinger Table highlights the sojourn of the Israelites in the Sinai desert, and the great world map at Hereford cathedral (ca. 1290–1300) clearly marks the winding trail of the Exodus. The father of church history, Eusebius of Caesarea (†ca. 340), is thought responsible for the depiction of Palestine divided among the twelve tribes of Israel; some believe its earliest surviving example is a famous sixth-century floor mosaic in Jordan. The twelve tribes appear in a thirteenth-century map by the English monk-historian Matthew Paris and, not long after, in a precociously fine rendering of the Holy Land by Pietro Vesconti. In what may be the earliest map for history in a Ptolemaic format, the tribal divisions of ancient Israel are shown in the third printed edition of Ptolemy. For much longer than one would think possible, geographical atlases made an exception of Palestine and presented it as Terra Sancta rather than in its modern configuration.[81]

Biblical maps gave early mapmakers a brief holiday from Ptolemy. Among the subjects available to them, none more vividly symbolizes the religious origins of Europe than the biblical Exodus. In the early sixteenth century, a much intensified version of this scene was drawn, retracing the route of the Chosen People in their flight from Egypt, the forty-year wandering in the Sinai desert, and the final attainment of and entrance into the Promised Land. With an authoritative text as guide, and forty or more stopping points to challenge the cartographer's skill, no hesitation was possible. The Exodus track was reproduced innumerable times, more than any other Bible map and sometimes with superb miniatures. Ortelius published it in the *Theatrum* simply as "Terra sancta," as though it were a current map, not history, but eventually moved it to the *Parergon*.[82] The winding trail of the migrants, often drawn at a width that our road maps assign to freeways, translates the myth of Judeo-Christian origins into an unforgettable image.

The Holy Land repertory had additional subjects. Maps of Paradise, encouraged by John Calvin, had been anticipated by miniatures of Adam and Eve in medieval manuscripts; the subject was entrenched by the end of the sixteenth century.[83] An alternative version, based on a new theory, was drawn by a front-rank French mapmaker in the eighteenth century and deemed important enough, in the Age of Enlightenment, to be finished and published well after his death. The threefold dispersion of peoples descending from Noah's sons, a featured subject in some medieval *mappaemundi,* was sometimes chosen even in the nineteenth century to signify the beginning of history.[84] Slowest to come into existence were depictions of the Crusades and crusading principalities. A forerunner in the early 1700s proved to be just that; the subject was not fully exploited until the late 1820s. Hesitantly or not, the inventory grew steadily and merged without interruption into present-day atlases of the Bible and church history.[85]

The English biologist Thomas Huxley (†1895) believed that familiar surroundings afforded the best geographical training; he ridiculed a curriculum that indoctrinated Victorian schoolboys in "Jewish history and Syrian geography."[86] Secularists like Huxley notwithstanding, biblical maps did not dwindle in popularity. Before, as well as after, the French Revolution, their place in atlases was more secure than ever. They continue to be in demand. Helped by the establishment of the state of Israel, their cartographic future seems better assured than that of maps of the classical world. The National Geographic Society's latest "Lands of the Bible Today" (1960) is faithful to tradition in its choice of inset maps—the Exodus, the travels of Saint Paul, and the Crusades.

The Paduan collection of 1699 went beyond Ortelius by supplementing the array of Bible maps with ecclesiastical geography—Christian, of course, and predominantly Roman Catholic. Hondius's and Sanson's precocious thematic maps depicting the religious branches of the world have no place in this grouping.[87] Ecclesiastical geography was concerned with the boundaries of patriarchates and bishoprics and with the locations of major councils, subjects that became routine in the mid–seventeenth century. A late edition of Lenglet Dufresnoy's *Méthode pour étudier l'histoire* affirms, "The table of organization of the church is assigned to medieval geography, since it was during this period (*intervalle*) that dioceses and ecclesiastical metropolises took shape." In the 1700s, such sensitivity to a *géographie du moyen âge* expressed a hope or suggestion, not a fact. In the early days of patriarchal and diocesan cartography, no one foresaw a specialty called "medieval geography," and if anyone in the eighteenth century, other than Lenglet Dufresnoy's reviser, pigeonholed ecclesiastical geography as medieval, he is not easy to find.[88] Regardless of the epoch documented by maps of ecclesiastical organization, they were for a time almost a French monopoly.

Bishoprics were a pervasive and fundamental administrative unit in Christian Europe. The first to be mapped in France was Le Mans in 1539;[89] others followed here and there, without a general plan. A comprehensive ecclesiastical geography was composed by Charles Vialart (or de Saint-Paul), bishop of Lisieux, with maps by Melchior Tavernier the Younger (1640–41).[90] Vialart's *Geographia sacra* makes a circuit of the five patriarchates, beginning with Rome, the largest. Within the patriarchal boundaries it moves from country to country, outlines the ecclesiastical provinces, and picks out the episcopal cities without attempting to fill in diocesan boundaries. The Sansons—Nicolas and son Guillaume—retraced Vialart's footsteps a few years later. The same order of presentation occurs in the Sanson-derived Paduan atlas. Other collections, such as Duval's *Monde chrétien* and Brion de la Tour's *Atlas ecclésiastique*, limit themselves to bishoprics and provinces.[91]

More specialized works tailored their ecclesiastical geography to the object at hand. The original *Gallia Christiana* of the brothers Sainte-Marthe, which catalogues all the bishops there had been in the Gallic and French church, has a single map demarcating the provinces of the various archbishops. The second, enlarged *Gallia Christiana* has one map per province, subdivided into bishoprics. Michel Le Quien's *Oriens Christianus* owes its maps of the four eastern patriarchates to the eminent cartographer Bourguignon d'Anville.[92]

Nicolas Sanson launched a project for mapping France in unprecedented detail. The plan, in large part carried out, involved tracing not only the boundaries of bishoprics but also those of infra-episcopal units, such as archdeaconries and *mandements*. However relevant such divisions were to local churches, anthologies of ancient geography had no place for anything so detailed.[93]

Church history was not represented exclusively by the outlines of patriarchates and bishoprics. Guillaume Sanson produced a successful two-sheet *Geographia synodalis;* his kinsman Pierre Duval had anticipated him in mapping the major councils. The sites of assemblies ranging in time from Nicaea in 325 to the recent Council at Trent (1545–67) were set out together on the same background; but the younger Sanson at least assigned separate sheets to the eastern and western parts of the Christian world. Church councils in Germany were mapped by the Jesuit Josef Hartzheim for Schannat's *Concilia Germanica* and also circulated separately.[94] These synoptic images, or pictorial catalogues, needed the support of chronological lists.

Whereas councils gained a firm place in the stock of ecclesiastical geography, maps of religious orders did not. The houses of Augustinian Hermits, Discalced Franciscans, Capuchins, Benedictines, Jesuits, and perhaps other congregations gave rise to more or less elaborate maps but did not appeal to compilers. Augustin Lubin's supplement to the *Roman Martyrology,* showing sites of martyrdom, also failed to tempt atlas-makers. Wilhelm Guppenburg's land-by-land survey of miraculous images of the Virgin did no better.[95] A map by Hubert Jaillot of the early desert hermits and monks proved more acceptable and moved actively among publishers.[96] Much later, diocesan geography dominates the three-volume *Orbe Cattolico* of Girolamo Petri, impressive mainly for its great weight. In keeping with an age of imperialism, its dedication to Pope Pius IX promises, among other things, a glimpse of "the field that is still open to [the Church's] triumphs."[97]

The "Ancient Atlas, Classical and Sacred," was the cartographic expression of ancient geography. Its four parts formed an impressive, if somewhat disjointed, collection; its core, lacking only patriarchates and bishoprics, existed already in Ortelius's *Parergon.* Collections directly descended from it exemplified ancient geography in a much more open—one might almost say, cheerful—way than the normally solemn Ptolemaic and comparative atlases. For a field of study that had been developing since the initial translation of Ptolemy, the *Parergon* was a great leap forward.

Ancient geography lost its preeminence in the nineteenth-century reorganization of geography, but its atlas—the classical and biblical maps whose cultivation Ortelius had done so much to foster—continues to prosper, gratifying steadfast, specialized clienteles. What had been a dual collection of profane and sacred maps emerged as separate atlases of ancient or biblical history. That was all the history they were able to accommodate. Whereas modern geography had no compunctions about poaching in the domain of its better-bred consort, "old" geography was poorly suited to reach beyond the fall of Rome. The comprehensive historical atlases of the future depended on bridging the ancient and modern branches of geography and on realizing that all the past, or all parts of it, were the subject needing to be mapped. Inhibited from yielding to these directives, the *Parergon* and its progeny retreated gradually into a niche.

HISTORY EVERYWHERE AND NOWHERE

Historical atlases like those we find on the shelves of the bookstores we frequent were not speedily and easily born. They certainly did not materialize in the 1580s in Ortelius's *Parergon* and proceed in a calm, stately fashion down to the present. Nothing like them existed as the seventeenth century drew to a close.

The main obstacle was that history seemed to be already present in abundance in geographical maps. Geography was useful to history; as Ortelius said, *historiae oculus geographia:* geography allowed history to be visualized on terra firma. For the modern period, special pains were superfluous. Geographical atlases were sought out by readers of history as well as by travelers, geographers, and others in search of the information they provided. The ancient period, before the fall of Rome, was privileged. An élite with humanistic training needed ancient geography; the making of such maps, when seriously done, demanded a high level of erudition and a taste for the distant past. Classical and biblical antiquity provoked more, and more varied, enterprise than the "modern" past. Comparative atlases, supplements to world atlases, as well as "Ancient Atlases, Classical and Sacred," all bore witness to the grandeur of ancient geography, a grandeur that retained luster well into the nineteenth century. But "ancient geography" was basically no different from "modern." The history dispensed by both kinds was diffuse and, with few exceptions, chronologically vague.

One map for "old" times and one for "new" were enough for all the past. This telescoping of time extended beyond geography. All the soldiers in John Speed's panorama of English battles, no matter how medieval, wear seventeenth-century battle dress, and the ships of Aeneas's storm-tossed fleet are of impeccably Elizabethan design. Precise historicity, scrupulous rendering of historical details, had not yet become the norm. The same tolerance of anachronism allowed a map of Louis XIV's France to be marked "Neustria" in its northwest quadrant and visualized as a part of the Merovingian kingdom.[98] The *Geschichtskarte,* or map for history as we know it, is a specialized artifact. "Europe at the Treaty of Westphalia, 1648," is for students of history, not for casual readers. Unusual persons down to 1700 thought that individual maps of this kind were needed and took pains to produce them. Their achievement was noteworthy but inconsequential. The maps they designed could hardly stand out among the mass of unspecialized cartography offering serviceable history, however hazy.

NOTES

1. Ortelius, 1579, 1570. Meurer, "Ortelius," 133, does not dwell on *geographia vetus.*

2. *Encyc. Brit.* 20:332; see also Eila M. J. Campbell, "The Early Development of the Atlas," *Geography* 34 (1949): 191; Black, *Maps and History,* 9 ("The first historical atlas arose as a result of the mapmaking of . . . Abraham Ortelius . . . "); Liliane Wellens-De Donder, "Un atlas historique: Le *Parergon* d'Ortelius," in *Abraham Ortelius (1527–1598), cartographe et humaniste,* ed. R. W. Karrow, Jr., et al. (Turnhout, 1998), 84 ("le premier atlas historique connu"); Hofmann, "Genèse de l'atlas historique, 14 ("auteur du premier recueil de cartes historiques"). For a recent, emphatic exposition of the theme, see Meurer, "Ortelius," 133–59, and n. 64, below.

3. Josef Schmithüsen, *Geschichte der geographischen Wissenschaft von den ersten Anfängen bis zum Ende des 18. Jahrhunderts* (Mannheim, 1970), 60: "Die Weiterentwicklung der Karten war eng gebunden an die Wiederbelebung der Gedanken des PTOLEMÄUS . . . , dessen Werk erst zu Anfang des 15. Jh. der abendländischen Wissenschaft bekannt geworden war." Similar judgments occur widely. For the twenty-seven Ptolemaic maps as the prototype of world atlases, see Campbell, "Early Development of the Atlas," 188. The relationship of Ptolemy to his maps is much disputed; so is their survival in Byzantium. See also Patrick Gautier Dalché, "Le souvenir de la *Géographie* de Ptolémée dans le monde latin médiéval (VIe–XIVe siècles)," *Euphrosyne, revista de filologia clássica,* n.s., 27 (1999): 73–106; the demonstration that medieval scholars recognized Ptolemy by name and occupation does not, it seems to me, detract from the enthusiasm with which his geography was greeted from the time when it became known.

4. See Leo Bagrow, *History of Cartography,* ed. R. A. Skelton (London, 1964), 61–73. François de Dainville, *La géographie des humanistes* (Paris, 1940), 12, notes that the geographical knowledge Ptolemy offered had been surpassed by the fifteenth century; his value resided in a mathematically based geography. (Dainville notwithstanding, some Ptolemaic teaching was still "state of the art.")

5. Ptolemy allows cartography to extend from the coasts inland: Max Eckert, *Die Kartenwissenschaft,* 2 vols. (Berlin, 1921), 1:42. On the rectification of the Caspian Sea, see Jean F. Bernard, publ., *Recueil des voyages au Nord,* new ed., 10 vols. (Amsterdam, 1731–38), 7:301.

6. Tony Campbell, *The Earliest Printed Maps, 1472–1500* (London, 1987), 124, 136–37; L. Gallois, *Les géographes allemands de la Renaissance* (Paris, 1890), 16–17; R. A. Skelton, in *Ptolemy, "Geographia," Strassburg, 1513* (Amsterdam, 1966), v–xx, takes pains to establish the contribution of the editor Martin Waldseemüller (who gave America its name). C. Koeman, *The History of Abraham Ortelius and His "Theatrum orbis terrarum"* (Lausanne, 1964), 21: in the Strasbourg Ptolemy and later, in such excellent works as those of Sebastian Münster, the text took precedence over the maps, whose role was primarily illustrative; "the maps lacked any spectacular or decorative function." Koeman stresses that ornament was essential to atlases from Ortelius onward. Numa Broc, *La géographie de la Renaissance (1420–1620)* (Paris, 1980), 10: the third edition of Ptolemy (Florence, 1480–82) shows the Holy Land divided among the twelve tribes, an exceptionally enduring feature of world atlases; unlike any of the other maps, "ancient" or "modern," it was oriented to the east (an orientation that was also exceptionally enduring). Ptolemy editions are listed in Moreland and Bannister, 288.

7. On Strabo, in brief, see *Oxford Classical Dictionary* (Oxford, 1949), 863. Strabo's text came from Byzantium to Florence in 1439. Strabo wrote *historicorum more;* the ideal was thought to be a combination of his discursiveness and Ptolemy's mathematical geography. See Marica Milanesi, *Tolomeo sostituito: Studi di storia delle conoscenze geografiche nel XVI secolo* (Milan, 1984), 10, 12. Strabo was a model but, though translated into Latin in the 1450s, long remained a rare text; Solinus, Isidore, and a few others were liked for their brevity (Gallois, *Géographes allemands,* 154–55). Sebastian Münster was complimented as a reborn Strabo (Broc, *Géographie de la Renaissance,* 15, 84).

8. H. J. Dacier, *Rapport historique sur le progrès de l'histoire et de la littérature ancienne depuis 1789* (Paris, 1810), 222; this report, commissioned by Napoleon, was presented to him in 1808. For a recent annotated edition, see *Rapports à l'empereur sur le progrès des sciences,* 5 vols., ed. François Hartog (Paris, 1989), 4:187–91, which contains notes by Christian Jacob on P. F. J. Gosselin (1751–1830), author of the part on geography. Jacob deprecates Gosselin's inflated view of ancient geography, Gosselin's own domain (189). That ancient geography was an auxiliary to history is less significant than Jacob thinks (191), since modern geography, arguably, was too. D'Anville, not long dead, exalted ancient geography as much as did Gosselin. About Lapie, see *Arch. biog. franç.,* fiche 598, frame 068.

9. Peter H. Meurer, *Fontes cartographici Orteliani: Das "Theatrum Orbis Terrarum" von Abraham Ortelius und seine Kartenquellen* (Weinheim, 1991), 19–20; this list appears under a different name in the first edition of the *Theatrum;* the name I cite was used in 1574 and became *Nomenclator Ptolemaicus* in 1579. Meurer, "Ortelius," 143, infers from this list that Ortelius started to work on the "old" maps as early as 1565.

10. Strasbourg Ptolemy, map no. 13 (see n. 6, above).

11. Ortelius, 1590, was recorded by Eduard Brandmair, *Bibliographische Untersuchungen über Entstehung und Entwicklung des Ortelianisches Kartenwerkes* (1914; repr., Amsterdam, 1954), 154, as among the fourteen maps in Additamentum IV to Ortelius, 1570 (no. 173 of the entire *Theatrum*). The legend on the map face (summarizing a longer account on the verso) reads: "En spectator, pilae totius terrae ichnographia, at veteribus, usque ad annum salutis nonagesimum secundum supra milles quadringenti cognitae, tantum geographiam" (Here, O onlooker, is a grid of the bounds of the entire earth, but a geography of the only limits known to the ancients down to the year of grace 1492). Owing to its direct visual message, admirably mixing topography and diagram, Ortelius, 1590, is one of the finest maps for history. (Robert Karrow, Newberry Library, very kindly provided me with information about

it.) Meurer, "Ortelius," 144, 153, claims that this map is of the "ancient world" or a "general map of the ancient world." He could not have taken account of Ortelius's message to the reader or even of what the map shows. See also James Romm, "Mythe, cartes et histoire: L'utilisation par Ortelius de l'Atlantide," in *Abraham Ortelius (1527–1598), cartographe et humaniste,* ed. R. W. Karrow, Jr., et al. (Turnhout, 1998),110, "carte du Vieux Monde."

12. Skelton, in *Ptolemy, "Geographia," Strassburg,* v; new maps came after old. Skelton is echoed by H. Elkhadem, "La naissance d'un concept," in *Abraham Ortelius (1527–1598), cartographe et humaniste,* ed. R. W. Karrow, Jr., et al. (Turnhout, 1998), 31– 42: the first part "est considérée comme un atlas historique" (32). We are not told by whom. Hermann Wagner, *Lehrbuch der Geographie,* 10th ed. (Hanover, 1920), vol. 1, pt. 1:20, credits Philip Clüver (†1622) with properly distinguishing modern geography from ancient. The accent here is on "properly"; the duality of the subject was well established before Clüver's time.

13. Bagrow, *History of Cartography,* 86: the ancient maps became a "traditional adjunct, . . . a kind of historical atlas," after the Ptolemy editions of 1513 and Münster's in 1540. But note the dignified title *Geographia universalis, vetus et nova* (Basle, 1545), and the existence of editions of Ptolemy wholly lacking new plates (they are called by Bagrow "tributes to tradition, of purely historical interest" [86]). *Catalogus mapparum,* 45, comments, "[Les] cartes modernisées, fondées sur la géographie ptoléméenne, facilitaient la vente des nombreuses éditions de la *Cosmographie* et accompagnaient le plus souvent les cartes antiques depuis l'extrème fin du XVᵉ siècle." The obsolescence of Ptolemy is certain; how soon it occurred and in what circumstances are less clear. Broc, *Géographie de la Renaissance,* 12–13, is a good example of the tendency, in the modern history of geography, to dismiss Ptolemy early, perhaps prematurely.

14. R. A. Skelton, "Early Atlases," *Geographical Magazine* 32 (April 1960): 530.

15. De Vrij, *World on Paper,* 18. (Maps of ancient, especially biblical, subjects occur occasionally in Lafréri atlases.) The "Atlas of Ancient and Modern Geography" compiled by Giovanni Antonio Magini (1596) advertised his Ptolemy as being "totally necessary for understanding ancient histories." There was a separate but equal geography whose mysteries Ptolemy continued to unlock.

16. Readers of history, for their own needs, might well prize maps of *geographia vetus* as geography rather than history. Meurer, "Ortelius," 147, maintains that Ortelius's topographical maps were the core of *vetus geographia.* This may be true, but topographical maps have not been central to atlases of medieval and modern history.

For the statistics, see Koeman 3:71; I include among scenic maps such panoramas as the Roman Empire. Excluded are Ortelius's Paradise maps (Daphne, etc.). Though important for the *Parergon,* they did not become permanent components of ancient geography. Meurer, *Fontes cartographici Orteliani,* 22: the *Parergon* obtained a title page only in 1592, more than a decade after becoming part of the *Theatrum.* For a convenient summary of the *Parergon* maps, see Marcel P. R. Vanden Broecke, *Ortelius Atlas Maps: An Illustrated Guide* (Westrenen, 1996), 229–80.

17. Rodney W. Shirley, *The Mapping of the World: Early Printed World Maps, 1472–1700* (London, 1983), 162, no. 39: this was part of Mercator's "grand scheme, occupying most of [his] life, to describe the whole ancient and modern world geographically and historically." See also J. Keuning, "The History of an Atlas," 37. Mercator's early plans for what would become his atlas coupled maps with description of the heavens, genealogy and the history of states, and chronology.

LC List 384, Mercator's Ptolemy maps, without text (1578): all but one of the twenty-seven maps have notes explaining Mercator's changes restoring the precise Ptolemaic appearance. Scholarship of this sort, later engaged in by others as well, more clearly separated modern geography from ancient.

18. J. Partsch, *Philipp Clüver, der Begründer der historischen Länderkunde: Ein Beitrag zur Geschichte der geographischen Wissenschaft* (Vienna and Olomouc, 1891), 23. Koeman, *Collections of Maps and Atlases*, 16, points out the large demand for Ptolemy's *Geography* among the learned, such that its last edition came only in 1730. A share of this demand was from collectors of classical texts unconcerned with geography (13).

19. Bertius, 1618/A; Bertius, 1628/A, is complementary. For more on Bertius, see ch. 2, n. 67, below. Quotations from Pastoureau, *Atlas français*, 66: "Arabia Eudaemon ex Plinio lib. 6 cap. 28"; "Orbis terrarum ex mente Pomponii Melae." Meurer, "Ortelius," 136, calls Bertius, 1628/A, a "miniature historical atlas." It's hard to tell what this far-from-diminutive work has to offer to students of history.

20. Sanson never compiled a dedicated "ancient" collection but made historical complements to his world atlas. The additions are listed in Pastoureau, *Atlas français*, 407, "Cartes générales de toutes les parties du monde" (1658 ed.), nos. 98–113 (corresponding to Ptolemy's districts); 410–11 (1665 ed.), nos. 157–71 (including the Trojan War and biblical history). For praise of d'Anville and ancient geography, and his publication of the first critically satisfactory atlas of the ancient world, see *Göttingische gelehrte Anzeiger*, 1811, 328; Christian Sandler, *Die Reformation der Kartographie um 1700* (Munich, 1905), 17. Delisle, like Clüver, prized modern geography as the springboard to ancient.

21. About Barbié, see below, ch. 5, n. 61. On the dialogue between old and new, see Koeman, *History of Ortelius*, 25.

22. E.g., Dufour, 1835, contains thirty-three modern maps and nine of "géographie ancienne et comparée." Five of the latter fall within the tradition of Ptolemaic ancient geography; the balance concern biblical history, Europe after the barbarian invasions, Charlemagne's empire, and Europe in 1789; precisely which of these are "géographie comparée" is hard to tell. For a straightforward example of this equivalence, see Hughes, 1869 (*A Popular Atlas of Comparative Geography*); we would say historical atlas. The belief that "classical atlas" was synonymous with "comparison of ancient and modern geography" is illustrated by Murphy, 1858/A. The one historical map out of a total of six in C. G. Nicolay and D. T. Ansted, *Atlas of Physical and Historical Geography* (London, 1852–59), is entitled, "Comparative chart of ancient and modern geography and geographical discovery."

23. Campbell, *Earliest Printed Maps*, 133–34 (about Berlinghieri): Spain, nos. 150–51; Gaul, nos. 152–53; Italy, nos. 156–57; modern Palestine and the Holy Land, nos. 170–71—the Ptolemy version is on the recto, rather than the verso, page, the recto being traditionally the privileged page in printing. For other editions, see *LGKartog.* 2:649–51; most, though not all, of the editions with *tavole moderne* place them on facing pages. I count at least six such editions between 1561 and 1598 (catalogue entries do not always indicate how the maps are set out). Venice, 1562, by Bruscelli, definitely invites comparisons of old and new.

24. For Fer, 1700/E, an astonishing comparison of the ancient and modern Holy Land, see Nebenzahl, 140–41. About the Thebaïd maps (not de Fer's invention), see ch. 2, nn. 116–20, below. On de Fer, see *Biog. univ.* 14:308. See Sandler, *Reformation*, 13, on de Fer's productivity.

25. Briet, 1647/A. On its popularity, see Mireille Pastoureau, "Les atlas imprimés en France avant 1700," *Imago Mundi* 32 (1980): 64. Sommervogel, 2:156–57, tells of unpublished volumes surviving at the BN. A fellow Jesuit is rumored to have prevented their pub-

lication. Nicolas Lenglet Dufresnoy, *Méthode pour étudier la géographie,* 7 vols. (Paris, 1716), 1:260–62, made much of Briet, but he became disillusioned and, in the third edition of *Méthode,* was content that Jean Hardouin (1646–1729, a learned, intellectually eccentric Jesuit) had disposed of Briet's unpublished volumes (he did not).

26. Lenglet Dufresnoy, *Méthode,* 3d ed., 1:260: Cellarius (see ch. 3, n. 27, below) had not provided ancient names with their modern equivalents; Lenglet Dufresnoy made up for his neglect. On the abbé Lenglet Dufresnoy (1674–1755), a colorful man of letters willing to pay a high price for the sake of personal independence, see *Biog univ.* 24:118–19; *Dictionnaire des lettres françaises: Le XVIIIᵉ siècle,* 2 vols. (Paris, 1960), 2:86–87.

27. Mentelle, 1778/A; on Mentelle, see Hofmann, "Genèse de l'atlas historique," 112–13. Philippe, 1787/A: Philippe's claim is on p. 1; he was not well informed about the history of school atlases.

28. Arrowsmith, 1828/A. The Arrowsmiths were a three-generation dynasty of esteemed map publishers: see *DNB* 1:595–97. Blair, 1853/A; cf. Patteson, 1825/A, whose subtitle is *Exhibiting in the Same Maps the Principal Features Both of Ancient and Modern* [*sic*] *Geography.* Mentelle combined ancient and modern on the same map in Mentelle, 1801/A, nos. 91c, 96. See also Anville, 1795, which was "materially improved by inserting the modern names of places under the ancient." Earlier still, Guillaume Delisle provided maps for young Louis XV, whom he was tutoring, showing the old and new place-names (*Mercure de France,* 1726, 485). Bilingual place-names on maps can be found even before Delisle; e.g., according to Milanesi, *Tolomeo sostituito,* 18, they were a way to keep the public buying. Blair, 1853/A, is the lithographed edition of Jones, 1830, according to Black, *Maps and History,* 28, 59.

Findlay, 1853/A. On Findlay, see *DNB* 7:23. Also in this vein, see Stackhouse, 1790/A; SDUK, 1851/A (especially the comparative atlas of 1844 entered under this heading).

29. Hughes, 1869; a title on the spine reads *Philips' Historical Atlas.* The quotation is from Hughes's preface. On Spruner, see ch. 5, 353–56 below.

30. LC List 4204–5: *Philips' Elementary Atlas of Comparative Geography* (London, [1914]); *Philips' Modern School Atlas of Comparative Geography* (London, 1913).

31. A. H. Robinson et al., *Elements of Cartography,* 5th ed. (New York, 1984), 29; cf. A. G. Hodgkiss, *Understanding Maps: A Systematic History of Their Use and Development* (Folkestone, Kent, 1981), 14.

32. Karl E. Fick, "Die kartographische Darstellung wirtschaftsgeographischer Sachverhälte im 18. Jahrhundert: Anmerkungen zu Arbeiten M. Suetters und des Geographischen Institutes in Weimar," *Geographische Zeitschrift* 59 (1971): 131.

33. R. A. Skelton, ed., *G. Braun and F. Hogenberg, "Civitates orbis terrarum," 1572–1618* (Amsterdam, 1965), xviii.

34. Vanden Broecke, *Ortelius Atlas Maps,* 121, no. 77 (Flanders under Gui de Dampierre); 125, no. 81 (ancient Frisia under the emperor Augustus); 127, no. 83 (eastern Frisia before 25 December 1277). Professor P. D. A. Harvey (Durham) very kindly drew these insets to my attention. On the first of these, see Henk A. M. Vander Heijden, "Ortelius and the Netherlands," in *Abraham Ortelius and the First Atlas: Essays Commemorating the Quadricentennial of His Death, 1598–1998,* ed. Marcel P. R. Vanden Broecke, Peter van den Krogt, and Peter Meurer (Westrenen, Netherlands, 1998), 283–86; the map belongs to the repertory of *vetus geographia* (*Germania inferior, Gallia Belgica*),

35. Vanden Broecke, *Ortelius Atlas Maps,* 213, no. 162 (Russia); 97, no. 56 (Germania).

36. Sanson, 1683 (one of the title pages has an alternative spelling of the title), contains historical notes. Lizars, 1831, includes a country-by-country historical commentary. Koeman,

Collections of Maps and Atlases, 69: the 1753 catalogue of Simon Emtinck's collection of 18,000 books classes atlases as "historia." Not all types were affected; books of nautical charts, itineraries, and city plans escaped. See also Hofmann, "Genèse de l'atlas historique," 100.

37. On the failure of Dacier, *Rapport historique,* to recognize any specialized "historical atlas," see ch. 5, n. 63, below.

38. Migeon, 1866; Johnston, 1846. A case similar to Migeon is Brué, 1822, which has nine maps of ancient geography out of a total of sixty-four.

39. Châtelain, 1705; for more about this atlas, see ch. 3, nn. 1–3, below. For the opinion that the Blaeu and Janssonius atlases did the same things better and more fully, see *Journal de Trévoux,* January 1716, 45–46 (referring to vol. 4 of Châtelain's *Atlas historique,* published in 1714).

The Châtelain collection tries by its format, layout, and comparative austerity—and name?—to make an appeal different from that of the luxury productions of the past. It retains a folio format but eschews the sumptuous map albums of Blaeu and Janssonius in favor of a plain, illustrated book (I have come across two or three colored sheets but no fully colored set or volume). The work is called a "French atlas" in *National Maritime Museum: Catalogue of the Library,* vol. 3, *Atlases and Cartography,* pt. 1 (London, 1971): 266–68, 366. Nothing seems to justify this, however, except language and the Huguenot origins of its publishers.

40. On the difficult and costly combination of engraved maps with letterpress, see Arthur H. Robinson, "Mapmaking and Map Printing: The Evolution of a Working Relationship," in *Five Centuries of Map Printing,* ed. David Woodward (Chicago, 1975), 8.

41. For Mercator's Ptolemy, see n. 17, above.

42. Guillaume Sanson, *Introduction à la géographie où sont la géographie astronomique . . . la géographie naturelle . . . la géographie historique, qui considère la Terre Par les Etats souverains, par l'Estendue des Religions, et par l'Estendue des principales langues* (Paris, 1693). Editions of 1681 and 1690 (and later) are at the BN. Nicolas's name (in the form "Sanson d'Abbeville") appears on the cover page, as though there had been an edition in his lifetime.

43. Ibid., 188.

44. P. Moulart-Sanson, *Introduction à la géographie* (Paris, 1719); this edition makes small additions to the earlier core and is fitted out with many maps. Still another "Sanson generation"—Robert de Vaugondy, father and son—carried forward the *Introduction.*

45. On the Exodus maps, see Catherine Delano-Smith and Elizabeth Morley Ingram, *Maps in Bibles, 1500–1600* (Geneva, 1991), 27; C. Delano-Smith, "Maps in Bibles in the Sixteenth Century," *Imago Mundi* 39 (1995): 3, 12; C. Delano-Smith, "Maps as Art and Science: Maps in Sixteenth-Century Bibles," *Imago Mundi* 42 (1998): 67, 70–71. On the eastern Mediterranean and Saint Paul, see Vanden Broecke, *Ortelius Atlas Maps,* 233, no. 181; Delano-Smith and Ingram, *Maps in Bibles,* 99 (apparently the earliest track of Paul's travels is in a Haarlem Bible of 1591); Delano-Smith, "Maps in Bibles," 9. Ortelius, 1579, like earlier maps, left Paul's course to the viewer's memory of the Book of Acts.

46. Vanden Broecke, *Ortelius Atlas Maps,* 276, no. 222. The omission of a track of Alexander's campaign is a conspicuous sign of Ortelius's austere handling of ancient geography.

47. The title of Mercator's map "Royaume d'Arles" is cited here as it appears in Le Long, 1:36, no. 420, although it is fuller on the map; see ch. 2, n. 57, below. Le Long, 1:35, no. 397, plausibly records 1642 for the date of the map I have listed in the catalogue as Tavernier, 1645

(Tavernier retired from business in 1644). For Duval, see ch. 2, n. 60, below. For maps mentioning "Austrasia," see Lenglet, 1768/B, 12:2; for Duval's four-sheet map, see Eberhard David Hauber, *Versuch einer umständlichen Historie der Land-Charten* (Ulm, 1724), 131.

48. Vanden Broecke, *Ortelius Atlas Maps*, 78, no. 37b; Koeman, 3:55, with no. 1a.

49. Hauber, *Versuch einer Historie*, 142.

50. Quotations from Ernst Bernleithner in *Wolfgang Lazius Austria Vienna, 1561* (Amsterdam, 1972), xiii; Florio Banfi, "Maps of Wolfgang Lazius in the Tall Tree Library in Jenkintown," *Imago Mundi* 15 (1960): 56. A twelfth map, of ancient Greece (supplementing Lazius's book on the subject), often circulates with the eleven maps of Austria. The *Typi* is in Eugen Oberhummer and Franz R. von Wieser, eds., *Wolfgang Lazius Karten der österreichischen Länder und des Königreichs Ungarn aus den Jahren 1545–1563* (Innsbruck, 1906). For a full facsimile, see Bernleithner, in *Lazius Austria*, xiii; Banfi, "Maps of Wolfgang Lazius," 152–65 (see a caution at 118, n. 51, below).

51. On the chorography genre in connection with Camden's *Britannia*, see F. J. Levy, "The Making of Camden's *Britannia*," *Bibliothèque d'humanisme et renaissance* 26 (1964): 77–78. In other connections, see ch. 2, nn. 87–88, below.

52. A. Horawitz, "Wolfgang Lazius," *ADBiog.* 18:89–93. Oberhummer and Wieser, *Wolfgang Lazius Karten*, 9–11 (Lazius's poor health; excellent portrait). See also *DBiog Archiv*, fiche 746, frames 259–76. Lazius won Ferdinand's favor after presenting to him *Rerum Viennensium commentarii* (Basle, 1546); it was much admired, won a civic prize in 1545, and made a great stir. Ferdinand, then designated successor to the empire (i.e., king of the Romans), became emperor in 1558.

53. For details of how Lazius made maps historical, see ch. 2, nn. 45–54, below.

54. Meurer, "Ortelius," 134, notes a map of the German empire of the time (Antwerp, 1547 and 1566) in which the settlement areas of the ancient Germanic tribes are shown, the presumed aim being, according to Meurer, to glorify the Holy Roman Empire. This map, not seen by me, may be similar to what Lazius was making, although on a more limited scale.

55. On Sanson and his descendants, see Mireille Pastoureau, in *LGKartog.* 2:699–701.

56. Sanson, 1651: *La France, l'Espagne, l'Italie, l'Allemagne et les isles Britanniques . . . Descrites en plusieurs cartes et differens Traittés de Géographie, & d'Histoire, suivant les plus belles & les principales distinctions, qui se peuvent remarquer dans tous les Autheurs anciens, & nouveaux.* The order of countries in the first edition was Britain, Spain, France, Italy, and Germany; Sanson relegated Britain to last place as a result of the execution of Charles I in 1649.

57. My quotations and summaries are from the copy in the BerlinPK. The German "Circles" were a major subdivision of the Holy Roman Empire since the sixteenth century.

58. Pastoureau, *Atlas français*, 430.

59. Nicolas Sanson, *Le Royaume de France et ses acquisitions* (Paris, 1665), 15 sheets: 1. bassins hydrographiques fluviaux; 2. France en 3 grandes parties; 3. France en deça de la Loire; 4. France en dessus de la Loire; 5. France au delà de la Loire; 6. provinces; 7. gouvernements généraux; 8. généralités; 9. chambres des comptes et cours des aydes; 10. Parlements; 11–15. Gaul (ancient geography). H. Jaillot 1695 (BN, Ge.DD.2987B [679]). I have not systematically recorded such maps. There must be many more than I touch on.

60. Cited in Shirley, *Maps of Britain to 1750*, 38–39 and pl. 21, with evidence of its popularity and use.

61. For the original version, and its spelling deliberately omitting accents, see Dangeau, 1697.

62. L. C. Desnos, *Catalogue des ouvrages tant anciens que modernes du Fond du Sr. Desnos* (Paris, 1765), 4–5, a collection of thematic maps (Desnos, 1765c). The form of an atlas tracing history by administrative boundaries seems to be espoused by Tomka, 1751, the first historical atlas of Hungary; see ch. 4, nn. 198–201, below.

63. Meurer, *Fontes cartographici Orteliani*, 21–22. See also n. 2, above.

64. Meurer, *Fontes cartographici Orteliani:* "Der *Parergon*-Teil des *Theatrum* als Atlas der antiken Welt ist der Beginn der modernen wissenschaftlichen Geschichtskartographie überhaupt" (1); "liegt seine grösste Bedeutung wahrscheinlich in der Erarbeitung von Landkarten mit historischen Inhalten" (21). Meurer, "Ortelius," 143, refers to "the epoch-making historical atlas *Parergon,* separating the early history of the discipline into [pre- and post-Ortelius]." By "discipline," Meurer means "historical cartography"—a problematic occupation. See Introduction, n. 20, above.

65. According to Emmanuel de Martonne, *Traité de géographie physique,* 4th ed., 3 vols. (Paris, 1925–27), 1:10, Ortelius considered the *Parergon* his capital work, reflecting the unanimous idea at the time (and for centuries after) that ancient geography was more serious and scientific than geography based on recent discoveries.

66. Koeman, *History of Ortelius,* 25; on Ortelius's lists, see n. 9, above.

67. Koeman, *History of Ortelius,* 44. My count of maps (1598 ed.) excludes "paradises," such as the Vale of Tempe, and includes scenes (Saint Paul) plus regions of the ancient world. The last edition of the *Theatrum* was 1612; the last edition of the *Parergon* was 1624. Brandmair, *Bibliographische Untersuchungen,* 152, notes that Ortelius always preferred ancient geography.

68. Mitchell, 1845. The first edition was 1844. On Mitchell, a major figure in early American map publication, see *Dictionary of American Biography* (New York, 1934), 7: 61–82.

69. W. J. Blaeu, *Theatrum orbis terrarum* (Amsterdam, 1639), pt. 1, no. 70; pt. 2, nos. 44–46. The first item of pt. 2 also is a map of "Gallia." Janssonius, 1652/A, ends the multivolume Janssonius atlas. For Beaurain, 1749, see ch. 3, nn. 58–59, below.

70. Cassini, 1792, 3: nos. 42–57. Vivien, 1825; maps 40–48, the last, are ancient, unadulterated by biblical or medieval subjects. Vivien de Saint-Martin (1802–97) was prominent in his day; he wrote an early history of cartography. See *Geog.: Biobib.* 6:133–38.

71. Padua, 1699. The maps, by Sanson and associates, are invariably attributed to their makers. See the useful discussion by Hofmann, "Genèse de l'atlas historique," 103. She calls the Padua atlas "une somme de la cartographie historique." Its own title, however, mentions only "ancient geography": *Tabulae geographicae quibus universa geographia vetus continetur.* Retaining the terminology of the time aids in understanding what was being done.

72. The same classification occurs in ClassAtl, 168-/A, a sixty-two-map *atlas factice* from the library of the Church of Paris. Its maps, of the seventeenth century, are in the same order as those of the Padua collection, right down to having "literary geography" bring up the rear.

73. The ninety-nine-map *Barrington Atlas,* the work of many contributors guided by Talbert, is published by Princeton University Press in CD as well as print versions. On the Ortelian background, see Meurer, "Ortelius," 137. The map of the Roman Empire has a liberal admixture of Ptolemy; but Ortelius is more remarkable for having largely substituted modern sources for Ptolemy in designing the *Parergon* maps.

74. Ortelius, 1579; Bertius, 1618/A; Spruner, 1850/A. Spruner distinguished "geographical" from "historical" sheets. The latter include such items as "XXV. Graecia a bello Pelo-

ponnesiace usque ad Philippum II," completely alien to the traditional "ancient geography" program but in line with the historical atlases of his and our time. Ancient geographers distinguished a "Great Germany" ("great" meaning "larger," as in "Great Britain") from the Roman provinces of "Germania" (Upper and Lower).

75. Meurer, "Ortelius," 147.

76. Vanden Broecke, *Ortelius Atlas Maps*, 253, no. 181 (Saint Paul); 254–55, nos. 182–83 (Abraham); 276, no. 222 (Alexander); 276, no. 223 (Aeneas); 278–79, nos. 224–25 (Ulysses, inset); 280, no. 226 (the Argonauts). Not everyone may agree on the precise extent of this selection.

Brandmair, *Bibliographische Untersuchungen*, 154, took the view (with little apparent evidence) that "all these historical maps were drawn for a particular purpose [e.g., a book] and only then were taken into the *Theatrum*." "Eneae navigatio" is in Jacobus Pontanus, S.J., *Symbolarum libri XVII quibus P. Virgilii Maronis Bucolica Georgica Aeneis . . . declarantur* (Augsburg, 1599). Cf. Dainville, *Géographie des humanistes,* 65, who has Ortelius acquire this map from Pontanus but gives no details. Meurer, *Fontes cartographici Orteliani,* 21–25, suggests that this is mistaken and may conceal Ortelius's originality. Vanden Broecke, *Ortelius Atlas Maps,* 277, no. 223, records 1594 for Ortelius's map, published in the *Parergon* in 1595.

77. Ortelius, 1579, marks the wanderings of the Israelites through the desert (he had many precursors), and in a very small inset of Vanden Broecke, *Ortelius Atlas Maps,* no. 182, he draws a line tracing Abraham from the Euphrates through Palestine toward Egypt. Other examples are *Atlas factice,* 1675/A, no. 68 (Philip Lea, "A New Map of Antient Europe to All Historiographers"); Philippe, 1787/A. On the Saint Paul track, see Hondius, 1607/A, and n. 45, above. The Aeneas map was annexed to history from the start by pictorial means— little ships, a storm, shipwrecks.

78. Duval, 1665; about him, see *Biog. univ.* 12:412. According to Didier Robert de Vaugondy, *Essai sur l'histoire de la géographie* (Paris, 1755), 225, every student had to have "his Duval." See also Dainville, *Géographie des humanistes,* 406–9. Pastoureau, "Atlas en France avant 1700," 63, maintains that Briet, 1647/A, was more important than Duval for pedagogy. The rich collection of "ancient geography" in Duval, 1667/A, listed in Pastoureau, *Atlas français,* 143–44, includes maps for, e.g., Herodotus, the Cyropaidea, Justin's *Epitome of Trogus Pompeius,* Eusebius (the church historian), Lucan's *Pharsalia,* and Caesar's *Civil Wars.*

79. Padua, 1699: Ulysses and Aeneas, 152; Alexander, 158; Argonauts through Saint Paul, 148, 156, 162, 164. Several Duval maps that might have had tracks do not (e.g., Pyrrhus, Demetrius Poliorcetes, Hannibal). Tracks were related to literature. They were also called for by the history of exploration and geographical discovery; for a very early set, see *Karten alter Meister, 24 ausgewählte Reproduktionen* (Gotha and Leipzig, 1954), no. 6.

80. For Spruner, 1855/A, see also n. 74, above. In addition to its main sheets the *Atlas antiquus* contains more than sixty insets, which incorporate much "literary" mapping.

81. Eusebius and the Madaba Mosaic: Nebenzahl, 19, 24–25. Two medieval maps with the twelve tribes, ibid., 30, 33; R. V. Tooley and C. Bricker, *Landmarks of Mapmaking: An Illustrated Survey of Maps and Mapmakers* (Amsterdam, 1968), 105. Vesconti's map of ca. 1320, an astonishing exploit, emphasizes both biblical features and current settlement. It illustrates Marino Sanudo's Crusade-oriented *Liber secretorum fidelium crucis:* Campbell, *Earliest Printed Maps,* 6; P. D. A. Harvey, *The History of Topographical Maps: Symbols, Pictures and Surveys* (London, 1980), 144, nos. 83, 84; J. B. Harley and David Woodward, eds., *History of*

Cartography, vol. 1, *Cartography in Prehistoric, Ancient, and Medieval Europe and the Mediterranean* (Chicago, 1987), 314 (Vesconti routinely signed and dated his works); Broc, *Géographie de la Renaissance,* 9–10. Palestine divided among twelve tribes is often combined with a map of the Exodus. The Peutinger Table: Nebenzahl, 32–33. On the Exodus track on the Hereford map (with a comprehensive account of other maps of the kind), see P. D. A. Harvey, *Mappa Mundi: The Hereford World Map* (London, 1996), 34; A. L. Moir, *The World Map in Hereford Cathedral* (Hereford, 1979), 27.

82. Delano-Smith, "Maps in Bibles," 3–5, 12; Nebenzahl, 72–77, 84–86, 94–97, 103, 105–6, 116–21, 123. On Ortelius, see Vanden Broecke, *Ortelius Atlas Maps,* 221–24, nos. 170–73 ("Terra sancta" in *Theatrum*), 232, no. 180 (the same in the *Parergon*). See also Brandmair, *Bibliographische Untersuchungen,* 89.

83. Delano-Smith, "Maps in Bibles," 9–10 (Calvin and the Garden of Eden); Delano-Smith and Ingram, *Maps in Bibles, 1500–1600,* 2–24 (as Eden rather than Paradise). Hondius, 1607, has a picture of Adam and Eve associated with his Paradise map. See also Fred Plaut, "Where Is Paradise? The Mapping of a Myth," *Map Collector* 29 (1984): 2–8.

84. For the alternative version of Paradise, see Delisle, 1764/E.

85. Not all subjects were pursued with equal zeal. The eighth-century *Commentary on the Apocalypse* by the Spanish Beatus of Liébana contains a world map showing the mission of the apostles across the world: Peter's head in Rome, Andrew's in Achaea, Thomas's in India. Described in Nebenzahl, 26–27. Maps of this apostolic diaspora are rare.

86. D. R. Stoddart, *On Geography and Its History* (Oxford, 1986), 190. Ewald Banse, *Entwicklung und Aufgabe der Geographie* (Stuttgart and Vienna, 1953), 49, expresses surprise at the imprisonment of early modern men in biblical credulity and ecclesiastical narrowness.

87. For maps of religious diversity, see ch. 3, n. 108, below. Note also the attention to this subject prescribed in the Sanson introduction to geography (see n. 42, above); and Nicolas Sanson, "Europe divisée par religions" (BN, Ge.DD.2907, 173, 174; NOT SEEN).

88. Lenglet, 1768/B, 1:284. Lenglet, 1740, vol. 3, pt. 2:27, is phrased more simply: whereas the civil history of the Middle Ages was poorly served, Sanson's maps of patriarchates and bishoprics were useful for those studying the church in the Middle Ages.

89. Mireille Pastoureau, "La France divulguée, évolution de la cartographie gravée du XVIᵉ au XVIIIᵉ siècle," in *Espace français: Vision et aménagement, XVIᵉ–XIXᵉ siècle,* exhibition catalogue, Archives nationales (Paris, 1987), 59. For a departure from French initiative, see Arn. Buchelius, *Dioecesis et ditio Traiectensis antiqua iuxta tabulas et diplomata formata* (Utrecht, 1642), cited by J. T. Bodel Nyenhuis and W. Eekhoff, *De algemeene Kaarten van de Provincie Friesland* (Leiden, 1846), 15, no. 22. Monographs on single bishoprics tend to precede collections embracing one or more lands.

90. For the subject in general, see François de Dainville, *Cartes anciennes des l'églises de France: Historique, répertoire, guide d'usage,* Bibliothèque de la Sociétée d'histoire ecclésiastique de la France (Paris, 1956), 114–15. Vialart, 1641/E; Vialart took the name "de Saint-Paul" when he joined the Theatine order. The maps appear in atlases with the date 1640 and Tavernier's name, but without an identified designer.

91. Duval, 1672/E. The balance of its title is *où sont les cartes des archeveschez et des eveschez de l'univers;* he draws a few provincial boundaries. Brion, 1766/E—an offshoot of the more comprehensive and frankly modern Brion, 1766a—is a world atlas adapted to a special purpose by using red ink to delineate provinces and underline bishoprics (the BN copy I saw was carelessly done); patriarchates are omitted.

92. Sainte-Marthe, 1656/E, vol. 4, frontispiece: "La France divisée en archeveschés,

eveschés, abbaies." Bishoprics and abbeys are marked, but not their boundaries. Nolin, 1715/E. Anville, 1732/E.

93. Dainville, *Cartes anciennes des églises de France,* 38–39; Pastoureau, *Atlas français,* 417–26 (N. Sanson, "Cartes particulières de la France"). The diocesan geography collection of Nolin, 1715/E, is commendable but too detailed for a general atlas.

94. Councils since Nicaea: Pierre Duval, "Tabula geographica locorum ubi habita sunt concilii tam generalia quam provincialia (La carte des conciles)" (Paris, 1660); see Pastoureau, *Atlas français,* 144, no. 43/113. Guillaume Sanson, "Geographia synodalis" (Paris, 1667); cited in Dainville, *Cartes anciennes des églises de France,* 115; Harms, *Themen,* 36. Hartzheim, 1758, 1762, includes sites of councils in Germany and bishoprics that were suppressed or transferred: "Episcopatuum ante primum millenarium qui vel desierunt vel in alia loca aut nomina transierunt."

95. Religious orders: Maximinus, 1649/A (Capuchins; a sixteenth-century order); Carl, 1732/E (Benedictines); Lubin, 1659/E (Augustinians); see Pastoureau, "Atlas en France avant 1700," 64. Martyrs and miraculous images: Lubin, 1660/E; see Dainville, *Cartes anciennes des églises de France,* 114 (usefully supplies the BN call numbers, Ge.FF.1141 and Imp. H.3397). Renner, 1823/E (Franciscans of Hungary), is a more modern example of the type. Guppenburg, 1657/E.

96. Jaillot, 1693/E; Rogg, 1740/E. See ch. 2, nn. 116–20, below.

97. Petri, 1858/E; about 2 × 3 feet. Dedication to Pius IX: "queste carte . . . presentano a colpo d'occhio le conqueste mirabili dell'Evangelo per le cure incessanti del Pontificato, la grandezza della Chiesa e del sacro suo regno, e il campo che resta ancora aperto a sui trionfi."

98. Speed, 1627b. For "Neustria" in an Ortelius map, see n. 48, above; in a map of Louis XIV's time (Duval), see ch. 2, n. 61, below.

CHAPTER 2

The Middle Ages in Sixteenth- and Seventeenth-Century Maps

A few maps of "ancient geography" representing scenes from history or literature were drawn in the sixteenth century. The most ambitious of them depicted the Exodus of the Israelites from Egypt. A somewhat smaller group of maps showed scenes of modern history and could be found here and there in Ortelius's *Theatrum*. Regardless of whether ancient or modern, historical scenes were far outnumbered by maps of countries or provinces—what might be called strict, timeless geography. In the seventeenth century, specialized maps for history were drawn in larger numbers. Some of them augmented the "literary" repertory of ancient geography, both profane and sacred. On the modern side, the motivation for historical scenes was less focused, and the maps rarely penetrated atlases. Nevertheless, these maps had the advantage of numbers: they far exceeded the additions to ancient geography. Their subject matter gravitated toward the kingdoms and principalities of the various European countries. In a few lands, the "national" past, as we would say, precociously called on cartographic skill. Paralleling ancient geography, these maps were "modern." In our terminology, however, their subjects were preeminently medieval.

Bygone Kingdoms

The Anglo-Saxon Heptarchy

The exploration and recovery of Anglo-Saxon England were salient achievements of English scholarship in the sixteenth century. Men of letters gave a lasting place in the history of England to the early English. As part of these activities, the lawyer and legal scholar William Lambarde (1536–1601) published a small map in 1568 that described "England, as it was devided in the Saxones tyme into VII kyngdomes" (fig. 2).[1] The map showed the kingdoms only, with boundaries between them. Their names, the only toponyms supplied, were written both in Latin and in characters considered Anglo-Saxon, which return intermittently in later maps of the Heptarchy.[2] Lambarde's map illustrates his *Archaionomia,* an edition and translation of the ancient laws of the English.[3]

The original of the "VII kyngdomes," simple, almost crude in appearance, is a woodcut, the normal medium in northern European lands that had not yet graduated to copperplate engraving.[4] Yet the map is noteworthy for more than English reasons. Lambarde is the first mapmaker—in the modern European tradition—to re-create a scene from medieval history, "modern" history in the terminology of the time. More than a half century passed until another map of this kind came along.

Although Lambarde moved in the same circles as the antiquary William Camden (1551–1623), there is no clear link between his pioneering map and the one that was eventually included in Camden's epoch-making survey of the British Isles, *Britannia.*[5] Camden's work ran through four editions before being finally enriched, in the fifth (1600), with a few maps by the Flemish-trained William Rogers, an initiator of copperplate engraving in England. The two history maps are called "Britannia. Provincia Romanorum," and "Englalond, Anglia, Anglosaxonum Heptarchia," one for each phase of British antiquity. The second has a selection of place-names in Anglo-Saxon characters, along with a box supplying the equivalents in Roman script. For the first folio-size *Britannia* in 1607, William Hole, another English pioneer of engraving, also redrew and enlarged the Anglo-Saxon map of 1600; "Major place names have all been transliterated into so-called Anglo-Saxon script, based on a key which appears in the top left hand corner. The seven Anglo-Saxon kingdoms are marked, and their emblem—a crown topped by an array of seven scepters—appears in the cartouche."[6]

As a result of the efforts of Lambarde and Camden, England became the first European country with a portrayal, however approximate and

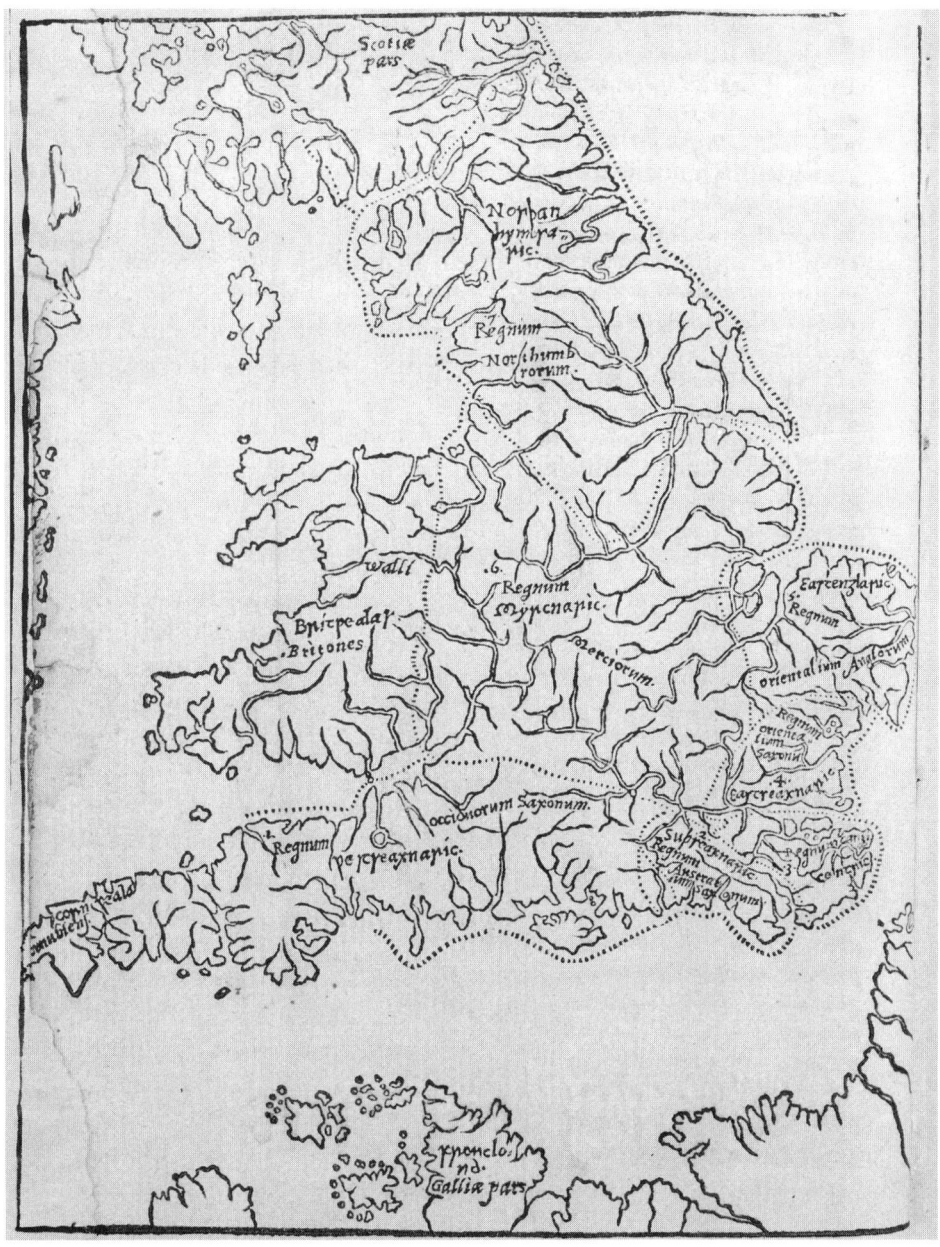

Figure 2 William Lambarde, Map of the Anglo-Saxon Heptarchy (1568). This appears to be the earliest map reconstructing a scene out of medieval history. It is at the head of Lambarde's book on the Anglo-Saxon laws, *Archaionomia, sive de priscis Anglorum legibus* (1568). The map title "England, as it was devided in the Saxones tyme into VII kyngdomes," is lacking in 1568 and comes from the reprint in John Fox's *Actes and Memorials* (1576). (Courtesy of the Beinecke Rare Book and Manuscript Library, Yale University)

incomplete, of its early political outlines. The Anglo-Saxon Heptarchy was the first feature of medieval history to be mapped.[7]

The seven kingdoms had a long, well-populated cartographic future. What came next involved mainly aesthetic and pictorial changes; the Anglo-Saxon letters went, and the topographical outlines were coupled with memorable images. The protagonist of this changed version was the historian and cartographer John Speed. Speed had originally been a tailor (and member of the corresponding London guild), then came to the attention of learned men through his private researches. By age forty-six he had won enough patronage to escape a manual occupation and engage in full-time scholarship (1598). In mapmaking Speed worked with Jodocus Hondius, a Flemish engraver of exceptional skill. Their collaboration, begun while Hondius was in London as a refugee from Ghent, extended over two decades and resulted in finished maps for Speed's *Theatre of the Empire of Great Britain* only a few months before Hondius's sudden death in 1612.[8]

With Hondius to carry out his designs, Speed forsook the pictorial modesty of Lambarde and Camden and produced a magnificently ornamental map of the Heptarchy, in which the seven kingdoms, trimmed with heraldic shields, are flanked by fourteen miniatures of early English history. The seven-sceptered crown, emblem of the Heptarchic monarch, adorns the cartouche, as it had in Camden's *Britannia*. Many localities are marked, without regard for when they existed, and the Anglo-Saxon characters are dispensed with.[9]

Splendid in its beginnings and widely diffused by Speed's atlas of Britain, the Heptarchy map attained its apotheosis when it was reengraved at the behest of Joan Blaeu, prince of Amsterdam atlas-makers, and added to the British volume of Blaeu's renowned atlas (1645). The new version derives directly from Speed/Hondius: the outline of Britain, the title, place-names, and all the figures and poses of the vignettes are copies. Nevertheless, the map was completely reworked, for aesthetic rather than topographic reasons. Rodney Shirley notes: "The unknown engraver for Jan Blaeu has re-created each of the 14 scenes as an unmistakable Dutch miniature in the dramatic style of the greater paintings of the time. When coloured in the best contemporary tradition, Blaeu's map is one of the finest available." Moreland and Bannister point out that "the copy is far finer than the original. . . . Arguably one of the finest maps published anywhere in the seventeenth century." These judgments are not exaggerated.[10]

Speed's original looks rough but not unappealing when compared with the elegance and artistry of Blaeu. The surrounding text in Blaeu's British volume is a new edition of Camden's *Britannia*. Spectacular miniatures in the

Dutch style of the day take the place of Hole's map and its Anglo-Saxon lettering; for the first time, the scholarly Camden is married to the pictorial and decorative Speed. Nothing is left of Camden's effort to re-create an Anglo-Saxon-looking landscape by choice of localities and by lettering in antique characters. Except for the outline kingdoms, one seeks in vain for anything discernibly Anglo-Saxon.[11]

The map of the Heptarchy gave unanticipated currency to obsolete kingdoms. It seemed, at least outside England, as though these hoary territories, far from being dissolved centuries ago, were integral to current English geography.[12] Atlas-makers who might have balked at mere Anglo-Saxon history assumed that the seven kingdoms were obligatory complements to the English counties.

A Heptarchy map circulated on the Continent even before the Dutch atlases multiplied Speed's design. The first version by the great Nicolas Sanson—colored boundaries on a mediocre modern base map—occurs in the collection now called *Les cinq royaumes,* discussed in chapter 1. Fourteen years later Sanson's new, improved Britain consisted of a British Isles in five sheets showing the counties and whichever of the kingdoms applies to the sheet in question. Sanson's large-scale Heptarchy transposed the Seven from history to established present-day boundaries.[13] This homogenization was in keeping with the practice of limiting coverage to a single map each for the modern and ancient periods.

Many more mapmakers depicted the Anglo-Saxon kingdoms: Hubert Jalliot in two sheets in 1673, "one of the most attractive maps of the late 17th century"; Vincenzo Coronelli, in a map dedicated to the newly deposed James II, as well as in the two-sheet Britain of his *Atlante Veneto;*[14] Pieter Schenk; Nicolas Visscher, who specified, "The Kingdom of England very precisely divided both into the seven ancient kingdoms of the Anglo-Saxons and into all the regions of today."[15] Visscher was the model for versions by J. B. Homann and Matthaeus Seutter in the next century.[16] Gilles Robert, heir to the Sanson map hoard, produced his first map of the Heptarchy in 1747, and another in two sheets a few years later. The misapprehension of the Heptarchy as current is best illustrated by the map in which Robert superimposes the main roads of contemporary England on the obsolete Anglo-Saxon kingdoms.[17] Generations of Europeans may have learned that England still had seven component realms.

British cartographers confined themselves to modern regional divisions. The exception is the information-packed wall map of England by Christopher Browne (1700). Incorporating almost everything that could be plotted, Browne's map also finds room for the merely historic Heptarchy.[18]

Some post-Camden maps of the seven kingdoms stand closer to history than those examined thus far. In a small-sized but well-furnished atlas, the ex-soldier Allain Manesson-Mallet (1630–1706) includes a few unusual items, such as reconstructions of the early kingdoms of Burgundy. His Heptarchy, following established models, portrays early England without including the later counties. The same holds for the Dutch publisher François Halma. Of the two maps of England in his pocket version of Sanson's *Cartes générales,* one "describes the Anglo-Saxon divisions of England without county boundaries (1700)." Halma later issued a map of English ecclesiastical divisions as part of a *Geographia sacra.* The Châtelain *Atlas historique* lives up to its name by offering a scattering of unprecedented scenes. The commonplace Heptarchy, though not overlooked, has modest dimensions as a mere appendage to Roman Britain. Cellarius and Köhler tried to be pioneers in medieval geography, but their versions of early England lack originality and fresh scholarship.[19]

All these would-be maps for history, except Manesson-Mallet's, might have been based on a much improved model, if the compilers had known of it. In the last decade of the seventeenth century, a new design for a map of Anglo-Saxon England gestated in a scholarly setting reminiscent of the one in which the first was born. After the Stuart kings were restored, Oxford witnessed a second blossoming of Anglo-Saxon studies, first at Lincoln College, then at Queen's, where Edmund Gibson (1670–1748) was a dynamic enthusiast. At the age of twenty-two he produced an estimable edition of the Anglo-Saxon Chronicle. Tipped into it is a newly compiled "Map accurately showing places mentioned in the Saxon Chronicle." Attention to the Heptarchy is limited, since its kingdoms had largely foundered by the time the chronicle began to be written. Gibson carefully analyzed the place-names in the chronicle, plotted them on a map, and transcribed them in Anglo-Saxon letters; *f* for *forsan,* "perhaps," indicates that the location of certain ancient places is doubtful. Here, none too soon, was a return to genuine scholarship.[20]

In time, Edmund Gibson was named bishop of Lincoln, then transferred to London. He earned lasting fame for assembling the still authoritative collection of Anglican canon law. Not long after his beginner's triumph, he planned to compile a larger map showing all old place-names in Anglo-Saxon texts. In the meantime, he organized a new, enlarged edition of Camden's *Britannia.* Only just approaching twenty-five, Gibson ably piloted this cooperative enterprise and enlisted Samuel Pepys among others as a collaborator. Responsibility for most maps was entrusted to the professional Robert Morden.[21] "Britannia Saxonica" in the 1695 edition was an enlarged, slightly simplified copy of Gibson's chronicle map, newly engraved by John Sturt. In a year that saw another bookseller dusting off Speed's almost centenarian portrayal of the Heptarchy in readiness for one more printing, Gibson's map of Saxon Britain was a desirable new point of departure.[22]

The main line of Heptarchy maps in the eighteenth century were in the tradition of Speed and the major atlases; Gibson's scholarly version remained peripheral. It occurs in an untitled early-eighteenth-century English atlas in Paris (possibly identical to the 1695 Camden) but not in one of 1720–31. Much reduced and simplified, it illustrates the *Histoire d'Angleterre* of Paul Rapin-Thoyras, a Huguenot exile from France.[23] Gibson's second edition of Camden appeared in 1722, but after this, eighteenth-century Anglo-Saxon studies may be said to have flagged. The chronicle map of 1692 makes its last notable outing at the end of the century in John Andrews's *Historical Atlas of England*. Along with the features it always had, the map acquires a track marking the invasion route of William, duke of Normandy. A complement is also there, under the title, "England divided according to the West Saxon, Danish and Mercian laws." At appropriate points, the names *Denelage, Merchenlage,* and *Westsaxenlage* (Danish law, Mercian law, West Saxon law) are lettered over a background consisting of the usual modern counties. This delimitation by "laws" was based on a late Anglo-Saxon document.[24]

As the Heptarchy or as merely Saxon, England outstrips any continental land in generating out of its early medieval past a widely accepted and reproduced subject for a map. Maps of Anglo-Saxon England regularly graced atlases, some narrowly English, others more broadly based, few expressly historical. The rare versions with place-names in a nonstandard alphabet were posters providing not only guides to landscape but also rudimentary instruction in a kind of Old English. From Camden's time, a portrayal of the Heptarchy was a natural complement to maps of Roman Britain descended from Ptolemy. Thanks to Edmund Gibson, at least one alternative "Saxon England" became available. It had the advantage of applying to the centuries of Wessex ascendancy as well as to those near the invasions from the Continent. John Andrews's version on the eve of the nineteenth century is simply a stepping stone in a by then very long tradition of maps of the Anglo-Saxon period. Modern narrative histories of England are sometimes thought to incline to the Normans and neglect the Anglo-Saxons. Thanks to Lambarde, Camden, and Speed, geographical depictions do not.

Where Is Austrasia?

It is not entirely clear which country's cartography deserves attention next. The contest is between Austria and France, but neither case is straightforward. Austria's claim to have given rise to maps of medieval subjects, and done so even earlier than Lambarde, has already been dealt with in connection with Wolfgang Lazius (who will be heard of again). Where France is concerned, medieval names as well as ancient ones were applied to some of its modern territories. Routinely, names from ancient geography were applied to modern lands without any intention that the maps in question should be accepted as re-creations of the past. Gaul stood for France, Pannonia for Hungary, Dacia for Denmark as well as Transylvania. The evidence needs examination, if only as another look at how maps became historical.

Le Long's *Bibliothèque historique de la France* (1768–78) attributes historicity to a series of maps of kingdoms that once occupied the territory of modern France. Hauber and Lenglet Dufresnoy do likewise. Mercator, Tavernier, and Duval are the main mapmakers involved.[25]

The name "Austrasia" occurs more often among these kingdoms and more widely than the others. Christophe Tassin, a Parisian atlas-maker, shows Austrasia, not explicitly called a kingdom, among his (nonhistorical) *Cartes générales des royaumes et provinces de la haute et basse Allemagne,* and two earlier atlases from Cologne do much the same. All three suggest that the term "Austrasia," whatever its past, was currently applied to lands comprising some or all of eastern Gaul outside the boundaries of the French kingdom of the time. It was a nonstandard name, probably old, for lands otherwise well known and usually mapped as Lorraine or Alsace or something else.[26] One additional reference to Austrasia is clearly different. Between 1545 and 1561, Wolfgang Lazius in his *Typi chorographici provinciarum Austriae* and in other work twice drew maps focusing on the Upper Rhine and, in a context of history, labeled them *Austrasia.* Royal portraits are included as an unequivocal sign that the cartographer wished to conjure up the early Frankish kingdom called Austrasia.

Austrasia is a familiar denizen of Frankish history. The root word "Auster" means the east and occurs, in reference to eastern districts, among the early Lombards and Saxons as well as the Franks. The Kingdom of Austrasia (with varied forms, such as "Auster") is first attested under that name, rather than that of a reigning king, in the late sixth century. It was part of the first, or Merovingian, Frankish kingdom and was distinct from Neustria, the western realm centered on Paris. One theme of Merovingian history, now somewhat muted, is the replacement in the seventh century of a Neustrian ascendancy by an Austrasian one, from which the second Frankish dynasty, the Carolingians, originated. After the dynasties changed (751), usage of "Austrasia" tended to be inconsistent and blurry, at least for a time. New eastern regions of the Carolingian dominions could be and were called "Austrasia," notably the lands of Frankish colonization beyond the Rhine. New Austrasia was only an extension of the old, which did not wholly lose its name, especially after 843, when the Carolingian Empire was divided into three pieces.[27] Another expansion of the Carolingian lands, from somewhat before 800, was a frontier district, or march, to the east of Bavaria in the direction of the future Hungary. The name applied to it was not "Austrasia"; from the late tenth century, not before, the region is attested as *Österreich*, "Austria," the name familiar to us.[28]

Aimoin of Fleury (ca. 1000), whom early scholars cherished as an authority on the Merovingians, broadly sketched the primitive layout of Francia: "When the Franks [originally] seized all these provinces, they called Austrasia the region that extends northward and lies between the Meuse and the Rhine, and Neustria the one that extends from the Meuse as far as the Loire." The seventeenth-century Jesuit Philippe Briet was more guarded: "It is not easy to determine the boundaries of this Austrasia, since they were sometimes curbed and sometimes enlarged." The past extent of this old province was uncertain, but its present extent was increasingly clear. From the late Middle Ages to the French Revolution and perhaps to the present, "Austrasia" served as the "ancient" name for the land commonly called Lorraine. In the Carolingian partition of 843, Charlemagne's senior grandson, Lothar I, obtained the "Middle" kingdom. Its northern third, the share inherited by his son Lothar II, early assumed the name of its second king and became Lotharingia. That territory included most of the former Austrasia.[29]

How soon and under what circumstances Austrasia came to be used as the old or poetic name for Lorraine needs investigation, but book titles from the sixteenth century to the end of the ancien régime confirm the existence of this hierarchy of names:

> Symphorien Champier, *Histoire des Royaumes d'Austrasie, ou Francie orientale, dite à present Lorraine* (Nancy, 1510)
>
> Richard de Wassebourg, *Antiquités du Royaume d'Austrasie et de Lorraine, jusqu'à François I, Roy de France* (Paris, 1549)
>
> Georges Aulbery, *Histoire de . . . saint Sigibert, roy d'Austrasie . . . contenant plusieurs singularitez du duché et de la ville de Nancy, capitale de Lorraine* (Nancy, 1616)
>
> Jean Sauvage, *Le Zodiaque sacré du grand soleil d'Austrasie, ou la Vie et mort heureuse de Henry II le Débonnaire, duc de Lorraine* (Nancy, 1626)
>
> Charles Hersent, *De la Souveraineté du roy à Mets . . . qui estoient de l'ancien royaume d'Austrasie ou Lorraine* (Paris, 1632)
>
> Du Bosc de Montandré, *Suite historique . . . où se voit l'établissement du Royaume d'Austrasie, son changement de nom en celui de Lorraine* (Paris, 1662)

Heinrich Vagedes (†1698), *Regnum Austrasiae seu Lotharingiae* (1682)

Affiches de Metz, renamed *Affiches d'Austrasie* (1766), then *de Lorraine* (1769)

Dom Jean François, *Vocabulaire austrasien* (Metz, 1773).[30]

The learned term "Austrasia" took its place alongside Noricum/Austria, Britannia/England, Pannonia/Hungary, and the other widely known "primitive" place-names. It differed mainly by being rooted in Frankish, rather than classical, antiquity.[31]

The association of Austrasia and Lorraine was not merely theoretical. In 1552, the French king Henry II "loudly announc[ed] the desire to take back the Kingdom of Austrasia, inheritance of the kings of France, and to press as far as the Rhine"; he allied with the Protestant princes of Germany and marched east with his army. This *voyage d'Austrasie* (as French historians still call the episode) netted him the three Lorraine bishoprics of Metz, Toul, and Verdun. In a duchy whose capital was Nancy, Austrasian identity allowed Metz to take pride in having been the capital under the Merovingians.[32]

French publicists were able to build Austrasia into their territorial claims because the name was not kindled from cold ashes but had a spirited, neutral, bookish life of its own. Authors and cartographers used it casually, without polemical intent. Moréri's historical dictionary (1681) mentions "the Lorraine of today which Latin authors sometimes call Austrasia"; acknowledging that Austrasia was a *pays d'Allemagne* within the existing Romano-Germanic Empire, the compiler innocently added "or rather of France (*ou plutôt de France*)," meaning "Francia," the Frankish ancestor to current France and neighboring lands. Although Tavernier's map of 1642 refers to Austrasia as the true inheritance of the French Crown, the lands given that name are not the large Frankish Austrasia but the modern Lorraine and its environs. A century later, Gilles Robert shows Germany in Carolingian and Ottonian times with "Austrasie ou R[oyau]me de Lorraine" in a suitable quarter.[33] Austrasia as a geographical name was well rooted regardless of what rights over it the kings of France might have had.

Those invoking Austrasia normally referred not to a historical entity, Merovingian or Carolingian, but to an existing territory, either limited to Lorraine or encompassing Lorraine and its immediate environs. Hauber cites Tavernier in the context of a map of Austrasia he was himself preparing and chides him for building ancient frontiers out of current ones. Hauber's own efforts, apparently never published, sought historical precision but by his own account seem to have focused exclusively on late Carolingian times: "I myself have in preparation a map of the *Regnum Lothariense* and the Kingdom of Austrasia in which the condition of this realm is displayed as it was initially, namely, a [territorial] portion of the Emperor Lothar, then also as it later was under his sons Lothar the Younger and Charles [of Provence], and after the latter's death as divided between the German king Louis and the Frankish, Charles the Bald, until under Otto it finally came in its entirety into the German Empire and developed into a twofold duchy; [in addition to these] the boundary lines of the main divisions among the sons of Louis the Pious."[34] While deploring Tavernier's roughness, Hauber agreed with him that the lands to be mapped centered on modern Lorraine. Disregarding the Merovingian subkingdom, he meant to show the changes undergone by Lotharingia/Lorraine since the ninth century. All the early maps labeled Austrasia—Lazius's excluded, as we shall see—fall within these specifications. The same is shown also by Bertius's map of the Carolingian Empire and d'Anville's eclectic portrayal of an "Intermediary Age between Ancient and Modern Geography."[35] Such maps document the history of a name; they do not focus on the Merovingian kingdom that our manuals define as Austrasia (often with myopic exclusiveness). Like the miscellaneous books just listed, the maps maximize the continuity between medieval Lorraine and its Frankish past and, in this way, illustrate the definition given to "Austrasia" in early modern times. There is some history in these portrayals, but the aim is, in general, to bridge and smooth vicissitudes rather than to commemorate them. Early France is farther from being depicted here than early England is in maps of the Heptarchy.

Austrasia Foreshadows Habsburg Austria

Wolfgang Lazius (1514–65) has already been mentioned. His Austrasia is Merovingian, and his intentions are to invoke a definite past. Lazius has an honorable place in the history of early cartography. Styled *medicus et historicus,* he was a physician and professor in the medical faculty at the University of Vienna, eight times dean of medicine, twice rector of the university. He was also a tireless humanist and a collector of coins and other antiquities. The Habsburg emperor Ferdinand I, besides naming Lazius historiographer royal, made him a privy councilor and otherwise rewarded him well for his researches and publications. Lazius assembled a notable library, traveled widely in search of records, and shortened his days by overwork; he died at fifty-one. He was much admired in life, especially in his homeland. Major scholars in the next generation, however—including Scaliger, Lipsius, and Vossius—thought poorly of the critical faculties displayed in his writings.[36]

Though richly rewarded and even ennobled by a grateful court, Lazius did not shrink from craftsmen's work. Austria lacked professional artists and engravers; undaunted, Lazius personally drew the illustrations for his books and even turned them into woodcuts and etchings for the printer's press. He attained proficiency as a draftsman and engraver without ever becoming totally expert. Lazius etched maps in steel, applying the technique used in chasing armor, and relied on type punches for lettering. The results, while perhaps not technically perfect, are nevertheless impressive and admirable.[37] Lazius goes far toward exemplifying the versatility prized in Renaissance men. The maps of Austria and of Hungary in the *Theatrum* of Ortelius are by him; he enjoys the esteem that comes from being listed in the Ortelian catalogue of cartographers. A few years after the original *Theatrum,* three more of his maps were welcomed into a supplement. The Lazius map of Austria continued to be reproduced by Hondius and Blaeu; an atlas by Covens and Mortier projected it into the eighteenth century. Lazius's handiwork can even be discerned in the map of Hungary of the Homann and Seutter collections.[38]

The eleven-map *Typi chorographici provinciarum Austriae* was his most important piece of cartography. Published in 1561, the *Typi* was fifteen years in preparation and had limited circulation. The collection was a chorography with historical content, like Camden's of Britain. Lazius, as befitted the historiographer royal, focused on the Habsburgs and their current possessions.[39] The work has documentary value for the time when Lazius composed it.

Lazius "did not limit his cartographic activity to the old Duchy of Austria but wished to portray all districts that . . . stood then under the scepter of the Habsburgs or were associated by descent or other historical relationships with the Habsburg house."[40] Each title of the *Typi* without exception spells out a dynastic connection, however faint: "Austria over the Enns by whose acquisition the distinguished Habsburg house won the archducal title"; "though not subject to the distinguished Habsburg family," Bavaria is included for the sake of the dioceses of Salzburg and Passau, which reach into Austria; "the Duchy of Carniola which defends the distinguished Austriadic family from its enemies"; "[t]he Kingdom of Austrasia, or Austria on the Rhine, instituted by the ancient kings of the Franks, from which the distinguished Habsburg family traces its origins"; "the County of Gorizia, from which the distinguished Austriadic house obtains many types of excellent wine."[41]

All Lazius's writings, theological items possibly excepted, are oriented to his patrons. In *De gentium aliquot migrationibus,* a major work, he turned the lands accumulated somewhat haphazardly by the dynasty he served into a panorama of ancient peoples marching from diverse origins and through diverse histories to association within the Habsburg fold.[42] The title page of the *Typi* is designed to show how his entire output is subordinated to a future *Commentarii rerum Austriacarum,* incomplete at his death (fig. 3). An early

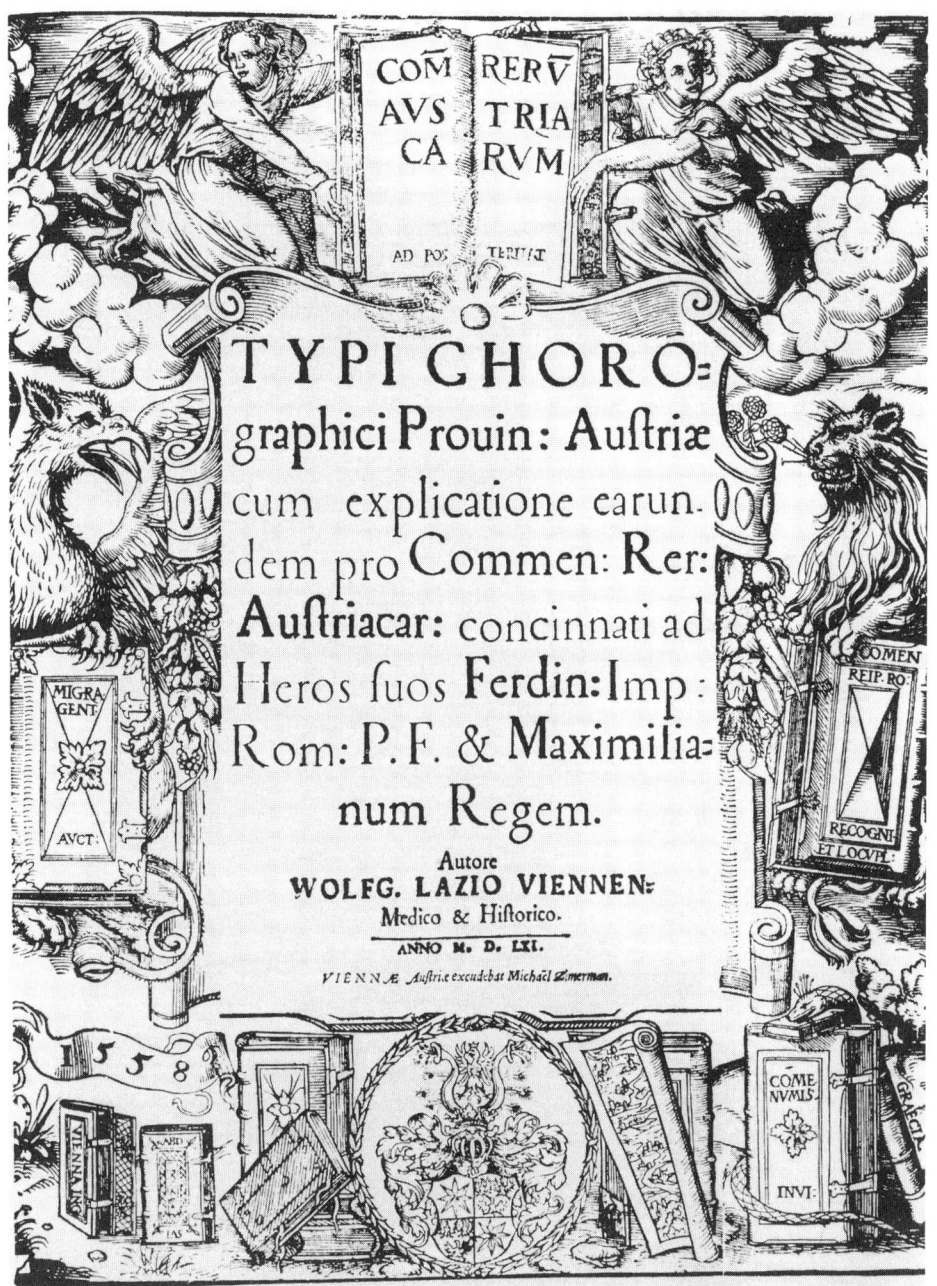

Figure 3 Wolfgang Lazius, Cover Page to the *Typi chorographici provinciarum Austriae* (1561). Lazius envisaged all his writings, including the *Typi chorographici*, as building up to a monumental *Commentarii rerum Austriacarum,* placed here in the topmost margin. Lazius proceeded as a skilled panegyricist, availing himself of historical information but avoiding the independence and detachment that historians traditionally claim. (Courtesy of the Map Collection, Yale University Library)

treatise of his on Roman government in conquered provinces is rendered on the *Typi* title page, balancing a depiction of his account of barbarian wanderings (*Migrationes gentium*). Even his *Commentarii rerum Grecorum,* also decorated with maps, was composed in the expectation that the "most serene Austrian house" would liberate Greece from the Turks, who cruelly oppressed it.[43] In maps as in written discourse, Lazius portrayed a past advancing, guided by destiny, toward the Habsburg present. This inclination, mutatis mutandis, was hardly unique with him. What unsettles us, but suits his epoch, is that he attached this ascending track to a dynasty rather than a country.[44]

Whether Lazius's map of Austrasia is more historical than the ones from Cologne and France just examined matters less than the claims that have been made for the *Typi:* that the work as a whole was a specialized historical collection. Lazius emphatically evokes the past. The first Frankish kings are not only traced through their genealogy but also pictured in an etching recycled from Lazius's *Migrationes gentium;* Charlemagne is featured as founder of a new "eastern kingdom," or improved Austrasia; the Eastern March comes next, bestowed by Otto I on the Babenberg counts; and, along the way, the Habsburgs are continually recalled as the highlight of all this past. The lands exhibited by the *Typi* maps are of the present day; the leaves dividing sections show images of contemporary inhabitants.[45] Five provinces are called by their contemporary names only, three more by decoratively "ancient" names. Lazius's maps are predominantly pictorial and include scattered historical notations, some even about Roman times. Similar notes occur in Camden's *Britannia* and Ortelius's *Theatrum.* Miscellaneous, as distinct from systematic, references to the past are not remarkable in comments about a district. Lazius, who generally provides no more than this, has as much or as little claim to having the *Typi* called historical as Camden and Ortelius do for their works.[46]

Lazius's chorography emphasizes the time of writing, as Camden would, but he meant unmistakably to dramatize the anciently entwined destiny of Austria and the Habsburg dynasty. This intention becomes particularly clear if his analytical table of contents (*Elenchus operis*) is coupled with the titles that appear either over the first four maps or in the commentaries to them:

1.   "Kingdom of Austrasia, or of Austria on the Rhine, instituted by the ancient kings of the Franks, from which the distinguished Habsburg family traces its beginnings." This section opens with a picture of the Merovingian *Rex Austriae* in his oxcart accompanied by a *maior domus Francorum.* There is also a programmatic text coupling the first with the second map: "About the double Austria, of which the one on the Rhine is the older,

whereas the latter extended from the Danube to the Italian Alps and the realms of the Hun [i.e., Hungarians]. In both of which the Habsburg princes gained glory."[47]

2. "Kingdom of the East or of new Austria placed by the guidance of Charlemagne on the Noric-Pannonian frontier, in which the distinguished Habsburg stock has obtained increase of fortune." A modern commentator explains, "According to this [inscription], the map shows the East March founded by Charlemagne between the Enns and the Raab, on whose territory under the Habsburgs there blossomed a new *Austria* (in contrast to the Frankish kingdom of *Auster* or *Austrasia*)."[48]

3. "The Ostmark placed by the Ottos in the charge of the Babenberg counts, with Saint Liutpold as shining intercessor."[49]

4. "The Bavarian March, by whose addition the Duchy of Austria took form."[50]

Lazius designed a one-sheet map in 1545 whose two side-by-side panels anticipate the opening maps of the *Typi*. The left panel is topped at the center by a heraldic shield with the legend "Austrasia dives"; the right is entitled "Archiducatus Austrie felix"; a center post dividing the halves is decorated from the bottom up with effigies of Clovis, first Christian king of the Franks, and eight early Frankish kings of Austrasia. Lazius was attached to the conception of "dual Austria," old and new.[51]

The presentation of the four opening maps of the *Typi* contains history, quite a lot of it, but of the kind associated with eulogy and homage. Much of the old kingdom, as well as Metz, its capital, is shown in Lazius's "Austrasia"; nevertheless, the map is better suited to portray "Austria on the Rhine" than the Austrasia of French and Lorraine cartographers. Long after the Merovingians (to whom the Habsburgs had no connection), this Rhenish land was intimately bound to the Habsburgs old and new: the family became prominent in eleventh-century upper Alsace, and it continued in Lazius's time to have extensive upper Rhenish possessions. For all the talk of early Franks, this is a modern map of a sector of the Habsburg dominions.[52] The same holds even more clearly for the map spoken of in connection with the Ottonian Ostmark, with its attention to the modern road net. As for Charlemagne, the connection suggested between him and Habsburg Austria seems purely topographical: behold the march he founded in which the Habsburgs, who came along four centuries later, have flourished.[53] The progression makes sense, is even admirable, as a celebration of the "Austriadic house"; the whole fits admirably within the design Lazius traced. As plain history, it lacks coherence, especially of a chronological kind, and presents little more than a set of unrelated coincidences. The four maps that seem historical are so closely bound to Lazius's argument that none may be carried over to anyone else's history.

Alsace-in-Austrasia had no tie to Carolingian and Ottonian eastern marches; these marches were irrelevant to Habsburg beginnings on the Upper Rhine; and none of the foregoing anticipated the thirteenth-century acquisition of the Duchy of Austria by Rudolf I. Lazius is a virtuoso in contriving to show that such connections existed. As in his *Gentium migrationes,* he did not reconstruct the past so much as adapt historical happenings to the glorification of his dynasty. The goal—commendable in eulogy—was to cheat time. Lazius's gifts for learned panegyric and for mapmaking are beyond question. They can be admired and applauded without classifying the *Typi* as a historical atlas.[54]

No matter how many authors identified Austrasia with Lorraine, no unanimity was attained about its territorial composition. Lazius went his way; so did Briet; and so did a pioneering work out of Austria that will be encountered among eighteenth-century productions.[55] Modern reference works rarely identify more than one of the possible Austrasias and are wholly silent about its grandfatherhood of Lorraine. It comes as no surprise that confusion reigned at an earlier time.

The Rest of France

Guillaume Delisle, during his brief period as tutor to the young King Louis XV (1717), twice drew Austrasia and its counterpart Neustria in an explicit setting of Merovingian history. That practice set a pattern for the future.[56] The original procedure when dealing with early kingdoms had been to label current territories with venerable names.

Le Long's *Bibliothèque* lists Gerard Mercator as producing a map called "The Kingdom of Arles." The full map title, in the *Gallia* volume of the original Mercator atlas, is less exclusive: "Southern Aquitaine. Kingdom of Arles." The map endured as long as the Mercator collection and was not superseded.[57] Mercator gave old names to contemporary lands to which no comparably collective terms currently applied. His Kingdom of Arles is no more historical, nor meant to be, than his Aquitaine—or than Gaul itself, the announced subject of the volume.

Mercator and others had at least one reason to remember the long defunct Arles. The district occurs as a kingdom in the *Parallela* of Philippe Briet and in the form "Royaume de Bourgogne et d'Arles" in Pierre Duval's four-sheet map of France. The historical Kingdom of Arles proceeded, like the nearby

Kingdom of Burgundy, from the fragmentation of the Carolingian Empire; the accepted date for its beginning is 933. It was brought into the Romano-Germanic Empire in 1032. By the 1370s, the Kingdom of France had largely swallowed Arles, whose regal status had lapsed. But it was not forgotten. In the sixteenth century an earnest polemic was initiated about the rights of the empire to this kingdom. Throughout the ancien régime, the theme of imperial rights to Arles continued to be obligatory in learned discussions of the prerogatives of the empire.[58] A map of the Kingdom of Arles showed lands that had meaning under that name for statesmen and scholars, even if not for the commoners of southeastern France. Neither Mercator's map of Arles nor those of Briet and Duval display medieval boundaries; the cartographers did not aspire to resurrect the past. They were satisfied to link the diverse dominions between the Rhône and the Alps with a place-name that had more academic significance than practical meaning.[59]

Duval's four-part map of 1671 has four entries in Le Long and was accepted by Hauber as an image of medieval France. Duval reprinted it several times. After his death, his brother-in-law, the Reverend Placide de Sainte-Hélène (1648–1734), had it reengraved by C. Insselin and reissued in an atlas of 1703–4.[60] The full map is called (in a cartouche on the lower-left sheet) "France with Its Ancient and New Boundaries," and modern France is what we are shown. Strips along the margins detail aspects of the land, such as the five cities of Limousin, with their names and the rivers they are on. One must look at the, admittedly conspicuous, titles on the upper and lower margins to find something historical: upper left, "The Kingdom of Western France, also called Neustria" (Le royaume de France occidentale dit autrement Neustrie); upper right, "The Kingdom of Eastern France, also called Austrasia, with a part of that of Neustria" (Le royaume de France orientale, dit autrement Austrasie, avec partie de celui de Neustrie); lower left, "The Kingdom of Aquitaine" (Le royaume d'Aquitaine); lower right, "The Kingdom of Burgundy and Arles, with the adjacent lands" (Le royaume de Bourgogne et d'Arles, avec les terres adjacentes). No attempt is made to delimit the entities in question or provide clarifications; these kingdoms existed at different times and had different durations. The names are simply there, looming over the general area occupied by the kingdoms in question at some past time or other.[61]

When Placide de Sainte-Hélène assembled these maps into a single, double-sheet map of France, he removed the margins containing the ancient names and left what was clearly there, namely, a large, handsome portrayal of current France and adjacent lands. Duval has already been encountered as a cartographer of classical history and literature. His most original works, according to a recent expert, are maps and small atlases celebrating the conquests of Louis XIV.[62] A similar concern for current events, rather than a learned interest in medieval history, may have informed his "France with Its Ancient and New Boundaries." The France of Louis XIV, enlarged by recent victories, still fell short of the extent of the ancient kingdoms composing the "France" of the Frankish past. Duval's map is well suited to suggest how far the borders of France still had to move, and in what directions.

One of Duval's kingdoms is "Burgundy and Arles." A preferable map of this region appears about a decade later in the small-format, five-volume *Description du monde* by Allain Manesson-Mallet (1630–1706). Manesson was a French military engineer long in the service of Spain and Portugal, where he designed and built many fortresses. On returning to France in the 1660s, he was employed as a mathematics teacher to a group of royal pages. Between 1671 and 1702, he published three multivolume works unusually profuse in illustrations. There are about 700 engravings in the geographical *Description du monde,* in which maps are sometimes repeated so as to save the reader from turning back a page for the image relevant to the text being read. Alongside the usual ancient and modern maps, Manesson includes town plans, ethnographic illustrations, ships, and famous buildings. Arabia leads into views of Medina and Mecca, as well as of the monastery of Saint Catherine at Mount Sinai. The *Description* could be labeled a "manual of general geography." The contemporary critic Pierre Bayle called it "an odd heap of a thousand things, crawling with errors and inaccuracies."[63]

Manesson has been mentioned above for a commendable map of the Saxon Heptarchy. Nevertheless, the focus of his *Description* is on the present and bypasses almost all the ancient French kingdoms that have concerned us thus far. His sudden, detailed interest in early Burgundy comes as a surprise: "Royaume de Bourgogne sous la race des Roys Bourguignons et sous la première race des roys de France"; "Royaume de Bourgogne sous la seconde race de roys de France"; and an unchanged repeat of the foregoing. Except for Lazius—a special case—no one who mapped Austrasia, Arles, or another French kingdom associated the land in question with a special set of kings. Manesson's maps lack eloquence; his successive Burgundian kingdoms include few place-names and no boundaries. Manesson uses words to supply history. The first map, his title implies, involves those early Burgundians whom the Roman government originally settled in Savoy and whose territory, enlarged westward to the Rhône and south to Provence, was conquered by the Merovingian Franks around 534. The conquest was integrated into the Frankish kingdom as one of the "three [constituent] kingdoms" (*tria regna*) that endured to the eighth century.[64] Manesson's second map concerns Burgundy under the Carolingian kings, in a much shorter time span. The second map adds little else to the first other than the district names Dauphiné, Transjurane Burgundy, and Cisjurane Burgundy.

Manesson's interest in this slice of the past may have been practical rather than antiquarian. In 1678, Franche Comté—the part of Burgundy in the empire rather than France—was awarded to Louis XIV by the Treaty of Nijmwegen and integrated into the French kingdom. Burgundy in the *Description* is grouped with lands "outside" France: Lorraine and its capital, Nancy; then Burgundy and its main city, Besançon. Manesson's maps of Burgundy fill in the past of Louis XIV's fresh acquisition, a district detached from the realm (as was believed) for eight centuries after the second Frankish dynasty "lost" it.[65]

From the late seventeenth century on, historians and geographers in France and Germany tended to be conscious of the Middle Ages. Writers on aids to history began to ask: what maps do we have of medieval geography? The answer for a long time was that the supply was inadequate, but not wholly barren. Patriarchates, church provinces, and many dioceses had been mapped; so had certain early kingdoms. Early maps of medieval territories were few and far between. The next one to be discussed is, since the maps of the Heptarchy, among the most obvious and deliberate re-creations of the past.[66]

The Empire of Charlemagne

In 1609, Pieter Berts, professor at the University of Leiden—the Petrus Bertius already encountered compiling atlases of ancient geography—delivered the funeral oration for Jacob Hermanns, a professorial colleague and old family friend better known as Jacobus Arminius, whose personal qualities Bertius greatly prized. But Arminius's ideas had aroused controversy in the Calvinist church. Ten years after his death, his teachings were conclusively condemned, and his friend Bertius was stripped of the Leiden professorship that, in one form or another, had been his since 1593. He was also denied other means of support. Bertius had, in 1618, dedicated to King Louis XIII of France a major work on ancient geography and had been honored with the title of king's cosmographer. Deprived of his livelihood by Dutch intolerance, he emigrated to France and, within a year, converted to Catholicism. His large family joined him in his new home and religion. Royal generosity toward Bertius brought him suitable employment, eventually a personal chair of mathematics at the Collège royal, now the Collège de France. In 1623, Paris printers published maps of "The Empire of Charles the Great and Neighboring Regions" in four-sheet and one-sheet versions. Dedicated to Louis XIII, the maps expressed Bertius's gratitude to his benefactor.[67]

Religious controversy had so major a part in Bertius's life up to 1619 that biographers often overlook his work with maps. Cartography ran in the family. He was the brother-in-law of both Jodocus Hondius and Pieter van der Keere, outstanding figures in map engraving, printing, and marketing. Protestants all, they had abandoned Ghent for permanent exile after Spanish forces captured it. Bertius first contributed to the map world by composing the Latin text for the enormously popular small-size atlas the *Caert thresoor* (1598), to which Bertius lent a more scholarly air in his translation, *Tabulae contractae* (1600).[68] As an alumnus of the University of Leiden and one-time travel companion of the scholarly celebrity Justus Lipsius, Bertius's historical horizons tended to focus on classical antiquity. His *Commentarii rerum Germanicarum . . . a Karolo magno ad nostra usque tempora* (1616) contains maps of ancient and modern Germany, together with maps of particular provinces and cities. The narrative, from which discussion of medieval Germany is absent, leapfrogs six centuries from the Frankish empire to the Empire-in-Circles as it stood in Bertius's time. His two-volume *Theatrum geographiae veteris,* already mentioned, was also his last work of this kind produced in the Netherlands, and his recommendation to the king of France. This collection of sources of ancient geography was complemented in Paris by a world atlas of ancient geography (1628). Bertius may have meant the latter to accompany the very brief introductory geography that he also produced in these years.[69]

The maps of Frankish-related kingdoms just examined labeled modern topography with old names; in the case of Bertius's "Empire of Charlemagne" map (fig. 4), there is no doubt that we are dealing with a history map, perhaps the first deliberate reconstruction after Lambarde's. Here and there, near the relevant spots, incidents of Carolingian history are briefly narrated, such as the destruction of a Muslim expedition against Corsica and Sardinia or Charlemagne's defensive measures against viking activity. On the spokes of a circle on the left side, we read "Frankish Names of the Winds." A Janssonius atlas at the New York Public Library (hand coloring by multiple colorists gives a distinctive character to each map) contains a copy of Bertius's map whose centerpiece is a "Western Francia" identified in bold lettering and delimited by a border west of the Rhine. A comparably vast, yellow-bordered "Eastern Francia" appears at the right. In the latter, "Eastern Francia" is modestly lettered as the equivalent of Franconia, a mere province. The lands bordered in yellow reach far to the east, encompassing Bohemia and districts to the north, as though Charlemagne's eastern border reached the Lower Vistula. "Western Francia" has its eastern edges at the Rhine and the Alps, far beyond the real limits of France in the 1620s but not wholly alien to French aspirations expressed in Bertius's lifetime. The standard of scholarship is unimpressive. The most startling trait of this side of the map is an ostensibly "internal" border running from the Rhine to the Channel at about the same points where (by a rough estimate) Louis XIII's France was divided from the Netherlands. Even to an indulgent spectator, such a line has no Carolingian relevance at all. Much the same distribution of tints occurs on the Charlemagne map in a Blaeu atlas at the Newberry Library. Its agreement with the Janssonius copy in New York is close enough to suggest that the latter is not unique or anomalous. Fortunately for Bertius's reputation, several examples of his map lack the anachronistic "Netherlands border," but they agree in placing Charlemagne's eastern borders fancifully far to the east.[70]

Other distortions invite attention. Spain is generally shown as a unit, with the Pyrenees as northern border and no trace of a Carolingian "march" in the area of modern Catalonia. The presence of Muslims is recorded, but only faintly; Christians are not much in evidence either. Bertius evidently copied a map of "Hispania," with the divisions appropriate mainly to Roman Spain. Italy, shown as a territorial entity from the Alps to Calabria, is much like "Hispania"; Charlemagne's conquest of 774 seems to be presumed, wrongly, to have encompassed all Italy. The full map extends to the Danube mouths and the Bosphorus, but what we would label as the Byzantine Empire is nowhere given that name. The colorists let fancy run free. In the Janssonius atlas just cited, three parts of Greece are picked out in contrasting tints, and a pink Byzantium-in-Europe faces a green Byzantium-in-Asia across the Straits, contradicting our idea of a united empire. Other copies of the map show greater restraint in coloring but no authorial control.

The accompanying letterpress, besides supplying a capsule Frankish history, discusses the coinage of the Franks, their language, and the extent of Charlemagne's empire. Bertius's program includes one question with political potential: "Was the Empire of Charlemagne German?" (*A sçavoir si l'Empire de Charlemagne a esté Germanique?*). Bertius marshals arguments, notably the uncontested equivalence of Franks and French, to demonstrate that this chapter of the past belonged to French history rather than to the Romano-Germanic Empire.[71]

How Charlemagne's realm related to modern France, a theme ostensibly relevant to France's eastern borders, was not topical when Bertius produced his map. The burning issues of the day were internal and religious. Louis XIII, aged twenty-two in 1623, was just starting his personal government; Richelieu had not yet entered the royal council. The Rhineland had been off the agenda for many years and would not return for many more. At other times, the question of the Germanness of the Carolingian Empire might have had inflammatory implications; Bertius, however, simply defended historical orthodoxy. As author of the *Commentarii rerum Germanicarum* of 1616 he might

Figure 4 Petrus Bertius, "Imperium Caroli Magni" (1623). Bertius's map, published in four sheets as well as one, is reproduced here from a one-sheet copy in a Hondius-Janssonius atlas of 1644. The next published Charlemagne map appeared more than a century later, in Germany (fig. 22). Despite this initial slow growth in interest, the "Empire of Charlemagne" became the scene of European medieval history most widely and frequently depicted from the nineteenth century onward. It still is. (Courtesy of the Beinecke Rare Book and Manuscript Library, Yale University)

have thought otherwise; now he was part of the crowd, in and out of France, committed to the idea that the Carolingians were French. No outcry arose when the Dutch atlas-makers decided that his Charlemagne map belonged in the volume assigned to France.[72]

Lenglet Dufresnoy, who attributes this map to Ortelius, calls it "belle et magnifique." He is right on aesthetic grounds: Bertius's map is a thing of beauty, especially in its four-sheet version. Even from the historical point of view, the map was far ahead of its time as a portrayal of a moment in the past of Europe as a whole. Hauber was familiar with Bertius's Charlemagne map, still not superseded a full century after its initial appearance, and considered it "very incomplete" (*noch sehr unvollkommen*)—mild criticism in view of what could be said if one wished to; he looked forward to Delisle's carrying out his plan to produce a new one. Better designs would see the light mere decades after Hauber: Georgisch, Hase, Rizzi-Zannoni, and others re-created the Carolingian Empire in maps unrelated to Bertius's and more attentive to history.[73] The long span in which his version stood alone is suggestive of how little sense Europeans had of a common past or desire to visualize it. Charlemagne was merely French, and too tainted by the German Empire to appeal to Frenchmen. Bertius—an exiled dissident ex-Calvinist geographer—had an understandable desire to show gratitude to his royal benefactor: this accident of fate is behind the only early commemoration of Charlemagne's "first Europe" that we have.

Mementos of a Shifting Province

"Bygone kingdoms" yield at this point to a province that is far from bygone and has a long past. Zeeland, a part of the Netherlands, is situated between Holland proper and Flanders; it consists mainly of islands forming the delta of the Scheldt and the Maas. More exposed to harsh seas than its dune-protected neighbors to the north and south, Zeeland is a battleground between nature and humans in the flooding and reclamation of living space. Its residents, like those of other water-threatened parts of the Netherlands, are at a meeting point of geography and history. They have reason to consider the ground under and around them to be a precarious environment subject to frequent change.[74] Mutability may be why maps of medieval Zeeland are plentiful in Dutch collections.

A modern account of Dutch cartography refers guardedly to early maps purporting to show Zeeland successively in 600, 1180, 1274, and 1288. This series can be augmented by a portrayal of Zeeland under the Batavians, thus pre-600, and by an inset in Ortelius's *Theatrum* showing the Scheldt mouth in about 1304.[75] These maps of Zeeland at six specific moments during the Middle Ages survive, for the most part, as single sheets drawn in the seventeenth and eighteenth centuries (they are dated), but one was printed by Ortelius, and several claim to have early sources. The monastery of Egmond near Alkmaar is mentioned several times; one map appeals to the Curia of Bruges, another to the archives of the Council of Flanders in Ghent.[76] Sources are sometimes mediated by histories, notably the mid-sixteenth-century Zeeland Chronicle of Reyzenbach. The name of Gui de Dampierre, count of Flanders (1280–1304), is often invoked; Zeeland is said to have been mapped in his time.[77] According to the history of Dutch cartography just cited, the many copies originate from historical reconstructions of the sixteenth century rather than from the Middle Ages. An earlier specialist, F. C. Wieder, posed the question whether the celebrated Jacob van Deventer was the first cartographer to map provinces of the Netherlands. In this connection Wieder surveyed the ostensibly medieval maps he had encountered. While admitting that incomplete evidence prevented certainty, he thought it likely that one or more makers of provincial maps had anticipated van Deventer.[78]

The physical survival of these maps attests to the willingness of seventeenth- and eighteenth-century patrons to pay for skilled reproductions. Did the other United Provinces receive comparable attention? None except Frisia has come to my notice. The proliferation of Zeeland maps is hard to explain. Did they meet a practical need, such as land claims, or were they merely collected as antiquities? The maps tend now to be listed chronologically, as I have listed them; catalogues, classifying by date, also do this. It is even suggested, "The series may have been produced as a form of regional historical atlas."[79] Intentions of this kind, which we intuitively assume, are not visible in the physical remains. The maps lack uniformities of design or cross-references. They do not form a recognizable series comparable to the sets by Alting and de Lamare discussed below. Only the maps for the years 600 and 1288 appeal to originals at Egmond (far from Zeeland, in north Holland). Several titles and brief explanations have chronological errors, such as the missionary Willibrord (†738/9) being placed in 600 rather than a century later, or Gui de Dampierre in the twelfth century (he died in 1304). There is no intimation of an atlas. Each map looks like a single, separate item, mainly related to other maps referring to the same historical moment, such as, for example, that of Saint Willibrord or of Gui de Dampierre.[80]

According to Fockema Andreae and van't Hoff, these maps are not derivatives of medieval originals but reconstructions. Their late date—if they are all late—is not disreputable in our context of maps for history. Neither the impulse to evoke past topography nor the apparent eagerness of later collectors to acquire the results can be scorned. What is open to question is the relative quality and rationale of the reconstructions: are these maps of Zeeland imaginative frauds, that is, deliberate counterfeits of ancient conditions, or are they scholarly or at least earnest, source-based efforts to recall the past?

Much depends on the sincerity of the source references. More of the Zeeland maps point in content toward Flanders than toward Holland. Derivations from Bruges and Ghent, which occur only twice, are more believable— though hard to verify—than originals from Egmond. The counts of Holland, who till 1323 fought with Flanders over Zeeland, had once been especially close to the abbey of Egmond, treasure house of the comital family. That intimacy ended in the twelfth century. Egmond was gutted by Calvinists in 1573.[81] Nevertheless, the chronicle of Reyzenbach, including maps invoking Egmond, dates from about 1550, when the monastery still stood. The circumstances are muddier than one would wish, and Reyzenbach is not considered trustworthy.

These maps of medieval Zeeland, an unavoidable group among early maps for history, are too poorly known to permit firm conclusions. In what sense they relate to the past remains a mystery. Whether earnest or fanciful, one or more persons thought them important enough to produce. Later on, other persons cared enough to have them copied and to collect them. A scholar with extensive knowledge of the history and geography of Zeeland would perform an estimable service by subjecting them to a comprehensive study.

Southern Netherlanders

The term "Southern Netherlands" designates the provinces of the Low Country that Spain held back in the Catholic fold, as distinct from the seven northern provinces that broke away. Several maps for history originating in the south apply to the early Middle Ages. The designers, though sharing the patriotic tendency to focus on their homeland, stand out for their interest in antiquities. Two out of four use maps to document ambitious and controversial theories about the early Frankish period. Another has maps accompany several folio tomes minutely investigating a district well supplied with sand dunes. Only one chooses a theme possibly appealing to a wide public.

The incomplete life work of the Jesuit Jacques Malbrancq (1579–1653) was a three-volume treatise *On the Morini and Morinian Affairs* (1639–52?). His subject lay near at hand. The Morini were the ancient residents of the Channel coast from Calais to Ostend; their principal town was the present Saint-Omer, where Malbrancq was born. (Geography had had an earlier outing in

Saint-Omer; the early-twelfth-century encyclopedist Lambert had included it in his wide-ranging *Liber Floridus*.) Volume 1 of Malbrancq's treatise has a large-scale map, oriented west, of what is now the Franco-Belgian coastline; it shows "[t]he extent of the Morini under Caesar, and place-names from 800 on." Straight lines are drawn between many points (their meaning is unclear); attractively drawn trees and watercourses define boundaries. Curious dotted lines mark the sea at a point offshore reminiscent of six-mile limit markings on our maps. The same map is in volume 2, "revised to 1647"; although a few names are changed, the main difference is the decorative addition of a large crucifixion scene with Charlemagne and his sister, Gisela, as mourners. This scene is the most medieval feature of the two maps.[82]

Whereas Malbrancq worked hard but with minimal critical sense (and training), the scholars who produced the next maps engaged in learned speculations about the early Franks, the dominant people in northern Gaul at the end of the Roman period. In doing so, however, both were working outside their areas of expertise.

Olivier Wrée, or Vredius (1580–1667), gained distinction as a legal administrator in Bruges, and was its burgomaster for a time. He saved Bruges from a raid by the prince of Orange just as his term as chief magistrate began. A talented Flemish poet, he spent his last twenty years in historical research. His excellent book on the seals of the counts of Flanders (1639), and a companion volume, are still considered worthy of consultation.[83] The Franks were a sidelight among Wrée's activities, but he was deeply attached to the theories he spun about their beginnings.

Godefroid Wendelin (1580–1667) was in even more alien territory than Wrée when reflecting on the origins of early Frankish law. His training was in astronomy, in which he had done estimable work.[84] As an antiquarian Wendelin lacked the skill evident in Wrée's books on Flemish seals. Both men were more imaginative than professionally rigorous in dealing with the very early Middle Ages.

Wrée's monographs "Pagan Flanders" and "Christian Flanders" (written in 1650) were prolegomena to a never-written major history of the counts of Flanders. They were published together soon after the author's death. His map, dated 1647, is called "The First Dwelling Places of the Franks." Spread over two pages, it reaches from Frisia south to Soissons, Reims, and Metz; one area of Frankish settlement is shown along the seacoast, a second along the Rhine from Xanten to Bonn. The map was designed to document Wrée's imprudent belief that the Flemings were the ancient Gauls and that "ancient Flanders was the first France."[85]

The early Franks were mapped again two years later. In "The Salic Laws

Exemplified: Their Native Soil Established," Wendelin argued a radical (and wrong) thesis about the date of the Frankish law code—placed very early, of course. Also discussed was the location of Dispargum, a town associated with Merovingian origins (it is featured again on an eighteenth-century map of early Frankish history). Wendelin identified Dispargum as current Diest, a town seven kilometers from his birthplace, Herk de Stad. Salic law, he maintained, had to have been composed in a land where money was plentiful, a condition that, in his view, fit Belgium in the fifth century. He also had ideas about the Arborychi, a people cryptically mentioned by the Byzantine historian Procopius; Wendelin associated them etymologically with forests (Lat., *arbores*, "trees").[86]

These notions are carried over onto a handsome, comparatively large-scale, and eye-catching map, oriented to the east and called "Second Salic Francia." (Maps oriented otherwise than north tend to look fresh and original when one expects a northern orientation.) Its perimeter extends north to Leiden, Utrecht, and Arnhem; south to Liège, Namur, and Bavay; east to Venlo and Maastricht; and west to Mons and Bergen-op-Zoom. Rivers, starting with the Sambre at lower right running up to join the Maas at Tournai, form a dramatic arc or semicircle on three sides. Very straight Roman roads run from eastern points to converge at Bavay; most place-names are given in both Latin and modern vernacular forms. In the northeast corner, on the right bank of the Rhine, one reads "First Francia" (*Francia prima*), as distinct from the "second" embodied by the map. Within or near two great patches of woodland, Arborychi of various kinds are marked. Wendelin's map is appealing until its historical content is taken into account. He and Wrée were right in one respect: the Franks were leading actors in the early Middle Ages; to be fascinated by them was no mistake.

The fourth map associated with the southern Netherlands has a very different context from the others. Like several exiled mapmakers we have met, Antoine Sanders (1586–1664) came from a Ghent family, the difference being that this Ghent native remained Catholic and even took holy orders. A priest and diligent man of letters, long striving to achieve success, he became a distant successor of the humanist Flavio Biondo and his influential chorographic survey, *Italia illustrata* (1453). In 1641, Sanders published the first of the two folio volumes of *Flandria illustrata sive descriptio comitatus istius*. Its place of publication, advertised as Cologne, was in fact Amsterdam. Hendrik Hondius had contracted with Sanders for *Flandria illustrata* in 1637 but ran out of patience. After long delays and costly alterations, Hondius sold the rights to the Blaeus, who thought the work so Catholic in its taste for relics and miracles that they omitted their name and substituted a Catholic city for the true place of publication.[87]

Sanders's text for *Flandria illustrata* is an unedifying patchwork of quotations. The value of the work, especially for archaeology, stems from a large complement of plans and views of cities, churches, and castles of contemporary Flanders, a feature derived from current Dutch models rather than Biondo. The glory Sanders has won was earned by his unending, often demeaning efforts to raise money to keep paying the artists responsible for this much prized pictorial record.[88]

Maps for history were little to Sanders's taste. Nevertheless, volume 1 contains a "New Geographic Map of Ancient Flanders as It Was . . . under Baldwin Iron (Arm) and Judith." Signed by Nicasius Fabius and adorned with no fewer than three cartouches—title, designer, acknowledgment of sources—it has the beauty and elegance associated with a Hondius product. Fabius was a canon of the cathedral of Tournai and famed as a local antiquarian. His map portrays the beginnings of the medieval county of Flanders. Judith was the daughter of Charles the Bald (reigned 840–877). Twice widowed by kings of Wessex when still very young, she eloped in 860 with Baldwin, count of several Flemish *pagi*. King Charles proclaimed that Judith was a "stolen widow," but after striving for several years to bring Baldwin to justice, he was reconciled with the pair and enlarged Baldwin's command (864). It was the starting point for a major principality of the French kingdom, perhaps the earliest to take form and one of the very few that modern France failed to attach directly to its Crown.[89] Fabius's map mainly shows the boundaries of Flanders during the founder's time.

Fabius's ninth-century subject was unequivocally medieval; except for him, these maps by southern Netherlanders might be said to pertain to ancient geography rather than to medieval. The dominant features in the learned ones come from late Roman sources; even Frankish history before Clovis belongs, technically, to the closing decades of the western empire rather than after its end. The authors, however, were all engaged in studies and arguments that looked to the medieval future. Sanders and Fabius, at least, were already there. Their map of Flanders under Baldwin and Judith stands alongside those of the Heptarchy and Bertius's Empire of Charlemagne as a pioneering portrayal of a medieval moment—provincial perhaps but nevertheless detailing events of significance in the short and the long run.

One more map fits into this section by date and subject, though only approximately. It is the work of Hermann Ewich, an obscure figure, associated with the town of Wesel, on the Rhine in the German Duchy of Cleves, not far

from the Dutch border. He published a history of Wesel in 1668. Cartography
does not specifically connect him with this locality or any other. His name
appears on a map called "The Ancient Fatherland between the Emperors
Julius [Caesar] and Charles the Great" ("Patria antiqua inter Iulium et Carolum Magnum, Caesares"). In choosing this subject Ewich takes the daring
step of bridging the ancient and modern eras. First seen in a Janssonius atlas
of 1652 and engraved by the Amsterdamer Salomon Savery, it was well liked
by compilers of atlases and often reproduced.[90]

Ewich's "Patria antiqua" catches the eye by being oriented west (not rare
in maps of the Netherlands). The Rhine, its affluents, and its mouths are the
central feature. The lands we are shown are reminiscent of a fan, extending to
Flanders at top right and to the Elbe mouth at lower left, bisected on a diagonal by the Rhine. The land on one side (west: left) of the Rhine is labeled "Part
of Gaul"; the other (east: right), "Part of Germany." The shifting contours of
the Netherlands do not detain Ewich; the Frisian Islands and Zuyder Zee have
their seventeenth-century shape. Ewich is engaged (as others were) in turning a present-day map into a map for history by means of an inscription, in
his case a catchy one. The names given most prominence identify the tribal
districts of the first century of Roman rule, those listed in Tacitus and similar
sources. The medieval period intrudes in little else than a message along the
lower edge: "Saxony in the time of Charles the Great."[91]

Antiquity is better served by Ewich than the Middle Ages. Even at that, the
main appeal of his map is an unusual layout inviting us to think seriously of
the Netherlands as the Rhine delta.[92] The southern Netherlanders with whom
I have grouped him here (owing to the subject of his map) could disregard the
physical contrast between the ancient and modern Lowlands in view of their
themes; Ewich should not have. It would be interesting to know what motivated his map, or what subtext underlies the words "ancient fatherland."

POSTER MAPS: ENGLISH BATTLES AND FAR-FLUNG FRENCH ROYALTY

John Speed's influential interpretation of the Anglo-Saxon Heptarchy was not
his sole contribution to maps depicting the English Middle Ages. The
"predilection" he shows in his depiction of the Heptarchy "for superimposing pictorial and anecdotal detail on a geographical map" is shown again in
his "The Invasions of England and Ireland with al their Civill Wars Since the
Conquest." Although Speed specified "invasions" and "civil wars" in the title,
the map displays many centuries of battles of all kinds, so that "battles map"

would be a more appropriate name (fig. 5). Almost ninety sites are marked down to 1600, each graced with a battle miniature, generously labeled by place-name and date, and consecutively numbered; sixty-four sites in England and seven in Ireland belong between 1066 and 1500 (the map omits Scotland except for a small southern part).[93] Several pages of text, keyed to the numbers, identify the contenders and give a brief summary. The formulaic but varied miniatures condense land battles as massed ranks of pikemen, whose tall weapons, sometimes crisscrossing, form a sort of roof; massed troops of this kind were routine on Italian military maps of the sixteenth century. Four of the battles Speed records feature cavalry engagements rather than pikemen, and one incident, in northern Ireland, occasions a unique scene. Occurrences off the coasts are shown somewhat differently; naval vessels, all of contemporary type, are set alongside explanations of the incidents in question.[94]

Speed's map lived on long after its century. Copies or adaptations were made in the French revolutionary era for purposes of encouragement or polemic. An unmistakable derivative was incorporated into the Lesage atlas (1801) and in at least one of its offshoots. The work is an early example of thematic mapping; no one could mistake its synchronic display for a comprehensive, let alone chronological, exposition. The ideal classification for it is as a "poster" or catalogue map.[95]

Clear references to the reigning queen show that Speed's map originated under Elizabeth I (†1603). Yet its wide dissemination stems from a late publication of Speed's, namely, *A Prospect of the Most Famous Parts of the World* (1627). The *Prospect*, a brief world atlas, augments the third edition of Speed's *Theatre of the Empire of Great Britain* (1st ed., 1611). A selection of foreign lands precedes and leads into the atlas of English counties. The "Invasions of England," inserted at the end of the world atlas (*Prospect*), and beyond its page numbering, is announced as belonging to the *Theatre* but was never fully integrated among the *Theatre* maps.[96]

The map in Speed's atlas of 1627 is the final link in a chain of "battles maps" that began in the very early 1600s. New works by Speed on this theme have recently come to light: his 1601 prototype for the *Prospect* map and a four-sheet wall hanging datable to 1603. The latter parallels a wall hanging of the same year by Hans Woutneel. Both acclaim the accession of James I and bear elaborate heraldry. There also survives at Oxford a small charred scrap showing ships and shoreline—all that remains of a sixteen-sheet map by the great cartographer John Norden, which, according to contemporaries, showed the battles fought in England from 1066 through Elizabeth's reign.

Figure 5 John Speed, "The Invasions of England and Ireland" (1601, 1627). Speed's map indicates the location of all battles in England and Ireland between the Norman Conquest and 1600. Capsule accounts of the battles are included; those for Ireland on the map face, the rest on separate sheets. Produced originally during Queen Elizabeth's reign (1601) and virtually uncirculated at the time, Speed's map came into its own twenty-five years later, newly engraved in Amsterdam by Cornelis Danckerts and included in the last of Speed's atlases, *A Prospect of the Most Famous Parts of the World* (1627, here ca. 1660). (Courtesy of the Beinecke Rare Book and Manuscript Library, Yale University)

Norden's enormous map has been destroyed and is undatable; its relationship to Speed's and Woutneel's work can no longer be puzzled out. The
maps of Speed and Woutneel have dates and distinctive traits: some refer only
to battles, others to "civil wars"; some mark battle sites by tactful tents,
whereas others show massed pikemen in conspicuous lines of battle. These
features allow us to link the Speed and Woutneel wall hangings of 1603
(battles, tents) and to differentiate them from the Speed single-sheet maps of
1601 and 1627 (civil wars, massed pikemen).

Speed's map of 1601 is the earliest datable item of the series. Its dedication
to "Oliver St John Knight" (1559–1630) and indication that Tyrone's rebellion
in Ireland was still in progress bring it into relation with an ongoing "civil
war." The next works on this theme, Speed's and Woutneel's wall-size maps
of English battles, soften the martial note and accentuate portrayals of the
British Isles. Land battles blend into the landscape.[97] Wall maps of Britain
identifying the clash of arms by means of sedate tents were suitable composi-
tions to salute the accession of the peace-loving King James, which is evidently what they were doing.[98]

Speed's "Invasions of England and Ireland" then dropped out of sight for
more than two decades. A collective portrayal of battles since 1066 would have
been superfluous in Speed's most successful atlas, *The Theatre of the Empire of
Great Britaine* (1611–12), since its county maps include plentiful representations of feats of arms. Then, surprisingly, Speed's small world atlas, the *Prospect*, revived the twenty-six-year-old battles map.[99] By then Speed was so old
that his personal participation in the *Prospect* cannot be taken for granted.

The *Prospect* version of the battles map is outstandingly handsome. Its engraver was the Amsterdamer Cornelis Danckerts, member of an eminent
mapmaking family. His version of the "Invasions of England and Ireland,"
copied from Speed's prototype and perhaps Danckerts's first production, easily surpasses in beauty every other battles map considered here. Speed's "Invasions of England and Ireland," long sidelined, its plate of 1601 gone, was resurrected under promising auspices.[100] Twenty years of peace under James I
were coming to an end. A change of course in foreign policy, prepared for several years, was completed by 1624. Sporadic armed intervention on the Continent took place and intensified under Charles I (1625). An atlas-maker could
anticipate that the public would have a taste for foreign lands and martial exploits.[101] The text accompanying Speed's rejuvenated map sketched British
exploits abroad in a folio column of text. The reorientation that commended
the world maps of Speed's *Prospect* also favored a wholly new edition of the
"Invasions of England and Ireland." Speed's map of 1601, however topical
when first drawn, was also the imposing, atlas-size display of "martial Britain"

suitable for the later 1620s. Danckerts's splendid engraving perfected an image in harmony with the times.

The closest France comes in the seventeenth century to equaling Speed's synoptic battles map is a grandly expansive survey, "Europe françoise," on a single sheet, of the lands governed at least once in their past by a member of the "royal and very illustrious family of France." No author's name is attached to this piece of patriotism, which appeared at the height of the régime of Cardinal Richelieu. The seller, Jean Boisseau, mainly called himself a map colorist but was also a publisher, notably of Bertius's "Empire de Charlemagne."[102]

Boisseau (or the nameless designer) was not committed to commemorating only lands held by members or offshoots of the royal dynasty. There is room for the commoners involved in the earliest and latest manifestations of French expansion. Attention is drawn to the Sea of Azov in about the Year of the World 3573, "from whose environs the first Frenchmen [i.e., Franks] departed . . . under the leadership of their chiefs, Suno and Pantenor"; and an inset at lower left indicates Quebec, "first dwelling of the Gentlemen of the Grande Compagnie." On the map, each applicable kingdom is outlined in green, and a framed box with the relevant name is attached to each. Medieval content is emphasized in the extract given here, in which I usually give only the first name in a list:

> Roys de Portugal de la Maison de France [= de la M. de Fr.], Alfonse Ier 1139
> Roi de Navarre [de la M. de Fr.], Philippe le Bel 1284
> Empereurs d'Allemagne [de la M. de Fr.] [Charlemagne to Conrad II]
> Rois de Hongrie [de la M. de Fr.], Charles dit le Martel 1290
> Roys de Naples et de Sicile [de la M. de Fr.], Charles de France, frère de St. Louis 1266
> Roys de Pologne [de la M. de Fr.], Charles dit Carobert 1315
> Empereurs de Constantinople [de la M. de Fr.], Pierre de Courtenay 1218
> [one box] Roys de Ierusalem et de Cypre [de la M. de Fr.] Foulques 5e Comte d'Anjou 1131 . . . Sybille 1185. Il y a plusieurs autres Princes de la mesme maison qui ont esté tiltrez Roys de Ierusalem, qui son icy obmis, pour n'avoir effectivement possedé le Royaume.[103]

In honor of the Crusades, all of Asia Minor and Syria are outlined in green, and "R[oyaume] de Ierusalem" is written alongside "Syrie," with several place-names but no attempt to provide the crusading principality with definite boundaries.

Boisseau's map is certainly synchronic; thanks to the legendary Franks out of south Russia, more than three millennia are spanned. But the relationship of the map to chronological history matters less than its function as a display of information. The same is true of Speed's "Invasions of England and Ireland." Boisseau offers no elaborate key and does not need one. The catalogue is adequately contained on the face of the map. As envisaged in "Europe françoise," all kings since the first Franks belong to a single "royal family," without subdivision into the traditional three "races" or dynasties. The roundabout connections of Peter of Courtenay and Fulk of Anjou to the Capetians are treated as if they were just as firm as the direct descent of Charles, brother to Saint Louis. Boisseau takes pains to be as comprehensive as possible. A post–World War I world map in the Putzger *Schul-Atlas* advertises all parts of the world where Germans had ever been or now were. Boisseau's "Europe françoise" illustrates an earlier moment of comparably imperial sentiment and display.[104]

An eighteenth-century commentator pointed out that medieval history was more directly relevant to the current condition of Europe than was ancient.[105] Speed's "Invasions of England and Ireland" and Boisseau's "Europe françoise," both free of any explicit reference to the Middle Ages, bore out in their maps the moral that had yet to be drawn.

OUTSIDE NORTHERN EUROPE

Toward the mid–seventeenth century, maps for history widen their range and reach from southern Italy to Armenia and Transoxania. In all but one case, however, the cartographers come from lands already heard from.

A story that may be *ben trovato* rather than true features the Neapolitan scholar Camillo Pellegrino (1598–1663). He resolved that his immense store of research material was not to survive him and made his wishes clear to his household. When he fell seriously ill and seemed to be at death's door, his servants hastened to reduce to ashes a lifetime's accumulation of notes. Pellegrino recovered and had five years in which to regret the alacrity with which he had been obeyed. The work on which Pellegrino's reputation rests, the two-part *Historia ducum Langobardorum,* was published when his notes were still intact. Part 2 starts with a map of the ancient Duchy of Benevento.[106]

Pellegrino's map of southern Italy places the Adriatic at the top edge and has the heel and toe of the peninsula form an almost precise right angle to the shank. The unusual oblong and angular design is attractive and visually arresting (it is not original with Pellegrino). Italy from the Tiber south—"Cis-Tiberine" in the author's terminology—is shown in firm and convincing outlines, with rather few place-names. Benevento is at the extreme left (north), and the Strait of Messina and a bit of Sicily are at the lower right. Pellegrino indicates a few medieval districts by name but does not draw boundaries.[107] The main part of the historical message is engraved in a large central box, of the kind already found on Ortelius's map of the Roman Empire. Pellegrino addresses the reader in tones of regret. Cis-Tiberine Italy, he says, has had one line after the other willfully drawn across it. The very extensive Duchy of Benevento that the Lombards had created was later split into the principalities of Benevento and of Salerno, from which, later still, the Campanian Principality was extracted. The Greeks—that is, the Byzantine Empire—disrupted many places originally called Calabria, then attached this name to Bruttium, and finally, after many disasters, established the two provinces (*themes*) of Calabria and Apulia. The Normans seized both Byzantine provinces and formed a Duchy of Apulia; to this they joined the principality of Campania, smaller dynastic lands, and especially the island of Sicily and turned them into a kingdom. The reader of these lines, with the map before him, is given a brief, estimable historical geography of southern Italy down to the twelfth century.[108] Pellegrino's model for a map—an eye-catching image combined with a brief, precise text—deserved imitation but reached a limited public.

For a long time, the western edge of the Byzantine Empire was situated just east of Benevento; a French mapmaker active soon after Pellegrino focused on a land to the east of Byzantium—Armenia in the sixth century. Philippe de La Ruë, personally obscure, has a modest, mainly biblical output. He was apparently connected to the Sanson enterprise, with the result that his maps had a wide circulation.[109]

La Ruë's work, "Ancient Armenia Divided into Four Parts at the Time of Justinian," is new and unexpected. It was the first depiction of any part of Justinian's empire and appears to be first, too, for the late Roman Empire and its alternative identity, early Byzantium. French atlases well into the eighteenth century reproduced this map, in which a "Great Armenia," presumably in the Persian Empire, is edged by three smaller Armenias within Justinian's Byzantium. La Ruë reveals that his sources are two laws from Justinian's codification. Special circumstances probably inspired this map, which held little appeal to ordinary map buyers. It may have been commissioned to illustrate a book.[110]

If sixth-century Armenia is exotic territory, the mapping of martyrdom throughout the Christian centuries is even farther off the beaten track. Martyrs—believers who were put to death for their faith—are the principal saints of Christianity. Intensely commemorated in the Middle Ages, they never lost their prominence in Catholic memories. Augustin Lubin (1624–95), a French member of the Augustinian order and compiler of *Geographia*

Augustiniana, undertook to map martyrdom in what he called an "Illustrated Roman Martyrology." Even though this book may be noteworthy in geography as an instance of thematic mapping, it proved a blind alley in the cult of saints. Marking the spots where martyrs were executed seems to have done little, in the eyes of the faithful, to enhance the fame of the saints in question. Lubin's small maps of martyrdom sites start in Spain and England and move eastward. It looks as though these arid illustrations never spread beyond the publication that initially housed them.[111]

The maps of La Ruë and Lubin, only marginally medieval, probably originated as chapters of Christian history and adjuncts to the Holy Land repertory. Pierre Duval, soon after them, takes us into securely medieval and thoroughly foreign territory. Duval did well with small atlases independently of his uncle, Nicolas Sanson. We have already encountered his three-part collection of 1665 called *Diverses cartes et tables pour la géographie ancienne.* More than most early publications, it sheds the constraints of ancient geography and modestly foreshadows a wider agenda.[112] Duval's third part, of maps to accompany accounts of travels, opens with "The Empire of the Saracens or of the Caliphs under Walid who Reigned in About the Year 700. Drawn from Abulfeda, Nassir-Eddin, Ulug Bey, and Other Arab Authors" (fig. 6). This broad panorama is complemented by one more map, supplying a single, unexpected detail: "The Regions beyond the Oxus, from Abulfeda Ismael, Prince of Hamah, the Year 1345."[113] At least one Arab geographer, Idrisi, had been available in translation since 1619, but Duval disregards him. The trio he mentions points clearly to translations recently published by John Greaves (†1652), sometime Savilian professor of astronomy at Oxford.[114]

Duval's very small scale portrayal of the caliphate in A.D. 700—the inscription is almost as large as the map itself—does little more than suggest the lands in question, from the Atlantic to the Indus, and provide a modest selection of Arab place-names. Sicily appears as "Sakaliyah" and the Caspian Sea as the "mer de Kilan"; some twenty names are listed to right and left of the title as a glossary with French equivalents. Neither boundaries nor other political complications are taken into account. This is a cursory geographic outline of the Muslim world, detached from its relations with Christian lands.

Figure 6 Pierre Duval, "La carte de l'empire des Sarrazins" (1665). Duval's small three-part atlas, *Diverses cartes et tables pour la géographie ancienne,* was not wholly committed to ancient geography and modestly foreshadowed a wider agenda. His third part, of maps to accompany accounts of travels, includes the one pictured here, "The Empire of the Saracens or of the Caliphs under Walid Who Reigned in About the Year 700." (Courtesy of the Beinecke Rare Book and Manuscript Library, Yale University)

LA CARTE de
L'EMPIRE DES SARRAZINS
ou
DES CALIPHES,

Sous VLII qui regnoit environ l'an 700.
Tiré d'Ulhusseda, de Nasir-Eddin, d'Uluy-Bei,
et d'autres Autheurs Arabes.
Par P. Dv Val Geographe Ordinaire du Roy.

Explication de quelques Noms Arabes.

Ammariyah, la Morée..
Athinah, Athenes..
Bugiah, Buge..
Buzantiyá que et
Constantinya, Constantinople..
Cortobach, Cordoüe..
Donkolah, Dankala.
Far Fez.
d'Frank, la Chrestienté..
Hend, l'Inde..
Magreb, la Barbarie..

Explication de quelques Noms Arabes.

makduiwah, la Macedoine..
al-Rum, la Grece.
Rumiyah magna, Rome.
Sakaliyah, la Sicile..
Seidiniah, Segelmesse..
Taniah, la Mauritanie Tingitane..
d'Torc, le Turquestan..
Trabolos in Occidente, Tripoli
de Barbarie..
Yaman, Arabie Heureuse..

Galice
ESPAGNE
AL-FRANK
MER MEDITERRANEE
Sardaigne
Corse
Mallorque
Valence
Sabr iyah
Candie
Bolgarie
ANATOLIE
PONT EVXIN
Ruslie
Comanie
ARMENIE
Caramanie
Damas
Jerusalem
Baghdad
ERAK
FARS
MER DE KILAN
Tabarstan
Mazanderan
Chuzistan
Kirman
Cherman
OMAN
ARABIE HIRIAZ Sarrasins
YAMAN
La Mecque
Mer de la Mecque
LE DESERT
MAGREB
Tropique de Cancer
Regaium Sudan in Barbaria
Alger
Tunis
Tripoli
Barca
Alexandrie
AL-TORC
Mavra Falnahr
INDE
HEND
Indus Fl.

The fragment of Abu'l-Fida's vast geography that Greaves edited concerns Khwarizm and Transoxania; Duval's map of the lands beyond the Oxus is a companion to this description. The region is fully shown, with district names, cities and towns, boundaries, and mountains in the usual pictorial style. One of the two boxes at the right margin lists the "four most delightful regions of the Mohammedans: the Soghd of Samarkand, Gautah of Damascus, Nahr Al-Ablah near Al-Basra, Shaab Bouuan in Persia." The other box comments: "It must be noted here that Abulfida counts ten more grades of longitude than does Nassir-Eddin." The better part of a century would pass until an attempt comparable to Duval's was made to draw a map of the premodern Islamic world.

The predominance of Frenchmen in all but the first of this series of maps is conspicuous; our last exotic subject is no exception. Hubert Jaillot (1632– 1712), a sculptor turned engraver, entered into association with Guillaume Sanson, son of Nicolas, with a view to keeping the great man's output attractive to the map-buying public. Jaillot is better known for enhancing the Sanson stock than for designing maps in his own right.[115] In 1692, however, he personally produced a dazzling map entitled "The Deserts of Egypt, of Thebaïd, of Arabia, of Syria, etc., on Which Are Exactly Marked the Places Inhabited by the Holy Fathers of the Desert." A faithful copy of Jaillot's "Holy Fathers of the Desert" was made, long after, by Gottfried Rogg and issued by Matthäus Seutter in 1745.Before this, the Jaillot version was given a northward orientation, labeled "Ancient Thebaïd," and coupled, in this form, with a detailed map of the so-called New Thebaïd, by which was meant the Cistercian monastery at La Trappe, refounded in 1664 on a very rigorous basis by Armand de Rancé, who had turned from a worldly abbé into an ultraascetic monk. The old and new Thebaïds were published as a pair by Nicolas de Fer in 1700, entered the atlases of Le Clerc and Châtelain, and enjoyed a wide circulation.[116]

Many early monks, including Anthony, Pachomius, Symeon Stylites, and other founding figures of Christian monasticism, are called "desert fathers." They not only exemplified the monk's discipline of austerity, asceticism, and prayer but also charted his spirituality—the pitfalls and blessings of the monastic state. Writings about these remote, eastern heroes, such as the collected *Lives of the Fathers* (*Vitae patrum*) and the works of John Cassian, were sources of inspiration to western monks throughout late antiquity and the Middle Ages and retained their vogue during the Catholic Reformation. Hermits in the desert father tradition were cherished by the pope who commissioned the Vatican Map Gallery and are conspicuous on its vaults.[117]

Jaillot's map is hard to explain unless somehow connected to Armand de

Rancé and the Trappist monastic reform (1664). Rancé, not content with the Rule of Saint Benedict even in a strict form, instituted much greater severity among his disciples: "the Trappist régime is probably the most penitential that has ever had any permanence in the Western Church." Two sixth-century desert heroes, John Climachus and Saint Dorotheus, were among de Rancé's sources of inspiration. He was at odds for much of his life with the leadership of French religious orders but had admirers at the court of Louis XIV. Because the early desert experience was a perennial model for extremes of asceticism and served to justify practices of drastic deprivation, it would not be wholly surprising if Jaillot, a prominent mapmaker, had vividly portrayed the heroic age of the founding fathers so as to evoke a past being reborn.[118]

Jaillot's design, combining features of old Holy Land maps, is arresting. The eastern end of the Mediterranean, drawn as a rectangle but in a startling diagonal direction, dominates the top, emphasized by Cyprus paralleling the shape of the sea basin; the map orientation is northwest. As a result, the Levantine coast forms the right side of a Y, the coast of Egypt and Libya the left, with the Red Sea as the descender, almost bisecting the lower part of the map. The rationale for this extraordinary disposition (originated in Mercator's "Palestine" of 1537) is that it permits maximum display of the arc of lands from Egypt to Syria settled by the desert fathers.[119]

Jaillot, indifferent to cartographic precision, designed a highly pictorial map. The main feature, other than boundaries and waters, is a series of stylized but nevertheless dramatic mountains, carefully shaded and often differentiated in height and orientation. Among them are disposed miniatures of the lives of the desert fathers, composed of tiny figures and identifying legends: Saint Anthony burying Saint Paul the Hermit, Saint Mary of Egypt found dead by Saint Zosimus, Saint Helenus sitting on a crocodile, and many more. A set of images concerns Saint Malchus; two (large) craft on the Mediterranean show Saint Hilarion causing pirates to withdraw; Arabs suspend a group of monks over a fire and suffocate them with its smoke. The map is basically a poster, meant to identify such early monastic sites as the deserts of Scete, Thebaïd, and Nitria in Egypt, and of Chalcis in Syria, and to show many scenes from the legends of the desert fathers. Jaillot transposed to early Christian monasticism the type of pictorial map long a part of Old Testament cartography.

An explicit connection between La Trappe and Jaillot's map comes in Nicolas de Fer's two related publications of 1700 or before. The map labeled "Ancient Thebaïd" is attributed to the authorship of "Religieux de la Trappe." This map, unmistakably derived from Jaillot's, was changed by the Trappist draftsmen as they copied. The Mediterranean is still angular but has regained a northward orientation—a more traditional design (though less fixed in Holy Land maps than in maps of the rest of the world). All land areas look more conventional. Margins to right and left are occupied by long "Avertissements" discussing important persons and places, such as the various desert centers of monasticism, with keys leading to marked points on the map. A didactic purpose obtrudes. Mountains, still very pictorial, are thinned out. The miniature scenes and identifications beneath them, also thinned, nevertheless remain important. Although it is impossible to be certain on merely internal grounds, the La Trappe "Thebaïd" may reprove the Jaillot prototype as being hard to decipher, and elegant rather than edifying. Another possibility is that the monk-draftsmen lacked the skill to do any better.[120]

The map by de la Salle of de Rancé's center of monastic reform—the "New Thebaïd"—is on a large scale. The abbey is shown in the midst of woods, ponds, wastes, and, at a distance, cultivated fields; no town is in sight. Finely engraved vignettes at the top show de Rancé at his writing desk gazing at a monk-apparition (Saint Bernard?) and, on the right, Saint Bernard also contemplating an apparition (a young Jesus Christ?). No monastic reform has left broader traces in cartography than this one.

The spectrum of history maps extending from Pellegrino's "Benevento" to Jaillot's "Holy Fathers of the Desert" is a promising indication that the repertory of map scenes could be widened. These were isolated efforts, however, sometimes buried in books that few would consult. None of them points toward an extensive program, let alone to the historical atlases of the future. Only their diversity anticipates maps to come.

Frisia: Tidal Flats and Historical Cartography

Friesland extends for a hundred and twenty miles from the Jutland peninsula of Denmark to Germany and the Netherlands—a string of slender islands paralleled by a line of coast and, in between, a confusion of shoals, banks, and channels. The Frisian Islands look as though they once were the continental shore of the North Sea and were torn from the mainland by the battering of the ocean. The widening estuaries of the Ems, Weser, and Eider, as well as the former Zuyder Zee, bear witness to fierce waters at work, hacking at the present shores and striving to break them, too, into islands.

The changes wrought by the sea are perceptible and memorable to those living near these threatened coasts. Twice daily, the tides uncover vast sandbanks that enlarge the land area many times over, then revert to open water a few hours later. Some of these drowned tracts, formerly settled and productive, succumbed to the sea within the memory of persons living on safe ground nearby, leaving visible traces of their former habitation. Written records log the ultimate flooding of other sandbanks, sometimes engulfing the inhabitants. The daily rise and fall of the tides expose lands and watercourses that once existed. Here, as in Dutch Zeeland, the residents live at a crossroads of geography and history.[121]

Husum, on the North Frisian coast of Schleswig, was the birthplace and lifelong home of Johannes Mejer (1606–74), a leader in early Danish cartography. Orphaned at eleven, rescued from herding livestock by a paternal relative, and trained in mathematics in Copenhagen, Mejer returned to Husum to live as best he could from his education. He taught, practiced as a consulting astrologer, published an annual almanac, and made maps. His cartography was in keeping with a local, Dutch-influenced surveying tradition, refined by what Mejer had learned in Copenhagen from pupils of Tycho Brahe.

In the 1630s, while mapping the marshes near his home, Mejer came to know Peter Sax, a Wittenberg-educated farmer and local historian from the Eidersted district just south of Husum, and was much influenced by him, notably in the graphic evocation of vanished landscapes. Mejer worked for Duke Frederick III of Gottorp in the next years, visited Holland in 1637 with his help, and carried out the duke's commission to survey and draw more than one hundred detailed local maps. The king of Denmark named him a royal mathematician in 1648, with a handsome annual retainer. Mejer was ready by then to begin issuing the collection that would be his only published monument—an atlas of the duchies of Schleswig and Holstein, finished in 1652. A number of its maps won wide circulation by being incorporated into the great Blaeu atlas of 1662. Mejer's career was only beginning. Under royal auspices he undertook a complete survey of Denmark, then a far-flung kingdom. Many traces of his activity are preserved in Copenhagen manuscripts and in a sumptuous 1940s publication, which reveal, among other things, that Mejer's excellent drawings were much coarsened when engraved. Our concern with him, however, ends with the published atlas.[122]

Mejer's collection of 1652 includes several maps with historical content, beginning with learned images forming an "ancient" prelude: first, the biblical dispersion of peoples, then ancient Germany in both northern (Scandinavian) and southern guises. These tributes to ancient geography are not related, except in spirit, to Mejer's depictions of medieval Friesland.[123] Mejer's history maps, though few, are more dynamic than any yet encountered. Their relation to present-day standards of historicity has been much discussed.

Regardless of whether they meet these standards or not, they enrich our collection of medieval scenes.

1. "Heligoland in A.D. 800, 1300, and 1649," a half page, beneath a much larger scale "Heligoland in the Year 1649"
2. "Map of Ancient North Frisia in the Year 1240: Ancient Cimbric Frisia," opposite "Map of North Friesland in the Duchy of Schleswig, in the Year 1651" (fig. 7)
3. "Northern Part of Ancient North Frisia until the Year 1240"
4. "Southern Part of Ancient North Frisia until the Year 1240"
5. Town plan, inset, "Oldenburg [Holstein] 1320," together with current "Oldenburg" (on a slightly smaller scale)
6. "Map of Ditmarschen in the Year 1559," opposite "Map of Ditmarschen in the Year 1651" [124]

In all these, Mejer invites the onlooker to compare the present to the not-very-distant past and to observe the differences that leap to the eye. The Heligoland of 1651 is a speck out of the great island of 800; that of 1300, whose outlines, though unlabeled, are clear enough, is a lamentable fraction of what was before. The years 800 and 1300 were not arbitrarily chosen for their round numbers: they were established moments of catastrophe. The same is true of 1240 in the maps of North Friesland, Mejer's own district. The storms of 1240 turned the sector called Nordstrand into an island that, in 1436, broke in two. On 11–12 October 1634, just as Mejer started surveying, a completely unexpected surge knocked the meager remnants of Nordstrand into four fragments. In Mejer's portrayal of current "North Frisia," none of the one-time sectors of Nordstrand remains habitable; it is marked as a vast submerged sandbank complete with signs and names of more than twenty hamlets with churches of their own, now underwater. Peter Sax, Mejer's historical adviser, wrote a work called *Nordstrand,* exhumed from manuscript and edited in 1910.[125]

Many more traces of losses to the sea are apparent when "North Frisia in 1240" is compared to the map of 1651: it seems as though all the underwater shoals were once dry land. Is this how Mejer traced past topography? There was no scientific method for reconstructing thirteenth-century topography. Yet if guesswork and art were involved, intelligence was too. The ancient estuary of the Eider is given a very plausible shape. Throughout, the relation of present to past is so convincingly loose that one needs to look hard to establish what it might be.

Ocean surges were not the only agents of change; Mejer's plan of fourteenth-century Oldenburg-in-Holstein illustrates another possibility. The town clearly had once been more prosperous; change for Mejer, in practice if not philosophy, was deterioration. A broad channel used to be directly accessible from the streets of Oldenburg; the waterway was now a narrow stream, and the current city had suffered accordingly.

Besides depicting economic decay, Mejer once relies on a source. Ditmarschen is the northern part of Holstein, just across the Eider from Husum and Eidersted. Mejer's two maps of the district are on the comparative pattern of "Heligoland" and "North Frisia"; the change in topography by loss of land to the sea is vividly illustrated on facing pages. The date 1559, absent from the annals of sea floods, points clearly to Mejer's informant: for the coastline of a century before, he used the woodcut *Holsatiae descriptio* (1559) of Mark Jørden (ca. 1531–95), the map incorporated in Ortelius's *Theatrum*. The bland commentary to the atlas, by Danckwerth, never mentions the sources of Mejer's maps or the particular circumstances they illustrate; fortunately, the passage of Jørden's map through Ortelius makes him easy to identify.[126]

Mejer's modern reputation stems from his work for the Danish government; for a long time his mapping was unequaled in Scandinavia. These activities were well removed from his reconstructions of islands and coasts overwhelmed by the sea. It was the latter, however, that made him a historian. "Here on the west coast [of Schleswig] the great flood of 1634 was the upsetting event"; and Mejer had been an eyewitness. Peter Sax was troubled by the effect of earlier floods; he reckoned up the "forsaken places" and acres of land of which North Frisia had been deprived, and he had firm beliefs about bygone topography. Sax's theories guided Mejer's portrayal of the earlier forms of Heligoland. Emotion, speculation, and, most of all, the lack of reliable information undermine Mejer's historical reconstructions—once the most discussed aspect of his work. The various localities that his maps indicate as formerly existing, now drowned parish villages in the waters between the shore and the outer islands (the Wattenmeer), "depend on vague conjectures, legendary tales, and imaginative interpretations of toponyms." "Questionable," "speculative," "misleading to uncritical users," "in keeping with the tastes of the times"—thus deprecated, Mejer's evocations of medieval North Frisia cannot be claimed for history.[127]

A recent commentator adds, unnecessarily, that Mejer's maps are not "forgeries." They were sincere efforts to show the authentic past, made without expectation of material profit. But the comparison is not wholly out of place. More than one forger has satisfied a perceived need that (under existing circumstances) could not be met except by invention.[128] Mejer and his

Figure 7 Johannes Mejer, "Frisia Borealis" ("Nortfrieslande") in 1651 and 1240 (1652). The small islands and mudflats along the north Friesland coast in Mejer's time (left map) are contrasted to the extensive dry land on the eve of ocean surges in the thirteenth century (right). The Friesland map was originally in Johannes Mejer and Caspar Danckwerth, *Newe*

Landesbeschreibung der zweij Hertzogthümer Schleswich und Holstein (Husum, 1652). Within a decade, most plates of this atlas were acquired by the great Blaeu firm in Amsterdam. The map shown is from the facsimile of a Blaeu atlas, 1663. (Courtesy of the Map Collection, Yale University Library)

adviser, Sax, shared this longing. In their case, it was for historical reconstructions of the local medieval past, with the special goal of illustrating its lamentable losses to the sea.

Now overshadowed by his "respectable" cartography of Denmark, Mejer's portrayals of the medieval Frisian coast are nevertheless deeply impressive. Had anyone before tried to chart historical transformations? By creating multiple frames, he forced on the eye incontrovertible proof of the ravages of the elements. Mejer lived in surroundings whose physical form might alter markedly within the span of a human life; the great flood of 1634 had been terrifying to behold. Comparably rapid mutations of political boundaries did not occur until the French Revolution. Mejer's maps, for all their limitations, mark an early moment when maps for history, better suited to rest than to motion, serve to illustrate change.

Dutch Zeeland and the Dano-German side of Friesland were not the only water-threatened lands to generate maps for history. In the Netherlands, Frisia, synonymous with Friesland, was not limited to the coasts, as in the east, but included a hinterland. It formed a definite province including barrier islands, threatened shores, and an interior boasting the fine cities of Leeuwarden and Groningen. Besides, the ancient Frisians were understood to have occupied most of the northern Netherlands. In geographical shorthand, Frisia might stand for what, from the sixteenth century, had become the seven United Provinces detached from the Catholic, Spanish south. Frisia symbolized the collective past of one of the most enviable countries of the age.[129]

Inset maps showing Friesland at historical moments had already appeared in Ortelius's *Theatrum*. Joachim Hopper (†1576) drew a map of Friesland under the Roman emperor Augustus; the territory in question embraces much of the northern Netherlands. A narrower East Friesland at the time of the great flood of 1277 was drawn by Jan Blommaerts, on the basis of a historical map by the famous Jacob van Deventer. Here, too, lands whose boundaries changed in historical, rather than geological, time encouraged cartography sensitive to history.[130]

Just before the end of the seventeenth century, Bernhard Schotanus à Sterringa compiled and published an extensive atlas of the Frisian province, formed of detailed maps drawn by technicians on the basis of accurate surveys. Some thirty years earlier, he had become well acquainted with the terrain and with land surveys when assisting his father, Christian Schotanus à Sterringa, in assembling a map-rich work called *Beschrijvinge van de Heerlijckheyd van Frieslandt* (*Description of the Splendors of Friesland*). Bernhard was then young, and fresh from classical studies; he supplied his father's book with the map "Ancient Frisia and the Islands of the Batavians . . . as They Were

When Controlled by the Romans and Then the Franks." The Frisia in question comprised the entire United Provinces.

Bernhard, grown in years and experience and a practicing physician, persuaded the States of Friesland in 1682 to let him retrace and improve his father's survey and to do so, like his father, at public expense. The new survey took sixteen years to complete. A copy of Bernhard's historical map, basically unchanged but enlarged, accompanied the atlas of Frisia that bears his name and was reprinted with it by the publisher Francis Halma. Halma used Schotanus's map again in his two-volume *Theater of the United Netherlands.* In both Halma reprints, the "Ancient Frisia" of Schotanus accompanies a sheaf of maps of Frisian history by Menso Alting (who will concern us presently). Halma provides all of them with wide margins, in which the Latin place-names are glossed in Dutch.[131]

Bernhard Schotanus's map of 1664 is so good that two mid-nineteenth-century experts considered it still usable. What impressed them most was his grasp of the terrain. While checking and correcting on his father's behalf, he had mastered the ground, at least in this province. In designing a map of greater Frisia through the age of Frankish rule, he faced a greater challenge than Mejer in Schleswig. By the 1250s or somewhat earlier, the Netherlands acquired a "Southern Sea" (Zuyder Zee)—an arm of the ocean that engulfed the formerly dominant freshwater lake and attained its full extent in 1421. Maps of ancient or early medieval Europe done in the nineteenth or twentieth century often show the Netherlands with its outlines of 1800 or 1900. The presence of, or attempt at, a pre–Zuyder Zee shape is a good test of attention to detail.[132]

Philip Clüver should not be overlooked in this context. A scholar rather than a native of the lands he studied (Danzig was his birthplace) he was a heroic exponent of "ancient geography" and won the equivalent of a research professorship at Leiden for his *Germania antiqua.* Where Frisia was concerned, Clüver's most important work was the "Commentary on the Three Branches and Mouths of the Rhine and the Five Peoples Formerly Inhabiting Them" (*Commentarius de tribus Rheni alvei et ostiis item de quinque populis quondam accolis*). The first of its three maps is fundamental for anyone trying to give the ancient northern Netherlands its "true" form.[133]

Aided, no doubt, by Clüver's work, Schotanus's "Ancient Frisia" portrays a topography unmistakably different from that of the modern Netherlands. There is a large lake, as well as several smaller ones, and the sea is still at a distance, held back by islands so ample as almost to form a coast. The northern and western districts are reminiscent of a loose-fitting picture puzzle. Whether Schotanus was guided by the forms of tidal sands, as Mejer may have been, is not clear. That resource was not available for all the regions relevant to his map. Comparison of past and present, so important to Mejer, interested him too: the old outlines are superimposed on faint but discernible traces of the present ones. It is a handsome and impressive map, perhaps not verifiably exact but a talented approximation.

In their survey of maps of Frisia, Bodel Nyenhuis and Eekhoff register surprise that Francis Halma, a learned publisher, published Schotanus's "Ancient Frisia" alongside the history maps of Menso Alting (1636–1713) without noticing that they clashed and that Schotanus's version was incomparably better. Halma was not alone in accepting inconsistency. Until the eve of the nineteenth century, Alting was held in high esteem and preferred to Schotanus, not least because he provided many maps, not just one, however admirable. Alting retains supporters; cartographers aware of the impossibility of accurately mapping the early medieval Netherlands commend him for daring.[134]

Like Schotanus, Menso Alting belonged to the Frisian establishment. The homonymous great-grandson of a pioneer Reformer, he was burgomaster of Groningen for close to thirty years (1686–1713), including those during which his maps were produced.[135] The first five appeared in Amsterdam in 1697 under the title *Description according to the Ancient Authors of the Batavian and Frisian Tract (ager), with the Adjoining Lands; or Account of Lower Germany on both sides of the Rhine now subject to the United Seven [Provinces].* ("Lower Germany," *Germania inferior,* was the name of a Roman province, comprising the lower course of the Rhine, from Mainz northward.) Alting led off his five maps with a version of Ptolemy's map of Germany. The balance of his atlas, nine maps in all, appeared four years later, as *Description of Frisia between the Old Port of the Scheldt and the Ems . . . according to the Medieval Authors.* Alting meant the two parts to be joined as a single volume.[136]

Thanks to Alting, the eighteenth century opened with an atlas of medieval geography—the first of its kind. Besides, Alting is responsible for the inaugural atlas of "national" history, and he almost equals the feat of Philippe de La Rüe, to be discussed presently, in producing a collection proceeding chronologically "from the origins to the present." Within narrower time limits than La Rüe, Alting chose a subject and illustrated its mutations in a sequence of maps. His exploit was matched soon after in Paris, by a set of plans of that city. In the eighteenth century, other mapmakers produced ample histories on a sequential plan.[137] Alting may have been too peripheral to be a model for anyone outside the Netherlands; nevertheless, he led the way. That he mapped Frisia is incidental. The design of his *Notitia Germaniae inferioris* is a milestone on the road to historical atlases of our kind (fig. 8).

Three stages of topographic change are acknowledged by Alting's atlas. A conspicuous feature of his maps is that strong black lines denoting established coastlines are clearly distinguished, in Zeeland as well as Frisia, from faint, indefinite sectors whose outlines are approximated and not vouched for. The maps range from general to particular, showing the same terrain on a progressively larger scale. Like Schotanus, Alting took account, throughout his first part, of the early freshwater Flevus Lacus, attested by ancient geographers. And he is well aware of the importance of later floods, since his coverage ends "after the thirteenth-century inundations (*eluviones*)." His geography seems most contestable as regards the Zuyder Zee. Alting shows it existing in the late Roman period, still separated from the North Sea but having replaced the freshwater lake. This is not a definite mistake. There were serious floods in late antiquity, especially through the western dunes. But the common opinion today is that, though the lake surface grew, the turn into an arm of the sea was very long delayed.[138] Right or wrong, Alting was aware of physical changes and made determined efforts to illustrate them.

Alting's fourteen maps remind one of a camera running over a fixed background: from frame to frame, the focus narrows and widens; the scene draws closer or recedes. Predictably, he starts wide and distant, and ends narrow and near.

The medieval part of Alting's atlas, running to the late thirteenth century, is concerned with a "Hereditary" (that is, dynastic or feudal) Frisia in the south, created by the Carolingians, and with a "Free" Frisia, between the Kennem and the Ems Rivers in the north. After general depictions of the late fifth century ("emigration of the Franks") and the eighth ("return of the Franks"), Alting maps smaller districts while continuing to advance through time: the whole of Hereditary Frisia under the Carolingians; part of the same under the Ottonians; another part under the Salians and Hohenstaufen (whom he calls Henrys and Swabians); finally the third part of Hereditary Frisia, again under the Salians.[139] Moving on from there to Free Frisia in the north, Alting uses nature as his clock. The map embracing the whole of Free Frisia runs from the eighth century to the great *eluviones,* whereas the two details illustrate conditions after the deluge.

Alting proceeded thoughtfully. Avoiding repetition and a mechanical time scale, he managed nevertheless to alight deftly on all the bases demanded by his theme. His topography and history may be faulty and are certainly imperfect; but his achievement commands respect. A modern commentator considers him to have "never been surpassed or even equaled." Close to a half century would pass until another country acquired a collection comparable

DESCRIPTIO
FRISIAE LIBERAE
inter
Kinnemum et Amisiam
sub Francorum reditum
usq: ad elationes sœculi XIII.

Auctore
M. ALTING.

Quincunx Cruciata.

GER

WISTRACHIA

AUSTRACHIA

OCEANVS

TEXEL

TEXLA Arx

COMITAT. STAVERIA

COMIT.

COMIT.

KINHEM

OESTERGO

WESTREGO

BORNEDA SINVS.

Franekera

Herlinge

Bolswardia

Walrichem

Stauria

GEMERCHO

WASIA

HANDACO.

DRECHTENSEd

MITATVS.

Winkelmeet

Melverde

Medemblec

Enchusa

Nyewendoren

Horna

Swollenchout

Widemse

BANCK ATRA

Stavr mer

FORKA

DE

FLEVS INFERIOR

Leona

Oosterzee

Urch

ZVTHER
ZEE.

FELII
PARS.

Olberch

Monikedam

Marchem

Slota

Amstelredamme

Cor. de Broen fecit Amstelodami.

53

54

Figure 8 Menso Alting, "Free Frisia before the Floods" (1701). Alting's fourteen-map atlas, *Descriptio agri Batavi et Frisii omnisque regionis quae hodie est in dicione VII. Foederatorum cis et ultra Rhenum* (1697–1701), achieved a double first: an atlas of medieval, as well as ancient, history and one of a "national" past. The map shown, from the second, medieval volume, distinguishes "free" Frisia from the Frisian lands in the German Empire. Alting's maps, reprinted several times, had a long-lasting effect on portrayals of the medieval Netherlands. (Courtesy of the Beinecke Rare Book and Manuscript Library, Yale University)

to Alting's atlas of the "United Seven." [140] That is only one of his claims to fame. For our purposes it matters more that his is the first collection in which ancient geography was combined with sustained portrayals of conditions in the Middle Ages.

THE FORESHADOWING OF HISTORICAL ATLASES TO 1700

The vignettes of the Vatican gallery, the Austrian chorography of Wolfgang Lazius, and Pierre Duval's map of France in four kingdoms are, in various ways, images aspiring to be maps for history; they need evaluation, however, and should not be fitted indiscriminately into an all-purpose category of maps about the past. Cartographers appealed insistently to history. Geography was the eye of history (Ortelius), the light and eye of history (Blaeu, 1662), and, with chronology, one of the two eyes of history (Jouvancy, ca. 1700).[141] War and exploration were "described" by geography, while history established their veracity.[142] These beliefs belong to their time and need not be argued with. Our problem is to appraise the material remains of cartography in the light of what we understand as history and maps for history. No deep musing is needed to decide that an atlas like Briet's *Parallela* does not look historical to us, and that the map of a seventeenth-century land labeled with the name of an old one lacks *acuratesse* for us just as it did for Hauber.

The two centuries just surveyed were not barren in specialized maps for history. Ortelius's *Aevum vetus,* with its Ptolemaic world map projected on a wider grid, splendidly illustrates historical change; Speed's "Invasions of England and Ireland" makes an impressive display; Pellegrino's "Ancient Duchy of Benevento" is not only visually arresting but also, thanks to the boxed-in commentary, an intelligent map for history; Lambarde and Fabius produced memorable pieces of medievalia; Jaillot's "Holy Fathers of the Desert" extends the Holy Land repertory into the early Christian centuries; Johannes Mejer's comparisons of the medieval and current Frisian coastline admirably combine geographic with historical themes. Individual mapmakers show that there was no lack of inventiveness and initiative in the creation of maps for history.

Chronology was conspicuously absent, however. As one leafs through Ortelius's *Parergon,* Paul the Apostle is found before Abraham the Patriarch, Alexander of Macedon before Ulysses; biblical heroes are separated from the Greeks by a map of the Roman Empire. The compiler treats his maps as isolated tableaux, without thought of grouping them into a meaningful aggregate. "Ancient atlases" slighted chronological order for a long time. In a handsome London collection of 1805, the travels of the apostles precede the dispersal of Noah's sons, the Roman Empire is followed by the route of the 10,000, from which a delicate jump leads to Saxon England.[143]

Ancient geography could dispense with chronology. The corpus of Ptolemaic maps (to say nothing of its early modern adapters) offered geography par excellence, not a historical atlas of the ancient world.[144] Once the Ptolemaic maps turned into illustrations of bygone times, they depicted a set of lands whose form presumably remained the same throughout antiquity; their message was timelessness. The predominance in ancient geography of maps of lands over historical scenes continually weighed in favor of geographical over chronological order. In the same way, ecclesiastical boundaries—once they started to be shown—were immune from mutability. The patriarchates, provinces, and bishoprics of the Christian church, impressively continuous with their beginnings in the administrative order of the Roman Empire, had not in fact been wholly spared shocks and fluctuations. Churches were overrun by Islam in North Africa and Spain; the Latin Empire of Constantinople (1204 – 61) subjected the patriarchate under its control to adjustments; and these were not the only disruptions. Even a comparatively stable country like France experienced intermittent alterations of provincial and diocesan borders. Change was disregarded in the early maps, though not necessarily in the accompanying clarifications. Tavernier, Sanson, and the others drew ecclesiastical geography with current boundaries, suggesting, consciously or not, that the contemporary layout was adequate for the past.[145]

For historical atlases like ours to come into existence, sensitivity to chronology was important. Nothing of the kind can be found in the sixteenth century. Lazius had historical personages in mind but ordered his maps by territories. John Speed, talented as he was in designing maps with historical content, never dealt with any ensemble more demanding than the English counties. By the mid–seventeenth century some rare atlas-makers moved toward successiveness. In *Cartes . . . pour bien entendre les historiens,* part 1 of Duval's *Diverses cartes et tables* (1665), he does not advertise the sequence in which its maps are disposed and did not set a fashion. Yet he kept to the order of time. Modifying Ortelius's motto about geography being the eye of history, the French Jesuit Joseph de Jouvancy gave history two eyes, geography and chronology.[146] The change suggested a deepening of reflection about the nature of history.

Even more remarkable than the initiatives of Duval and Jouvancy is the six-map *La terre sainte* published in 1651 by Philippe de La Ruë, the unsung cartographer whose map of Byzantine Armenia has been discussed above. Not only does La Rue announce a unified theme, but also the maps of his small atlas move in a deliberate sequence from the land of Canaan and the Exodus, to the Promised Land in twelve tribes, Solomon's kingdom, the land of the Jews in the time of Christ, the Christian Patriarchate of Jerusalem, and modern Syria under Ottoman rule. The maps have a narrative purpose both singly and as a collection (fig. 9).[147] By comparison with the lack of chronology prevalent in "old geography," this group is astonishingly perfect. La Rue's atlas "from the earliest times to the present," with balanced and sequential representation of all epochs, seems to be the very first work cast in a pattern that has long ceased to surprise us.[148] "Sacred" in content, La Rue's maps had frequent reprintings. The admirable example they set for historical collections was noticed only by a far-sighted commentator, Augustin Lubin (1624–95).

Lubin, an Augustinian friar and mapmaker, foreshadowed historical atlases, but in words alone. In 1678, he published *Mercure géographique, ou le guide du curieux des cartes géographiques*. It was a glossary—the first one—of the terms that someone consulting maps would encounter in Latin and need guidance with. Lubin classifies geography into three chronological parts:

> Ancient geography, *vetus*, the finest and most learned . . . , must be divided into sacred and profane. . . . The Second *Geography*, which is called of the middle age, *medii temporis*, deals with the condition of the world as it was in approximately the seventh, the eighth, and the ninth centuries. It is distinguished from Ancient [geography] because of the notable change undergone by the face of the world. This Geography has been dealt with hitherto only by discourses without maps. Nevertheless, P. Bertius supplied a map of the Empire of Charlemagne.
>
> The Third *Geography*, called new (*nouvelle*), represents the state of the kingdoms of recent times, especially of those [lands] needed by persons who, by their occupations, are obliged to deal with foreigners. [This geography] is useful to those who travel and agreeable to lovers of contemporary history and those curious about new things.[149]

Between a first and third geography that Lubin describes in conventional terms, a medieval geography makes its appearance, limited to three centuries—a periodization that will be encountered again in connection with Christoph Cellarius (1638–1707) and other contemporaries. This "middle" period, Lubin said, had been only written about, except for Bertius's "Charlemagne," which appeared more than a half century before Lubin's *Mercure*.

Lubin does not stop there: "These divisions [into three periods] are needed because very considerable changes have taken place in kingdoms and the divisions of their provinces. Some authors have muddled these times and made big mistakes." Someone portraying a kingdom or region should, Lubin says, show it at the three times (ancient, medieval, and modern), "so that the reader or spectator of these three different maps might be able all at once (*en un moment*) to observe the changes it experienced." Lubin continues in a manner prophetic of historical atlases:

> It seems to me that this enterprise might even be pushed further. For, as wars and various other distressing [*fâcheux*] happenings have caused notable changes in the lands invaded from time to time, and inhabited by various foreign peoples, the whole appearance [*face*] of the government, the names of countries, cities, and rivers, [have] changed. . . . It would therefore be very relevant to observe, through history, the considerable changes of the states whose polished [geographical] description one wishes to make, and engrave as many maps, indicating the chronology, that is to say, to which century this map applies. This is what Ph. de la Rüe did with much success, whereas on the other hand the splendid work of Adrichomius [a historical map of the Holy Land] is spoiled to some extent by the intermingling of these different geographies.[150]

Lubin seems more conventional in the next extract, but his call for specialized maps for history is unmistakable:

> It is to be desired that learned geographers should give us maps that would be called "historical," on which the sites of great deeds, which histories speak of, and notably battles, would be marked with two or three words [indicating] the date and the outcome. And for individual histories one or two maps might be drafted, in keeping with the subject. They would furnish the diagram or the theater of the history [in question], in which, so as to avoid confusion, only those places would be engraved that the historian mentions, so that the reader might have before his eyes the location of places in which all the great deeds took place [that are told by] the history he is studying. That is the best way to remember history.[151]

Lubin is a remarkable forerunner; his suggestions anticipated by at least twenty-five years the most pioneering mapmakers who carried out the projects suggested in his pages. His special value is as an intermediary or

Figure 9 Philippe de La Rüe, "Sourie ou Terre sainte moderne" (1651). La Rüe was a Holy Land specialist associated with the mapmaking enterprise of Nicolas Sanson, patriarch of French cartography. Hardly anything is known of La Rüe. His small atlas, *La Terre sainte en six cartes géographiques* (Paris, 1651), is the first collection of maps laid out "from the origins to the present," in a systematically chronological order. This sequence, now commonplace, was so far from customary in La Rüe's time that no one imitated it for about fifty years. (Harvard Map Collection, Harvard College Library)

stepping stone, who makes it less astonishing that Alting, Lamare, and the other early-eighteenth-century makers of maps for history showed the originality they did.

Many years passed before atlases adopted the structure of La Ruë's *La terre sainte*. At the turn of the eighteenth century, as we have seen, Menso Alting became the uncontested winner of several prizes that were not yet being awarded: he compiled the first "national" atlas as well as the first atlas including the Middle Ages as the continuation of antiquity. Alting also laid out a chronological, unmechanical sequence of fourteen maps highlighting the Batavians of antiquity and the medieval Frisians.[152] Maps ordered in this way embodied a narrative purpose—precisely the design that ancient geography eschewed. Even narrower in scope than Alting's atlas and visually more compelling as history are the eight plans by Nicolas de Lamare charting the successive enlargements of Paris from Roman times to the modern city (fig. 19). Often reprinted and copied, they were first published in the initial volume of Lamare's *Traité de la police* (1705).[153] Chronological order allied with narrative form is a comforting signpost of a historical scheme.

Maps of ancient geography, rarely ordered by time, came early to synchronic displays—a single background showing different moments separated by a more or less long span of years—and have never renounced them. This is what Lubin complained about in the otherwise admirable map of Adrichomius. In Tilleman Stella's great Holy Land map of 1557, the Exodus, the territories of the twelve tribes, and the Roman provincial borders coexist. (Lucas Cranach's much copied Holy Land map of 1510 depicts the Exodus not far from the ship that, in 1498, bore the elector of Saxony on his Holy Land pilgrimage, which Cranach was commemorating.) The Duchy of Milan panel of the Vatican Map Gallery includes four incidents extending from the third century B.C. to the sixteenth century A.D., all clustered near each other. In Speed's map of battles between 1066 and 1600, the English fire ships that were launched against the Spanish Armada share the same sector of the Channel as the fleet of William the Conqueror.[154] Just so, in the figurative arts, two, three, or more successive but related scenes were and had long been routinely combined in a single composition. The practice reached back to pharaonic Egypt or farther. Synchronic maps are as common today as in the seventeenth century. All of World War I, for example, can be found pictured as a single image. An esteemed German historical atlas of 1956 shows the tracks of Attila's Huns passing over those of Alaric's Visigoths as though the actors were not separated by a half century.[155] Such maps will concern us again.

Alting's historical reconstructions of the Netherlands and that of Paris by Lamare are criticized for being too imaginative. The charge, however merited, is too limited in its targets. Johannes Mejer was imaginative when ambitiously depicting the Frisian shoreline at various epochs; so was the Jesuit Athanasius Kircher when he "conjectured" the transformation of the world after the Flood.[156] Even practitioners of ancient geography could be tempted to exceed factuality: the poetic heroes Ulysses and Aeneas were equipped with precise tracks of their travels. The Exodus also called for invention; forty-odd stops in the desert had to be somehow fixed to the ground in the absence of authoritative guides.[157] Well into the nineteenth century, serious atlases chose the dispersion of peoples after the Flood as the first event in history. It seemed preferable to supplement inadequate sources by imagination than to give up a particular historical theme as impossible to depict in a map.

Maps illustrating history were not alone in needing imaginative license. Depictions of North America sometimes showed beavers engaging in cooperative enterprise, complete with an exemplary division of labor. Africa was so richly populated with fictional physical features and animals that d'Anville won lavish praise for wiping the slate clean.[158] The share of imagination bore some relation to a standard of geographic accuracy that, for extensive parts of the world, had barely advanced beyond Ptolemy: the Caspian Sea was circular, the Sea of Aral nonexistent, and until about 1700 the Mediterranean had a much wider east-to-west expanse than medieval mariners had known it to. Historical accuracy and geographic accuracy were not perceptibly discordant. Early designers of maps for history usually deserve more applause than reproach for their appeals to imagination.

During the seventeenth century, individual designers took admirable individual initiatives in the making of maps for history. But the relation of geography to history changed little from what it had been in the days of Ortelius. World atlases continued to appear with the same premise as the *Theatrum orbis terrarum:* that their primary value was for readers of history and that they were basically historical. Barely disturbed by the intermediate "medieval geography" known to Lubin, the duo of "ancient geography" and "modern geography" remained dominant, and while the first gloried in its learned prestige in offering entrée to the *antiqui,* "modern geography" was more straightforward about offering its ample historical resources. Because history pervaded most corners of cartography, it was hard for any type of map collection to be historical par excellence. Historical atlases today are a small, distinctive fraction in a mass of nonhistorical cartography. Before 1700, the special characteristics of "maps for history" had been explored by a few mapmakers but had not yet taken firm shape.

NOTES

1. Lambarde, 1568, opp. fol. Ej. The commentary accompanying this map came late to my notice, and I regret not having seen it. On the revival of Old English studies, see Robin Flower, "Laurence Nowell and the Discovery of England in Tudor Times," *Proceedings of the British Academy* (1935), reprinted in Eric Stanley, ed., *British Academy Papers on Anglo-Saxon England* (Oxford, 1990), 1–27; David Douglas, *English Scholars, 1660–1730,* 2d ed. (London, 1951), 52, 69; Stuart Piggott, "William Camden and the *Britannia*," *Proceedings of the British Academy* 37 (1951): 201–2; Levy, "Making of Camden's *Britannia*," 77–78 and, on the ecclesiastical aspect of the Anglo-Saxon revival, 79–80; May McKisack, *Medieval History in the Tudor Age* (Oxford, 1971), 79. I give another account of this subject in "The First Venture into 'Medieval Geography': Lambarde's Map of the Saxon Heptarchy (1568)," in *Alfred the Wise: Studies in Honour of Janet Bately,* ed. Jane Roberts, Janet L. Nelson, and Malcolm Godden (Cambridge, 1997), 53–60.

2. Shirley, *Maps of Britain to 1650,* 48, no. 106.

3. Lambarde, 1568. Cf. *Landmarks in Learning about the Anglo-Saxons, 1566–1853,* Durham University Library, exhibition catalogue (in honor of the International Society of Anglo-Saxonists conference), (Durham, 1989), 3, 5, no. 3. Shirley, *Maps of Britain to 1650,* 41, no. 83b, and pl. 48, Lambarde's map from John Foxe, *Actes and Monuments* (1576 and later). It was also reprinted in Foxe's *Book of Martyrs* (Shirley, *Maps of Britain to 1650,* no. 115a).

4. The map is undecorated but resembles its descendants. The kingdom names are discreetly lettered. Each kingdom is numbered, as a reference to the explanation on the verso; boundaries are marked.

5. Ifor M. Evans and Heather Lawrence, *Christopher Saxton, Elizabethan Map-Maker* (London, 1979), 36–38, 59–65. On Lambarde's notes (published in 1730) and Camden, see Levy, "Making of Camden's *Britannia*," 78.

6. Camden, 1600. Boundaries indicated by dotted lines. (The 1st ed. of the *Britannia* is 1586.) Shirley, *Maps of Britain to 1650,* 95, nos. 231–32, 113, no. 280 (source of the quotation), 308, with pls. 86 and 99; Douglas, *English Scholars,* 20–21, 25; *Landmarks in Learning,* 5, no. 6. On Rogers and Hole, see *DNB* 17:145 (Rogers said to be the first English copperplate engraver) and 9:1022.

7. Lambarde's claim to priority is not undermined by the pre-Ortelian maps for history listed by Meurer, "Ortelius," 134.

8. On Speed, see *DNB* 18:726–28; R. A. Skelton, in *John Speed, "A Prospect of the Most Famous Parts of the World," London, 1627* (Amsterdam, 1966), vii; he was in the Merchant Taylors Company like his father. His main work is a *Historie of Great Britaine* (London, 1610), much praised but also known to consist of old material and to retain the old blunders. On Speed, 1627, see nn. 93–101, below. On Dutch engravers in England, see Skelton, *John Speed, "A Prospect,"* 5, 7. On Hondius, see Koeman, 2:136. Hondius came to London after the duke of Parma captured Ghent; he set up shop as a mapmaker and married the sister of a future mapmaker. After the business was moved to Amsterdam in 1593, the acquisition of the plates of Mercator's atlas in 1604 was the decisive step ensuring the firm's prosperity. Hondius died after a short illness, leaving a family able to carry on the business. Hondius engraved the whole cycle of county maps for Speed, 1611.

9. "Britain as it was devided in the tyme of the Englishe-Saxons especially during their Heptarchy," in Speed, 1611; cf. Thomas Chubb, *The Printed Maps in the Atlases of Great Britain and Ireland: A Bibliography, 1579–1870* (repr., London, 1966). For later editions, see the

next note. The twelfth-century *Historia Anglorum* by Henry of Huntingdon (Goffart, "First Venture," 55 n. 10), carefully separating secular from ecclesiastical affairs, was well suited to be the source of Speed's fourteen Anglo-Saxon scenes.

10. Willem Blaeu and Joan Blaeu, *Theatrum orbis terrarum sive Atlas novus,* Pars 4[a] (Amsterdam, 1645), no. 1a. The quotations are from Shirley, *Maps of Britain to 1650,* 188, no. 549, with pl. 151, and from Moreland and Bannister, 221, with illustration. For a recent price, see Faupel's *Catalogue* 107, no. 27 (May 1995): $1,850. There is an excellent reproduction of the Blaeu map in John Goss, ed., *Blaeu's "The Grand Atlas of the 17th Century World"* (New York, 1991), 72–73.

11. The Old English lettering, though not accurate by our standards, was an honest approximation of a scribal hand of the tenth or eleventh century; it is more evocative of Anglo-Saxon characters than Roman ones.

As soon as the Blaeu version of Speed's map appeared, the Janssonius firm hastened to have the new version copied and marketed it on its own account (1646). The Janssonius copy, all agree, does not equal its model, though it is close to literal. See Shirley, *Maps of Britain to 1650,* 193, no. 577. Moreland and Bannister, 222, comment that, by comparison with its model, the Janssonius map is relatively rare (though it was reprinted down to the 1740s, much later than Blaeu's).

12. The elimination of all the kingdoms except Wessex is illustrated by a time chart in David Hill, *An Atlas of Anglo-Saxon England* (Oxford, 1981), nos. 39–40.

13. On Sanson, *Les cinq royaumes,* see ch. 1, nn. 56–57, above. Sanson, 1651: the English maps are marked "Tavernier [the publisher], 1641"; see Pastoureau, *Atlas français,* 430, no. 7. Shirley, *Maps of Britain to 1650,* 173, no. 506, disparages the base map. Sanson, 1654: the Heptarchic kingdoms are given the main, colored boundaries; the counties are also shown. See also Schenk 2 in n. 15, below. Cf. Shirley, *Maps of Britain to 1750,* 122. Cf. Sanson, 1665 (Pastoureau, *Atlas français,* 408: *Cartes générales,* no. 117). For the particular maps (1654) gathered in *Cartes générales,* 1658, nos. 23–26, see Pastoureau, *Atlas français,* 403.

14. For Jaillot, see Shirley, *Maps of Britain to 1750,* 74 (Jaillot 1): two large sheets, "showing physical features, county boundaries," as well as the seven kingdoms and Hadrian's Wall; shown, pl. 47. Shirley, *Maps of Britain to 1750,* 45–47: Vincenzo Coronelli, 1689, includes a heavily symbolic cartouche concerning James II by a noted artist; counties are combined with the Heptarchy. "Historical information is distinguished by a small star"; shown, pl. 26. In another Coronelli version, in two sheets, in *Atlante Veneto* (Venice, 1691), "counties are grouped within the boundaries of the seven ancient Anglo-Saxon kingdoms."

15. Shirley, *Maps of Britain to 1750,* 122; Schenk 2, "Anglia in Septem Anglo-Saxonum Regna distincta" (Amsterdam, ca. 1690); see also Koeman, 3:119, no. 93, counties with the Heptarchy, "following the practice of the time." Shirley, *Maps of Britain to 1750,* 143, draws attention to the similar plate by Visscher: Nicolas Visscher 3, "Angliae Regnum Tam in Septem Antiqua Anglo-Saxonum Regna quam in omnes Hodiernas Regiones accuratissime distinctum" (Amsterdam, ca. 1695), which marks the Heptarchy "although the normal county boundaries are shown as well."

16. Shirley, *Maps of Britain to 1750,* 69–71 (pl. 44), Homann 3, "Magnae Brittaniae pars meridionalis in qua Regnum Angliae tam in septem antiquos Anglo-Saxonum regna quam in omnes hodiernas regiones ostenditur" (Nuremberg, ca. 1715), with acknowledgment of the Visscher "archetype"; the seven kingdoms "are typically accentuated by distinctive wash colouring." It would be hard to make them out otherwise, given the abundance of other places. Shirley, *Maps of Britain to 1750,* 131, Seutter 2, "Britanniae sive Angliae regnum tam

secundum priscae Anglo-Saxonum Imperia quam recentiores provinciarum divisiones (Augsburg, after 1733); based on the foregoing; "wash colouring typically applied to accentuate the kingdoms" that are "superimposed on the normal county boundaries."

17. Robert, 1753, two sheets. For the modern road map of England plotted on a Heptarchic background, see RVaugondy, 1757, *Atlas universel,* no. 15.

18. *A New Mapp Of The Kingdom of England* (1700); cited above, ch. 1, n. 60. Shirley, in *Maps of Britain to 1650* and in *Maps of Britain to 1750,* documents the (judicious) neglect of the Heptarchy by English mapmakers.

19. Manesson-Mallet, 1683/A, vol. 5 (Europe, ancient and modern), one of five maps of Britain. On Manesson, see n. 63, below. Shirley, *Maps of Britain to 1750,* 87. Shirley, *Maps of Britain to 1750,* 66, Halma 1 (Amsterdam, 1700), in a pocket-size edition of Sanson's *Cartes générales;* the first of two British maps. Shirley, *Maps of Britain to 1750,* 66, Halma 2, bishopric boundaries, from Halma's *Geographia sacra* (1704). Shirley, *Maps of Britain to 1750,* 42, Châtelain 1, "Carte pour l'Introduction à l'Histoire d'Angleterre où l'on voit son premier gouvernement," in Châtelain, 1705; shown, pl. 23. See ch. 3, n. 1, below. Köhler, 1730, 2:109. Cellarius, 1776, no. 5, "Saxonia Transmarina" (a Heptarchy map). This Latinate name for England is used in Bertius, 1623. The sources of Köhler's and Cellarius's maps are not obvious.

20. On the Oxford "Saxonists" and Gibson, see Douglas, *English Scholars,* 64–71. For Gibson's relations with George Hickes, see Richard L. Harris, ed., *A Chorus of Grammars: The Correspondence of George Hickes and His Collaborators* (Toronto, 1992), 44–45. E. Gibson, ed., *Chronicon Saxonicum ex MSS Codicibus* (Oxford, 1692). Unpaged *Praefatio:* very candid about the genesis of the edition. At the end of the preface, Gibson states that he has included nothing in the map but what he has firmly established in the included *Explanation of Place-Names* (*Explicatio nominum locorum*). A comparison of his map to Camden's shows that there is genuine improvement. On Gibson's edition, "a most remarkable performance," see Charles Plummer, ed., *Two of the Saxon Chronicles Parallel* (Oxford, 1892), 2:cxxiv, cxxix–cxxxi.

21. On Gibson, see *DNB* 7:1153–54. Piggott, "Camden and the *Britannia,*" 209–13, gives an engaging account of Gibson's leadership. See also Douglas, *English Scholars,* 257–58. On Morden, see *DNB* 13:851–52. Gibson never carried out his larger Anglo-Saxon map.

22. Camden, 1689, no. 3, "Britannia Saxonica." Dimensions: 293 × 368 mm (i.e., 163 percent of the size of the *Chronicle* map). The cartouches are simplified and more elegant. Watling Street is the most conspicuous feature of the *Chronicle* map omitted in Sturt's enlargement. Shirley, *Maps of Britain to 1750,* 99, attributes this map to Morden; its difference in style from all the others in the volume, and Gibson as source, might have been noted. On the engraver Sturt (1658–1730), see *DNB* 19:138–39; he specialized in miniature calligraphy.

23. *Atlas of England,* 1720, no. 2; *Magna Britannia et Hibernia, Antiqua et nova* (London, 1720–21), no. 5, "Britannia Saxonica." Paul Rapin-Thoyras, *Histoire d'Angleterre,* 10 vols. (The Hague, 1724), 1:91, "Carte d'Angleterre sous les Saxons" (Gibson acknowledged as source). On Rapin and the Heptarchy, see Henry Hallam, *Supplemental Notes to the View of the State of Europe during the Middle Ages* (London, 1848), 203; Simon Keynes, "Rædwald the Bretwalda," in *Voyage to the Other World: The Legacy of Sutton Hoo,* ed. C. B. Kendall and P. S. Wells (Minneapolis, 1992), n. 66.

24. Andrews, 1797, no. 10, "Saxon Britain according to the Saxon chronicle," and no. 11. For the three laws, see Hill, *Atlas of Anglo-Saxon England,* 198, no. 174. Hill marks borders. Andrews's map is comparable to the final map of Rizzi-Zannoni, 1764, showing the division

of France by "coûtumes" (see ch. 3, n. 93, below). Andrews is more fully discussed in ch. 4 on 278–80.

25. For Mercator's *Arles,* Tavernier, and Duval, see Le Long, 1:35–35, nos. 397–99, 415, 419–20. Lenglet, 1768/B, 12:2; Hauber, *Versuch einer Historie,* 131.

26. Tassin, 1633, no. 34, "Austrasie," shows part of the Rhine (flowing west of Frankfurt), the Moselle (with Trier), the archbishopric of Trier, the duchies of Luxembourg and Zweibrücken, and, at bottom, parts of Lorraine and Alsace. The maps flanking Austrasia are "Limbourg" (no. 33) and "Franconia" (no. 35). Metellus, 1594: see Meurer, *Atlantes Colonienses,* 168. Metellus's general account of France refers to "vetus Austrasiae regnum," but not as something to be laid out on the map; one map is called Lorraine, another Alsace, none Austrasia. Conrad Leo (Löw), *Italia, Hispania, Francia, Austrasia, Helvetia, tabulis aeneis incisae* (Cologne, 1598) (NOT SEEN).

27. A. Huguenin, *Histoire du royaume mérovingien d'Austrasie* (Paris, 1862), 42; T. Zotz, "Austrien," *Lexikon des Mittelalters,* ed. Robert Auty et al., 10 vols. (Munich and Zurich, 1977–), 1:1258–59; H. A. Anton, "Austrasia/Austria," in *Reallexikon der germanischen Altertumskunde,* ed. Johannes Hoops, 2d ed. (Berlin, 1967–), 1:512–13; Paul Kretschmer, "Austria und Neustria: Eine Studie über spätlateinische Ländernamen," *Glotta* 26 (1938): 207–40; Margret Lugge, *"Gallia" und "Francia" im Mittelalter: Untersuchungen über den Zusammenhang zwischen geographisch-historischer Terminologie und politischem Denken vom 6.–15. Jahrhundert* (Bonn, 1960), 32–37 and passim. These authors suggest that the subject is a minefield. J. Moreau, *Dictionnaire de géographie historique de la Gaule et de la France* (Paris, 1972), 31 (s.v. "Austrasie"), is detailed about the Merovingian period, more general about the Carolingian ("ne désigne plus que les pays d'outre-Rhin"), and silent about the future. Michel Mourre, *Dictionnaire encyclopédique d'histoire* (Paris, 1978), s.v. "Austrasie," is exclusively Merovingian.

28. For one summary account out of many, see Alfons Dopsch, "La naissance et la formation de l'État autrichien," *Revue historique* 177 (1938): 34–35. In addition to the Eastern March, the Carolingians established a southern one in this region, extending from Frioul. Earliest occurrence of "Osterrîchi": A.D. 996. Louis Moréri, *Le grand dictionnaire historique* (Paris, 1743), 2:1150: "Austria, est un nom Latin qui convient également à deux pays très différens et fort éloignez l'un de l'autre, je veux dire à l'*Austrasie* et à l'Autriche" (the same in earlier editions of Moréri).

29. Aimoin, *Historia Francorum,* bk. 1, ch. 4: "Has omnes provincias dum Franci occupavissent, illam regionem, quae Septentrionem versus tenditur, et inter Mosam et Rhenum est, Austriam: illam autem, quae a Mosa ad Ligerim usque pertingit, Neustriam vocaverunt." Cited by Charles Dufresne de Du Cange, *Glossarium mediae et infimae Latinitatis,* ed. L. Favre (Niort, 1883–87; repr., Graz, 1954), s.v. "Austria" ("Austrasia" cited as a variant). For a recent map showing the boundaries of Austrasia and Lotharingia to 870, see Michel Parisse, ed., *Histoire de Lorraine* (Toulouse, 1978), 105 (ch. 4, "De l'Austrasie à la Lorraine," implies that the idea that today's Lorraine used to be called Austrasia retains currency). Briet, 1647/A, 2:64, "Porro de terminis huius Austrasiae non est facile statuere, cum aliquanto restricta sit aliquando dilatata." While recognizing that Lorraine was "composed of a not inconsiderable part of Austrasia" (75), Briet defines Austrasia for the purposes of his exposition as containing only the Rhenish Palatinate, the bishopric of Speyer, and the Duchy of Zweibrücken (64–66). Briet's Austrasia is definitely absent from his list of districts composing France; he considered it a part of the German Empire.

30. All but the last of these are cited in Burchard Gotthelf Struve, *Bibliotheca historica,*

ed. J. G. Meusel, 11 vols. (each in two parts) (Leipzig, 1800), 10, pt. 1:125–28, 130–31. Philippus Ferarius and M. A. Baudrand, *Lexicon geographicus,* 2d ed. (Paris, 1670), 1:162: "Lotharingia passim Latine Austrasia nominatur." Parisse, *Histoire de Lorraine,* 350, provides an additional eighteenth-century citation; at the outbreak of the French Revolution, the provincial estates of Lorraine were sometimes referred to as "Estates of Austrasia" (353). The claims of Metz to be capital of the province can be traced to ca. 566, when it became chief city of the Frankish subkingdom.

31. These equivalences were not necessarily erudite. In connection with Gaul/France, Dainville, *Géographie des humanistes,* 347–49, concludes that by Richelieu's time (ca. 1640) everyone who mattered was familiar with Caesar and knew from him that, because the Rhine was the frontier of Gaul, it should be that of France.

32. Auguste Longnon, *Géographie de la Gaule au VIe siècle* (Paris, 1878), 193, 370–71; Fredegar, *Chronicon,* bk. 3, ch. 32, ed. A. Kusternig, *Quellen zur Geschichte des 7. und 8. Jahrhunderts* (Darmstadt, 1982), 114, with n. 94. The capital of modern Lorraine was Nancy. Auguste Digot, *Histoire de Lorraine,* 6 vols. (Nancy, 1856), 5:250, in the early seventeenth century, "Depuis nombre d'années, on faisait en Lorraine de fréquentes allusions à la résurrection du royaume d'Austrasie"; 4:285, it was believed that Charles III of Lorraine (ca. 1593) "wished . . . to obtain later the title of king of Austrasia." On Henry II, see Auguste Longnon, *La formation de l'unité française,* ed. H. F. Delaborde (Paris, 1922), 307. The campaign netted Metz, Toul, and Verdun for Henry before resistance thwarted the advertised goal of reaching the Rhine. Recently, Henry's historical claim is played down; Jean Daniel Pariset, *Les relations entre la France et l'Allemagne au milieu du XVIe siècle d'après des documents inédits* (Strasbourg, 1981); Frederick J. Baumgartner, *Henry II, King of France, 1547– 1554* (Durham, 1988).

33. Concerning Tavernier's patriotic attitude toward Austrasia, see Boisseau, 1641, below nn. 102–3. On historians able to resurrect Austrasia, Rosamond McKitterick, "The Study of Frankish History in France and Germany in the Sixteenth and Seventeenth Century," *Francia* 8 (1980): 556–72. Moréri, *Grand dictionnaire historique,* 2:1149, "la Lorraine d'aujourd'hui, que les Auteurs Latins appellent quelquefois Austrasie"; the usage was not limited to Latin authors. Robert's map of Germany, Robert, 1745.

34. Hauber, *Versuch einer Historie,* 142: "Ich habe selbst eine Charte von dem *Regno Lothariensi* und dem *Königreich Austrasien* under Händen, in welcher so wohl der Zustand dieses Reichs, wie es anfänglich eine *Portion* Käysers Lotharii ware, als auch wie es nachgehends under seinen Söhnen Lothario dem Jüngern, und Carolo, nach dieser Todt aber unter dem Teutschen König Ludovico, und dem Fränckischen, Carolo Calvo zertheilet worden, biß es endlich unter Ottone gantz und gar an das Teutsche Reich gekommen, und zu einem gedoppelten Hertzogthum erwachsen ist, neben denen Gräntz-Lineen deren Haupt-Theilungen derer Söhne Ludovici Pii vorgestellet wird." By this account, Hauber's map would display the developments of about one century.

35. Bertius, 1623. Anville, 1771. Bertius's Austrasia coincides with d'Anville's in embracing Upper and Lower Lotharingia (both spelled out by d'Anville). D'Anville clearly excludes Alsace—part of Merovingian Austrasia—and attributes it to Alamannia (historically appropriate for ca. 700).

36. For biographical references, see ch. 1, n. 52, above. Lazius's father, a Swabian, was a professor of medicine at Vienna. Lazius the son became uncommonly rich thanks to his historical appointment. His library entered the emperor's; much survives at ViennaÖNB.

Amidst religious turmoil in which Vienna shared, Lazius stood with Ferdinand I as an unwavering Catholic. He died after taking extraordinary pains over the *Commentarii in genealogiam Austriacam* (Basel, 1564). Oberhummer and Wieser, *Wolfgang Lazius Karten,* 13 (judgments on his work).

37. For an assessment of Lazius's graphic accomplishment, see Andreas Andresen (1865), repr in *DBiogArchiv,* fiche 746, frames 266–76. On Lazius as etcher of his maps, see Banfi, "Maps of Wolfgang Lazius," 56–57, 65. About the map in Banfi, see the caution at n. 51, below. Rather negative, Bernleithner, *Lazius Austria,* xiii. Essential reading about his maps is Leo Bagrow, *A. Ortelii Catalogus cartographorum, Pt. 1 (A–L),* repr., *Acta cartographica* 27 (1981): 191–97. Bagrow notes the lack of medical writings in Lazius's output; history (mainly ancient) and cosmography were the foci of his scholarship (128).

38. Bagrow, *Ortelii Catalogus,* 128–30, nos. 3, 6, 132, nos. 20, 21, 24. Koeman, 3:35, nos. 27, 42; 38, nos. 9, 14 (each number includes several Lazius maps from the *Typi;* see next note). For Hondius, see Koeman, 2:347, no. 27, 350, no. 34, etc. (always the same map). For Blaeu and Covens and Mortier, see "Lazius" in Koeman's index of cartographers. For the map of Hungary, see Oberhummer and Wieser, *Wolfgang Lazius Karten,* 55.

39. About the *Typi,* see ch. 1, nn. 50–54, above. On its limited circulation, see Oberhummer and Wieser, *Wolfgang Lazius Karten,* 28. On the *Typi* and history, see Bernleithner, *Lazius Austria,* xiii; Banfi, "Maps of Wolfgang Lazius," 56. On its relation to the chorography genre, see ch. 1, n. 51, above.

40. Oberhummer and Wieser, *Wolfgang Lazius Karten,* 26, 15.

41. Bernleithner, *Lazius Austria,* xix, xxv; Oberhummer and Wieser, *Wolfgang Lazius Karten,* 35, 31.

42. Banfi, "Maps of Wolfgang Lazius," 54, 56. Wolfgang Lazius, *De gentium aliquot migrationibus* (Basel, 1557). About this work, see W. Goffart, "The Theme of 'The Barbarian Invasions' in Late Antique and Modern Historiography," in *Rome's Fall and After, by W. Goffart* (London, 1989), 122. On the twenty-one woodcuts by Lazius providing "phantastische Kostümbilder" of the *gentes,* see Andresen in *DBiogArchiv,* fiche 746, frames 274– 76. Andresen calls the book, too summarily, a "Geschichte der deutschen Völkerwanderung"; it bears little resemblance to later works on this theme.

43. Wolfgang Lazius, *Reipublicae Romanae in exteris provinciis, bello acquisitis, constitutae, commentariorum libri 12* (Basel, 1551; 2d ed., Frankfurt-am-Main, 1598); a loose translation of this title is "Discussion in 12 books of the Roman government established in the foreign provinces it gained by war." For the very interesting title page of the *Typi,* see fig. 3; also Oberhummer and Wieser, *Wolfgang Lazius Karten,* 27 and 12: all Lazius's writings as preparations for the great work that he did not live to write. A sketch of the major work survives in manuscript at the ViennaÖNB: M. Kratochwill, in *LGKartog.* 1:443–44. About the work on Greece, see Oberhummer and Wieser, *Wolfgang Lazius Karten,* 16–17. According to Meurer, "Ortelius," 134, the maps of Greece are genuinely historical (he does not elaborate). Lazius's main (not unique) theological work is the first printed edition of a New Testament apocryphon: Abdias, *De historia certaminis apostolici* (with other hagiographica) (Paris, 1560)—possibly reprinting an original Basel, 1552 edition whose existence I cannot adequately verify.

44. James Vann, "Mapping under the Austrian Habsburgs," in *Monarchs, Ministers, and Maps,* ed. D. Buisseret (Chicago, 1992), 153–57, gives an excellent account of Habsburg ambivalence toward maps, reflecting a situation in which the particularism of each Habs-

burg province was stressed in obedience to a Habsburg emperor heading the (highly decentralized) Romano-Germanic Empire. These conditions militated against maps that suggested that the Habsburg dominions formed, or should form, a unified state.

45. Oberhummer and Wieser, *Wolfgang Lazius Karten,* 33, 31, 34.

46. On modern and ancient district names, see Banfi, "Maps of Wolfgang Lazius," 55 (from Bagrow). Bernleithner, *Lazius Austria,* xv, notes the pictorial, rather than mathematical, nature of Lazius's maps. Banfi, "Maps of Wolfgang Lazius," 56, remarks that Lazius, with a view to his major *Commentarii rerum Austriacarum,* "fills the maps with references to past history. The numerous inscriptions, legends and directions abound in such references, their purpose being to record historical events." Banfi distinguishes these historical inscriptions (on all the maps) from maps "purely historical in character," such as "the Austria of Charlemagne." Oberhummer and Wieser, *Wolfgang Lazius Karten,* 15–16, point out that Lazius's map of the Schmalkaldic War (1546) sets ancient tribal names (in very large letters) alongside modern ones. On references to the Roman past, see Oberhummer and Wieser, *Wolfgang Lazius Karten,* 29; to early Christian times, Bernleithner, *Lazius Austria,* xix.

47. Oberhummer and Wieser, *Wolfgang Lazius Karten,* 35: "Regnum Austrasiae sive Austriae ad Rhenum a Francorum regibus antiquis inceptae, unde inclita Habspurgensis gens initia repetit." (a description of boundaries follows); "De duplici Austria, quarum ille ad Rhaenum vetustior, haec a Danubio ad Italiae Alpes Hunnorumque regna extenta fuerat. In quarum utraque Habsburgenses Principes ad famam venerunt." Another title for this map is "Regnum Austrasiae ad Rhenum cum Edellsassia et ducatus Alemaniae" (Banfi, "Maps of Wolfgang Lazius," 55). (The reference to Alsace, associated with Habsburg origins, is noteworthy. Although Alsace had been in Frankish Austrasia, it had no place in the equivalence of Austrasia with Lorraine current in the sixteenth century.) The royal image is recycled, from Lazius's *Migrationes gentium,* with rearrangement of the figures and new labels.

48. Oberhummer and Wieser, *Wolfgang Lazius Karten,* 30: "Regnum orientale sive Austriae novae Caroli Magni auspicio in finibus Norici-Pannoniae positae, in quo inclyta Habsburgensis propago fortunae incrementa accepit." The boundaries are then spelled out. The modern gloss is from the same page.

49. From the *Elenchus operis,* unpaged, in the facsimile: "Marcha orientalis in qua Babenbergenses Comites rerum potiri ab Ottonis caepere, Beato Liupoldo intercessor clari." The investiture of the Babenbergs occurred in the 970s. The Babenberg saint, patron of Austria and father of the historian Otto of Freising, is Liutpold III, duke in 1096.

50. From the *Elenchus;* see previous note.

51. The map of 1545 is reproduced by Banfi, "Maps of Wolfgang Lazius," opposite 65. For the date, see Oberhummer and Wieser, *Wolfgang Lazius Karten,* 35. At the Forty-sixth Wolfenbütteler Symposion (26–29 October 1999), Dr. Franz Wawrik (Kartensammlung, ViennaÖNB) pointed out that Lazius's authorship of this map is now questioned. Lazius's authorship, very plausible at first sight, is maintained here, until the question of authorship is settled.

52. Richard Vaughan, *Charles the Bold, the Last Valois Duke of Burgundy* (London, 1973), 85: "Since the end of the fourteenth century the lands of the house of Habsburg, which stretched discontinuously across Europe from Vienna to Basel, had been divided between two, or even three, princes who described themselves as 'duke of Austria.'" On the Habsburgs in sixteenth-century Alsace, see Philippe Dollinger, ed., *Histoire de l'Alsace* (Toulouse, 1970), 264 (with map), 267–68. Their power base was the Sundgau and, on the right bank, the Breisgau, the latter of which accounts for Lazius's inclusion of the Duchy of

Alamannia. Oberhummer and Weiser, *Wolfgang Lazius Karten,* 35, forthrightly label the map as being of "Vorder-Österreich," the established term for these western possessions. They comment about the map of 1545, "Sie steht offenbar in Zusammenhang mit seinen Studien über die Genealogie und Geschichte des Hauses Habsburg, dessen Ursprung und älteste Besitzungen bekanntlich im obern Elsass zu suchen sind."

53. Banfi, "Maps of Wolfgang Lazius," 61–62, points out that the area shown by the first map is much larger than the Carolingian Eastern March; and the second is smaller than its advertised borders. He claims paradoxically that these maps, with their faulty historical content, are "purely historical." For a good map of the medieval development of Austria, see J. B. Freed, "Austria," in *Dictionary of the Middle Ages,* ed. Joseph R. Strayer, 13 vols. (New York, 1982–89), 2:4–9.

54. They are definitely "historical maps," in the definition of Witt, in *Lexikon der Kartographie,* 201, but not *Geschichtskarten,* that is, earnest reconstructions of past conditions.

55. Bessel, 1732. See ch. 4, nn. 21–22.

56. Delisle, 1717; Le Long, no. 393 (division of the kingdom at the deaths of Chlothar I and Dagobert).

57. Mercator, 1585: "Aquitania australior. Arelatense regnum /Aquitania australis. Regnum Arelatense cum confiniis." The map was a very stable component of the *Atlas* before and after its acquisition by Hondius. This volume for France first appeared ten years before the full Mercator *Atlas,* with which it is integrated.

58. For Briet and Duval, see next notes. On Arles, see *Encyc. Brit.* 2:557–58; Eugene L. Cox, "Dauphiné," in *Dictionary of the Middle Ages,* 4:108–9. Jürgen Voss, *Universität, Geschichtswissenschaft und Diplomatie im Zeitalter der Aufklärung: Johann Daniel Schöpflin* (Munich, 1979), 174: Schöpflin's lectures on the claims of European sovereigns (1750s) contained a section on the rights of the empire to the Kingdom of Arles. The qualifier "Holy Roman" was not applied to the medieval German Empire until the twelfth century.

59. Briet, 1647/A, 1:415, explains that he would not argue whether there was a Kingdom of Arles or what its boundaries were; what mattered for his purposes was that it was a useful toponym for grouping the four regions he would now describe: Dauphiné, Provence, Savoy, and the Prefecture of Lyons.

60. Pastoureau, *Atlas français,* 367–69.

61. Hauber, *Versuch einer Historie,* 131, and Le Long, 1:35, nos. 398, 399, 1:36, nos. 415, 419, agree on 1671 as the initial date of these maps, which were also published in many later years. Duval, 1679, nos. 42–45 (Newberry; different order in other copies): for no apparent reason, the sheets are ordered counterclockwise from Aquitaine (with the cartouche), then Burgundy-Arles, Austrasia, and Neustria.

62. Placide, *Cartes de géographie,* no. 7 (for which see Pastoureau, *Atlas français*). About Duval's works on current affairs, see Pastoureau, "Atlas en France avant 1700," 63.

63. Manesson-Mallet, 1683/A. About him, see Pastoureau, *Atlas français,* 309; *Biog. univ.* 26:256 (with the quotation from Bayle); Quérard, *France littéraire,* 5:487. Other Manesson-Mallet works are *Les travaux de Mars ou l'Art de la guerre* (1671); *La géométrie pratique* (1702). He was teacher to the pages of the Petite Écurie.

Both the Sinai monastery and Mecca are in the great Jerusalem map of Bernhard von Breydenbach's Holy Land narrative; see Nebenzahl, 63–66 (with illustration).

64. Karl Ferdinand Werner, *Les origines (avant l'an mil),* in *Histoire de France,* ed. Jean Favier (Paris, 1984), 1:323–25.

65. Manesson-Mallet, 1683/A, 5:231, "De l'origine des Bourguignons . . . " (the main

point is to fill in the past of newly acquired Franche Comté); 233, first map (no boundaries; Rhône Valley at center; Aquitaine left, Lombardy right; West Franks to northwest, East Franks to northeast); 235, second map (to the foregoing adds Cis- and Transjurane Burgundies, plus Dauphiné and other novelties); 237, second map repeated. Franche Comté, long a Spanish possession, was conquered by the French in 1668 but restored the same year at the peace treaty. A second conquest, in 1674 (the start of continuous French control), was confirmed by the peace of Nijmwegen: *Encyc. Brit.* 10:931–32.

66. Lenglet, 1768/B, 1:284; Hauber, *Versuch einer Historie,* 157. Tavernier, 1645.

67. On Bertius (Dutch "Berts"; French "Bert"; 1565–1629), *Biog. univ.* 4:170–71; A. J. Vander Aa, *Biographisch Woordenboek der Nederlanden,* 7 vols. (Haarlem, 1852–78) , 1:137–38; A. Thijm, in *ADBiog.* 2:509–10 (Arminius and Bertius's family); Jules de Saint-Genois, in *Biographie nationale de Belgique* (Brussels, 1866), 1:292–98; Knipsch, in *Nieuw Nederlandsch Biografisch Woordenboek,* ed. P. C. Molhuysen et al., 10 vols. (Leiden, 1911–37), 1:320–23; Pastoureau, *Atlas français,* 65. Born in Flanders, Bertius as a child was a refugee first in a London suburb, then in Rotterdam, where his father was named minister. Several countries claim him. Early supporters of Arminius are called Remonstrants; many were more harshly treated than Bertius. Discourses justifying his emigration and conversion are at the BN and the BL. The capsule biography in Koeman, 1:60, says nothing of his troubles and exile.

Charlemagne map: Bertius, 1623. The Paris versions were probably marketed by several sellers. The first atlas printing was in Blaeu, 1635; see Koeman, 1:100. One would have thought that, owing to family connections, the map would have gone straight to the Hondius firm. In fact, the Blaeu version precedes Hondius-Janssonius by three years (see n. 72, below).

68. Koeman, 1:60. See also Koeman, 2:216–18 (van den Keere). The triangle Ghent, England, northern Netherlands was common to Bertius, Hondius, and van den Keere. Type cutting was the specialty of van den Keere's family. On the *Caert thresoor,* see Koeman, 2:252–55, 146. The maps of the Latin edition are sometimes (wrongly) attributed to Bertius.

69. *Commentarii rerum Germanicarum* (Amsterdam, 1616; repr., 1632). Bertius implies that the Romano-Germanic Empire then existing began with Charlemagne. Bertius, 1618/A: for its contents, see Koeman, 1:63. Louis XIII, only seventeen years old in 1618, cannot have had a very personal part in awarding the honor of king's cosmographer to Bertius. The second atlas is Bertius, 1628/A. *Biographie nationale de Belgique,* 1:295, credits Bertius with *Variae . . . tabulae XX geographiae ex antiquis geographis et historicis* (1602), possibly the same as Bertius, 1628/A. Bertius's *Breviarium totius orbis terrarum* (Paris, 1624) was incorporated into George Horn's best-selling *Introductio ad geographiam antiquam* (Amsterdam, 1653, and many later editions); see Janssonius, 1652/A.

70. Details in Bertius, 1623.

71. The Janssonius copy (see next note) has four pages of letterpress on the back of the map and after.

72. Pierre Chevallier, *Louis XIII, roi cornélien* (Paris, 1979), 233–49; Victor-L. Tapié, *France in the Age of Louis XIII and Richelieu,* ed. and tr. D. M. Lockie (London, 1974), 95–129; Bertold Baustaedt, *Richelieu und Deutschland* (Berlin, 1936), 17–25: Louis campaigned against Huguenots in 1620–22; France was impotent when the Thirty Years' War began. Richelieu entered Louis's council in 1624.

Blaeu (see n. 67, above): Koeman, 1:100, no. 215 (no Charlemagne map in Blaeu's fullest atlas). Janssonius (1638): Koeman, 2:399 (opening the French section, in a historical introduction headed by the Gaul of Julius Caesar, then of Strabo). The exception is Janssonius, 1652/A (his atlas of ancient geography; Koeman, 2:499–502).

73. Lenglet, 1965/B: 96; Hauber, *Versuch einer Historie*, 142 n; Delisle's project, see ch. 3, n. 36, below. For Charlemagne maps in the eighteenth century, see ch. 4, nn. 156–59.

74. "Zeeland," *Encyc. Brit.* 28:965; "Flanders," Encyc. Brit. 10:478–80; "Holland, County of," Encyc. Brit. 13:606–9. "Holland proper" distinguishes the county, meant here, from the informal usage of Holland as equivalent to the northern Netherlands as a whole.

75. S. J. Fockema Andreae and B. van't Hoff, *Geschiedenis der Kartografie van Neder-land* (The Hague, 1947), 12. For Zeeland under the Batavians, see W. S. Unger, *Catalogus van den historisch-topografischen Atlas van het zeeuwsch genootschap der wetenschappen: 1st Deel, Gedrukte kaarten en plattengronden* (Middelburg, 1931), 9, no. 41. I learned of the exis-tence of these maps from Black, *Maps and History*, 7. For the inset in Ortelius's *Theatrum* referring to Gui de Dampierre, see ch. 1, n. 33, above.

76. *Inventaris der verzamling Kaarten beustende in het Rijks-Archiev* (Collectie Hing-man), pt. 2 (The Hague, 1871), 2:299, no. 2761 (1737), no. 2762 (1666), no. 1233 (1770), no. 2764 (1540?); Unger, *Catalogus van den historisch-topografischen Atlas*, nos. 43, 47, 49; C. de Waard, *Rijksarchief in Zeeland: Inventaris van Kaarte en Teekeningen* (Middelburg, 1926), 1– 6: nos. 3 (1718), 4–8 (1650), 9 (1727), 10 (1729), 11 (1773), 12 (1617), 13–14 (1701), 15 (1731), 16 (1785), 17 (1805), 18 (1763), 19 (1763), 20 (1727), 21–24 (1649), 25 (1848), 26 (late eighteenth century), 27 (1807), 28 (ca. 1550), 29 (1610, 1625, 1667), 30 (1727), 31. Leiden University Li-brary, Map Collection, Museum Bodellianum, Portfolio 33, nos. 2 (eighteenth century), 3 (ca. 1663), 4–5 (1715), 6 (1726), 7 (1727), 8 (1760s), 9 (1551), 10, Zeeland map "in den eersten druk der Chronyk van Zeeland door Jan Jañs Reyzersberch" (1550). Other Dutch collections (e.g., Amsterdam) should be checked for maps of this kind. References to Egmond: Waard, *Rijksarchief in Zeeland*, nos. 3, 4; Unger, *Catalogus van den historisch-topografischen Atlas*, no. 43 (derived from Gargon's chronicle, see next note); Leiden, nos. 2, 3, 21, 22. Leiden, no. 5, mentions the Curie of Bruges; and *Inventaris . . . Rijks-archief*, no. 1233, mentions the Council of Flanders, Ghent.

77. Jan Reyzenbach (Reigersbergh), *Chronyk van Zeeland* (Antwerp, 1551) (the name Reigersbergh has many spellings; see *NUC*); M. Smallegange, *Nieuwe cronijk van Zeeland: Eerste deel* (Middelburg, 1696), contains chronicles of Jac. Eyndius and J. Reigersbergh. Mattheus Gargon (1661–1728), *Walchersche Arkadia* (Leiden, 1715–17). Fockema Andreae and van't Hoff, *Geschiedenis der Kartografie,* dismiss the Zeeland map of Reigersbergh's and Smallegange's chronicles (it is often invoked in the maps). In these chronicles, Gargon's *Walchersche Arkadia* (1715), also mentioned in this context, is very probably derivative. Ref-erences to Gui de Dampierre: Leiden University Library, Map Collection, Museum Bodel-lianum, Portfolio 33, nos. 3–8; *Inventaris . . . Rijks-Archief,* nos. 2762, 2764, 1233; Waard, *Rijksarchief in Zeeland,* nos. 9–20, 29–31. About Gui de Dampierre, see *Encyc. Brit.* 10:479; Henri Pirenne, *Histoire de Belgique,* 2d ed. (Brussels, 1902): 1:239, 365, 374, 377, 380, 398. Gui's last years were troubled, but he ruled prosperously for many decades.

78. Friedrick Caspar Wieder, *Nederlandsche historisch-geographische Documenten in Spania,* Koninklijk Nederlandsch Aardrijkskundig Genootschap te Nederland (Leiden, 1915), 304–6. Wieder's context is the mapping of provinces, not the making of maps for history or a historical atlas of Zeeland—a possibility that does not occur to him.

79. Fockema Andreae and van't Hoff, *Geschiedenis der Kartografie,* 12; for the cata-logues, see n. 77, above (the chronological sequence is in keeping with normal methods of classification; no one argues that the maps form a progression). The quotation is from

Black, *Maps and History*, 7, wrongly suggesting that the maps are a series and all stem from Egmond.

80. The late copies I consulted at Leiden (see n. 76, above) are not linked decoratively. Catalogue entries and physical remains indicate single items, not parts of an atlas. The maps do not always show the same part of Zeeland, any more than the same date, yet no pattern of coverage leaps to the eye. For Alting, see nn. 135–40, below; for Lamare, see ch. 4, nn. 44–45, below. On Egmond, see A. C. F. Koch, "Egmond," in *Dictionnaire d'histoire et de géographie ecclésiastique*, 27 vols (Paris, 1900–), 4:23–27; N. N. Huyghebaert, in *New Catholic Encyclopedia*, 15 vols. (New York, 1967), 5:191; J. Hof, *De abdij van Egmond van de aanvang tot 1573*, Hollandse Studiën 5 (The Hague and Haarlem, 1973).

81. Most of the maps in Waard, *Rijksarchief in Zeeland*, are listed as being of Zeeland and Flanders; a scattering of the others are of Zeeland and a part of Flanders. None appears to reach northward instead. About Egmond, see the previous note.

82. Malbrancq, 1639. Coverage extended from 309 B.C. to A.D. 1313. A. Beeckman, in *Biographie nationale de Belgique*, 13:212–14, notes Malbrancq's lack of critical discrimination. Gisela was abbess of Chelles, a prominent monastery for women. On Lambert and the *Liber Floridus*, see C. Hümmörder, "Lambert von St-Omer," in *Lexikon des Mittelalters* (Munich and Zurich, 1991), 5:1626; A. Derolez, *Lambertus qui librum fecit*, Verhandlingen van der Konigklijke Akademie voor Wettenschappen, Klasse der Letteren, 40th year, no. 89 (Brussels, 1978).

83. A. Vander Meersch, in *Biographie nationale de Belgique*, 6:22–24. Wrée entered the Jesuit order but withdrew as a novice. His *Sigilla comitum Flandriae* is complemented by a work on the genealogy of the counts (1642); see R. L. van Caeneghem, *Kurze Quellenkunde des westeuropäischen Mittelalters*, tr. Maurits Gysseling (Göttingen, 1964), 322. Wrée has been called the first scholarly sigillographer.

84. L. Godeaux, in *Biographie nationale de Belgique*, 27:180–84. After some years in Provence, Wendelin returned, was ordained in 1619, and had, for a time, a canon's prebend.

85. Wrée, 1647; repr., Aa, 1729a.

86. Wendelinus, 1649. Wendelin's book, a Plantin product, is very attractive. Part 1 is a text of the *Pactus legis Salicae*, presumed to be the earliest state of the Frankish law code, *Lex Salica*. Procopius of Caesarea is a well-known Greek historian contemporary with Justinian (527–65).

87. Fabius, 1641. On Biondo, briefly, see *Enciclopedia Italiana*, 7:56. Sanders is likely to have modeled his work on Zander van Boxhorn (see next note) rather than Biondo. On Sanders and his enterprise, see the very full account of V. Fris, "Sanderus (Antoine Sanders, dit)," in *Biographie nationale de Belgique*, 21:317–67; on his attachment to listing relics and miracles, 338. Robert L. Volz, Chapin Library, Williams College, very helpfully responded to my inquiries about *Flandria illustrata* (18 October 1990).

88. *Flandria illustrata* was never completed. It occasioned a new flurry of interest a century later; the resulting reprints presumably explain how the map came to the attention of Köhler, 1730. On Sanders's financial difficulties, see Fris, in *Biographie nationale de Belgique*, 21:335–39; the value of his work, 344–45; 344, his inspiration for illustrations of the volumes came from Marc Zander van Boxhorn, *Theatrum sive Hollandiae comitatus et urbium nova descriptio* (Amsterdam, 1632).

89. Fabius, 1641. Fabius acknowledges debts to Philip Clüver, Aegius Bucher (Gilles Bouchier, a Belgian Jesuit), and J. J. Chifflet, a celebrated scholar. On Fabius, see A. Vander Meersch, in *Biographie nationale de Belgique*, 6:818–19. Note also Henry Stevens, *Bibli-*

ographia geographica et historica (London, 1872), pt. 1: no. 1058, "Flanders [A Map of, with the following Inscription,] Carolus Calvus Imp. Balduino Ferreo et Judithae Filiae suae Conjugibus Somonae Scaldi Oceanoque inclusam Flandriae nomine appellatam in Dotem tradit," with the fanciful date 861. This map (NOT SEEN) is probably a single-sheet derivative from Fabius's. On Charles, Judith, and Baldwin I, see *The Annals of St-Bertin,* tr. and ed. Janet L. Nelson (Manchester and New York, 1991), 97, 103–4, 106, 110; Janet L. Nelson, *Charles the Bald* (London, 1992), 203–4. Rosamond McKitterick, *The Frankish Kingdom under the Carolingians, 751–987* (London and New York, 1983), 249–51, points out that medieval Flanders is better thought to have originated with Baldwin II (879–918) than with his father.

90. Ewich, 1652. Part of the map title is catchy, as though to say "between Jules and Charles Caesars." On Ewich, see *DBiogArchiv,* fiche 299, frame 134 (thin). The name of the engraver Savery (1594–after 1665) appears elsewhere in the form "Savry": *Allgemeines Lexikon der bildenden Künstler,* ed. V. Thieme and E. Becker, 57 vols. (Leipzig, 1907–55), 29:505.

91. Several map titles in Philip Clüver's *Germania antiqua* (Leiden, 1616; 2d ed., 1631), a famous book, anticipate Ewich's catchy phrasing: no. 3,"Cisrhenan Germany about the time of J. Caesar"; no. 4, "The same as it was between J. Caesar and Trajan"; no. 7, "The tribes of Germany between Rhine and Elbe between Caesar and Trajan," facing no. 8, "The same, as they lived about the time of J. Caesar"; no. 9, "The tribes of Germany between Rhine and Elbe, where they lived between the emperorship of Trajan and about the age of Ammianus [Marcellinus]."

92. A map in Ortelius's *Theatrum* shows the Netherlands with a westward orientation. Another example is Doncker's famous *Leo Belgicus,* near the time of Ewich's map: Jan W. H. Werner, *Leo Belgicus gedrukt door Hendrick Doncker: Een beknopte toelichting* (Amsterdam, n.d. [1990?]).

93. The difference between seventy-one medieval battles and the total of ninety is made up by the battles between 1500 and 1600.

94. Speed, 1627b, is my basic reference. Quotation from Skelton, in *John Speed, "A Prospect,"* x. See also Shirley, *Maps of Britain to 1650,* 104, no. 255, 106, no. 261. Medieval battles in Speed's list, in England, nos. 1–64 (A.D. 1066–1497); in Ireland, nos. 1–7 (1170–1399). Postmedieval battles are 14 percent of the total in England, 50 percent in Ireland, a contrast reflecting the chronology of British-Irish history. Earlier Italian portrayals of pikemen: Novacco, 19—/O, no. 13 (1559), cf. no. 30; AtlFact, 16—/O, nos. 26, 27 (1560s); Lafréri, 1980/B, no. 143 and passim. I shall give a fuller account of this map in a separate article.

95. For Speed's map in the French revolutionary era, see ch. 4, nn. 39–42, below. For use in the *Atlas Lesage,* see ch. 5, n. 42; the offshoot is Lévi-Alvarès, 1840. The name "poster" map is suggested by a comment by Johannes Dörflinger, cited in ch. 5, n. 42, below.

96. My separate article will deal with the relationship of *Prospect,* Speed's map of battles, and *Theatre.*

97. For massed pikemen, see n. 94, above.

98. Michael McCormick, *Eternal Victory: Triumphal Rulership in Late Antiquity, Byzantium and the Early Medieval West* (Cambridge, 1986), 386–96.

99. Nigel Nicholson and Alasdair Hawkyard, *The Counties of Britain: A Tudor Atlas by John Speed* (London, 1988), a reproduction of *Theatre;* battles appear in Bedfordshire, Berkshire, Buckinghamshire, Cumberland, Durham, Gloucestershire, Herefordshire, Hertfordshire, Leicester (where Speed expresses typical regrets that the civil dissension of York and

Lancaster "spent England more blode than twice had done the winning of France"), Lincoln, Norfolk, Northamptonshire, Shropshire, Stafford, Sussex, Warwick, and Worcestershire (plus two more in Wales). The vignettes are large enough for triangular pennons to peep out over the massed pikes.

100. On the Danckerts, see J. Keuning, "Cornelis Danckerts and His 'Nieuw Aerdsch Pleyn,'" *Imago Mundi* 12 (1955): 136–39; Koeman, 2:88–90; R. V. Tooley, *Dictionary of Mapmakers* (New York, 1979), 145–46. Only Tooley identifies Cornelis II Danckerts (1603–56) with the engraver of Speed's map. Keuning dates Cornelis's first appearance to 1628; Koeman mentions 1633. All dated maps in *Prospect* are marked 1626; "Corn. Danckertsz sculpsit" has to be Cornelis II. He was apparently a beginner. Koeman underscores the quality of his work.

101. On the political dimension, see Thomas Cogswell, *The Blessed Revolution: English Politics and the Coming of War, 1621–1624* (Cambridge, 1989), 1, 13–14; Roger Lockyer, *The Early Stuarts: A Political History of England, 1603–1642* (London, 1982), 23, 205–6; Roger Lockyer, *Buckingham: The Life and Political Career of George Villiers, First Duke of Buckingham, 1592–1628* (London, 1981), 168. I am very grateful to Dr. Neil Cuddy, my sometime colleague, for these and other references.

102. Boisseau, 1641. See also Pastoureau, *Atlas français*, 67.

103. Boisseau, 1641. The legend of Franks proceeding out of south Russia first occurs in an eighth-century history. Concerning Jerusalem, Boisseau states that several persons called kings are not named because they never effectively controlled the kingdom.

104. The three French "races," dynasties to us, are the Merovingians, Carolingians, and Capetians. In its appeal to history to corroborate contemporary self-esteem, Boisseau's map is reminiscent of Lazius, 1561. For the post–World War I edition of Putzger, 1877, see F. W. *Putzgers historischer Schul-Atlas*, rev. and ed. by Max Pehle and Hans Silberborth, Grosse Ausgabe 50. Jubiläums-auflage (Bielefeld and Leipzig, 1931), no. 134, "Das Deutschtum" (BAV, call no. Geog. III, 42); commentary, Wolf, "100 Jahre Putzger," 708–11.

105. See ch. 3, n. 13, below.

106. Pellegrino, 1644. About Pellegrino, *Biog. univ*, 32:396 (this source, unconfirmed, is not wholly reliable for a story like this). Aa, 1729a, vol. 17, no. 2, reproduces a second, but modern, map by Pellegrino from a different book.

107. Pellegrino, 1644. For maps of Italy with similar designs, see Novacco, 19—/O, nos. 67, 69, 70 (1557–91); there is one in Ortelius's *Theatrum*.

108. "Lector. Typum vides universae *Italiae Cis-Tyberinna*, in qua . . . alia, et alia pro arbitrio formant schemata ac reformant; *Langobardi* latissimum olim instituere Ducatum *Beneventi*, quem postea in *Beneventanum* pariter principatum et *Salernitanum* dispertivere, ac rursus Principatum ex iis extraxere *Campanum*. Graeci quoque plura, disjunctaque antiquo sub nomine *Calabriae* primitus servantes loca, notisque dein Brutiis eiusmodi vocabulum ascribentes, post varios postremum casus bina Themata *Calabriae* et *Apuliae* statuere: *Nortmanni* vero ditiones utrisque ademptas Ducatum appellarunt *Apuliae*: qui hunc tandem *Campano* simul Principatu minoribusque aliquot aliis *Cis-Tyberini* etiam sitis Dynastiis, et praeterea *Siciliae* Insulae ei adjunctis, in unius regni formam transmutarunt."

109. Pastoureau, *Atlas français*, 293. Duval, 1667/A, contains a wide selection of La Rüe maps.

110. The Holy Land atlas, La Rüe, 1651/E. For Armenia, see La Rüe, 1653. As sources, La Rüe cites *Codex Iustinianus*, bk. 1, ch. 29, ¶5 (adjusting to our standard edition), and *Novella* 31. About Justinian and Armenia, briefly, see H. Gelzer, *Die Genesis der byzantinischen Themenverfassung* (Leipzig, 1899; repr., Amsterdam, 1966), 23–24.

111. Lubin, 1659/E (*Orbis Augustinianus*), 1660/E (martyrs). Mentioned in ch. 1, n. 95, above.

112. Duval, 1665. For an earlier discussion, see ch. 1, n. 78, above.

113. Duval, 1665, pt. 3, no. 1, "La carte de l'empire des Sarrazins ou des caliphes sous Vlit qui regnoit environ l'an 700. Tiré d'Abulfeda, de Nassir-Eddin, d'Ulug-Bei et d'autres auteurs Arabes"; no. 2, "Regionum ultra Oxum tabula geographica ex Abulfeda Ismaele. Principe Hama, anno 1345." On Nasir al-Din al-Tusi (1201–74), see Joachim Lelewel, *Géographie du moyen âge,* 4 vols. (Brussels, 1852; repr., Amsterdam, 1966–67), 1:116–23; Heinrich Suter, *Die Mathematiker und Astronomen der Araber und ihre Werke* (Leipzig, 1900), 146–53; J. Ruska, "al-Tusi . . . ," *Encyclopedia of Islam,* new ed., vol. 4, pt. 2:980–82; George Sarton, *Introduction to the History of Science* (repr., Huntington, N.Y., 1975), 2:1001–13. On Abu'l-Fida (1273–1331), see H. A. R. Gibb, "Abu'l-Fida," *Encyclopedia of Islam,* new ed., 1:118–19; Lelewel, 1:147–52; Suter, 160; Sarton, 3:200, 793–99 (Abu'l-Fida was prince of Hamah, a town on the Orontes, about halfway between Aleppo and Damascus, recently leveled to the ground). On Ulug Bey (†1459), see Lelewel, 1:155; Sarton, 3:1120. Nasir al-Din, astrological counselor to the Mongol Hulagu (scourge of the Abbasid Caliphate), is the most distinguished scientist of the three.

114. On John Greaves (1602–52), see *DNB* 8:481–82; he lost his professorship in 1648 because of his loyalty to Charles I. His editions: *Epochae celebriores . . . ex traditione Ulug Beigi,* with *Chorasmiae et Mawaralnahrae, hoc est, regionum extra fluvium Oxum descriptio, ex tabulis Abulfedae Ismaelis, Principis Hamah* (London, 1650); *Binae tabulae geographicae, una Nassir Eddini Persae, altera Ulug Beigi Tartari* (London, 1652). Greaves's *Insigniorum aliquot stellarum longitudines . . . ex astronomicis observationibus Ulug Beigi* was published in his professorial predecessor John Bainbridge's *Canicularia* (Oxford, 1648). The three geographers partly edited by Greaves, and no others, are mentioned by Duval. One would like to know what led Duval to situate his map of the caliphate in the time of Walid ("Ulid")—not that any details of the map point specifically to this time.

115. About Jaillot, see *LGKartog.* 1:353–54.

116. Jaillot, 1693/E. On its publication, see Pastoureaux, *Atlas français,* 140–41, 250 (Jaillot, I D, no. 107). The date is a terminus ante quem; Jaillot's map probably circulated before being incorporated into the Sanson atlas. Handsome reproduction with commentary in Nebenzahl, 136–37. Rogg, 1740/E is a reengraved copy rather than a reprint. Altered version published with map of the New Thebaïd (La Trappe): Fer, 1700/E. Reprintings: Le Clerc, 1705/A, nos. 37–38; Châtelain, 1705, vol. 7, no. 35 (reduced).

117. Desert fathers and their literature (including John Cassian) are explained in *New Catholic Encyclopedia,* 4:793. For their relevance to Pope Gregory XIII, see Goffart, "Christian Pessimism," 794 n. 22.

118. Cuthbert Butler, "Rancé, Armand de (1626–1700)," in *Encyc. Brit.* 22:885, and "Trappists," in *Encyc. Brit.* 27:213–15 (quotation, 214). L. J. Lekai, in *New Catholic Encyclopedia,* 12:78–79, gives a scathing sketch of Rancé's régime. The reform was instituted at the monastery of Notre-Dame de La Trappe (the Trappe is a stream). On the influence of the desert fathers on Rancé, see Gérard Michaux, in *Dictionnaire du Grand Siècle* (Paris, 1990), 1300–1301. For background on John Climacus, see *New Catholic Encyclopedia,* 7:1045; on Dorotheus, whose main work Rancé translated into French, see *Bibliotheca sanctorum,* 12 vols. (Rome, 1961–69), 4:826–28.

119. Nebenzahl, 72–73 (Mercator's map is on such a large scale that the filiation is not immediately apparent); for other models, see Nebenzahl, 136.

120. The de Fer map was, I gather, first published in 1696, only a few years after Jaillot's (the one I record is marked 1700). According to Johann Hübner, *Museum geographicum: Das ist Ein Verzeichniss der besten Land-Charten* (Hamburg, 1726), 203, the monks requested it. La Trappe, though perhaps isolated, was certainly not cut off from centers of population. Many visitors came to attend retreats (and acquire maps as souvenirs?).

121. My opening sentence owes much to Erskine Childers, *The Riddle of the Sands* (London, 1903; repr., Penguin, 1978), 84 and ch. 7; a matchless evocation of Frisia. "Of the shoal spaces which lie between [the islands] and the mainland, two-thirds dry at low-water, and the remaining third becomes a system of lagoons" (153). For details, see *Encyc. Brit.* 11:233, "Frisian Islands": 1065 sq. mi. in 1250, 105 sq. mi. in 1850; they mark "the outer fringe of the former continental coast line." Despite attempts at defense they are bound to crumble "under the persistent attacks of storm and flood." "Many . . . Frisian legends and folksongs deal with . . . submerged villages and hamlets." For an improved account of this coastline (and the actual, secondary nature of the islands), see F. J. Monkhouse, *A Regional Geography of Western Europe*, 4th ed. (London, 1974), 29–32, 47; there is a certain equilibrium of land and water, 27. For annals of "[s]ea floods which caused losses of land and/or cost many human lives" (off the North Sea), see H. H. Lamb, *Climate: Past, Present and Future*, vol. 2, *Climatic History and the Future* (London and New York, 1977), 120–26, table 13.3: the Jadebusen (Weser Estuary) was formed in 1218 and 1511; the Dollart (Ems Estuary) was flooded in 1277 (see also Monkhouse, *Regional Geography*, 31); the Zuyder Zee was created from 1219 to 1421. See also Mark Bailey, "*Per impetum maris:* Natural Disaster and Economic Decline in Eastern England, 1275–1350," in *Before the Black Death: Studies in the "Crisis" of the Early Fourteenth Century*, ed. B. M. S. Campbell(Manchester and New York, 1991), 184–208, esp. 191–96 (destruction of houses and towns, soil erosion). Also interesting is Harms, *Themen*, 127: a contemporary map of the great sea surge of 1717 in "Nieder-Teutschland."

122. On Mejer (sometimes spelled Meier or Meyer), see *ADBiog.* 21:200–202; *LGKartog.* 1:151; H. A. Hens, in *Dansk Biografisk Leksikon*, vol. 9 (1981), 498, and "Sax, Peter," in *Dansk Biografisk Leksikon*, vol. 12 (1962), 635–66. Concerning Mejer's cartography, see Christian Degn, *Schleswig-Holstein, eine Landesgeschichte* (Neumünster, 1994), 69; he stresses his stay in Holland. A prospectus for the atlas of the Danish kingdom appeared in 1657 (one volume of the seven was destined for historical maps); war interfered with the project. See now Mejer, 1942/O. In the 1933 exhibition recorded in *Catalogus mapparum*, 21–22, no. 35, Mejer represented Denmark. Was the decision to feature this Dane from (lost) Schleswig based only on scholarly considerations?

Mejer, 1652; Danckwerth, 1652. Danckwerth, a medical doctor and burgomaster of Husum, wrote the text (for a positive appreciation of him, see Degn, *Schleswig-Holstein*, 138). Mejer judged it very unsatisfactory in political orientation and ignorance of the land and so wrote an 1,100-folio revision, never published. Danckwerth and he fell out publicly and sought legal remedies. The atlas plates, sold by Danckwerth's widow for a derisory 360 thaler, were soon acquired by Joan Blaeu (for the Blaeu publication, see n. 124, below).

123. The three maps situate the microcosm of Schleswig-Holstein in the macrocosm. "Orbis vetus cum origine in eo gentium a filiis et nepotibus Noe": Mejer, 1652: 28; "Germania antiqua australis": 32; "Germania antiqua septentrionalis": 44. Both "Germanies" are subdivided into tribal districts (not an innovation). Scandinavia is also treated as "Germanic" in Clüver, 1616/A; the connection was not based, as with us, on primarily philological grounds.

124. (1) "Helgelandia A° 1649. Helgelandt in annis Christi 800 1300 and 1649," in facsimile in *Géographie Blaviane* (Amsterdam, 1663), 1:xxxviii. The same pair also appear with
an alternative name for the first map: "Newe Landtcarte von der Insull Helgelandt Anno
1649." (2) "Frisia Borealis in ducatu Sleswicensi sive Frisia cimbrica Anno 1651" and "Frisia
Borealis in ducatus Sleswicensi Anno jz40 [= 1240] Frisia Cimbrica antique" (both on the
same scale), facsimiles in *Géographie Blaviane*, 1:xxxvii. They also appear under the titles
"Landtcarte von dem Nortfrieslande in dem Hertzogthumbe Slesswieg. Anno 1651" and
"Landtcarte von dem alten Nortfrieslande Anno 1240. Frisia Cimbrica antiqua." "Cimbria,"
which occurs in ancient writers, was the term used in refined Latin for Denmark. (5) Oldenburg, in *Géographie Blaviane*, 3:61a. Note, too, *Géographie Blaviane*, 3:64: plan of Oldeschloh in 1380, without modern comparison (Oldeschloh is east of Lübeck). (6) "Ditmarsiae
tabula ann. 1559" and "Ditmarsiae tabula ann. 1651" (both on the same scale), facsimiles in
Géographie Blaviane, 3:68a. Cf. Camden, *Britannia* (London, 1607), 230–31, map of Kent;
outside Rye and Winchelsey, "Vindelis Insula. Olde Winchelsey whose Ruyns lurk unseen
under the sea waves."

125. Lamb, *Climate: Past, Present and Future*, 120–22 (Heligoland; 800 is marked "and/
or 806"; 1216 is also noted as catastrophic for Heligoland), 123–25, cf. 433 n. 1 (North Frisia);
word was sent to the Council of Basel (1431–37) "that 60 parishes accounting for over half
the agricultural income of the then Danish diocese of Slesvig (Schleswig) had been 'swallowed by the salt sea'" (123). Heligoland appears on the edge of Mejer's maps of North
Friesland, as well as on its own page. For Sax's *Nordstrand,* see the biographical article cited
in n. 122, above. Degn, *Schleswig-Holstein,* 69, makes much of a vastly destructive surge in
1362—not, I think, mentioned by Mejer; 139, the heavy casualties of 1634.

126. One copy of Jørden's map (Hamburg, 1559) survives at Leiden; a later edition of
1580 is wholly lost. About Jørden, see Bagrow, *Ortelii Catalogus,* reprinted in *Acta cartographica* 27:187–88.

127. My main guide (quotations included) is Christian Degn, unpaged introduction to
K. Domeier and M. Haack, eds., *Die Landkarten von Johannes Mejer, Husum* (Hamburg and
Bergedorf, 1963). Also on collaboration with Sax, see *Catalogus mapparum,* 21. Peter Sax, *De
praecipuis rebus gestis Frisiorum septentrionalium . . . libri sex* (1656), bk. 6, ch. 4, "De locis
desolatis in Friside septentrionali," in E. J. von Westphalen, *Monumenta inedita rerum Germanicarum,* 4 vols. (Leipzig, 1739–45), 1:1377–82. Sax's language is steeped in the terminology of Roman law. Mejer was not alone in daring to attempt bold historical reconstructions;
e.g., in a map of ancient Italy, Giacomo Gastaldi "endeavoured to reconstruct the conditions [of the seacoast] as they existed in Strabo's epoch"; see Roberto Almagià, "An Historical Map by Giacomo Gastaldi," *Imago Mundi* 5 (1948): 14–15.

128. John Wain, "Alternative Poetry," *Encounter,* 42, no. 6 (June 1974), 26–38, here 34–
35, with special reference to James MacPherson and his Ossian fabrications.

129. Dutch Friesland is composed of the provinces of Friesland (capital, Leeuwarden)
and Groningen. Some districts and islands on the western side of the IJssel Meer (ex–Zuyder
Zee) may also count as Frisian but do not concern us. The wider sense of Frisia was recognized in the eighteenth century, e.g., by Le Long, 1:35, nos. 405, 406: "Cette grande Frise
répond aux Provinces-Unies d'aujourdhui, dont la partie méridionale, bornée par l'ancien
Rhin, étoit seule de l'ancienne Gaule; mais le reste fut du domaine de nos Rois, à la fin de la
première Race, et sous une partie de la seconde."

130. For these early maps, see Vanden Broecke, *Ortelius Atlas Maps,* 127, no. 83, 121, no.
77; Meurer, *Fontes cartographici Orteliani,* 175, 146.

131. Schotanus, 1664, 1698. The thirty-five maps of the latter are larger than and entirely different from those of 1664. For the title, see Koeman, 3:122, where the 1664 atlas is mistakenly attributed to Bernhard rather than his father. The *Friesche Atlas* was reprinted by Francis Halma as *Uitbeelding der Heerlijkheit Friesland* (Leeuwarden, 1718), with Bernhard named as author (unlike 1698). There is a facsimile of the Halma reprint of *Uitbeelding* (Amsterdam and Leeuwarden, 1979) listing Schotanus as author. See Koeman, 3:122–23. Francis Halma, *Toneel der Vereenighde Nederlanden* (Leeuwarden, 1725), is basically a dictionary, with images or portraits of major figures in Dutch history, as well as maps. On the origins of Schotanus's survey, see Bodel Nyenhuis and Eekhoff, *Kaarten van Friesland,* 55. See also J. Keuning, "Bernardus Schotanus à Sterringa, zijn leven en zijn kartografisch oeuvre," *Vrije Fries* 42 (1955), 37–87 (particularly detailed on the difficult financing of the atlas).

Christian Schotanus (1603–71), but not Bernhard, has a large entry in Vander Aa, *Biographisch Woordenboek der Nederlanden,* 6:136–37; and in Bergmans, in *Nieuw Nederlandsche Biografisch Woordenboek,* 5:700–701. A minister and educator, he is noted as an early historian of Frisia.

132. Bodel Nyenhuis and Eekhoff, *Kaarten van Friesland,* 6–7, consider Schotanus's 1664 map still usable. Excellent care is taken to portray an "ancient" form of the Netherlands in Dubos, 1742; the map is of Gaul in 407. Early Netherlands in recent atlases: *Anchor Atlas,* some maps (notably of ancient times) are pre–Zuyder Zee, many are more casually anachronistic; *Times Atlas of World History,* modern outlines from the tenth century at least; *Harper/Collins* (initially French), modern outlines exclusively; Putzger, 1877, 42d ed. (1920), shows the lake in the first century (European provinces of Rome), then the Zuyder Zee from the tenth or eleventh century. Early maps for history are often more scrupulous. About the Zuyder Zee, briefly, see *Encyc. Brit.* 18:1049; *Encyc. Brit.,* 15th ed. (1974), 12:940 (the central lake apparently widened in late Roman times, but water levels were lower from 700 to 1200); Monkhouse, *Regional Geography,* 31–32.

133. Clüver, 1611 (Rhine branches), 1616/A. *ADBiog.* 4:353–54; better, *LGKartog.* 1:144; Stephen A. Bromberg, "Philipp Clüver and the 'Incomparable' *Italia antiqua,*" *Map Collector* 11 (June 1980): 20–25. Clüver's adventurous life took him widely over Europe; he held that ancient geography must trace early conditions from modern ones.

134. See introductory remarks by J. J. Kalma, in *Uitbeelding* facsimile (1979), 13–14 (n. 131, above); Bodel Nyenhuis and Eekhoff, *Kaarten van Friesland,* 6–10. Earlier disparagement: J. A. de Chalmont, 1798. L. P. C. van den Bergh, *Handboek der Middel-Nederlandsche Geographie naar de bronnen bewerkt* (Leiden, 1852; 2d ed., 1872; 3d ed., 1949), ix, despite many defects, "Alting's work has never been surpassed or even equaled."

135. *ADBiog.* 1:369 treats the burgomaster as a footnote to the Reformer. Also Vander Aa, *Biographisch Woordenboek der Nederlanden,* 1:69, "a man of great ability" (the Reformer is Menso 3.; the burgomaster, Menso 6., said here to be grandson to the foregoing); *Nieuw Nederlandsch Biografisch Woordenboek,* 1:98. The earliest Menso Alting recorded by biographers was a counselor to the duke of Guelders and died in 1325; the mapmaker was a direct descendant, trained in law and its administration.

136. Alting, 1701. Cf. Muller, 1863/B, 4:14. On the Alting maps in *Aloude Holland,* 1745, see ch. 4, n. 141.

137. For La Rüe, see nn. 148–49, below; for Lamare and Paris, see ch. 4, nn. 44–47, below.

138. Lamb, *Climate: Past, Present and Future,* 433, 450 (Zuyder Zee as a thirteenth-

century creation); see also n. 128, above. Schotanus knew better, and Alting almost certainly was acquainted with his map.

139. The "emigration" and "return" of the Franks—formulas unfamiliar to our history books—refer to the Franks' setting out to conquer Gaul and, three centuries later, bringing the (pagan) Kingdom of Frisia under their rule.

140. For quotation, see n. 134, above. The next national atlas is Hase, 1750b.

141. "Chronology and geography are the two eyes of history": Joseph de Jouvancy, S.J. (1643–1719), is identified as the originator of this saying by Ildefons Stegmann, *Anselm Desing, Abt von Ensdorf, 1699–1772: Ein Beitrag zur Geschichte der Aufklärung in Bayern,* Studien und Mitteilungen Ordinis sancti Benedicti, Ergänzungsheft 4 (Munich, 1929), 132. Monique Pelletier, "Les géographes et l'histoire, de la Renaissance au siècle des Lumières," in *Apologie pour la géographie: Mélanges offerts à Alice Saunier-Seïté,* ed. J. R. Pitte (n.p., 1997), 146, names Cicero as presumed author, but Hugo Merguet, *Handlexikon zu Cicero* (Leipzig, 1908), does not corroborate either at "historia" or "geographia," and "chronologia" is unattested. On Jouvancy, see also Sommervogel, 4:830–59, whose bibliography does not confirm (or disprove) Jouvancy's authorship of this dictum.

142. Quoted by Alpers, *Art of Describing,* 159.

143. Wilkinson, 1797/A (1805 ed.): iii-3, iv-2, vii-1, vii-4.

144. Nebenzahl, 14–17.

145. Sanson's great project (see ch. 1, n. 94, above) illustrated the continuity of the present with the late Roman *notitia Galliarum,* dramatizing the utility of modern ecclesiastical boundaries for antiquarian research.

146. On Jouvancy, see n. 141, above.

147. Duval, 1665. La Ruë, 1651/E: "1. Terra Chanaan ad Abrahami Tempora Per Populos XI Item per Toparchias Idumeae totidem et Stationes XLV ad Moisis Tempora; 2. Terra Promissa in Sortes seu Tribus XII; 3. Regnum Salomonicum; 4. Regnum Iudeorum in Filios Herodis Magni Per Tetrarchias divisum Ad Tempora Christi Domini; 5. Pinax Geographicus Patriarchatus Hierosolymitani [553–1250]; 6. Sourie ou Terre Saincte Moderne [political divisions under Ottoman rule]; Addition. Assyria Vetus Divisa in Syriam, Mesopotamiam, Babyloniam, et Assyriam Proprie dictam."

148. La Ruë's modern Syria was modern in name rather than content. See Pastoureau in *Nicolas Sanson, Atlas du monde, 1665,* ed. Mireille Pastoureau (Paris, 1988), 57; she emphasizes that the map is mainly the "œuvre pieuse" of a "naiveté toute médiévale," showing Abraham's oak, Adam's birthplace, etc. The combination of naive content and forward-looking atlas design does not have to be contradictory.

149. Augustin Lubin, *Mercure géographique ou le guide du curieux des cartes géographiques* (Paris, 1678), 8–10. I applaud Hofmann, "Genèse de l'atlas historique," 103, for having given prominence to Lubin's book.

150. Lubin, *Mercure géographique,* 10–11.

151. Lubin, *Mercure géographique,* 356 (in the explication of "Mappa").

152. Alting, 1701. Part 1 is ancient (five maps); the medieval part 2 (nine maps) is, in fact, the earliest atlas of medieval history. See nn. 136–40, above.

153. See ch. 4, nn. 44–45, below. Very near Lamare's time, Jacques de La Feuille drew five maps of the development of Amsterdam; they flank a current map of the city, "Amstelodami veteris et novissimi delineatio," reproduced in W. F. Heinemejer et al., eds., *Amsterdam in Kaarten* (Ede, 1987), 46–47 (Hoofdstuk 10).

154. Tilleman Stella, in Nebenzahl, 76–77. Speed, 1627; discussed in nn. 93–101, above. On the Cranach Holy Land map, see Delano-Smith, "Maps as Art and Science," 67–68. Vignettes on the Vatican map of the Duchy of Milan: Gambi and Pinelli, *Galleria,* 3, map 19.

155. Narrative painting "in which a story is illustrated by showing several stages of a narrative side by side within the same frame" has a long history: see David Woodward, "Reality, Symbolism, Time, and Space in Medieval World Maps," *Annals of the Association of American Geographers* 75 (1985): 510–21 (quotation, 514). World War I in one map, plus four insets: *Times Atlas of World History,* no. 252. Alaric's tracks cross Attila's: *Westermann's Atlas zur Weltgeschichte,* Teil 2, *Mittelalter,* ed. H. Quirin and W. Trillmich (Braunschweig, 1956), no. 48, map 1.

156. About Lamare, see ch. 4, nn. 44–45, below. About Alting, see Bodel Nyenhuis and Eekhoff, *Kaarten van Friesland,* 7–10; Mees, 19. About the much respected Philip Clüver, Hauber comments (*Versuch einer Historie,* 131n), "Unterweilen trauet er seiner Phantasie allzuweit zu, allein ein Leser [as distinct from a viewer] kan ja unterscheiden, was nur eingebildet, und was bewiesen seye." For Mejer, see n. 128, above. Kircher, 1675: 192; about Kircher, see Anthony Grafton, *Defenders of the Text: The Traditions of Scholarship in an Age of Science, 1450–1800* (Cambridge, Mass., 1991), 159, 184.

157. The stopping points are probably less certain to us than they were then: Y. Aharoni and M. Avi-Yonah, *The Macmillan Bible Atlas,* 3d ed. by A. Reimy and Z. Safrai (New York, 1993), no. 48; *Harper Atlas of the Bible,* ed. James B. Pritchard (New York, ca. 1987), 57.

158. Edward H. Dahl, "The Original Beaver Map—De Fer's 1698 Wall Map of America," *Map Collector* 29 (1984): 22–28. On d'Anville and Africa, see Siegmund Günther, *Geschichte der Erdkunde* (Leipzig and Vienna, 1904), 172, 183.

CHAPTER 3

From 1700, New Departures

Three quarters of the way through the eighteenth century, Johann Christoph Gatterer, professor of history at the University of Göttingen, prepared a map collection in forty-four sheets, of which twenty-four formed a historical atlas from approximately A.D. 100 to 1500. This atlas was meant as a *historische Hilfswissenschaft,* an aid to history, for the course in geography that Gatterer was about to undertake. A full list of Gatterer's twenty-four subjects survives, although most of the maps do not. His subjects bear a recognizable resemblance to what a comparable collection would contain today.

Gatterer's atlas for use in classrooms indicates how far maps for history had advanced toward respectability. At the beginning of the eighteenth century, anyone who wished to visualize modern history, since the fall of Rome, could resort only to modern maps, unmodified for the purpose. Even atlases of ancient geography offered many more maps of ancient lands than of moments or scenes of ancient history. Since the days of Ortelius, specialized maps for history had been produced from time to time in very small numbers. The eighteenth century experienced a livelier pace of change: from its earliest decade, maps were published, often in groups of two or three, detailing moments of medieval history. Somewhat later, complete historical atlases structured by chronological, rather than geographical, order were compiled. Innovations were intermittent, suggesting that mapmakers worked with little competition or cooperation among themselves, at least in this part of their labors. Nevertheless, cumulative changes were not without effect.

ATLASES CALLED "HISTORICAL"

The first atlas explicitly called "historical" appeared in 1705 and took more than fifteen years to be fully published. Its title perfectly illustrates the idea current since Ortelius that general atlases were primarily of historical value — the "eye of history." The Amsterdam *Atlas historique* is anonymous, ascribed on its title pages to "M. C*****," and is now associated with the name of its Huguenot publisher, Zacharias Châtelain. Catalogues and bibliographies more commonly cite as author a different Châtelain, a shadowy figure named Henri Abraham C., but his participation lacks corroboration and this ascription is gradually being eradicated. A prudent recent commentator restores the *Atlas* to anonymity. Yet there is a plausible, if not conclusive, account attributing the work to the publisher Zacharias Châtelain. A reviewer at the time affirms without qualification that Zacharias was the author and that, for several decades, he had left the publishing firm to the care of his wife and children while he engaged in learned studies.[1]

Zacharias Châtelain presumably oversaw the general design of the *Atlas* and provided it with maps, illustrations, chronological charts, and other historical materials. The "dissertations" accompanying the illustrative material were delegated to the literary jack-of-all-trades Nicolas Gueudeville, a runaway to Holland from a monastery in Normandy. Gueudeville's participation brought controversy to the *Atlas*. An exile from France and monastic life, he made no secret of his hostility to Louis XIV and the Roman Catholic church. Most French reviewers, while praising the main compiler's share, found Gueudeville's style incongruous and his grasp of fact insecure. Gueudeville's failings attracted more attention than Châtelain's learning. Far from damaging the enterprise, however, criticism had a positive effect on sales and Gueudeville's reputation. The seven folio volumes issued by Zacharias Châtelain and his son-in-law François L'Honoré were enthusiastically acquired by buyers and are plentiful in map collections today.[2]

The Amsterdam *Atlas historique* is unoriginal in design and composition. Its subtitle declares that it was "[a] new introduction to history, chronology, and ancient and modern geography, set out in new maps in which notice is given of the establishment of the states and empires of the world, their duration, their fall, etc." The promised product was an introduction to history and geography, and contemporaries certainly thought that it contained a welcome complement of chronological, genealogical, and other historical materials. The end result nevertheless was a conventional survey of the countries of the world. If the subtitle just quoted advertises a new and different product (which need not be the case), the compilers fell short of fulfilling their pledge.[3]

In portraying the past, the Châtelain collection draws mainly on the repertory of the "Ancient Atlas, Classical and Sacred," adding finely engraved chronological and genealogical charts of ancient kings and emperors. Historical time is more telescoped than ever: the map of the four ancient world empires singles out the homelands of the invading peoples held responsible for the fall of the Roman Empire; the Greek empire indiscriminately includes the tracks of the Argonauts, Xenophon's 10,000, Alexander the Great, and Saint Paul; the map of the Roman Empire at its height comes complete with traces of Julius Caesar's campaign against Pompey as well as comments concerning— again—the dangerous barbarians to the north and south. The seventh volume, a supplement, fills gaps in the coverage of ancient geography. One of its maps combines Noah's Mount Ararat, Alexander's campaigns, Leonidas at Thermopylae, Mount Olympus, and the battle between the Romans and Hannibal at Cannae. In the same spirit, the explanatory notes sometimes anticipate the dizzying shortcuts found in future summaries of the barbarian invasions. Chronological and thematic clarity was sacrificed to compression and thrift.[4]

Even before the last Châtelain volume left the printer, there appeared in Frankfurt a work bearing the same name, the *Atlas historicus* by Johann Georg Hagelgans.[5] The Hagelgans "atlas" was not a competitor to the Amsterdam volumes; his creation features pictures rather than maps. It is a world chronicle in little drawings, complete with the parallel columns found in the earliest Christian chronicles (those of Eusebius and Jerome). In 1699, the Nuremberg printer Christoph Weigel published a visually oriented history, *Sculptura historiarum,* featuring ten small engravings per century.[6] Hagelgans, not content just to illustrate historical highlights, sought to speak in pictures. His *Atlas historicus* was more ambitious than Weigel's and also less effective, in that the "little pictures" neither illustrated the past nor achieved eloquence.[7] This did not prevent the book from being twice reprinted and presumably finding buyers till midcentury. The works published by Weigel and Hagelgans suggest that, in the early 1700s, visual history called mainly for pictures of scenes and persons. Maps as a medium of historical narration were hardly even born.

Hagelgans's *Atlas,* however, is not devoid of maps. The four he supplies, each occupying half a leaf, seem to say something about the course of events between the Roman Empire of Augustus and the era after Charlemagne. They have to be reshuffled in order to make a collective point. If the letters A-B-C-D stand for chronological order, Hagelgans resorts to the puzzling sequence C-D-B-A.[8] However that may be, one of the maps is noteworthy as the earliest attempt to chart the *migratio gentium,* or "barbarian invasions." This initiative earns a longer look in the next chapter. Hagelgans's barbarian map is an early attempt to depict movement on a map.

A prominent book dealer in nineteenth-century Holland, Frederick Muller, outlines a wider setting for what, alongside Weigel, Hagelgans may have been doing: "A 'historical atlas,'" Muller says, "is a chronologically ordered collection of prints, maps, and portraits that represent the occurrences of one country and whose plates of persons illustrate those who were prominent in the events."[9] Muller spoke for local, rather than general, conditions. The atlases he described were portfolios of individual leaves rather than published books. In the eighteenth century, picture-history was a Dutch speciality meant for domestic consumption. It was nourished by an extraordinary demand for engravings of Netherlandish subjects, matched by an extraordinary output.[10] Collectors of Dutch history had even less use for maps than Hagelgans. He and Weigel of Nuremberg addressed wide stretches of the past and kept their pictures small so as to provide balanced coverage of all ages; the Dutch piled up images of normal print size about a small country, especially during the short, glorious span since its revolt against Spain.

The Dutch were not alone in these tastes. Contemporary Frenchmen also assembled scenes and portraits of national history; major collections were given to the Bibliothèque royale.[11] Like Weigel (and no doubt others), Dutch and French collectors provided history with a monumental visual element composed of representational pictures and little else. "Historical atlases" of this kind, suggestive of the historical picture books widely available today, are a side branch, detached from collections in which maps of historical subjects predominate.

Anyone tempted to think of the historical atlas as a sixteenth-century creation has reason to ponder the confusion of terminology prevailing in the early 1700s. Châtelain's basically modern atlas, Hagelgans's chronicle in little pictures, and the Dutch and French assemblages of historical prints were all "historical atlases" to their makers and the public; none of them was a work of the kind we would recognize under that label. The term was not yet restricted. A century later, some learned men still considered world atlases synonymous with historical ones. They were behind the times, unaware of noteworthy developments. The posthumous *Atlas historicus* of Johann Matthias Hase in 1750 marks the birth of the technical term we are looking for. Hase's maps also come nearer than any earlier ones to forming the sort of historical collection we expect.

MOVING ON FROM ANCIENT GEOGRAPHY

During the age of Louis XIV (1648–1715), France became the homeland of progressive cartography. As Weigel and Hagelgans pursued their activities beyond the Rhine, the Sansons were into their third generation and retained luster less by making new departures in cartography than by exploiting the massive and continually reprinted stock of maps bequeathed by the patriarch, Nicolas. The leading French mapmakers of the early 1700s were Guillaume Delisle and the somewhat younger Bourguignon d'Anville, who eventually became more celebrated. Unlike the established French school, they were receptive to the data that the Paris Observatoire had accumulated since the 1670s.[12] Delisle markedly raised the scientific level of mapmaking; a student of the elder Cassini at the Observatoire, he availed himself of new astronomical observations (especially for determining longitude), strove to bridge the differences among diverse measures of distance, and applied the tables of magnetic declination to compass readings.[13] He, like d'Anville after him, worked in many branches of cartography. Delisle and his father, Claude, a teacher of history, designed highly praised maps of contemporary New France (largely Claude's work); Guillaume's assistant (and later son-in-law), Philippe Buache, gave impetus to physical geography; and Guillaume's brother, Joseph, gained fame as a geographer of Russia. However significant their achievements were for the history of geography, our attention here is focused on Guillaume Delisle's and Bourguignon d'Anville's maps for history.[14]

The inaugural year of the Châtelain atlas saw Delisle publishing a very well received, two-sheet "Historical Vista for A.D. 400" ("Theatrum historicum ad annum Christi quadringentesimum"); it will concern us in connection with his wider plans (fig. 12).[15] In 1711, Delisle drew two ambitious maps for Anselmo Banduri's *Imperium orientale,* largely inspired by the learned tenth-century Byzantine emperor Constantine VII Porphyrogenitus. Delisle showed the eastern empire divided into "themes" (the Middle Byzantine equivalent of provinces) in the eighth century (fig. 10) and again under Constantine VII in the tenth. These were the first comprehensive maps of the Byzantine Empire. Another illustration of his in these years was for a history of Genghis Khan.[16]

Somewhat later in the decade, the young Bourguignon d'Anville was commissioned by his mentor, the abbé Dufour de Longuerue, to design nine

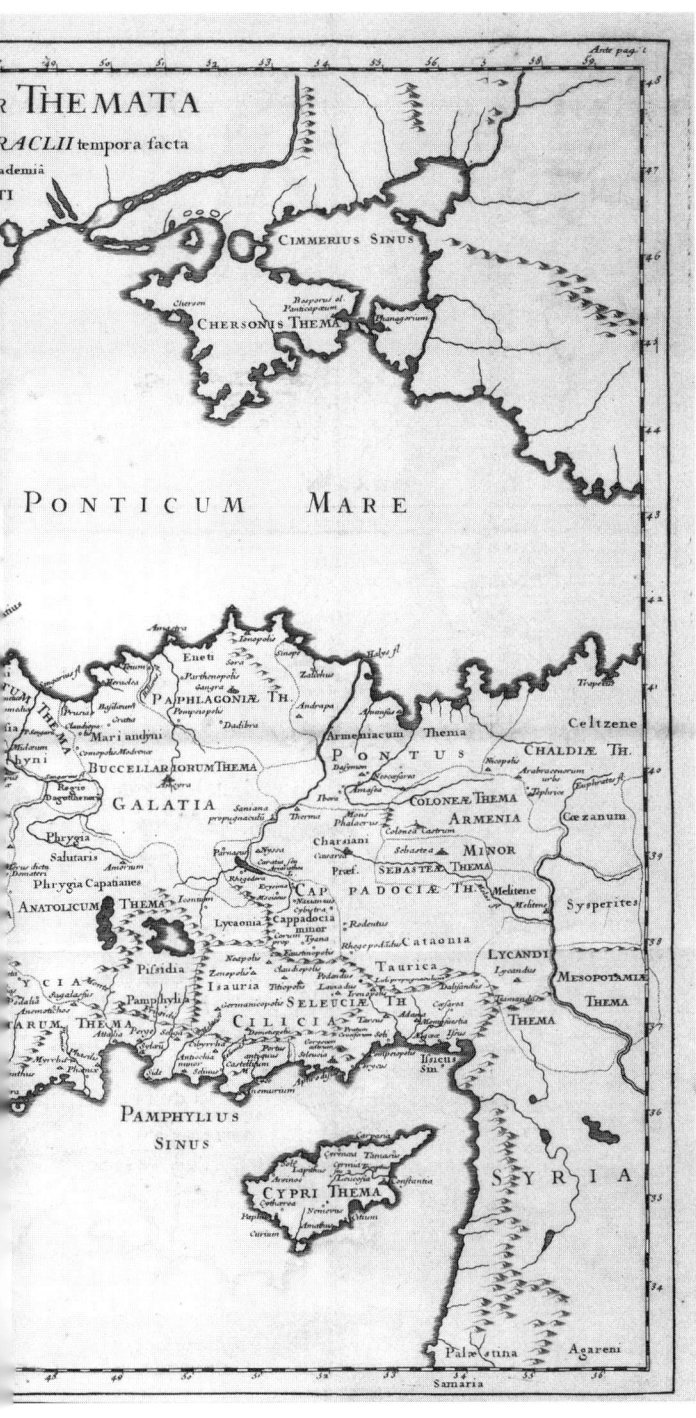

Figure 10 Guillaume Delisle, the Byzantine Empire in the Eighth Century (1711). Whereas maps for the history of Germany, Hungary, and many other European lands were initially produced long after 1711, the eastern Roman, or Byzantine, Empire had the advantage not only of being portrayed comparatively early but even more of having the very talented Guillaume Delisle as its mapmaker. The map was occasioned by the volume *Imperium orientale,* which the Benedictine Anselmo Banduri contributed to the officially sponsored French collection of Byzantine sources. Delisle's posthumous catalogue of works contains a heading "Pour l'histoire du moyen âge"—the first "medieval" collection that a professional cartographer admitted to. The map shown here is one of Delisle's four acknowledged medieval productions. (Courtesy of Houghton Library, Harvard University)

maps of "ancient" France to accompany Longuerue's *Description historique et géographique de la France ancienne et moderne*. D'Anville's maps, admirably clean and uncluttered, are less accomplished historically than Delisle's. Only map 2, called "La France ancienne," to contrast with the opening "La Gaule ancienne," is supplied with distinctive, though jumbled, features suggestive of the medieval period. The novel goal of mapping France between Roman and modern times was not attained.[17]

Delisle is heard from again in 1726 in a set of three maps of lands associated with the Knights of Malta and drawn for the abbé René Vertot's *Histoire des chevaliers hospitaliers de S. Jean de Jérusalem*. Unexpectedly, they were the cartographer's swan song.[18] Two years later, Henri Liébaux, who sometimes engraved Delisle's designs, published a set of maps illustrating the conquests of Clovis and the establishment of the Franks. Though not the earliest depictions of Gaul under the "first dynasty," they were original as a consecutive group of three. Liébaux had been anticipated a decade before, when Delisle, as geography tutor to young Louis XV, had illustrated the early history of France in five maps, from Clovis's death to the partition of Charlemagne's empire. Though the maps were never published, careful, finished manuscript copies survive, including a handsomely executed one by Buache, Delisle's assistant.[19]

At least one German historian nearly kept pace with the French in venturing into uncharted territory. Jakob Karl Spener was the son of Philipp Jakob Spener (†1705), the founder of Pietism. In 1717, as a professor of public law at the University of Halle, Jakob Spener published *Notitia Germaniae antiquae*, illustrated by four maps. "Ancient" Germany was not a new subject. Ortelius and the short-lived Philip Clüver, among others, had expressed their attraction to this topic in books.[20] Spener broke cartographic ground in a section concerned with "middle [i.e., medieval] Germany as it was in the sixth century and a little after"; two of the maps he supplied were said to be "in accordance with the systems of antiquity and the Middle Ages (*ad veteris mediique aevi rationes*)," whereas a third depicted an undiluted "Germany adjusted to the conditions of the first centuries of the Middle Ages." The contents of the last of these best reflects the mid–ninth century but has traces of earlier times, as the announced program implies. By our standards, there is too much intermingling of ancient and medieval in the two maps given this dual role. Spener at least acknowledges that one age merged into the other and made a first attempt to outline how the German lands looked in the centuries separating Ptolemy's "Magna Germania" from the modern Romano-Germanic (or Holy Roman) Empire.[21]

The next decade witnessed Germans making further additions to the stock of maps for history. One of the earliest, drawn twice within three years, portrayed Charlemagne's unfinished canal linking the Danube and the Rhine in the Bavarian Nordgau.[22] Somewhat later and in a more conventional vein, Friedrich Zollmann, court archivist in Weimar, produced two attractive maps portraying "ancient" Saxony down to A.D. 1000 and medieval Saxony from the eleventh to the fourteenth century; the pair was esteemed by contemporaries and continued to be well regarded thereafter. Saxony, as we will see below, had an unusual affinity to maps reconstructing history; Zollmann led off. So different from ours were the chronological sensitivities of the day that the viewer is led backward in two steps from modern to "ancient" Saxony. Zollmann's maps are entered in reverse order of time in the Homann atlas of Germany (1753) and are so listed in his bibliography.[23]

Zollmann's medieval Saxony appeared in 1732, the same year as a remarkable book published at the Bavarian abbey of Tegernsee—the *Chronicon Gotwicense* by Gottfried Bessel and Franz Joseph von Hahn. The title seems to promise a medieval source but in fact applies to a history of the Austrian monastery of Göttweig, whose abbot Bessel was. He wished the *Chronicon* to equal the high levels of historical scholarship reached by the Benedictines in France, notably Dom Jean Mabillon. The published volume contains a large body of material preparatory to the history of Göttweig. This history itself was never published, though it is believed to survive in manuscript. The preparatory material contained in the published *Chronicon* is a "noteworthy encyclopedia of sciences auxiliary to history," in the spirit of Mabillon's epoch-making and French-centered *De re diplomatica* (1681). Its focal point was the charters of the medieval German emperors. Göttweig lay in the land whose Habsburg ruler was the current descendant of Charlemagne and Otto I.[24]

The main contribution of the *Chronicon* to geography is a collection of everything known about those districts of the empire called *gaue*. Three large, not very legible maps are presented: the palaces and royal estates of the East Frankish kingdom (fig. 18), medieval Germany (with *gau* names but few boundaries), and Eastern Francia. They are meant to portray the situation prevailing before the German Interregnum (1254). Though criticized for merely assembling data, the maps of Bessel and Hahn are more ambitious ventures into medieval geography than anyone else would make for many years to come. No atlas-makers paid them any attention, however, perhaps because Bessel's maps, like Hagelgans's, lacked decorative appeal or legibility.[25] A more humdrum subject, J. H. von Falckenstein's portrayal of the (Bavarian) Nordgau in the twelfth and thirteenth centuries (1733), was welcomed alongside Zollmann's maps in the *Particular Atlas of Germany* produced by the firm of Homann Heirs in Nuremberg (1753).[26]

Innovative but Unrealized Atlases

The small sets of published maps just surveyed would be less noteworthy if Guillaume Delisle and two German scholars, Christoph Cellarius and J. D. Köhler, had been able to carry out the ambitious plans for maps of the medieval period that they are known to have had. Cellarius (1638–1707)—in scholarship the Latin form of his name was preferred to the German "Keller"—held a chair at the University of Halle and played a many-sided role in German humanistic studies from the later 1670s to his death. His appetite for unremitting toil was proverbial; rumor had it that in his years at Halle he was glimpsed only once taking a stroll. He is sometimes credited with the extension of the name "Middle Ages" from a type of Latin to a period of history.[27] His *Notitia orbis antiqui,* an austere work in two volumes and thirty-four maps, was a scholarly breakthrough, "the first complete and systematic treatise on [ancient geography]."[28] Cellarius intended to continue beyond antiquity and provide an equally large work on the geography of the Middle Ages. He took the period to last only from Constantine through Charlemagne.[29] Eighteen plates for this project were engraved by the talented J. B. Homann before Cellarius's death cut it short (fig. 11). Two generations later, the original Leipzig publisher dusted off the plates and issued them as part of a "Threefold Appendix to Cellarius's *Notitia orbis antiqui.*" Owing to the long delay in publication, Cellarius's plan is less interesting for the maps it gave rise to than for the awareness it showed that the medieval period was a geographical gap needing to be filled.[30]

Others shared this belief. The *Mercure de France* for May 1720 reported the death of Claude Delisle, father of Guillaume. The announcement leads rather oddly into an account of the son's plans for three ambitious sequences of historical maps. Guillaume, it seems, had allotted one set of eight items to the Middle Ages; a second of nine items, largely ancient, was to resemble his "Historical Vista for A.D. 400" (fig. 12) and show how the world looked at the nine moments in question; and a third, of six items concerned with the Holy Land down to the present, would form a multiple "Vista of Ecclesiastical History" ("Theatrum historicum ecclesiasticum").[31] Two or three of these twenty-odd items were already in print; none of the others was destined to appear, except three biblical ones, posthumously published in 1764 by Joseph Delisle.[32]

Guillaume's three projects called for versions of many items in the traditional "Ancient Atlas, Classical and Sacred"; the two announced sets of "historical vistas" would have been composed mainly of familiar subjects, newly rendered and improved.[33] But Delisle's "Historical Vista for A.D. 400" had been a novel conception, expressing a wish to advance out of antiquity into uncharted territory, and the schemes described in 1720 retained this impetus. There was to be a "Historical Vista for A.D. 600" and another for the Islamic Empire at its height (ca. 750). The Orient would claim more than half the medieval series: the world known to "the Nubian Geographer" (al-Idrisi), the Levant during the Crusades, the Latin Empire of Constantinople, the Empire of the Tartars, and that of Tamerlane. Delisle's plan was symptomatic of his epoch: eighteenth-century map sets of the medieval period almost invariably gave a salient place to the Orient.[34] Delisle, like Cellarius, had little idea of what to make of high medieval Europe; he proposed only one map, "France, Germany, and Lorraine in the Middle Ages."[35] Charlemagne was more familiar territory. His empire was slated for a "historical vista"; and the divisions of Charles's empire among his grandsons and great-grandsons were to have a map of their own.[36] Almost half the subjects that Delisle projected were either rare or unprecedented.

In 1764 the *Journal des sçavans* recalled that Delisle had meant to supply each "principal historical epoch," seven in all, with a map on the model of his *Theatrum historicum*.[37] This reminiscence undervalues Delisle's plans: they went well beyond seven items. Punctilious about chronological order, Delisle proposed a threefold series that, if realized, would have provided a coherent progression from the biblical dispersion of peoples, through the early empires and Rome, into the Middle Ages, and, via the Orient rather than the West, to Tamerlane in the early fifteenth century.[38] Only recent history, after 1500, was left out of account. Delisle would have renovated even familiar subjects by converting them from isolated scenes into a chronicle-in-maps. Alongside the main sequence, the cycle of biblical history could either exist by itself or be fitted into the interstices, down to modern Syria—the one niche allowed to contemporary history, as it had been by La Rue.

Delisle was forty-five and vigorous when his father died. It might have been expected that he would carry out the twenty-odd maps of his historical project, but in fact, his life ended within six years. His projected atlas would have been a remarkable step in the direction of our collections of maps for history.[39]

Johann David Köhler (1684–1755) is alone among these innovators to have seen most of his new maps into print. An esteemed professor of history while at Altdorf, near Nuremberg, he was called to be the first incumbent of the chair of history at the new University of Göttingen (1735). Besides being a specialist in the medieval Romano-Germanic Empire, Köhler lectured on the

Figure 11 Christoph Cellarius, "Germania Medii Aevi" (before 1707; published in 1776). Cellarius undertook a medieval sequel to his famous *Notitia orbis antiqui* (1701–6). He provided the young J. B. Homann with eighteen maps to engrave but died before writing the commentaries that would have completed the project. The map plates, kept by the publisher, were finally printed without commentary almost seventy years after Cellarius's death.

The sample map shown here illustrates the confused blend of times and places that characterizes most of Cellarius's maps for this project. Moving from "ancient" into "medieval geography" was more difficult than Cellarius anticipated. (Courtesy of the Beinecke Rare Book and Manuscript Library, Yale University)

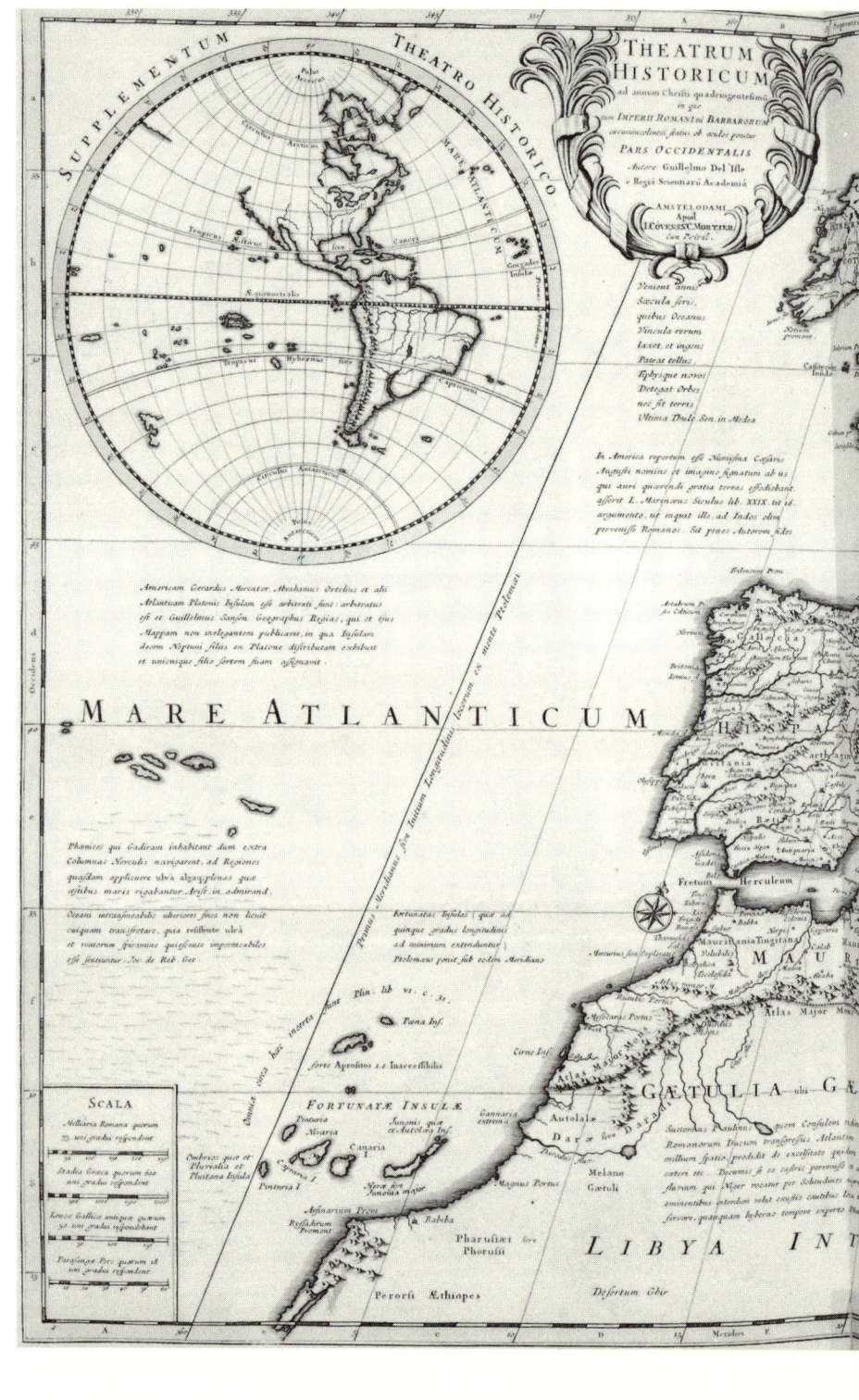

Figure 12 Guillaume Delisle, "Theatrum historicum ad annum Christi quadringentesimum, pars occidentalis" (1705). The subtitle in the cartouche reads: "in which the condition of both the Roman Empire and the barbarians living around it is placed before the eyes." Delisle's "Historical Vista," in two sheets, documents a desire to extend the scope of maps for history beyond classical antiquity into the Middle Ages. A heightened sense of chronology is also shown. Delisle, unlike his predecessors, attached a firm date to the vista being portrayed. The "Historical Vista" immediately had a warm reception and influ-enced the universal, sequential atlases from the mid–eighteenth century onward. A resemblance to fig. 15, below, is plain. From a Covens and Mortier reprint of 1733. (Courtesy of the Beinecke Rare Book and Manuscript Library, Yale University)

"auxiliary sciences," especially numismatics, on which, over the years, he published a work in twenty-two parts—probably his main contribution to scholarship.[40]

Köhler taught geography and, in cooperation with his publisher, Weigel of Nuremberg, compiled atlases; he did not make maps himself. His most attractive product, a classical atlas called *Descriptio orbis antiqui* (1720), is an anthology in which the original designers are often identified. Its predictable repertory, descended from Ptolemy and Ortelius's *Parergon*, includes many maps exemplifying eighteenth-century improvements, such as Adriaan Reland's widely acclaimed maps of Palestine.[41] Köhler's earliest compilation, *School and Travel Atlas for Learning Ancient, Medieval, and Modern Geography*, suggests by its appeal to a threefold geography that buyers would escape the entrenched dualism of the discipline. "Medieval geography," mentioned by Lubin as early as 1678, gave rise in 1712 to a book that was rather narrowly focused on medieval German charters. Among Köhler's 140 maps "for learning ancient, medieval, and modern geography," the only hint of medieval geography is a reduced and unattributed copy of Delisle's "Historical Vista for A.D. 400." The same two-sheet reduced map, still anonymous, reappears in Köhler's *Descriptio orbis antiqui*.[42]

Köhler had no sooner completed his classical atlas than he began, like Cellarius, to contemplate a medieval continuation. After arranging with Weigel for its production, with a suitable commentary to accompany the maps, he dropped the project: medieval geography was more demanding than he had anticipated. Cellarius had left eighteen maps; Köhler's efforts resulted in nine, most of them printed while he lived. They were lodged in his well-received *Brief and Thorough Introduction to Ancient and Medieval Geography*, published in two parts in 1730 and 1737, and completed with a third in 1765, ten years after Köhler's death. Each part contains nine ancient maps and three medieval. Buyers patient enough to wait for thirty-five years gradually acquired a nine-piece medieval atlas. The six maps available by 1737, when rivals were still rare, gave better value than the full set.[43]

Either way, the selection was patchy; Köhler readily admitted in the foreword that he offered only a sampling of medieval geography. All the maps are small, and many are not new. The second installment includes "Britain under the Anglo-Saxons," a subject that reaches back via Sanson and many others to Lambarde in 1568. The opening volume contains "Gaul in the Fifth Century," whose like had just been published by Liébaux, and, more interestingly, a "Lombard Italy" that—as Köhler appreciatively acknowledges—came from a recent volume of Ludovico Muratori's great source collection, *Rerum Italicarum scriptores*.[44] No acknowledgment is given for the map "Flanders in the

Ninth Century," a surprising and important subject, reprinted in the 1730s from Antoine Sanderus's century-old *Flandria illustrata*.[45] Köhler's posthumous third part reproduces "Charlemagne's Canal in the [Bavarian] Nordgau" and a fifty-year-old "Constantinople Divided into Fourteen Regions in the Fifth Century." Georg Andreas Will, editor of part 3, regretted that the medieval maps he was left with were too specialized.[46]

Among Köhler's maps that are (presumably) new, three contribute little to history or geography. His atlas charts Britain, Gaul, Germany, Spain, and Italy one by one at moments that tended to be before the end of antiquity rather than after. Typically, "Spain in the Fifth Century" outlines the Roman provinces of Spain, then makes them medieval by scattering barbarian names in likely spots. The next millennium is more frugally handled, all of it squeezed into a diminutive "Gaul, Germany, and Italy in the Middle Ages (*mittleren Zeiten*)." Köhler was as unnerved as Cellarius and Delisle had been by the High Middle Ages.[47]

ATLASES REFLECTING THE COURSE OF WRITTEN HISTORY

The planned historical atlases of Cellarius and Delisle, innovative as they may have been, were not carried out, and Köhler's offered little more than scattered and nondescript additions to the "Ancient Atlas, Classical and Sacred." What the opening decades of the century did not accomplish began to be fulfilled in the 1740s. A French reviewer, looking back from the 1760s to recent developments in historical maps, knew nothing of the projects of Delisle or Cellarius. As far as he was concerned, marked improvements began with the German "Hasius."

Johann Matthias Hase, called Haas in German and Hasius in Latin (1684–1742), was professor of "lower" mathematics at the University of Wittenberg and a remarkably wide-ranging and productive scholar. A leading collaborator of the Homann firm, he practiced cartography as the scientific equal of Delisle and d'Anville, the latter of whom lived long enough to profit from his work. Hase chose the astute projection normally used in Homann maps.[48] History was no more than a second or third string in his bow, yet he stands with Johann Christoph Gatterer as preeminent in the making of maps for history in eighteenth-century Germany. Hase's small historical atlas is the first visible anticipation of what such works would become.

In 1750, almost a decade after Hase's death, Homann Heirs issued the major *Atlas historicus* that became his monument and was even reprinted in 1813.[49] This retrospective collection assembles Hase's historical maps from the later part of his career, distributed in four groups that had been separately published in previous years. Only two groups are relevant to our concerns.[50]

As dean of the Wittenberg faculty of philosophy in 1728, Hase devoted a commencement address to outlining his plan for a set of maps illustrating "the greatest empires" (*imperia maxima*). His discourse was later printed in a book called *Phosphorus historiarum,* designed as a companion piece to the maps. Hase explains in the preface that the maps had been ready five years earlier. Without naming Delisle, he stresses that they were deservedly called "Historical Vistas" (*Theatri historici*). But the (unnamed) publisher, he added, had inexplicably drawn back and induced long delays. Now, thanks to the firm of J. F. Gleditsch in Leipzig, which had kindly agreed to publish the *Phosphorus* and forced the hand of the original printer (*excusor*), the enterprise was back on track.[51] Hase's days, however, were running out. The *Imperia maxima* was seen into print in 1743 by an editor, A. G. Böhme (Boehmius).[52]

This quarto atlas of twenty-eight maps contains Hase's most notable innovations. The work, in three parts, opens with a narrative entitled "Universal Political History," followed by a color-coded time chart ("chronology of the greatest monarchies and empires"). The maps, forming part 3 and small in size, are concerned with the greatest empires from early antiquity to the present. Hase meant the work to be a university textbook.[53] Besides designing many new medieval maps, he produced maps of the ancient world foreign to the stock of the "Ancient Atlas, Classical and Sacred" and extended the historical spectrum into modern times.

Chronology mattered to Hase. His small historical maps stray twice from the order of time for the sake of topical cohesion, but the empires are set out in a deliberate progression. His ancient maps depart from the traditional repertory by multiplication. Instead of packing the early empires into a single display, he supplies four maps from Sesostris to Darius, another four from Alexander to the Parthians, and three for the Roman Empire, including an unprecedented "Empire under Justinian I" (fig. 13).[54] Delisle's project had given a large place to Asia; almost two-fifths of Hase's maps—especially of the Middle Ages—are allotted to the Orient: five for the "Arab Empire" (including one at the time of the Crusades), three for the Mongols from Genghis Khan to the Moghuls in India, and three for the Ottoman Empire. The Moghuls and Ottomans account for three of the maps of modern history. The Russian Empire, also modern and straddling east and west, brings the atlas to a close. Hase acknowledged the western Middle Ages in five maps tucked between the Arabs and the Mongols and dedicated to the Romano-Germanic Empire, from Charlemagne to the death of Charles VI (1740).[55] The scope of the 1743 Hase atlas, its orientation to the classroom, and its proximity to narrative history more than faintly anticipate familiar recent collections such as the *Anchor Atlas* of 1974.

Hase's other creation that concerns us is *Seven Geographical Maps to Illustrate as Many Periods of the History of Germany*.[56] It is a slightly extended and larger-scale version of his small-sized Romano-Germanic set. As there, the group includes both medieval and modern reigns. Copies of a single base map are colored in such a way as to portray, successively, the empires ruled by Charlemagne's descendants Lothar I and Louis II in 851; Otto I at his death (973); Conrad II in 1031; Frederick II on the eve of his death (1249); Frederick III at his death (1493); Charles V on his retirement (1556); and the recently deceased Charles VI (1740). Special care was taken in choosing the moments to be portrayed. The years 851, 973, and 1249 are not weighty in themselves, but all are perceptively chosen to result in a map useful for historical study. Each of the seven maps fills three-fifths of a leaf alongside a legend occupying the remaining space and elaborately explicating the colors, lines, and other graphic details shown opposite. As a result of Hase's efforts, the German Empire came closer than any other European country, except the Netherlands, to having by the mid-1700s a historical collection of the kind that the biblical lands had long had. The next atlas of German history lay years in the future.[57]

The effect of the first half of the eighteenth century on the organization of historical maps reached beyond Hase. A year before his *Atlas historicus* of 1750 was marketed, there appeared in Paris the last of a lavish, sixteen-volume production resembling the great Dutch atlases of the previous century and possibly exceeding their size. Its publisher was Jean Beaurain, a minor figure in the Paris cartography of his day. This final volume is a historical appendix in the *Parergon* tradition, called *Geographical Atlas Containing the Maps Helping to Understand Sacred and Profane Geography*. Beaurain, an earnest map collector, gathered into this volume a rich retrospective of ancient and biblical maps, including even an old favorite from the sixteenth century.

Beaurain's anthology did not reflect only tradition. Historical sensitivity had grown more acute. No trace appears in his atlas of "literary geography" (such as scenes from the *Odyssey* and the *Aeneid*). Presumably it was omitted on grounds of being frivolous. Also dispensed with are "Illyricum," "Magna Germania," and the other ancient regions descended from Ptolemy's atlas. The subjects selected are historical entities, such as "the Persian Empire," and the maps are set out in chronological order.[58]

Beaurain did not limit history to classical and biblical antiquity; he felt bound to add a section "Pour l'histoire du Moïen Age." The section bearing this title in a Covens and Mortier atlas of 1733 housed only the four maps by Delisle classed as "medieval" in the ordering of his works. Beaurain, however,

Figure 13 Johann Matthias Hase, "Imperium Romanum sub Iustiniano I" (1743). Hase's geographical survey of the "Greatest Empires" (*Tabulae geographicae . . . de summis imperiis*) was long in preparation, suffered delays, and appeared only after his death (1743). No early atlas more clearly breaks with tradition and anticipates the scope and design of future historical collections. The portrayal of Justinian's empire, shown here, is one of several moments that Hase was the first to map. (Harvard Map Collection, Harvard College Library)

Tab.XI. vel III. Imper. Romani

EVROPA

50   55   60   65   70   75   80

Chou asmiæ Lac

Bulga

PONTVS EVXINVS Phasis

Charracarta
Zariaja

CONSTANTINO

Alani

Albania

40

Bactris
Nisaea Par-
thaumsa

35

Armenia

Dahæ

Smyrna

Per armenia

Antrop

Hyrcania

Nisibis   Susia Ari:

Rhodus I.

IMP.

Cyprus I.

PER

SARVM   Iaba
Deserta   30

RANEVM

Apadana   Itatiche

ALEXANDRIA

Regiostrato

Ctesiphon

Cyropolis
Pasagarda

Hierosolyma

Sin Persicus

ARABIA

Sin armusia   25

DES   ERTA

Hamonis

Gerrha

Macæ   Emporium
Omanum Moscha

Libyæ

A

Sachalitæ
OMANA

NILVS

R   Minnæi   20

Sabbatha

Mecca Makka

Mara Mare
aba

AETHIOPIA

Thapsa Thebæ

Rubrum   Nagara   F   Sappar

MEROE Ins.

Homeritæ   Charomoticæ

15

Sembrita   Æthiopes

Arabia felix
Adana

AV   XV   MIT   ANI

Christiani et foe   derati Romanor.

10

50   55   60   65   70

has eighteen maps for medieval history. Half of them are d'Anville's set for Longuerue's *Description,* and only one depicts a land other than France.[59] Even a lackluster medieval selection was an improvement.

The figures whom we have been considering, such as Köhler, Hase, and Beaurain, operated in the capitals of cartography. There was also activity on the periphery. Menso Alting, for example, worked in provincial Groningen. Less concentrated than Alting's work was the nineteen-map *Small Atlas of Hungary* by János Tomka Szászky, which appeared in Pozsony (now Bratislava) soon after Hase's posthumous *Atlas historicus.* Not exclusively historical, Tomka's work contains many ancient and medieval scenes alongside maps of current Hungary. A strictly historical six-map abridgment was published two decades after his death. Even in the middle Danube Valley, the design of a historical atlas was no longer the mystery it had still been to Longuerue and d'Anville.[60]

UNIVERSAL HISTORY IN SIXTY-SIX MAPS

When Beaurain's sixteenth volume reached the market, an atlas radically different from all earlier historical collections had been recently drawn or was about to be. The holdings of the Bibliothèque nationale include an unsigned, undated, and unpublished "Complete Atlas of the Vicissitudes That the Terrestrial Globe Has Experienced from the Beginning of the World to the Present." The unpublished "Atlas complet" is composed of sixty-six maps, each of them preceded by a page or more of handwritten historical explanations ("Observations"). The card catalogue at BN Cartes et plans proposes a date of 1747, but the source for this information is unknown. Throughout the atlas, a single horizontal two-folio engraving is used, portraying Eurasia from Spain to Korea; each spread gives an impression of immensity. On each of the sixty-six identical maps, a bold hand supplied the changing colored boundaries of world history. The illuminations are usually rough and include cancellations and second thoughts. This is not a finished text ready for duplication, coloring, and distribution.[61]

Responsibility for the composition of this bulky draft is harder to determine than one would like. The entry of the *Biographie universelle* for the prominent Paris cartographer Gilles Robert (de Vaugondy) refers to the sixty-six map collection as being his "least known work" and lists its imperfections, then traces the fate of the sole copy down to a publisher's auction in 1813, at which it fetched sixty francs. A handwritten note at the head of the BN copy declares, "Ce Receuil concernant les révolutions des Empires . . . appartiens à

M. Robert de Vaugondy," and a BN acquisitions register reports, "L'Atlas avait été adressé par Robert de Vaugondy à M. Boudet, libraire, le 29. Mars 1773." Because Gilles Robert died in 1766, the Robert conveying the atlas to the bookseller-publisher Antoine Boudet must be Gilles's son, Didier Robert de Vaugondy. The two notes together might confirm authorship, but only if the evidence were explicit. The two existing documents show only that Didier Robert de Vaugondy once owned the draft atlas and disposed of it to Boudet, a long-time business associate.[62] Neither Gilles nor Didier claims responsibility for creating the atlas, which is also absent from the various catalogues of their maps. The lengthy handwritten "Avertissement" and "Observations" in the sole copy are not thought to be in the hand of either. The Roberts' latest biographer breaks with tradition and absolves Gilles Robert of responsibility for the "Atlas des révolutions."[63] No alternative candidate or secure date has yet been proposed.

The sixty-six map atlas is astonishing and noteworthy even if it exists only in a single copy of unknown authorship (fig. 14). Its faults are many and conspicuous. Not only does the underlying double plate show the boundaries of modern European states, but also, when the illuminator came to the plates for 1660 B.C. and many equally remote years, he colored in the outlines of Scotland, England, France, and Spain in the shapes familiar to us. The small scale of the maps is an obstacle to legibility; far from getting a vivid visual impression from the unfolding of map after map, the reader needs to study the coloring at a slow pace, often consulting the "Observations." The underlying historical narrative, which, for example, situates the origins of Prussia and Lithuania in the seventh century, is more defective by our standards than by those of its time.

These and other blemishes are overshadowed by uncommon virtues. No collection of historical maps on this plan or vast scale had ever been attempted. From China to Europe (but excluding the New World), almost all lands with known histories are given a place; the full span of chronicled time is there, carefully grouped into intervals ("divisions"), some long, some short, rarely of as much as one century. Each entry is introduced by a handwritten exposition establishing the historical limits of the "division"—from such-and-such an event up to but not including such-and-such a somewhat later event—and summarizing intervening occurrences. Every so often, a map and its narrative introduction concern only "new states," briefly delaying the onward course of years (e.g., divisions 36, 48, 51, 59). Antiquity, from the "Dispersal of Peoples after the Flood," is assigned thirty-two or thirty-three maps; modern times, to the union of Lorraine with France (1738), rate only three or four. The medieval period appears to run from division 32, "From the First Dismemberment of the Western Empire [A.D. 407]," to

Figure 14 Universal History in Sixty-Six Maps (ca. 1747). "Complete Atlas of the Vicissitudes That the Terrestrial Globe Has Experienced from the Beginning of the World to the Present," in 66 identical engraved maps ("Atlas complet des révolutions que le globe de la terre a éprouvées depuis le commencement du monde jusqu'à présent," en 66 cartes gravées toutes pareilles). This large map for A.D. 1102–57—two folio pages glued horizontally to form a very wide surface—is no. 56 of the total atlas and has sixty-five companion pieces that differ from it only by colored lines applied by hand. The anonymous and unpublished work in question inaugurates the universal, sequential atlases that long were the preferred format for historical collections (e.g., see fig. 28). The anonymous Paris atlas starts with the dispersal of the sons of Noah and reaches 1738, when Lorraine was annexed to France. (Département de cartes et plans, Bibliothèque nationale de France)

divisions 62 or 63, "The Union of Sweden and Denmark" or their division (A.D. 1521). The "Avertissements" accompanying the maps for the Middle Ages often dwell on the Far East or the Islamic lands.

The goal of Hase's *Imperia maxima* had been virtually the same. The Paris draft took a relentlessly chronological approach and sought comprehensiveness by projecting history on a uniform Eurasian background. The anonymous author condensed the equivalent of a universal history into colored

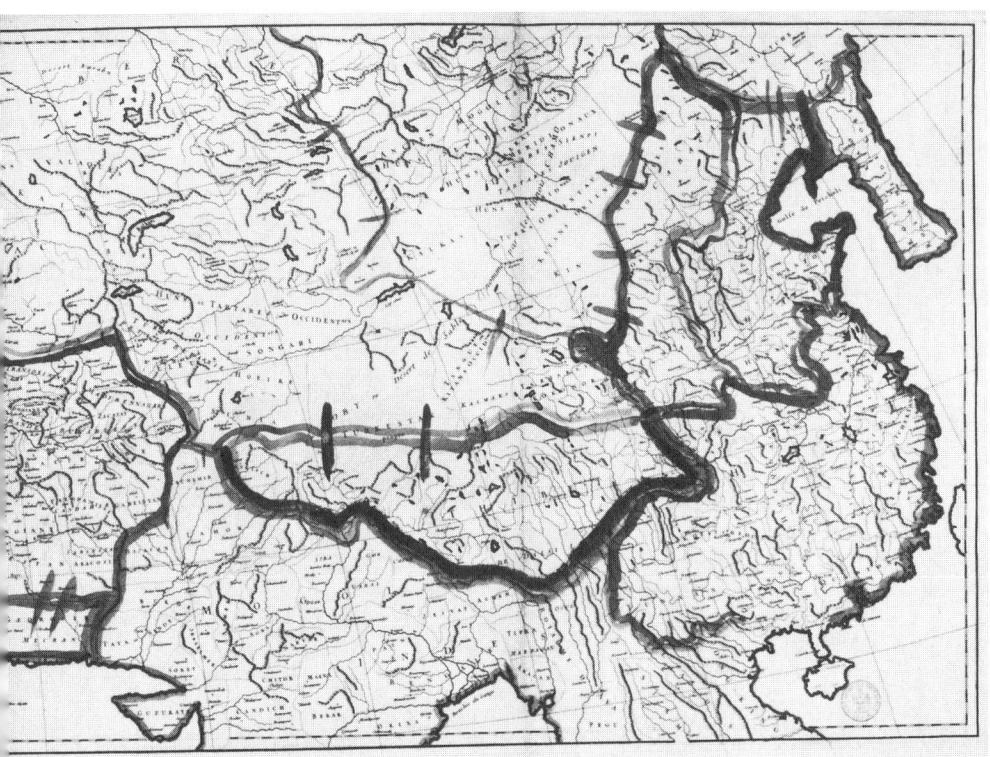

lines on maps. Two complementary forms of discourse, words and illustrations, proceed in parallel. No one had tried this before, and that, indeed, may have been the reason that the work was unfinished. Maps were often commissioned to illustrate books. For the "Atlas complet," however, the equivalent of a small book was composed in order to generate maps; the full span of world history was adjusted to amplify features suited to mapping. Even more than Hase's *summa imperia*, the Paris draft adopts the ideal of arraying maps alongside a narrative of universal history. For most of the next century, this universal, sequential format was favored for historical atlases of the kind that the eighteenth century had devised.

A DISCONNECTED FRAGMENT

After decades in which "atlas historique" denoted a variety of works, the scope of the term had, by 1750, attained relative stability. It differed from the "Ancient Atlas, Classical and Sacred," by including medieval and sometimes modern history and, more often than not, by proceeding in company with a specially composed narrative.[64] These traits became fully established in three Paris publications of the early 1760s.

The first was barely begun when discontinued. In January 1761 the *Journal de Trévoux* announced that Pierre Luneau de Boisgermain was taking subscriptions for a twenty-map *atlas historique* subtitled "Maps of the Principal Parts of the Terrestrial Globe Subject to the Periodic Turmoil (*révolutions*) That It Has Experienced." The first map, available for inspection, was considered by the journal to be "well executed and pleasing to the eye"; the other nineteen would be issued at the rate of one every three months. Luneau promised to set out the progress of the successive populations of the world and to reveal to the eye what chronology and history imparted to the intellect.[65]

His venture did not last long. Only three maps were produced, extending to the Exodus and the classical Flood of Deucalion; they were issued by themselves and, a few years later, as an annex to Luneau's *Cours d'histoire universelle et de géographie* (1765–68). They are exceedingly rare. The sole trace to have yet surfaced is a large sheet showing "the principal parts of the terrestrial globe to illustrate the history of the two first centuries from the creation of the world"—a barren outline of Eurasia and the Americas, enlivened only by details about the Earthly Paradise. The surviving black-and-white sheet was presumably meant to be colored. A historical interval extending somewhat beyond Adam and Eve might have afforded a livelier scene.[66]

Luneau was a dedicated teacher, especially of languages and literature, and a dogged foe of the printers' publishing monopoly. History, geography, and maps occupied him for only a few years. The premature interruption of the atlas is probably not a major loss to the genre. Could a universal history whose third map reached only the Exodus have been completed in twenty steps? Regardless of what the answer might be, Luneau's title contains an echo of the large draft atlas, whose plan it clearly resembles: the major vicissitudes of world history from the start to the present, more or less. The classical and biblical past, without being discarded, was merged in a larger historical aggregate, subject to "revolutions" and inhabited by populations experiencing "progress." Rather than being an aid to literary study, Luneau's maps aspired to project to the eye the course of universal history.

THE FIRST FULLY PUBLISHED SEQUENTIAL ATLAS

No sooner had Luneau called off his project than there appeared an atlas called *Les révolutions de l'univers,* "presenting the political divisions of the various regions . . . from the dispersion of the Children of Noah until the re-union of Lorraine to France, divided into 30 intervals and shown on 60 maps." The designer did not venture to show the whole of Eurasia in the daring way it was shown in the Paris draft of 1747; in the published atlas, each period had one folio for the West and one for the East. The background design used throughout consists of two engravings set out two by two. As the title says, history is shown on sixty maps but divided into only thirty periods of time. Modern frontiers make discreet appearances on the base outline; historical boundaries, changing thirty times, are supplied by coloring. This voluminous collection was neither an isolated draft nor an interrupted fragment: it could be bought in Paris for sixty livres (appropriately enough). In the year of its publication it was listed alongside Hase's small atlas, and at four times the price, in the printed catalogue of Roch-Joseph Julien. Copies survive in several public collections (fig. 15).[67]

The individual associated by name with the *Révolutions de l'univers* is Michel Picaud of Nantes, but he is credited only with drawing the base "map for general history, on which all the vicissitudes (révolutions) of the universe may be shown." A column of separately printed comments, about six inches wide, is glued to the left edge of each "interval" of the atlas.[68] These explanations, also available as a separate book, are attributed to "M. D. P." on the title page. In two copies of the atlas surviving with title pages, the authorship is claimed by a "M. Dupré," presumably a pseudonym; D. P. and Dupré seem linked.[69] Barbier's repertory of anonymous works ascribes the atlas to E. A. Philippe de Prétot, a known educator and atlas-maker (in this case D. P. would stand for "de Prétot"). The identification fails to satisfy. Philippe's signed work lacks originality and shows no trace of the Dupré scheme. No real Dupré has been identified as potentially responsible for this atlas, nor has anyone else materialized behind the pseudonym. Nevertheless, "Dupré" as compiler of this atlas is preferable to the improbable Philippe de Prétot. In many reference works, the *Révolutions de l'univers* is ascribed to Picaud, who undoubtedly took part in the enterprise, but only as a draftsman, unknown to biographical dictionaries. "Dupré," advertised on the title page, comes closer to representing the author than any other name.[70]

The marginal comments of this atlas refer almost continually to its sources, such as Delisle, Hase, and Robert de Vaugondy, specifying in every case one or more published maps. No reference is made, however, to the Paris draft. The compiler may have thought it pointless to draw attention to a source existing in a single copy inaccessible to the public. In spite of this lack of acknowledgment, the "Atlas complet des révolutions" almost certainly lies behind the Dupré atlas. That the *Révolutions de l'univers* should be independent

Figure 15 A Sequential Atlas of World History in Published Form (1763). Dupré (pseud.), *The Vicissitudes of the Universe, Presenting the Political Divisions of the Sundry Regions . . . from the Dispersion of the Sons of Noah to the Reunion of Lorraine to France, Divided into 30 Intervals and Shown in 60 Maps* (*Les révolutions de l'univers offrant les divisions politiques des differentes régions . . . depuis la dispersion des Enfans de Noé jusqu'à la réunion de la Lorraine à la France, divisées en 30 intervales et representées en Soixante cartes*) (Paris, 1763, repr 1775). Shown is one-half of "Intervalle XVI." Dividing the world in two and using one map for each part resulted in a less cumbersome atlas than that of 1747. The scope of the "Dupré" atlas is clearly similar to that of the Paris anonymous (fig. 14). (By permission of the Syndics of Cambridge University Library)

seems much less likely than that the two were related. The universal or Eurasian breadth of their base maps is identical; the sons of Noah as starting point and the union of Lorraine to France in 1738 as ending precisely correspond. The Orient, near and far, claims a large share of attention in both. The distribution of emphasis among the ancient, medieval, and modern periods is very similar, especially in neglecting recent times.[71]

The later work has improvements. Picaud's design, though too small to please students, is visually and dramatically impressive, more so than that of the Paris draft. It is better engraved.[72] The base map is projected and curved to suggest a globe; northern Europe is noticeably foreshortened. Boundaries are colored only after the lands in question enter history. France, Spain, and other modern countries are not shown (as they are in the Paris draft) as though they already existed in the 1000s B.C. As in the draft atlas and other maps for history multiplying a single prototype, place-names are supplied once and for all, no matter how anachronistically. The colorist, responsible for highlighting the places relevant to each interval, did not always fully carry out this part of his task. A reviewer thought that users would have found it easier to distinguish ancient from modern names if more than one base map had been supplied.[73]

Obviously, new plates for each "interval" would have been ideal. And why stop at thirty (or sixty-six) frames? Accustomed as we are to film animation, the idea underlying the design of the *Révolutions de l'univers* is familiar. If the technical resources available to Dupré (and Picaud) had equaled ours, the map of Eurasia would conceivably have served as a stable background for the continual gyrations of animated political boundaries dramatizing historical change from the Atlantic to the shores of Japan across the millennia of recorded time.[74] Forced to be content with more modest means, Dupré sought unmistakably to express the changes of universal history in graphic form.

Ending the *Révolutions de l'univers,* the author of the marginal matter regretted that he had had to dwell on only those "revolutions" that could be portrayed to the eye. Even at that, he said, errors were unavoidable in drawing most boundaries except those of the few lands, such as the Roman Empire, whose frontiers were certain.[75] The Dupré thirty-map collection and its unpublished antecedent, without establishing a fashion, took the boldest steps yet seen toward an atlas that did not merely help the study of books but narrated history through maps.

The History of France in Sequential Maps

The Paris draft and the Dupré atlas each has a single background and repeats it with changing colored lines at studied intervals in a long sequence of years. Two years after the *Révolutions de l'univers* appeared, the same pattern was applied with variations to the much narrower dimensions of a single country. French history was the subject. In sixty maps of largely uniform projection, the coastal outlines and waterways of France and its near neighbors stay the same, but the lettering changes from page to page and so do the location and size of woods and mountains.[76] To achieve this result each map needed a plate of its own, no more than quarto size in this case. The publisher, Louis Charles Desnos, was a prominent geography specialist in Paris, with the mildly derogatory reputation of publishing everything that was brought to him.[77]

Desnos's catalogue of 1765 advertises three different, newly compiled historical atlases of France. First cited, and in greatest detail, is the *Atlas historique et géographique de la France ancienne et moderne,* "adapted to the understanding of the History of Messrs. Velly and Villaret. . . . It has been prepared by Mr. Rizzi-Zannoni, of the Royal Academy of Sciences and Letters of Göttingen." Desnos presents the Rizzi-Zannoni atlas as "Part 1," immediately complemented by "Part 2"—a sixty-map *Tableau analytique* of current France. Each part cost thirty-three livres, bound.[78] Two more historical atlases are then listed, sharing the quarto format of Rizzi-Zannoni's; neither of them is attributed by name to a compiler other than Desnos himself: first, *Atlas historique, géographique et chronologique de la France ancienne et moderne,* "designed for understanding the *Abrégé de l'histoire de France* by President Hénault"; then, in a shortened citation, an "other atlas adapted to the history of France of the Rev. Daniel and of Mézerai." The Hénault companion is quoted at twenty-one livres; that for Daniel and Mézerai at twenty-four.[79] Desnos, with an eye to sales, saw to it that each of his atlases was connected to well-known, even best-selling narrative histories. The *Journal de Trévoux* drew the intended conclusion: just as the history of Velly and Villaret is in everyone's hands, so Rizzi-Zannoni's atlas should be, too.[80]

This cluster of Desnos publications includes the earliest historical atlas of France. Toward 1820 Adrien Hubert Brué, the next compiler of such a work, cites Rizzi-Zannoni as his one predecessor, then belittles and dismisses him.[81] Brué is close to correct in his bibliographical information. A plan had appeared in 1750 for a fifteen-map historical geography of France, together with an affirmation that nothing of the kind had yet been done (the project was called off). In 1763 the dealer Roch-Joseph Julien listed the historical atlases of Hase and Dupré but could not supplement them with one of France. Soon after, however, the *Bibliothèque historique de la France* of Le Long records two of the three atlases announced by Desnos: Rizzi-Zannoni's as well as the Hénault companion. The abbé Boullemier, reviser of Le Long's *Bibliothèque*, restrained his enthusiasm. He regarded Rizzi-Zannoni's atlas as being at best a trial run (*essai*). Boullemier's disapproval was welcomed by Brué in 1820; it meant that his own atlas properly inaugurated French history in maps.[82]

Giovanni Antonio Bartolomeo Rizzi-Zannoni (1736–1814) became in time a major cartographer. When the historical atlas bearing his name appeared, he was still in his twenties. He is surprisingly discreet about his part in this project. In a third-person autobiographical fragment dated 1774, he associates his atlas with someone prominently mentioned in the Desnos catalogue, namely, Gaspard Moise de Fontanieu, an *érudit*, charter collector, and lofty royal official to whom the atlas is dedicated: "In 1765, M. de Fontanieu, whose special avocation was French history, asked Zannoni to carry out a work intended to show the extent of the great fiefs, their dismemberment, and their reunion with the Crown, as well as the losses and successes of the nation under the three dynasties of the French monarchy." Fontanieu's library was open to him while compiling the maps.[83]

The themes mentioned by Rizzi-Zannoni match the atlas he produced for Desnos, but the year "1765" does not. Most of the maps bearing Rizzi-Zannoni's name are dated 1764, so that if Fontanieu approached Rizzi-Zannoni with his request only in 1765, his wishes would have been fulfilled even before they were expressed. In addition, the *Journal de Trévoux* for January 1765 announced the atlas as forthcoming and, in its July issue, provided a full review, as did the *Journal des sçavans* in August. Desnos is not mentioned in Rizzi-Zannoni's autobiographical fragment, yet Rizzi-Zannoni worked for or with him in these years, and the publisher's signature is beside his own in the "Avertissement" to the *Atlas historique* of 1764. Rizzi-Zannoni made a habit of fudging his life story; we often cannot tell why.[84]

A Paduan by birth, Rizzi-Zannoni came to Paris from Nuremberg in late 1759, seconded by the firm of Homann Heirs to work with a French mapmaker especially on maps of the war then in progress. Rizzi-Zannoni's activities in the next few years included compiling a large variety of maps of France, a small-size circuit of its coasts, and the pioneering historical atlases. Nevertheless, the *Atlas historique de la France* did not glitter either in his own estimation or, over the years, in anyone else's. Military maps aside, it was his solitary venture into maps for history.[85] His career truly prospered only after a clouded departure from Paris and return to Italy (1776). Invited to Naples, he was charged with mapping the Kingdom of the Two Sicilies on a large scale and made a success of this multiyear enterprise. Wolkenhauer calls him the "foremost leader" of Italian cartography in his day. He combined great talents as a mapmaker with energy, restless exertion, and puzzling character flaws.[86]

The same issue of the *Journal de Trévoux* that advertises Rizzi-Zannoni's work also mentions the atlas serving as companion to Hénault's *Abrégé* (fig. 16). Le Long's *Bibliothèque historique,* which assigns both to the year 1764, specifies that "[Rizzi-Zannoni's] collection contains the same maps as [the Hénault companion]."[87] The "atlas designed for understanding Hénault" is rare; it embodies a first state of Rizzi-Zannoni's historical maps of France, complete with his name as designer on almost every sheet (the alternative is anonymity). This first state occurs not only in the Hénault volume but also in the 1764 version of the Velly and Villaret atlas, whose title page mentions Rizzi-Zannoni. The choice of moments of French history, the cartouches, and the placement and form of woods and mountains are all identical to those found in the Rizzi-Zannoni atlas of 1765 and after. In the Desnos manner, each map, deftly tinted in discreet, pleasingly assorted colors *à la mode hollandaise,* appears on a page adorned with an engraved frame or border printed from a second plate and distinctive for each atlas.[88]

The map frame of the Hénault companion differs from the Rizzi-Zannoni frame of whatever year, but the atlases of 1764 underwent extensive changes before becoming the editions of 1765 and after. Someone examined the historical details with a critical eye, found them too skimpy, and caused notable improvements and additions to be incorporated in the collection bearing Rizzi-Zannoni's name.[89] In the Charlemagne map, even the cartouche is moved, from lower to upper left, so as to uncover Pyrenean Spain and permit the insertion of the Carolingian Spanish March. Occasionally, in the cartouches, special provision is also made for mentioning additions to the domain of the French kings, Fontanieu's special interest.

Until a copy of the atlas for the histories of Mézerai and Daniel surfaces, the genesis of Desnos's precocious aids to French history will be incompletely known.[90] Rizzi-Zannoni was deeply involved in the two we have (and in a third survivor as well, absent from the Desnos catalogue). The one bearing his name clearly went through a process of improvement. It is alone among the Desnos atlases of French history to be widely represented in public map collections.[91]

Rizzi-Zannoni's *Atlas historique* is an elegant and attractive volume. Planned, like Dupré's *Révolutions* and its predecessors, to illustrate history by using color to indicate changes over a long span on one repeated map outline,

Figure 16 Giovanni Antonio Rizzi-Zannoni, *Atlas historique de la France* (1764, 1765). These maps document revisions of Rizzi-Zannoni's map no. 6, "Empire de Charlemagne, VIIIᵉ siècle," in his *Atlas historique de la France.* The publisher, Desnos, made a point of associating the atlas with best-selling histories of France. The upper map comes from the atlas "pour l'intelligence de l'Abrégé chronologique . . . de M. le président Hénault" (1764); the lower, from the version "pour faciliter l'intelligence de l'Histoire de MM. Velly & Villaret" (1765). The Velly-Villaret companion was the atlas most commonly advertised and sold. The 1764

printing is rare. In the later version, the cartouche was conspicuously relocated to uncover the Carolingian frontier province called the Spanish March. (1764 printing, courtesy of the University of Illinois Library, Urbana; 1765 printing, courtesy of the Beinecke Rare Book and Manuscript Library, Yale University)

Rizzi-Zannoni's France is remarkable for recording historical variations not only by colors but also by a changing base map. The result is not wholly admirable. The form and size of the land stay the same, but as one display yields to the next, the forests appear to wax, wane, and wax again with alarming speed and ease, and even to change location altogether. Mountains, shown as schematically as forests, also prove astonishingly migratory. The publisher's provision of multiple plates avoids monotony at the cost of fanciful variation. The positive effect is that place-names and boundaries are not set down once and for all on a single plate, as in Dupré's atlas and elsewhere. The presence in the Desnos atlases of political, regional, and place names and boundaries adapted to the depicted moment partly offsets the whimsical travels of forests and mountains.

Rizzi-Zannoni's main enhancement of the atlas signed by him was a group of twenty-four "repeating maps" (*cartes de répétition*) documenting the widening over nine centuries of the lands directly ruled by the French kings—the so-called royal domain. Desnos published these thematic maps separately, as usual without Rizzi-Zannoni's name, as a companion atlas to P. N. Brunet's *Abrégé chronologique des grands fiefs*.[92] The repeating maps are of the same economical kind as those of Dupré and his predecessors: place-names, basic territorial divisions, and mountains are identical throughout; each sheet has its engraved title but no decorative cartouche; a uniform blue outline, with pale blue-white infill, marks the king's gradually widening lands. Rizzi-Zannoni thought it proper to qualify them apologetically as "repeating," a product inferior to the main, changing series. The atlas could be bought without them.

The sixty maps begin with a foldout of late Roman Gaul, acknowledged as deriving from d'Anville. They end with Louis XV and, from a different standpoint and again as a foldout, with a "unique (*singulière*) map worthy of attention," namely, of France divided according to the ancient customs (or customary laws) followed in the various provinces: "this map seems to show a prodigious number of separate small states enclosed in a single one."[93] The Charlemagne map has a smaller scale than the others so as to include more territory. Otherwise, every page exhibits the same familiar outline, varying often but only in particulars. Lands other than France document seventeenth-century wars and Louis XIV's colonies. A characteristic blue-and-white perimeter periodically announces the next chapter in the unfolding saga of the royal domain. In choice of subjects, though perhaps not in correctness of boundary lines or graphic appearance, Rizzi-Zannoni's progression through French history does not differ very much from its many successors in the next century.

The abbé Boullemier complained about the size of Rizzi-Zannoni's collection: "It seems that the maps of this little atlas have been excessively multiplied." He probably meant that a preferable design, such as the one proposed in 1750 but not executed, might call for no more than fifteen well-chosen moments of French history. If all history since Noah's sons could fit into sixty-six or thirty maps, then France since the fifth century A.D. ought to be manageable in fewer than Rizzi-Zannoni's sixty.[94] Regardless of Boullemier's criticism, the Desnos atlases creatively modify the sequential plan of Dupré, Luneau, and the Paris draft. The design is adapted to a single country and avoids a limitation of past models—the use throughout of an unchanging map.

Contemporary reactions were positive. The *Journal des sçavans* considered the work eminently useful for understanding French history and placed the undertaking in context:

> [Hase] conceived and carried out a similar project for the general history of the world, but his maps, too few and too small, present only the main epochs. His work is much less extensive than [Rizzi-Zannoni's], whose subject is only France. Hase's project was very well received, and several persons have aspired to imitate it in greater detail, but these undertakings have not been carried to complete perfection. This one, however, limited to France, is more complete; nothing seems to have been overlooked to make it pleasing; all the maps are accompanied by cartouches. Another aspect matters more, namely, the precision of geographical details; users of this Atlas will reveal, book in hand, whether the authors devoted themselves to this essential point [i.e., precise details] and omitted nothing that might be desired. The project is fine, deserves praise, and, if it should contain defects, might be corrected.[95]

The reviewer leaves us in no doubt about the place of Rizzi-Zannoni's atlas in cartographic developments descending from Hase through various works typified by the unnamed Dupré. The multiple maps for France alone are deemed a special merit rather than a cause for complaint.

Praiseworthy or not, the Rizzi-Zannoni/Desnos atlas fell short of commercial reward. A volume of the Velly and Villaret *Histoire* recommended that it be consulted. Reprints of the atlas in 1777 and 1782 survive. Successive Desnos catalogues advertise it as well as the Hénault companion. These traces document that it had a modest place in the book trade but not the wide sale and financial returns that so unprecedented a collection might have deserved. Author and publisher, instead of being acclaimed as pacesetters, were all but ignored by later compilers of historical atlases of France and other lands.[96]

The public may not have been ready for their innovative product. More than half a century passed before something comparable was tried again.

THE LOST ATLAS OF JOHANN CHRISTOPH GATTERER

By the middle of the eighteenth century, two models for a historical atlas had been devised: a topical plan in chronological order, exemplified by Hase's "Greatest Empires," and a succession of time intervals projected on a uniform, usually universal background, exemplified by the Paris draft, Dupré, and others. The topical scheme, which had a future, lagged in development. Meanwhile, the alternative design, preferring universality to the national focus of Rizzi-Zannoni, showed astonishing vigor and attracted skilled designers. It was the main line of development for historical atlases down to 1830. Its last example in the eighteenth century has affinities to the "Atlas complet des révolutions," but not without marked variations. Perhaps for the first time, a historian not wholly forgotten today was involved.

Johann Christoph Gatterer (1727–99) succeeded Köhler as professor of history at the University of Göttingen and held the chair for forty years. He has a starring role in Herbert Butterfield's account of the German historical school. Gatterer never wrote a major history; his style was thought better adapted to articles than to longer works. Like Köhler, he was very active at Göttingen in teaching the auxiliary disciplines, such as genealogy and geography; his manual of diplomatic—the science of charters—was singled out for special praise.[97]

Another major concern of Gatterer's was universal history. He published eight works on this subject; almost all are incomplete, and each differs from its predecessor because further research kept swaying him. They were primarily for his Göttingen students, to accompany his courses of lectures; but there was public demand for them, too, and some sold very well. His time line, however, which we know had six engraved and colored tables (and was independent of his maps), has yet to be discovered.[98]

Gatterer's geographical collection was potentially as perishable as his time line. Toward 1776, in connection with his *Abriss der Geographie* and the resumption of his lectures on this subject, he caused a large set of maps to be engraved at his own cost in Göttingen and Nuremberg and to be colored according to the needs of his teaching. He undertook to lecture on "comprehensive geography following [the *Abriss* and] a set of maps designed in an entirely new way." The maps were not expected to instruct by themselves even though hardly any written matter accompanied them. In students' hands, they were to provide the visual basis for the lecturer's oral commentary and a ground for the student's notes. Some sale outside the university was expected; the appropriate privilege (equivalent to our copyright) was secured. But Gatterer's maps were school outline maps, without names and legends (*methodische* was the German term for the type); they did not have the look of a finished commercial atlas.[99]

The main consequence of these origins is that Gatterer's maps are rarities. His long bibliography in a who's who of then living German scholars says nothing about his mapmaking. One atlas of his survives in Berlin, another at Harvard; their contents, duly "methodical," are in turn comparative, modern, or physical.[100] Gatterer's historical atlas, however, cannot be found. Until a copy surfaces, it is knowable only indirectly through his own description and those of three early witnesses, one of whom attaches a sample page. In 1805 Christian Kruse, who had been bringing out a historical atlas of his own since 1802, published in the Weimar journal, *Allgemeine geographische Ephemeriden,* a lengthy "Analysis (*Probe*) of Gatterer's Maps of the Migration of Peoples." Kruse paid homage to the not-long-deceased Göttingen professor. He did not relish examining Gatterer's twenty-four maps for flaws; but certain persons, even some unfamiliar with the Göttingen atlas or his own, had alleged that Kruse's atlas was like Gatterer's or depended on it without acknowledgment. The main culprit, unnamed, was Gabriel Gottfried Bredow (1773–1814), prospective editor of the minor Greek geographers. Kruse felt forced to show in detail that the charge was false. As further proof of how independent the Kruse atlas was, Gatterer's eighteenth map was reproduced and tipped into the journal (fig. 17).[101] Kruse, Bredow, and presumably others were familiar with Gatterer's historical atlas; it had circulated widely enough to reach interested parties outside Göttingen. Gatterer's extensive summary of its contents in the *Abriss der Geographie* was more a prospectus than the description of an atlas already on sale. Kruse's article establishes that publication had taken place. The Göttingen *Historisches Journal* announced the appearance of Gatterer's maps in 1776, and the collection reappeared forty years later in the comprehensive bibliography of Woltersdorf.[102]

Woltersdorf's three-part entry for Gatterer presents a corpus of the latter's mapmaking and notes that Gatterer's historical atlas was obtainable as a whole for twelve reichsthaler (an amount then equivalent to roughly thirty shillings sterling, inexpensive for an atlas).[103] Only the second part concerns us:

> [Part 2] the two hemispheres, one sheet octavo, which by coloring in various ways produces the following forty-four sheets: 1–2) according to rivers; 3–5) the lower and upper hemispheres according to countries (*Länder*), these

XVIII

Geographia
Migrationis Gentium:
Periodus XVIII, a conjunctione Regni Burgundici
cum Germania et ab initio Imperii Seldschukidarum
usque ad initium Regnorum Siciliae utriusque ae
Portugalliae et Imperii Mohadarum h.e ab A.1037
_1127 et 1146.

Figure 17 Surviving Sheet of a Twenty-four-Map Teaching Atlas (1775). Johann Christoph Gatterer prepared maps in 1775 for the geography course he was to give at the University of Göttingen. With wide margins for student notes and almost no writing, the maps were meant to accompany his lectures and to be explicated by them. Twenty-four of the forty-four maps he compiled formed a sequential historical atlas, called *Maps for the History of the Migration of Peoples* (*Charten zur Geschichte der Völkerwanderung*). The atlas is lost, but four separate descriptions exist. Map no. 18 (showing mainly the eleventh century) was reproduced in 1805 to document an article in *Allgemeine geographische Ephemeriden* by Christian Kruse. (University of California Library, Berkeley)

according to ancient and modern geography; 6–12) according to the ancient
states; 13–36) to illustrate the history of the Migration of Peoples and the
Middle Ages; 37–38) according to today's constitution; 39–40) according to
languages; 41–42) according to religions; 43–44) according to the conditions
of commerce.[104]

Kruse's critique addresses the items numbered 13–36, those overlapping the
period of his own atlas. Gatterer's description in the *Abriss der Geographie* includes numbers 6–12 as well as 13–36. Nothing fully prepares us for the
breadth of the atlas that Woltersdorf records.[105]

Its least auspicious feature is the use, in this small size, of a base map featuring hemispheres. The absence of names helps the coloring to stand out,
but so limited an expanse drastically restricts detail. However that may be,
much of Gatterer's work reflects a tradition by now familiar to us. The twenty-
four maps Kruse worried about—which Gatterer entitled "of the *Völkerwanderung*" and we call medieval—are in keeping with the format of the Paris
draft and Dupré. Gatterer's hemisphere as the canvas for universal history was
a suitable counterpart to Dupré's sweep of Eurasia. Gatterer explained: "Because, in the science of states (*Staatenkunde*), one must always survey the entire known world, the subject cannot be served by parts of the earth or maps
of countries. Hemispheres are necessary here, always identical of course, but
colored in contrasting ways, so that one may, every time, be immediately oriented on the earth's surface and able to compare one age with another."[106]
The surviving sample (fig. 17), appended to Kruse's article, shows a quarter
sphere rather than a half, centering on Europe and ranging from Africa to the
North Pole. The coloring is attractive, but the selection of place-names—surprising to find at all in outline maps—is predictably meager and unfinished.

The group of maps heading Wolterdorf's list (nos. 1–5) is hard to visualize on the basis of his words alone. If fundamentally physical, they would fore-
shadow nineteenth-century atlas designs in which physical geography was set
before political. Gatterer, a disciple of Philippe Buache's theories about
mountain chains and watersheds, was unusually concerned with the physical
side of geography and is credited with producing the first physical atlas.[107]

The closing maps are more certainly innovative. Gatterer, like the Paris
draft and Dupré, slighted major modern events. Subjects commonplace today, such as "Europe at the time of the Peace of Westphalia (1648)" are left out.
Gatterer passed from 1517 ("The Reformation") directly to the present. His
originality consisted in not limiting modern coverage to the political boundaries of "the current constitution"; he added maps of the languages, religions,
and commerce of the modern world. The first two had been charted before—

languages in the past generation, world religions as early as the sixteenth century—but had never yet penetrated a historical collection. His attention to commerce is the more remarkable because the subject had gone virtually unnoticed in previous work. The *Tableau analytique de la France* that Desnos offered as a complement to Rizzi-Zannoni's atlas somewhat anticipated Gatterer's three thematic maps; Gatterer differed from the Paris publisher in much more clearly integrating this feature of his collection with the maps for history.[108]

Gatterer taught *Staatenkunde*, "the science of states," a cross between political science and political history, at the very time when statistics was on its honeymoon, developed by a Göttingen professor into a celebrated course on contemporary lands. Gatterer's colleagues included Anton Friedrich Büsching (1724–93), whose enormously influential, multivolume *Neue Erdbeschreibung* did much to define political geography and provide statistics with a home. Gatterer's atlas showed that the modern world could and should be portrayed in more dimensions than political boundaries.[109]

The universal, sequential Paris draft and its equally comprehensive descendant, Dupré, continued to exert a profound influence. Consciously or not, Gatterer espoused their universality by projecting his atlas on hemispheres and mapping his twenty-four "scenes" of the medieval period from interval to interval. His collection anticipated that of Christian Kruse (1802–18) closely enough for an overeager critic to believe in a filiation, and differed from it enough for Kruse, a temperate man, to resent the rapprochement. Not Gatterer's atlas but Kruse's—European rather than universal in scope—is the culminating point of the developments that these pages have traced from the Paris draft onward.

The sixty-six map "Atlas complet des révolutions"—an effort to chart the course of universal history by intervals in a single sequence from start to present—turned maps for history in a direction to which they clung for close to a century. As far as we can tell, the sequential atlas materialized abruptly and without explanation (except for Hase's influence), and stayed unpublished, yet marks a starting point that cannot be overlooked. In their early incarnations, map collections of this type paid exceptional attention to the affairs of the Orient, near and far. This tendency debouched in the next century in the independent Asiatic atlas of Klaproth and, to a surprising extent, in the first historical atlas of Russia. With these exceptions, interpretations of the sequential kind focused increasingly on Europe and Europeans. Another offshoot was Rizzi-Zannoni's French atlas, not the pioneer among "national" atlases but certainly the fullest yet produced. The unrealized atlas project of Guillaume Delisle, though splendid, did not point toward a design of the sequential, universal kind, and Hase's excellent collection of twenty-eight maps did so only ambiguously. Nevertheless, the infancy of historical atlases is exemplified by the single-line, universal design in its French and German forms.

The lesson was well enough assimilated at the dawn of the nineteenth century to be embraced by Conrad Malte-Brun in his *Précis de la géographie universelle*. The historical atlas, he affirmed, called for "a series of geographies [i.e., maps], each of which, very different from the one before or after, is nevertheless true, accurate, and complete for the year, or even the century, to which it belongs." A contemporary German set out brief, straightforward specifications: many maps in chronological order, so as to illustrate gradual changes.[110] In this form the emergent historical atlas found its spine, but only provisionally. Less than a century after the Paris draft, historical maps cut loose from the world chronicle of literary tradition and drew closer to the main line of geographical atlases, those all-purpose collections deemed invaluable to students of history since the days of Ortelius.

NOTES

1. Châtelain, 1705. *Journal de Trévoux,* March 1705, 544 (Slatkine repr., Geneva, 1968, 5:149): "D'Amsterdam. L'Auteur de l'Atlas historique est un nommé Chastelain, qui s'est associé d'un libraire, et qui veut être auteur à quelque prix que ce soit." This is disobliging rather than helpful but bears out the early anonymity. H. A. Châtelain is rejected as author by Aubrey Rosenberg, *Nicolas Gueudeville and His Work (1652–172?)* (The Hague and Boston, 1982), 79. He is taken most seriously by Leo Bagrow, *Die Geschichte der Kartographie* (Berlin, 1951), 337, where he is dated 1684–1743 (catalogues credit him with *Sermons sur divers textes de l'Écriture sainte*). His association with this atlas seems to be now abandoned by everyone. For Zacharias Châtelain as *auctor,* see *Acta eruditorum* (Leipzig), July 1709, 294, with details. Koeman, 2:33, classes this atlas under Zacharias Châtelain—probably as publisher rather than author (the senior Zacharias had a son of the same name involved in the family publishing business). *Le grand théâtre généalogique* (Amsterdam, 1719), signed by the "author of the *Atlas historique*" (*NUC* 104:504), is compatible with what we know about Zacharias Châtelain, author of *Histoire abrégée des Provinces unies des Pays-Bas* (Amsterdam, 1701), incorporated into the *Atlas historique.* See also Isabella Henriette van Eeghen, *De Amsterdamse Boekhandel, 1680–1725,* 4 vols. (Amsterdam, 1960–66), 3:67–71.

2. Review of the atlas: *Histoire des ouvrages des savants,* November 1704, art. VII, 484–99; *Journal de Trévoux,* January 1716, 45–97 (review of vol. 4, published in 1714); *Journal des sçavans* 43 (1708): 497–513; *Journal des sçavans* (Amsterdam) 58 (July 1715): 106–13; *Journal des sçavans* (Amsterdam) 65 (December 1719): 616–27 (review of vols. 5–6, published in 1719). Review of *Supplément à l'Atlas historique* (1720): *Journal des sçavans* 68 (July–December 1720): 183–90: fanciers of Gueudeville's howlers (*buffoneries*) will be disappointed be-

cause the letterpress is now by Limiers. *Journal des sçavans* 117 (1739): 280 comments on the new edition of *Atlas historique,* "Ce grand Ouvrage est trop connu pour qu'il soit necessaire d'en parler." Some libraries with copies are listed by Rosenberg, *Gueudeville,* 165–66; add LC, UIll, University of Iowa, Harvard, UMich-Clements, Wayne State.

3. Woltersdorf, 119 (the *Atlas historique* is listed under Gueudeville): "All the maps are handsomely engraved and cleanly colored [see ch. 1, n. 38, above], but many are too small, because the larger share of space is taken up by the accompanying historical, geographic, and genealogical notes and tables"; the value of many essays in the first volumes is very uneven. The beauty of some of the nongeographic plates is pointed out.

4. Châtelain, 1705, 1: nos. 4, 7, 10 (world empires, Greek and Roman Empires; Caesar's civil war was "literary" cartography, relevant to reading Lucan's *Pharsalia*); Supp., 7: no. 34. For the capsule accounts of Goths and Vandals, see ibid., vol. 2: nos. 3, 4.

5. Hagelgans, 1718. About editions, see W. Goffart, "The Map of the Barbarian Invasions: A Longer Look," in *The Culture of Christendom: Essays in Medieval History in Memory of Denis L. T. Bethell,* ed. Marc A. Meyer (London, 1993), 1–2.

6. On chronicles: Brian Croke, "The Origin of the Christian World Chronicle," in *History and Historians in Late Antiquity,* ed. B. Croke and A. M. Emmett (Sydney, 1983), 116–31. On aspects of history-in-pictures (to 1697): Christoph Weigel (1654–1725), *Sculptura historiarum et temporum memoratrix* (Nuremberg, n.d. [1699?]; and later eds.); Gregor Andreas Schmidt and Samuel Faber were responsible for the historical content. Weigel also published *Historiae celebriores veteris et novi Testamenti iconibus repraesentatae* (Nuremberg, 1700; and later eds.), and other collections of biblical pictures. Reusing the pictures of *Sculptura historiarum,* but in coarser form, Weigel cooperated with J. D. Köhler (see n. 41, below) in the more popular *Die Welt in Ein Nuss* (Nuremberg, 1722; many translations). Each century in this abridged world history belongs to some collectivity: the fourth century to the Arians, the sixth to the Goths, the eighth to the Germans, etc. Weigel and Köhler also produced a picture album of memorable eighteenth-century events.

7. P. Richter, "Ueber Johann Georg Hagelgans," *Nassauische Heimatsblätter: Mitteilungen des Vereins für nassauische Altertumskunde und Geschichtsforschung* 3 (1899–1900): 35–49, at 37, may overrate the influence of Weigel's *Sculptura;* his collection and Hagelgans's use pictures in very different ways.

8. For the details, see W. Goffart, "The Map of the Barbarian Invasions: A Preliminary Report," *Nottingham Medieval Studies* 32 (1988): 51–53. Chronological order (and narrative sense) can be gained by reading the maps counterclockwise from lower right. Though reshuffling the four map plates should have been possible, the order was unchanged in the reprintings.

9. Muller, 1863/B, 1:xiii. See also the quotation by Koeman, Collections, 67.

10. Koeman, Collections, 67–69. G. van Rijn and C. van Ommeren, *Atlas van Stolk: Katalogus der Historie-, Spot- en Zinneprenten betrekkelijk de Geschiedenis van Nederland verzameld door A. van Stolk,* 10 vols. (Amsterdam and The Hague, 1895–1931). The Dutch type of "historical atlas," consisting of large collections of historical prints (including maps), reached massive proportions, and several of these hoards survive in public libraries. One, in Rotterdam, has become an endowed foundation in its own right.

11. Le Long, 4:11ff. (catalogue of Fevret de Fontette's historical engravings, organized in chronological order, much like Muller's future Dutch catalogue), 110ff. (Gaignère's collection of portraits drawn from monuments). Portraits were privileged: René Vertot's *Histoire des chevaliers hospitaliers de S. Jean de Jerusalem: appellez depuis chevaliers de Rhodes, et*

aujourd'hui chevaliers de Malthe, 4 vols. (Paris, 1726) has three unremarkable maps but an exceptionally fine series of portraits of the Grand Masters of the Order of Malta. Nicholas de Fer, *Histoire des rois de France* (Paris, 1722), offers more than sixty portraits of French kings. Muller, 1863/B, 1:xiv, suggests that the English, too, were somehow involved in the enterprise of history-in-pictures. I have not found out what he is referring to.

12. On French ascendancy, see De Vrij, *World on Paper,* 15, 92; see also G. Alinhac, *Historique de la cartographie,* 2 vols. (Paris, 1965). Delisle, vigorous and theatrically handsome in the portrait of Harms, *Künstler,* 149, died at age fifty-one. His name, not spelled consistently in his lifetime, occurs in at least three forms in the library catalogues of today: Delisle, Isle (de l'), and Lisle (de). *BN Impr.* differs from BN, Cartes et plans, in its choice. "Delisle" is used here.

D'Anville did not let frail health affect his fifteen-hour working days. He reached his eighties, outliving Delisle by half a century. His intensely critical spirit, which broke cartographic ground by leaving blank the undocumented parts of the world, was especially famous. He had a predilection for ancient geography and was the dominant figure of his time in this branch. Gibbon owned more of his works than of any other geographer: *LGKartog.* 1:18–21; R. V. Tooley, *Maps and Map-Makers,* 6th ed. (New York, 1978), 43; Geoffrey Keynes, *The Library of Edward Gibbon: A Catalogue of His Books* (London, 1940). A mid-nineteenth-century commentator regarded d'Anville as still unequaled in France: Louis Vivien de Saint-Martin, "De l'état actuel de la cartographie en Europe et particulièrement en France," *Bulletin de la Société de géographie,* ser. 4, 10 (October–November 1855): 12–19.

13. Alinhac, *Historique de la cartographie,* 1:61. For a contemporary account of these refinements, see [Nicolas Fréret], "Lettre à M. de L. R. sur les ouvrages géographiques de M. De Lisle . . . et sur sa mort," *Mercure de France,* March 1726, 471–72; phrased as criticism of the Sanson maps that were still being marketed without overhaul. See also J. P. Niceron, *Mémoires pour servir à l'histoire des hommes illustres* (Paris) 10, pt. 2 (1731): 9–54 (probably by Fréret). The astronomer Jean-Dominique Cassini, first of the dynasty and teacher to Delisle (Fréret, "Lettre," 475–76), died in 1712.

14. On Delisle, father and son, see Jean Delanglez, "The Sources of the Delisle Map of America, 1703," *Mid-America* 25, no. 3 (1943): 275–98. For Buache and physical geography, see Buache, 1767/O, 1768/O; Schmithüsen, *Geographischen Wissenschaft,* 142–43; and n. 107, below. Guillaume Delisle was Buache's patron; three years after Guillaume's death, his only child married Buache and barely survived to the first wedding anniversary. Delisle is therefore called Buache's father-in-law, as he is, typically, in Delisle, 1754 (reprinting of Delisle maps). Joseph (1688–1768), who was younger than Guillaume, spent more than twenty years in Russia and was long-lived (*LGKartog.* 1:160–61). For his publications from Guillaume's *Nachlass,* see ch. 4, n. 89, below.

15. See ch. 4, nn. 1–4, below.

16. The canon of Delisle's maps drawn up by Buache after his death (Delisle, 1730) includes a section for the Middle Ages, consisting of two of the Byzantine Empire plus Toul and Dauphiné: Delisle, 1707, 1710b, 1711. For Genghis Khan's empire, see Delisle, 1710a.

17. Anville, 1719. On Longuerue, see *Biog. univ.* 25:79–80. The three general maps head vol. 1 of Longuerue's book; the six particular ones, vol. 2. The book offended the court; withdrawn for revision, it was reissued in 1722. D'Anville's eulogist stresses that merit, not friendship, convinced Longuerue to have d'Anville draw the maps for his book. The terms "Middle Ages" and "medieval" do not appear on the maps.

18. Delisle, 1726. Fréret, "Lettre," 487, says that Delisle had finished the maps just before his death on 25 January 1726. They were published in the abbé Vertot's *Histoire des chevaliers hospitaliers.*

19. Liébaux, 1728. Le Long, 1:34–35, nos. 394–95; Fréret, "Lettre," 487, notes that before Liébaux began to teach geography in Paris, he engraved some of Delisle's works. For the maps for Louis XV, see Delisle, 1717; Le Long, 1:34, nos. 393, 396, 414. Fréret, "Lettre," was convinced that they were ready for publication. I am not sure this could be rightly said of the gaily colored surviving maps, but they are highly finished. A faint pencil inscription in the cartouche of BN, Res.Ge.D.7830, makes the attribution to Buache. He was thought to have copies of many of Delisle's manuscript maps. Delisle's widow published Guillaume's best-known map for young Louis XV, the "Expédition d'Alexandre," shown in Harms, *Themen,* 146–47, no. 65.

20. Abraham Ortelius, *Aurei saeculi imago, sive Germanorum veterum vita, mores, ritus, et religio iconibus delineati et commentariis ex utrumque linguae auctoribus descriptae* (Antwerp, 1596); Clüver, *Germania antiqua,* which made his reputation. Philipp Jakob Spener, besides founding Pietism, inspired the establishment of the University of Halle: *Encyc. Brit.* 25:638–39.

21. Ptolemy's map 5 is of "Magna Germania"; for its sense, see ch. 1, n. 74, above.

22. For details, see ch. 4, nn. 17–22, below.

23. Zollmann, 1732.

24. Bessel, 1732. Quotation from Fritz Curschmann, "Die Entwicklung der historisch-geographischen Forschung in Deutschland durch zwei Jahrhunderte," *Archiv für Kulturgeschichte* 12 (1914): 138.

25. Curschmann, "Die Entwicklung der historisch-geographischen Forschung," 138–39, stresses the limitations of the work on *gaue.* For the maps' being pre-Interregnum, see P. Emmeram Ritter, "Gottfried Bessel—der 'deutsche Mabillon,'" in *Gottfried Bessel, 1672–1749: Diplomat in Kurmainz—Abt von Göttweig,* ed. Franz R. Reichert (Mainz, 1972), 314. Le Long, 1:35, nos. 400–402, records the maps.

26. Falckenstein, 1733 (has an inset of Charlemagne's canal). Johann Heinrich von Falckenstein (1682–1760) published *Antiquitates Nordgavienses* (Frankfurt, 1733) and, in the same year, *Codex diplomaticus antiquitatum Nordgaviensium* (Frankfurt and Leipzig, 1733). His scholarship is not highly esteemed (*ADBiog.* 6:555).

27. A handsomely ferocious picture of Cellarius is reprinted in Harms, *Künstler,* 77; it is the frontispiece of Cellarius, 1701b/A. According to the *Grand dictionnaire universel (Larousse) du 19ᵉ siècle,* 17 vols. (Paris, 1865–90), 3:682, "Toute sa carrière fut consacrée à relever les études classiques. . . . Les progrès réalisés au XVIIIᵉ siècle par la science allemande furent préparés par les nombreux ouvrages de Cellarius et par son activité incessante." *ADBiog.* 4:80–81 complains that his many text editions lack acuity of judgment and are unsystematic. His reputation came from writing practical textbooks, promoting the improvement of Latin style, and supplying popular, small-size editions of the classics. *Biog. univ.* 7:309–10. Without "inventing the Middle Ages," Cellarius transposed to history the philologists' familiar concept of *Latinitas mediae aetatis.* See George Gordon, "*Medium aevum* and the Middle Ages," *Society for Pure English, Tract* (Oxford) 19 (1925): 3–28; Lucie Varga, *Das Schlagwort vom Finsteren Mittelalter* (Baden bei Wien, 1932; repr., Aalen, 1978), the most comprehensive study; Butterfield, *Man on His Past,* 45–46.

28. Cellarius,1701b/A. *Grand dictionnaire universel (Larousse),* 3:682. On the small

maps, see Harms, *Künstler*, 76. The second edition of *Notitia orbis antiqui*, published by
L. Io. Conrad Schwartz (Leipzig, 1731), is preferred.

29. Review in *Journal de Trévoux*, January 1702, 78–102; "On en promet une suite pour
la Géographie du moyen âge surtout depuis Constantin jusqu'à Charlemagne" (81–82).
Grand dictionnaire universel (Larousse), 3:682: "Cellarius avait commencé un traité analogue
sur la géographie du moyen âge." The titles of Cellarius's historical compendia suggest that
he regarded the Middle Ages as lasting to 1453 only in the East, whereas the western Middle
Ages ended with Charlemagne. This periodization is implied by Hagelgans's sequence of
maps (see n. 8, above). It may also have influenced Augustin Lubin's *Mercure géographique*
of 1678 (see ch. 2, n. 149).

30. Cellarius, 1776. On Homann as engraver, see J. G. Hager, *Geographischer Büchersaal
zum Nutzen und Vergnügen eröfnet*, 3 vols. (Chemnitz, 1766), 1:380. Cellarius, 1701b/A, had
an enriched edition in Rome, 1774 (collated in *Map Collector* 10 [1980]: 50). This edition encouraged production of the threefold appendix, including Appendix 3, "Tabulae quaedam
geographicae a Cellario in usum geographiae medii aevi descriptae." The anonymous editor
was well aware that, in 1776, Cellarius's unfinished medieval maps were mainly a memento
of the great man. See also ch. 4, nn. 104–6, below.

31. *LC Bibliog. of Cartog.* 2:147 leads to the obituary in *Mercure de France*, May 1720,
127–33; the author's interest turns toward Guillaume by the second page. Nicolas Fréret
(anonymously) wrote Delisle's obituary in the *Mercure de France* for March 1726. He may
be responsible for the father's as well (whose student he had been). In the three series of historical maps Guillaume had been projecting, the last two maps of the second series were
also to be the opening items of the first series.

32. Definitely in print: Delisle, 1705. The projected "Empire des Tartares" may be related to Delisle, 1710a. For the posthumous publications, see ch. 4, nn. 87, 89, below.

33. Fréret, "Lettre," 481–82, 485: in 1723, Delisle produced an innovative map of
Xenophon's 10,000 for the use of young Louis XV; he intended to produce several more on
the same scale, notably the Persian Empire under Darius, the Macedonian Empire under
Alexander, and the Roman Empire at its greatest extent. The planned "Theatrum historicum ecclesiasticum" is very similar in selection to La Ruë, 1651/E. Both end with maps of
modern Syria, but Delisle proposed a Crusade age map in place of La Ruë's fifth map, "The
Patriarchate of Jerusalem."

34. Concerning al-Idrisi as "the Nubian Geographer," see C. F. Seybold, "Al-Idrisi,"
Encyclopedia of Islam (Leyden and London, 1927), 2:451; Lelewel, *Géographie du moyen âge*,
1:94–95 n. 226. Hagelgans, 1718, gives generous coverage to Asia. Duval, 1665, has a map of
the Islamic Empire (fig. 6), but in a context of geographical discovery rather than history.

35. Comparable maps of this indefinite kind occur in Köhler, 1730 (the posthumous
third installment), and Anville, 1771.

36. Bertius, 1623. The two Carolingian maps in Delisle, 1717, in a separate file folder
from the other three, as well as the map for Charlemagne, show signs of extension beyond
the established frame. They may have been being turned into "théatres historiques." Another fragment of this project may be ArchNat, Marine, Serv. hydrog., 6JJ/70 no. 76.

37. Review of Dupré, 1763, in *Journal des sçavans*, April 1764, 393.

38. The dovetailing of the "historical vistas" with the medieval set is particularly impressive (the last two of the former were to be the first of the latter). Fréret, "Lettre," 484:
for young Louis XV, Delisle drew several maps with modern and ancient names of the same

places; "[leurs] divisions étoient relatives à certaines époques déterminées, afin d'éclaircir entièrement l'Histoire des temps ausquels elle avoit apport." Such sensitivity was still rare.

39. The circumstances of Delisle's death are presented by Fréret in suitably dignified colors. Dr. Mary Pedley (Clements Library, University of Michigan) tells me there is a different account in the letters of the Delisle sister Angélique to Joseph in Russia.

R. V. Tooley, "The De l'Isle, Buache, Dezauche Succession (1700–1830)," in his *The Mapping of America* (London, 1980), 3, states that Guillaume's father, Claude, produced an *atlas historique* in 1684. That year, Claude published *Tableau historique du Siam.* Some early reference works give the title *Atlas historique* to Delisle, 1718/B, without intending a geographic sense; it contains sixty-one nongeographical engraved charts. Claude has no historical atlas to his credit.

40. *ADBiog.* 16:442–43. The University of Altdorf lasted from 1623 to 1809. Köhler, named professor of logic in 1710, exchanged this chair for history four years later; he was also university librarian. He supervised many more theses there than at Göttingen. Picture in Harms, *Künstler,* 155. Köhler's *Kurtzgefaste und gründliche Teutsche Reichs-Historie* (Frankfurt and Leipzig, 1736) is more a chronicle than an interpretive essay. J. D. Köhler, *(Wochentliche) Historische Munz-Belustigung,* 22 vols. (Nuremberg, 1729–50 [1756]); each weekly issue features one or two coins or medals (sixteenth to eighteenth centuries), with an engraving, description, and historical explanation. The final, posthumous volume is by Köhler's successor at Göttingen, J. C. Gatterer, with a full account of Köhler's life and works (see Köhler, 1730). See also Arthur Kühn, *Die Neugestaltung der deutschen Geographie im 18. Jahrhundert* (Leipzig, 1939), 20.

41. Harms, *Künstler,* 155, calls Köhler Weigel's collaborator and counselor in matters geographical. His classical atlas is Köhler, 1720/A. Adriaan Reland, *Palaestina ex monumentis veteribus illustrata* (Utrecht, 1714), 9 maps; about his reputation, see, e.g., *Encyclopaedia Judaica,* 16 vols. (Jerusalem, 1971–72), 14:65; Nebenzahl, 142–43, stresses the modernity of Reland's maps.

42. On medieval geography: Christian Juncker, *Anleitung zu der Geographie der mittleren Zeiten* (Jena, 1712). The compilation of Köhler's friend Samuel Faber, gymnasium rector in Nuremberg, is the core of Köhler, 1718, and anticipates its title. On Faber and Weigel, see n. 6, above, and Harms, *Künstler,* 108.

43. Köhler, 1730: *Kurtze und gründliche Anleitung zu der alten und mittleren Geographie nebst XII. Land Chärtgen (Compendium geographiae antiquae et mediae);* 12 maps per installment.

44. For the Anglo-Saxons, see ch. 2, nn. 1–24, above. Liébaux, 1728 (n. 19, above). Muratori, 1727; see also Le Long, 1:36, no. 422.

45. On Sanderus and Fabius, see ch. 2, nn. 88–89, above.

46. For the canal, see n. 22, above. The plan of Constantinople is one of two in Banduri, 1711.

47. See n. 35, above.

48. A fuller account of Hase occurs in ch. 4, nn. 160–73. See C. Sandler, "Die homännische Erben," *Zeitschrift für wissenschaftliche Geographie* 7 (1890): 333–55, 418–48; reprinted in *Acta Cartographica* 5 (1969): 370–428. See especially the forthcoming study by Johannes Dörflinger, "Das geschichtskartographische Werk des Johann Matthias Hase," in *Geschichtsdeutung (Archäologie und Geschichte) auf alten Karten* (proceedings of the Fortysixth Wolfenbütteler Symposion, 26–29 October 1999). (I am very grateful to Professor

Dörflinger for supplying me with an advance copy of his contribution.) In view of the importance of Africa in d'Anville's reputation (Günther, *Geschichte der Erdkunde,* 183), it is high praise of Hase when Sandler comments, "The first maps of Africa drawn on a strictly scientific basis were by Hase, not d'Anville" (416).

49. Sandler, "Die homännische Erben," 423: Chris. Fembo, successor to the Homann Heirs, reissued Hase, 1750a, in 1813. It was still remembered and thought relevant in the mid–nineteenth century (Köhler was too, for that matter); "Die damals [in 1820, 1824] bekannten Charten [of the Middle Ages] von Hase und Köhler hatten sich überlebt." (unpaged introduction to Wedell, 1849); see ch. 5, n. 158, below.

50. The latest information on publication dates is in Dörflinger, "Werk des Johann Matthias Hase."

51. Hase, 1739/B: the preface and commencement address are unpaginated. He had readied nine maps from which thirty-three or -four *schematae* could be produced. Gleditsch of Leipzig was the publisher who eventually issued Cellarius, 1776.

52. Wilhelm Bonacker, "Johann Matthias Haas (1684–1724), sein Leben, seine Schriften und Karten," *Zeitschrift des historischen Vereins für Schwaben* 59–60 (1967): 278. Hase announced his historical atlas in the university lecture list for the summer term of 1742. On Böhme, see ch. 4, n. 166, below.

53. Hase, 1743. The narrative part 1 is described as being for *lectiones academicae.* In the summary of the contents of part 3, the *imperia* are subdivided into *antiquiora* and *recentiora;* the medieval empires are classified as "recent."

54. The closest competitor is La Ruë, 1653.

55. Two of the maps of the medieval empire are not assigned to a definite time but give a "General view of all the regions that were ever attached to the Franco-Roman or Romano-Germanic Empire" (*Conspectus generalis omnium regionum quae ad Franco-Romanum vel Romano Germanicum Imperium unquam pertinuerunt*), one each for West and East.

56. Hase, 1750b.

57. See the comment of Wegele, in ch. 4, n. 171, below.

58. Beaurain, 1749. Vol. 15 is devoted to profane antiquity, vol. 16 to sacred history, followed by the Middle Ages. Vol. 16, nos. 1–45, cover successively the ancient empires, biblical history, and ecclesiastical geography—three of the four categories of Padua, 1699, but with the empires and biblical history each in chronological order. Only nos. 20 and 22 are not by French cartographers.

59. For the Covens and Mortier atlas, see Delisle, 1733. Beaurain, vol. 16, nos. 48–51, are Bertius, 1623 (Charlemagne's empire in four maps). Beaurain, who does not mention an author, probably assumed these four maps to be French, like all the other medieval maps.

60. Alting, 1701. For Tomka's *Parvus atlas Hungariae* (Bratislava [Pozsony, Pressburg], 1750–51), see ch. 4, n. 198, below. The abridged, posthumous atlas is Tomka, 1781; maps dated 1750 and 1751.

61. "Atlas complet," 1747. BN, Cartes et plans, Journal des entrées des cartes, Registre C 14094 à 14970, no. 14526. The Journal indicates that the atlas had a title page and six pages of Avertissements. If these still exist, they are elsewhere than with the volume. Hofmann, "Genèse de l'atlas historique," 117 n. 68, considers them lost.

62. *Biog. univ.* 36:138–39. In the first quotation (which, with a flourish, closes the Premier avertissement of "Atlas complet," 1747), "appartiens" has the ambiguity of "belongs": either "owned by" or "the intellectual property of." The second quotation is from the BN

Journal des entrées. For the long association of the Roberts with Boudet, see Pedley, *Vaugondy*, 51–53 and passim.

63. Pedley, *Vaugondy*, 233. A systematic comparison of Gilles Robert's handwriting with that in the "Atlas complet" would be useful. Hofmann, "Genèse de l'atlas historique," 117, regrets that there is no handwriting sample from Gilles Robert at BN, Cartes et plans.

64. Neither then nor afterward did "historical atlas" become a technical term. *Atlas historique, géographique et topographique de l'Empire d'Allemagne,* 2 vols. (Paris, 1758–59), confines its historical content to the printed text: see *Journal des sçavans,* March 1759, 52–57. An atlas of current Lower Saxony (Paris, 1757) has *Atlas historique* as main title: *Journal des sçavans,* March 1758, 131–32. Reischl, 1758/B, called *Atlas historicus,* is a small octavo without maps.

65. Luneau, 1760. "Revolution" is a well liked but not a strong word before 1789 (1776?). I am tempted to translate it simply as "event," provided "event" is understood as an occurrence worthy of historical commemoration and not just a domestic incident. "Vicissitudes," which I use, may be preferable. About Luneau's projected atlas, see *Journal de Trévoux,* January 1761, 175–76. About Pierre Joseph François Luneau de Boisjermain (1732–1801), see Quérard, *France littéraire,* 5:394–95, "instituteur zélé et littérateur médiocre." *Biog. univ.* 25:482–84: Luneau trained as a Jesuit and dropped out in good standing. Also see Hofmann, "Genèse de l'atlas historique," 118–19.

66. Luneau, 1760, has markings: upper right, "Première feuille"; inset lower left, "Supplément à la carte du Paradis terrestre," Palestine *and* Armenia shown as alternatives, favored by some, to Mesopotamia as the site of Paradise. Much attention to Adam, Eve, and Paradise; hardly any place-names. Writing lower-left center (alongside the inset): "On présente le Globe Terrestre [Mercator projection] tel qu'il est aujourd'hui, quoique probablement il a été altéré par le Déluge."

67. Dupré, 1763. The copies at Cambridge and Göttingen have title pages. Julien, *Nouveau catalogue,* 65–66, spells out the titles of the thirty divisions but gives no author or compiler.

68. *LC Bibliog. of Cartog.,* s.v. Picaud: "[Map] Dressée sur les mémoires de M. D. P. La carte a été dressée la même année [as the accompanying text] par M. Michel Picaud de Nantes." Picaud is unknown to biographical dictionaries. The accompanying text is *Les révolutions de l'univers ou remarques et observations sur une carte géographique destinée à l'étude de l'histoire générale* (Paris, 1763), 174 pp. in 12mo.

69. Dupré (without initials) is on the surviving title pages (see previous note). Dupré is cited there as though it were a full and adequate name, which just possibly it is; but library catalogues and biographical dictionaries produce no Dupré who might be the author.

70. A. A. Barbier, *Dictionnaire des ouvrages anonymes,* 4 vols., 3d ed. (Paris, 1872–79), 2:353. *Biog. univ.* 33:125–27 (on Philippe), ascribes to him the small book elucidating and paralleling Picaud's maps. Nothing confirms this conjecture. Philippe, 1768, 1787, are very run-of-the-mill compilations; Velly, 1787, in which Philippe was probably involved, is even worse. Basically a teacher of the classics who edited classical texts, Philippe branched successfully into classes in geography and history and took part in drawing up the curriculum for the military schools.

71. *Biog. univ.* 36:139 attests to an early, undocumented belief in the descent of Dupré's atlas from the sixty-six-map draft. Although our manuals do not date the acquisition of Lorraine to 1738, the atlas-makers had a basis for doing so. For oriental content, see *Journal*

de Trévoux 64 (1764): 200; only Delisle, 1705, and Hase, 1743, can bear comparison to Dupré's attention to the East, and they had severe limitations "parce qu'on ignoroit alors l'histoire des Nations plus orientales" (an exaggeration: Hagelgans, 1718, shows how attentive a contemporary of Delisle's already was to Chinese history); the summary of Dupré's "intervalles" bears out the prominence of Chinese and Islamic history. The atlas (and the foregoing Paris draft) can hardly be called Eurocentric. Distribution of emphasis: (Paris draft) ancient, thirty-two; medieval, thirty; modern, four; (Dupré) thirteen, fifteen, two. By comparison with the sixty-six maps of the Paris draft, the thirty of Dupré's atlas inflict the greatest losses on antiquity. Philippe de Prétot, in his certain works, showed little interest in the Middle Ages.

72. Pedley, *Vaugondy*, 233.

73. *Journal des sçavans* (April 1764), 397.

74. Some of this result is achieved by the computer program Centennia, a historical atlas of Europe and the Middle East, A.D. 1000 to 1993. (The Gold Bug, Alamo, California). The goal of this program, available in the early 1990s, was to trace frontier changes in a way basically similar to that of the Paris atlases.

75. Interval 30. French historians of the time, notably Hénault and the abbé Velly (see nn. 78–79, below), were well aware that readers deplored history with a one-sidedly political and military focus. They took pains to satisfy the demand for more uplifting or entertaining fare; contemporary testimony: *Journal des sçavans*, October 1761, 463. But political boundaries were still the chief implement in the historical geographer's tool kit.

76. The sixty maps are subdivided into thirty-six tracing the history of France and another twenty-four recording the possessions of the domain of the kings of France. Rizzi-Zannoni called them "cartes de répétition."

77. On Desnos's standing and reputation, see Lenglet, 1768/B, 1:340–41. His equals as map publishers were Julien and Lattré. About his receptivity to all maps offered to him: "dans le nombre il s'en trouve de fort bonne, ou du moins dont l'idée ne demande qu'à être étendue ou perfectionnée."

78. L. C. Desnos, *Catalogue des ouvrages tant anciens que modernes du Fond du Sr. Desnos* (Paris, 1765), 4–5; concerning Rizzi-Zannoni's atlas, "il est le premier qui est paru dans ce genre. On peut le regarder comme très-utile, nécessaire même à la Nation et à l'instruction de la jeunesse" (the *Catalogue* is bound in with these atlases at UIll, Princeton, and other collections). The prose history referred to is the commercially very successful work of P. F. Velly (1709–59) and Claude Villaret (†1766), *Histoire de France depuis l'établissement de la monarchie,* 16 vols. (Paris, 1755–65) (a third collaborator, J.-J. Garnier, started at vol. 16). Velly and Villaret's history is no longer esteemed; see Eduard Fueter, *Geschichte der neueren Historiographie* (Munich and Berlin, 1911), 146. Velly, 1787, is a map supplement. On Desnos's second volume for the atlas, see ch. 1, n. 62, above.

79. Desnos, *Catalogue des ouvrages,* 5–6. Desnos, 1765b, relates to Jean François Hénault, *(Nouvel) abrégé chronologique de l'histoire de France* (Paris, 1744; 7th ed., 1765)—a set of annals, not a sustained narrative. Fueter, *Geschichte der neueren Historiographie,* charges Hénault's *Abrégé chronologique* with harboring political speculation like Montesquieu's rather than history. Its very successful publication history extends to 1855. Announcement of the atlas: *Journal de Trévoux,* January 1765, 199. The histories of Mézerai (1668; new ed., 1755) and Gabriel Daniel, S.J. (1696, 1721; new ed. 1755–60), are considered much more serious than that of Velly and Villaret (Fueter, *Geschichte der neueren Historiographie,* 145–46). Desnos's "autre atlas" appears to be a companion to both of them, perhaps inspired by their

reprinting in 1755. Le Long, 1:3, no. 3, lists the Hénault companion as having thirty-two leaves (identical to the UIll copy) and has no knowledge of Desnos's companion atlas to Daniel and Mézerai. Lenglet, 1768/B, though concerned with maps, takes no account of the Desnos atlases.

80. *Journal de Trévoux*, July 1765, 213–22, on 222. A fourth Desnos atlas is paired with a narrative; see n. 92, below. A later Desnos list, *Catalogue des ouvrages de géographie . . . dont est composé le fonds du sieur Desnos* (Paris, 1775), 1, lists the historical atlas of France without reference to Rizzi-Zannoni and points out that it is endorsed in the "seizième volume de M. Villaret, p. 386."

81. Brué, 1820: 3 n. 3 (spelling, "Zanoni").

82. Le Long, 1:3, plan of 1750 (published in *Journal des sçavans*); the Desnos atlases are listed with the comment that Rizzi-Zannoni's "ne peut être considéré que comme un essai," echoed by Brué, 1820: 3 no. 3. See also Julien, *Nouveau catalogue*. On the other hand, the *Journal de Trévoux* and *Journal des sçavans* warmly welcomed the Desnos atlas.

Le Long died early in the eighteenth century. The abbé Charles Boullemier (1725–1803) published a much enlarged edition of Le Long's *Bibliothèque historique* (1768–78), with indexes by Fevret de Fontettes and Barbaud de la Bruyère: *Dictionnaire des lettres françaises,* vol. 2, *Le XVIIIᵉ siècle,* ed. Georges Grente, new ed. (Paris 1994–96), 82.

83. Aldo Blessich, "Un geografo italiano del secolo XVIII: Giovanni Antonio Rizzi Zannoni (1736–1814)," *Bolletino della Società geografica italiana,* ser. 3, vol. 11 (Yr. 32, vol. 35) (1898): 62 (Rizzi-Zannoni's own account of the historical atlas). On Fontanieu (1693–1767), a councilor of state and intendant, see *Biog. univ.* 16:332. His rich charter collection, given to the royal library, is summarily catalogued in Le Long, vol. 5.

84. In this case, Rizzi-Zannoni may simply have misremembered the year. The maps in Desnos, 1765c (see 46n. 62, above), document Rizzi-Zannoni's work for Desnos in these years. Vladimiro Valerio, *Società uomini e istituzioni cartografiche nel mezzogiorno d'Italia* (Florence, 1993), 107: "la scarsa attendibilità del Rizzi Zannoni 'biografo.'" Valerio, a sympathetic observer, evokes "the enigmatic and sometimes disquieting aspects of his complex personality" (85).

85. On Rizzi-Zannoni's beginnings in Paris, see Valerio, *Istituzioni cartografiche,* 85–87, and ch. 4, n. 185, below. Typically, the manifold activity documented by his publications is wholly disregarded in his stately memoir of 1774. The energy we prize in Rizzi-Zannoni does not seem to have won his own approval. For Rizzi-Zannoni's early story of coming to Paris to draw war maps, see BN, MS français 22120 (57), September 1763. I owe this information and what follows to the generosity of Dr. Mary Pedley (Clements Library, University of Michigan).

86. W. Wolkenhauer, *Leitfaden zur Geschichte der Kartographie* (Breslau, 1895), 60; Rizzi-Zannoni is mentioned honorably by Günther, *Geschichte der Erdkunde,* 187, though many worthies are not. Wolkenhauer disregards his atlas of French history. Harms, *Themen,* 106, reproduces Rizzi-Zannoni's "Carta del littorale di Napoli e de luoghi antichi piu rimarchevoli" (1794). His Neapolitan atlas is at PhiladAmPhil. On this part of his life, see Giancarlo Alisio and Vladimiro Valerio, eds., *Cartografia napolitana dal 1781 al 1889: Il Regno, Napoli, la Terra di Bari* (n.p. [Naples], 1983), and esp. Valerio, *Istituzioni cartografiche.*

87. *Journal de Trévoux,* January 1765, 195 (Rizzi-Zannoni's atlas), 199 (Hénault's atlas). Desnos is featured in both announcements; no mention of Rizzi-Zannoni. Le Long, 1:3.

88. The only copy of the Hénault companion known to me is at UIll. I saw it in 1989. It

appears to be lost now. The first state of the Rizzi-Zannoni atlas is illustrated by BL, Maps 24.d.6. Harms, *Themen*, 101, reproduces a colored example of Desnos map printing, identical to the appearance of the historical atlases (and Desnos, 1765c, for that matter); described as "Colored copper-etching (*Radierung*) by 2 plates"—one for the frame, another for the map. The delicacy and precision of the coloring give greater charm to Desnos maps than the frame does. The reviewer for the *Journal des sçavans* was impressed (August 1765, 564). The Desnos map style was proper to his firm and was not connected to Rizzi-Zannoni. On the other hand, W. W. Jervis, *The World in Maps: A Study in Map Evolution* (New York, 1937), 41, credits Rizzi-Zannoni with some of the "finest examples" of decorative mapmaking and of achieving a good general effect "in spite of the minuteness of the decorative design."

89. Desnos, 1765b (the Hénault companion), was probably not denied a share in these changes.

90. Possibly, the Daniel/Mézerai atlas never existed. An atlas in this form could have been advertised on the chance that a demand would develop: if there were customers, existing maps could be assembled and provided with a title page. Echoing the silence of Le Long, Desnos fails to list such a work in his later catalogues, in which the two surviving atlases duly appear to the end of the century (see n. 96, below). See also *Journal des sçavans,* August 1765, 562: though the Rizzi-Zannoni atlas is mainly oriented to the history of Velly and Villaret, "on s'est proposé de ne pas borner son utilité à cet ouvrage seul, il peut servir pour la lecture de Mezeray, du P. Daniel, et de tous les autres Historiens." Although a special Desnos atlas for Daniel and Mézerai may yet turn up, its contents are unlikely to be surprising. Desnos, *Catalogue des ouvrages,* 13, also lists three single-sheet maps of Gaul and France by Rizzi-Zannoni.

91. Desnos seems to have been very active in the 1760s, notably in collaboration with Brion de la Tour. Paris, BN, *Inventaire de la Collection Anisson sur l'histoire de l'imprimerie* (Paris, 1900), 2: no. 22149/9–10 (ca. 1761, Desnos seeks permission to promote the stock of Nicholas de Fer and heirs). Collections in which the Rizzi-Zannoni French history atlas survives are too many to list. Pastoureau, "La France divulguée," 62, traces Desnos's business career.

92. Desnos, 1766. Catalogued at Harvard only under Desnos's name.

93. *Journal des sçavans,* August 1765, 563. See 114 n. 24, above.

94. Le Long, 1:3. Another interpretation of Boullemier's comment is possible, but I don't think he meant that too many copies of these maps had been placed in circulation.

95. *Journal des sçavans,* August 1765, 563–64.

96. For mention of the atlas in Velly and Villaret, *Histoire de France,* see n. 78, above. Reprints: 1777 at Oxford, 1782 at BN (possibly new title pages for old stock). Desnos, *Catalogue des ouvrages;* L. C. Desnos, *Catalogue raisonné* (Paris, n.d. [post-1783]) (BN, Ge.FF.7181); L. C. Desnos, *Catalogue des atlas historiques et cartes géographiques . . . dont est composé le fonds du sieur Desnos* (Paris, n.d. [ca. 1790]) (BN, Ge.FF.7182). For Brué, see Brué, 1820. Another symptom of neglect is that Rizzi-Zannoni/Desnos is disregarded in Léon Mirot and Albert Mirot, *Manuel de géographie historique de la France,* 2d ed. (Paris, 1947).

97. Butterfield, *Man on His Past,* 42–44. Gatterer's life and work: *ADBiog.* 8:410–13. Princeton University Library, for example, has nothing of Gatterer's except three major works on diplomatic (i.e., the study of early charters). I comment further on Gatterer in "The Plot of Gatterer's 'Charten zur Geschichte der Völkerwanderung,'" to appear in *Geschichtsdeutung (Archäologie und Geschichte) auf alten Karten* (proceedings of the Forty-sixth Wolfenbütteler Symposion, 26–29 October 1999) (forthcoming).

98. *Biog. univ.* 15:646–47, says seven; so does *GV* 44:6–8. Butterfield, *Man on His Past,* 48, counts eight and calls the first epoch making. *NUC* 192:412–13 records four of the seven listed in *Biog. univ.* Gatterer's works on universal history: *Handbuch der Universalhistorie nach ihrem ganzen Umfang,* 2 vols. (Göttingen, 1761–64); *Abriss der Universalhistorie* (Göttingen, 1765); *Abriss der Universalhistorie* (Göttingen, 1773); *Einleitung in die synchronistische Universalhistorie,* 2 vols. (Göttingen, 1771); *Weltgeschichte in ihrem ganzen Umfang,* 2 vols. (Göttingen, 1785–97); *Kurzer Begriff der Weltgeschichte* (Göttingen, 1785); *Versuch einer allgemeinen Weltgeschichte* (to 1492) (Göttingen, 1792). Entries in library catalogues add many qualifications: partly finished second volumes, reprints, etc. Gatterer's time line: *Synopsis historiae universalis sex tabulis comprehensa* (1766; new ed., 1769), folio.

99. On the conception and goals of Gatterer's maps, see *Historisches Journal von Mitgliedern des königlichen historischen Instituts zu Göttingen* 8 (1776), IV. Stück: 17; the lecture announcement of 1775 is quoted in Kühn, *Neugestaltung der deutschen Geographie,* 114. For early student maps, see Dangeau, 1697; J. B. Homann, *Atlas methodicus* (Nuremberg, 1719) (sometimes called the first school atlas); Gaspari, 1811/O. On outline ("methodical") maps, see *LGKartog.* 2:718.

100. G. C. Hamberger and J. G. Meusel, *Das gelehrte Teutschland oder Lexicon der jetzt lebenden teutschen Schriftsteller,* 5th ed. (Lemgo, 1796; repr., Hildesheim, 1965), 2:490–94 (an impressively full entry). Gatterer, 1775/O, 1780/O. The earlier of the two is hard to reconcile with the listing in Woltersdorf, 144–45; but see also the next note.

101. C. Kruse, "Probe der Gattererschen Charten zur Geschichte der Völkerwanderung, mit Anmerkungen für diejenigen welche diese Charte mit meinem historischen Atlas zu vergleichen wünschen," *AGEphem.* 16, no. 4 (April 1805): 377–99; Kruse calls the maps "an essay but by an expert"; Gatterer is "our most famous scholar"; Kruse believed Gatterer had Hase's atlas before him and wished to correct it, and he singled out Gatterer's work on the eastern peoples for special praise (382–83). In light of the forty-four-map atlas listed by Woltersdorf (see previous note), it is noteworthy that Kruse's numbering, including map 18, relates to a twenty-four-map collection only. Gatterer's published map differs very much from any map of Kruse's. On Kruse, see ch. 5 nn. 46–59, below. About Bredow, see J. S. Ersch and J. G. Gruber, eds., *Allgemeine Encyclopädie der Wissenschaften und Künste,* sec. 1, 12 (1824), 334; *ADBiog.* 3:282–83.

102. Gatterer, 1775/B, xiii–xvii. Half the *Abriss* was ready in 1775, the rest (including the pages describing the atlas) three years later. There is no overlap with Gatterer, 1789/O, a country-by-country survey. I am thankful to Jürgen Furgyik, of the Niedersächsische Staats- und Universitätsbibliothek, Göttingen, for his generous help in dealing with this question, on which we do not wholly agree.

103. Kruse, "Probe der Gattererschen Charten," 382, despite his reservations about Gatterer's atlas, believed that, because of its modest cost, it gave much better value than the very expensive atlases coming from England and France. He was probably thinking of the pricey Lesage atlas, issued in London in 1801 and Paris in 1803–4; see ch. 5, nn. 4, 5, and 13, below.

104. Woltersdorf, 144–45. "Conditions of commerce" translates "Zustand der Handlung"; *Handlung* where we would expect to see *Handel* is recorded as an obsolete form by German dictionaries of today. Gatterer seems to have produced maps only in the mid-1770s and never again in his career. The collections surviving at Harvard and Berlin incorporate some of the items listed by Woltersdorf.

105. Woltersdorf's description is foreshadowed by the *Historisches Journal.* Kruse says nothing about the other twenty maps of Gatterer's collection.

106. Gatterer, 1775, xiii–xiv. Some maps of hemispheres, but not historical ones, survive at Göttingen: Gatterer, 1789.

107. Gerhard Engelmann, *Heinrich Berghaus: Der Kartograph von Potsdam* (Halle, 1977), 26, mentions Gatterer, 1775/O, as an aid to Gatterer, 1775/B, and as the oldest physical atlas ever compiled. Most of the maps are, in fact, concerned with comparative ancient and modern geography. Responsibility for precocious physical atlases might rest with Buache himself (or his disciples). See Elizabeth Rogers, "An Eighteenth-Century Collection of Maps Connected with Philippe Buache," *Bodleian Library Record* 7 (1962–67): 99; a Buache publication of 1752 marks the revival of interest in physical geography. On Buache inspiring Gatterer, see Oscar F. Peschel, *Geschichte der Erdkunde bis auf Alexander von Humboldt und Carl Ritter,* ed. Sophus Ruge, 2d ed. (Amsterdam, 1961), 806. See also Buache, 1767/O, 1768/O. The originality of Buache's physical atlas is stressed in *Journal des sçavans,* July 1757, 74–79, 392–93.

108. For language maps, see Harms, *Themen,* nos. 17 (1593, a very limited district), 50 (1723), 76 (1741, comprehensive). No. 99 is a map of commerce about a decade after Gatterer's. Hondius, 1607, "De religionum diversitate per universum Orbem" (BSB copy: 760–61). Nicolas Sanson has such a map, but I have not seen it ("Europe divisée par religions," BN, Ge.DD.2907, 173, 174). *Map Collector* 50 (1990): 41, reports a hitherto unknown world map by Franciscus Haraeius (Verhaer) (1614?) showing the distribution of religions. See also Desnos, 1765c. Religious diversity is given prominence in G. Sanson, *Introduction à la géographie,* discussed in ch. 1, 24–25.

109. On G. Achenwall and *Statistik,* see Butterfield, *Man on His Past,* 51; on Büsching, *ADBiog.* 3:644–45. Gatterer's thematic maps are as lost as the others.

110. Conrad Malte-Brun, *Précis de la géographie universelle,* 8 vols. (Paris, 1810–29), 1:7, quoted by Brué, 1820: 2; *Allgemeine Litteratur-Zeitung* (Halle), no. 346, 4 (December 1804): 522. Compare the statement of William Hughes (1869) quoted in ch. 1, at n. 29.

CHAPTER 4

Eighteenth-Century Maps of the Middle Ages

Single Maps and Small Sets

From the beginning of the eighteenth century onward, maps of medieval scenes increased markedly in numbers, and so did the variety of ways in which such scenes were depicted. Some mapmakers chose ways to design medieval moments that proved to be blind alleys, while others took directions that were convincingly right. One senses the advent of a "medieval geography," fascinated like its "ancient" precursor with turning the data of medieval sources into maps of historic districts and territories. Maps were drawn when needed, especially when called for by the production of a book. Certain scenes were repeated; others, oddly neglected. Isolated maps include the worst and best of the century's medieval mapping. However that may be, by the beginning of the nineteenth century, the efforts of atlas-makers presented devotees of history again and again with graphic surveys of medieval Europe.

Isolated Scenes

Barbarian Invasions

If Menso Alting, little known outside the Netherlands, was the first person to produce a true medieval atlas, the few but choice maps by Guillaume Delisle that are classed as "géographie du

187

moyen âge" were symptomatic of an activity that, with him, attracted an internationally famed cartographer and acquired respectability. Delisle's medieval maps are widely dispersed, remarkably numerous, and usually created as book illustrations. One unpublished set, introduced in chapter 3, was drawn by Delisle to help teach King Louis XV, in his minority, about his predecessors. In this chapter, Delisle's many medieval maps will be presented under several subheadings rather than all at once, so that they may be discussed in the context appropriate to each one.

Delisle's farthest-reaching historical map is officially classed as "ancient" but is deliberately poised at the borderline between antiquity and the Middle Ages. This is his "Historical Vista for A.D. 400 ("Theatrum historicum ad annum Christi quadringentesimum") in which the condition of both the Roman Empire and the barbarians living around it is presented to the eyes." First published in 1705, the map is in one sheet each for the eastern and western parts of the empire. Copies, attributed or not, were made very soon; it was printed in London by Herman Moll in elephant folio and by Weigel of Nuremberg in a reduced format. The public seemed to have been waiting for a map like this and welcomed it.[1] Delisle, the son of a history teacher with whom he also collaborated, meant it to be the first of a series including additional "vistas" of both ancient and medieval history—a plan checked by his premature death.[2]

The comments, or "petites remarques," that accompany the two-part map were available as a small book and are sometimes glued to the edges of the map. They reveal Delisle's eagerness to depict the moment when the Roman Empire was about to yield its western half to new management: "I chose this period so that those studying history might better know the movements of the barbarians who, soon after, settled on the lands of the empire; and I began [at this time] because the irruption of the barbarians in 406 and 407, which is the beginning so to speak of the ruin of that empire, is also [the starting point] of all modern histories, and because many people are not very interested in much earlier antiquity." Histories tended still to be only ancient or modern; Delisle shows no familiarity with a moyen âge. But the concept of an intermediary period was far from unknown to the friendly commentator, who, in connection with the death of Delisle père, hailed the maps to be expected from the son: "He has also undertaken the production of a series of maps for the Middle Ages. How useful a corpus of such maps would be is well known, because they have much greater bearing on the current condition [of the world] than do maps of antiquity."[3] As maps gained value for understanding history, bridging the cartographic gap between Rome and modern times seemed worth doing.

Contemporaries noted disparagingly that Delisle's "Theatrum" much resembled Sanson's "Roman Empire"; the projection is reminiscent of Ortelius's "Roman Empire" as well. Justifying himself, Delisle stressed among other things Sanson's neglect of the barbarians: "I have felt obliged to give them a part of my cares, with a view to being as useful as possible to those who read history. What the barbarians did on the lands of the empire once they were established there is well enough known; but everyone does not know where they were before, and many persons have often asked me. I have marked their dwelling places as well as I could so that the route that they followed might be better seen." Addressing problems that later designers of maps of the barbarian invasions would also face, Delisle abstained from hazardous cartography. He thought that the dynamics of history exceeded the capacities of a single portrayal. Because maps could not be infinitely multiplied, and only so many markings could appear on a single sheet before muddling everything, the map should be left simple, while the intricacies of history were consigned to the written word. His reasoning was sensible and lucid:

> I found it difficult to determine the localization that I should give to certain of these peoples because they did not have fixed homes, they were in continual movement, and they were found at the same time in different places and in the employ of different sponsors. It occurred to me to write their names two or three times and to accompany [the names] with various dates, so as to show that they were here at such-and-such a time, and there at another. But, besides the confusion occasioned by this repetition of names, I still could not, nor could anyone, have shown all these changes of homes without drawing a special map for almost each one of these peoples. It is these changes among the barbarians, as much as those that occurred in the empire, that convinced me to supply these brief comments.[4]

Netherlanders had charted early Franks several times, as we have seen, and traced the passage of the Lower Rhine from ancient to medieval times; and many atlases showed the divisions of Britain after the Saxon conquest. Delisle's map for general history at the end of the Roman Empire was ambitious and new, well suited to be the platform for a sequence of historical maps embracing the slice of the world in which the best-known events had taken place.

The "Theatrum" displays the world between 25° and 60° north latitude, dividing the two halves (with overlaps) at 45° east longitude. If joined together, the halves would form a long strip embracing the traditionally inhabited lands, excluding China. Delisle allots to each part an expanse strictly equal to the other's. Because the *pars orientalis* embraces Persia and extends somewhat beyond the Oxus, *occidens* reaches south to include a vast tract of African desert, and far west into the Atlantic, yielding one-third of its breadth to empty water. An inset hemisphere in *occidens* portrays the Americas as then known; another in *oriens* presents the three continents, and traces of Australia, in their entirety.

By comparison with other of Delisle's medieval maps, the "Theatrum" seems unremarkable in content. Content matters less in this case than the example Delisle set, borrowing the widely copied "theatrum" terminology of Ortelius, in defining the stage for ambitious general history. When Hase, near midcentury, came close to carrying out the extensive historical collection that Delisle had planned, he insisted that his work was truly a *theatrum historicum*. The anonymous cartographer who strove to outdo Hase in the Paris draft of 1747 adopted a stage whose oblong form linked Delisle's two sheets into one and adjusted its breadth so as to include more of the inhabited world.[5] The "Theatrum" of 1705 had a future.

Apparently by coincidence, rather than out of a common impulse, the early-eighteenth-century maps concerned with the Middle Ages seem inclined, like Delisle, to focus on the barbarian neighbors of the Roman Empire. An unexpected setting for such notions is the *Atlas historique* of Zacharias Châtelain.

This atlas, as was noted in chapter 3, should not be judged by its title. It is primarily a survey of contemporary geography, with intermittent glances at historical backgrounds. It starts with the world, then engages in a country-by-country circuit: the Romano-Germanic Empire, France, England, northern and eastern Europe, and so forth. Its plan is conventional. The compiler, no historian, had little interest in the Middle Ages. Only one medieval episode unconnected to the barbarian invasions of late antiquity rates a map.

Along with geographical tables, genealogical ones are provided for purposes of historical instruction. Charts of Roman emperors from Augustus via Charlemagne to the present, and of kings of Spain, England, and other lands, all have sizable barbarian components. Also appreciative of barbarians is the very decorative (and imaginative) illustration "Trophy raised to the glory of the heroes of Germanic liberty," complete with portraits of heroes.[6]

The maps convey a less amiable impression of the invaders. An early one, showing the four providential world empires, is bent on identifying the homelands of those peoples who "in their times were the scourges of Europe." Not limited to late antiquity, the invaders run the gamut from Scythians of the seventh century B.C. to Arabs of the eighth century A.D. A comparably extravagant time span is evoked in a "Historical, Chronological, and Geographic

Map of the Roman Empire." The "scourges of Europe" look more pleasing in
the "Map of Germania and of the Various States to Which It Bore Its Weapons" (vol. 2). The narrative of Gothic migrations in the map of "Ancient Germania" is drastically telescoped: "Goths, a people from the neighborhood of
the Baltic Sea, come to settle near the Danube and then pass into Italy and
Spain." A similar capsule is given about the Vandals: "Vandals, having left the
vicinity of the Baltic Sea, pass into Africa, where they establish a kingdom that
lasted about 105 years." [7] The triumphant reference to Germanic weapons has
no graphic elaboration; enlightenment comes only from the very brief section
"Remarks about the Conquests of the Ancient Germans."

The *Atlas historique*'s contradictory and inconsequential treatment of barbarians suggests an insecure grasp of the past. New images for medieval history nevertheless appear in its pages. They occur in the volume concerned
with England and may include the first map with a track of barbarian movements. Beneath a title saying that the Britons appealed in 443 to the Saxons for
help against the Picts, we are shown four embarkation points on the Continent and as many tracks of ships crossing the North Sea; the tracks converge
on the sea or at least pass each other's wakes; some ships come ashore at the
Humber; others, farther south, at Thanet. This image condenses more drama
than all the other barbarian evocations of the Châtelain collection. However
humble as cartography, the Saxons crossing to England inaugurates the depiction of barbarian peoples storming the Roman Empire.

Equally surprising is the other piece of medievalia in the *Atlas:* Richard I
(Lionheart) is traced on his Crusade to the Holy Land (1190–92). This inset
map is little else than an outline of the Mediterranean with a track of some
kind running through it. So unimpressive an image might easily be overlooked. Yet, the scene was without precedent not just for Richard or England
but for the crusading epoch as a whole.

These moments of remarkable originality in the Châtelain atlas come out
of, and lead, nowhere. The *Atlas* from then on neglects any aspect of history
except the biblical-classical repertory. Its unprecedented images featuring invasion and Crusade tracks remain isolated forerunners of scenes independently elaborated a century later. [8]

The *Atlas historicus* of Johann Georg Hagelgans (1718) has already been introduced. Its four maps are subsidiary to the real contents—a world chronicle in little pictures, an astonishing design for a history book. Of its four
maps, the one called "Migrations of Peoples" ("Migrationes gentium") expresses the same apparent fascination with barbarians found in Delisle and
Châtelain. Still quite young, Hagelgans had recently worked for the Swedish
diplomatic mission to the Imperial Diet in Frankfurt. He had presumably

been exposed to the pride modern Swedes took in having launched the Goths and many other peoples upon Europe. Hagelgans's map is the inaugural attempt, except for Châtelain's Anglo-Saxon crossing, at deploying flowing lines and other graphic resources to illustrate the movements of barbarian tribes.[9]

Hagelgans's "Migrations of Peoples" shows a background of Europe and the Near East and crisscrosses it with tracks, complete with arrows at the start and end of each one. The dotted lines (one careless exception aside) are identified as "route of [a named] people": "route of the Gauls," "of the Normans," etc. One peculiarity distinguishes this migration map from most of its unwitting descendants: Hagelgans's *gentes* are selected out of a time span of 2,500 years. The usual barbarians of the Germanic invasions—Goths, Vandals, Angles, Lombards, etc.—are present but so are the Greeks who colonized Sicily (eighth century B.C.), the Gauls and Cimbri who, at different times B.C., attacked early Rome, and the Tartars who terrified and mauled Islam and eastern Europe (thirteenth to fifteenth centuries A.D.). The archaic Greek colonists of southern Italy come face-to-face with the Norman colonists of the 1040s A.D.[10]

This millennial *longue durée* might be shrugged off if Hagelgans's map were a synchronic hodgepodge like those of the Châtelain collection, but its title seems to denote a conventional historical moment, the *migrationes gentium,* rendered in German as *Völkerwanderung,* the events of two or three centuries that our books call the (Great) Barbarian Invasions. A look at the cartouche reveals a different intention.[11] Hagelgans's 2,500-year scene maps all the "migrations" mentioned in his picture-chronicle, without privileging the conventional *Völkerwanderung* coinciding with Rome's fall. Hagelgans's portrayal inadvertently undermines the idea that there was a *migratio* par excellence, belonging specifically to late antiquity. These considerations are unimportant. The originality and internal complexities of Hagelgans's map neither enhanced nor harmed his reputation; he seems to have remained entirely unknown to cartography.[12]

Whereas Châtelain evoked "scourges" spanning the fifteen centuries from the Scythians to Islam, Hagelgans evoked (presumably harmless) "migrants" and added one more millennium to Châtelain's. Diluted in the immensity of time, a phenomenon generally thought historical ("barbarian invasions") turned into generalized sociology. Hagelgans jettisoned a long but limited process, such as Delisle had in mind, and substituted an age-old human proclivity to movement. That he intended to propose such a "mild" interpretation is improbable. At least one cartouche suggests that "migration" was a destructive occupation. Hagelgans never explained himself in words; it is up to readers to guess his reasoning.

More in the spirit of Delisle than of Hagelgans, the Benedictine Urbain Plancher took pains, in his *Histoire générale et particulière de Bourgogne* (1739), to chart two stages in the advance of the people called "Burgundiones" toward the lands that would become Burgundy. His main map, "The Old Kingdom of Burgundy," encompasses all the lands that, in the early Middle Ages, lay within one of the various, often regrouped kingdoms of Burgundy. Plancher's rendering is much more of a composite than the Burgundy maps of Manesson-Mallet. At the very start, he presents an experimental portrayal, whose eastward orientation catches the eye: "Map of the Lands of Germania Inhabited by the Vandals and Alamanni, and in Which the Ancient Burgundians Made Their Home before Crossing the Rhine to Come Settle in Gaul." Plancher, like Delisle, took an interest in early tribal homelands, which he believed he could plot cartographically. An undefined area running eastward from the Vistula and deep into modern Russia is called "First major dwelling place of the Burgundians"; a second Burgundian home is shown having its center from the Sudetenland through Bohemia. First and second homelands for the Vandals and Alamanni are also marked. Unlike Hagelgans, Plancher stops short of tracing barbarian movements, but the resting points he provides for fixed and homogeneous peoples continuously living under the same name are a step in the direction of linking together homes with lines defining movements.[13]

In late summer 1759, the *Journal des sçavans* drew its readers' attention to a map recently published by Nicolas Bellin (1703–71). Bellin was *ingénieur-hydrographe* to the king's navy and a celebrated figure in French cartography. His two-sheet "Map of a Part of the World for Understanding the Migration of Peoples" was essential, the journal said, to elucidate the just finished magnum opus of the distinguished Orientalist Joseph de Guignes (1721–1800). Bellin's map appears to be lost. According to the *Journal,* its two sheets, extending from the Sea of Japan to the English Channel, indicated the main places mentioned in de Guignes's study, and for easy reference, they might be placed at the head of volume 1 of de Guignes's *Histoire générale des Huns, des Turcs, des Mogoles et des autres Tartares occidentaux.* Bellin apparently showed, among other things, all the various colonies that the Huns had established in Italy and Europe.[14] Those familiar with the history of the Huns in the west may well wonder what was meant by their "colonies."

Ever since the writings of de Guignes, based on Chinese records, scholarship has had to reckon with a theory according to which the fall of the Roman Empire was ultimately brought about by the stiffening of Chinese defense against its barbarians, the Hiung-nu, or Huns, and the consequent push of the latter westward toward softer prey. No convincing proof of this theory has ever been given. Scholars usually reject it as improbable, but it refuses to go away. Owing to its graphic possibilities, it is often illustrated in recent atlases, much more often than it is expounded in books. Bellin, guided by de Guignes, appears to have inaugurated the mapping of this alternative theory. One hopes that Bellin's two-sheet map will be rediscovered.[15]

Among eighteenth-century mapmakers, Delisle and Plancher focused single-mindedly on the barbarians of late antiquity; Bellin drew a cartographic aid for a book that considered Asiatic vicissitudes responsible for the fate of the Roman Empire; Châtelain and Hagelgans embraced wider and wider time spans, and ever more numerous and diverse peoples. Like Delisle, they were certain that barbarians were relevant to the end of Rome, but they lapsed into generalized epochs comparable to de Guignes's. The efforts of these mapmakers are a chapter in the historiography of the barbarian invasions.

Invasions of Christian Europe, in general, did not stop in the sixth century. Hagelgans and some later mapmakers had no chronological compunctions about including Scandinavians in their maps, but viking attacks were not a favored subject in this century. One exception comes out of the Netherlands—an untitled, elegantly engraved map, probably out of a book. The University of Amsterdam library catalogue identifies it as "The Netherlands in the 10th–11th century"; a more precise title would be "Viking Devastations in the Netherlands, 9th–11th Centuries." The map illustrates viking attacks by means of dates pointing to distinct territories: 839, two pointers in Frisia; 889, pointing northwest of Leiden; 923, north of Haarlem; and 1018, west of Dordrecht. It is too bad that the book in question eludes identification. From the little one may tell, the map is, for its century, an exceptional illustration.[16]

Carolingian Sidelights

A corner of the Bavarian Nordgau, near the town of Weissenburg, still contains impressive traces of the canal by which Charlemagne tried to link the Atlantic and the Black Sea or, to put it more modestly, the Rhine and Danube Rivers. This canal may have been the most ambitious engineering project of the early Middle Ages (it was not in fact completed until 1992).[17] Whereas Bertius's Charlemagne map of 1623 stood unrevised for more than a century, several mapmakers in the early eighteenth century were moved to chart Charles's canal. The cartographic theme they launched has had enduring appeal.

The existence of the canal, locally called "the Ditch" (*Graben*), has been known since at least the sixteenth century, but it was not charted until about 1717, when J. G. Vetter showed it in some detail as part of a map of the district of Nuremberg. A more evocative map was provided a few years later by the Jesuit historian of France Gabriel Daniel. Daniel's treatment of the canal occurs in *Histoire de la milice française*. His depiction owed little to archaeology or to a knowledge of the site, but he did illustrate the link between the North and Black Seas that the canal implied.[18]

The subject then became and stayed German. Almost simultaneously with Daniel—the exact date is unknown—an impressive outline of the remains of the canal was included by J. B. Homann in a map entitled "Accurate Prospect and Outline of the Region of the Imperial Free City of Weissemburg in the Nordgau, with All Its Existing Antiquities" (ca. 1720). The canal itself, though, was not precisely delineated for a long time. Vetter and Homann illustrated a prominent feature of the districts they were mapping, whereas Daniel evoked Charlemagne's wider vision rather than the surviving traces of the canal. In 1726, G. Z. Haas finally supplied a detailed, large-scale map focusing on Charles's enterprise; it illustrated Haas's history thesis, *De Danubii et Rheni coniunctione*. Haas's survey is careful and accurate; extensive notes indicate dimensions and mark out the parts full of water and fish or marshy or dry. Elaborate drawings suggest imaginatively how Carolingian boats were to be lifted and lowered (locks, a late medieval invention, are avoided as anachronistic). Haas did not have the last word; the eminent Strasbourg professor Daniel Schöpflin wrote on the same subject, with a map similar to Daniel's, for the Paris Académie des inscriptions in 1753. But it was Haas who properly introduced Charlemagne's canal into the repertory of historical maps.[19]

A few other versions had claims to originality. Doederlein, who wrote a creditable account of the canal early in the century, provided a careful image (not equaling Haas's) as an illustration for J. G. von Eckhardt's source collection *Commentarii de rebus Franciae Orientalis*. Another rendition, far from satisfactory, appears as an inset in Falckenstein's widely circulated "Bavarian Nordgau in the Middle Ages." Köhler, ever on the lookout for medieval scenes, did not neglect Charles's canal. A small map taken from one of the available models appears in the posthumous part of the *Kurtze Anleitung* (1765). A trail of added maps leads through the rest of the century and into the next.[20]

These maps were not inspired by Charlemagne's fame or some other historical phenomenon. Then, as today, the large and impressive remains of the canal were visible, a marvel for the population of the Nordgau (and others) to be proud of, a fitting subject for pictures to hang on the wall. The scenic glamour of the canal, rather than its relevance to medieval history, is the probable basis for the unusual interest it aroused.

Royal residences or palaces were a second Carolingian feature to be mapped and, from a historical standpoint, seem more vital than the canal. The already mentioned *Chronicon Gotwicense* of Bessel and Hahn is an exploit of eighteenth-century scholarship. Its three maps are preoccupied with *gaue* (believed to be the ancient territorial unit par excellence) rather than with moments in history. But its strictly businesslike, undecorated "Map of the Palaces or Royal Estates in the German Kingdom of Eastern Francia, Adjusted to the System of *Gaue* out of Medieval Charters and Diplomas" has a more definite chronology. All the maps are said to date from before the Interregnum (1254 and after); the palaces are attributed to the "Teutonic Realm of the East Frankish Kingdom." Whether this involves the same pre-1254 time span as the others or a shorter one is not entirely clear (fig. 18).[21]

The *Chronicon* follows the lead of Mabillon's *De re diplomatica,* the beacon of research in medieval history. Early kings were often in their *palatia* when issuing diplomas; whether the ruler was present or not, royal dwellings were foci of early medieval government. Erudition profited from the charting and identification of royal residences. Scholarly value, however, was not guaranteed to reconcile laymen to the austerity of unornamented maps.[22]

Europe as Seen By . . .

Queen Anne of England (1702–14) was the second daughter of the Catholic Stuart James II (1685–88). Reared as a Protestant, she followed her like-minded sister, Mary, as Stuart sovereign after James was expelled in 1688. Four years before her death, Anne appointed a Tory ministry whose sympathy for the son of James II—the Jacobite pretender—occasioned widespread alarm about the return of "popery." Thomas Bray (1656–1730), a divine famed for missions in America and for founding the two oldest societies in the Church of England (including that for promoting Christian knowledge), did what he could to avert the looming danger. The collection he edited under the title *Papal Usurpations and Persecution . . . Designed as Supplemental to the Book of Martyrs* is composed of "choice and learned treatises of celebrated authors, ranged and digested into a regular history." Part 2, devoted to papal oppression of the rank and file, contains the English translation of a seventeenth-century *History of the Old Waldenses and Albigenses* by Jean Paul Perrin. The two maps provided as illustrations are by John Senex (†1740), an established engraver and publisher: first, "The Valleys of Piedmont and France which were the seat of the Waldenses or Vaudois both Ancient and Modern"; then, "The Seat of the Albigenses so famous for their Sufferings by the papal croisados and Massacres in the 13th centurie."[23]

The subjects are mainly medieval, but neither Bray nor Senex ventured anything more historically ambitious than modern maps of the lands in which the Waldenses and Albigensians were persecuted. Typically, walled towns are shown ringed with fortifications of modern design. Despite the absence of historical color, Bray's effort is not contemptible. Without actually mapping thirteenth-century persecutions, he showed that the subject was there to be done.

The accomplishments of the Middle Ages in geography and mapmaking were little known until the nineteenth century. Two exceptions in the 1770s came, fittingly enough, from the northern edge of medieval Europe, the same region whose defective portrayal by Ptolemy resulted in early supplements to the Ptolemaic atlas. The first was a product of Anglo-Saxon England. King Alfred (†899) sponsored the translation from Latin into Old English of a small library of basic books for Christians. They included the *Histories* of Orosius, which opens with a geographical survey of the world. The Old English Orosius contains an addition reporting the northern journeys of two ninth-century voyagers, Ohthere and Wulfstan. It now occasioned a map.

Daines Barrington, editor of *King Ælfred's Orosius* (1773), offset what he called the tedium of its first chapter by annexing a map "which contains the names of most of the European places named in this geographical chapter, and [also traces] the voyage of Ohthere and Wulfstan in these Northern Seas." During Barrington's preparations, a widely accomplished and ill-tempered German scholar named Johann Reinhold Forster (newly elected to the Royal Society) needed any employment he could find; the map was his work. Barrington acknowledges only Forster's counsel and announces that Forster was, at the time of publication, "with the vessels sent upon discoveries in the southern hemisphere." Forster had been rescued from ill-paid hack work by a call for him and his son to be naturalists for Captain Cook's second voyage (1772–75).

Meanwhile, Forster's map appeared in Barrington's edition. Immediately preceding it are Forster's long and careful "Notes on the First Chapter of the First Book of Ælfred's Anglo-Saxon version of Orosius." The map and notes have been called "the first direct continental contribution to English scholarship of Anglo-Saxon." The map uses contrasting lines to mark the travels of Ohthere and Wulfstan; Roman numerals, explained in the narrative, indicate the number of days' sail between stopping points; and all names of countries and towns shown by the map north of the Mediterranean—a moderate number of localities—are written in an Old English typeface (the rest are in Latin). Perhaps the most astonishing feature of the map, said to derive from the underlying text, is a Germany extending over a vast expanse from the Rhine to the White Sea and the mouth of the Don.[24]

Figure 18 Reconstructions of Medieval Features and Districts (1732). J. G. Bessel and F. J. Hahn, *Chronicon Gotwicense.* Bessel and Hahn's undertaking documents the eighteenth-century enthusiasm for scholarly reconstructions of medieval districts and administrative units, in part to facilitate study of early charters. Illustrated here is their "Map of the Royal Palaces or Manors in the German Kingdom of Eastern Francia, from Medieval Charters and Diplomas. . . ." (Courtesy of Yale University Library)

Palatiorum antiquorum, Romani et Franco-Theotisci,
in Germania adhuc extantium reliquiæ.

LIBER III.
DE ANTIQUORUM REGUM
AC IMPERATORUM TEUTONI-
CORUM PALATIIS, VILLIS, ET
CURTIBUS REGIIS.
CAPUT I.
De Imperatorum ac Regum Teuto-
nicorum Palatijs, Villis, ac Curtibus
Regijs in communi.

Xtremam Rei Diplomaticæ perio-
dum absolvit penitior locorum illorum con-
sideratio, quorum mentionem in ipsis Di-
plomatum *Datis* invenimus. Cùm verò
plurima in ijs occurrant Palatia, Villæque
& Curtes Regiæ, è quibus tanquam è Sedi-
bus, ubi ad tempus aliquod morabantur, sua
concesserunt diplomata Reges & Imperato-
res, rem benevolo Lectori haud ingratam
facturos nos esse arbitrati sumus, si peculia-
rem de Palatijs Imperatorum nostrorum
Teutonicorum commentationem huc adjí-
ceremus: non minùs enim ad plenam argumenti diplomatici perfectionem,
quàm

Scopus hujus libri.

Necessitas Et Splendor hu-jus materiæ.

Tom. I. Kkk

Forster reprinted his map a decade and a half later in the long and widely translated *History of the Voyages and Discoveries Made in the North*. It bears his signature (absent from the Orosius edition) and the title "A Map of Europe for the Illustration of the Geography of the Middle Ages and King Alfred's Anglo-Saxon Translation of Orosius." Forster took this occasion to counteract Barrington's quiet attempts to treat the map as his.[25]

Only a few years later, a prominent Norwegian scholar, Gerhard Schöning (1722–80), was officially encouraged to travel through Norway gathering materials for its history. When rector of the cathedral school at Trondheim a decade earlier, he had designed a pair of maps showing how the North was envisaged by writers of classical antiquity.[26] He returned to mapmaking in the decade of his historical travels and, this time, was concerned with the Middle Ages. After portraying "The Appearance of the Three Northern Kingdoms of Europe according to the Precepts of the Old Writers," he also projected the place-names found in medieval Norse and Icelandic writings onto a map of Europe and neighboring lands. The Latin title of this final effort elaborates its program; the simpler German version is "Geography of the Middle Ages, Especially of the Ninth and Tenth Century." The lands shown run from Iceland as far as the Caspian and a little beyond; all the Mediterranean is included, along with the northern edge of Libya; many place-names are marked.[27] The marriage of modern cartography and medieval writers might have proved useful to readers of Old Norse–Icelandic literature. Whether the map really showed the "appearance of Europe . . . as conceived" by early Scandinavians is less certain.

In the late 1720s, a group of scholars in England undertook one of the more ambitious historical enterprises of the century: a work composed of suitable parts (ancient Rome, France, China, etc.) and entitled *Universal History from the Earliest Times*. Its opening set of volumes appeared between 1730 and 1736. The venture was so well received that Dutch, French, German, and Italian translations followed. Maps and other images graced the various editions, but not in lockstep with the English progenitor, the less so when (as with the German edition) a new, improved text was preferred to mere translation of the original.[28]

Precisely how the maps for these multiple volumes were secured is hard to know. They are often taken from an existing source, scarcely ever acknowledged. In England, estimable map publishers issued a separate atlas, *Maps and Charts to the Modern Part of the Universal History,* and the same was done in the Netherlands. The German edition, initially dependent on the Dutch for its maps, became more enterprising as time went on. The "modern" maps of the English edition lack any taint of the Middle Ages. Some continental translations contain more than a few medieval images, including items not (yet) found elsewhere. The map to be presently discussed is the first of several from this collection to concern us.[29]

After Anglo-Saxons and Norsemen, it was time for Slavs to have an impact on maps, the more so as the treatment of northern Europe by the English *Universal History* was deemed in Germany to be particularly defective.[30] The early Slavs, who had not expressed their geographical ideas in writing, could not provide a written witness like that of the Anglo-Saxons and the Norse. An ethnographical map seemed possible; some general ones had been tried, such as that by Robert de Vaugondy, which was limited to skin color.[31] At a time when ethnographic maps had yet to come into their own, an effort was made to depict "all the lands settled by Wends and Slavs."[32]

The anonymous map is centered on Russia, with the Black Sea a little right of center on the lower edge. The left side reaches some of the Adriatic and the North Sea; the right falls short of the Caspian. There is much writing and little chronological discrimination. "Rhine Franks" (*Rhein Franken*) is written at the left edge, and "Russian Empire" appears prominently on the right half, although nearly a millennium separates these entities from each other. The indicated Bulgarian Empire seems to be the short-lived one centered in Macedonia that, from its foundation, was locked in war with Byzantium (early eleventh century). The cities shown were not all contemporary to each other; names of regions (such as "Weis Serbien") outnumber names of peoples ("Lächische Völcker"). A winding river of letters indicates the "most ancient land of the 'Slavinen'"; and not far to the east, in echelon, the "most ancient land of the Antes." Other peoples besides Slavs are mentioned; the name "Petschenegen" marches across southern Russia, and Baltic tribes are abundant to the north. There is no inclination to minimize the Slavonic diaspora; names of Slavonic peoples appear in numbers far to the west. This early ethnographic display is at best a composite, synchronic work of reference to be consulted by those who, for example, came across "Volhinia" and "White Russia" and wished to know their location. It does not aspire to show Slavonic settlement in a precise epoch. Despite pronounced shortcomings, the map is a creditable effort by an author who, one hopes, will some day be identified.

Poster or Catalogue Maps

John Speed's "Invasions of England and Ireland" illustrates a type of historical map composed of a numbered list of dated incidents keyed to a depiction of where these incidents occurred. It might be called a "catalogue map" but fits equally well into the more dignified category of a "poster."[33] In 1667,

Guillaume Sanson issued a "Geographia synodalis," setting all church councils from Nicaea I to Trent on one background each for East and West. It resembles Speed's map by being accompanied by conspicuous lists.[34] Maps of this kind with emphatic visual appeal are more posterlike than others. Catalogue maps, including those qualified "posters," are few and far between. Eighteenth-century ones concerned with general history—no more, apparently, than three in number—are a Bavarian monopoly.

In 1731, the lyceum of Freising first issued a map of universal history that was reprinted intermittently, by various presses, until as late as 1808. This very original work, presenting all history as "a picture or a comedy," was designed by Anselm Desing, a Benedictine then teaching in the secondary school of the episcopal city of Freising. Desing (1699–1773) had a brilliant career before him as a leading Catholic educator. He was responsible, among other things, for anchoring geography in the curriculum. A resourceful instructor, he may have inaugurated glass mounted slides as a teaching aid, and he even used a camera obscura to project slides of pictures like those on his map, but better drawn.[35]

By itself, Desing's map, which lacks his name, gives a modest, even shabby impression, but it was soon improved by a professional engraver. It reduces universal history to 215 items and plots them geographically down to the foundation of the Ostend, or Austrian East India Company, in 1719. The mere fifty-five items between Attila (no. 124) and Columbus's discovery of America (no. 178) suggest that this Benedictine paid little attention to the Middle Ages. Still, Desing's interests were wide enough to include "The withdrawal of Latin, origin of the French language" (no. 135), among noteworthy events.[36] Desing favored geography for the lower school grades and history only for the higher ones. His very elementary map evidently exposed geography students to a controlled dose of universal history.

A decorative appearance, highly suitable for atlases and public display, characterizes the two catalogue maps issued by the firm of Mattäus Seutter of Augsburg toward 1745: "History of the Romano-Germanic Empire" and "History of the Bavarian Circle and Neighboring Regions." A specific circumstance, such as the brief, ill-fated intrusion of a Bavarian emperor, Charles VII (1742–45), or at least an inventive designer, is likely to lie behind this unusual publication and needs clarification.[37]

The maps record incidents from the entire history of the Romano-Germanic Empire and, in greater detail, of the Bavarian Circle of the empire. Both have a long numbered list of events, chronologically ordered, in German rather than in the Latin of the title: "Key to the illustrations found on the historical map of Germany and its neighboring lands." The corresponding number, with coordinates, leads to the appropriate site on the map, as, for example, "218.B.e. The brave hero Prince Eugene is born in Soissons A.D. 1663." Dispensing with symbols, the two Seutter maps are frankly pictorial, full of carriages, tents, jousting, troops in ranks firing rifles, marching men, and other little figures in great variety. Several arrays of soldiers are reminiscent of Speed's.[38] The concern of these maps for the Middle Ages was subordinated to a main theme: incidents of imperial and Bavarian history. Desing's map has a plainly pedagogic look; with Seutter as with Speed, however, the desire to instruct and inform seems coupled with a more diffuse subtext, a statement of some kind rooted in current affairs. Nothing more profound need be involved than a sense that the public was ready for such catalogues and that the maps would sell.

A final cluster of catalogues, in the poster category like Seutter's, reaches into the nineteenth century and is conspicuously related to momentary circumstances. Best of all, they involve a tacit but obvious rejuvenation of old material. John Speed's "Invasions of England and Ireland" was topical at least three times during the wars of the French Revolution and Napoleon, years sometimes qualified in England as the "Great Terror" or fear of French invasion. France had succeeded by 1797 in bringing the Dutch and Spanish navies under its control. Combined, the three fleets promised a numerical superiority adequate to sweep aside the British navy and open the way to a cross-Channel invasion. For England, 1797 was marked not only by this critical naval imbalance but also by mutinies aboard its outnumbered ships. In September of that year the firm of Laurie and Whittle issued John Enouy's "The Invasions of England and Ireland." It was incorporated into Kitchin's *New International Atlas* in 1798 and reprinted for many years thereafter.[39] From an English standpoint, the lesson of this map was hearteningly plain: after William the Conqueror, invasions had failed.

Enouy's adaptation adds no more than seven events to Speed's list. Fifty or more of the battles are still medieval, if such details rather than mere numbers interested the public. Of the seven added occurrences, the most topical involves an abortive French landing at Bantry Bay in Ireland (1796). The incidents of the Spanish Armada are still there, with ships modernized to the current standard. On land there is an obtrusive change: Speed's large, picturesque battle symbols have been removed (at a great saving in skilled engraving). His numbered boxes keyed to the legend are retained; each battle is identified by name in its box without an accompanying symbol. The legends, one each for England and Ireland, are placed on the map face to facilitate consultation. Enouy's "Invasions" is a single self-sufficient sheet.

A copy of Enouy's map presumably reached Paris in short order. In Year VI of the Revolutionary calendar (September 1797–August 1798), the well-known mapmaker P. F. Tardieu issued "Map of the landings made in England and Ireland since William the Conqueror." The extensive accompanying letterpress explains that, ever since the Directory (the then governing body of France) had resolved upon an expedition against England, all citizens hastened to support this vast undertaking. Tardieu's contribution was the attached map showing that, out of forty-five landings in England, forty-one had succeeded (the names are listed); Bonaparte would ensure victory by the French troops. History proved, at least to Tardieu, that the obstacles were trivial, that control of the sea was unnecessary, and that, once the troops had landed, the English campaign would be a walkover because there were no fortifications or strong points.[40] The relevance of the map to current enterprises, left implicit by Enouy, is spelled out by Tardieu, who boldly reverses Speed's message of English security.

The Directory, reasoning differently from the optimistic cartographer, shelved the attack. Bonaparte turned his back on England and took a French army east—to Egypt. But Tardieu's map retained appeal to a country whose main enemy was England.[41] A few years later, after the breakdown of the Peace of Amiens (1803), Napoleon tried again. He bent every effort to launch a massive invasion of England, and the updated, reinterpreted Speed map came back into circulation.

It was now called "Map of the British Isles and of the Neighboring Coasts, to Assist Understanding of the History of Landings Made in These Isles from [but not including] the Romans Down to the Present" and was signed by P. G. Chanlaire, an established cartographer. Some changes had been made. The number of events rose to eighty, most of them, according to a contemporary reviewer, insignificant and inconsequential. The French naval preparation for the invasion failed in 1805, and Napoleon led his troops east, to victories against Austria. For this reason, perhaps, Chanlaire's map is lost, attested only by a contemporary journal. After the defeat of Trafalgar, few customers may have been ready to acquire a map that interpreted Speed's "Invasions of England and Ireland" as a pledge of success for the invader.[42]

This Revolution age cluster of catalogue maps includes medieval events but has no bearing on mapping the Middle Ages. Speed's "Invasions" was a source of the Lesage atlas and was copied into at least one more historical collection. In these cases, pedagogy outweighed propaganda. Enouy's revival of the "Invasions" map in 1797 may, but need not, have won Speed this more tranquil afterlife.

Divisions and Limits of Medieval Lands

A new direction taken in eighteenth-century maps of the medieval period, and exemplified many times throughout the course of the century, was the reconstruction of limited districts on the basis of available documentation. Credit and acclaim for being first with maps of this sort are sometimes given to Karl Spruner and his historical atlas of Bavaria (1838); in France, credit—even longer delayed—tends to go to Auguste Longnon, for a standard-setting historical atlas of France (1885). This chronology is too limited. From the days of Guillaume Delisle, a remarkable series of maps anticipates nineteenth-century achievements with more or less adequacy and success.[43]

The eight plans of Paris, five of them medieval, accompanying the anonymous *Traité de la police* in 1705, have already attracted our attention as being among the early attempts to use maps to illustrate historical change. These city maps are known to be by Nicolas de Lamare, or Delamare (1639–1723), a Paris magistrate under Louis XIV, always called *commissaire,* who specialized in the whole spectrum of urban problems and services.[44] His four-volume *Traité,* a pioneering work on the administration of a big city, opens with a historical discourse setting out the factual underpinnings for eight plans, from Roman Lutetia to Louis XIV's Paris in its twenty wards. The medieval plans have the following descriptions:

> Lutetia conquered by the Franks from the Romans, or second map of the city of Paris.
> Third map of the city of Paris, its extent and the boroughs neighboring it in the reign of Louis the VIIth of that name [fig. 19].
> Fourth map of the city of Paris. Its growth and its condition in the reign of Philip Augustus, who died in the year 1223 [fig. 19].
> Fifth map of the city of Paris. Its growth and its fourth enclosure, begun under Charles V in the year 1367 and completed under Charles VI in the year 1383.
> Sixth map of the city of Paris [1422–1589].[45]

With two modern maps at the end and an ancient one at the start, Lamare comes as close as La Ruë to compiling a (small, specialized) atlas "from the earliest times to the present."

Plans of Paris have been much studied, yet it is not entirely clear which engraver-publisher brought Lamare's set into print, or whether the maps are fanciful or historical. Notwithstanding dissent, the first seven maps are likely

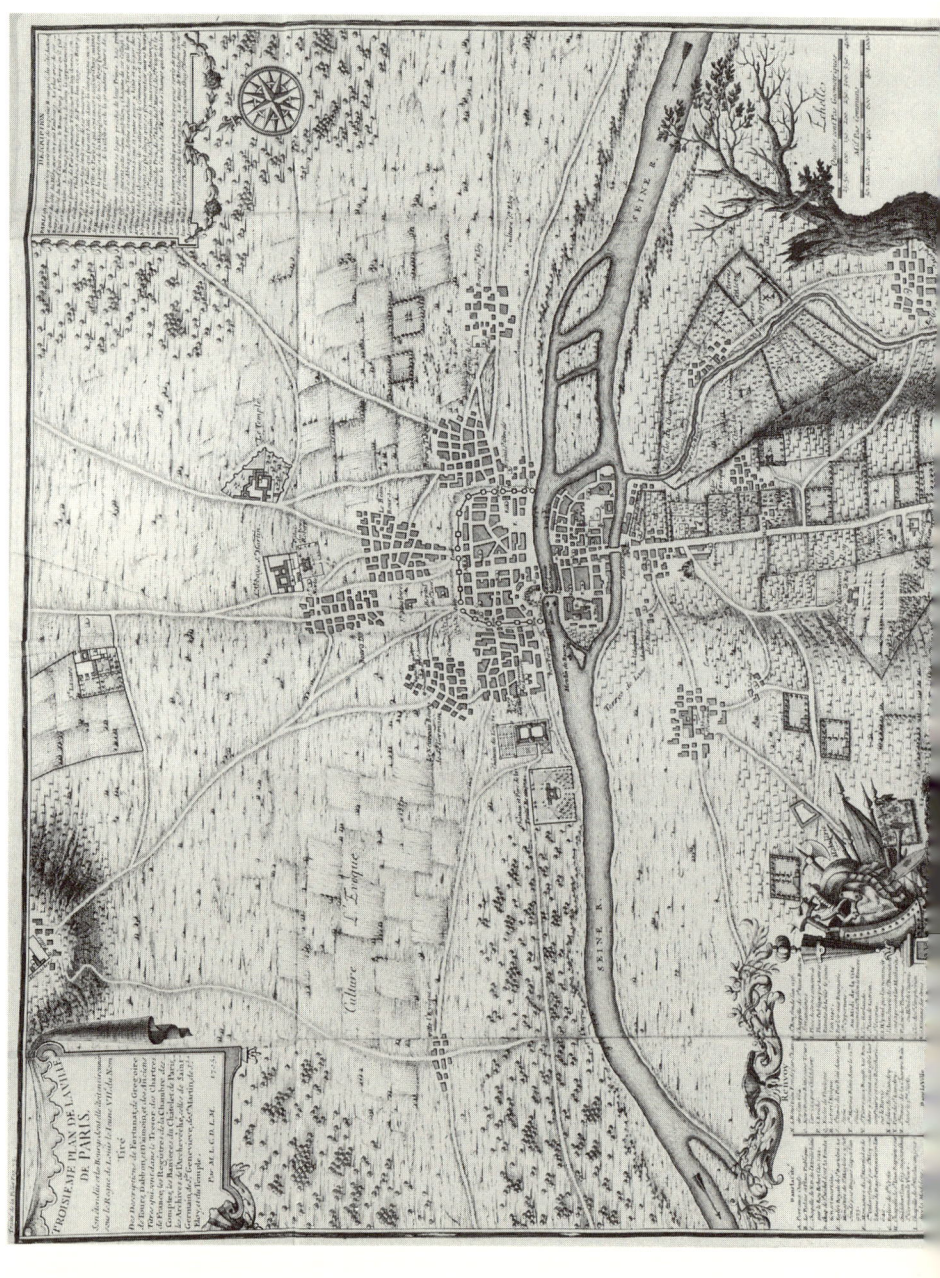

Figure 19 Third and Fourth Plans of Paris in Nicolas de Lamare, *Traité de la police* (1705). Eight maps illustrating the development of Paris accompany Lamare's multivolume work on city administration. Maps 3 and 4 show Paris under Louis VII (†1180) and Philip Augustus (†1223). City plans are well adapted to charting change on maps. (Courtesy of the Beinecke Rare Book and Manuscript Library, Yale University)

to have been designed by A. Coquart and the eighth by Nicolas de Fer. Co-quart, using the materials supplied by Lamare, created seven original maps; whereas de Fer's eighth map is an off-the-shelf contemporary city map. De Fer, however, definitely helped to popularize Lamare's plans by making and selling a full series of his own that was long marketed by Paris map sellers, alongside Coquart's or separately.[46] Nineteenth-century opinion was decid-edly against the historicity of Lamare's maps; a reference mentions them as "[e]ight imaginary maps" in Lamare's work. More recent judges are silent. If Lamare's maps are imaginary, he took considerable pains to give them a sem-blance of truth; authorities are cited, and explanations supplied. Lamare's reach may have exceeded his grasp, but his success in conveying images of ur-ban growth are as worthy of applause as Johannes Mejer's historical maps of northern Frisia.[47]

None of Guillaume Delisle's many and varied contributions to the "géo-graphie du moyen âge" are more central than the detailed maps of medieval Toul, an episcopal principality in Lorraine, and of the province of Dauphiné. Both, predictably, were occasioned by learned books based on documents: Benoît Picart's *Histoire ecclésiastique et politique de la ville et du diocèse de Toul*, and the anonymous *Mémoires pour servir à l'histoire de Dauphiné sous les Dauphins de la Maison de la Tour du Pin*, known to be by J.-P. Moret de Bourchenne de Valbonnais.[48] Delisle's title and legends do not say that these districts are being shown as they were in the Middle Ages. In the portrayal of Toul, all place-names are written in French and Latin. The two maps are on a large scale and outline the ancient territorial subdivisions. In *Dauphiné*, these are "principalities, counties, baronies." Delisle states that his sources are "an-cient charters under the princes of Dauphiné," that is to say, prior to the four-teenth-century incorporation of the land into France. The map of Dauphiné is more clearly medieval at first glance than that of Toul, but Toul is certainly not shown in its modern divisions; it is usable for medieval history.

The relationship of Delisle's maps to local histories or antiquarian studies of limited districts is unsurprising. Meanwhile, an early German example of such mapping was produced by Frederick Schannat in connection with the gift charters (*traditiones*) of the great abbey of Fulda: "New map of ancient Buchonia." Schannat's edition of the *traditiones* includes a reconstruction of the "Patrimony of St. Boniface, or ancient Buchonia with its bordering lands, extracted from the gift charters of Fulda . . . from the times of King Pepin [751] to the beginning of the fourteenth century." The map of the forest of Bucho-nia, late in the book, accompanies this analysis of the Fulda patrimony. The boundaries of a series of *gaue* are shown, with the names of the villages in them; woods (very plentiful, as one would expect) hills, and other topographical features are indicated by pictorial symbols.[49] One could hardly ask for more from a historical standpoint than this reconstruction based on contemporary charters.

Ludovico Muratori's *Rerum Italicarum scriptores* is a famed eighteenth-century national collection of medieval sources. Its opening preface (1723) promises three maps: for the current volume, "in the time of the aging Roman Empire"; next, "in the, as they say, Middle Ages, [including] of course the Lombard kingdom"; finally, "the restored fortunes of the Italians under the Emperors and the other Italian princes."[50] This announcement is an interesting echo of the periodization found in Lubin, Hagelgans, and probably Cellarius, which envisages the *medium aevum* as a brief, though troubled, interval starting with the decline of Rome and ending with Charlemagne.

The map called medieval is the frontispiece to volume 10 of Muratori's *Scriptores.* Anonymous, but known to be by the Benedictine Gasparo Beretti, the map was included (the introduction announces) so that one might know at a glance what appearance Italy had under the Lombard and Frankish kings; for, although the earth was the same as under the Romans, new provinces and cities had been created, and ancient castles, towns, and cities had been blotted out. Beretti's map is complemented by a "*Dissertatio chorographica* about medieval Italy for consultation of the map of Greco-Lombardo-Frankish Italy, as it was transferred from the Greeks and Lombards to Charles [the Great]," which takes up more than 300 columns.[51]

Beretti's map is a large, handsome foldout with room for much detail. It mirrors many moments, rather than just one, of the period it is designed to portray but is rich in information and guidance about the boundaries separating the diverse entities present in early medieval Italy. Lombard and Byzantine duchies are there, along with the eastern and western parts of the Lombard kingdom and the Byzantine Exarchate and south Italian possessions. The Paris dealer Roch-Joseph Julien advertised this map as showing Italy divided into gastaldates, the units of Lombard royal government. Strictly speaking this is not true—the divisions are not uniform—but several gastaldates are shown. The map invites close study, especially in company with Beretti's massive commentary.

Almost as soon as it was published, Muratori's map of Lombard Italy caught the eye of Köhler. Volume 1 of his *Kurtze Anleitung* includes a reduced copy that nevertheless retains many details of the original. More unusual is that Köhler identified the source. Another copy, also acknowledged, was incorporated much later in the relevant volume of the *Algemeine Welt-Historie.* As is too often true in this collection, the copy is reduced, coarsely done, and printed on flimsy paper.[52] A map as comprehensive as Muratori's Lombard Italy inevitably won more attention than Delisle's Toul or Schannat's Buchonia.

A two-year period during the 1730s markedly enriched the mapping of medieval Germany. One contribution, already encountered, is the *Chronicon Gotwicense* of Bessel and Hahn. Its main cartographic goal was to portray the medieval *gaue* as they had been reconstructed to that time. Without engaging in original research, Bessel collected the results attained by earlier scholarship and, including other divisions as well as *gaue,* plotted the boundaries of Germany in two maps: one general, the other more limited and detailed. The titles are elaborately phrased:

> Germany, carefully distributed in its ancient provinces, duchies, major and
> minor *gaue* (*pagi*), with the place-names stated in the medieval manner,
> drawn from medieval diplomas, charters, and plans.
> Duchy of the Austrasians or of Eastern Francia [i.e., Franconia], with the *gau*
> of Eastern Thuringia, drawn in its individual *gaue,* under the Franconian
> and Saxon emperors, from various medieval diplomas, charters, and doc-
> uments.[53]

The first, estimable for being the earliest attempt at a major map of German *gaue,* has serious shortcomings: "innumerable names are entered . . . but few attempts made to draw boundaries; only with the greatest effort can one decipher the regional and *gau* names that criss-cross the sheet."[54] The second map looks like a magnified fraction of the first. It shows a "duchy of Austrasia" largely coinciding with the later duchy of Franconia; the western edge reaches into Lorraine but includes only a small triangle of land to the left of the Rhine.[55] Bessel did not situate Austrasia in an idiosyncratic place so much as reflect the loose way in which the term had been used after Merovingian times. In the text of the *Chronicon,* the *pagus Ripuaria* is said to comprise the larger part "of the Frankish kingdom of Austrasia and later of the Kingdom of Lothair." Bessel may or may not have endorsed the idea that Austrasia was the "old" name of Lorraine, but he must have assumed that the Austrasia associated by the text with the *pagus Ripuaria* lay elsewhere than the Franconian Austrasia of the map.[56] From a broader perspective than *gau* geography, both maps shine with authenticity and practical value by comparison with the Ptolemy-derived experiments, in Köhler's atlas and elsewhere, that claim to be related to medieval Germany.[57]

The year of the appearance of the Göttweig volumes saw two detailed maps of early Upper Saxony, along with one of the modern duchy. They were the work of Friedrich Zollmann (1690–1762), a career civil servant in Weimar. In 1728, when the very small, fragmented principality of Saxe-Weimar came under the rule of Duke Ernst August, the territory was restored to a single ruler, and primogeniture of the ducal office was instituted. This was also the year of Zollmann's appointment as head archivist of the duchy, the post he held to the end of the reign in 1748. He wrote much history and published most of it. The work for which he is best known focuses on the main heraldic arms of Saxony—a subject well suited to gratify Duke Ernst August, who headed the most senior branch of the entire Saxon house.[58]

In their order of publication, Zollmann's three maps of Upper Saxony are deliberately numbered in reverse order of time. They proceed from "The Electoral Duchy of Saxony" to "The Duchy of Upper Saxony . . . between the Tenth and the Fifteenth Century, Drawn Principally from Medieval History," and end with "The Duchy of Upper Saxony in Its Most Ancient State . . . from the Birth of Christ to A.D. 1000, Compiled from Sources of Saxon History." Whether Zollmann actually designed his maps from present-day Saxony backward is not recorded, but he, and his Nuremberg publisher, saw nothing wrong in setting them out counter-chronologically. They are considered his finest work: "All these maps are characterized by great precision and were often the basis for later cartography of the lands concerned."[59] The date and context suggest that their appearance had something to do with the hopeful expectations for the future aroused by the accession of Duke Ernst August.

Zollmann's maps are more attached to the Middle Ages than their order and titles suggest. The centuries from the Incarnation through the Carolingians are all but disregarded. With effort it is possible to make out the names of tribal territories of Roman times—"Hermunduri," "Cherusci," "Longobardi"—but these are mere nods toward the distant past. The Upper Saxony we really see is preoccupied with Slavs. Its prominent divisions from west to east are Thuringia, the Sorabian March, and the district of the Slavs (*Duringia vel Germania specialis; Sorabicus limes; Sclavorum regio*). The *gaue*, too, are carefully outlined. At lower right, the cartouche conveys a message of forcible conversion nearer to A.D. 1000 than earlier. To the rear, mounted troops seem to be fighting. A cut-stone platform stands in the foreground topped by three idols marked Radagart, Swantwith, and Flins. At the base, men break an idol with large mallets, while another statuette is readied for destruction and a third, inscribed Zernibog, is dragged off the pedestal. Charlemagne's persecution of Saxon paganism goes unnoticed; the proscribed idols are Slavic, a scene from the next wave of Christianization.[60]

"Medieval Upper Saxony" is even more decorative than the "ancient" model. The upper edge of the map boasts a line of coins, then a series of heraldic shields. Figures profusely adorn the lower edge. The main scene shows the great tournament at Nordhausen held by Henry III the Illustrious in 1263, the year that the Landgraviate of Thuringia finally became his secure possession. Unidentified figures to the right begin with a knight in full plate armor and his standard bearer. Farther right, as a separate scene, stand a richly dressed lord and lady; she cradles the model of a church. Behind both groups, a large river flows through the landscape. The main territorial divisions on this very beautiful map are smaller than those of the earlier period, and their outlines seem unrelated to the past. They include Thuringia (or Upper Saxony), the County of Wettin (from which the ruling family took its name), the Palatinate of Saxony, the County of Brene, and a series of *marchiae*. Many smaller districts, and bordering ones, are marked.[61]

To say that Zollmann's maps are of high quality and may have guided later cartography is not quite enough. One senses that his choice of territories and images was underpinned by a historical program aspiring to link past and present. His reverse chronology and unusual periodization should probably be played down by comparison with the more immediate dynastic circumstances in which his maps were created. Someone familiar with Saxon history will, one hopes, take the trouble to decipher Zollmann's scheme in its details. Among the many mapmakers intent on tracing the historico-genealogical vicissitudes of Saxony, Zollmann is first in line.

The last of this two-year collection of regional maps of medieval Germany has been mentioned in connection with Charlemagne's canal. In 1733, Johannes Heinrich von Falckenstein (1682–1760) published two works of erudition about the Bavarian Nordgau: one (in two volumes) called *Antiquitates Nordgavienses,* the other a complementary collection of charters, the *Codex diplomaticus antiquitatum Nordgaviensium.* They are not thought to be of high quality. The accompanying "Map of the Old Nordgau as It Was in the Eleventh and Twelfth Centuries" shows the *gaue* and major lordships of this large district north of the Danube. Its scale, extent of detail, and general appearance resemble Zollmann's "The Duchy of Upper Saxony . . . between the Tenth and the Fifteenth Century" and "The Duchy of Upper Saxony in Its Most Ancient State" of the previous year. The Homann atlas for Germany welcomed Falckenstein's endeavor under the summary title "The Medieval Nordgau." It joined Zollmann's pair for Upper Saxony as the only maps in this atlas attempting to portray the topography of medieval German regions.[62]

France, without finding anyone to rival the quality of Delisle's reconstructions of Toul and Dauphiné, witnessed new efforts to map medieval territories. Falckenstein's big books coincided with the appearance in Paris of the multivolume *Histoire générale du Languedoc* by the Benedictines Claude de Vic and Joseph Vaissette, one of the most respected works of eighteenth-century French erudition. It continues down to the present day to be much consulted by scholars. Jean-Baptiste Nolin Jr. (1686–1762), whose maps of French bishoprics illustrate *Gallia Christiana,* provided de Vic and Vaissette with a "Roiaume et duché de Septimanie" to place at the head of their second volume.[63]

Nolin's map, as usual for the age, specifies no date but shows the major districts that Septimania contained in the Middle Ages: Marquisat de Toulouse, Duché de Gascogne, Duché d'Aquitaine, Marquisat de Barcelone ou d'Espagne (the Spanish March created by the Carolingians), Duchez de Septimanie, Royaume et Duché de Provence. Although all the districts, as drawn, look as though they coexisted, the coloring (if one finds a colored copy) introduces a measure of complexity, suggesting that units bordered in dark green did not originate simultaneously with those in pink; a special color betrays that Provence is separate, outside Septimania. The map exists as a reference aid for the *Histoire générale* and was not meant to speak for itself. The lettering is suitably graded in order of importance, but much of it is in long, straight, oblique lines, too much so to meet normal standards of grace and elegance.

An abridgment made the monumental history of Languedoc accessible to a wider public. Dom Vaissette, whose writings include general cartographic work, not only carried out the condensation but also provided a substitute for Nolin's map. An elaborately descriptive title is provided, along with the precise date 8 March 1749: "Carte du Languedoc avec les provinces voisinnes où l'on à [*sic*] marqué la division du Royaume et Duché de Septimanie et celles des Trois Senechaussés anciennement comprises sous le nom de Languedoc." The seneschalsies, which in the full edition have a map to themselves, are here combined with the kingdom and duchy. There is little originality in this.[64]

The encouragement that Bessel, abbot of Göttweig, gave in the German sphere to the study of charters and the mapping of their contents was taken to heart in the 1740s by Joachim Berward Lauenstein, pastor of Saint Michael's church in Hildesheim, in Saxony. In a work that almost subordinates charters to the production of a map, he focuses strictly on the lands he knew best: "Map of the Medieval Diocese of Hildesheim, in Which [the author] Collected the Names of Villages, Castles, Towns, etc., from Various Charters of the Middle Ages, and Carefully Delimited the *Gaue,* Both Major and Minor, after the Guidance of Charters, and Couched the Names of Rivers, Woods, etc., in the Medieval Dialect." The long title, attached to an atlas-sized map, nearly serves as an author's introduction. Unannounced, a scattering of historical notes about early rulers of Saxony are written at suitable sites. They show, for example, the spot where the Saxon duke Henry (later king and progenitor of the Ottonian dynasty) obtained his surname "the Fowler." Another note indicates that Pope Gregory V (996–99), also born an Ottonian, originated from Melverode, between Braunschweig and Wolfenbüttel. The outlines given of the bishopric of Hildesheim are those at the time of publication.[65]

Lauenstein's program, rather too grandiloquently announced, is not obviously less satisfactory than others we have looked at. There were flaws enough, however, for the map to be taken fiercely to task in the next years by Johann Friedrich Falcke (1699–1756). The polemical relationship of Falcke to Lauenstein justifies a brief departure in the narrative from strict chronological order, as well as the addition of one more German reconstructing the medieval map of his homeland. We then move to France.

Falcke, a merchant's son who took the unusual step of attending university and eventually became pastor of a church in the Hildesheim district, worked at historical labors for several years before obtaining his pastoral living. Once secure, he shouldered the ambitious task, comparable to Schannat's at Fulda, of publishing an edition of the charters of Corvey. The abbey of Corvey had been founded in the early ninth century as a projection into the frontier territory of Saxony of the prominent West Frankish monastery of Corbie. Falcke's *Codex traditionum Corbeiensium,* published in 1752, was equipped with no fewer than five maps, portraying the wide territories in which the monastery owned lands.

An index map is needed but not supplied. The titles of the first four convey an adequate sense of overlap:

1. Part of Old Saxony or Angaria in the Eastern Region Divided in *Gaue* [vertical]
2. Part of Old Saxony in the Eastern Region Divided in *Gaue* [horizontal]
3. Part of Old Saxony or Westphalia as well as Angaria in the Western Region Divided in *Gaue* [horizontal]
4. Part of Old Saxony or Angaria in the Western Region Divided in *Gaue* [vertical]
5. Old Frisia and Part of Old Brabant Divided in *Gaue*

The first four titles have a disconcerting sameness, as though Falcke could not think of ways to differentiate them. With patience, the pattern can be sorted out. The right side of number 1 overlaps the left side of number 2. The second pair has a different relationship: number 3 is the general map, whereas number 4 enlarges a part of it. The rationale for this visual confusion, if provided at all, is tucked away out of reach of any but a very meticulous reader. The *gaue* have the lion's share of attention; they are basic to medieval land grants and are the main reason for having maps in a collection of *traditiones*. Some localities and other features are noted. Falcke generally avoids historical notes of the type inserted by Lauenstein but cannot resist one about Duke Henry being called the Fowler (no. 2).

Falcke's introduction to the *Codex* devotes several disappointing pages to the maps. He starts cogently enough by asserting that his maps conform, not to the observations of astronomers, but to the location of watercourses, which are shown in the best current maps. Whether this is said with pride or in apology might have been made clearer. His remarks then decline into several pages of polemic about the worthlessness of Lauenstein's historical map of Hildesheim. Time has not been kind to his own competence; later scholars have been displeased with the text of the *traditiones* as well as with the copious annotation. Even though an assistant is said to share responsibility for these flaws, Falcke has a poor reputation. His five maps do not redress the balance.[66]

There can be little doubt that, by midcentury, the charting of narrow districts in their medieval shape had become an established scholarly occupation in Germany. A few years after Falcke, the court librarian of the Principality of Anhalt, Johann Sigmund Strebel (1702–64), published a map of the medieval Rangowe, in preparation for his major book on Franconia, of which, in the event, only the first volume appeared. The program of the map is set out on its face: "The Rangowe district restored out of the Middle Ages with the places mentioned in old charters, in accordance with the pronunciation and spelling of the old Franks, [and] with the addition of a brief historical explanation." The map, which celebrates Strebel's recent promotion to the rank of privy councilor (1757), makes a poor first impression. Every bit of it is covered, except for watercourses, by completely uniform little trees; only the clumps of grass on which they stand vary, and only slightly, from tree to tree. Anhalt (*Onoldus*) is almost at the center. Neither the Rangowe nor neighboring *gaue* are delimited. A moderate number of localities are indicated. The advertised historical explanation, which might have made up for the mediocre visual effect, is nowhere to be found.[67]

Strebel's ambitious *Franconia illustrata,* "or Essay to Elucidate the History of Franconia from Trustworthy Archival Documents," was published four years later, in 1761. It opens with a complaint, frequent since the start of the century, about the lack of medieval geography: "It is a known thing that we Germans lack an adequate geography in the Middle Ages, though one is indispensable for the real understanding of old charters." Maps of the Middle Ages are envisaged from the narrow, though important, standpoint of diplomatic, in the tradition of Mabillon and Bessel. Strebel does little with his own map, however; the promised explanation does not even occur in the more expansive confines of his book. Disappointing in our eyes, the work was admired in its time as local history and was long consulted.[68]

In the French kingdom, unlike fragmented Germany, peripheral territories like Toul, Dauphiné, or Septimania were better served with medieval maps than the heartland. In 1751, Daniel Schöpflin published the first volume of *Alsatia illustrata,* his masterpiece, concerned with the most outlying province of France. Its frontispiece is a map of early medieval Alsace.

Schöpflin (1694–1771) stands out among eighteenth-century academics not so much for scholarship or intellect, though he lacked neither, as for the exceptional regard in which he was held from an early age at the courts of Versailles, Vienna, Saint Petersburg, and other European capitals. Professor at the (Protestant) University of Strasbourg, Schöpflin chose to remain there, on suitable terms, in spite of many flattering invitations to accept a position in a more illustrious center. Strasbourg declared a public holiday to honor his fiftieth year of teaching (1770). He was "fond of titles and successfully collected them." About the time *Alsatia illustrata* appeared, Schöpflin created, within the university, an "academy" of diplomacy and statesmanship later headed by Guillaume Koch, who was also destined to be important to maps for history. The academy attracted a glittering, international group of young noblemen. In Schöpflin's many-sided life, mapmaking has a minor part.[69]

Yet maps directly benefited from his prominence and wide contacts. J. B. Nolin, the second-generation professional who drew Septimania for de Vic and Vaissette, won little more favor as a cartographer than his father; Schöpflin, however, was able to secure no less an expert than d'Anville to give the maps for *Alsatia illustrata* their final polish. A special feature of the two volumes, the very fine "Geographic Map of the Duchy of Frankish Alsatia, Divided into *Gaue* and Counties, with Towns, Castles, Palaces, Monasteries, [and] Villages," is oriented west, so that the course of the Rhine might be its main axis. The names are given, usually in Latin, as found in the documentary sources, whose intensive use in the preparation of this work is one of Schöpflin's most certain titles to scholarly fame.[70]

A decade passed between volume 1 of *Alsatia illustrata* and its conclusion. Schöpflin provided the sequel with another interesting piece of cartography, done in cooperation with J. Stiedbeck. The expansion of Paris had been shown by Lamare's eight-map cycle; the noteworthy map by Schöpflin and Stiedbeck is called "Strasbourg et ses agrandissements" and couples a main map of modern Strasbourg with subsidiary insets showing the growth of the city. Dainville comments, citing the Lamare set, "This ingenious portrayal of the development of Strasbourg has more than local interest . . . [after the Paris cycle] it is one of the rarest items of historical [city] plan-making."[71] The idea of mapping a sequence of historical moments, less singular in 1762 than it had been in 1705, was still far from standard practice.

Schöpflin's impact on maps extended beyond Alsace. Much earlier in the century, the then Elector Palatine of the Rhine had moved his capital from the religious tensions of Heidelberg to the friendlier city of Mannheim (1720). His successor, Karl Theodor (1742–99), preferred literature, art, and music to martial pursuits. Schöpflin aided him in establishing a learned academy in the new capital, the first of a series of societies that, for a time, made Mannheim a cultural beacon. Schöpflin accepted the honorary presidency of the new "Theodoro-Palatine" Academy, attended its biennial open sessions to the end of his life, and took a sustained interest in its development. He also offered the Elector his prize pupil and disciple, Andreas Lamey (1726–1802), to be the academy's executive secretary.[72]

Lamey, a thoroughly trained antiquarian, guided the historical section of the Mannheim academy as an institute of local antiquities and history. He had closely assisted Schöpflin in preparing *Alsatia illustrata,* was schooled in his methods, and eventually published its complement, a collection of Alsatian charters. Elector Karl Theodor was very keen to have a "Palatinatus illustratus"; he had invited Schöpflin to undertake this task and, when Schöpflin pleaded age, followed his suggestion to create the academy to carry it out in his place.[73] Lamey drew up plans and prospectuses for "Palatinatus illustratus"; collaborators were named to the historical class of the Theodoro-Palatine Academy; and the academy made the prolegomena to the big history its first group project. Lamey personally contributed an edition of the medieval charters of the ancient abbey of Lorsch (*Codex Laureshamensis diplomaticus*), as well as seven articles, which directly concern us, on the *gau*-geography of the Palatinate. His academy associates worked in the same direction, exemplifying the freshly conceived belief that *gau* boundaries coincided with diocesan ones. The Mannheim doctrine about *gau* limits held sway until strongly challenged in the 1830s and survived even that challenge.[74]

Lamey's *gau* studies, distributed among the academy *Acta* between 1766 and 1794, form a notable part of this activity. They were meant, among other things, to be geographical complements to the newly edited Lorsch cartulary. Each article is accompanied by a large-scale map of the *gau* in question. "F. Denis, Lt. of Engineers," presumably a French technician, drew the first six (the engravers, one from as far as Nuremberg, change often); Lamey himself signed the seventh. The maps are "very well disposed, [and] noteworthy for their time." They are a considerable advance in the serious mapping of medieval territories:

1. "The Lobodengau from Charters of the Eighth and Ninth Centuries"
2. "The Wormsgau from Charters of the Eighth, Ninth, and Tenth Centuries"
3. "The Rheingau from Ancient Charters, Mainly of the Eighth and Ninth Centuries" (fig. 20)
4. "The Speyergau from Charters of the Middle Ages"
5. "The Medieval Craichagau Drawn from Ancient Charters"
6. "The Navagau from Ancient Charters"
7. "The *Gau* of Elzengau in Rhenish Francia as It Was Mainly in the Middle Ages"
8. "Weingarten from Ancient Charters"[75]

One map is oriented northwest and three to the west, like Schöpflin's Alsace, for the same reason of alignment with the dominant physical feature. A special symbol considerately identifies "More recent localities not yet attested in the Carolingian age." The large scale helps maintain a sense of distance between population centers; cities like Worms and Frankfurt are marked modestly enough to suggest overgrown villages rather than urban agglomerations. The studies of local geography Lamey and his coworkers engaged in followed methods and attained results that have still not been wholly superseded.[76]

One more contribution to this complex can be mentioned. Christoph Jacob Kremer was one of Lamey's closest associates. His life's work was *History of Rhenish Francia under the Merovingian and Carolingian Kings Down to 843*, that is, to the eve of the major partition of the Frankish kingdom. The subtitle, *As a Foundation for Palatine Political History*, recalls some at least of the goals of the Mannheim enterprise. Posthumously published in 1778 with Lamey as editor, the work is accompanied by an admirably drawn and engraved map: "The Duchy of Rhine Francia Divided into Its *Gaue*, with the Neighboring Territories." It centers on Worms, Mainz, and Frankfurt, with appropriate extensions all around—a bridging point between older districts of Frankish rule and eastward extensions. Although *gaue* are carefully bounded and named, larger districts lack territorial boundaries and are named only. It was a worthy contribution to Mannheim history.[77]

The *Catalogus mapparum* of 1933 has a special favorite in German regional historical mapping prior to Spruner, namely, Franz Güssefeld's "Geographic Examination of the Territorial Divisions That Took Place in the Ernestine

Line of the House of Ducal Saxony, and Presentation [of these divisions] in an Accurate, Newly Designed Five-Part Detail Map." The *Catalogus* calls it the first attempt, for its time highly successful, to compile a particular historical map on a scientific basis, with reference to unpublished charters. While everyone else was immersed in *gaue*, Güssefeld addressed intricate dynastic transactions of 1485 and after, well suited to geographical exposition. He was very productive as a collaborator of the Weimar Geographical Institute, which published his *geographische Übersicht* of the Ernestine divisions. Important or no, the survival of this exemplary work of historical geography appears to hang on the thread of a single copy in Berlin.[78]

How many more maps of medieval German districts are dispersed in eighteenth-century books and academy publications is hard to tell; items not recorded here are certain to turn up. One last sample, handsomely colored, impressive in design, and partly overlapping with Lamey's territory, is Wilhelm Theodor Schmidt's "Map of Hesse and the Wetterau with the Neighboring Lands, after the Medieval Geography of the Eighth to the Twelfth Century." It illustrates H. B. Wenck's *Hessische Landesgeschichte*, a work similar to Schöpflin's on Alsace or the unwritten "Palatinatus illustratus." Well before the end of the eighteenth century, the German territories, in their cultural vigor and passion for local history, were already the promised land of that cartography of medieval landscapes that was applauded in 1933 as though it were a recent achievement.[79]

A final group of maps, attested only at second hand, takes us outside Germany. Two unexpected medieval scenes occur in the *Algemeine Welt-Historie*, probably reprinted from earlier works and almost certainly from outside Germany. Their workmanship does not equal the best of the maps already encountered, but they are noteworthy all the same. The more rudimentary of them portrays "the localities near Auxerre named in Nithard['s *History of the Sons of Louis the Pious*] in connection with the Battle of Fonten[oy-en-Puisaye in 841], and other localities." Fontenoy was a decisive moment in the civil war among the Carolingian kings; its outcome decided that the empire would be divided among the sons of Louis the Pious. The very small map of the *Welt-Historie* shows Auxerre in the upper-right corner and the Upper Yonne River circling south and east. Despite a paucity of names and features and the lack of a scale of distance, its role is clear. The volume concerns Charlemagne; its main map portrays his empire, and the battle detail holds the place that, before long, would often be assigned to an illustration of the three-way division of 843.[80]

Figure 20 Reconstructions of Medieval Features and Districts (1768). Andreas Lamey's eight maps of medieval *gaue* are still praised. The map shown here (from "Pagi Rhenensis qualis sub regibus Carolingis maxime fuit descriptio," *Acta Academiae Theodoro Palatinae* 2 [1770]: 253) bears an earlier date than the academy proceedings containing Lamey's article. Lamey's maps were prompted by his edition of the Cartulary of Lorsch, the collected charters of a major early abbey, and by the preparation of a major work on the Rhenish Palatinate—the collective goal of the Mannheim academy, whose executive head Lamey was. (Courtesy of Yale University Library)

A later volume of the same *Welt-Historie* includes a large and, as always, unattributed map entitled "Territory of Milan in the Middle Ages." The coverage reaches west to Bergamo, south to Pavia (Ticinum), and east to Novara; main roads are shown. To the north, three Alpine lakes are seen, along with Bellinzona and convincingly high mountains. Both the Po and the Adda Rivers appear with wide banks and islands. Towns are named in large, not overwhelming numbers, and the legend includes symbols for monasteries, forts, parishes, hospitals, and colleges of canons regular. Counties, profuse and delimited by dotted lines, are the sole territorial divisions. Only an expert could tell whether this countryside was really medieval. Even if the map falls short of precision, the approximation it achieves is convincing. Where did the *Welt-Historie* editor find it? Muratori's collection, drawn upon in the previous volume, is not the source. Until someone has the good fortune to happen upon it, the German reprint has to suffice.

The finest and most ambitious depictions of medieval scenes known to the eighteenth century are among the maps just surveyed. The detailed, often archival research underlying them yields more convincing results than the more summary work of big names in cartography. How respectable Delisle's Toul, Zollmann's Saxony, Beretti's Lombard Italy, Schöpflin's Alsace, and the others are by modern standards are interesting questions, which some may have the time and skill to verify. Our century can probably do better. But it may hardly be doubted, after the works just seen, that scientifically creditable maps of medieval history were already being produced—and in fairly sizable numbers—during the supposedly antimedieval Age of Enlightenment.

"Outside Europe"

Byzantines, Crusaders

The volumes of Karl Spruner's famous *Hand-Atlas* on the ancient world and Europe are complemented by a third volume whose binding is stamped with the brief title *Ausser Europa* (1851). Spruner's usual mastery is evident in the contents, but "Outside Europe" is very thin, especially by comparison with the European volume. A less grudging attitude prevailed in the previous century.[81] The stock of maps of non-European history, almost nonexistent before 1700, swelled to respectable proportions thereafter. It is in the historical atlases (discussed later in this chapter) rather than in the single maps now being examined that the desire of eighteenth-century cartographers to reconstruct the Middle Ages on a Eurasian scale is best assessed.

The Byzantine Empire, in its many configurations, was always rooted in

Europe, and the Crusades were European enterprises. These two subjects afford a gradual passage toward the Orient. Where isolated maps of Byzantium are concerned, only one cluster mattered in the eighteenth century. Köhler appropriated a city plan from this cluster for the *Kurze Anleitung;* and d'Anville designed an original map of northern Persia and adjacent regions to accompany his study of the great war of the emperor Heraclius (1762).[82] There appear to be no independent forays other than these. Köhler's map has broader appeal than d'Anville's very learned one, but neither is memorable.

The France of Louis XIV stimulated Byzantine studies by producing the twenty-seven folios of original sources long known to scholars as the "Byzantine du Louvre." To this collection the Benedictine Anselmo Banduri (1671–1743) contributed a volume called *Imperium Orientale sive antiquitates Constantinopolitanae.* Last of the parent series, it was underwritten by Banduri's patron, Grand Duke Cosimo III of Tuscany, and dedicated to him.[83] Its twin foci are the scholar-emperor Constantine VII Porphyrogenitus (912–59) and a sheaf of writings about the city of Constantinople, notably the *Patria sive origines urbis Constantinopolitanae,* which Banduri uncovered in a Paris manuscript. In connection with the latter, Banduri drew two city plans. The writings of Constantine Porphyrogenitus gave rise to more ambitious cartography—the earliest depictions of the Byzantine Empire worthy of notice. The maps are by Guillaume Delisle and, after "Toul" and "Dauphiné," complete his canon of "géographie du moyen âge."

Banduri's two city plans have nothing geographic about them except an outline of the peninsula of Byzantium, its water boundaries, and part of the lands beyond the Golden Horn on which Galata stands. The outlines, slightly different in scale, are identical. The first image is labeled "Constantinople, Divided into Fourteen Regions, as It Was under Honorius and Arcadius [395–423], as well as Its Suburbs" (395–408); the second, "Constantinople Divided into Three Parts, as Described by an Anonymous Author Who Lived in the Time of Alexius Comnenus" (1081–1118). Ostensibly separated by six centuries, both show an interchangeable network of roads or trenches outside the great land walls. The content of the maps is written rather than drawn. In the first, the fifth-century wards, schematically rendered, form a frame for the names of the main monuments of each of the fourteen wards as listed by a contemporary source. The end product is not so much a map as a decorated diagram. Banduri's Comnenan Constantinople is even less visual; the monuments are not listed or plotted but written in, as near to where they stood as the means for showing them allowed. Presumably, an eye-catching historical plan of the city was unattainable; no draftsman felt able to venture a reconstruction of medieval Constantinople. Banduri's diagrams at least suggest relative locations and give modestly visual assistance to the rich documentation of the volume.[84]

Delisle supplied one map to illustrate the treatise "On the Themes" (*De thematibus*) by Constantine VII, and a second to accompany the emperor's advice on "how to run the empire" (*De administrando imperio*).[85] Scholar-emperors were as rare in Byzantium as anywhere else; Constantine VII, grandson of Basil I, founder of the Macedonian dynasty, took up scholarship while waiting thirty years to acquire effective leadership of the state. The seventh- and eighth-century reorganization of the much diminished empire into "themes" marks the passage of Byzantium from antiquity into the Middle Ages. The provincial units central to Delisle's first map, and prominent in the second, were the specifically medieval divisions of Byzantine provincial government. Delisle, thanks to the subject assigned to him, gave visual form to a Byzantine Empire that had managed to survive vast losses of territory and population and regain stability and power. His handsome, carefully researched maps do not disappoint readers.

The outlines Delisle presents are closer in time to each other than the map title of the first suggests. Whereas we are only told "after the days of Heraclius [610 – 41]," the divisions drawn on the map are derived predominantly from Constantine VII's *De thematibus;* the themes in question are of the tenth century, not the seventh. Predictably, the Byzantine Empire in Delisle's second map is divided along the same lines as the first, or almost. As Constantine VII's *De administrando imperio* is mainly about foreign affairs, so its map is much concerned with the wider world that Constantinople dealt with. Foreign peoples are plotted with as much as was known of their boundaries. Petchenegs are there, with Uzes, Russians, and many others whose names assure us that the denizens of late antiquity (as identified, for example, by Delisle's "Theatrum historicum") were long gone and superseded by a new cast of characters. Later cartographers have improved Delisle's geographical, administrative, and ethnographic outlines, but tenth-century Byzantium can hardly be given a more impressive and monumental look than it has in his maps.

The Crusades are ubiquitous in our experience of medieval history. They belong not only to Europe but also to the Holy Land, with all its popularity for pious audiences. Yet maps of the Crusades, late bloomers, were almost wholly neglected in the eighteenth century. Modern models are most closely anticipated in a map of the Châtelain *Atlas historique,* already mentioned, showing the track of Richard I (Lionheart) to Palestine. This meager delineation is noteworthy only as a foretaste of the future; where Châtelain traces a single route, later maps of crusading expeditions would highlight many tracks. The small start in Châtelain was not gradually enlarged with one more track after the next; no stepping stones connect the Châtelain début to the developed Crusades map of a century later (1829).[86]

Meanwhile, the big names of Delisle and d'Anville were at least enlisted as illustrators of specialized narratives involving the Crusades. The commission Delisle completed on the day of his sudden death involved three maps for the abbé Vertot's *Histoire des chevaliers . . . de Malte:* geographical views of Cyprus and Rhodes, and, as a frontispiece, a "Map of the Lands to Which the Knights of Malta Bore Their Weapons." This wide, very horizontal map extends from Spain to the edge of India, with lots of place-names but no specifics. It is the most decorative of the three, but very small and serves only the limited needs of Vertot's readers.[87]

D'Anville's scope is, if anything, narrower than Delisle's. He provided two maps each for Claude Marin's *Histoire de Saladin, Sulthan d'Égypte et de Syrie* and for an ambitious edition of Joinville's *Histoire de St. Louis.* All four are narrowly addressed to the books in question. The Joinville edition, for example, is equipped with a general map of the eastern Mediterranean and a "Special Map for St. Louis's Egyptian Expedition." Excellent, comprehensive Crusade maps might conceivably be found under such titles, but this is not the case here. The cartographers had specialized commissions and carried them out. They were not asked to illustrate thorough histories of the Crusades.[88]

The closest one comes in the eighteenth century to a comprehensive map is Delisle's posthumous "Carte générale de la Syrie, de la Palestine et de l'Isle de Chypre . . . pour servir à l'histoire des croisades et . . . des royaumes de Jérusalem du Temps des croisés." Probably drawn in connection with Vertot's book, it was not published until 1764, in circumstances about which too little is known. Guillaume was not the only gifted Delisle. His long-lived brother Joseph-Nicolas (1688–1768) was an astronomer who worked in Russia for twenty years and had an honorable scientific career there and in Paris. In his midseventies, Joseph began to publish items from the long-dead Guillaume's scholarly legacy. The Dépôt de la Marine, in which Joseph's own learned collection had recently been deposited, held more than twenty fat portfolios of Guillaume's unfinished work, comprising not only drafts but maps in near-publishable form. Joseph made known the existence of these treasures so that they might be gradually published.[89]

The only items that Joseph withdrew from this rich and varied collection concerned the Holy Land. Guillaume's treatment of this edifying region had been emphasized before Joseph. Soon after the great cartographer's death, his widow presented to the king a memorandum on Guillaume's new, unpublished map of the Holy Land. In the 1760s Joseph drew attention to a wider spectrum of subjects than the widow had; there were 117 pieces in the Holy Land portfolios, not all publishable, of course. The map that aroused greatest interest showed the site of Paradise; its publication was sponsored by the duc de Choiseul, then dominant in the government. Two years later Choiseul's cousin, the Navy minister Choiseul-Praslin, lent his patronage to an astonishing "Map of Babylonia Now Called Hierac-Arab, with Its Names Both Ancient and Modern, and the Tracks of the Expeditions of Cyrus and Julian the Apostate, as well as Those of Teixeira, Benjamin [of Tudela], and Other More Recent Travelers." This visually unappealing map of Iraq displays impetuous historical exuberance. It is covered with notes about every age (Tower of Babel, battle of Kadesiah), mixes ancient and modern personages (Cyrus, Teixeira) to an extent that Guillaume does nowhere else, and draws tracks that defy one's ability to trace their course in full. Either Joseph "improved" a piece from the heritage, or Guillaume left a draft never meant for publication.[90]

The Holy Land maps that Joseph chose as posthumous tributes to his great sibling were uncommon. Such, too, was the large rendering of the Levant meant to accompany "histories of the Crusades and of the Kingdoms of Jerusalem at the time of the crusaders."[91] Guillaume was concerned with the bridgehead the crusaders made, not with the routes they took to get there. Later atlases often assign an inset map to the Kingdom of Jerusalem and adjacent principalities. Anticipating such images, Guillaume's is larger in size and more reticent in information. It is also handsomer and better designed than those of his other cartographic vestiges published by Joseph. Without foreshadowing the Crusade maps of the still rather distant future, Guillaume supplies geography and a scattering of places. On this austere outline, free of territorial boundaries, castle sites, battlefields, and other historical paraphernalia, the user is free to add whatever he needs. Guillaume's evocation of Iraq, historically jumbled though it is, looks as though it may have conveyed more instruction.

Asiatic Peoples

The first maps of farther Asia with medieval content are associated with a dynasty of Orientalists at the French royal court. François Pétis (1622–95), interpreter for Arabic and Turkish during much of Louis XIV's reign, had been drawn into this profession by his maternal uncle, the interpreter for Turkish in the navy ministry. At the urging of the great Colbert, Pétis undertook a history of Genghis Khan from Arabic and Persian sources. Unfinished at his death, it was seen to publication by his son, also named François. The second François (1653–1713) was responsible for adding "de la Crois" to the family name. Starting in his midteens he spent a decade in the Near East at government expense and became a remarkable linguist. He was named to the chair of Arabic at the Collège royal (now Collège de France) in 1692 and succeeded his father in 1695 as court interpreter. Literature knows him as translator of several collections of oriental tales, notably the "Thousand and One Days." Besides ensuring publication of his father's book on Genghis Khan, Pétis de la Croix complemented it with a fifteenth-century Persian history of Tamerlane, whose translation he did not live to finish. Very soon after he died, his sixteen-year-old son, Alexandre Pétis de la Croix (1698–1751), sailed to the Ottoman Empire for his decade of language training at state expense. The completed Tamerlane translation came off the presses a few months after Alexandre's return.[92]

These books needed maps. It is no surprise to find that Guillaume Delisle drew northern Asia for François Pétis's *Histoire du grand Genghizcan, premier empereur des anciens Mogols et Tartares . . . contenant la vie de ce grand can . . . traduite et compilée de plusieurs auteurs orientaux et de voyageurs européens.* Most of what is said about this map—"Asie septentrionale"—is engraved onto its surface: "A Map of Northern Asia to Complement the *Histoire de Genghiz-Can* Composed by M. Pétis de la Croix, Interpreter to the King for Oriental Languages, by Guillaume Delisle, of the Royal Academy of Sciences, 1710." Delisle is introduced by the publisher as "one of the ablest geographers of the age," who drew the map from information supplied by Pétis and son. Delisle himself appeals only to oriental geographers and the itineraries of several travelers and adds that any discrepancies between his longitudes and those given in the book result from his reliance on the most exact information.[93]

References to scientific accuracy sound inflated in connection with this small and modest map. The publisher seems not to have been ready to pay for Delisle's best effort. Only the area shown, running from the eastern half of the Black Sea all the way to Korea, is expansive; the coverage is "northern" in as much as the peninsular half of India and all Southeast Asia are excluded. Many places and lands are named, but the map is far from crowded. Dotted lines indicate various boundaries. The campaigns of Genghis Khan or his armies are not marked. At most, the reader is given a rough idea of the situation of the various lands—not localities—mentioned in the narrative. Delisle's portrayal, based on the materials supplied by François Pétis, is not likely to be strictly true to thirteenth-century Asia. Imperfect or not, it inaugurates the maps of medieval Mongol conquests. A much improved but recognizable descendant bears the name "Mongolenreiche" in the standard *Putzger's historischer Schul-Atlas.*[94]

François père's *Ghengizcan* is a composite of several Islamic sources. His son's work on Tamerlane translates the narrative of a single fifteenth-century Persian author and as a result is catalogued everywhere under the name Sharaf al-Dīn 'Alī Yazdi, a well-regarded historian whose florid language the translator took care to adapt to French usage. Four volumes long, in contrast to one for Genghis, the *Histoire de Timur-Bec* is also more fully illustrated. One map is provided for each volume and an extra for volume 3. Half the size, at best, of Delisle's "Asie septentrionale," they show more limited districts and do so in greater detail. Only the last, comprehensive map bears a cartographer's name. All five, identical in style, are likely to be by the one person named; this is J. B. Nolin, already encountered as designer of the map of Septimania for the *Histoire générale du Languedoc.*[95]

Nolin's maps are shoddy as well as small. In the first, for example, the Black Sea is out of proportion with the Caspian, and the outlines of both are, by our standards, pathetically inaccurate.[96] The general map in volume 4 reaches from the Arctic to Australia ("N[ouvelle] Hollande") and, amplifying Delisle's claim, speaks of reliance on the "Gentlemen of the Academy (of Sciences)" to rectify the information supplied by the person who ordered the map ("M. de la Croix"). These are brave words in a context of lackluster cartography. The titles indicate what subjects were thought worthy of mapping:

> (1:3) Carte du Capchac, Partie du royaume de Gete, de la Transoxiane, de la Moscovie Georgie
> (2:1) Carte du Mogolistan, d'une partie du royaume de Gete, et des pays voisins
> (3:1) Carte de l'Expedition de Tamerlan dans les Indes;
> (3:193) Carte de l'Expedition de Tamerlan dans l'Irac Agemi, l'Irac Arabi, le Courdistan, l'Anatolie &c par M^r. de la Croix
> (4:1) Asie divisee en ses principaux Estats lors de l'expedition de Tamerlan sur les memoires de Monsieur de la Croix rectifiée sur les observations de Messieurs de l'Academie par J. B. Nolin[97]

Although "expeditions" are mentioned, their routes are not shown on the maps.

The two centuries of Mongol conquest and ascendancy begun by Genghis Khan and ended by Tamerlane are conspicuous in medieval history, though sometimes undervalued by modern narratives that look little farther than western Europe. The one major thrust of Mongol armies into Europe (1241–42) crushed Hungary and adjacent lands but was stopped by the death in China of the supreme khan and the departure of his generals to attend an election that, in the event, was long delayed. Eastern Europeans were spared further attacks, and the west was never touched. In Russia and the Islamic principalities of central and western Asia, the Mongols had a devastating and fateful impact. At the same time, they opened up a secure overland route to the Far East and made Asia accessible to Christian merchants and missionaries. These circumstances and much else give the epoch of Mongol power compelling interest. Soon after Delisle and Nolin, historical atlases would include improved maps of this epoch. Until they did, the unprepossessing illustrations for the translations of François Pétis and son attested to the share Asia had in the dawning notice given to the medieval past.

Bellin's map for de Guignes's history of the central Asiatic nomads, mentioned above under the heading "Barbarian Invasions," would not be out of place here, as a delineation of Asiatic lands and peoples. Midcentury, the time of Bellin's map, witnessed a return to the subjects pioneered by Delisle and Nolin. The setting, now, was the *Algemeine Welt-Historie,* improving the English cooperative *Universal History.* Successive volumes of the *Welt-Historie* include an anonymous "Portrayal of the Onetime Turko-Tartar Empire of the Middle Ages, Drawn in 1760" and, right afterward, "The Empire of Timur-Beg or Khan, or the Tamerlane, ca. A.D. 1405, Presented after the Pattern of a Small Map by the Late J. M. Hase, 1761." The latter was based, as the title indicates, on Hase's small-scale atlas of great empires. It would not be surprising if the former were also based on the same collection of 1743.[98]

Whatever their sources, both maps are anonymous and seem to be by the same hand. The first reaches a little farther east than Delisle's; the second, in keeping with its subject, extends from Serbia to the Gobi. Much larger than the French pacesetters, they embody improvements, notably a credible outline of the Caspian Sea. Nevertheless, these are unattractive, amateurish, hard-to-read maps, in keeping with the generally poor visual quality of illustration in the *Welt-Historie.* Despite their mediocrity, they are evidence of the attention then paid to the Mongol period of the Middle Ages.

European Travelers to Asia

In its map of "Mongolenreiche," *Putzgers historischer Schul-Atlas* lays aside its normal aversion for tracks and draws three thin, black-and-white, differentiated lines across Asia: one stands for the Silk Road; another for the Franciscan William of Rubruck (1253–55); and a third for Marco Polo (by internal evidence 1271–95). Cartographers and their patrons have never agreed when and how often maps for general history should commemorate incidents from the history of travel and discoveries. Most explorations are confined to specialized records; a small assortment wins the favor of historians. Sometimes, viking colonization of Iceland, Greenland, and points west is picked out. The "great discoveries" on both sides of 1500 have even better claims to inclusion. By the time the Putzger atlas displayed both the Mongol Empire and two prominent Europeans who penetrated it, quite a few other historical atlases had made the same choice. It was not in history books, however, but in the literature of travel that maps of these noteworthy medieval voyagers had their original home.

Pieter van der Aa, to whom these travelers lead, is deplored by cartobibliographers. An Amsterdam bookseller, van der Aa published enormously between 1682 and 1733, much preferring quantity to quality. The borrowed maps that passed through his hands were tortured whenever commercial pressures counseled surgery, and though he sponsored new maps to illustrate his publications, his goal was to sell books rather than to win cartographic blue ribbons. Van der Aa's reprintings have proved useful here as a stepping-stone to earlier maps, such as Pellegrino's of southern Italy. His atlases are under a cloud not for being derivative but for making free with the originals.[99]

A van der Aa publication of 1729, *Recueil de divers voyages curieux,* reprints earlier editions of the four major accounts of medieval travelers to Asia apparently available at that time; "I have also improved this edition," the publisher announces. "The maps and figures which were wanting in the old editions are added here." The travelers whom van der Aa equipped with maps are the twelfth-century Spanish Jew Benjamin of Tudela, who journeyed as far as Baghdad long before there were Mongols in sight (1159–73); Giovanni di Plano Carpini (1246–47), a Franciscan sent by the Council of Lyons; another Franciscan, William of Rubruck, journeying privately; and the Venetian merchant Marco Polo, today the best known of the four.[100]

Among the maps, the pair concerning the two Franciscans has priority in time; Carpini and Rubruck had long been on van der Aa's list. In 1706 he had published a pamphlet-size translation of Carpini into Dutch, the work of a clergyman, Solomon Bor. The next year van der Aa issued a travel collection in Dutch and, soon after, took only its maps into the two-volume *Cartes des itinéraires et voyages modernes . . . dans toutes les parties du monde.* The track of Carpini's journey, already available in 1706, was added to Bor's translation. It was incorporated together with Rubruck's on a pair of small, very distorted, but decoratively framed maps published in 1707.[101] These illustrations seem to have been specially ordered by van der Aa. Little recommends them as cartography. Nevertheless, this publication of 1707 may mark the origin of maps registering European travelers to Asia.[102]

What van der Aa did on behalf of Carpini and Rubruck in 1706–7 he repeated in 1729 for Benjamin of Tudela and Marco Polo. He claims to have enriched their narratives with maps; earlier editions of Benjamin and Polo appear to lack geographic illustrations. The maps of 1729 differ considerably from the earlier pair; they look more dignified and are larger by more than three-quarters. Stylistically similar to each other, and free of traveler's tracks, they are consistent with van der Aa's mapmaking. Much of Central Asia is shown in the map for Benjamin, which credits him hyperbolically with a "voyage . . . around the world." The Marco Polo map, presenting the lands of his "very uncommon and quite remarkable voyages carried out through all Asia, Tartary, Mongolia, Japan, the East Indies, adjacent islands, and Africa (*voyages très-curieux & fort remarquables achevées par toute l'Asie, Tartarie, Mong[olie], Japon, les Indes orientales, Isles adjacentes, et l'Afrique*)," is on a smaller scale, so as to include the East Asian sea. The legends on both of them are in French only. Since van der Aa tended to reprint rather than originate, prototypes are likely to turn up.[103]

In Hagelgans's picture-chronicle (1718), already examined, distinct columns are assigned to Arabs, Muslims, and Persians, and also, very fully, to China. Medieval history "outside Europe" proved only modestly interesting to mapmakers, whose depictions of these regions rarely equaled the overall quality of their output. Yet there is a contrast with the previous century. The single maps dealt with thus far need to be complemented by relevant atlases. When assembled, the evidence confirms that historical collections managed to attain a Eurasian scope even before midcentury.

Getting It Wrong

The wish to evoke the geography of the Middle Ages was not necessarily accompanied by adequate means for doing so. Accuracy is a minor consideration in judging the output; high standards cannot be expected. Most designers of maps for history, early or late, may be faulted in one way or another; and some flights of imagination, such as Mejer's in North Frisia, merit applause for vision and daring. But a number of eighteenth-century maps claiming to be of the Middle Ages illustrate, at best, the author's inability to give the epoch he aspired to portray an adequately historical dimension.

The medieval maps of Christoph Cellarius would hardly deserve a glance if his name were not attached to them. Designed in the early 1700s and engraved by J. B. Homann, the plates moldered for seven decades in the hopes that someone would take up the original plan and avail himself of ready-made illustrations. No one did. Eventually, the eighteen plates were printed as one of three appendices to Cellarius's widely circulated *Notitia orbis antiqui* and gathered into a supplement, *Appendix triplex,* to the *Notitia.*[104]

Cellarius enjoyed great posthumous fame. His major geography of the ancient world, the just mentioned *Notitia,* was a standard work of scholarly reference. His smaller comparative geography, *Geographia antiqua iuxta et nova,* was turned by an English schoolmaster, Samuel Patrick (1684–1748), into a map-rich *Geographia antiqua* (1731) whose periodic reissues reached into the next century. The editor of Cellarius's *Appendix triplex* was well aware that its medieval maps were outdated, but the author's reputation gave value to his scholarly remnants even if only as a monument to him, so the three very dated appendices were brought to market. The longing of late-eighteenth-century readers for Cellarius's unpublished fragments may have been overestimated: the *Appendix triplex* is a rare book.[105]

If Cellarius's medieval maps had been accompanied by a text comparable to his ancient geography, and if they had been published before the great man's death, the work would have had the very special distinction of being the inaugural general atlas of medieval geography. The extension of the *Notitia* into the Middle Ages was promised in Cellarius's lifetime and considered noteworthy by reviewers. The stabs at comprehensive medieval atlases in the early eighteenth century—Delisle's and Köhler's, as well as Cellarius's—have been sketched. The ill-starred *Appendix triplex* of 1776 gives us an idea of what the earliest of these never-finished atlases would have included.

The trouble with Cellarius's maps is that almost all are, at best, late Roman and feature little else than the boundaries of Roman provinces. For example, the map of "Western Illyricum" has no conceivable connection to the Middle Ages other than a reference to lands beyond the imperial border as the "seat of the Carpi and Goths before they crossed the Danube," in the very marginally medieval third century! The maps coincide with the Roman Empire and its frontiers, and they bear impeccably classical titles. Four of the eighteen— "Gallia, Francia et Gothia," "Saxonia Transmarina," "Belgica Medii Aevi," and "Germania Medii Aevi"—allow the Middle Ages to overshadow the Roman past. "Overseas Saxony" is none other than the English Heptarchy, very familiar to cartography by Cellarius's day. But the four-map foray beyond Roman times is hesitant and confused.[106]

Cellarius was at a loss about how to relate geography to chronology. Only twice did he explicitly distinguish late Roman from medieval times by allowing each period to have a map of its own; and even when medieval space was carved out, he could not resist filling it with anachronistic antiquities. In

"Medieval Belgica," the ancient "Island of the Batavians" stands cheek by jowl with "Brachbantum," a twelfth-century duchy, as well as the Kingdom of Lothar II (860s) and the—again ancient—tribe of Chatti. In three of the four maps with discernible medieval content (the Heptarchy is the exception), these ill-assorted chronological mixtures abound, and nowhere more so than in "Germania Medii Aevi" (fig. 11). The fourth-century Salian Franks and the even earlier Lombards-on-the-Elbe keep company with the ninth-century Moravians and tenth-century Hungarians; "Wandalia" on the Baltic, presumed home of the Vandals before their equally presumed first- or second-century migration, sits just to the east of the kingdom of the Slavic Abodrites, Charlemagne's northeastern neighbors. If these were large maps, with many names from many periods, the mingling might be excusable, as though on an encyclopedic poster; but the maps are small, with a limited selection of very legible names. No comprehensible principle, or history, guided Cellarius's choices. A hero of ancient geography, set in his thinking, came to grief in his invasion of the Middle Ages.

The likelihood that Jakob Karl Spener also interwove antiquity into the Middle Ages to the detriment of both is suggested by the title of his treatise of 1717: "Description (*Notitia*) of Ancient Germany, from the Beginning of the Commonwealth (*respublica* [*sic*]) to the Establishment of the Germanic Kingdoms in the Roman Provinces. . . . A Survey Is Added of Medieval Germany as It Was in the Sixth and the Next Few Centuries."[107] Both Gallia Belgica (with a part of Britain) and Rhaëtia and Noricum are mapped "according to the systems of antiquity and the Middle Ages (*ad veteris, mediique aevi rationes*)." Spener tries manfully enough to carry out this program by writing ancient names in Roman capital letters and medieval ones in Italic capitals. Yet the Frankish *Austrasiae regnum* is in the same Roman capitals as the ancient *Germania inferior,* and the names are an almost arbitrary selection, arduous to read and interpret. The juxtaposition of typefaces is even more complex in the map of Rhaëtia. The medieval tier shares space with at least two layers of Roman provincial organization, and observers are offered inscriptions on the pattern "Seat of the Lombards, then of the Huns [Avars? Hungarians?]."[108]

Spener's map of Germany claims to be exclusively medieval. It invites comparison with Cellarius's, then unpublished. The resemblance is astonishing in the outlines of land and water (the maps of Belgica, also comparable, differ much more). The selection of names is also surprisingly similar, with predictable differences: the ancient names that clutter and confuse Cellarius's map are cleared away from Spener's by his concentration on the Middle Ages. It would not be surprising if Spener, a professor at Cellarius's old university, had been able to consult his late colleague's papers; his title *Notitia* echoes and honors his predecessor's most famous work. Spener's portrayal of Germany keeps clear of the more distant past and may have a span extending from the eighth century to the tenth. There are no boundaries or chronological markings. Was the past foremost in Spener's program? The choice of names makes one suspect that the map was conceived with current Germany in mind. "The Eastern March, Later Austria" ("Marchia Orientalis quae deinde Austria"), is only the most conspicuous example of a tendency to choose and emphasize those medieval names directly foreshadowing the present. He may have pioneered the reconstruction of (early) medieval Germany, but he was as distant from strict history as Duval had been in his 1670s evocation of medieval France.[109]

Spener's book had barely appeared when the twenty-two-year-old Bourguignon d'Anville, already named "king's geographer," furnished nine maps for the *Description historique et géographique de la France ancienne et moderne* by his mentor, the abbé de Longuerue. In the 1850s, Vivien de Saint-Martin surveyed cartography in France and found no one to equal d'Anville; in Vivien's eyes his brilliance and rigor put to shame the mapmakers of the present. D'Anville's star has never dimmed. He seems to have been the last cartographer—perhaps also the first—who could establish that ancient observations, correctly interpreted, produced more accurate results than modern science. He identified himself wholeheartedly with ancient geography, but with the understanding that modern geography had to be mastered for the sake of ancient. His achievements in both areas were prodigious.[110]

What of the Middle Ages? Although d'Anville's study of the Persian wars of the emperor Heraclius relates to the early 600s, he surely regarded Heraclius as an antique personage rather than medieval. The effort expended on the emperor is characteristic of d'Anville's meticulous critical investigation of Greek and Latin sources. None of his medieval products already examined here equals the study of Heraclius. The Beaurain atlas of 1749 introduces his nine maps for de Longuerue's *Description* as though they had general importance; they were "to assist the reading of histories of France and, among others, that composed by the abbé de Longuerue."[111] So matters may have seemed in 1749. Three decades earlier, d'Anville's cartography is more likely to have been at the exclusive service of his mentor.

Neither de Longuerue nor d'Anville can have given much thought to medieval France. The opening sheet is assigned to ancient Gaul, a land that had been mapped often before and that later occasioned one of d'Anville's most celebrated works. The third through ninth items, under the collective title "France and Neighboring Lands to the Boundaries of Ancient Gaul," display the equivalent of *modern* France and parts of its neighbors, in general and in regional detail. The second map alone, "la France ancienne," to contrast with "la Gaule ancienne," is charged with the burden of linking Roman Gaul to the France of Louis XIV.

In 1719 as well as in his second foray into this alien epoch, d'Anville avoided the label "moyen âge." The term could be considered superfluous or redundant in application to France. Because Clovis and his Franks—"Français" in the then current usage that blended Frank with French—were deemed to inaugurate France at the close of the fifth century, Roman Gaul had an immediate, unequivocal, and lasting successor; the discontinuity implied by "middle age" was avoided. The medieval map for de Longuerue's *Description* is an improvement on Duval's four-sheet precursor, but with only one sheet for a thousand years, chronological confusion prevails. Names of widely separated periods are written in. Features applicable to the two Frankish dynasties, some obsolete by 700, others by 900, amicably coexist with principalities of the tenth and thirteenth centuries. The linguistic divisions "Linguadeoui" and "Linguadeoc" are shown as though they were territorial names. The few boundaries marked by dotted lines are left for the reader to interpret. Possibly, d'Anville supplied only a base map, allowing the buyer to clarify the history he was interested in by applying color highlights according to individual needs; "la France ancienne" was well adapted to this settled practice. If copies of the map with varied historical coloring came to light, it would be easier to believe that the initiative was left to buyers and readers.

Avoided by d'Anville, "medium aevum" is conspicuous in an anonymous, undated single-sheet map entitled *Gallia medii aevi,* catalogued by the British Library as "London 1720?" English names are given in English; no boundaries are drawn; the very limited selection of identified sites point to the Frankish, Carolingian period. Like Spener's Germany and d'Anville's "France ancienne," this anonymous "Gaul" avoids mixing ancient with medieval, but we are left wondering what it might be portraying or what clientele it served.[112]

Johann David Köhler, as noted before, was the third after Cellarius and Delisle to plan a medieval atlas that never saw the light. In lieu of an atlas, Köhler issued a three-part *Introduction to Ancient and Medieval Geography* (*Anleitung zu der alten und mittleren Geographie*). On completion, ten years after Köhler's death, it comprised a nine-piece anthology of medieval scenes. Köhler probably derived "Gaul in the Fifth Century A.D." from an illustration by Henri Liébaux. "Germania in Seculo V P.C.N.," possibly from Spener's medieval Germany, features many names, such as "Francia orientalis," "Frankfurt," and "Marchia orientalis," that, however suitable for Spener, have no business in Köhler's advertised fifth century.[113]

When trying to depict aspects of the Middle Ages, few early mapmakers could bear to cut loose from antiquity and plunge farther than the ninth century. (It did not help that Cellarius, Hagelgans, and others considered the western Middle Ages to end with Charlemagne.) To Köhler, whose most attractive and successful atlas was *Descriptio orbis antiqui,* even the fifth century seemed daring. His map "Spain in the Fifth Century A.D."—source untraced—is immediately betrayed as Roman by its covering of place-names. The lone concession to the Middle Ages is a scattering of names of invading barbarians written across various provinces: *Silingi* (Vandals) in Baetica, *Suevi* and *Vandali* across Galicia and Lusitania, and *Gothi* on the boundary of Tarraconensis and Carthaginensis. For Spain, a moment out of the first half of the fifth century was probably singled out because a source explicitly documents which barbarians (briefly) controlled what areas ca. 411.[114] Köhler's only effort to cut loose from the earliest medieval centuries is called "Gaul, Germany and Italy in the Middle Ages" (*mittlern Zeiten*) and appears in the final installment of the *Anleitung.* In the absence of an identified source, Köhler is the surmised designer. The subject is reminiscent of a map Delisle planned, and it precedes d'Anville's large-scale map of a similar subject by only a few years. What should such a map contain? What features might be representative of, for example, the thirteenth century? Köhler, or his unidentified source, was clearly at a loss for answers. To go by this map, medieval Gaul, Germany, and Italy had little or nothing to recommend them. D'Anville did not do much better.[115]

In the long interval between the first and last installments of Köhler's *Anleitung,* Gilles Robert drew a large, handsome map to illustrate Juan de Ferreras's *Histoire générale d'Espagne* in its French edition. The map is entitled "The Monarchy of the Goths in Both Gaul and Spain" ("La monarchie des Gots tant dans les Gaules qu'en Espagne"). Robert, an assiduous worker, was the successor and continuator of the Sanson dynasty of mapmakers. He and his son Didier Robert de Vaugondy have about six times as many maps to their credit as d'Anville, their contemporary. Although the Roberts did not make a specialty of the past, Gilles is responsible for many maps for history, along with everything else, and several are the first of their kind. Such is the case of his illustration for Ferreras's *Histoire générale.*[116]

Not long after, Johann Matthias Hase acknowledged the originality of Robert's Visigothic map and withheld further comment. The kingdom of the Visigoths was founded in the decade after Alaric sacked Rome and was conquered by the Muslims three centuries later (418–711); it deserved a cartographer. Gilles Robert fell short of the challenge. He provides generous borders, presenting the kingdom as it may have been in the few years of its widest extent. The limits shown may outline the "space" of the kingdom rather than delimit its boundaries at a definite time. Problems begin with the choice of names to inscribe. Something recognizably Visigothic is hard to find. "Septimania," the conspicuous Visigothic enclave in southern Gaul, goes unmentioned; in its place there is a (Roman) "Province de Narbonoise." Everything else is heard from, at least as a sample: pre-Roman peoples (Celtiberia); Roman provinces, of course (Baetica, Lusitania, Cartaginensis); fifth-century kingdoms (Sueves); medieval kingdoms (Leon, Aragon, Portugal); Muslim kingdoms (Algarve, Granada, Murcia); medieval provinces (Estremadura, Old Castile, Catalonia, Viscaya). Readers of Ferraras's history might avail themselves of Robert's map to some extent in many periods of late Roman and medieval Spain; but the Goths, of all people—three centuries of them and a monarchy—are virtually left out.

The same lesson is often repeated. In a map of medieval Belgium, "Gallia Belgica ad historiam medii aevi concinnata," the Jesuit Charles Wastelain seems as ineffective as others in determining a time period. This is one of three maps illustrating a "description of Gallia-Belgica according to the three ages of history, the ancient, the medieval, and the modern." Medieval Belgica is accompanied by ancient and modern maps (the latter by Rizzi-Zannoni). Despite the presence of "ancient Belgica" to give the past its due, the medieval map contains quite a few Roman names—even the tribe of Sicambri from the days of Augustus. Few place-names are characteristically medieval. There are no boundaries or delimitation of principalities. "Medieval" tends, as usual, to mean earlyish, but "when?" stays vague.[117]

The steady diet dispensed here of books and authors acknowledging the existence of the Middle Ages is not representative of the daily fare of the eighteenth-century reading public. Bourguignon d'Anville—in the footsteps of Cellarius and Köhler, but sublimely ignoring them—felt obliged after completing an ancient geography to address himself to the "interval" between ancient and modern times. The compulsion to proceed beyond antiquity had been often expressed before; d'Anville speaks as though venturing into an uncharted wilderness:

> It is very common in geography to consider only two widely separated objects—antiquity and current conditions. But to examine geography under only these two points of view involves neglecting a very lengthy interval and to pass brusquely without connection from the first to the second. . . . [Antiquity extends at most to the end of the fifth century.] One feels the need in the intervening period for an image of the state of things in this great passage, and an intermediate age (*âge mitoyen*) from which the current state of affairs directly

descends. Current geographies do not offer this. After doing [an ancient geography] I believe I may hardly serve the public any better than by following with a work in which one would see the formation in Europe of the states that arose on the ruin of the western empire.[118]

The work in whose preface these lines appear, "States Formed in Europe after the Fall of the Roman Empire," was well received. It was among the eighteen d'Anville items in Gibbon's library and was cited prominently thereafter in contexts of medieval geography.[119] In the half century since contributing to de Longuerue's *Description,* d'Anville had still not admitted "moyen âge" into his vocabulary. His sense of history, also shaped by his earlier experience, did not go beyond tracing the direct background to current conditions. Nevertheless, he composed an early historical geography of western Europe.

The year of "États formés en Europe" is also that of d'Anville's large and beautiful "Germanie, France, Italie, Espagne, Isles brittaniques dans un âge intermédiaire de l'ancienne géographie et de la moderne"—the map supplementing his book. Reprintings continued for more than half a century. D'Anville achieves remarkable grace and beauty with great simplicity of line and total avoidance of decorative devices. So close are the outlines of his map to those seen today that an untrained eye might easily take it for a twentieth-century product. There are better reasons to commend it as art and as contemporary cartography than as a depiction of the past.[120]

A single sheet, even if large, for an undefined but millennial "intermediary age" is a tall order. The best to be hoped is that the map will identify the locations of many places and peoples regardless of when and how long they existed. D'Anville shows good sense in paying no attention to the barbarian invaders of the Roman Empire. His omission of the Caliphate of Cordova and other Islamic entities in Spain seems more arbitrary and facile. Dutch Frisia, in the northwest of the continent, looks as though its islands formed the North Sea coast. Though its inner waters are given the modest dimensions of Lake Flevus, they are identified as "Zuthera Zea" (Zuyder Zee). To the northeast, the Baltic bears the unhistorical name "Sea of Barbarians." In southern Spain, "Vandalitia" (for Andalusia) in large capitals is also hazardous.[121]

The map does allow one to locate many medieval places, but notable names left out surpass those included. There may be method in this. England, for example, shows only its Heptarchic kingdoms, not its counties; presumably the latter could be sought in maps of modern England. So, too, elsewhere. D'Anville's "Francia" is not Capetian France but an earlier Frankish entity, or even Roman Gaul; it features the historical and long vanished sub-kingdoms of Neustria, Austrasia, and Burgundia. Beyond the Rhine, Carolingian "Francia Orientalis" occupies the space to which later medieval "Franconia" has an equal claim; the Carolingian "Marchia Hispaniæ" [sic] is marked where Catalonia would be. All names are in Latin. D'Anville may have sought to emphasize obsolete, outgrown, and vanished place-names—the ones absent from current maps but found in histories. Yet an effort to emphasize the forgotten past is doubtful since localities familiar to modern maps are present in abundance. What seems certain is that d'Anville's handsome map should not be mistaken for a reconstruction of medieval western Europe. With him as with Cellarius and others, the vagueness and confusion of times imply a much less ambitious goal. "Germanie, France . . . " suggests a timeless, very incomplete geographic aid to the reading of medieval history.

Late in the century the hardworking Edme Mentelle designed a map of France under Clovis. It deserves to be mentioned here because it shares the various mishaps just surveyed. In most of his large output Mentelle succeeded in steering clear of the Middle Ages. Comparisons of ancient and modern geography were his favorite activity. What occasioned the Clovis map is unknown, perhaps a book, but the circumstances clearly did not call for limiting the contents strictly to Clovis's lifetime (†511). A few Merovingian divisions are included, about half of them anachronistic for Clovis's day. Other names come out of the distant future, such as the Spanish March (late eighth century), Upper and Lower Lorraine (tenth century), Dauphiné (thirteenth century). The result, cramped and lacking beauty, treats its subject much as d'Anville handled "La France ancienne." [122]

Trained medievalists were rare in the eighteenth century, and none (with the exception of Gottfried Bessel) is likely to be found among those engaged in cartography. Everyone was more or less self-taught, groping in territory much less familiar than antiquity. Some mapmakers, like Delisle and Hase, took well to medieval subjects. Others were less effective.

Beginnings without Continuations: The History of France

Even talented cartographers like Hase felt more comfortable when working in the earlier medieval centuries (ca. 500–1050) than in the segment we call "High." French history, whose medieval period is normally associated primarily with the Capetian dynasty (967–1314), suffered conspicuously from the inclination toward maps of very early medieval scenes. Not until the French atlas of Rizzi-Zannoni was there a printed map portraying any age later than that of Charlemagne's grandchildren.

The competition to supply the High Middle Ages with a map barely even started until the mid–eighteenth century. Guillaume Delisle, the most talented contestant, qualifies only on an honorary basis since his maps remained unique manuscript sheets. For early depictions of French history, the unexpectedly central figure is the Jesuit Gabriel Daniel (1649–1728), whose map of Charlemagne's Rhine to Danube canal has already been described. The first installment of his *Histoire de France* (1696) aroused fierce objections owing to its critical approach to the ancestors of Clovis. Almost two decades passed before Daniel published the balance of his history, with a revised opening but also with a happy outcome: the complete *Histoire* became a standard work, ranking with the narratives of Mézerai and (later) Hénault. It was reprinted several times.[123]

Daniel's original installment of 1696 contains what seems to be the earliest map of French history. Without identifying words connecting it to Daniel's work, the map is called "Depiction of France in Relation to the Reign of Clovis and His Children" ("Description de la France par rapport au Regne de Clovis et de ses Enfans") and marked "Gravé par Berey." A Paris cataloguer uncertain about its origin and age associated it with Nicolas Berey *père* (†1665); Nicolas, a map colorist, was the father-in-law of Hubert Jaillot. The date of Daniel's *Histoire* implies that the engraver is almost certainly Berey *fils*, whose clients included not only Delisle but also Henri Liébaux, an engraver and mapmaker who eventually had a direct connection to Daniel's history.[124]

Despite references to Clovis and sons, the Daniel/Berey illustration shrinks from actually portraying French history. Gaul is shown at an ostensible date of ca. A.D. 480, distributed among a kingdom of the Visigoths, one of the Burgundians (wrongly called "de Bourgogne"), and a Roman Empire, occupying the north. The map is a "description of France" only by anticipation, since nothing Frankish is discernible, with the possible exception of a "France Germanique" on the east bank of the Rhine. The eastern part of the map is marked "Germanie" and aligns five peoples from north to south. The clearest anticipation of France is a sheaf of provincial names anticipating a distant future, such as Languedoc. The map has an old-fashioned outline of Gaul. It missed by a few years the radical change that the contours of France underwent near the turn of the century, the change that produced the profile familiar to us. Daniel/Berey retains the common seventeenth-century silhouette, awkwardly broad (to modern eyes) and sporting a decidedly drooping Brittany.[125]

Daniel's full *Histoire de France*, published in 1713, did not have its complement of maps changed until the posthumous reprinting of 1729. In the interval, Guillaume Delisle worked for a short time on the beginnings of France. He provided his pupil, young Louis XV, with maps of French history that the market evidently could not supply. None of Delisle's maps for the king was printed. A friendly obituary writer anticipated speedy publication: "[Delisle] had drawn up several maps to serve the History of France; they are bounded (*divisées*) according to the various partitions of the Monarchy, among the descendants of Clovis and among those of Charlemagne. I believe they are absolutely finished and in condition to see the light." Forty years later, Le Long's *Bibliothèque historique* recorded the maps as drafts ("en minute") stored at the Dépôt de la Marine; five are listed by individual titles, as though historians should take them into account even in manuscript. The Paris Archives nationales hold copious, and much consulted, portfolios of Delisle material, including maps in bright colors closely approximating the Le Long titles.[126]

The draft maps of French history among Delisle's papers are not limited to the Le Long five. There is a barely begun map of France under Philip IV the Fair (1285–1314) and another, much more developed "Map for the Reign of Louis XI [1461–83]." A third item is marked in a very neat hand, "Mlle Delisle m'a donné cette carte en avril 1734," the writer presumably being Philippe Buache, Guillaume's posthumous son-in-law. The map in question, of the drooping Brittany type and first saved by the spinster sister—perhaps a piece of Guillaume's juvenilia—is called "Map of the French Kingdom in the 10th Cent., Divided into Its Duchies and Subdivided into Their Counties" ("Carte du Royaume de France au Xe s. divisé en ses Duches et sudivisés en ses comtés"). The rigid system of this title—kingdom/duchies/counties—is of the artificial kind that constructs feudal pyramids. These are archival items, of uncertain attribution and, even if by Guillaume, of value mainly to a biographer.[127] What gives them relevance to our concerns is that they venture into a later segment of French history than the Carolingian dynasty. Maps of the reigns of Philip the Fair and Louis XI did not reach print until Rizzi-Zannoni's atlas, yet cartographers early in the century did give a little thought to the possibility that France after the Carolingians might be portrayed.

Delisle tutored Louis XV in 1718, the year in which he was promoted to full member of the Académie des sciences and given the newly created title of *premier géographe du roi*. Louis emerged from Delisle's teaching with a publication to his credit. The maps to assist his education in French history were finished in bright, lively colors that remind one of a gaily colored teaching atlas done by Louis de Dangeau in the late 1690s.[128]

Delisle's program seems clear: to illustrate the history of the first two dynasties of the kings of France, with particular attention to the divisions of the kingdom. The three maps treating the reigns of the Merovingian kings underscore the practice of inheritance partitions: "France Shared among the Sons of Clovis [511]," "France after the Division of the Sons of Chlothar I [561]," "France Divided at the Death of Dagobert into Neustria and Austrasia

[629]." Delisle's theme reflects a preoccupation with partitions of the Frankish kingdoms shared by many generations of historians when dealing with the
Merovingian dynasty; these partitions are prominent whenever the Merovingians are mapped.[129] According to the list in Le Long, the last map of Delisle's
set was "The Empire of Charlemagne, with the Division of His Grandsons
and Great-Grandsons." A draft answering to this description may have once
existed, but in the surviving papers the "Map of Charlemagne's Empire" is
separate from a "Division of the Empire," apparently portraying the situation
at the death of Lothar I (855) rather than at the more familiar Treaty of Verdun (843). One item seems not to survive from the five listed in Le Long,
namely, the opening "Map of the First Establishment of the Franks (*François*)
in Gaul." Its subject comes closest to overlapping with the printed Daniel /
Berey map.[130]

Delisle's French set stands out for overall quality and systematic unfolding. Its large size is unrivaled in the century; its completeness for the early history of the kingdom was not equaled until the Rizzi-Zannoni atlas.[131] Typically, however, even Delisle did not trace the vicissitudes of France later than
the Carolingians. French history could be begun, especially in its severance
from Roman Gaul, but there was no call for continuing beyond the "second
race," at least in cartography. The limitation of Delisle's unpublished maps
was shared by the few printed maps that, for several decades, monopolized
the field.

The prominent name in this connection is Henri Liébaux, who engraved
many of Delisle's maps and worked alongside him on at least one project. The
British Library has loose copies of the two maps of French history by Liébaux
that Le Long's *Bibliothèque* catalogues, whereas the Bibliothèque nationale has
one. Le Long lists them individually with the date 1728, which the loose copies lack. The maps originated in Daniel's *Histoire de France*. Daniel's introduction to his full *Histoire* describes a single map corresponding precisely to
the illustration engraved by Berey. Daniel died in 1728. The reprinting of the
Histoire one year later has introductory remarks by the publisher, indicating
new illustrations: one map "of France under Charlemagne, and as it was in
the middle times; another for the latest reigns, within two hundred years approximately." The volume introduced in this way contains four maps, of
which three are signed by Liébaux. The first is of current France—early modern, we would say—as specified by the publisher. The others evidently fulfill
his promise about "ces moïens tems," even though Charlemagne's France is
nowhere to be seen.[132] These are the maps found as loose copies in London
and Paris. They also reappear, sometimes improved and reengraved, in the
German and French editions of the *Universal History,* as well as in other col-

lections. While Delisle's teaching aids slept in the Dépôt de la Marine, Liébaux supplied the long-dominant portrayals of medieval France.

Only two maps for Daniel's *Histoire* are of interest here. France for 1713–14 is dispensable, and so is the Daniel/Berey map, surprisingly included despite its no longer being needed; it retains the obsolete geographical outline of France but displays the new, more accurate title "Situation of Gaul Containing the Three Monarchies That Shared It When Clovis Carried Out His Conquest." [133] Liébaux was not content with this improvement. Inhibited from moving briskly forward, he positioned the earlier of his own medieval additions contemporaneously with Daniel/Berey: "Map of Gaul Showing the Dominations to Which It Was Subject When CLOVIS Came to Lay the Foundations of the French Monarchy." The title is almost identical to the renamed Daniel/Berey map, and the contents are a virtual copy. One difference stands out: a heavy line, suggestive of a causeway, runs directly from a locality named Dispargum (on the Upper Weser in Germany) westward across the Rhine, through the Ardennes, and on to Soissons. The line is marked "Route de Clovis" (fig. 21). The Châtelain *Atlas historique* and Hagelgans's picture-history contain precocious barbarian invasion tracks. Here, in Liébaux's development of the Daniel/Berey illustration, is a third example of the practice. Clovis never took such a route; the bold line thrusting into Gaul misinterprets a passage of Gregory of Tours. The designer superimposed an eye-catching feature on the Daniel/Berey core—drama based on a misunderstanding. [134] In this way he rendered Clovis's Gallic conquest graphically memorable.

Somewhat new ground is broken by Liébaux's "Map of France for the End of Clovis's Reign and for the Division of His Territories among His Children." Because Clovis died in his forties, no great leap into the future is involved, but this map hazards, however cautiously, into the Merovingian age instead of pausing apologetically on its threshold. [135] For the twelve hundred years from 511 until 1713, readers of Daniel's narrative were on their own for geographical assistance.

Daniel's *Histoire* circulated widely. In it—or in the German and French editions of the *Universal History* or as loose copies—Liébaux's maps cut a wide swath. For their time they are the sole portrayals of French history. If words rather than images were in question, a possible contender would be Jean-Baptiste Dubos's *Critical History of the Establishment of the Franks in Gaul*—a respected book that retains historiographic importance. Its introductory map, however, cannot rival the Daniel/Liébaux maps. Dubos's design was evidently inspired by Delisle's "Theatrum historicum" of 1705; we are back in ancient history: "Vista of the Gauls and Neighboring Regions toward A.D. 407." Commendable pains are taken to avoid anachronism in depicting

HIBERNIE

Lieues de 25. au degré
5 15 25   50

GRANDE
BRETAGNE

Londres

RIPUAIRES
Tournay
Somme R.   Arras
Cambray
Amiens
Rouen   Noyon
Bauvais
Seine R.   Soisso
Baïeux   Mar
Evreux   PARIS   Meaux
Coutance   Séez   Melun
Avranches   Chartres
PETITE BRETAGNE   EMPIRE
ou   Le Mans   Orleans   Sens
ARMORIQUE   Rennes   Maine
Vannes   Anjou
Nantes   Angers Tours   Amboise
Touraine   Bourges   Ne

OCÉAN   Loire R.   ROYAUME
Poitou
Vouille
Sof
Poitiers
Vienne R.   AQUITAINE
Garonne R.   Saintonge   Limoges   Auvergne
Saintes   Angouleme   Limosin   ou Clerm
Perigueux   Auveron
Perigord   Dordogne R.
Bourdeaux   DE S   Cahors Lot R. M
Rodez
Bazas   Agen   Rouergue
Eause   Garonne R.   Alby
Aire   Auch   Septimanie
Lescar   Tarbe   Toulouse Bezie
Comminges   Carcassone
Monts   Languedoc
Pyrennées   Elne

ESPAGNE

CARTE
DES GAULES
où l'on voit les Dominations ausquelles
elles étoient soumises, lorsque CLOVIS
vint y jetter les Fondemens de la
Monarchie Françoise.

Par HENRI LIÉBAUX Géographe
1728.

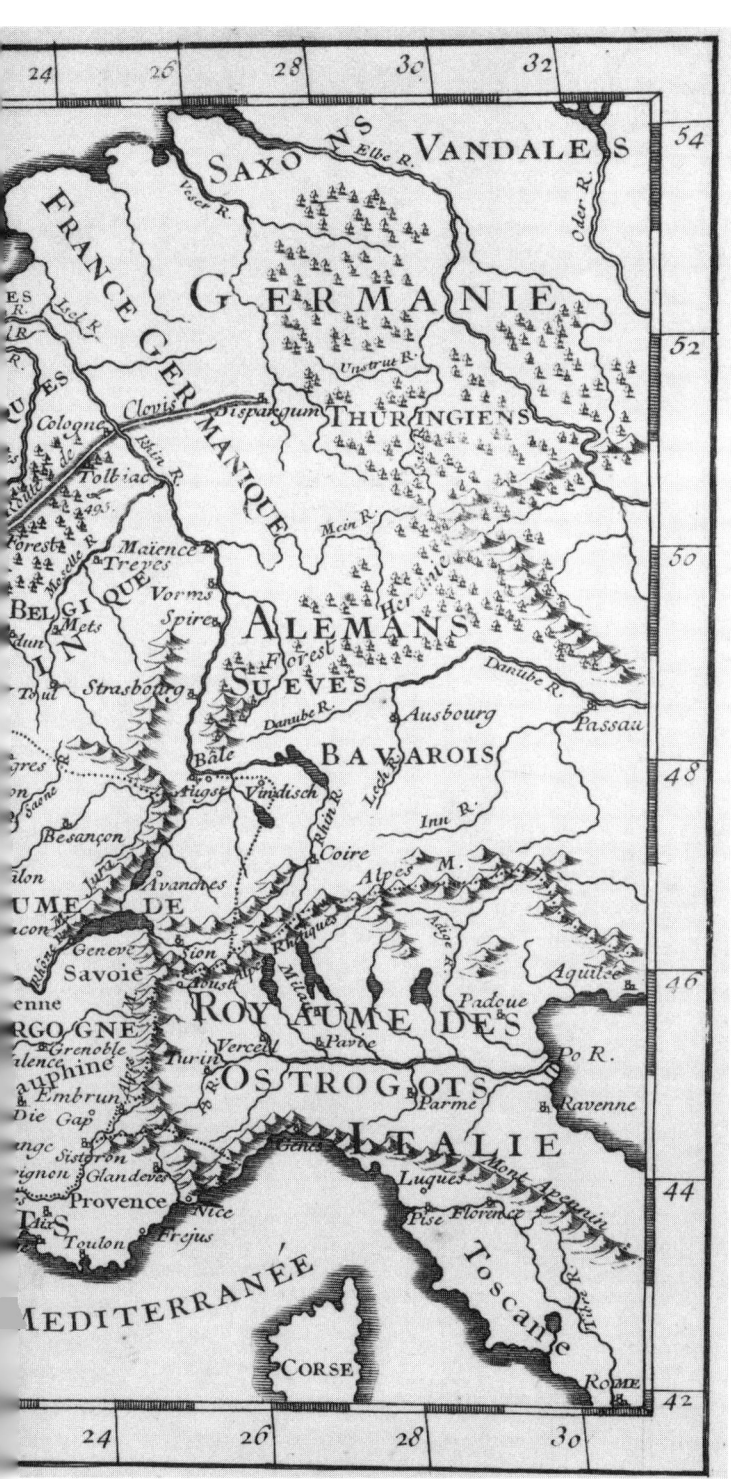

Figure 21 Henri Liébaux, Gaul at the Time of Its Conquest by Clovis (1728). Gabriel Daniel's much appreciated *Histoire de France* long had only one map, showing the situation of Gaul ca. 482. The first posthumous edition of the *Histoire* increased the maps to four, three of them by Henri Liébaux. The one featuring Clovis, shown here, replaces the original map for ca. 482, but with a difference: an emphatic line marks the "route of Clovis" from the edge of Thuringia to the city of Soissons. Beginners might interpret this track as the course of Frankish conquest, but the track simply interprets Clovis's early life as told in the *Histories* of Gregory of Tours. These illustrations of Daniel's narrative were the only maps for French history until the 1760s. (Rare Books/Special Collections, Library of Congress)

the Netherlands, a minor corner of the map. The shores of Frisia and Zeeland are shown in an exemplary reconstruction of late Roman outlines. The heart of the map is less remarkable. Its theme is the early-fifth-century invasion of Gaul that divides the peace of Roman rule from the rapid changes thereafter. Even though an important moment is featured, the resulting image marks no advance in the mapping of medieval France.[136]

Maps for French history improved in quality in the 1740s, but remained as cramped as ever. A collection of the narrative sources of French history, meant to pick up where an earlier effort had foundered, had been under discussion from the 1670s on. Half a century later, the head of the Benedictine monks of Saint-Maur, the congregation long preeminent in French scholarship, accepted the task of editing the early French historians and placed Dom Martin Bouquet in charge of the project. The resulting multivolume *Recueil des historiens des Gaules et de la France*—the first six guided personally by Bouquet—is now antiquated in editing methods but remains the fullest French source collection ever compiled; in the words of David Knowles, "in a sense [it is] the grandmother of all the great national collections."[137] Gilles Robert, no stranger to us, was commissioned to provide Bouquet's collection with large foldout maps; for volume 3, "Status of France under the Kings of the First Dynasty"; and volume 5, "Empire of Charlemagne, Emperor of the West." Here, too, perhaps with better justification, maps were supplied for the beginnings of France but not for the sequel.[138]

Robert's Merovingian map takes its stand toward the close of Clovis's life. Whereas Aquitaine and Novempopulania had been seized from the Visigoth Alaric II, the Burgundian kingdom in the Rhône valley remained unconquered. The underlying divisions are still Roman—the provinces of late antiquity and many *civitas* districts. Robert's table of symbols, sharing the concerns of Delisle and Liébaux, carefully analyzes the partition of 511 among Clovis's sons and that of 561 among Chlothar I's. Although two centuries of Merovingian history followed this date, Bouquet seemed content that they should be left unillustrated.

However limited the coverage was, Bouquet did ask Robert for a Charlemagne map. Neither of them realized in the 1740s that, a decade earlier, a new, careful delineation of the Frankish empire had been carried out in Saxony by Peter Georgisch. Robert must have regarded himself as taking up where Bertius had left off in 1623. Georgisch's work was unknown, probably because enclosed in a book that made no impression. Yet his two maps for the Carolingians are reminiscent of Delisle's unpublished pair. Earlier than Robert's map, they are also preferable as history.

Peter Georgisch (1698–1746) appears to have worked placidly in the civil service of Electoral Saxony. He was eventually named archivist in Dresden and given the rank of privy councilor. As a scholar, he specialized in editions of texts, notably a school version of the early Germanic laws and, more ambitiously, a chronologically ordered calendar of every kind of charter he could secure.[139]

Georgisch's mapmaking was destined for a somewhat different kind of book, namely, *Provisional Introduction to Romano-German History and Geography in Chronological Order, along with Relevant Maps of Antiquity and the Middle Ages.* Of the seven maps accompanying this work, only the last two concern the Middle Ages: "The Merovingico-Carolingian Kingdom and Empire" and "Division of the Kingdom of Lothar of the Year 870 . . . Taking Account of the Earlier Divisions in 843, 855, and 858." Georgisch's mapmaking emphasizes the earlier period; like his counterparts farther west, he abstained from venturing into later medieval centuries.[140]

The Carolingian maps are good, a major step forward from Bertius, even though lacking professional quality and finish. Georgisch's reference to the Merovingians is intended only to connect the two Frankish dynasties; his map is strictly Carolingian, on the premise that Charlemagne built on Merovingian foundations (fig. 22). Bertius's "Imperium Caroli Magni" reached to North Africa and the Black Sea; Georgisch's is more tightly focused. The land outlines are a little odd, as though stretched horizontally. England is blanked out by the cartouche. Dutch waters are given an approximately historical form. Italy is shown as being under Frankish rule down to and including the Duchy of Benevento. Most of all, Charles's empire is carved into relevant provinces, whose boundaries, though not perfect, are respectable reconstructions for the time. Georgisch draws a firm eastern frontier and writes in the names of the Slavic peoples beyond them, without indicating whether they were tributary to the Franks or not—these are foreigners, not a second category of empire beyond the first. Georgisch's map takes a long step in the direction of modern representations of the Carolingian Empire. There is much to improve, but one need not, as with Bertius, start all over again.

The *Versuch einer Einleitung* was not meant for professionals. Georgisch, not a teacher himself, was nevertheless trying to launch a book for schools. Ending his preface, he explains that the last three maps (including the Carolingian ones) include faint outlines of modern provincial divisions so as to orient young readers—a novel and shrewd idea, whatever one might ultimately think of such a superposition of past on present. These lines photograph poorly; they are so discreetly drawn that even in the originals they

Figure 22 Peter Georgisch, "Regnum et imperium Merovingo-Carolingicum" (1732). Georgisch's map appeared in a work that had little circulation: *Versuch einer Einleitung zur römisch-teutschen Historie und Geographie in chronologischer Ordnung nebst zugehörigen Land-Charten der alten u. mittleren Zeiten (Introductory Essay to Roman-German History and Geography in Chronological Order, Together with Relevant Maps of Ancient and Medieval Times)* (Halle, 1732). Unknown to mapmakers (who would have profited from it), this map documents the lack of communication among cartographers. Although Georgisch availed himself of an obsolete geographical outline of France, he paid greater attention to historical detail than Bertius, and his Charlemagne map compares well with the much more monumental version of Gilles Robert de Vaugondy. (Niedersächsische Staats- und Universitätsbibliothek, Göttingen)

remain unobtrusive until looked for. Georgisch was not afraid to take his readers into account and construct maps that might actually involve them.

Georgisch also maps the divisions of the Carolingian realms. Although he builds his depiction on that of Charles's empire, he simplifies many details in order to focus on the main subject of partitions among Charlemagne's descendants. Georgisch emphasized 870 because that agreement (the Treaty of Meersen) came near to establishing the enduring line between the Romano-Germanic Empire and the France of the future. He also tried to mark the intervening divisions from the tripartition of 843 onward; the map takes in the vicissitudes of twenty-seven years. Many later mapmakers combined events in the same way. Such a map, somewhat muddled in appearance, has to be carefully deciphered with help from a written guide. Historians had long recognized that these successive partitions brought into being an important frontier in European history, one basic to many future quarrels between France and the various German regimes. Georgisch translated this awareness into a map. His *Versuch einer Einleitung* circulated almost as little as Delisle's manuscript map of the same phenomenon and was probably less known. It deserved a better fate.

When compared to Georgisch's, Gilles Robert's map of Charlemagne's empire for Dom Bouquet takes us halfway back to Bertius. The geographical outlines differ markedly from those of the seventeenth century and outdo Georgisch as well. Other than that there is little improvement. Bertius's wind rose with the Frankish names for the winds is back, and coloring is applied in such a way as to suggest contemporary Italy, Spain, Germany, and other lands, rather than the divisions of the Carolingian world. Hardly any feature other than the title evokes the year 800. Robert drew a handsome map derived from Bertius's equally attractive model. The professionalism of his design casts a shadow over Georgisch's school-oriented cartography. But history was not Robert's forte here or elsewhere. An enlarged version, with a scientifically improved base map, was issued in 1752 and included in the widely circulated Vaugondy *Atlas universel,* as the item of medieval geography best suited—so says the compiler, Gilles's son Didier—to rouse the curiosity of Frenchmen.[141]

Except in unfinished drafts, mapmakers and their patrons seemed unable to illustrate French history later than the Frankish kings. The third "race"— Hugh Capet and sons—remained an unattainable shadow. France was not alone in being so treated. Everywhere, straying far from the comforting support of ancient geography was frightening. Few except the Germans who dealt with early charters in a regional setting could feel that they had a secure, agreed-upon geography of the Middle Ages to operate in.

Other Countries at Medieval Moments

Thanks to Menso Alting, the Netherlands was better equipped than any other country with historical maps. In 1745, the lackluster firm of Pieter de Hondt in The Hague issued a nine-map *Atlas of Ancient Holland* in Dutch and French editions. According to Koeman, "The maps have been copied from Menso Alting's work. They are newly engraved in a poorish way." Additional work, perhaps only editorial, entered into the atlas. All the titles have been changed, and the "Map of Gaul at the Time When Clovis Made Himself Its Master" derives from Liébaux's illustration for Daniel rather than from Alting's collection. This is one of the two medieval items. The period coinciding with Roman history claims two-thirds of the total, and one item concerns the fifth century. The last of the set shows the Duchy of Frisia as granted by the emperor Louis the Pious to his senior son, Lothar I, in 839. This map, which ventures no farther than the Carolingian age, is more limited than Alting's work, which had dealt with Frisian history as late as the thirteenth-century floods.[142]

In the year of the de Hondt Dutch atlas, the industrious Gilles Robert published maps for Joseph Barre's *Histoire générale de l'Allemagne.* Typically, Robert's "ancient Germany" is on an equal footing with its medieval counterpart, and the latter is pictured at as early a moment as possible. Both maps look old-fashioned, perhaps because they are related to the Nicolas Sanson map hoard in Robert's possession. A Ptolemaic "Magna Germania" is the recognizable background of Robert's "Ancient Germany"; the map focuses narrowly on Europe from the Alps to the Baltic. On the other hand, "Germany [the ancient name, *Germanie,* not the vernacular *Allemagne*] under the Carolingian and Saxon Emperors" is set in a context consisting of the larger part of Europe, taking in all of Spain and reaching the Black Sea.[143]

The design of Robert's map of the early German empire is very unassuming by comparison with portrayals in the next century. He emphasizes regional names rather than larger units. "Grande Bretagne" is marked, but dismissively, as the sole name in the British Isles. In Spain, the Caliphate of Cordoba *may* be delimited, but is not named. The prominent units are districts like Lusitania, Valentia, Asturias, Navarre, Catalonia—all given borders and identified. The Byzantine Empire is visible, at least in part, but anonymous. Instead, we have Greece (and districts), Thrace, Dalmatia, and Bulgaria (marked as though a Byzantine province). The same holds for France and Germany, whose various components are seen at an indeterminate time, part eighth century, part tenth. Italy shows no sign of having been penetrated by Otto I. Correctly understood, Robert's map (like others seen here) does not pretend to depict its subject; it aspires only to be a geographical companion, useful to readers of the accompanying book and of others like it.

Except for Speed's battles map and other catalogues, battles are not featured in the maps of medieval scenes that concern us. There is a single exception, listed by the 1763 catalogue of the Paris dealer Roch-Joseph Julien; he calls it a battle near Basel when Charles VII was king of France. The map was published by the Homann Heirs in 1748. Its own title develops into a heroic description: "Authentic Portrayal of the Battle and Environs of St. Jacob before Basel"; the prose continues: "where on the 26th of August 1444, 1,200 Swiss (*Eidgenossen*) fought from early morning until late evening against 60,000 French, English, and German [troops] led by Louis, son of Charles VII, king of France; and after they had slain more than 8,000 of the enemy, exhausted by victory, all but ten of them fell where they stood."[144] The statistics are shakier than the heroism. The battle of Saint Jacob—known to every schoolchild in today's Switzerland—was part of a virtual civil war in which Zurich stood with the emperor Frederick III in opposition to the Confederacy. Meanwhile, a lull in the Franco-English (Hundred Years') War created military redundancies. Charles VII dearly wished to nudge this dangerous element out of France: Zurich and its imperial ally had use for these troops. A battle took place near Basel, with the Swiss cantons on the defensive. More than 2,000 casualties were inflicted on the French side. Among the Swiss, fully five-sixths of the force engaged was lost—1,268 dead out of 1,500. Although the invaders overcame the Confederates, the war was cut short. Those concerned, especially the French, were deeply impressed by Swiss valor. At Frederick III's court in Nuremberg, the humanist and future pope Aeneas Silvius Piccolomini coined the heroic sentiment about the Confederates that comes to us on the map as their being "exhausted by victory": not defeated so much as overcome by their feat (*nicht sowohl besiegt, als von Siegen entkräftet*).[145]

The map of 1748, pleasingly drawn, shows troop movements as well as the countryside. It occupies half the sheet; the lower half has a keyed "Erklärung" of troop movements and other occurrences.[146]

The circumstances occasioning this attractive image are suggested by its date. During the War of the Austrian Succession (1740–48) and just before that of the Polish succession, the Swiss Confederacy thought its neutrality endangered by the troop movements of the belligerents. Basel was strongly reinforced in 1743; soon after, French forces advanced (across south Germany) to Lake Constance and, in response, the Swiss placed the whole northern frontier in a state of defense. No fighting took place and, in the traditional manner, Swiss recruits flocked by the thousands to all sides in the war. Nevertheless, the possibility that the neutrality of the Confederacy would have to be defended by armed force was thoroughly dramatized. It would not be surprising if these dangers were a favorable moment to remind the Swiss of their vaunted valor. Besides, the third centenary of the battle of Saint Jacob fell in 1744.[147]

The early past of the Netherlands is again commemorated by a map accompanying Jan Wagenaar's *Vaderlandsche Historie*. Wagenaar (1701–73) has a high reputation among early historians of Holland, enhanced in his own time by his being a champion of particularist liberties and opposing the risk of domination by the House of Orange. A constitutional crisis in 1747–48 gave special weight to Wagenaar's history. Part 1 is graced by a portrayal of "Ancient Holland"; part 2 features "Old Map of the Now United Netherlands to Elucidate the *Vaderlandsche Historie* in the Middle Ages." Though no one is named as cartographer, Wagenaar is credited with the underlying design.[148]

Precisely what we are shown as the medieval Netherlands is hard to say. The Zuyder Zee is fully formed. "Friesland" is lettered very large in a diagonal from the Scheldt to the Weser. Counties are numerous and conspicuous, but even more so are three parts of Friesland—a different Friesland from the one just mentioned, extending in its three parts from the Scheldt no farther north than the Lek and Rhine. Each bears the same date, 870, the only date given on the map. The commentators Bodel Nyenhuis and Eekhoff were exasperated: "What does this mean?" The Treaty of Meersen among the Carolingian heirs occurred in 870; Wagenaar seems to have believed it divided southern Frisia into three slices, but Bodel Nyenhuis and Eekhoff refused to agree. They saw that, however original Wagenaar wished to be, his design had been arrived at "with an eye to Alting's maps," from which he borrowed a locality that did not exist. Nevertheless, Wagenaar's map looks impressive; it is well drawn and engraved and free of conspicuous anachronisms (except the considerable one of the Zuyder Zee). The reproof of Bodel Nyenhuis and Eekhoff gains in its universality when one recalls how poorly medieval France was provided with maps at the time.

The London mapmaker Thomas Kitchin has marginal associations with maps for history. Nevertheless, his name as engraver, and none other, is attached to an isolated, small size "Map of the Antient Dominions of the Kings of England in France with some adjacent countries." France is shown as extending as far as the Rhône with, mainly, the boundaries of its principalities from Artois to Roussillon. This design, though unexciting, adds to the repertory of medieval subjects and falls in the "high," or late, part of the Middle Ages that other cartographers shunned. The map is very likely to be a book illustration. Its date prompts one to check Tobias Smollett's *Compleat History of England*, but this guess is unproductive. The year suggests another possibility. There was an "invasion scare" in the months immediately preceding the outbreak of the Seven Years' War. On the French coast energetic preparations were made to launch a cross-Channel assault. Kitchin's map might be a defiant response. But in the absence of some kind of confirmation, this proposal is merely speculation.[149]

Ireland was a more troublesome but enduring English "dominion." Between 1691 and 1745, close to a half million Irishmen are said to have scattered abroad, mainly as recruits in Continental armies. It was to a second generation of "Irish troops in the service of France," and presumably to their French-speaking dependents as well, that, in 1758, James MacGeoghegan, a chaplain in Paris, dedicated a three-volume *Histoire de l'Irlande ancienne et moderne.* One commentator dismisses it as relating secondhand history and adding nothing; another prizes MacGeoghegan's freedom to speak truths that were banned from Ireland at the time.[150] The publisher, Boudet, had close connections to Gilles and Didier Robert de Vaugondy. MacGeoghegan's work is ambitiously illustrated: two comprehensive maps and four regional ones — astonishingly thorough for a history. In drawing them Gilles Robert further extended his historical repertory. The general portrayals are "Ancient Ireland to accompany the history of this country" and "The Kingdom of Ireland . . . 12th to 17th century." The counties of Connaught, Leinster, Munster, and Ulster are then surveyed; each one is said to be "divided into dynasties for the first centuries of Christianity."[151]

The issues occasioning the four regional maps are more than geographical. MacGeoghegan's history pays careful attention, early on, to "the Different Divisions of Ireland" and expounds the connection between current families and the primitive holders of the land. Neither a scholar nor an innovator, he was guided along this risky trail by Roderic O'Flaherty's *Ogygia, seu rerum Hibernicarum chronologia* (1685), which he cites repeatedly.[152] Other sources are also mentioned. Thanks to them, it seemed possible to trace "dynasties" and their territories from pagan times to the present. Gilles Robert plotted the historical details supplied to him.

With a tidy — if unplanned — symmetry, Dutch cartography, which began this section, also ends it. An anonymous author in 1792 availed himself of a map of the Netherlands in Roman times to develop what he called "A Geographic Table of the Middle Ages of Holland, Zeeland, and Friesland." It looks as though special attention is paid to the frontiers of Holland proper and to river mouths. Yet another exercise in impersonating ancient geography as medieval has a soporific effect. Unsigned, undated maps, lacking any indication of where they come from or hope to go, sometimes repay the effort of identifying them. This one looks as though it may not.[153]

HISTORICAL ATLASES

Single maps of medieval subjects and small, related ensembles such as Falcke's for the charters of Corvey or Liébaux's for early French history were necessarily conceived in a more limited way than historical atlases taking account of long spans of time. The eighteenth century opened with Menso Alting's pioneering atlas of the United Netherlands down to the thirteenth century. Fifty years passed before additional national collections of this sort were attempted. In the interval, three atlases paying special attention to the Middle Ages were planned and abandoned. Those of Cellarius and Köhler have already been adequately discussed. They document a dawning interest in the medieval past without offering any improvement in re-creating the past on maps.

Delisle's Unrealized Outline of History

Guillaume Delisle's intended collection of historical maps, forestalled by his early death, anticipated the directions that later atlas-makers, such as Hase, took. How firm Delisle's plans were is not entirely clear. The anonymous source, author of the obituary for Delisle's father, Claude, in the *Mercure de France,* may be Guillaume's talented friend Nicolas Fréret and may have been well informed; but he could convey Guillaume's plans only as they stood at the time of writing. The scheme spelled out in the *Mercure* consists of three parts: a medieval series; a set of "Historical Vistas," mainly classical in content, two of whose maps would do double duty with the first series; finally, a sequence of largely biblical "Vistas of Ecclesiastical History." The preference for three groups, instead of a consolidated one, subordinates chronology to thematic dispersion. The dispersal may reflect the order in which Guillaume wished to proceed, or it may betray commercial considerations: a market more receptive to small collections than to a combined, unified atlas.[154]

The *Mercure* writer was especially gratified that Delisle was taking maps in a medieval direction. He praised Guillaume for already endowing the Middle Ages with two maps of the Byzantine Empire and those of Toul and Dauphiné and drew special attention to the painstaking attention he had paid there to small districts. Further work of this kind was particularly useful inasmuch as maps for the Middle Ages "are far more relevant to the current state of things than are the maps of classical antiquity."[155]

Utility and bearing on the present seem somewhat remote from the subjects Delisle proposed to address in his three-part collection of historical maps. Of the possibly eleven maps of medieval interest, only one concerns

Europe after the Carolingian ninth century. Most of Delisle's attention is turned to Asia. To indicate the unity of conception together with the three projected parts, the classifying letters M, T, and TE are used in the table that follows (M, medieval series; T, historical vistas; TE, vistas of ecclesiastical history). Each entry closes with a brief indication of surviving traces of the map, if any:

> T6.   Historical Vista for A.D. 400 (published)
>
> T7.   Historical Vista for A.D. 600
>
> T9.   Empire of Charlemagne (MS, for Louis XV)
>
> M1.   Empire of Charlemagne Divided among His Sons and Grandsons into France, Germany, and Lorraine (MS, for Louis XV)
>
> M2.   France, Germany, and Lorraine in the Middle Ages
>
> T8.   The Islamic Empire at Its Greatest Extent, 750
>
> M3.   The Geography of "the Arab of Nubia" (i.e., al-Idrisi)
>
> M4, TE5.   Lands of the Levant for Understanding the History of the Crusades (Vertot; ed. J. N. Delisle, 1764)
>
> M5.   Latin empire of Constantinople
>
> M6.   Empire of the Tartars (F. Pétis *père, Histoire du grand Genghizcan*)
>
> M7.   Empire of Tamerlane [156]

The *Mercure* obituary contains unintentional ambiguities. "Charlemagne's Empire Divided" (M1) may have been meant to be merged with "Charlemagne's Empire" (T9); the case for distinguishing them is that two drafts survive in manuscript.[157] "The Islamic Empire at Its Greatest Extent" (T8) is supposed to overlap with a map of the medieval set. "The Geography of 'the Arab of Nubia'" seems meant, but this conclusion is the result of a process of elimination and the need for an overlap rather than from firm evidence. A friend's report of Delisle's plans cannot be assumed to be a reliable version of his final decisions.

Out of the three series listed in the *Mercure,* quite a few items are classical and biblical, and a single one, in the set of "Vistas of Ecclesiastical History," is outrightly modern. Like La Ruë, Delisle meant to end his biblical scenes with current Syria. The medieval program, besides leaning toward the Orient, anticipates the reticence of Köhler and d'Anville when faced with designing a map for the High Middle Ages. Delisle, whose talent for historical maps was exceptional, would probably have depicted "France, Germany, and Lorraine in the Middle Ages" in a way that outdid his successors. He had evidently decided to show the lands that had proceeded from the Carolingian partition of 843. But he, too, preferred origins to continuations.

Charlemagne and the division of his empire, at the start of Delisle's medieval plan, are worthy but unremarkable choices; they are almost obligatory in early medieval history. Among the subjects in this subsection, the "Historical Vista for A.D. 600" is a measure of Delisle's sensitivity to the past. Later atlases, especially after 1800, abound in maps situated in A.D. 500 more or less—not a contemptible selection. Nevertheless, A.D. 500 calls for re-creating conditions soon overturned by Frankish expansion and Justinian's conquests.[158] Delisle's preference for A.D. 600 was similar to his choice of A.D. 400 for his pacesetting "Historical Vista"; he fastened on a moment soon to be followed by marked discontinuity, notably the near collapse of Byzantium, the expansion of the Slavs, and the onset of Islam.

Delisle's six maps of Asia subdivide into two for Islamic lands, two for the Crusades, and two for the Mongols. A chronology of medieval periods is implicit in these: the Islamic pair apparently representing the eighth through eleventh centuries; the sites of the Crusades, the early twelfth to the thirteenth; and the Mongols, the mid-thirteenth to the fifteenth. One assumes that Delisle's goal for these maps would have been primarily geographic, showing relevant frontiers without attempting to illustrate particular historical changes or conquests. Delisle's map of Tartary for Pétis's book and of Crusade lands for Vertot, as well as the larger, posthumously published map of the Levant in the crusading era, all point in this direction. These would be maps to assist the reading of histories, at best including brief commentaries attached to the map edges, as in the published "Vista for A.D. 400."

The traces we have of Delisle's handling of these subjects are rough guides to the major maps he had in mind, whose scope is better represented by fully finished productions, such as the two Byzantine images. What matters more than the map details, since Delisle's atlas was never assembled, is the historical conception guiding the choice of subjects. His Middle Ages are more Asiatic than European. The same is true of all eighteenth-century atlases placing the Middle Ages in a comprehensive setting. The perspective is no doubt European, but the maps reflect narratives of the medieval centuries in which the largest share, especially in what we call the "high" period, is assigned to activities in the East.

The Succession of Great Empires

Johann Matthias Hase (1684–1742) is already known to us. Born into the tolerated Protestant minority of Augsburg, he spent his working life in arch-Lutheran Wittenberg and was a leading associate of the Homann Heirs in Nuremberg. Well regarded in learned circles ever since his Leipzig student days, he had an agonizingly long wait before paid employment came his way. At thirty-five he was appointed to a professorship of mathematics at the University of Wittenberg (1719). Rotating terms as dean of the faculty of philosophy and university rector punctuated his career. He married in 1724 and sired a daughter and four sons.[159]

Hase was involved with maps, even historical maps, well before obtaining the Wittenberg appointment. He published an ambitious work "on the construction of maps of all kinds" in 1717 and equipped a friend's edition of Quintus Curtius's history of Alexander the Great with several illustrations. Cartography competed for his attention with his teaching responsibilities, but it was not wholly unrelated to his mathematical appointment. Mapmaking had an extraordinary place in his life. Nonhistorical maps of the Ottoman Empire, China, Africa, Russia and Tartary, Silesia, the Swabian Circle (of the Empire), the war theater in Bohemia, Asia Minor, and other places proceeded from his worktable, all of them planned with impressive care and critical intensity. Several projects he was much involved in with the Homann Heirs, notably a multisheet map of Germany, absorbed great effort only to prove unrealizable. Most of Hase's maps, like his atlases, appeared posthumously. Recognition and esteem did not wholly elude him in life, but the expenses of his cartography ate into meager resources. In 1731–32, he petitioned the king of Saxony for a pension to supplement his academic stipend and, in 1732, with compliments for "his widely praised assiduity," was granted 100 thalers per year, a modest reward.[160]

Hase cared enough about history both to want and to make major improvements in the maps accompanying historical writings. Nevertheless, cartography had priority over history in his outlook. His article "Notes to His Maps of the Great Empires" (the "difficult and vexing" project that had taken much of his time, rivaling with many others) appeared in the same year as the posthumous omnibus of his maps for history, the earlier mentioned *Atlas historicus*. Throughout the "Notes," practical issues of cartography are stressed. Correct projections, he says, matter even for historical maps, and so do surveying techniques, especially in portraying wide expanses of terrain, such as inner Asia and Russia. Unlike such luminaries as Cellarius, he made his own maps and strove to acquire the latest information. His Caspian Sea was up to date; it was interesting to observe, he said, that its much reformed (non-Ptolemaic) outline agreed remarkably with the data of the Arab cartographers. He was well aware that geographical knowledge constantly increased, but he could not continually disturb the engraver of his drafts and had to leave unchanged some details needing amendment in the light of new information. Hase's comments continue in the same practical vein: his maps are proportionate to the size of the empires they represent; eschewing the convention of coloring the main subject fully and leaving neighboring lands lightly tinted, he has, for the sake of history, colored everything, at no small effort. Why, he asks, did he not adopt the size of standard topographic maps? Because, in such dimensions, his empires, when gathered and bound together, would have formed an irregular, awkward volume. Why does he not have more place-names? Because he did not just lift place-names from a source but precisely located them one by one. Why did the maps include places that did not exist at the same time? "It is easy to see that I made use of a single plate to show many lands that have to be tinted with separate colors; [I did so] simply so that the costs of this already expensive work might be limited." He longs for a rich patron who would finance a historical collection with a separate plate for each scene.[161] Hase eloquently addresses the theoretical and practical problems then involved in designing maps for history.

According to opinions reported (but not documented) by Bonacker, Hase leaned to the historical rather than the geographical aspect of mapmaking: "[Hase's] opinion, far in advance of the times, was that it was inadequate to attach to historical writings any old maps that expressed nothing about the course of historical events. [Historical] maps should rather be designed in such a way that not only the boundaries between individual states but also the course that the peoples had covered might be gathered from them. The changing outlines of states and their development should be presented to the eyes through special maps, serving only this purpose."[162] Menso Alting and Delisle had also been uncommonly aware of what they were doing, but their explanations largely escape us. Hase both reconstructed past geography and articulated his goal and methods.

The "great empires" of his most important atlas evoke an established theme in historiography. Even in pre-Christian times, the idea of organizing the past as a succession of rising and falling empires had had advocates. The scheme is most familiar in the Christian form worked out before 415 by the Spanish priest Orosius and propagated, with variations and amendments, by many successors. Orosius's *Histories against the Pagans* remained enormously influential for more than a millennium.[163] Much happened to historical methods and thinking in the early modern period; nevertheless, several German universities, especially if close to their religious roots, continued to cultivate the Orosian scheme. Its central idea was that, throughout historical time, God had chosen an empire as his instrument of social order on earth, fostering it for some time, until, for sufficient reason, he chose another more capable and worthier for the same task. For Orosius and most of his descendants, God's final choice was the Roman Empire, into which the Son of God was incarnated; it was destined to last till the end of the world. Thanks to Charlemagne's revival of a Roman empire and its rarely interrupted continuation, the Orosian scheme held its own through the Middle Ages. It retained credit in Hase's day and was not yet wholly obsolete.[164]

Despite the currency of such notions, Hase chose empires in such a way as to suggest that no theological scheme controlled his atlas. His selection includes entities alien to Orosian specifications, such as the Parthians and the Sassanian Persians, as well as the Ottomans, Russians, and Chinese.[165] Hase's goal seems to have been to focus on the profane "great powers," as they would later be called in a narrowly European context. Hase's basis for choice was not a fascination with dominators. He needed a framework for the "universal political history" (Historia universalis politica) announced by his general title. The empires appeared with "their more important modifications." Although the atlas was published posthumously by Hase's pupil A. G. Böhme, its plan had been laid out in a university address in 1728. It was, as Hase said, a difficult and vexing project.[166]

Delisle (only two years dead when Hase outlined his project) had planned for his Asiatic maps to take up six out of eleven medieval items in his threefold collection. Hase's ratio, if the synthetic items 19–20 are disqualified, is eight of eleven. The "Arab Empire" alone, a newer subject to historical maps than the Mongols, accounts for five, the largest single group. Hase, who never expresses special concern for the Middle Ages, produced the collection in which for the first time the medieval centuries obtained slightly more than equal billing with classical antiquity.[167]

11. The Roman Empire under Justinian ca. 565
12. The Arab Empire (early Abbasid Caliphate)
13. The eastern part of the Arab Empire
14. The western part of the Arab Empire
15. The Arab Empire at the time of the Crusades
16. The Arab Empire (western North Africa) under the Almoravids
17. The (Frankish or Romano-Germanic) Empire of Charlemagne
18. The Empire of Otto I and Conrad II (962–1031)
[19. All the parts once attached to the Frankish or Romano-Germanic Empire—west
20. All the parts once attached to the Frankish or Romano-Germanic Empire—east]
22. The Mongol Empire or Ghengis Khan and successors ca. 1257
23. The Empire of Timur Beg (Tamerlane) ca. 1405
25. The Empire of the Ottoman Turks ca. 1453.[168]

Hase was blamed for not attaining the uniformity of scale and orientation that, ideally, atlases should have. Those following in his tracks took pains to conform to the ideal but were able to do so only by adopting dimensions that sacrificed convenience of consultation.[169]

A remarkable proportion of Hase's medieval items are new designs. Charlemagne's empire had been mapped by Bertius and, recently, by Georgisch (whose maps Hase may not have known). Gilles Robert's Charlemagne, of the same year as Hase's small atlas, is by definition independent. Hase's Empire of Otto and Conrad is also independent of Robert's "Empire des empereurs carlovingiens et ottoniens" (1745); their subjects only roughly overlap. Where the Mongols and Tamerlane are concerned, Delisle and Nolin have priority, but Hase's maps represent a leap in quality and historical articulation. Byzantium, pioneered by Delisle, is a noticeable absentee; Hase must have considered it not to be a *summum imperium*. For seven subjects, beginning with the Empire of Justinian, he has a completely clear field. The "parts pertaining to the Romano-Germanic Empire" (the bracketed nos. 19–20), original as well, are not historical in the same sense as the others. Do they express patriotic irredentism? These maps seem comparable to those of a Gaul larger than France that, in the not very distant past, had anticipated "reunions" to the French kingdom of imperial territories such as Alsace.

Hase shared the embarrassment of Delisle, Köhler, and d'Anville when faced with the "High" Middle Ages in Europe. The prominence he accords to Muslims, Mongols, and Ottoman Turks stems only in part from a pondered appreciation of their place in history. A mapmaker needs subjects that lend themselves to the means at his disposal; Hase's craft was best served by prominently shifting boundaries. Muslim states could be grasped as imposing blocks, the vast eighth-century caliphate fragmenting into large pieces. Islam may have been emphasized as much because it offered Hase preferable subjects for small-scale mapping as because he specially valued Islamic history.

His partiality for Muslim and Asiatic affairs did not imply blanking out all other lands. The maps centering on the Islamic world reach north to the Loire and into much of central Europe, whose vicissitudes are not ignored. More of even "high" medieval Europe is dealt with here, though on the margins, than had been before.

The *Greatest Empires* was not the only work in which Hase dealt with the Middle Ages. His more sustained treatment, not avoiding "high" centuries, occurs in the *Seven Maps to Illustrate as Many Periods of the History of the German Empire*. In what year Hase finished this series, published only in the *Atlas historicus* of 1750, is a perplexing question, unanswered here.[170] Two of the seven "periods" are modern (Charles V, 1556, and Charles VI, †1740). The others offer a fair sample of the medieval centuries: Charlemagne's empire in 851, after the death of his grandson Lothar I; the empire at the death of Otto I (973), next at that of Conrad II (1031), then Frederick II's (1249), finally Frederick III's (1493). The duplication or overlap with the *Greatest Empires* is largely outweighed by its larger size, greater detail, and timely clarification.[171]

The Romano-Germanic set, unlike the *Imperia maxima,* develops out of a single base map, almost twice as large as those of the small atlas. All place-names are present and fixed from the start. What changes is the coloring. A very prominent legend, to the right of each map and occupying one-third of the total framed space, identifies the colors and is detailed enough to be almost a commentary. The historical circumstances on the maps are identified and color coded; anyone keeping track of the mutations in lines, tints, and words is exposed to a narrative of a special kind complete with the immediacy of visual impression.

This effect is not accidental. Hase maintained that maps should be drawn to present to the eyes "the changing outlines of states and their development"; they would do this if truly historical. To achieve such an effect in a single frame is close to impossible. Almost all the maps of the small atlas are in sets, some of three, some of as many as five. The comparison of identical geographical forms in their changing coloring and political configurations was the reader's task, his main way of profiting from the mapmaker's efforts.[172]

The *Seven Maps* carried this program to perfection. The fixed background leaves the reader undistracted, able to focus from a stable platform exclusively on the changing lines and colors. Hase has a story for the reader, one that filters, if not leaps, to the eye. We are given to understand that the empire was at its maximum extent under Charlemagne, after whom the first blow to its fabric was dealt by his grandsons. The tale of shrinkage, not necessarily dramatic but nevertheless plain, carries on thereafter. Flat colors covering the whole of a territory signify lands integrated in the empire; they degenerate into border colors, merely highlighting frontiers and leaving the interior transparently neutral; the legends spell out a loosening sequence of subjection, clientage, independence. The story is not catastrophic; it would not lend itself well to Hase's design if it were. Almost a millennium after Charlemagne, much remained for Charles VI to be emperor of. But there had been gradual transformations of a kind that could be expressed in graphic form. Hase realized his ambition to present historical change to the eye. With Alting and Lamare, he tried to dramatize history in maps rather than just offer a geographical companion to the written word.

No great leap of the imagination is needed, when looking at Hase's small atlas, to see in it a direct forerunner of the historical atlases to come, even ones of the later twentieth century like the *Anchor Atlas* of Kinder and Hilgemann. One reason is the small, handy size of his maps: not immediately influential, they eventually became (for better or worse) the basic format of school texts. The program of the "Ancient Atlas, Classical and Sacred," though not abandoned, is treated with cartographic rigor and incorporated in a scheme of universal history from the origins to the present, a scheme wholly new by comparison with the collections of "ancient geography" descending from the *Parergon*. Hase was noticed and praised; he may not have been praised enough.

World Chronicles in Maps

Hase was a point of departure. The creative force in historical mapping for several decades, first in Paris, then in Germany, was universal history in sequence, so designed as to convey historical change in a visual way; or so it was hoped. This format was more closely anticipated by Hase's *Seven Maps* than by his *Greatest Empires*. About one decade after the Hase/Homann *Atlas historicus*, a reviewer for the *Journal des sçavans* drew a balance sheet: "[Hase's] maps [for the general history of the world], too few and too small, present only the main epochs. . . . [His] project was very well received, and several persons have aspired to imitate it in greater detail, but these undertakings have not been carried to complete perfection."[173] The details laid out are scanty, but the reviewer knew what he was talking about. Two sizable traces survive of these "imitations of Hase in greater detail." One is the sixty-six-sheet single-copy atlas at the Bibliothèque nationale, of unknown date; the second, called *Les révolutions de l'univers*, bears the pseudonym Dupré and was published by Lattré.

The Paris draft might be called a "manuscript" if it did not resemble all other atlases of its time in combining an engraved plate with hand coloring; it differs from comparable atlases by having clear traces of tentativeness and never being published. Its unfinished state testifies to someone's dissatisfaction. Yet the incomplete work impresses by its monumentality and brave ambition. Hase's small, particular maps intimate their (near) universal scope only by cumulative effect. The Paris draft pioneers a base map of Eurasian breadth announcing universality in an emphatic, evident way. That base map, a forceful step beyond Hase, is adopted with variations in several later atlases. The unchanging background in the Paris draft, however anachronistic in projecting the borders of modern states backward into the 1000s B.C., has the same use as Hase's *Seven Maps:* it draws attention away from the fixed elements and toward the shifting colored lines.

The draft atlas assigns thirty-two double sheets to the medieval period (assuming, for this purpose, that the period begins in A.D. 407–19 and ends in 1483–1521). The historical coverage forms an ostensibly continuous series, the time span of each map leading into the next. This system works most of the time. There is a forty-year gap between sections 33 and 34, five missing years (almost trifling) between 38 and 39, thirty-seven between 53 and 54, and about thirty between 55 and 56. Conversely, sections 52 and 53 largely overlap in time. The exceptions are too sporadic to suggest a plan. The irregularities in continuity pale by comparison with the long periods chosen for two sections: almost the whole twelfth century, 1102–93, is assigned to the single section 55, and an even longer span, 1228–1358, belongs to section 58. Most centuries are normally allotted about three sections, and the ninth century has no fewer than seven to itself. The anomalously long durations crammed into sections 55 and 58 suggests that the anonymous designer also had problems with the High Middle Ages.

However eloquent the maps were expected to be, especially when compared to each other, the compiler realized that explanations were needed; each section is headed by "Observations." A general introduction six folios long, now unfortunately missing, may have spelled out his intentions.[174] The "Observations" come as pointers to the important features of each map and also as intermittent listings of "New States"; whole maps are occasionally reserved for new states, rather than time periods. One example of "Observations" is representative, though it does not speak for all:

> 56th Division, addressed to the time that passed from the year 1226, period of the first conquests of Genghis Khan, prince of the Tatars, up to and not including the division of his empire carried out in the year 1228. This map differs from the foregoing in that one sees in it . . . the conquests of Genghis Khan, prince of the Tatars, who united to his empire Persia, the terra firma of India or rather the peninsula beyond the Ganges in our direction, and Shensi, one of the western provinces of the kingdom of China.

> New States [1187–1218]

> The kingdom of Jerusalem united to Egypt 1187. The kingdom of Bulgaria. The kingdom of Cyprus 1191. Khorasan united to Dergigian. The principality of the Seljuk Turks of Iran united to that of the Mongols 1201. The Empire of Nicaea and Adrianople 1204. Principality and empire of Trebizond. Principality of Thessalonica. Duchy of Athens. Principality of Achaia. Principality of Epirus and Albania. Union of Lower Navarre to the kingdom of Navarre. Union of the kingdom of the Karaites to that of the Mongols 1205. Union of

the principality of Akhlat to the kingdom of Egypt 1206. Duchy of Naxos and
the islands of the Archipelago 1207. Union of the kingdom of Si-hia to the empire of the Mongols 1209. Union of Tangut to the same. Kingdom of Tunis
1210. Kingdom of Granada 1212. Union of the kingdom of Persia to that of
Khwarizm. Union of the kingdom of Kipchak to the empire of the Mongols.
The kingdoms of Estremadura and Andalusia. Union of the kingdom of Leon
and that of Castile. Union of the lordship of Ravenna to the States of the
Church 1218. Principality of Mesopotamia 1218.

The author's intention is that map no. 56 should be about the growth of the
empire of Genghis Khan. The list of new states invites readers to observe
other changes on the map, changes that, as it happens, fall within a gap in coverage. But the contents of the list make the gap look weightier than a count of
years would suggest. There are minor events, to be sure, and total inclusiveness should not be expected. Still, the names beginning with the kingdom of
Bulgaria and the Empire of Nicaea and continuing to the principality of
Epirus and Albania (plus the duchy of Naxos) collectively imply the adven-
tures of the Angelus dynasty in Byzantium, the Fourth Crusade, and the ensuing partitions of the empire. Most histories would not treat these as incidental happenings. Another aspect of the "new states" leaves in doubt how
the unions said to be "to the empire of the Mongols" relate to the main subject. The scale of values seems askew. However virtuous brevity might be,
more features of the maps need explanation than get it. The observer/reader,
unless otherwise knowledgeable, is forced to resort to historical handbooks,
and the atlas returns to being a companion to reading.

An atlas presenting the world as a whole, or almost, creates problems for
itself that are avoided in a more fragmented format. A succession of maps like
that of the Paris draft, all reaching from Spain to Korea, implies a synchronic
world history, one in which, for example, Scandinavian or Moroccan happenings compete for attention with Chinese or Mongolian ones. The author's
"Observations to the 56th Division" shows his response to this problem: he
singles out one, presumably dominant, "revolution" and relegates other
changes to a catalogue. By comparison, the prototypical world chronicle of
Eusebius arrayed its information in parallel columns, one each for the states
involved; and the collaborative English *Universal History,* begun in 1730, dealt
with countries one by one, much like the Cambridge medieval and modern
histories familiar to us. The universal coverage offered by narratives was
formed of separate parts.[175] The attempt of the Paris atlas to show world history as a whole was daring.

It was also an explorer's foray into territory prudently avoided by most historians. The taking of Constantinople by the Fourth Crusade in 1204 struggles to be noticed if its gravity has to be weighed against Genghis Khan's unification of the Mongol tribes—or vice versa. Oftentimes, too, an interval of general history proves barren of excitement, so that the featured "revolution," needed to keep the maps going, seems less than world shaking. An example in the Paris atlas involves the union of Sweden and Denmark (62: A.D. 1404–83) and their divorce (63: A.D. 1483–1521)—events apt to be profoundly uninteresting to most readers. The Paris draft was an effort to do better than Hase, an ambitious experiment more memorable for the exhaustiveness and uniformity it aspired to than for the ways it went wrong. Its main defect perhaps was that mapmaking outran history. A massive succession of enormous maps could be and was differentiated by colored boundaries, but no one apparently asked whether narratives existed that these maps could complement.

A reviewer early in the next century explained that historical atlases called for many maps in chronological order; he considered this to be the standard form of such collections.[176] He was not thinking of venerable *antiqui* such as Ortelius, Briet, or Cellarius. Hase, especially in the *Seven Maps,* had taken a large step toward what would be the new standard. The Paris draft turned Hase's initiative into a system. Closely in its tracks (after a few years) came the atlas signed by a veiled "Dupré," using a map signed by an otherwise unknown Michel Picaud of Nantes. This collaboration succeeded, where its forerunner had not, in crafting a product that Lattré, a leading Paris map specialist, was willing to publish in 1763 and reprint a decade later.

The Dupré atlas is somewhat less than half the length of the Paris draft, a likely reason for its greater commercial promise; and it is worked out on a more attractive and better engraved base map, on a smaller scale.[177] Modern frontiers are faintly traced on the unchanging background but not colored until they become "real." Dupré's printed commentary, glued to one edge and also available as a small separate book, is more circumstantial than the sketchy "Observations" of its forerunner. Dupré takes the unusual step of citing sources. His research is not exhaustive; Köhler and Menso Alting are unknown but offset by Delisle, Hase, Robert, and a few others. Reviewers considered the fullness of Dupré's coverage of the Orient a major asset. Delisle's "Theatrum" had the defect, a reviewer said, of reaching only to 98° east longitude; Hase, too, did not push far enough to the east. According to Dupré, de Guignes's recent writings on eastern peoples had facilitated this aspect of his work.[178]

The Middle Ages were bound to be important in a chronological sequence whose years beckoned to be completely filled in. Dupré had a traditional idea of the course of events: the Roman Empire, like previous empires, perished under its own weight; "Several barbarian peoples, after having long troubled the borders, penetrated the provinces and . . . it is in this epoch that it is customary to place the Geography of the Middle Ages . . . [; it is] ordinarily carried forward to the mid–fifteenth century, that is, to the capture of Constantinople by the Turks and the discovery of America." "Geographers have very much worked over the Roman Empire . . . ; they have too much neglected the geography of the times following the ruin of this empire." Dupré allots a little more than half his maps to medieval geography—sixteen "intervals" out of thirty; the sixteen cover as few as 13 years and as many as 135.[179]

This special regard for medieval geography is carried over to the maps in the same way as in the Paris draft. The salient trait of both is homogeneity or evenhandedness; every epoch, from biblical to modern, looks the same, has the same visible features, and sounds in commentary like every other. There is much to be said in principle for such an approach. What one experiences, however, is not so much a laudably egalitarian past as the jolt of a history in which events from Korea to the Pyrenees lose any distinctiveness by stewing in the same pot.[180] The flavor of this sameness needs sampling:

> [Sample, Classical Antiquity] VIII. *From the changes that took place in the Seleucid Kingdom up to the conquests of the Huns over the Yve-Chi or Getae.*
>
> The Kingdoms of Paphlagonia, of Pergamum, of Galatia, of Bactria, of the Parthians, of Armenia Minor, and of Judea took shape within the bounds of the Kingdom of Syria, and greatly diminished its breadth. While the Seleucid Empire was truncated, the Romans, who had taken nearly five centuries to become masters of Italy and Sicily, conquered the island of Corcyra [Corfu], New Epirus, and Macedonia. In China, Stam-Vam, king of Tsin, conquered the state of the Tche-ou; his successors seized the kingdoms of the Han, of the Goei, of Tçou, of Hien, of Tchao, of Tien-tçi, of the Mangi or southern China, except for the provinces of Quang-tung and of Quang-si, which submitted to the Chinese only under the reign of Venti, toward A.D. 177. The history of the Huns since 210 B.C. tells us that, at the time, Tartary was possessed by six main peoples, namely, the Topa, the Huns, the Yve-chi, the Ou-siun, the Su, and the Tim-lim. Various other states of these times are known to us by their connection with Roman history; such are those of the Japydes and the Liburnians, the kingdom of the Dardanians, that of the Gauls of Thrace, those of Mauritania, of the Massaesylian Numidians, of the Dalmatians, of the Cottian Alps, and the Republic of the Lycians and of the Carians. The course of the period also includes the expeditions of Pyrrhus and of Hannibal in Italy.

[Sample, Middle Ages, 1329–79] XXVI. *From the establishment of several new states in the kingdoms of Iran and of Zagathay up to the conquests of Tamerlane.*

In addition to the divisions of the Empire of Genghis Khan already mentioned, there arise within it the principalities of Kirman, of Segesta, of the Mazauderan, of Lorestan, of Giorgian, of a part of Khorasan: these were established in Iran. A new kingdom of Kwarismi, and the principalities of Kerch and of Balkh, took shape in Zagathay. China was removed from the main branch of the Mongols and took the name of kingdom of Kalkas. Toward the same time, Little Bokharia and Tibet regained their independence. Also visible are the birth of a kingdom of Bulgaria in Asia and the kingdom of Schirvan. The sultans of Egypt extend their domination as far as Cilicia. In Europe France loses Navarre and gains Dauphiné. The Kingdom of Majorca is united to that of Aragon, and Viscaya passes to the king of Castile. Wallachia and Moldavia wrest their independence from the Hungarians. Casimir, king of Poland, subjects the Russians of Halicz.

[Sample, Modern Times, 1498–1640] XXIX. *From the conquests of Sçhaïbek to the reestablishment of the Portuguese monarchy by John, duke of Braganza.*

Sçhaïbek, descendant of Genghis Khan, is seen snuffing out the power of the Timurids and seizing Andekan. Babar, who reigned in the latter principality, flees to India, conquers the Kingdom of Delhi, and founds the Moghul Empire. The Sophis, founders of the Kingdom of Persia, add Ghila and the island of Hormuz to their principality of Azerbaijan. Cochinchina separates from Tokin. Selim seizes Aladulia and dispossesses the Mamluks of Egypt. John IV, prince of the Russians, conquers the Kingdoms of Kazan and Astrakhan; his successors become masters of Touran or Siberia, extend their domination to the Glacial Sea and to Kamtchatka. The Nogaïs Tartars submit to the khans of Astrakan; the principality of the Circassians subsists as a Russian dependency. Poland acquires Lithuania, Volhina, Samogina, and Livonia. Aragon, separated from Castile, is reunited to it twelve years later; while separate, it seizes Naples and the part of Navarre south of the Pyrenees; the part north of the mountains becomes French at the accession of Henry IV. Philip II takes possession of Portugal. Bologna is reunited to the Holy See; Parma and Piacenza are separated from it. The Medici bring the Republics of Siena and Lucca under the power of the Florentines. The Frisians submit to Charles V. The Netherlands throw off the yoke of Spanish power. Scotland and England unite in a single kingdom. Geneva becomes a republic. Lorraine is declared independent of the empire of Germany. The Swedes detach themselves from Denmark. Transylvania separates from Hungary. In Africa, the Sherifs of Morocco take the Kingdom of Fez from the Oatazes; the Kingdom of Tafilet

> revolts and thirty-seven years later takes the name of Kingdom of Morocco. The Spaniards conquer Mexico and Peru.[181]

A well-known jibe at medieval chronicles is that they relate "just one damned thing after the other." Dupré outdoes the chroniclers. Every now and then, knowledgeable readers can pick up a recognizable thread, but much of what they are told, and see on the maps, is equivalent to the popping up of "new states" in the Paris draft: a few or many of the entities present in the previous period are wiped away, and another set materializes. Succession is all. On this modest, thoroughly egalitarian basis, Tibet blends in well enough with Navarre.

The main contemporary reviewer did not object: "We know nothing more useful [than this atlas] for the study of Universal History; the facts are made perceptible and, provided one gains familiarity with this map, imprint themselves on the memory without confusion"; the atlas merits the serious attention of "everyone wishing to fasten the most important events (*faits*) to their memory."[182] Dupré's work is not spoken of as an independent form of historical exposition; it earns praise as a study aid, teaching educated readers how to distinguish places and times and situate salient events among them.

Dupré, while disclaiming the intention of writing history, thought he was doing something more than supplying a study aid. He is at his most reflective in the comments to the final Interval. His goal has been to furnish a precise enumeration of revolutions. External ones have been preferred because they, unlike internal revolutions, can be expressed in visual form. Dupré articulated the limitations of historical maps as then conceived. Maps could be set in motion by little else than historical events that altered frontiers—the one "speaking" variable, but one not guaranteed to excel in importance. Atlases designed like Dupré's and its predecessors illustrated the onset, swelling, and contraction of states, great and small; "revolutions" so conceived were their repertory. This limitation of the genre has still not been wholly overcome.[183]

The faults of the Paris draft and the Dupré atlas were not obvious to contemporaries. Universal history was prized, and there was reason to think that a novel method had been found for expressing this difficult subject and imprinting it indelibly on the memory. Chronologically successive maps seemed a sound method. What to put on them was less certain.

The History of France Graphically Unfolded

One may guess, though there is no evidence, that the cluster of historical atlases of France published by Louis Charles Desnos in the mid-1760s was inspired by the example of Dupré or the Paris draft. Reviewers saw the connection; it could hardly have escaped the compiler or compilers. In a promising modification, the multimap format reserved up to now for "general" history was adapted to a single country and applied to a "Tableau varié des Révolutions qu'a subies la Monarchie Française."

The Desnos atlases trace, in boundaries and colors, the shifting development of the kingdom in thirty-six intervals. Owing to the limited quarto size, there are no subsidiary insets; whatever information is supplied takes up a full map. From the first (or 1764) state of the plates to the second, many details were changed for the better thanks to someone's study or knowledge of French history. Whether this was Rizzi-Zannoni or someone else, the knowledge was gained in the library of Gaspard Moise de Fontanieu, an official of high rank, a charter collector, and the author of unpublished books on aspects of French history.[184]

Giovanni Antonio Rizzi-Zannoni was too fully engaged in a variety of projects to devote lengthy research to this one. A designer of exceptional facility and skill, he was well chosen to execute those parts of the maps that went beyond the reiterated outlines of France. Rizzi-Zannoni speaks of doing the atlas at the request of Fontanieu, whose contribution must have been considerable.[185]

Within the next decade, Rizzi-Zannoni worked twice for distinguished patrons who, personally or through agents, assembled the materials for major maps. The Neapolitan abbé Ferdinand Galiani engaged him to produce an ambitious map of the Kingdom of the Two Sicilies (1769); and Prince J. A. Jabłonowsky, Palatine of Nowogródek, supplied him with the information out of which came a twenty-four-sheet map of Poland (1772). Rizzi-Zannoni signed his name to both, without complaint from either sponsor that he had taken undue credit (later, Galiani was midwife to Rizzi-Zannoni's appointment as cartographer of the Neapolitan kingdom). Rizzi-Zannoni had good claim to featuring his name. Galiani and Jabłonowsky had the resources and means to gather materials but needed the expertise and craftsmanship of a trained cartographer to turn their material into finished maps. Whoever signed the finished product incurred the risk of a poor critical reception. Presumably, Fontanieu's relations to Rizzi-Zannoni were similar.[186]

The Rizzi-Zannoni atlas, belittled or ignored initiator of the myriad atlases of French history, distributes its sixty maps among twenty-four items concerned exclusively with the royal domain and thirty-six addressing the main theme. Medieval France takes up a generous share of the total and, for a change, more maps are devoted to the later Middle Ages than to the earlier centuries (see table).

1. État de la Gaule sous Pharamond et ses successeurs [later Roman Empire]
2. Établissement de la monarchie française dans la Gaule transalpine sous le règne de Clovis et de ses Enfants
3. Carte pour l'histoire de France au VIIᵉ siècle
4. France au VIIᵉ siècle [in fact, to the end of the Merovingians, 751]
5. Empire de Charlemagne, VIIIᵉ siècle
6. Couronne de France (Louis I à Odon)

7. France sous Charles IV, Raoul, Louis IV . . .
8. France sous les 1ᵉʳ Capétiens

9. Conquêtes britanniques en Normandie et en Aquitaine, XIIᵉ siècle
10. France sous Philippe-Auguste

11. Monarchie françoise sous Louis VIII et St. Louis
12. France sous Philippe III

13. France sous Philippe IV

14. France au début du 14ᵉ siècle
15. France sous les Valois jusqu'à 1356 (Poitiers)

16. Sous Jean II and Charles V (jusqu'à 1380)
17. France sous Charles VI (jusqu'à 1422)

18. France sous Charles VII

19. France vers la fin du XVᵉ siècle

1. Tableau du Domaine royal isolé de toutes les souverainetés qui subsistoient après la mort de Charlemagne jusqu'en 898
2. Tableau du domaine royal à la fin de la 2ᵉ race
3. Domaine royal sous Philippe I aggrandi par plusieurs conquêtes
4. Pertes et acquisitions du Domaine sous Louis VII
5. Accroissement du Domaine sous Philippe-Auguste
6. Le Domaine de la couronne s'accroit considérablement sous St. Louis
7. Tableau du Domaine Royal augmenté par la réunion du Marquisat de Provence, des Comtés de Toulouse, d'Alençon, et de Chartres sous Philippe III
8. État du Domaine sous Philippe le Bel accrû de la Navarre, Angoumois, Marche, Lionnois, et partie de la Brie
9. Le Domaine sous Charles IV
10. Le Dauphiné et la Champagne sont unis au Domaine sous Philippe de Valois
11. Élargissement prodigieux du Domaine sous Charles V
12. Le Domaine diminue à cause des invasions des Anglais
13. Après la bataille de Formigni le Domaine royal s'accroit de nouveau par la réunion de toute l'Isle de France, de la Normandie, du Valentinois et d'une grande partie d'Aquitaine sous Charles VII
14. La Provence et la Bourgogne sont incorporés au Domaine Royal sous Louis XI

French history was eminently mappable. Frontiers were numerous, and their frequent reshuffling provided the draftsman with the opportunities he needed. Even Hase's atlas of the German Empire (*Seven Maps*), with its larger dimensions, did not attempt to show the complexity of medieval German territories. Rizzi-Zannoni's display of major French principalities initiates the mapping of medieval fragmentation, sometimes considered characteristic of the epoch. The relatively large units of the Merovingian and Carolingian periods give way to a patchwork of large and small dominations that not only undergo rearrangements but also, little by little, fall to the Crown and are incorporated to its sphere of direct rule, the Domaine. The scale of Rizzi-Zannoni's atlas is too small to allow precision in mapping even the major principalities, but it offers a suggestive approximation. In a unique variation, the map relating to the earliest Capetian kings salutes William of Normandy's conquest of England by drawing a few ships in the Channel, some apparently sinking. Adrien Hubert Brué, Rizzi-Zannoni's unappreciative successor, followed his lead in allotting a small illustration to this event.[187]

France had established heroes, villains, and themes; its past was well-plowed territory. Gabriel Daniel had dared to contest the proto-Frankish king Pharamond; Pharamond is present in Rizzi-Zannoni/Desnos, even if only as a pretext for a map of late Roman Gaul. Everyone agreed that the divorce of Louis VII from Eleanor of Aquitaine was deplorable because, as a result, much land was lost to the monarchy. This view is endorsed here as in later atlases. Well into the twentieth century, the growth of the Domaine supplies French history, particularly in the Middle Ages, with its triumphal, ascending theme.[188] It was as well suited to graphic presentation as to expression in words.

The Rizzi-Zannoni/Desnos atlas is not markedly inferior to its successors in historical content. It can even be praised for avoiding the morbid fascination of later collections with the successive divisions of the Merovingian kingdom. The thematic maps of the Domaine, identical in outline, plain and unequivocal in substance, are creditable. The longer group errs by prettiness. The interior of France is graced by movable forests and mountains. Only waters stay put. The borders and cartouches have the graceful Louis XV map style of the Desnos firm, which sometimes wins praise. Was it well suited, even then, for a teaching manual? Rizzi-Zannoni's French atlas benefited from Fontanieu's solicitude. Its subject was dealt with as competently as any other contemporary national collection and much more fully. For more than fifty years it had the market to itself. It merited a better reception than the ambivalent one it got. Might its elegance have detracted from the subject?

A Geography of the Migration Age

The origins and disappearance of Gatterer's historical atlas have already been discussed. Although this work and its prose accompaniment first became available in 1776, near the time when Gatterer published *Abriss der Geographie,* the date found on the surviving map relates to a third edition of 1789.[189] Bredow, Kruse, Woltersdorf, and the reviewers of the Göttingen *Historisches Journal* were familiar with Gatterer's *Geography of the Völkerwanderung;* yet all sheets but one have vanished from sight. The existence of Gatterer's maps is not even acknowledged by his bibliographers. Possibly, the unfinished form of the maps—predominantly "methodical," colored-in outlines, almost entirely devoid of place-names and other writing—jeopardized their survival. The loss of Gatterer's maps for history is suggestive of the low esteem in which such productions have been held.

Gatterer worked in the tradition of the Paris draft and Dupré and is likely to have been acquainted at least with the published atlas of 1763. Whereas Dupré drew a Eurasia so projected as to suggest the curvature of the earth, Gatterer took the final step of adopting a hemispheric design, though only a quarter of the globe is shown in the one piece we can consult. Several examples of his full hemispheres survive, fitted out for other than historical purposes.[190]

In the verbal summary of the *Abriss,* Gatterer sorted the twenty-four maps into five acts ranging from four to six scenes each. The time span is A.D. 93 to 1517, and the acts range in duration from 164 to 383 years. The theatrical articulation makes one hope that the scene summaries will be tinged with drama, but austerity prevails. Gatterer's titles are stepping-stones from event to event, of the same "interval" type as in the Paris draft and in Dupré. They are no more informative when grouped in theatrical acts and scenes than when listed, as by Kruse, in twenty-four consecutive entries.[191]

The title *Geography of the Völkerwanderung* is unique to Gatterer and compatible with his outlines of universal history. Narrow definitions of the *Völkerwanderung,* reflected in Delisle's "Theatrum historicum" and other works, and acknowledged by Gatterer himself (act 1, scenes 3–4), situate the start of the great migration on one side or the other of A.D. 400.[192] Gatterer basically espouses a longer view. He does not reach quite so far back as the invasion of Roman space by the Cimbri and Teutones in the second century B.C., but he does stage the first scene of his atlas toward the turn of A.D. 100. The Romans and Germans, as well as the Huns and Chinese, are seen at rest, at their respective "starting lines." Gatterer notifies us, besides, that the Arabs will long be dormant and that the Persians have no protracted future. Such was the lineup of peoples "before the great changes began out of which, little by little, the states of today's world originated."

In books, Gatterer gives a more expeditious account of the "great changes." The Germanic invasions take place, the barbarian kingdoms are established on the lands of the one-time empire, and with them the states of the modern world begin. Gatterer's geography, though not inconsistent, was more complex. The *Wanderung* was not distinctively Germanic; migrations went on, involving Slavs, Hungarians, and others as well as Germans, until the "states of today's world" came into existence. These states, by implication, were stable. Only the events, notably migrations, leading up to them were volatile. Prerevolutionary Europe seemed solid. Gatterer was spared the troubling thought that "today's" states might be no more than way stations in an unending flux of peoples.

Gatterer's interest in migrating peoples was not limited to his ancestors. Kruse praises him for doing much for the history of the oriental peoples.[193] The end of Gatterer's prospectus gives an overwhelmingly Asiatic impression: "act V" belongs wholly to Mongols, Timurids, Ottomans, and Persians. The period that, in our manuals, runs from the Fourth Crusade to Luther's posting of his theses (1204–1517) is populated, not by popes, merchant cities, and mendicant friars, but by descendants of Genghis Khan and focuses on central Asia. On closer scrutiny, though, Gatterer's summary lags behind Dupré in its attention to Asia. Until the late Middle Ages—terrain that commonly led to cartographic perplexity—Gatterer's Asiatic interest is restrained; Europe predominates. He points out that the Chinese put an end to the kingdom of the eastern Huns (and set them in motion toward the west); he shows concern for the Arabs, notably their seizure of Spain; and he takes account of the Seljuk Turks. His real highlights, however, are the development of the Romano-Germanic Empire, the settlement of the Slavs, and the emergence of the Hungarian kingdom. His theme of *Völkerwanderung* proceeds from the neighbors of the Roman Empire to the making of what seemed to be a durable modern Europe. Once the descendants of the migrating barbarians acquired their permanent space, Gatterer could pay more attention to the Orient.[194]

Maps that cannot be examined, but only read about, limit what may be said about them. Kruse, who closely studied Gatterer's collection and spoke approvingly of it, criticized the individual pieces for encapsulating fifty or even a hundred years. Kruse was arguing in favor of his own method; Gatterer's procedure was that of his predecessors. Although Kruse pointed out many faults, he commended Gatterer's inexpensive work: it was much better, despite its flaws, than the high-priced atlases now finding their way to Germany from England and France.[195] By 1805, when Kruse's article appeared, the Lesage (Las Cases) atlas was being welcomed in the marketplace and disdained by German critics. Gatterer, who carried forward the type of serious historical atlas-making that had prevailed for half a century, offered too unfinished a product to win a following and make a difference outside the lecture hall. In a few decades Hase's much older atlas would be remembered, while Gatterer's was forgotten. Later universal, sequential atlases, however, were not wholly free of the latter's influence.

Isolated Initiatives: Beaurain, Tomka Szászky, Andrews

Not all eighteenth-century atlases containing historical material shared in the ambitions that have been traced here from Delisle to Gatterer. The margins of cartographic activity were discreetly colonized.

Beaurain was an atlas-maker in one of the established meanings of the word—a collector, assembler, and publisher of maps by other cartographers. A sixteen-volume atlas of his at the Bibliothèque nationale handsomely exemplifies the *atlas de choix* amassed for collectors.[196] As Ortelius joined the small *Parergon* to the large *Theatrum,* so Beaurain supplemented fourteen volumes of modern maps with two of ancient geography. Biblical and classical subjects had the lion's share, but Beaurain was modern enough to include a section "Pour l'histoire du Moïen Age."

Only France, Beaurain implies, had produced medieval maps; Köhler, if consulted, would have taught him a thing or two about how to assemble a medieval collection. The most recent productions in Beaurain's miscellany are by Gilles Robert: France under its first dynasty; the kingdom of the Visigoths in Spain and southern France; and Charlemagne's Empire. A new outing is also given to all four sheets of Bertius's aged "Empire de Charlemagne." Famous old maps, such as Adrichom's sixteenth-century portrayal of the Holy Land, were welcome in Beaurain's atlas. Delisle was drawn upon for medieval Toul and Dauphiné but, typically, not for his Byzantine reconstructions (Byzantium was not in France). Nine slots, half the total, are allotted to d'Anville's largely unmedieval maps for de Longerue's *Description historique.*[197]

An *atlas de choix* is unlikely terrain for innovation. Even the skimpiness of Beaurain's selection is only to be expected. What mattered more was that, in a "supplement" whose prototype contained only ancient geography, medieval history was acknowledged to merit a place, however limited and narrowly patriotic.

János Tomka Szászky's *Parvus atlas Hungariae,* and the six-map abridgment extracted from it, are rare works that may not have traveled far outside central Europe.[198] The *Parvus atlas,* if taken as a "national" historical collection, was preceded only by Menso Alting on the Netherlands (1697–1701) and Hase on the Romano-Germanic Empire (1750). Its contents were more comprehensive than strictly historical and were oddly irregular.

Tomka (ca. 1692–1762)—often catalogued as "Szászky"—was a Lutheran educated as a minister at Jena. Returning to Hungary, he held several religious appointments before moving in 1732 to Pressburg (Pozsony in Hungarian, now Bratislava) to fill a position combining church duties with the direction of the Protestant school. Pressburg, not far down the Danube from Vienna, had been the capital of Hungary since 1541, complete with royal palace, coronation church, and assembly hall for the Hungarian Diet. These circumstances seem less important to Tomka's mapmaking than his association there with circles wishing to improve local Lutheran schooling. A major influence was Matthias Bel (1684–1749), whom the emperor Charles VI had encouraged to compose an extensive *Notitia Hungariae novae;* another stimulus was Samuel Mikoviny, a fellow Jena alumnus of Tomka's, who provided Bel's *Notitia* with eleven excellent county maps. So inspired, Tomka turned to geographical, as well as historical, studies. His atlas of Hungary, in a handy format, was designed to serve as a teaching aid in secondary schools.[199]

Tomka's *Parvus atlas* combines scenes evoking contemporary, as well as historical, conditions. The contemporary kingdom is mapped in general and by circles (an administrative unit); "ancient" Hungary, reaching back to the Scythians and into the Roman period, is conscientiously delineated; the relevant sections of the Peutinger Table are reproduced; and the Middle Ages are carefully included. One senses that these maps are for a curriculum thoroughly grounded in classical antiquity but meant also to emphasize the Hungarian past. The maps have long, thorough titles, some even specifying the applicable times.[200]

A generation after the original *Parvus atlas* and after Tomka's death, a selection was published to illustrate *Introductio in geographiam Hungariae antique et medii aevi,* apparently by István Jeszenák. (Neither his name nor Tomka's is cited.)

The original *Parvus atlas* appears to have been designed as a source book that teachers would use according to their preferences, rather than as an ordered work like Hase's *Imperia maxima* and *Seven Maps.* The four copies seen by me disagree markedly in order of presentation. In the following list, Bu = Budapest, 18 maps; Brno = Brno, 13 maps; Gö = Göttingen [1781], 6 maps; Bu81 = Budapest [1781], 11 maps. The number joined to each atlas abbreviation indicates where in that atlas the map is situated. The Budapest copy

(Bu) is fullest and comes nearest to chronological order; it is therefore listed first:

> Bu1/Brno6: "Scythia Europaea et Asiatica cum Chersoneso Taurica ante et post Christum natum ad seculum V" [1750]
>
> Bu2/Brno8/Bu81, 9: "Cumania Magna, olim Hunnia Orientalis" [1750]
>
> Bu3/Brno5/Bu81, 8: "Dacia atque Moesia" [1750]
>
> Bu4/Brno4/Bu81, 10: "Illyricum in Dalmatiam, Chrobatiam, Slavoniam et Serbiam divisum" [1751]
>
> Bu5/Brno7/Gö1/Bu81, 3: "Pannonia Illyricum et Quadorum Iazygumque regiones" [1751]
>
> Bu6/Brno10/Gö4/Bu81, 1: "Hunnia occidentalis per Scythiam sive Gepidiam Daciae et Ostrogothiam Pannoniae Sec. V, VI extensa" [1751]
>
> Bu7/Brno11/Gö5/Bu81, 7: "Hunnia occ. Abarica per Pannoniam atque Daciam, sec. VII, VIII prorogata" [1751]
>
> Bu8/Brno9/Gö2/Bu81, 4: "Magnae Moraviae pars per Hunniam occidentalem extensa" [1751]
>
> Bu9/Brno12/Gö3/Bu81, 5: "Hungaria seu Turcia in octo capitaneatus divisa" [1750]
>
> Bu10/Brno1/Bu81, 11: "Hungaria Magna in regna et provincias divisa" [1750]
>
> Bu11/Brno13/Gö6/Bu81, 6: "Hungaria aevi medii . . . in Megas [counties] LXXII distincta" [1751]
>
> Bu12/Brno2: "Hungaria hodierni temporis in partes II, et circulos IV divisa" [1750]
>
> Bu13, 14, 15, 16: each of the four Circles of modern Hungary
>
> Bu17/Brno 3a/Bu81, 2: (a) [Peutinger Table] "Itinerarium Norici, Pannoniae, Moesiae Superioris, Liburniae et Dalmatiae; Secundum tabulae Peutingerianae segmentum III, IV a Marcomannis ad Sarmatas usque"
>
> Bu18/Brno3b: "Itinerarum Sarmatiae Scythiae. . . . secundum tabulae Peutingerianae segmentum V, VI [1751]"[201]

Alting, Lamare, Delisle (in his atlas project and maps for Louis XV), and Hase all organized their maps in a narrative sequence. Something similar is found in the Budapest copy of Tomka's *Parvus atlas:* ancient "Hungary" (1–4, 17–18), migration age "Hungary" (5–8), the first Hungary (9–11), present-day Hungary (12–16). The one anomaly is that the copies of the Roman itinerary (Peutinger Table) are placed, inexplicably, at the end. Every other set of Tomka's maps considered here follows its own fancy, that of Brno being perhaps the most random of all. Little can be inferred from these variations (who can tell in what form the maps were distributed?). The contents suggest that

Tomka's conception of a historical atlas resembled Sanson's, Dangeau's, and other French mapmakers' attention to administrative boundaries, rather than a narrator's concern with chronology and a developing story.

Tomka went lovingly about his task and was particularly determined to individualize each map by delineating internal divisions. The sixth-century historian Jordanes vaguely indicates where in Pannonia the Goth Valamer and his two brothers established themselves after Attila's domination collapsed. Tomka sharply demarcates at least two of the three divisions, treading firmly where today's historians admit ignorance. A very colorful map of the set concerns the Avars, representing nine Avar fortified enclosures, or "rings"; in one corner, the ring type is shown both in perspective and in plan. Imagination is at work. Perhaps the most resourceful idea—unlikely to be Tomka's invention—involves the belief that, for centuries after the brief passage of the real Huns, a "Western Hunnia" existed as a stable territory perpetuating their name. Only a specialist can unravel how far Tomka ventured beyond established fact. Did Bel serve as guide? Alting is criticized for having left his maps too bare; Tomka made them too full. However that may be, he endowed Hungary with a precocious, visually attractive, and ambitious historical atlas.[202]

The "Chronological Map of Europe describing the Revolutions of Its Principal States" published by the prominent London map publisher William Faden in December 1783 has an original form, but not one destined to set an example. It attracts attention for being a type of historical compendium that had not been tried before. Faden's map shows a large outline of Europe, reaching far to the east; each country has its current borders; a brief historical synopsis is written within these borders, which serve as margins. The dimensions of the script vary with the size of the country and the amount said. Each thumbnail sketch begins with an identification of the capital. Ireland affords a brief, representative example (the italics are Faden's): "*Capital*, Dublin. *This country, it is supposed, was peopled from the West of Scotland and Danes seiz'd upon part of it 795 A.D. Henry II subdued it 1172 and settled a Colony from Bristol in Dublin. There were several rebellions, & a dreadful Massacre but they were reduced to obedience by Cromwell 1653.*" In the entries of other lands, history usually starts with the Romans. Only Iceland and Sicily are barren of writing. The quality of information is not high: the Polish synopsis omits the recent partition; France's stops with Henry IV (†1610); Hungary is said to have been conquered by Charlemagne, then have become independent; the Turks apparently "reduced" the Saracens and afterward turned Muslim (not wholly wrong but inappropriate as background to the Ottoman Empire).[203]

The novel conception of this map matters more than its amateurish contents. Faden's "geochronology" clearly intends to teach by a synthesis of geography and history. Users learn the geographical position and shape of the European states together with a smattering of historical information about each one. Presumably the designer—in an act of novel invention—hoped that the integration of script with geography would make both more memorable. The experiment is an instructive reminder that mapping of the sort considered in this book is mainly sustained as an aid to the assimilation of history, not as an outlet for scholarship.

The *Historical Atlas of England* by John Andrews was issued sheet by sheet for a considerable span of time. The map dates, not always provided, range from 1790 to 1797, the latter being the official date of publication and found on more maps than any other. Andrews's atlas, a luxury product in its time, is very rare. Of its twelve sheets, the British Library copy lacks three, very important ones; they are at Oxford. I am not aware of other locations for this work.[204]

Andrews's atlas is the oddest and in some sense the least historical work we have encountered, yet not so strange as to be disregarded. The Weimar journal *Allgemeine geographische Ephemeriden* greeted the early sheets with praise and enthusiasm, even though the maps espoused Philippe Buache's notions of physical geography, with which the reviewer disagreed. Later issues out of Weimar record a growing disillusionment, motivated by the slow pace of publication and particularly by a map of English fisheries and spas.[205] The reviewer was still waiting for history.

As the Weimar journal suggests, Andrews had an eclectic conception of the past. He would have helped matters by explaining and defending his plan. Instead he simply announced what would come. Continental historical atlases had no influence on him. He unwittingly foreshadowed a distant future of dominant thematic maps and appears to subscribe to an *histoire (presque) totale,* extending from the physical formation of England to its modern deaneries and healing waters, not forgetting the Saxon and Norman invasions.[206]

Andrews opens memorably with a presentation of how England looked when the waters of the Deluge receded—the dry land uncovered when the waters had subsided to so-and-so many feet of altitude. Assuming the Deluge is history, this map showing drainage through the river basins of England is a creditable alternative to the elsewhere approved scene of the dispersal of Noah's sons. Andrews's advertised first item was to have been "A Physical Map of the Terrestrial Globe" along the lines of Buache's ideas; it is absent from the London and Oxford copies. The Deluge map is followed by three more items of physical geography, peculiar in this context but integral to Andrews's conception. Only then does "normal history" begin.[207]

Andrews's collection closes with very conventional maps of religious and civil boundaries. Aside from a small nonmedieval exception, no history after 1066 is recorded. If "historical" had been dropped from the title we would have been left with an adequate physical and political atlas of England. Even four of the six historical maps are thoroughly tame. Andrews thought he should show the Britons before the Romans came and did so on the basis of ancient sources, superimposing the results on an outline of modern county names and boundaries ("Stoleme" for Ptolemy is obtrusive, but trivial). Roman Britain is then given a double-size display, followed by two maps for the Anglo-Saxon period: one of Britain according to the Anglo-Saxon Chronicle (a descendant of Edmund Gibson's map of 1692), augmented by a track of Duke William's invasion; the other, also double-sized and based on a medieval document, of England divided according to the West Saxon, Danish, and Mercian laws.[208] These scenes build on earlier maps of British history and have a future in later atlases, whether copied from Andrews or not.

Andrews is most original, even eccentric, in a map that contains historical information, yet is well designed to disconcert readers of history. Polar projections are familiar for certain kinds of information and had long been drawn. The same does not hold for a projection in which London takes the place of a pole: "A political, historical, astronomical and commercial chart of Europe, To shew relatively the situation of places to the rays of the sun, the rumb of the wind, and hours of the day, from the meridian and parallel of London, on which are delineated the zones and climes with the tracks of the Carthaginians, Phoenicians, Romans, Saxons, Danes & Swedes, with rectiline distances in circles of 100 miles each." The map, an exceptionally large foldout, radiates out of London; it reaches the North Pole and the Sahara and includes part of Greece and the Black Sea, as well as Scandinavia and Iceland. Much navigational information is recorded, and many localities are noted, including negligible ones in Iceland. The various tracks—discrete, labeled lines—are Andrews's bow to conventional historical events. The lines trace the more or less permanent visitors to Britain, some as tin merchants, others as raiders and settlers; Saxons are distinguished from Anglo-Saxons. The unannounced, least expected track is of the Stuart Pretender, both coming and going, in 1745. Much is fitted into this large map. Andrews outdid himself in striving to couple the cartography of navigation to some at least of the early navigators who reached British shores.

A seagoing spirit also informs Andrews's "Map of the position of South Britain with the Points of the wind and distances from the meridian and parallel of London." Although the design has a more familiar appearance, navigation retains prominence. Shallow banks at sea are carefully marked. Tracks again appear, fewer than in the previous map and drawn with greater care. The Saxon and viking invasions alone are considered and much simplified. Track 1 shows Danes and Swedes coming to the Humber; track 2 has Anglo-Saxons going from Rypen in Jutland to Margate; track 3 has Saxons sailing from the Elbe mouths to about the same destination as the Anglo-Saxons of track 2. The pattern, Scandinavians aside, is reminiscent of the precocious tracks in the Châtelain atlas.

Andrews's atlas is odd. Not lacking virtues, it combines many disparate things, from the extraordinary experiment of the sixth map to the parochialism of modern English spas. The other-than-geological history that he displays is almost exclusively medieval and even pre-1100. As usual, the High Middle Ages seem to lack anything worth mapping. Andrews did not set a fashion. His association of physical and historical cartography does not lack merit. The improbable coupling of navigation and history might have been worth developing but proved to have no future. The work has the poignant appeal of a historical atlas done by someone wholly detached from a historical turn of mind.

INNOVATIVE ATLASES AND PERPLEXING MAPS

Eighteenth-century Europeans, sometimes credited with an enlightened aversion to a past (as they thought) sunk in superstition, in fact paid much attention to the Middle Ages and made an effort to give parts of the period geographic form. Major cartographers had a hand in this extension of maps for history. The masters involved were not outsiders or rebels against convention. Delisle, Nolin, Hase, Rizzi-Zannoni, Robert, and d'Anville were leaders of contemporary cartography. Medieval subjects were dealt with in keeping with tradition: medieval geography joined the established twosome of ancient and modern geography. History was not the primary consideration. Much of the mapmaking, especially that involving medieval territories, was based on charters and other primary sources. Each segment of geography addressed the sources proper to it. Some excellent eighteenth-century mapmakers proved clumsy in dealing with medieval subject matter; others were at a loss about what to show after ca. A.D. 900. If re-creating medieval scenes had been uniformly easy, the successes would not merit applause.

In the material surveyed, timeless geographic representations of medieval territories, "passive" maps designed as companions to reading, were the most common. Delisle's "Dauphiné," Beretti's "Lombard Italy," Hase's *Imperia maxima*, Lamey's *gaue* of the Palatinate, d'Anville's "Age intermédiaire,"

Rizzi-Zannoni's atlas of France, and many more largely outnumber any other type among the eighteenth-century maps examined. Maps for history did not differ radically from geographical maps; they highlighted boundaries and provided names of places and regions. The utility of such reference maps— common to the "ancient geography" of Ortelius and the historical atlases of today—is beyond doubt.

Poster, or catalogue, maps continued to be intermittent and rare, but estimable when they turned up. The two by Seutter in 1745 are the salient examples. The type was not trouble-free. A map of councils and one of Benedictine monasteries might be similar in design, yet have very different chronological implications. The first posed no problem: viewers saw places where there had been a flash of activity, much as in Speed's map of battles, followed by quiet. The multiplicity of instants was obvious. Other subjects were more troublesome. Maps of enduring establishments, such as monastic houses, could hardly avoid generating chronological muddles. Founded at different times, some houses survived, others dissolved; a collective map homogenized them timelessly. A poster of monastic houses might be useful enough as a geographical locator, provided time was disregarded.

A type of map for history, apparently new and unprecedented, came into existence in association with the Paris draft of 1747 and its descendants. In these atlases each map encompassed an announced period of time, more or less long. Each atlas was composed of a succession of these maps. A typical example, already cited, declares that the span of the map is "the time that passed from the year 1226, period of the first conquests of Genghis Khan, prince of the Tatars, up to and not including the division of his empire carried out in the year 1228." Maps incorporating an announced set of years were created for the needs of these innovative atlases. They have been an enduring addition to the repertory of maps for history and continue in the historical atlases today. Their shortcomings are evident. If the map incorporates individual features selected from the entire announced time period, a chronological hodgepodge ensues. If, instead, the cartographer decides to show only a cross-section of events at the end or the beginning of the announced time period, he risks being charged with deceptive labeling, but is likely to produce a more firmly anchored map. Writing a legend saying that a map displays so-and-so many years, decades, or centuries is easier to do than designing a convincing depiction of a long period.

Rarest among eighteenth-century maps for history were the attempts to suggest movement. The Châtelain *Atlas historique* has tracks of Saxon ships landing in Britain, as well as of Richard I (Lionheart) going on crusade. The timid Châtelain lines are surpassed by Hagelgans's map of all Europe crisscrossed by "migrations of peoples," twenty of them scattered over 2,500 years. These efforts, anticipated by Duval's and other maps of ancient geography, were made within a few years at the beginning of the century. They were matched a decade later by Liébaux's portrayal of Clovis conquering Gaul, complete with a very conspicuous arrow-like route from inner Germany to Soissons. In all cases, except Richard I's Crusade, barbarians are the subject. Liébaux's Clovis is deplorable from a factual standpoint but much more forthright as a conquest than the others. These isolated attempts to impart dynamism to a map can hardly be called the beginning of a trend, but they suggest the effects mapmakers occasionally wanted to achieve. That aspiration did not fade.

The eighteenth century was much more important than Ortelius's sixteenth for the origins of historical atlases. First as aspiration, then as accomplishment, there came into being atlases no longer limited to ancient geography but open to historical periods and themes. The atlases common to our bookshelves, with a nonclassical repertory and a concern for chronology, truly begin. The designers took a sustained interest in medieval times, cared about chronological order, and subdivided the past into intervals that allowed historical "revolutions" to be narrated in images as well as words. No lasting shift took place, however, in the spectrum of maps available to the public. Buyers, more apathetic than acquisitive, saw to it that historical collections of the new type remained peripheral, still overshadowed by the continually reprinted compendia of ancient geography.

NOTES

1. Delisle's alternative title, in ArchNat, Marine, Serv. hydrog. 6JJ/59, no. 14, is "Partie orientale du théâtre des invasions barbares au IVe siècle"; along the same lines, Hübner, *Museum geographicum*, 207: "One sees on them how it looked in the west and the east at the start of the fifth century when the great *migratio gentium* occurred." The "Theatrum historicum" occurs often in the very numerous Delisle atlases. The parallel transmission is more varied: John Senex, *Modern Geography* (London, 1708–25), nos. 30–31; Moll, 1709; Faber, 1711 [1740], nos. 46 ("Scena historiarum orientalis quinti seculi P.N. Chr.") and 47 (". . . occidentalis"); Köhler, 1720, nos. 43, 44. The "Scena" printed by Weigel is never attributed to Delisle; its Caspian Sea differs greatly in outline from his. Hauber, *Versuch einer Historie*, 135, notes the delight of students that Moll copied Delisle, 1705, in large format and Weigel in small.

2. On Delisle's planned atlas, see ch. 3, nn. 31–39, above, and nn. 154–58, below. On his father, see ch. 3, n. 14, above.

3. *Mercure de France*, April 1720, 127–32: "Il a entrepris de donner aussi une suite de

Cartes pour le moyen âge. On sçait de quelle utilité seroit un corps de ces Cartes, parce qu'elles influent beaucoup plus sur l'état present, que ne font les Cartes de l'antiquité" (129). Cf. Lenglet, 1752/O, 1:272: the geography of the Middle Ages "est proprement relative à l'histoire de chaque nation."

4. The quotations are from Delisle, *Remarques sur le Théatre,* attached to the maps (LC) or a separate publication (BL). The second and third are from the "Pars orientalis" sheet. Some nineteenth-century mapmakers wrote names of barbarian peoples several times at different spots, not wholly without success despite Delisle's misgivings.

5. Hase, 1739/B, (unpaginated) *praefatio,* says he has deservedly called his maps THE-ATRI HISTORICI. For the anonymous cartographer, see n. 174, below. Dupré, 1763, appropriates the inset hemispheres of Delisle's "Theatrum historicum."

6. Châtelain, 1705, 1: nos. 15, 31; 2: nos. 1, 5, 42; 7: no. 7. The Germanic heroes are associated, in this image, with the coats of arms of prominent current German families (dynasties); "tum signa veterum militaria, tum hodiernarum familiarum in Imperio illustrium insignia" (*Acta eruditorum* [Leipzig], July 1709, 297). Germanic antiquity was much admired by Dutch mapmakers (see ch. 3, n. 20, above). Book titles, such as P. van den Keere's *Germania inferior,* remind us that the Netherlands were in "Lower Germany."

7. For thumbnail sketches of barbarians, see Châtelain, 1705, 2: nos. 3, 4.

8. Châtelain, 1705, vol. 2: nos. 42 ("Carte pour l'Introduction à l'histoire d'Angleterre"; the Saxon crossing is one component), 45 ("Carte pour l'intelligence de l'Histoire d'Angleterrre"; inset, Richard's Crusade journey; the map also has various items concerning the English in France).

9. Hagelgans, 1718. The author's preface (in the 1718 ed. only) refers to the maps as being only of ancient history and illustrating the places that the picture-chronicle mentions. The same uninformative (and somewhat inaccurate) account is in a separate publication: Goffart, "Longer Look," 2–3. On Hagelgans's years as a junior diplomat in Swedish service (1711 to 1719/21), see Otto Renkhoff, "Johann Georg Hagelgans, 1687–1762," *Nassauischen Lebensbilden* 5 (1955): 56–69, at 57.

10. Goffart, "Preliminary Report," 51–53. Hagelgans's Greeks include both the colonists of Sicily and the troops of Alexander of Macedon; his Normans are both the vikings and the adventurers from Normandy active in southern Italy. The traditional, mainly Germanic denizens of the *Völkerwanderung* account for more tracks than the peoples who muddle the chronology. For a similarly all-inclusive later account, see Alfred C. Haddon, *The Wanderings of Peoples* (Cambridge, 1911; 2d ed., 1927).

11. The Latin phrase, apparently coined by Lazius, is said to have been translated into German only in the last quarter of the eighteenth century: Herwig Wolfram, "Gothic History and Historical Ethnography," *Journal of Medieval History* 7 (1981): 312. This late date is improbable; see, e.g., Johann Jacob Mascov, *Geschichte der Teutschen* [to 751], 2 vols. (Leipzig, 1726–37), where the term is duly translated. I have not compared the migrations of Hagelgans's picture-chronicle to those of the map so as to verify whether they correspond.

12. The cartouche title says, "wanderings of the peoples mentioned in the Atlas historicus" (*migrationes gentium in Atlante historico memoratorum*). Presumably (though this should not be taken for granted), the tracks Hagelgans traces on the map correspond precisely to the *gentes* shown in his picture-chronicle. If this is what he meant, the conventional *Völkerwanderung* is diluted in a much vaster process.

13. Plancher, 1739; Le Long, 1:36, no. 418. Plancher provides the Upper Rhine with a different source from the Lower (1:1). This was surely not a widely shared misconception at

the time. "Grande première demeure des Bourguignons" stretches from the Vistula to deep into Russia; the first Vandal dwelling place, from the base of Jutland to the Vistula; the second Burgundian dwelling place, from the Sudetenland through Bohemia; "Allemagne ou seconde demeure des Allemans," from the Elbe to the Rhine.

14. Bellin, 1759; *Journal des sçavans,* August 1759, 129–30; not found by me. ArchNat, Marine, Serv. hydrog. 6JJ/43, nos. 5A and 5B, is "Carte d'une partie du globe pour l'intelligence des migrations des peuples suivant l'histoire des Huns, des Turc, des Mogols &c de M. de Guignes 1758." The map is anonymous, in two parts, black and white: (1) from Spain to east of the Sea of Aral; (2) from the edge of the Caspian to Kamchatka (detached from the mainland). The geographical outline is relieved by a few names but nothing more. Any trace of colonies established by Huns is lacking. Would coloring have made a difference? The map may be an unfinished proof. On de Guignes, see *Encyc. Brit.* 12:690. His major book is dated 1756–58.

15. De Guignes's most prominent continuator was Julius Klaproth (Klaproth, 1826). An influential convert was Edward Gibbon, *History of the Decline and Fall of the Roman Empire,* ed. J. B. Bury, 7 vols. (London, 1896–1900), 3:75, 261 [ch. 41, 1776]. A recent map of this kind, with commentary, is "Les migrations des peuples en Eurasie du IVe au VIe siècle," in Georges Duby, ed., *Atlas historique Larousse* (Paris, 1978), 180. J. F. Horrabin, *Atlas of European History from the 2nd to the 20th Century* (London, 1935), 17 (about his map 3): "the gradual drying up of the pasture land of Western Asia . . . was the direct cause of a vast movement and displacement of peoples which resulted in the break-up of the Roman power." Along the same lines, Giorgio Bombi et al., *Atlas of World History,* tr. T. Touioli and C. Tite (London, 1987), 87: "Asiatic Peoples and Empire between the Fourth and Sixth Centuries A.D." Potent refutations: Ferdinand Lot, *Les invasions germaniques: La pénétration mutuelle du monde barbare et du monde romain* (Paris, 1935; repr., 1945), 52–54; Lucien Musset, *Les invasions: Les vagues germaniques* (Paris, 1965), 60–61 with n. 1; L. Vajda, "Zur Frage der Völkerwanderungen," *Paideuma* 19 (1973): 47 n. 12. See also Walter Goffart, "What's Wrong with the Map of the Barbarian Invasions?" in *Minorities and Barbarians in Medieval Life and Thought,* ed. S. J. Ridyard and R. G. Benson (Sewanee, Tenn., 1997), 176–77 n. 42.

16. Netherlands, 17–. Another copy is at the University of Leiden, Port. 3, no. 25, "Kaart van Nederland in de 10te eeuw." W. Bilandijk was the draftsman.

17. "North Sea Linked to Black Sea," *Manchester Guardian Weekly,* 11 October 1992, 12, reprinted from *Le Monde* (Paris), 26 September 1992: on 25 September, "Charlemagne's dream . . . came true." The claim is contestable, though the new canal is real enough. A postcard of the Fossa Carolina, thoughtfully sent to me by my sister-in-law, Lilo Frank, states, "Erst im 19. Jahrhundert . . . wurde durch den Ludwig-Donau-Main Kanal dieser geniale Plan Kaiser Karls des Großen Wirklichkeit." The Ludwig canal, among the foremost early German works of this kind, was sponsored by Ludwig I of Bavaria (1825–48).

18. On the historiography of the canal, the standard work is Friedrich Beck, *Der Karlsgraben: Eine historisch, topographische und kritische Abhandlung* (Nuremberg, 1911) (on detection of the canal in the 1550s, see 10–12). Daniel, 1721, shows an (imaginary) reconstruction of the track of the full canal, reproduced in Harms, *Themen,* 140–41, no. 62.

19. Beck, *Karlsgraben,* 35–37. The relevant detail from Homann's map is reproduced by Beck; for Haas's map, see Beck, *Karlsgraben,* 36. G. Z. Haas, *De Danubii et Rheni conjunctione* (Regensburg, 1726); this Haas is different from J. M. Haas or Hase. Daniel Schöpflin, "Sur la jonction du Danube avec le Rhin projetée par Charlemagne," *Histoire de l'Académie des inscriptions et belles-lettres* 18 (1753): 256–60, with map facing 256 (shows the area in

which the junction took place, with the Main at the top and the Danube at the bottom; the actual canal is very small and presented pictorially). On Schöpflin, see n. 69, below.

20. J. O. Doederlein, "Conspectus Fossae Carolinae pro coniungendo Danubio et Rheno," in *Commentarii de rebus Franciae Orientalis,* by J. G. von Eckhardt, 2 vols. (Würzburg, 1729), 1:750 (occupies part of the page with no reference to the designer); reduced copy in Beck, *Karlsgraben,* 37. Hans Hubert Hoffmann, *Kaiser Karls Kanalbau* (Sigmaringen, 1969; 2d ed., 1976), ranks Haas's map over Doederlein's. The latter is reprinted in Falckenstein, 1733. Köhler, 1730, pt. 3, no. 10 (a rare omission by Beck). Hoffmann, *Kaiser Karls Kanalbau,* is very helpful for identifications. For later maps, see Beck, *Karlsgraben,* 38, 40; his list includes every map in which the canal appears, regardless of whether in detail or incidentally to a larger map, as in Bessel, 1732. Predictably, the canal attracted the attention of Karl Spruner; see ch. 5, n. 152, below.

21. "Palatiorum sive villarum Regiarum in regno Franciae orientalis Teutonio tabula, ex chartis atque diplomatibus medii aevi, ad rationes pagorum adtemperata" (the placement of the map in various copies of the work is not fixed).

22. Bessel, 1732; list of the maps (441). Bessel is sometimes called "the German Mabillon"; see ch. 3, nn. 24–25, above. For current teaching about *palatia,* see the trailblazing work of Carlrichard Brühl, *Palatium und civitas: Studien zur Profantopographie spätantiker civitates vom 3. bis zum 13. Jahrhundert,* 2 vols. (Cologne, 1975–90).

23. Bray, 1711 (the quotation is part of Bray's subtitle). About Bray, see *Encyc. Brit.* 4:438; *DNB* 2:1148–49, with the general assessment that hardly anyone performed "more real and enduring service to the church." On Senex, see *LGKartog.* 2:738. For a list of Perrin's works, see *NUC* 451:362–63. For maps of later persecution of Waldensians, see Harms, *Themen,* 75; Nolin, 1746. The fear that Queen Anne's leading ministers favored a Jacobite (let alone a papist) reaction had weak foundations (*DNB* 17:623).

24. Barringon, 1773; the first Barrington quotation, xxiv; Forster's "Notes," 241–59 (followed by the map); Forster in the South Seas, 241 n.; about Germany, 244. On Forster (a descendant of seventeenth-century English emigrants to East Prussia), see *ADBiog.* 7:166–72. On his scholarship, see E. G. Stanley, "The Continental Contribution to the Study of Anglo-Saxon Writings Up to and Including That of the Grimms," reprinted in Stanley, *A Collection of Papers with Emphasis on Old English Literature* (Toronto, 1987), 54. Professor Scott Westrem (CUNY) kindly tells me that the Hereford Map has a similarly vast idea of Germany. The problem is unaffected, since the Hereford Map is unlikely to have been Forster's source.

25. Forster, 1788. The map is unchanged from that of 1773 except in title (identical in all editions). On Forster and Barrington, see Stanley, "The Continental Contribution to the Study of Anglo-Saxon Writings"; Forster asserts that Barrington tried to take credit for his map, yet the geographical "Notes" of the Orosius edition are plainly credited to him. Forster's son had a role in the geographical education of the famous geographer Alexander von Humboldt.

26. On Schöning's activities, see *Catalogus mapparum,* 199, in which he is the featured Norwegian. Norway belonged to Denmark in his time; ceded to Sweden in 1814, it became independent in 1905. Schöning, 1763: maps for classical antiquity accompanying his much esteemed account of Scandinavia in this period, *Kiøbenhavnske Selskab Skrifter* 9 (1765): 151–360, tr. A. L. Schlözer, in *Algemeine Welt-Historie* (Halle, 1771), 31:1–206 (on this multivolume work, see n. 28, below). Schöning's map titles all begin with "aspect" or "appearance," literally "face" (*facies*).

27. Schöning, 1777 (Schöning, 1771, may be a preparatory version). Schöning, 1779; Schöning's "national" role in the *Catalogus mapparum* explains why his map of Europe is disregarded there.

28. George Sale and T. Salmon, eds., *An Universal History from the Earliest Times* (London, 1730–36); Butterfield, *Man on His Past,* 47–48. The translations are sometimes hard to locate; the Dutch has eluded me. *Histoire universelle depuis le commencement du monde jusqu'à présent . . .* (Amsterdam, Paris, 1742–92), NOT SEEN or located (WorldCat reports an edition at the Bibliothèque nationale du Québec); nouv tr., 126 vols. (Paris, 1779–89). For the first French translation, see Barbier, *Dictionnaire des ouvrages anonymes,* 2:836–37; for the second, see LC: D20.U595. Siegmund Jacob Baumgarten, ed., *Algemeine Welt-Historie von Anbeginn der Welt bis auf gegenwärtige Zeit,* 57 vols. (Halle, 1744–91); *Storia universale del principio,* 4 vols. (Venice, 1772).

29. Bowen, 1766/O; Bachiene, 1785. On the initial source of the maps of the German edition, see Ludewig Albrecht Gebhardi, *Vorrede,* in *Algemeine Welt-Historie* (Halle), 50:vi.

30. Butterfield, *Man on His Past,* 47–48 n.

31. Pedley, *Vaugondy,* 136, no. 15, cartobibliographic listing; 98–99, reproduction; 103, comments. Vaugondy's map was anticipated by Moulart-Sanson's edition of the Sanson *Introduction à la géographie* (see ch. 2, n. 44, above).

32. "Charte aller von Wenden und Slavinen besessenen Länder," in *Algemeine Welt-Historie* (Halle), 51: opposite 670.

33. The term "poster map" was introduced in connection with Speed's battles map; "catalogue map" is an alternative (see ch. 2, n. 95, above).

34. For councils and religious orders, see ch. 1, nn. 94–95, above. Part of Speed's list or legend is on the map. This is probably not obligatory: it suffices that the written catalogue should accompany the map, without necessarily being inscribed on its face.

35. Stegmann, *Anselm Desing,* xvii (no. 6), 131–37, 171 (quotation), 172. Desing, 1731; the author's name does not appear on the first edition; place and date in pencil. Engraved by Desing himself at the Freising lyceum; I have not seen the improved version engraved by a professional. Stegmann, *Anselm Desing,* xvii, and *DBiogArchiv,* fiche 231, frames 120–55, use a title different from the first edition: "Kürziste Universal-Historie nach der Geographia auf der Land-Karte zu erlernen." See, now, Johannes Dörflinger, "Die Karten und Globen von Anselm Desing," in *Anselm Desing (1699–1772): Ein benediktinischer Universalgelehrter im Zeitalter der Aufklärung,* ed. Manfred Knedlik and Georg Schrott (Kallmünz, Germany, 1999), 197–207, 322–27, 419–21; includes a meticulous bibliography of Desing's geographical publications.

36. Desing, 1731, shows Europe, Africa, and Asia (to Sumatra), with modern place-names and many numbers keyed to events, such as 113, Council of Nicaea; 122, the empress Eudocia is unlucky with the apple (this calls for knowledge of the reign of Theodosius II); 130, Ostrogothic kingdom; 139, Mohamet; 177, expulsion of the Moors from Spain, 1483; 178, Columbus discovers America.

37. Seutter, 1745c and b. "Circles" were subdivisions of the Holy Roman Empire since ca. 1512. The two maps were issued at the same time; that of Bavarian history is advertised on the imperial one. In the War of the Austrian Succession (1740–48), Charles Albert of Bavaria was elected emperor in opposition to the Habsburg candidate, Maria Theresa's husband. Charles VII was promptly driven from Bavaria, restored in 1743, again ousted, restored in 1744; he died in January 1745. Events in both maps reach into 1745, implying publication after Charles VII's death, but only just.

38. Seutter, 1745c: second sheet, "Erklaerung deren Figure, so sich auf der historischen Land-Charten von Teutschland, und derent Angraenzenden Laendern befinden" (provides the quotation). The "Germany" referred to is a geographical, not political, term. Some symbols (such as a sword for a battle) appear on the maps, probably to make room for pictorial markings. Seutter, among others, published decorated lists, without maps, of popes (with symbols for the character of each), emperors, European kings, imperial electors, etc.: Seutter, 1720, 1770.

39. Enouy, 1797. On the events, see Georges Lefebvre, *La révolution française* (Paris, 1951), 473–76, 493–99; H. F. B. Wheeler and A. M. Broadley, *Napoleon and the Invasion of England: The Story of the Great Terror,* 2 vols. (London, 1908), includes many contemporary cartoons and medals but no maps.

40. Tardieu, 1798. In this handsome map, with the usual Speed coverage (i.e., England and Ireland, almost no Scotland), the counties are border-colored.

41. Enouy, 1798, was reprinted in 1801. This reflects a moment when, on the French side, right after the Peace of Lunéville on the Continent, Napoleon organized the Camp of Boulogne and contemplated a cross-Channel attack: Georges Lefebvre, *Napoléon* (Paris, 1965), 103; Wheeler and Broadley, *Napoleon and the Invasion of England,* 1:159–94 (the month when the map was reprinted was the climactic moment of tension).

42. P. G. Chanlaire, *Carte des Isles Brittanique et des côtes, qui les avoisinent, servant à l'intelligence de l'Histoire des descentes faittes* [*sic*] *dans ces Isles depuis les Romains jusqu'à présent, avec la date et la notice de chacune des ces descentes,* revue et augmentée en l'an XII (Paris, 1803–4); NOT SEEN. My information is from the review in *AGEphem.* 18 (1805): 214– 16. The topical nature of the map was obvious to the reviewer. Why, he asks, did Chanlaire leave out the important Roman landing? No reference to Speed (or Enouy). About Napoleon's efforts to invade England (1803–5), see *Encyc. Brit.* 19:234–35 (Napoleonic Campaigns); Lefebvre, *Napoléon,* 175–77; Wheeler and Broadley, *Napoleon and the Invasion of England,* 2:1–196.

43. *Catalogus mapparum,* 154–55; Spruner, 1838a.

44. Coquart, 1705; [Nicolas de Lamare], *Traité de la police, où l'on trouve l'histoire de son établissement: On y a joint une description historique et topographique de Paris et huis plans gravés,* vol. 1 (Paris, 1705). Quérard, *France littéraire,* 4:476, provides a table of contents and eulogistic biographical sketch; see also François Monnier, *Dictionnaire du Grand Siècle* (Paris, 1990), 453–54. *Commissaire* was the title of assistants to the provost of Paris. For the sense of "police," see P. Robert, *Dictionnaire de la langue française* (1975), 5:311: "La *police* consiste à assurer le repos du public et des particuliers, à purger la ville de ce qui peut causer des désordres, à procurer l'abondance et à faire vivre chacun selon sa condition" (royal edict of 1669). There is a 261-volume Lamare archive: Collection de Lamare, Département des imprimés, BN, *Catalogue de l'histoire de France* (Paris, 1855), 1:xiii n. 1.

For a similar and contemporaneous map of Amsterdam, see ch. 2, n. 153, above; but Amsterdam is only marginally "medieval" (first walls, 1482).

45. These maps have been often reproduced (nos. 3–6 are in my "Breaking the Ortelian Pattern," fig. 2-1). Fer, 1705, may be a cruder rendering issued in the year of Lamare's book. Two of the latter are in Pastoureaux, *Atlas français,* figs. 69, 70. NYPL records an *atlas factice* called Collection of Maps of Paris, nos. 1–7 (1714); 8 (1705); two leaves of descriptive letterpress with each map, plus a folded "carte de la ville par Incelin," a ten-leaf survey of the wonders of Paris, and fifteen illustrations. The bookseller Desnos advertised "Les huits plans de Paris des differens ages" in a broadside attached to an atlas of 1764 (BL, Maps c.24.d.22).

46. Loose copies at MilwaukeeAGS are signed by Coquart (nos. 1–7) and by de Fer (no. 8). *A Catalogue of Maps and Charts in the Library of Harvard University in Cambridge, Massachusetts* (Cambridge, 1831), 114–15, attributes 1–7 to Coquart, 8 to de Fer. Pastoureau (*Atlas français,* 179–80, nos. 188–92, see also nos. 193, 195–96) suggests that de Fer did the whole set. Léon Vallée, *BN, Catalogue des plans de Paris* (Paris, 1908), nos. 2458–61, names both Coquart and de Fer without clarifying the relationship; reproduces source citations from the maps. A photo collection at BN, Cartes et plans, reference shelves, vol. 42, *France: Paris des origines au XVIIe s.,* shows separate designs by Coquart and de Fer, the latter different from Coquart and coarser. Entries for de Fer in NYPL, *Dictionary Catalog of the Map Division,* 10 vols. (Boston, 1971), 3:902, are helpful: one entry indicates a de Fer Collection of Maps of Paris, eight maps, the last one dated 1705, and the others, 1714; a similar set mentioned in the next column is alleged to be derived from Lamare's book and is dated 1705–35, but the entry says nothing suggesting that de Fer produced the eight maps in 1705. De Fer probably supplied Lamare's last map, then almost a decade later copied Coquart's maps. It may be that, as the years passed, by dint of merchandising, he came to be regarded as the designer of the full set. But more study is needed.

47. *NUC* card for "Fer, 1705," cites an encyclopedia: "Huit plans de fantaisie intercalés, par le commissaire Delamarre." Another plan of medieval Paris, Dheulland, 1756, circulated under alternative headings, either fourteenth-century Paris or the time of Charles V and VI (1364–1422). See also Julien, *Nouveau catalogue,* 44. The attribution to the fourteenth century was rejected in *Journal des sçavans,* February 1757, 225–27; *Journal de Trévoux,* April 1757, 1131–34: the correct date was Charles IX (1560–74). Dheulland's plan was apparently based on a contemporary wall map at the Abbey of Saint Victor, obviously postmedieval. Didier Robert de Vaugondy intimates (1760) that the history of Paris was a recent interest: Pedley, *Vaugondy,* 108.

48. Delisle, 1707; I have not seen the thirteen-page accompanying "Avertissement." Delisle, 1710b. Both are in the official list of Delisle's medieval maps; copies are abundant. Picart is often cited as Père Benoît. Moret de Bourchenne's work on Dauphiné, a thick folio with many documents and dissertations, has the subtitle *où l'on trouve tous les actes du transport de cette province à la couronne de France.* For identification of the author, see Barbier, *Dictionnaire des ouvrages anonymes,* 3:236d. Reprinted with additions as *Histoire de Dauphiné* (Paris, 1722). *AGEphem.* 10 (1802): 180, wrongly attaches "medieval" to Delisle's modern "Carte d'Artois et des environs" (Paris, 1704), in Adrien Mallart, ed., *Coutûme générale d'Artois* (Paris, 1704).

49. Schannat, 1724.

50. L. A. Muratori, ed., *Rerum Italicarum scriptores,* 25 vols. (Milan, 1723–51), 1, pt. 1: preface, unpaged (fol. g2v); the speaker appears not to be Muratori): states reasons for giving maps of Italy at various times; (1) "Senescentis Imperii Romani aevo" (= vol. 1, pt. 1, facing p. 1); (2) "medio, ut aiunt, aevo, nempe Langobardico regno"; (3) "restauratae sub Imperatoribus, atque Italis aliis Principibus Italorum fortunae." No. 3 should be medieval too; I have not found it and wonder whether it exists.

51. Muratori, 1727. The identification of Beretti is in Le Long, 1:36, no. 422.

52. Julien, *Nouveau catalogue,* 64. Köhler, 1730, "12. Italia Longobardica et Graeca in Seculo VI, VII, VIII"; source acknowledged at the end of the unpaged *Vorwort. Algemeine Welt-Historie,* vol. 40 (Halle, 1778), in the unpaged *Vorrede,* "Tabula Italiae medii aevi."

53. Bessel, 1732.

54. Curschmann, "Die Entwicklung der historisch-geographischen Forschung," 138–39 with n. 5.

55. Bessel, 1732. See the discussion of the *Chronicon Gottwicense* at ch. 3, nn. 24–25. The title of map 3 is "Austrasiorum sive Franciae Orientalis ducatus, cum pago Thuringiae Australis, in suos pagos singulares, sub imperatoribus Francicis et Saxonicis, ex variis medii aevi diplomatibus, chartis ac documentis descriptus."

56. The second map is entitled "Germania in priscas suas provincias, ducatus, pagos tam majores quam minores, curate divisa, nominibus locorum ad medii aevi dialectum expressis, ex diplomatibus, chartis & tabulis medii aevi descripta"; margin label "T. 1. L. 4. P. 527." It faces p. 526 in the copy consulted. P. 749, *pagus Ripuaria*, a large district comprising the principal *portio* "Francorum Regni Austrasiae et postmodum Regni Lothariensis." Austrasia, often mentioned, is not in the index. Casual reference in relation to a palace in Alsace (491); about Metz, the kingdom of Metz, or Austrasia (493). Both items refer to Merovingian Austrasia, a different entity from the one mapped.

57. For the impugned items, see nn. 104–12, below (notably Cellarius and Spener).

58. Zollmann, 1732. On Saxe-Weimar, see *Encyc. Brit.* 24:261–63. On Zollmann and his historical writings, see *DBiogArchiv,* fiche 1418, frames 308–11; *ADBiog.* 45:428–29 (his father was a court official in Gotha). The book on Saxon *Hauptwappen* stands out among Zollmann's works in library catalogues. The division of the Wettin family into Ernestine and Albertine branches occurred in 1485 (*Lexicon der deutschen Geschichte* [Stuttgart, 1979], s.v. "Wettiner"); Zollmann's duke was of the Ernestine branch.

59. Zollmann, 1732; the quotation, *ADBiog.* 45:429. About chronology, see also ch. 3, n. 23 above.

60. See Marija Gimbutas, "Slavic Religion," in *Encyclopedia of Religion,* ed. Mircea Eliade (New York, 1987), 13:353–61. The one deity firmly identified as Slavic is Swantwith = Svetovith; Radagart resembles another name given by Gimbutas. The two others, definitely not Germanic, must have been considered Slavic deities by Zollmann.

61. On Henry the Illustrious, see *ADBiog.* 11:544–46; Volker Mertens, "Markgraf Heinrich III von Meissen," *Die deutsche Literatur des Mittelalters: Verfasserlexikon* (Berlin) 3 (1981): 786–87. He held famous tournaments: Meissen, 1241; Nordhausen, 1263 (tree with gold and silver leaves); Meissen, 1265; Merseburg, 1286. He was a *Minnesänger,* responsible for major pieces of lay and church music. About the Nordhausen tournament, see Richard Barber and Juliet Barker, *Tournaments: Jousts, Chivalry and Pageants in the Middle Ages* (Woodbridge, 1989), 54. Henry's rival for Thuringia was the combative daughter of Saint Elizabeth, Sophia of Brabant; see Hans Patze and Walter Schlesinger, eds., *Geschichte Thüringens* (Cologne, 1974), 2, pt. 1:43. In his last decade, Henry made Dresden his capital and fostered the city's development; is this relevant to the river background? I have not identified the four figures on lower right.

62. Falckenstein, 1733; Homann, 1753.

63. Nolin, 1733. Cf. Le Long, 1:36, no. 416. About Nolin, see *Biog. univ.* 21:13; Pastoureau, *Atlas français,* 357 (esp. Nolin Sr.; see also *Biog. univ.* 10:334); the Nolins lack esteem. Diocese maps, Nolin, 1715/E; see ch. 1, nn. 88–93, above.

64. Vaissete, 1749.

65. Lauenstein, 1745.

66. Falcke, 1752. The Latin original of the titles translated above is in the Catalogue of Maps and Atlases. About Falcke, see *ADBiog.* 6:546–47. His assistant was named Spangenberg.

67. Strebel, 1761. He styles himself *Consiliaris intimus Onoldinus.* See *ADBiog.* 36:551.

68. *ADBiog.* 36:551.

69. Schöpflin, 1751. On Schöpflin, Baden-born but otherwise a lifelong Alsatian, see *Biogr. univ.* 38:407–9; *ADBiog.* 32:359–68; esp. Voss, *Universität, Geschichtswissenschaft.* For his fondness for titles, see Chantal Grell, "J. D. Schoepflin et l'historiographie française des lumières," in *Strasbourg, Schoepflin et l'Europe au XVIIIᵉ siècle,* ed. B. Vogler and J. Voss, Pariser historische Studien 42 (Bonn, 1996), 255. Biographers stress Schöpflin's gregariousness and skill as a conversationalist. A lifelong bachelor, his aversion to women was notorious. His skill with words won him appointment as university orator, including the task of delivering an annual Latin panegyric of the king of France.

70. On d'Anville's contribution, see Voss, *Universität, Geschichtswissenschaft,* 249–50 (based on Schöpflin's correspondence); I do not know whether d'Anville had anything to do with the Strasbourg items of the next paragraph. The map of Alsace is relatively narrow; it reaches well south of Basle; its limits are 47° to 49°15′ north latitude, 35°40′ to 36°15′ east longitude. Two handsome cartouches are signed as being drawn and engraved by Weis, Strasbourg.

71. Imaginary views of Gallic and Roman Strasbourg precede the map of Strasbourg with insets. I have no information about Stiedbeck. François de Dainville, "L'Alsace comme la voyaient les cartes anciennes," in Dainville, *La cartographie reflet de l'histoire,* ed. M. Mollat de Jourdain et al. (Geneva, 1986), 426–36 (with illustrations); quotation, 436. Dainville explains what each of the plans of Strasbourg portrays. Was he aware of the Amsterdam historical map (ch. 2, n. 153, above)?

72. On religious tension and the move to Mannheim, see *ADBiog.* 15:331. On Schöpflin and Mannheim, see Voss, *Universität, Geschichtswissenschaft,* 204–21. On Karl Theodor, Elector Palatine and, later, Elector of Bavaria (1742–99), see *ADBiog.* 15:250–58; inclinations to poetry, etc., 250; cultural societies in Mannheim, 252. In 1777, the death of a childless relative made Karl Theodor head of the Wittelsbach house and Elector of Bavaria (252), in which roles he was apparently deplorable. On Lamey, see *Biog. univ.* 23:89; *ADBiog.* 17:568.

73. Voss, *Universität, Geschichtswissenschaft,* 284–89 (about *Alsatia diplomatica*), 272–81 (Schöpflin provided his native state of Baden with a history), 205–6 (local history central to the establishment of the Mannheim academy).

74. Ibid., 209–10, 212, 216–19 (about the *Codex Laureshamensis diplomaticus*), 215–16 with n. 81 (*gau* research). On the title "N. illustratus," see ch. 2, nn. 87–88, above; it did not imply a fixed program for the contents. Also on Mannheim and *gau* research, see Curschmann, "Die Entwicklung der historisch-geographischen Forschung," 141–46; Curschmann believed that the theory of *gau* origins associated with the Mannheim academy was "in the air" at the time and that neither Lamey nor anyone else claimed to have originated it but only applied and exemplified it. For opposition in 1831, see ch. 5, n. 154, below (Spruner). For the Lorsch cartulary, see Lamey, 1766.

75. Quotation, Curschmann, "Die Entwicklung der historisch-geographischen Forschung," 142 n. 2. "Map (*descriptio*) of" has been omitted from all the translated titles. On the roundabout way these maps came to my attention, see Introduction, n. 15, above.

76. Orientation west, nos. 2–4, northwest, 7. Voss, *Universität, Geschichtswissenschaft,* 216 n. 81, cites a work of 1973 showing the continuing currency of Lamey's *gau* geography.

77. Kremer, 1778.

78. Güssefeld, 1796; NOT SEEN. Curschmann, in *Catalogus mapparum,* 181. However worthy Güssefeld was, he comes somewhat late to have authored the first historical map on

a scientific basis. A good selection of maps by Güssefeld, especially of parts of Germany, is in an *atlas factice* at MilwaukeeAGS.

79. Schmidt, 1800.

80. *Algemeine Welt-Historie* (Halle, 1744), 36:108. The same volume has a small, unclear map of Charlemagne's empire (opp. 15). About Nithard, see Janet L. Nelson, "Public *Histories* and Private History in the Work of Nithard," *Speculum* 60 (1985): 251–93. Delisle, 1717 (lost MS map, see Le Long, no. 413), and Georgisch, 1732, had already complemented the empire of Charlemagne with the partition of 843.

81. Spruner, 1850/A, 1837. To be fair, Spruner's European atlas includes quite a few non-European subjects, such as the Islamic caliphates and successor states.

82. Köhler, 1730, pt. 3, no. 12 (last in the total collection). The surprising omission of Delisle's Byzantine maps (see n. 85, below) is probably explained by their previous incorporation into Faber, 1711. Anville, 1762; very special in purpose, it was never reprinted outside the proceedings of the Paris academy.

83. On the Louvre collection (officially called *Corpus Byzantinae historiae,* Paris, 1645–1711), see Louis Bréhier, *Dictionnaire d'histoire et de géographie ecclésiastique* (1938), 10:1512–14, explaining several complications. See also Georg Ostrogorsky, *History of the Byzantine State,* tr. Joan Hussey, 2d ed. (Oxford, 1968), 3–5.

Banduri, 1711. Older accounts of Banduri are superseded by S. Impellizzeri and S. Rotta, "Bandur, Matteo (Banduri Anselmo Maria)," in *Dizionario biografico degli Italiani* (Rome, 1963), 5:739–50. Born in Ragusa (Dubrovnik) and won to Byzantine studies by a local antiquary, he joined a nearby Benedictine house. The celebrated French Benedictine Hellenist Bernard de Montfaucon "discovered" him in Florence and recruited him for the enterprises of the Benedictines of St. Maur. Banduri's first two decades in Paris—as a guest at St.-Germain-des-Prés from 1702 to 1724—were financed by a meager pension from Grand Duke Cosimo. The French regent (in Louis XV's minority) then became his patron. Banduri never returned to Italy. His later studies were mainly numismatic. He was denied permission to publish the notable finds of Italian humanist writings (Petrarch, Salutati) he had made in Paris.

84. Reduced copy of the first of these, Köhler, 1730, pt. 3, no. 12. About the *Patria* and associated treatises, see Karl Krumbacher, *Geschichte der byzantinischen Litteratur,* 2d ed. (Munich, 1897), 423–24, 426. Banduri's technique somewhat resembles that of Faden, 1783.

85. Delisle, 1711. For a very large map study that may have been done in preparation of Delisle's second map, see ArchNat, Marine, Serv. hydrog. 6JJ/70, no. 18 (eastern Europe and western Asia). The *De administrando imperio* has a Greek/English edition, ed. G. Moravcsik, tr. R. J. H. Jenkins (Budapest, 1949).

86. For Richard I's track, see n. 8, above. Multiple tracks of Crusaders, see ch. 6, nn. 56, 69, 70, below.

87. Delisle, 1726. Contemporary notices (not reviews) are in *Mercure de France,* March 1726, 487; *Journal des sçavans* 102 (1734): 503–5. The layout of Delisle's frontispiece resembles Duval, 1665 (map of the caliphate in pt. 3). Georges Lefebvre, *La naissance de l'historiographie moderne* (Paris, 1971), 110, reports a famous anecdote about Vertot: shown new sources about a siege of Malta he had recently described, he pushed them aside with annoyance, saying, "Mon siège est fait." Vertot's remark, now proverbial, is taken to typify how distant eighteenth-century historians were from tireless erudition.

88. Saladin: Anville, 1758. Joinville: Anville, 1757 (it was ready long before the text it illustrated). Each map is about half atlas size.

89. Delisle, 1764. See also Le Long, 1:36, no. 423 (variant title). On Joseph-Nicolas Delisle, see *Biog. univ.*, 10:335–36 (the king bought his scientific collection in 1754); *Dict. biog. franç.* 10:40–42. He drew maps but did not specialize in cartography. He, not Guillaume, is in the *Encyc. Brit.* (7:963–64). On the Delisle brothers (one went to Russia with Joseph), see *Biog. univ.* 10:336; *Dict. biog. franç.* 10:840–41, 839. A valuable guide to these collections is Étienne Taillemite, "Les cartes anciennes du service hydrographique de la Marine conservées aux Archives nationales," in *La carte manuscrite et imprimée du XVIᵉ au XIXᵉ siècle: Journée d'étude sur l'histoire du livre, Valenciennes, 1981,* ed. Frédéric Barbier (Munich, 1983), 19–32.

Joseph Delisle, *Mémoire sur la carte de l'ancienne Palestine ou de la terre sainte* (Paris, 1763), 9–11, discusses Guillaume's papers and MS maps at the Dépôt de la Marine. Joseph Delisle, *Mémoires sur trois cartes nouvellement publiées en juillet 1764* (Paris, 1764); the maps in question are (1) "Carte particulière de la Syrie, ou seulement des partis septentrionale à la Palestine" (one of the maps for Vertot's book; Joseph indicates he retouched it); (2) "Une carte plus générale de la Syrie, qui comprend en même tems [*sic*] la Palestine avec les Pays voisins, pour servir à l'histoire des Croisades, etc." (= Delisle, 1764); (3) "Une carte encore plus générale et plus étendue pour la recherche de la situation du Paradis terrestre" (see next note). Contemporary comments: *Journal des sçavans,* July 1763, 239–42; ibid., October 1763, 331–35; ibid., January 1765, 127–28 (Lattré bundled Joseph's *Mémoires* with the maps).

90. About the MS presented to the king by Veuve Delisle, see Philippe Buache, *Idée générale de la carte nouvelle de la Terre Sainte dressée par Guill. Delisle, et comparée avec les cartes du même pays* (Paris, 1729). Guillaume's unpublished map is possibly related to G. Delisle, "Terrae sanctae tabula," *opus postumus* (1763), in L. C. Desnos, *Atlas moderne* (Paris, [1771–83]).

Delisle, 1764/E. About the duc de Choiseul, see *Biog. univ.* 8:182–86; *Dict. biog. franç.* 8:1219–22. His support of the philosophes and the *Encyclopédie* is well known. The interest of Choiseul and his cousin (and political associate) in these Holy Land maps may need explanation. Possibly, though, only routine support for the arts and sciences was involved, without special concern for the subject.

Delisle, 1766. Kadesiah is the battle of 637 at which the Muslims overcame the Sassanian (Persian) Empire; Benjamin of Tudela, a Jew from Spain, is a famous twelfth-century traveler (see *Encyclopaedia Judaica* 4:535–38); Pedro Teixeira (*Biog. univ.* 41:206–7) crossed from the Persian Gulf to Aleppo in 1604–5 and wrote *Relaciones de Pedro Teixeira* (1610).

91. Delisle, 1764.

92. On the Pétis family (including the maternal uncle of François Sr.), see *Biog. univ.* 31:585–87. Alexandre supplied a biographical notice of his father at the opening of the Tamerlane volume; see Nolin, 1722. Alexandre was named to the Collège royal in 1744; he had no son. The long entry of François Jr.'s writings in *NUC* bears out the success of his literary translations.

93. Delisle, 1710a.

94. Delisle's preparation for the Pétis map was a modern map of "Tartarie" in 1706. Putzger, 1877: 36th ed. (1913), no. 16b; cf. Shepherd, 1929/O, no. 92; a family resemblance continues in Kinder and Hilgemann, *Anchor Atlas,* 1:178, despite the introduction of campaign tracks.

95. Main entry under Nolin, 1722.

96. The Caspian in Delisle's Mongol map is no better than Nolin's, and both maps lack

a Sea of Aral. The fault stems from the legacy of Ptolemaic cartography and is paralleled in contemporary atlases. For a new map of the Caspian, commissioned by the czar (1719–21), with an outline much more familiar to our eyes, see Bernard, publ., *Recueil des voyages au Nord*, 7:301.

97. The lettering of the original omits accents. "Royaume de Gete," also in Delisle's map, corresponds to the Khanate of Chagatai, not the ancient Getae of Europe.

98. Bellin, 1759. *Algemeine Welt-Historie* (Halle, 1760), 21:475; (1761), 22:294.

99. Van der Aa was encountered above, ch. 2, n. 85 (he is abbreviated Aa in the Catalogue of Maps and Atlases). For disobliging comments about him, see Koeman, 1:1–2. About Pellegrino, see ch. 2, n. 106, above. Van der Aa's most useful publication for detecting earlier maps is Aa, 1729a.

100. Aa, 1729b. A connection is claimed with Bergeron, 1634, which, however, lacks maps.

101. Details in Aa, 1707.

102. For whatever negative evidence may be worth, Duval, 1665, takes no notice of these travelers.

103. Aa, 1729b.

104. See ch. 3, nn. 27–30, above. Cellarius, 1776: 25 (I abridge and paraphrase the comments of the editor of 1776): This is what is left of a vast work that the great man was planning. He proposed to write about medieval geography on the same scale as ancient, and to this end quite a few copperplates were engraved. But death cut off the project. Since that time the publisher, Gleditsch of Leipzig, carefully saved the plates, hoping that someone else would take up and finish the task. Long disappointed, he was told that, if the job were done today, new plates would be needed. So he decided to release them as they stand rather than have them forgotten. Let the reader now have these remains, unfinished as they are, as a memento of Cellarius.

105. Cellarius, 1701b/A, the major geography. Cellarius, 1701a/A is his best-seller. On Samuel Patrick, see *DNB* 44:44 (for a time second master at Charterhouse). I have come across Cellarius, 1776, only at BerlinPK and Yale. The few copies surviving may suggest how little of the print run was sold.

106. Cellarius, 1776. Cellarius's normal inspiration is typified by the inept legends (4) "Medieval Roman Britain" and (2) "Medieval Gaul under the Romans." The maps that I mark with an asterisk contain some name or other feature genuinely stemming from the Middle Ages: (1) Italia Medii Aevi; (2) Galliae medio aevo conspectus sub Romanis;* (3) Gallia, Francia et Gothia; (4) Britannia medii aevi Romana;* (5) Saxonia Transmarina;* (6) Belgica medii aevi;* (7) Germania medii aevi; (8) Rhaetia and Noricum; (9) Illyricum occidentale; (10) Illyricum orientale; (11) Thraciae sex Provinciae; (12) Asia Minor; (13) Armeniarum et Syriarum medio aevo conspectus; (14) Palaestina; (15) Aegyptus; (16) Africa occidentalis medii aevi; (17) Libya medii aevi; (18) title not noted. Cellarius apparently reasoned that Constantine or Diocletian marked the end of antiquity and that any later time, such as the fifth century, was medieval.

107. Spener, 1717. See ch. 3, nn. 20–21, above.

108. Spener, 1717, 1:245, 271 (Spener's first map is exclusively "ancient"). The brief description of the maps is on the frontispiece, a handsome composition (including a map). The maps are attractive, with decorative drawing along the lower edges. In the map of Rhaëtia, the reference to the Lombards excludes the possibility that the "Huns" should be the fifth-century invaders normally so called. The Asiatic Avars are the likely people once Lom-

bards are mentioned (later sixth to late eighth century). If "Huns" is used in a wholly loose way, then the Hungarians (from the very end of the ninth century) are possible.

109. Hauber, *Versuch einer Historie*, 133, clearly affirms that Spener used Cellarius's maps (that is why Spener's maps of old Germany were so good, he added). Hauber does not make it clear whether the maps he meant were those of the published *Notitia* or the medieval ones deposited with the publisher, of whose existence he was aware (137).

110. Anville, 1719; see ch. 3, n. 12, above (biographical details).

111. Scattered medieval maps: Anville, 1757, 1758, 1762. Beaurain, 1749, nos. 54–62 (cited ch. 3, n. 58, above). Beaurain mistakenly classes all nine maps as "géographie du moyen âge."

112. *Gallia,* 1720. Probably the sheet is from a book not yet identified.

113. Köhler, 1730. The maps cited are 1:10 and 1:11; the subjects of the others are, summarily, Italy, Spain, Anglo-Saxon Britain, and Flanders (1:12; 2:10, 11, 12); western Europe in the Middle Ages, Charlemagne's canal, and the city of Constantinople (3:10, 11, 12).

114. Monumenta Germaniae historica, *Auctores antiquissimi,* 11:18, no. 49 (the Chronicle of Hydatius). Köhler is faithful to a part of the Chronicle, but if he had read further, he would have learned that within twenty years all *gentes* but the Sueves had left the stage, and the Goths came later.

115. Köhler, 1730, 2:10, 3:10. For Delisle's projected map, see ch. 3, n. 35, above.

116. Robert, 1742.

117. Wastelain, 1761: "Gallia Belgica ex antiquis auctoribus descripta," 4; medieval, 24; modern by "Rizzi Zannoni de la Société Cosmographique de Nuremberg 1760," 74.

118. Anville, 1771. *Journal des sçavans,* October 1771, 327: "L'intervalle [between ancient and modern] doit être également étudiée, puisque c'est dans cette intervalle que les différens états qui existent actuellement en Europe ont pris naissance."

119. Keynes, *Library of Edward Gibbon,* 21.

120. Anville, 1771. For a less critical view than mine about d'Anville and the Middle Ages, see Hofmann, "Genèse de l'atlas historique," 107.

121. Early commentators withheld praise; Dacier, *Rapport historique,* 249: (about Anville, 1771) "[d'Anville] was able to trace, often even with an uncertain hand, a feeble sketch of a tableau that he did not dare or did not wish to undertake to complete."

122. Mentelle, 1782. In size and style the map could fit into one of Mentelle's atlases. The BN copy is black-and-white; colors would help. Other regions named include (possibly Merovingian date) Austrasia, Neustria, Bourgogne, Aquitaine, Gascogne; (Carolingian and after) Flanders, Normandy, Marche de Bretagne, (Spanish) March, Lonbardie [*sic*], Dauphiné, Marche de Provence. On Mentelle, see Hofmann, "Genèse de l'atlas historique," 112–13, 123.

123. Daniel, 1696. See editions and reviews in Sommervogel, 2:1795, 1806–9; critical reviews, including the 1696 map, 2:1809–10. A reviewer in *Journal des sçavans* (Amsterdam ed.), March 1730, 313–15, urges that Daniel's hypercriticism about the first Merovingian kings be offset by Bernard de Montfaucon's *Monuments de la monarchie française;* see also G. P. Gooch, *History and Historians in the Nineteenth Century* (1913; repr., Boston, 1959), 13. On Mézerai and Hénault, see ch. 3, n. 79, above.

124. For the attribution to Daniel's *Histoire de France,* see Le Long, 1:34, no. 392, who does not mention Berey.

125. Pastoureau, "La France divulguée," 61 (with an illustration of this silhouette, for which Nicolas Sanson seems to have been responsible).

126. Delisle, 1717: the ArchNat call numbers are in my main entry (Catalogue of Maps and Atlases). Le Long, 1:34–36, nos. 393, 396 (three maps), 414. Quotation from *Mercure de France*, March 1726, 485.

127. Delisle, 1721. Philip IV the Fair, BN, Ge.C.6189, with the note, "Papillon ms sura-jouté sur l'édition ordinaire de la carte de France, avril 1721." Whatever project this may have been, it was barely begun. Louis XI, ArchNat, Marine, Serv. hydrog., 6JJ/71/1, no. 19: an undated MS historical map of France mainly in the fifteenth century; historical notes on the map and beside it. The reference to Louis XI is on the back. Tenth-century France, BN, Ge.D.15788, one-sheet MS, with border coloring; the system imagined by the mapmaker is suggestive of the schoolroom. Guillaume's spinster sister was named Angélique. The note suggests Buache as receiver. The younger brother, Joseph, another candidate for being given this draft, was in Russia in 1734.

128. Louis XV, *Cours des principaux fleuves et rivières de l'Europe* (Paris, 1718); three copies at BN. Delisle's maps for Louis XV: Delisle, 1717. See also Dangeau, 1697; his atlas contains many *cartes muettes* with gaily colored boundaries.

129. For example, Auguste Longnon, sometimes considered the creator of the first real historical atlas of France, initially won fame with his *Géographie de la Gaule au sixième siècle*, which included detailed maps of these partitions.

130. Delisle, 1717. Daniel, 1696. The Le Long references are in n. 126, above.

131. Rizzi-Zannoni, 1764.

132. Loose copies of the French history maps, Liébaux, 1728. Daniel's publisher comments in Daniel, 1729, 1:xci, xcvi. Three maps are at 1:1; the fourth at 1:77. The original French history map is Daniel, 1696. Liébaux as engraver for Delisle is attested particularly well in *Catalogue of Maps and Charts of Harvard* (1831). He and Delisle made maps for F. Catrou, S.J., and P. J. Rouillé, *Histoire romaine,* 20 vols. (Paris, 1725–48); Delisle, vols. 1–4; Liébaux, vols. 6–7, 9–10, 12–16, 19. On Liébaux, see Pastoureau, *Atlas français,* 146 (nos. 64, 66), 148 (Duval), 159 (no. 1, de Fer XIIA 1697), 424 (no. 41, 1703); Tooley, *Mapmakers,* 392. He lacks even a brief biographical entry in works about cartographers.

133. Liébaux, 1728, with details and derivatives.

134. Liébaux, 1728. Le Long, 1:34, no. 394. Gregory of Tours, *Historiae,* bk. 2, ch. 9 (at the end of the chapter), is the source for this causeway. Not necessarily consulted by Lié-baux, it traces Clovis's *ancestors,* not Clovis himself, from Dispargum in "Thoringia" (on the *left* side of the Rhine) to Cambrai. The location of Dispargum is uncertain, as befits a possibly imaginary place. No historian (known to me) traces a "route de Clovis" from Dis-pargum to Soissons. (Duisburg *is* identified with Dispargum, complete with an account of Frankish origins, in Braun and Hogenberg, *Civitates orbis terrarum,* 2:34.) For early barbar-ian tracks, see nn. 8–10, above. Notable anachronisms among the names on Liébaux's map involve Languedoc and Dauphiné in the south and "Vandales" to the north, between the Elbe and Oder.

135. See n. 132, above. Le Long, 1:35, no. 395, "pour la fin du règne de Clovis." Clovis died in 511, aged forty-five (as far as we know).

136. Dubos, 1742. No cartographer is named.

137. For the *Recueil* (abbreviated *RHGF*), permanently associated with Bouquet's name, see "Bouquet, Martin," in *Biog. univ.* 5:254, with a good account of its beginnings and early history; also David Knowles, *Great Historical Enterprises: Problems in Monastic History* (London, 1963), 55.

138. Robert, 1740, 1743; Le Long, 1:34–35, nos. 391, 411. Both maps are dated a year earlier than the volumes containing them. The reprints of Bouquet's *Recueil* do not include the
Robert maps.

139. *DBiogArchiv*, fiche 381, frames 44–47; two of three obituaries omit the book containing the Carolingian maps. The entries in *GV* 45:223 are abridged and inconsistent.

140. Georgisch, 1732 (with the map titles in the original Latin). Map no. 5 fits into ancient geography: "Gallia et Germania ultimorum Romani Imperii temporum aequalis."
From within the Romano-Germanic Empire—that is, from Georgisch's perspective—maps
of the ancient Roman Empire represented the "old" as compared to the "new," not something of merely historical interest.

141. RVaugondy, 1752; Le Long, 1:36, no. 412 (the map title differs slightly from Robert,
1743, but is comparably long); RVaugondy, 1756, no. 12 (atlas). Didier claims to incorporate
his father's map in the latter, but the atlas version is signed "Didier" and differs in details
from its source. A probable derivative is Blair, 1768, no. 8, "Imperium Caroli Magni occidentis imperatoris ad finem saeculi post Christum VIII." Charlemagne is the single subject
by which Blair links biblical and classical antiquity to modern maps; so, too, RVaugondy,
1756.

Hugier, 1756—two MS maps of early French history at the BN—are too unfinished to
permit discussion.

142. *Aloude Holland,* 1745. Abbreviated map titles in Koeman, 2:148, dH-1/2; the titles
or subjects do not consistently correspond to Alting's; one map is by Liébaux. The later
compiler took some pains to reshuffle the maps and avoid the appearance of a direct copy.
The atlas is disregarded (without comment) by Bodel Nyenhuis and Eekhoff, *Algemeene
Kaarten van Friesland.*

143. Robert, 1745; RVaugondy, 1756, is in many Robert atlases.

144. Julien, *Nouveau catalogue,* 65: Basel, 1748. After the title: "da den 26. Augustom.
1444. 1200 Eidesgenossen wider 60 000 von Karls VII Königs in Frankreich Sohne Ludwig
angeführte Franzosen, Engelländer und Deutschen von frühen Morgen an bis in den spaten
Abend gestritten, und, nachdem sie über 8000. der Feinde erschlagen, endlich von Sige [*sic*]
ermüdet alle bis auf 10. Mann auf dem Platze geblieben." Louis, in charge of this expedition,
was the future Louis XI. For the circumstances, see James Murray Luck, *A History of Switzerland* (Palo Alto, Calif., 1985), 88–91; Johannes Dierauer, *Geschichte der Schweizerische
Eidgenossenschaft,* 2d ed., 6 vols. (Gotha, 1920–31), 2:108–10.

145. On the casualties, see Dierauer, *Geschichte der Schweizerische Eidgenossenschaft,*
2:108–9. On the fame of the exploit, see ibid. 2:110 (source of the quoted paraphrase);
Luck, *History of Switzerland,* 122 n. 2 (the dauphin Louis was so impressed that, when king,
he worked hard to win Swiss friendship and eventually obtained it). Peter Dürrenmatt,
Schweizer Geschichte, 2 vols. (Zurich, 1976), 1:174, notes that the battle took place during
the Council of Basel.

146. The map proper occupies less than 23 cm × 46 cm.

147. On threatened neutrality, see Ulrich im Hofe in *Handbuch der Schweizer Geschichte,* 2 vols. (Zurich, 1977), 2:705; Dierauer, *Geschichte der Schweizerische Eidgenossenschaft,*
4:256–57 (including recruitment at the time).

148. Wagenaar, 1749. See also Muller, 1863/B, 4:4–5. On Wagenaar, see *Biog. univ.*
44:189–90. The map is reprinted in Velly, 1787, 1: no. 24. For commentary and attribution
of the design to Wagenaar, see Bodel Nyenhuis and Eekhoff, *Kaarten van Friesland,* 11, 15,
nos. 15, 25.

149. For Kitchin and Bowen, see *Algemeine Welt-Historie,* vol. 1 (Halle, 1744). The map of France is Kitchin, 1756. A connection seemed promising since Smollett's *Compleat History* was dated two years later (1758). On the invasion scare in the spring of 1756, see Julian S. Corbett, *England in the Seven Years' War: A Study in Combined Strategy,* 2 vols. (London, 1918), 1:86–95, 139–40.

150. On emigration after 1691, see *Encyc. Brit.* 14:779 (Ireland). For recent works, see Thomas Bartlett, *Times Literary Supplement* 4897 (7 February 1997): 29. James Ma-Geoghegan (MacGeoghegan), *Histoire de l'Irlande ancienne et moderne,* 3 vols. (Paris, 1758–62); tr. Patrick O'Kelly (Dublin, 1844; New York, 1851; repr., 1870). *DNB* 12:530–31, is unimpressed. MacGeoghegan says, "Europe . . . was surprised to see your fathers . . . following the fortunes of a fugitive king [i.e., James II]."

151. On Boudet and the Vaugondys, see ch. 3, n. 62, above. The maps: Robert, 1757.

152. MacGeoghegan, *History of Ireland,* tr. O'Kelly (1851), 117–29. On O'Flaherty, see *Encyc. Brit.* 14:756; *DNB* 14:903–4 (full of praise); *Biog. univ.* 14:182–83 (s.v., Flaherty), less laudatory but with fuller analysis and description.

153. Netherlands, 1792: "Geographische Tafel der Midden-Eeuwe van Holland, Zeeland en Friesland."

154. For the obituaries, see ch. 3, n. 31, above. For Guillaume's father and chronology, see Delisle, 1718/B. Also see Hofmann, "Genèse de l'atlas historique," 106–7.

155. Quoted in full with French text, n. 3, above.

156. T6, Delisle, 1705; T9, M1, Delisle, 1717; M4, TE5, Delisle, 1726, 1764; M6, Delisle, 1710a. See also ch. 3, nn. 31–34.

157. The *Mercure de France* description (M1, given above) suggests a combined map, such as offered by Hase; see below. For the manuscript maps in question, see nn. 126–27, above (Delisle, 1717).

158. See 387–88, below. Europe in A.D. 500, or the equivalent, is in dozens of atlases.

159. Bonacker, "Johann Matthias Haas," 273–80, is based on much archival and bibliographic research. Sandler, "Die homännische Erbe," 412–17, adds appreciative notes. The University of Wittenberg was amalgamated with Halle in 1815. On Augsburg, see Étienne François, *Protestants et catholiques en Allemagne: Identité et pluralisme, Augsburg, 1648–1803* (Paris, 1993): the tolerated status of Lutherans dated from the Peace of Westphalia (1648); their specialties included engraving and graphics. Hase's last-born also obtained a position at Wittenberg but died in his twenties. The study of Hase is rejuvenated by Johannes Dörflinger, "Werk des Johann Matthias Hase."

160. See Bonacker, "Johann Matthias Haas": maps illustrating editions of several classical texts (289–91), territorial maps (292–93), map of Germany (301–3), royal pension (Hase pleads the cost of needed instruments) (278). In brief, *Neue deutsche Biographie* (Berlin, 1968), 8:21–22; *ADBiog.* 10:743–46 (s.v. "Hasius").

161. J. M. Hase, "Anmerkungen über seine Landkarten von den grossen Weltreichen" (tr. from Latin), in *Kosmographische Nachrichten und Sammlungen auf das Jahr 1748 zum Wachsthume der Weltbeschreibungswissenschaft von den Mitgliedern der kosmographischen Gesellschaft zusammengetragen* (Vienna and Nuremberg, 1750), 7:329–47. Censuring existing historical maps, Hase refers in particular to a work he calls THEATRUM HISTORICUM, "considered splendid in other respects by booksellers and the inexperienced" (329). Its map of Alexander's conquests is cited as conspicuously bad. The work whose title is closest to this is Delisle, 1705 (from which Alexander is absent), but it is not what Hase had in mind.

Book catalogues offer nothing appropriate, other than reprints of Janssonius, 1652/A. For lack of an alternative, I think Hase meant Châtelain, 1705, which fits his specifications in all but title (*Theatrum historicum* vs. *Atlas historique*).

162. Bonacker, "Johann Matthias Haas," 288. He adds in commentary, "We do not dare to say that, by these opinions, Hase was the founder of the historical map, though much speaks in its favor. All is still in flux in the history of cartography and much still uncertain. One is on very shaky ground in trying to claim priority for a particular cartographer in the various branches of thematic mapping." Hase's comments, though admirable, apply only to the sort of map he was designing. Scenes like Delisle, 1707, or Schannat, 1724, are not taken into account. Nevertheless, Hase, when evoking the genre, envisaged the main existing type. (An astonishing amount of Hase's message was anticipated by Lubin, *Mercure géographique*.)

163. Ernst Breisach, *Historiography: Ancient, Medieval, and Modern* (Chicago, 1983), 86–88, see also 59, 83–84; C. A. Patrides, *The Grand Design of God: The Literary Form of the Christian View of History* (London, 1972), 16–20. Orosius won much popularity from being thought to be the historical mouthpiece of Saint Augustine of Hippo, whom Orosius mentioned as sponsor. Augustine's imprimatur was a seal of sound doctrine. Whether Orosius did or did not speak for Augustine does not matter here; Augustine's influence on him is beyond question.

164. Breisach, *Historiography*, 103–5, 143–44, 164, 179–81, 185, and see previous note.

165. Hase, 1743 (pt. 3): "Tabulae geographicae . . . de summis imperiis eorumque mutationibus potioribus: quarum membrum I exhibet Imperia Summa potissimum antiquiora: Aegyptium, Assyriacum et Assyrio-Medo-Chaldaeum, Persicum prius, Graecum, Romanum prius et Partho-Persicum, vel Persicum posterius; membrum II vero Imperia potissimum recentiora: Arabicum, Francicum et Romanum posterius; Mugalicum prius, Timuricum vel Mugalicum posterius, Ottomanno-Turcicum, Russicum, tandemque Sinense. Specimen exhibentes Geographiae alicuius absolutae Antiquae, Mediae et Novae." The announced map of the Chinese Empire, lacking in the copies I have seen and their tables of contents, may not have been drawn or engraved.

The Parthian and Sassanian realms are hard to mistake for "world empires," or divinely established imperial powers. While they existed, both were overshadowed by the Roman Empire. They are good illustrations of Hase's freedom from the Orosian scheme. From a perspective of ancient political history, there were impeccable reasons for including them.

166. Böhme (1719–97) later taught mathematics in the Saxon corps of military engineers: *DBiogArchiv*, N.F., fiche 145, frame 292; Bonacker, "Johann Matthias Haas," 294 n. 73. His part in the publication need not have been large; Hase, when still alive, announced the imminent publication of the atlas: Bonacker, "Johann Matthias Haas," 278.

167. Alting, 1701, is not an exception, since Alting was concerned with only one country.

168. Hase, 1743. A relative latecomer to maps of the Middle Ages, Hase is virtually the first to publish maps of modern history: (no. 21) "The Romano-Germanic Empire, 1740"; (24) "The Moghul Empire after Timur (Asia)"; (26) "The Ottoman Empire before the Siege of Vienna, 1683"; (27) "The Ottoman Empire after the Treaty of Passarowitz, 1718"; (28) "The Russian Empire (Peter the Great)." Possibly, nos. 24 and 28 do double duty as geographic maps of the lands in question. The others, however, are parts of series implying historical changes. Hase's no. 10, "Imperium Romanum post divisionem in Occidentale et Orientale," belongs to the fourth century and may be safely classed as "ancient."

169. Hase, 1739/B, unnumbered preface, refers to maps nos. 33–34. Yet Hase, 1743, has

only twenty-eight maps. The difference is probably accounted for by maps never completed, such as the Chinese Empire.

170. Bonacker, "Johann Matthias Haas," 296, lists the item within the posthumous omnibus of 1750; so does Dörflinger, "Werk des Johann Matthias Hase." Julien, *Nouveau catalogue,* 64, says 1749, also posthumous.

171. Hase, 1750b. The subtitle is baffling: "[this] is a fragment of Hase's posthumous major work about the greatest empires, whose remaining maps we promise for another time" (fragmentum est ex opere maiori posthumo Hasiano de summis imperiis cuius reliquas tabulas alio tempore promittimus). From our perspective, the "opus de summis imperiis" was published first, too close to Hase's death to allow the *Seven Maps* to precede it.

Franz X. van Wegele, *Geschichte der deutschen Historiographie seit dem Auftreten des Humanismus* (Munich, 1885), 562: "[H]e is the first who made an attempt at a historical atlas of German history."

172. Hase's *Imperia maxima* was accompanied by a narrative and by a (sketchy) time chart. The relevant title page of the *Atlas historicus* introduces section 3 (the empires) as being "pro illustratione doctrinae *in Sect. I. de* Idea Hist. Vnivers. polit. *expositae*" (from a photograph of the BSB copy).

173. For the full quotation, see ch. 3, n. 95, above. The work under review is Rizzi-Zannoni, 1764.

174. Hofmann, "Genèse de l'atlas historique,"117 n. 68, conjectures that these missing pages were the basis for the BN catalogue date of 1747 for this atlas.

175. On Eusebius, very briefly, see Breisach, *Historiography,* 80–81; on the *Universal History,* see Butterfield, *Man on His Past,* 47; *Cambridge Medieval History,* ed. H. M. Gwatkin et al., 8 vols. (Cambridge, 1911–36); *Cambridge Modern History,* ed. A. W. Ward et al., 13 vols. (Cambridge, 1907–11).

176. See ch. 5, n. 36, below.

177. See ch. 3, nn. 67–75, above.

178. *Journal de Trévoux* 64 (1764): 199–200. De Guignes is mentioned at the final "Intervale" of Dupré, 1763.

179. Quotations from Intervale 12. The medieval intervals are distributed as follows: (13) 409–54, (14) 454–534, (15) 534–68, (16) 568–81, (17) 581–746, (18) 746–827, (19) 827–907, (20) 907–51, (21) 951–1035, (22) 1035–91, (23) 1091–1169, (24) 1169–1226, (25) 1226–1328, (26) 1328–79, (27) 1379–1404, (28) 1404–98.

180. Criticism along these same lines by Edme Mentelle is discussed by Hofmann, "Genèse de l'atlas historique," 123–24. Mentelle preferred geographic to chronological order.

181. My quotations are meant to convey the flavor of the atlas, not to provide an accurate account of the vicissitudes of kingdoms and principalities. The quotations are from the very favorable review in *Journal de Trévoux* 64 (1764): 215–17, 244–45, 248–50. Some but not all the proper names used by Dupré have been turned into the current equivalents. A wholly consistent translation of name forms has not been attempted.

182. *Journal de Trévoux* 64 (1764): 252, 259.

183. Ch. 3, n. 74, above, calls attention to this sequential mode in atlases for computers.

184. Details above, ch. 3, nn. 75–96.

185. Incomplete listing of Rizzi-Zannoni's output after reaching Paris: 1759, four-sheet war maps; continuation, five sheets; map of Oldenburg, 1759 (Seutter); map of Belgium 1760

(repr., Wastelain, 1761); Rizzi-Zannoni, 1761; Bonne, 1762 (with Bonne and R. Janvier); Rizzi-Zannoni, 1762; Lake Geneva, Lake Constance, 1766 (Homann Heirs); multisheet map of Sicily, 1769. See also Valerio, *Istituzioni cartografiche,* with special reference to a major atlas of Germany. On appreciation of Rizzi-Zannoni's ability, see Jervis, *World in Maps,* 44; Alisio and Valerio, *Cartografia napoletana,* 123: "le sue capacità topografiche di grande sintesi tanto geometrica quanto figurativa." On Fontanieu, see ch. 3, n. 83, above.

186. On Rizzi-Zannoni and Galiani, see Valerio, *Istituzioni cartografiche,* 84, 90–98. About the Polish project, see Karol Buczek, *The History of Polish Cartography from the Fifteenth to the Eighteenth Century,* tr. Andrzej Potocki (Amsterdam, 1966; repr., 1982), 93, 95; map of Poland, 100–101. On the basis of Rizzi-Zannoni's own acknowledgments, Buczek blames him (sometimes intemperately) for taking undue credit for the Polish map. It was really Jabłonowsky's, Buczek claims. Rizzi-Zannoni should be spared sins he did not commit. Actual cartography differs very much from collecting the materials for a map (I gratefully acknowledge the counsel of Dr. Mary Pedley, who bears no responsibility for my conclusion).

187. See ch. 6, n. 82, below.

188. A recent, respectable exemplification of the theme is Robert Fawtier, *Les Capétiens et la France: Leur rôle dans sa construction* (Paris, 1942); tr., *The Capetian Kings of France: Monarchy and Nation, 987–1328* (London, 1960).

189. The reprinting may have been associated with the publication in that year of Gatterer, 1789/B.

190. Gatterer, 1789; no title or binding; the title given is a library label. Nine hemispheres: (1, 3) rivers, (2–3) current political divisions, (5–6) current empires, (7–8) "Neue Länder," (9) world known to the ancients.

191. Kruse, "Probe der Gattererschen Charten," 379–81.

192. The long and short perspectives on the barbarian Invasions are outlined in W. Goffart, *Barbarians and Romans: The Techniques of Accommodation* (Princeton, 1980), 4 with n. 2. An extreme form of the "short perspective" is argued by Peter Heather, "The Huns and the End of the Roman Empire in Western Europe," *English Historical Review* 110 (1995): 4–41.

193. Kruse, "Probe der Gattererschen Charten," 382. Gatterer, *Universalhistorie,* vol. 2, pt. 1 (no pt. 2 appeared), is wholly devoted to East Asian history (China, Japan, Korea, and Tibet). Different in form from vol. 1, it looks like Gatterer's reading notes of a basically annalistic digest.

194. As was typical before the French Revolution, the "states of today's world" did not have to be traced beyond 1517.

195. Kruse's criticisms, in "Probe der Gattererschen Charten," 385–99: the Kingdom of Jerusalem never looked as Gatterer shows it; Chazaria and Poland have arguable boundaries; Denmark, Norway, Sweden, Scotland, and Iceland are not colored because they do not illustrate the theme of *Völkerwanderung,* yet there were migrations in and out of these lands, as witness the colonization of Iceland. The lack of writing means that the Petchenegs, very important in the eleventh century, are supposed to convey their importance by color alone, which is impossible; the Serbs, Croats, and Bulgarians are lumped together as Slavs, yet the early Bulgarians were not Slavs.

196. The definition of an *atlas de choix* runs the risk of overlapping with that of an *atlas factice.* The former might be taken to be an atlas composed of maps chosen by a seller for

quality or other virtues, whereas an *atlas factice* is one assembled by a collector for reasons of his own, which may include just tidying up an eclectic group of maps. (With help from Dr. Mary Pedley, thankfully received.)

197. Robert, 1740, 1742, 1743; Bertius, 1623; Christian van Adrichom, in Nebenzahl, 90– 91, 94–97; Delisle, 1707, 1710b; Anville, 1719.

198. Tomka, 1751, is cited as a work in nineteen maps by Johannes Dörflinger, *Die öster-reichische Kartographie im 18. und zu Beginn des 19. Jahrhunderts: Unter besonderer Berück-sichtigung der Privatkartographie zwischen 1780 and 1820*, 2 vols. (Vienna, 1984–88), 1:60. The authoritative work of Purgina (see next note) confirms nineteen and supplies a list of titles. Eighteen are in the Budapest copy seen by me; thirteen maps are listed by title in Karel Kuchař, *Mapová Sbírka B. P. Molla v Universitní Knihovně v Brně* (Prague, 1959), 316, referred to here as the "Brno catalogue." For Tomka, 1781, see ch. 3, n. 60, above.

199. Works in English call him Tomka; others call him Szászky, the name adopted by his grandfather from the locality to which the family moved. See Constant von Wurzbach, *Biographisches Lexikon des Kaiserthums Oesterreich*, 60 vols. (Vienna, 1856–91), 40:201–2 (on Tomka, "Szászky, Johannes"), 1:235–36 (on Bel). The *Biographisches Lexikon* cites only the six-item *Introductio*. Other works by Tomka include a very long *Introductio in orbis hodierni geographiam* (Bratislava, 1749), 2d ed. by J. Severini (1777). Tomka's historical re-search is documented mainly by a book on ordeals and other aspects of early Hungarian criminal justice (1740) and what seems to be an ethnographic supplement to Bel's *Notitia* (1759). On Pressburg, see *Encyc. Brit.* 22:299. The frontispiece to the *Parvus atlas* features Maria Theresa surrounded by Hungarian notables.

I have made what use I could of the extensive account of Ján Purgina, *Tvorcovia Kar-tografie Slovenska do Pol. 18. Storočia* (Creators of Slovakian Cartography to the Middle of the 18th Century) (Bratislava, 1972), 45–75; German summary, 81–82; black and white maps from the *Parvus atlas,* nos. 36–38. Kindly supplied by Professor Johannes Dörflinger.

200. Here are the first seven titles in Kuchař, *Mapová Sbírka Molla,* listed by the num-ber-order they have in the main list of the *Parvus atlas* maps (see next note): 10, 12, 14–15, 4, 3, 1, 2. Individuals clearly were free to group Tomka's maps as suited their wishes.

201. The titles of the 1781 abridgment (Göttingen copy) have the following numbers in the main list of *Parvus atlas* maps: 5, 6, 7, 8, 9, 11. The list reconstructed in Purgina, *Tvor-covia Kartografie Slovenska,* 62–63, can be compared to the one here as follows: 1 (2 in Purgina, since his 1 is the frontispiece), 5, 6, 7, 8, 9, 2, 11, 12, 10, extra ("Hungaria in partes II et cominatus III divisa"), 13–16, 4, 3, 14–15.

202. For an ambitious map of tenth-century Hungary, see Hell, 1771. According to György Györffy, *Az Árpád-kori Magyarország Történeti Földrajza (Geographia historica Hun-gariae tempore stirpis Arpadianae)* (Amsterdam and Budapest, 1966), 5, Hungarian study of the historical geography of medieval Hungary was begun by Martinus Georgius Kovachich at the very start of the nineteenth century. The terms of reference need to be qualified; Tomka deserves better than exclusion.

203. Faden, 1783. Presumably, Faden was only the publisher; the author or designer is unidentified.

204. Andrews, 1797, twelve maps. Not in *NUC.* The BL copy lacks nos. 6, 7, 8. The long, but not very informative, introduction promises maps of peerages, noblemen, and gentle-men's country seats, but there is no trace of this. Item 1 is supposed to be "A physical map of the terrestrial globe to show the connection of the great chains of mountains, seas and rivers

with those of Britain"; it was not supplied. As a result, the twelve maps are numbered 2 to 13. Andrews's maps come in a variety of sheet sizes, including foldouts; see next note for the possibility that map 1, which must have been particularly large, existed in the original issue.

205. *AGEphem.* 1 (1798): 97–105 (closes with a list of Andrews's other maps of England); *AGEphem.* 2 (1798): 147–50; *AGEphem.* 4 (1799): 38–39. Andrews's no. 5 charts mineral waters and bathing places; fisheries are mentioned in the letterpress, not shown on the map. The reviewer may have received the first map, missing from the BL and Oxford copies.

206. Andrews has a long introduction and says what he will do without reference to what he might have been expected to do. Owing to the rarity of historical atlases—Andrews's was the first produced in England other than classical/biblical ones—he was in no position to compare his selection of subjects to those of others and justify it. His initiative is not wholly unprecedented if one recalls that Rizzi-Zannoni's historical atlas was marketed together with thematic maps of contemporary France (Desnos, 1765c).

207. Andrews, 1797: (3) "England Physical and Natural, division into watersheds"; (4) "Rivers"; (5) "Mineral waters and bathing places," drawn on an ordinary map of England with county divisions. Nos. 2–5 are colored in the Oxford copy.

208. Maps of religious and civil boundaries in Andrews, 1797: (12) "England civil and ecclesiastical divided into provinces and subdivided by dioceses and counties," and (13) "The Dioceses, rural deaneries and chapters in England." Conventional history maps: (8) "The Ancient Britons of South Britain according to Stolome" (ancient and modern names superimposed, and modern county names and boundaries written in); (9, double size) "Roman Britain"; (10) "Saxon Britain according to the Saxon chronicle" (names given in a variety of forms; track of William, duke of Normandy); (11, double size) "England divided according to the West Saxon, Danish and Mercian laws" (an ordinary county map with *Denelage, Merchenlage,* and *Westsaxenlage* lettered in at appropriate points without specific boundaries; see ch. 2, n. 24, above).

CHAPTER 5

Historical Atlases Come of Age

The century whose opening Napoleon dominated also witnessed at its start two makers of maps for history who achieved remarkable commercial success and, in doing so, set an example that their peers could hardly avoid noticing. In his best-selling atlas Christian Kruse adopts and modifies the eighteenth-century sequential plan. Émmanuel de Las Cases, author of the other publishing success, applied the layout of geography textbooks or comprehensive atlases of the world to his historical collection. Dissimilar in design and quality, both works proved memorable in the decades to come. Each of them merits extended discussion, starting with the more innovative French specimen.

The Midas Touch of Émmanuel de Las Cases

In Stendahl's *Le rouge et le noir* (1830), the young hero first turns up dazed by a blow from his father and lamenting the loss to a millstream of his most cherished book, an admiring commemoration of Napoleon called *Le mémorial de Sainte-Hélène*. The *Mémorial*, first published in 1823, was the work of a diminutive, ancien régime aristocrat named Émmanuel de Las Cases.[1] A marquis by hereditary right, and a count and chamberlain by Napoleonic patent, Las Cases attached himself to the defeated emperor after

Waterloo and spent eighteen months in the imperial entourage on Saint Helena before being expelled back to Europe. The *Mémorial* was published when
the banished emperor was two years dead; it launched the Napoleonic legend
and had an enormous success. Already somewhat padded in its first edition,
it grew longer in later ones and had a profusion of printings. Las Cases's place
in French publishing history is secure.[2]

When the *Mémorial* came onto the market, Las Cases already had one
best-seller to his credit. As an exile to London from revolutionary France, he
had compiled a historical atlas that appeared in English in 1801. Carried forward by Las Cases and nursed for many years, its frankly utilitarian program
was rewarded by astonishing popularity in every corner of Europe.

None of Las Cases's formative experiences, as known to us, laid a basis for
the *Atlas*. He was twenty-nine in 1795 when he turned seriously to authorship.
In July of that year, he narrowly avoided death in a landing in Brittany promoted by England that not only failed but took the life of his closest friend
and fellow exile, Jean-Henri de Lage de Volude. After that, Las Cases renounced politics and strove to break out of his poverty-stricken condition.[3]
He had occasionally tutored before; now he became a thriving teacher who
had to subcontract to others the excess of pupils flocking to him. His atlas,
possibly conceived and begun with Volude before 1795, was mainly elaborated
during these years. Las Cases kept the details of its genesis to himself. Frequent allusions in its pages to pedagogy suggest that he decided to transpose
his tutoring method to graphic and written form.[4]

Much or most of the atlas seems to have been completed in the later 1790s,
when Las Cases lived in Bloomsbury. His British Museum pass was renewed for
six months in July 1796. In 1799, he left London and slipped secretly into Brittany to marry the childhood friend to whom he had become engaged on the eve
of his emigration.[5] There was much to be done when Las Cases came back and
took up residence at 26 South Street, Manchester Square, London. A team of
helpers for drawing and coloring was formed; a subscription list of many hun-
dreds was gathered; and in 1801 there appeared the first incarnation of his
brainchild.[6] It was called *Genealogical, Chronological, Historical, and Geographical Atlas*, dedicated to the duke of Gloucester, and signed with the bland
pseudonym A. Lesage, which already graced preparatory publications (fig. 23).
The *Atlas* was associated with this homely name throughout its complex history. Even after the author's identity became public, "Comte de Las Cases" was
normally recorded only as a gloss to the trademark Lesage.[7] The London first
edition, complete in twenty-five "maps," sold for the high price of £4 14s.
Within a year, Las Cases widened the potential clientele by making available, at
£3 10s., eighteen *Select Maps out of Lesage's Complete Historical Atlas*.[8]

GENEALOGICAL,

Chronological, Historical, and Geographical

ATLAS,

EXHIBITING

ALL THE ROYAL FAMILIES

IN EUROPE,

THEIR ORIGIN, DESCENDENCY, MARRIAGES, &c.

TOGETHER WITH

VARIOUS POSSESSIONS, FOREIGN WARS, CIVIL COMMOTIONS, FAMOUS BATTLES, RELIGIOUS TROUBLES,
MINORITIES, TITLES AND ORDERS, COURTS OF LAW, REMARKABLE EVENTS, &c.

OF EACH KINGDOM.

By Mr. LE SAGE, No. 26, South-Street, Manchester-Square.

Dedicated, by Permission, to His Royal Highness the DUKE of GLOUCESTER.

TITLES AND ORDER OF THE MAPS,

The PLAN and CONTENTS of which are detailed in the next Sheet.

1. Genealogical, Chronological, and Historical Map of England.
* 2. Genealogical, Chronological, and Historical Map of France.
* 3. Geographical Map of France; *Two Maps on the same Sheet.*
4. Genealogical and Chronological Map of Spain and Portugal.
* 5. Geographical Map of Europe.
6. Genealogical, &c. Map of Denmark, Russia, and Sweden.
* 7. Geographical Map of Germany; *Two Maps on the same Sheet.*
8. Genealogical Map of the House of Habsburg, or Old Austria.
9. Genealogical Map of Prussia.
10. Genealogical Map of the House of Lorrain, or New Austria.

* 11. Map of Italy, *with the Campaigns of Buonaparté and Swarrow.*
* 12. Genealogical Map of Sardinia, Naples, Tuscany, &c.
* 13. Map of the Ancient World, with the Four Great Monarchies.
* 14. Ancient Greece, for the Illustration of Fabulous History, &c.
* 15. Map of the Roman Empire, marking its gradual Increase, &c.
* 16. Map, exhibiting the Transmigration of the Barbarians, &c. &c.
* 17. Map of Africa, *with the European Settlements, &c.*
* 18. Map of America, *with the European Settlements, &c.*
* 19. Map of Asia, *with the European Settlements, &c.*
20. The Turkish Empire, and the great Revolutions of Asia.

N. B. *Each Number is* 3s. 6d. *Only those marked* (*) *may be had separate, and at the same Price.*

Just Published, being a Supplement to the Atlas,

A new Edition of the *Geography of History*, or detailed View of general Modern History, from the Christian Æra to the present Time; giving the Cotemporary Princes, &c. in *Two Maps, 3s. 6d. each.*

General Picture of *Universal History*, being a true and easy Key to the Knowledge of all that passed in the World before and after Jesus Christ; in *Two Maps, 7s. the Set.*

London:

PRINTED BY J. BARFIELD, WARDOUR-STREET, SOHO,
PRINTER TO HIS ROYAL HIGHNESS THE PRINCE OF WALES.

1801.

Figure 23 A. Lesage (Émmanuel de Las Cases), *Genealogical, Chronological, Historical, and Geographical Atlas* (1801). The origins of the *Atlas Lesage* were muddied in Las Cases's own lifetime. Priority was commonly given to the Paris edition of 1803–4, so that the work might seem to have emanated from France. In whatever language, the Lesage atlas sold very well. It was perhaps the first historical collection to win great favor with the public and to spawn profuse editions, translations, and adaptations. Until 1802, Las Cases was in England as a refugee from the French Revolution. The first edition of his atlas, whose cover page is shown here, was dedicated to the duke of Gloucester and published in English by a printer in Soho. (Boston Public Library/Rare Books Department—Courtesy of the Trustees)

In one form or the other, the *Atlas Lesage* was an immediate success. Favorable circumstances in 1802 allowed Las Cases to return to France.[9] The once destitute refugee was now the possessor of 2,500 guineas (£2,625), earned from publications that had been on the market for only a few months. Las Cases later remarked that the English version of the atlas was no more than a start, a shapeless preliminary draft.[10] He was right that his work, further developed, had a great future. Nevertheless, quite a few original features were final; throughout its later permutations the work retained the (only slightly altered) layout and typography of the first London edition.[11]

Back in Paris Las Cases took cramped mezzanine rooms on the rue St. Florentin.[12] He kept his wife and child in Brittany and set in motion the French edition of his work. A subscription list of 900 was secured. In eight installments from 1803 to 1804 the *Atlas historique, généalogique, chronologique et géographique de A. Lesage* was delivered to the public. Las Cases was his own distributor, in association with the printer Jules Didot, nearby at the Louvre.[13] Subscribers had a choice of editions at 80 or 120 francs.[14]

The atlas was an immediate hit in Paris, as it had been in London.[15] From the distance of Oldenburg, Christian Kruse, slowly publishing his own historical maps, caustically compared the high-priced products arriving from England and France to Gatterer's inexpensive but, as he believed, much more valuable historical atlas.[16] An anonymous reviewer in *Allgemeine geographische Ephemeriden* raked the *Atlas Lesage* over the coals: the French, adept at so much, should leave history to the Germans. In Halle, the *Allgemeine Literatur-Zeitung* was emphatically negative. Such criticism, however cogent, had little effect. Wide sales of the atlas spoke loudly for its acceptance by the public and soon outweighed all disparagement. Within a decade or so, Las Cases's *Atlas* could not be mentioned even in Germany other than with admiring respect.[17]

In an introduction to a twentieth-century edition of the *Mémorial de Sainte-Hélène*, André Maurois comments that Napoleon's Boswell was "naïvely vain, [and] obsessed by his pathetic atlas."[18] Maurois spoke too confidently from surface impressions. Las Cases, for all his blue blood, was a self-made man, and the merchandise on which his first fortune was based can seem pathetic only to those who know nothing about it. The Paris launch of his *Atlas* was followed by a new, more leisurely edition in 1806, the version that comes nearest to being authoritative for the inaugural state of the *Atlas*.[19] In the same year, the first pirated edition was published in Florence, allegedly improved "with corrections and additions." Sustained by another Paris printing, Las Cases was able in 1807 to leave his narrow mezzanine, reclaim his wife and son from Brittany, and start family life. When he applied in 1809 to be named an honorary chamberlain to Napoleon—one in a horde of successful applicants—his authorship of the *Atlas* was public knowledge.[20]

The history of the *Atlas Lesage* in France involves continual reprintings, supplementary sheets, abridged elementary editions, and augmented student ones. Las Cases, not resting on his laurels, continually worked at the contents and commercial exploitation of his work. His daughter took over these tasks when his energies were claimed by the *Mémorial* and a seat in the Chamber of Deputies.[21]

Translations began to appear in 1809, with the first installment of a Russian version, and were regularly undertaken: Italian (1813–14, from the Florence counterfeit), Spanish (about 1826), German (1825–28), and Polish (a selection, 1844). Mme Coindé, one of the original London subscribers and a former neighbor of Las Cases's on South Street, was in charge of the English versions of 1813 and 1818. A reviewer wished she wrote more idiomatically.[22]

More complex editions followed. An "improved" but unauthorized *Atlas Lesage* was undertaken in England by C. V. Lavoisne († ca. 1807) and especially Charles Gros, a London schoolteacher; Las Cases considered it counterfeit. Published by Barfield, printer of the London edition of 1801 (but not of the Coindé translation), the atlas normally called after the deceased Lavoisne might be described as an evolutionary, improved version of the *Atlas Lesage*. It had special success in the United States, as the many copies surviving in public libraries attest. Influenced by it, the Philadelphia publishers H. C. Carey and Isaac Lea compiled a historical atlas of the United States in the manner of Lesage, promptly translated into French by J. A. Buchon, who, while echoing the reference to Las Cases, disregarded Carey and Lea.[23] The "method and plan" of Lesage were also invoked by Adrien Jarry de Mancy (1796–1862) in three or four atlases and tableaux, wholly without maps, concerning a variety of subjects, notably ancient and modern literatures (1826–30s). Another derivative, this time unavowed, was the *Atlas politique de la France . . . depuis 1789 jusqu'au règne de Charles X* by Weiss de la Richerie. An ambitious Dutch encyclopedia chose for its initial installment an enhanced version of Las Cases's map of the barbarian invasions.[24] In 1827, Joseph Marchal in Brussels produced an improved, probably unauthorized edition of the *Atlas Lesage*. It was reprinted several times till 1853.[25] A Venetian omnibus collection, incorporating an improved Italian translation and augmented with the American material of Buchon and the literary history of Jarry de Mancy (also translated), formed an impressive volume, destined for wide sale. A more modestly expanded and enriched French version went no farther than a manuscript now in the Library of Congress.[26] The *Atlas Lesage* fathered a numerous progeny in direct and collateral lines.

Las Cases died in 1842. For all his success in the marketplace, he seemed unconcerned with money, spent freely, and was something of a gambler.[27] From the 1820s on, he was preoccupied with the *Mémorial* as well as with political activity as an unreconstructed Bonapartist. Nevertheless, his *Atlas* held its own against competitors throughout his life and affected the design of future collections.

The *Atlas Lesage* is largely detached from the historical atlases of its time. Many of its "maps" consist entirely of letterpress. There is no sequence of geographic images at successive intervals of history. Its passage from English into French brought the qualifier "historical" from third to first position; the primacy of this adjective suited a comprehensive package of aids to history. Las Cases had the same multidisciplinary ambitions as the compilers of the Châtelain atlas of a century before: "This work presents the complete union of history, geography, chronology, and genealogy; these four sciences whose relations are so intimate are here constantly blended together."[28] Time charts, genealogical tables, "geographies" of history were, as he saw them, on an equal footing with topography. Many are color coded by washes and set out in such a way that location on the page might facilitate memorization.[29] Some features of the work had appeared in earlier atlases. Claude Buy de Mornas's *Atlas méthodique et élémentaire,* in 1761–62, was flanked or otherwise accompanied by writing, even in ample quantities. Abbé Courtalon's *Atlas élémentaire de l'empire d'Allemagne* (1774), updated by Mentelle in 1798, is like the *Atlas Lesage* in giving geography only a fractional share among the other aids to history (fig. 24).[30]

Las Cases notes in the *Mémorial de Sainte-Hélène* that geography was the weak part of his atlas and that the maps (the pages of the atlas that first fascinated Napoleon) embodied much less work and depth than the historical and genealogical tables.[31] The last-place position of "geographical" among the adjectives in the title makes the same point. Geographical maps are not only few but shoddy: coarse outlines, sporadic place-names, no grid. An early reviewer observed that the absence of fixed coordinates of any kind in the map of Hannibal's expedition made it impossible to check any positions.[32] Las Cases in London cannot have intended to equal or even approximate the cartographic standards of his day, nor was he concerned, later on, to improve this aspect of the atlas.[33] Others were, led by such adapters of his work as Gros and Marchal, but not he.

The early editions of the *Atlas Lesage* in London and Paris appeared in an unsystematic rush that affected their layout. Las Cases, hastily issuing installments to his Paris subscribers, notified them with a touch of exasperation that each buyer was free to assemble the sheets as seemed best to him.[34]

The original pattern, where discernible, seems to involve, first, a succession of current European countries, sometimes with medieval backgrounds; next, maps of classical antiquity ending with the barbarian "Transmigration"; finally, the rest of the world. By 1809, when publication of the *Atlas* had become more systematic, the "ancient" section moved to the beginning, the European countries occupied the central position, and the other continents stayed at the end. This change was better adapted to the chronological bias of historical readers and was also consistent with a basic structure that moved from the general to the particular.[35]

The layout of the entire *Atlas Lesage* stemmed from the then-standard geographic atlas of the world rather than from the collections of "sequential" historical maps. A contemporary German well versed in developments since 1750 spelled out what readers were entitled to expect of a work with Lesage's title: "A historical atlas must . . . supply many topographical maps (*Landkarten*) and they must follow each other in chronological order, so as to present to the eye the gradual changes in the setting of events." A like-minded commentator observed that the German public asked for much more from a historical atlas than Lesage offered; not content with fragments, they wished "to trace the course of world history in its continuity." Such comments imply that the sequential format of Dupré, Rizzi-Zannoni, Gatterer, and those like them had become compulsory.[36] Not for Las Cases, however. He had no commitment to quantity and continuity. Each of his maps stood by itself, as the focus for letterpress amplifications, and not in a series with other maps.

Las Cases took a different course from what others considered normative: "[Lesage] believes that a historical atlas has to be organized in the way our geographical atlases are generally organized. In these one finds first a map of the globe, then Europe with individual maps of the individual states and their provinces, then the remaining parts of the world with the individual maps pertaining to them." The reviewer showed how the *Atlas Lesage* followed this scheme, partly in its tables. Lesage goes astray, he continues, by not realizing that particular topographic maps—those, that is, of a geography—have to be much more detailed than historical ones: "It is enough to put in our heads one, to some extent complete, general historical map of Europe. . . . [A] particular history of Italy or of any other country interests us only in a moral or aesthetic respect, or in connection with the history of the whole." In other words, a general panorama of European history was enough; Lesage's particular maps of France, Italy, and the other lands mistakenly transposed to a historical atlas a design suited only to geographic ones.[37]

In criticizing Las Cases's scheme, the anonymous reviewer tacitly advocated the design of the rival atlas of Christian Kruse, which did well (as we

D'ALLEMAGNE.

gnes, sous Charles-le-Gros, ...
magne. A la mort de ce prince, ...
... à Rome, les peuples élurent ...
... que sept ans : celui-ci mourut ...
... gréenne allemande.
... se croire dégagés de toute sujé-
... ique indépendants ; l'extinction de ...
... ... roient sans doute jouir chacun ...
... ... clamé ; cependant ils ne le firent ...
... ... e craindre tint lieu de justice.
... e Charlemagne, avoient brisé les ...
... ... dévastatrices jusque dans le cœur ...
... ... Vénètes et les Danois désoloient ...
... ... porta les Germans à resserrer ...
... ... Toute la nation s'assembla, et ...
... ... à cause de son grand âge, et ...
... ... peu de temps, et mourut avant ...
... tchie. Les peuples, assemblés de ...
... ... et présentèrent la couronne au ...
... pre, l'accepta.

DE SAXE.

les vertus militaires aux talents du ...
... ... et bien rare dans des temps ...
... ... semblaient avoir effacées, ...
... il repoussa les Hongrois sur leur ...
... à Conrad son prédécesseur ; il ...
... France ; il recouvra la souverai- ...
... renferma enfin les Danois dans ...
... ... guerriers de Henri ; voici en ...
... ... l'ordre parmi ses vassaux, et les ...
... milices régulières, qu'il stationna ...
... ... abattit, rebâtit les villes détrui- ...
... ... enfin une place parmi les plus ...

... l'éclat de ses armes et la gloire ...
... maître. Les commencements de ...
... ... ; il les soumit, les dépouilla, ...
... battit aussi le roi de Danemark, ...
... chrétienne. Tant de victoires rem- ...
... ne dans le midi. Alix ou Adélaïde ...
... ore, gémissoit dans les fers d'un ...

... de sa prison : elle lui fit proposer ...
... verse les Alpes, vole auprès d'elle, ...

... avoient eu le loisir de s'occuper ...
... Charlemagne avoit imprimé tant de ...
... lie.

... e. Comme ce titre d'empereur ap- ...
... in, on crut que c'étoit à Rome ex- ...
... seul à le donner. Aussi les prin- ...
... pire de Charlemagne, profitèrent ...
... confusion qui désoloit la France et ...
... ne s'occupèrent plus dès-lors qu'à ...
... ... lambeau sanglant. C'est ainsi que ...
... les Bérenger de Frioul, les Gui de ...
... à de simples marquis d'Ivrée ; ils ...
... nt plusieurs le portent à la fois. ...
... imperial. Il s'élance dans l'arène ...
... Alors il marche fièrement à Rome, ...
... ps, avilie par la foiblesse, mais de ...
... ndeur sur le front de Charlemagne. ...
... noissent à jamais la souveraineté de ...
... nommer à tous les bénéfices de leurs ...
... ire personnelle, attacha pour tou- ...
... ... propriété de l'Italie à ses succes- ...
... glise ; dons funestes, qui, nous dès- ...
... calamités, et feront couler le sang ...
... it les palatins provinciaux, et finit ...
... ent et de Capoue sur les empereurs ...

... t ans. Il employa tout son règne à ...
... à réprimer l'audace des Romains, ...
... au moment d'une révolte générale ...
... des Bohémiens, conduits par le célè- ...
... ard'Ihui. ...
... pere. Il lui succéda néanmoins, ...
... ... que des révoltes apaisées en Alle- ...
... Théodora, remplissait Rome de ...

... sordres et de confusion. Jusque-là ...
... le leurs peres. Othon n'en laissoit ...
... trône. Enfin le duc de Baviere l'em- ...
... regna sous le nom de Henri II. ...
... le clergé de factieux dangereuses. ...
... dance par des surveillans militaires ...
... ... de dévotion, supprima ces officiers ...
... ... ta-fait indépendants. Il fut occupé ...
... x en Italie, toujours troublée par les ...
... nvahir par le midi. ...

FRANCONIE.

... rit de la maison de Saxe, quoiqu'il ...

... Othon-le-Grand par les femmes, fut ...
... ... successeur. Jamais élection ne fut ...
... mp par les vœux des évêques, des ...
... ... demeurèrent campés pour cette élec- ...
... ... dans les belles plaines de Worms ...
... ... ognes à la mort de Raoul ou Rodol- ...
... ... ient en grand nombre, Conrad étoit ...
... ... traités, et il les soumit par la force. ...
... ... es entre l'Aar et le Reuss en Suisse. ...
... vaste pays, profitant des troubles de ...
... ubra presque aussitôt. Les seigneurs ...
... onserva qu'un vain titre d'archiduc. ...

... ... aint, aimant les lettres et les culti- ...
... ... : il mit de l'ordre dans son empire ...
... l'Italie sans troubles, fit renouveler ...
... e sans le consentement de l'empe- ...
... ... , qu'il les nomma tous. C'est sous ...
... important pour l'empire en ce qu'il ...

... ... cèda à l'âge de six ans. Ce règne est ...
... ... par les infortunes du prince, et les ...
... et les italiens d'une faute à laquelle on ...
... ... ministres qui indisposent les peu- ...
... soulèvements et facilite l'audace des ...

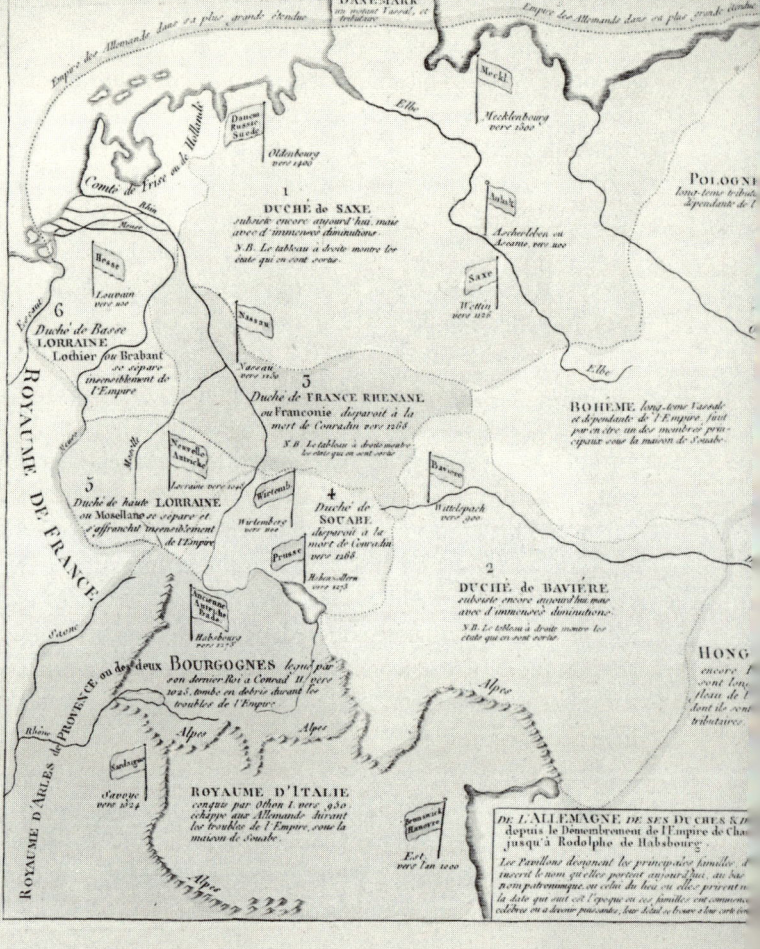

DE L'ALLEMAGNE, DE SES DUCHÉS, ET DE SON HISTOIRE JUSQU'À RODOLPHE DE HABSB...

OBSERVATIONS GÉNÉRALES.

Plan du Tableau et des Matières.—L'Allemagne depuis le démem-brement de l'empire de Charlemagne jusqu'à l'accession de Rodolphe de Habsbourg présente un intervalle de plus de 500 ans, divisé en quatre périodes bien distinctes ; celles des maisons de Saxe, de Fran-conie, de Souabe, et la longue anarchie.

Nous traçons à côté ces quatre périodes dans des colonnes verti-cales qui renferment d'une manière correspondante entre elles leur durée, les empereurs qu'elles ont fournis, les prérogatives et l'auto-rité de ceux-ci, enfin l'état politique de l'empire.

La carte ci-dessus présente l'Allemagne, ses duchés, et dépendances durant les premiers succès de cet empire. On n'attend pas que nous en avons tracé les contours avec une exactitude parfaitement rigoureuse ; c'êtoit difficile, pour ne pas dire impossible ; puisque dans ces temps le nombre et l'étendue de ces provinces varia souvent. Nous nous sommes contentés d'une exactitude suffisante pour l'accomplissement de notre objet, qui est seulement de faciliter l'intelligence de la con-stitution germanique, d'attacher pour ainsi dire à l'entrée de ce dédale politique, et surtout de présenter, à l'aide de la carte à droite, d'une manière simple et facile l'origine de tous les états actuels d'Al-lemagne, sortis de ces provinces primitives.

Une bande ou ligne circulaire coloriée embrasse dans la carte ci-dessus, à gauche, l'empire germanique dans sa plus grande étendue quand il comprenoit toute l'Allemagne et les royaumes d'Italie et des deux Bourgognes. Une bande ou ligne circulaire semblable marque sur la carte, à droite, l'étendue de ce même empire avant les dernières changements ; et l'on voit qu'à cette époque il avoit diminué de la Hollande, du Brabant, de la Lorraine, de l'Alsace, des deux Bour-gognes, et de l'Italie ; car, quoiqu'il conservât encore sur divers états de ce dernier pays quelque souveraineté féodale, elle étoit purement

Période de Saxe.	Période de Franconie.	Période de Souabe.	La longue a...
Dure 124 ans, et donne 5 empereurs.	Dure 101 ans, et donne 4 empereurs.	Dure 116 ans, et donne 7 empereurs.	Après les trois cedent arrive un... confusion et d'an... allemands, ne c... leur indépendance ...
Ils conquièrent l'Ita-lie, et la gouvernent, ainsi que l'Allemagne en vrais souverains.	Ils héritent du royaume des deux Bourgognes, dont ils ne peuvent pro-fiter.	Ils conquièrent Na-ples, qui augmente leur embarras, en augmen-tant la haine des papes.	donnent des chef... laires dans les pro... de Cornouaille... tille. Durant ce te...
Ils nomment ou con-firment les papes, et dis-posent de tous les béné-fices dans leurs états.	Les papes secouent le joug, et commencent la guerre du sacerdoce, qui dure près de 500 ans.	La guerre du sacerdoce continue toujours avec fureur, et finit par la des-truction des empereurs.	désordre monte au ... noud s e t déchires ... convenance et la f ... du mal même un...
Ils enrichissent le cler-gé pour l'opposer aux grands vassaux, et éta-blissent les palatins pro-vinciaux pour restrein-dre l'autorité des ducs.	Le clergé, les ducs, et les princes, s'unissent contre les empereurs, et profitent de leurs em-barras pour élever leur indépendance.	L'Italie échappe aux empereurs, et les prin-ces allemands secouent le joug ; se rendent in-dépendants, et détrui-sent la monarchie.	les princes en vien... chef, et choisi... Habsbourg ; épo... importante, par... le système mona... et qu'elle place au... une maison dont...
Les fiefs, qui n'étoient à l'abord qu'à vie, com-mencent à devenir hé-réditaires.	Tous les fiefs devien-nent héréditaires, et les titres même se trans-mettent sans office.	La plupart des fiefs originaires ne sont plus que de véritables sou-verainetés.	lintérieure dans l'h... De toutes les inst... gements auxquels ... donne naissance, ...
Sous cette période les empereurs confèrent tous les fiefs vacants suivant leur bon plaisir.	Sous cette période les états gagnent de con-courir à la collation des fiefs majeurs.	Sous cette période les empereurs ne confèrent plus de fiefs ni de titres sans les états.	le la plus considér... cedent avant de... vient le siège de l... time hérédité ; p... jusqu'alors avoit e...

Figure 24 A. Lesage (Émmanuel de Las Cases), Section on Germany from *Atlas historique* (Paris edition). Critics gave unusually good marks for accuracy and comprehensiveness to the pages concerning the history of Germany, shown here. They had been in the 1801 core of the *Atlas Lesage*. Little flags and other decorations highlight the maps, which are sur-

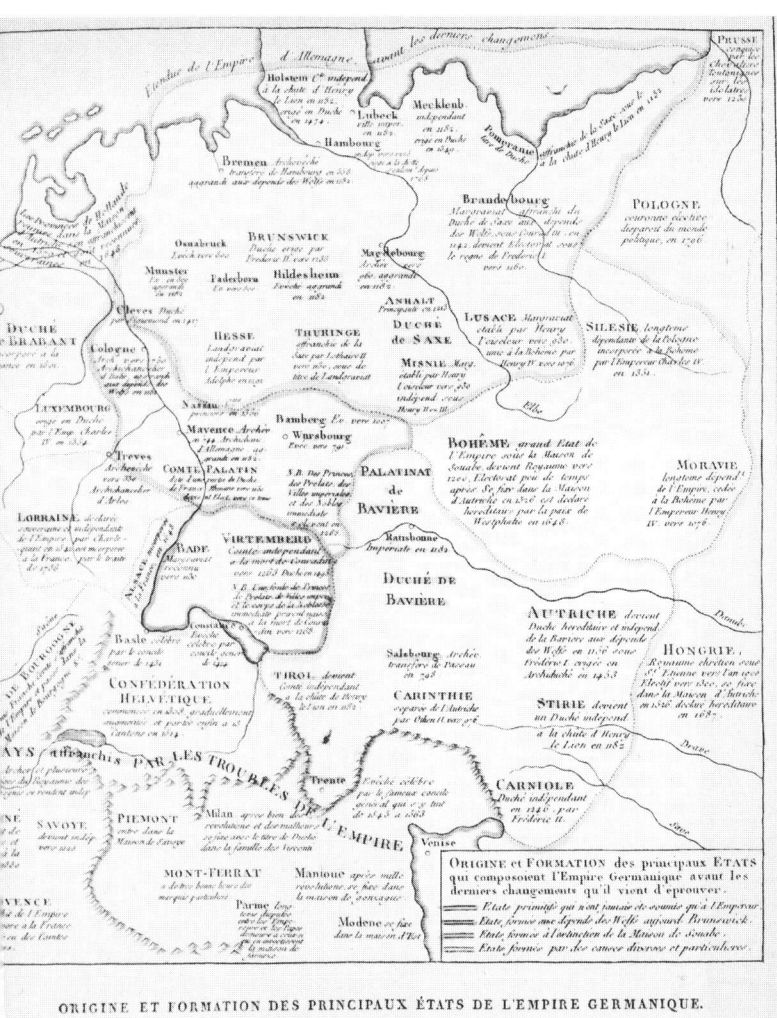

ORIGINE et FORMATION des principaux États qui composoient l'Empire Germanique avant les derniers changemens qu'il vient d'éprouver.

— États primitifs qui ont subsisté et se sont conservés qu'à l'Empereur
— États formés aux dépens des Welfs aujourd'hui Brunswick
— États formés à l'extinction de la Maison de Souabe
— États formés par des causes diverses et particulières

ORIGINE ET FORMATION DES PRINCIPAUX ÉTATS DE L'EMPIRE GERMANIQUE.

ÉTATS PRIMITIFS.	ÉTATS FORMÉS	ÉTATS FORMÉS	ÉTATS FORMÉS
PLAN DU TABLEAU.	AUX DÉPENS DES WELFS.	À L'EXTINCTION DE LA MAISON DE SOUABE.	À DIFFÉRENTES ÉPOQUES
Observations générales.	Chûte de Henri-le-Lion.	Fin tragique de Conradin.	ET PAR DES CAUSES PARTICULIÈRES.

rounded by extensive columns of print. A deliberate method was embedded in this layout. Las Cases's prize pupil, Napoleon (heading for Saint Helena), finally learned how he must proceed: "he began to pore over Asia, harmonizing the margins with the picture." (Courtesy of the Map Collection, Yale University Library)

shall see) with sequential maps of Europe. But the implied preference of this reviewer—Kruse himself might be speaking—does not make his comments less instructive. He thought Las Cases went astray not only by presenting particular maps disconnected from one another but also by being arbitrary in what he portrayed. Garish march-routes stand out in the maps: Charles XII of Sweden in that of Europe, Napoleon and Suvaroff in that of Italy, Gustavus Adolphus in one of Germany, others too. Even supposing that these tracks were beyond reproach, readers are left wondering why some campaigners are favored while many other qualified ones are left out. These maps with casually selected highlights typify a general failing of the *Atlas Lesage,* namely, the piling up of disconnected fragments: its columns of letterpress are like the odds and ends that fill out the columns of newspapers. None of the sections for countries, except, surprisingly, that for Germany, provides a consecutive narrative history of the land in question. The atlas gave the impression of being an aggregate of particles lacking coherence.[38]

As a naval lieutenant in exile, Las Cases cannot have been familiar enough with the design of atlases for the past half century to be wholly deliberate in how he went about his own, but he evidently did something right. He found his model in the standard world atlas, kept its order of presentation, gutted it of geographical content, and transformed it into an album of world history, illustrated with intermittent maps.[39] In his words: "the difficult study of history, chronology and genealogies [is reduced] to the positive and simple study of geography."[40] The continuity of historical time was displaced by a scattering of historical and ancillary information ordered primarily on a country-by-country basis. "The geographical canvases (*tableaux*) are for the most part new in their composition," Las Cases announced.[41] Interval-to-interval historical changes were not for him. He provided images that no one had yet thought of placing in an atlas and, most inventive of all, maps highlighting the history of recent centuries, hardly ever found in atlases before.

Johannes Dörflinger points out that the Lesage maps "manifest a new (more poster-like) form of representation." This observation holds true for his portrayal of the "Transmigration of the Barbarians" as well as for modern campaigns dramatized by ribbon-shaped tracks. Las Cases's ribbons, much more freely used than in earlier atlases, were also gaudier than the norm. He used most of them in maps of modern history, a commodity that he had (almost) sole title to until Kruse marketed more austerely scientific versions in 1818.[42]

Las Cases preferred synchronic maps to sequential ones. The *Atlas Lesage* reproduces Speed's battles map in an attenuated form, indicating the sites of events in a pleasing manner but also adding tracks for campaigns by Charles I,

Charles II, and the Stuart Pretender. The eclecticism that, here, combines William the Conqueror (1066), the Spanish Armada (1588), and Bonnie Prince Charlie (1745) is common to the Lesage maps. The tracks of Hannibal's invasion of Italy in the third century B.C. are drawn on a Roman Empire divided by the administrative boundaries of the fourth century A.D. The one map of Italy displays the march-routes of the latest French conquests and, in the margins, a table of the most famous Italian painters, their styles, their masterpieces, and additional aesthetic details.[43] The maps nonchalantly combine disparate times; they are scrappy, as hostile reviewers stressed; each has its particular title and object.

Content to divert and please, the *Atlas Lesage* remained, most of all, unpretentiously pedagogic. Las Cases was fifteen when he left school for the navy. He said of himself that, when a tutor in England, he taught one day what he had learned the day before. He wished to stimulate his pupils' retentive capacities. How could the multiplicity of historical facts be durably inculcated? In a context of anxious schooling, the intrusion of Hannibal into the fourth-century Christian Roman Empire mattered just as little as the seven-century convergence on a single map of William the Conqueror and the Stuart Pretender. Users of the atlas were not expected to reflect thoughtfully on a total image, only to extract separate pieces of information quickly, painlessly, productively. The barbarian invasion map, telescoping five centuries, did not intentionally project the picture of a period or collective phenomenon. What mattered was the relationship of boxes and ribbons to topographical outlines. The map was a chart, rudimentary in its geography, visually aiding anyone forced to learn seriatim the origin of the various peoples, their travels, and ending points. The *Atlas Lesage* was a tool for cramming, useful for those, young and old, who desired to equip themselves hastily with useful particles of knowledge.

Las Cases's best-seller exasperated reviewers: its vast omissions, elementary errors, and inferior maps stared out at them. Its redeeming virtues were hard to see. Yet the *Atlas Lesage* went from success to success. No comparable printing histories rewarded the pioneering historical collections of the eighteenth century. Fembo, successor to the Homann Heirs, reissued Hase's *Atlas historicus* in the days of Lesage. A map of Napoleon's empire was now added, but Hase's atlas in its half-century existence had never been improved or translated. The Rizzi-Zannoni/Desnos atlas of French history, visually appealing and original in content, had been coolly received and was presumably forgotten.[44] Against all reason, the *Atlas Lesage* managed to avert the curse that had kept specialized historical atlases from winning public favor.

Quality was not the key; the critics were right about this. Instead, Las

Cases was original in the structure of his *Atlas,* equally original in the design of many maps, and resourceful in putting together a winning package. Proper world atlases had recently begun to suggest that geography was independent from history. Las Cases, without contesting, let alone threatening this division of competence, led the way to an atlas identical in plan to its geographical counterpart—a continent-by-continent, country-by-country basis—yet so resolutely historical in content as to be only marginally geographical. Histories without maps, like Jarry's, could profess to follow in its tracks. The *Atlas Lesage* claimed to contain everything needed by young and old, students and laymen, to learn and understand history on an elementary and skeletal level. Las Cases had a flair for pitchman's prose, and the product he offered was credible: "it forms a whole library in itself; it is the *vade mecum* of the merchant, of the schoolmaster, of the scholar, and of the man of the world."[45] Comparable works, more or less extensive, had long been available. Las Cases, no expert in the material he handled, had the idea of combining an anthology of historical aids with a geographical framework not notably rich in maps but more geographical than rival handbooks; and he proposed, besides, to integrate knowledge through geography. A compelling blend of ingredients may have been Las Cases's secret, the recipe that enabled him, by grace of his first best-seller, to cast off titled penury and to join with success in the scramble for fortune and rank in Napoleon's empire.

Kruse's Europe at 100-Year Intervals

When the *Atlas Lesage* was on the eve of its London début, an unsuspecting Christian Kruse sent to the *Allgemeine geographische Ephemeriden* a long communiqué describing the work he was ready to offer to the public. He invited subscriptions to an "Atlas for the history of all European states from the origins to the year 1800." A later variant of the title is "Atlas and tables for surveying the history of all European lands and states from their first settlement to the most recent times."[46] Except for Gatterer's modest student workbooks, no comparably thorough atlas had appeared in Germany since Hase (1743, 1750) or elsewhere since Dupré (1763). Kruse wrote from the Duchy of Oldenburg in northwest Germany; he signed himself as tutor to its serene princes, that is, the sons of the reigning duke. Whenever, in atlases of the next 150 years, we find panoramas of Europe in or near a double-zero year, as we often do, they have a good chance of descending from the work Kruse announced. It lent itself more easily to discreet imitation than Las Cases's.

Kruse's announcement appeared in the Weimar journal when his atlas was not yet ready for instant delivery. An interested clientele was won by the initial installments, in 1802 and 1804, but completion of Kruse's *Atlas und Tabellen* lay almost twenty years in the future. Six editions were eventually published down to 1841, in association after 1816 with Kruse's scholarly son Friedrich (1790–1862), professor of history at the University of Halle, then at Dorpat (Tartu), who shared the title page from then on. The long-delayed sheets for 1600–1816 were distributed in 1818.[47]

Kruse senior, who expected readers to have a little Latin, planned to issue his work for both the German and the French markets and provided the maps with a French title in the upper margins, as the firm of Homann Heirs had long done on its sheet maps. The French version lapsed after a single install- ment. A full and more ambitious French translation appeared in 1836, com- plete with a pointed remark that the "beau livre" of Las Cases had shown but not satisfied the need for maps for history. The translator Félix Ansart re- cycled Kruse's maps, slightly altered, under his own name in 1840 and 1846.[48] Meanwhile, the German version was guided steadily by Friedrich Kruse to its final edition. The original array of maps, somewhat improved and refreshed in the third edition, could still count on sales forty years after having been first placed in circulation.[49]

Kruse's *Atlas und Tabellen* has more to commend it than commercial suc- cess. Though not universal in scope, it probably marks the culminating point of the sequential atlases that have long had a starring role in my narrative. Kruse's concentration on Europe may have contributed markedly to its favor- able reception. The Paris draft and Dupré were weighed down by a base map reaching to the Sea of Japan. Europe, focal point of the clientele's interest, was unavoidably small in a universal design. Kruse escaped both these drawbacks. Along with the somewhat younger Las Cases, he made the changes in scope that herald the coming wave of historical atlases. The prize for popularity goes to Las Cases, that for cartography and scholarship to Kruse.

A native Oldenburger, Christian, or Karsten, Kruse (1753–1827) was born into poverty so dire that his parents could not afford to rear him. Family friends arranged for the boy to be taken at age ten into the Halle orphanage.[50] His hardships were not at an end:

> So as to complete his studies . . . at the University of Halle [a French biogra-
> pher writes], he had to engage in every kind of work that the courage of poor
> students from across the Rhine resorts to when they are determined to learn.
> Back in Oldenburg, Kruse combined the paltry functions of under-cantor at
> the Church of St. Nicolas and under-corrector at the Gymnasium without
> finding enough pay in both positions to make ends meet, even after joining

to them an evening class for young women. Marriage into modest affluence enabled him to win notice.[51]

A good match saved Kruse from the indigence of his beginnings. He established a family, became tutor to the sons of a court official, and settled into a strict and unvarying work routine. Undersized (like Las Cases) and physically frail, he long outlived the expectations of the anonymous guest who, at his wedding, said to his bride, "But how long will you have him?" He was worth having; undamaged in spirit by early hardships, he was liked, respected, and esteemed in the community. The ex-orphanage boy was named by the duke in 1788 to be tutor to his sons. Kruse accompanied them to the University of Leipzig and saw them through their studies, while himself obtaining the equivalent of a Ph.D. (1803–5). He was appointed head of the Oldenburg gymnasium on his return.

By then, the ambitious atlas project, long in preparation, had begun publication, not without subsidy from the duke. The medieval sheets, from A.D. 400 to 1500, were distributed in three installments from 1802 to 1810 and proved more than satisfactorily lucrative (fig. 25). In 1811, another of Napoleon's many reorganizations of Germany suppressed the duchy, drove the ruling house into exile, and turned Oldenburg into the Département des Bouches-du-Weser in Greater France. Kruse, who had been named ducal *Hofrat* on the eve of these changes, renounced much of his fortune and moved east into exile. Almost at once, the University of Leipzig appointed him professor of the auxiliary sciences of history. Kruse remained there contentedly to the end of his life.[52]

Kruse's atlas is simple but rigorous in design. The ancient world, classical and sacred, is omitted because amply provided for already (Kruse thought the normal "ancient atlas" rather poor in technique). The maps he offers do not pretend, he says, to vie with those of Cellarius or Köhler in contributing to a learned *geographia antiqua et media*. His work, he insists, is merely a "geography of history" (*Geographie der Geschichte*)—to us, a historical atlas. The notion of a work of this type still familiar today is thoughtfully articulated by Kruse's pen, who distinguishes it from its main competitor, the classical atlas. He spoke as a schoolmaster sympathetic to the needs of the audience: "In view of the mass of languages, sciences, and arts that the young student of today has to learn all at once (*zu gleicher Zeit*), instruction in this important part of history [medieval and modern Europe] can be only brief and hurried (*gedrängt*)." Something has to make up for the dominance of students' time by other subjects. Unassumingly, Kruse offers a complete survey of European history, intending only to be useful to young scholars and other lovers of history. The map sequence begins in A.D. 400 and continues to 1816, showing how Europe stood on 31 December at the turn of each century.[53]

Kruse was aware that the times were revolutionary: "in the present [eighteenth] century, the geographic appearance of our part of the world has been markedly altered by fourteen wars and other occurrences"; "the incessant fluctuations in the geography of all states" were much on his mind. Critical of classical atlases that bring together in the same display provinces and cities that never existed at the same time, he stresses that "no map can pass for completely true for more than a single year (and often only for months)." Periods of fifty and even a hundred years, he notes, were embraced in Gatterer's maps of the *Völkerwanderung,* but recent history shows the fatal defect of this method: can a map of Europe be colored in such a way as to portray simultaneously and correctly the countries of 1804 and those of 1794? An obvious impossibility. Even medieval examples can illustrate the error of crowding decades into a single map.[54]

Accordingly, Kruse makes a point of having his maps record only the final moment of each century—that is, the eve of each following one—leaving out all the alterations of the intervening ninety-nine years. The accompanying chronological tables (commended for their excellence by contemporary critics) record the conspicuous changes of the century. In this way, each map is associated with written modifications that allow users to trace whatever adjustments are needed for their purposes. The tables, in the same folio format as the maps, provide not only these supplements but also historical summaries and genealogical charts.[55] Where letterpress is concerned, Kruse's atlas bears an obvious and coincidental resemblance to Las Cases's: any atlas aspiring to assist historical study could not be limited to geography. In Kruse, however, text and image travel separately; geographical maps count for much more than in the *Atlas Lesage.*

The maps are large, approximating Homann products, and aesthetically appealing. Kruse's choice avoided the enormous format of Dupré and the small one of Hase's *Imperia maxima* and Rizzi-Zannoni's "France." The likeness to Homann maps was Kruse's nod in the direction of geography. The inclusion of Iceland, the North Cape, and the Urals results in a smaller scale for the heart of Europe than if the periphery had been more severely trimmed. But Kruse's draftsmanship is precise and delicate, and the coloring, whose execution Kruse promised to supervise personally in the copies destined for subscribers, is judicious, informative, and, in the original edition, exceptionally gratifying. Without deliberate efforts to be decorative, Kruse's maps are a pleasure to behold.[56]

Kruse's plate of Europe changes with every interval. Minor differences

D. *Ducatus, Duché,* Pr. *Principatus,* M. *Marchionatus* C. *Comitatus*

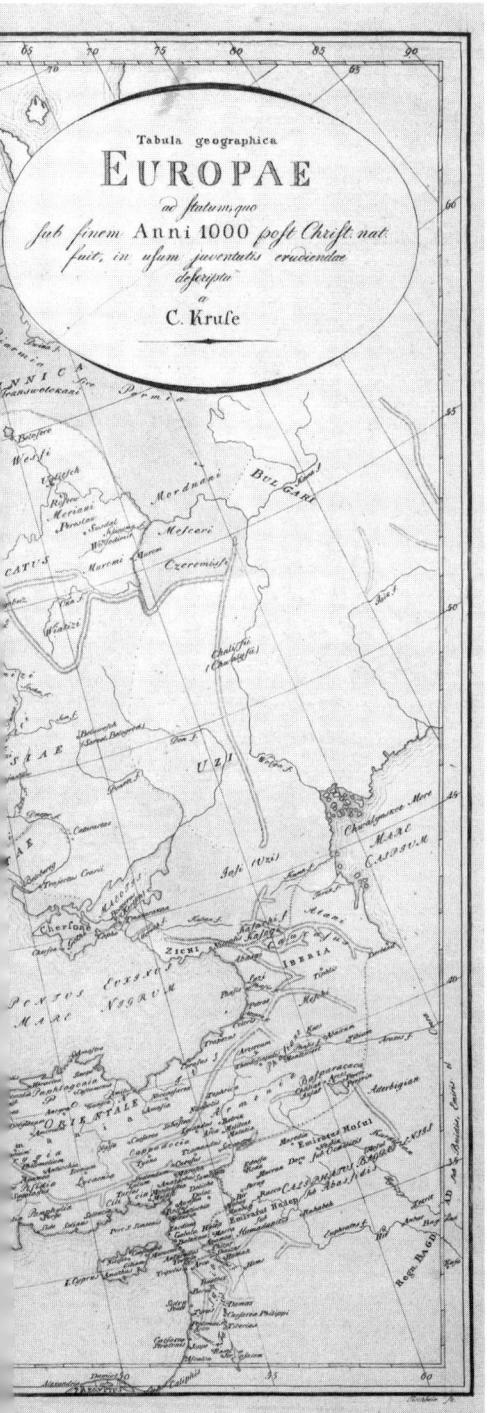

Figure 25 Christian Kruse, Europe in A.D. 1000 (1802–10). Kruse's *Atlas for Surveying the History of All the European States* (*Atlas zur Übersicht der Geschichte aller europäischen Staaten*) (Oldenburg and Halle, 1802–10) dispensed with the universal scope of earlier sequential atlases and even cut away ancient geography. Each century was allotted one map, starting in A.D. 400. The most recent times were somewhat more generously provided for. Kruse's most disputed initiative was austerely scientific: he mapped political frontiers as they stood in the year ending with two zeros. He expected that users, aided by these foundations, would trace changes within centuries from written information. Critics maintained that the intervals of such maps should, instead, be determined by historically eventful years. (Harvard Map Collection, Harvard College Library)

from map to map are numerous and easy to discern, such as the lettering of *Oceanus Atlanticus* or the drawing of Iceland and Greenland from 800 onward. Wedge-shaped segments of the oval cartouche are rolled apart in trompe l'oeil on the two occasions when, otherwise, something needing to be seen would be concealed.[57] No games are played with woods and mountains; the latter form a discreet and unvarying pattern of "hairy caterpillars," and the former are left out. Names of provinces and places, thoroughly overhauled from plate to plate, are Kruse's greatest triumph. Well aware of the need to adapt the assortment to each century and, even more, to be selective so as not to overload the plates, he created an atlas that leapt ahead of all predecessors and contemporaries in its careful handling of place-names.

Kruse was not a penetrating scholar, and he had somewhat exaggerated ideas of what could be demonstrated by means of maps. His collection is nevertheless admirable; he knew how a historical atlas should be made. The European frame he chose included as much of the world as could be dealt with on a scale permitting adequate detail, especially in the inclusion of place-names; and it was much better adapted than the broad span of Eurasia to the arena of French revolutionary and Napoleonic turmoil.[58]

A map should show only the entities that were there at one moment, Kruse maintained. Departures from this course earned his firm, though courteous, censure. Unsurprisingly, therefore, he himself portrayed frozen time. Before and after publication, critics argued with him that intervals should coincide with momentous events rather than with arbitrary moments on the calendar. Kruse's austerely scientific choice of a single day per century proved to be the most controversial feature of the *Atlas,* one that had few defenders and no imitators. He responded, with his usual simplicity, that no other course was realistic: "If, for example, I had wished to treat the most notable peace treaties as periods, ten or more maps would be needed for the present century alone."[59] His maps provided secure, thoroughly attested baselines; this standard was their primary benefit. The baselines could then be supplemented by the accompanying letterpress tables, carefully listing changes through the century. Visually impressive though the *Atlas und Tabellen* was, it was a set of images to be studied with written text in hand.

Kruse differed from Las Cases in being able to talk about his atlas. If necessary, he discoursed about subscription prices, conditions of mailing, and the need not to be hasty in engraving and coloring maps. But he was thoroughly articulate, as Las Cases most definitely was not, in discussing the sources he used, the problems of place-names, and the other details of composition. Maps following each other in succession (*auf einander folgende Charten*) mattered to him; he belonged to the tradition of the Paris draft,

Dupré, and Gatterer. In ideal circumstances, the universal scope of Kruse's forerunners was preferable to the European frame he adopted and more erudite; but other considerations mattered, too, such as public demand. Kruse's atlas, and probably that of Edward Quin (soon to be encountered), are the works in that tradition known to have sold well; and though Kruse's choice of arbitrary 100-year periods rather than momentous events failed to set a fashion, he persuaded many future atlas-makers to include maps of Europe on the eve of the next century.

The Historical Atlas Still Unknown and Disavowed

By the early 1800s, there existed a repertory of specialized maps for history detached from that of ancient geography. The habit had been acquired of assembling collections to illustrate the course of events after the fall of Rome. The maps of Kruse's *Atlas und Tabellen* are exclusively medieval and modern; Las Cases's attention to antiquity is overshadowed by his mainly modern maps. Even in atlases continuing the tradition of the Paris draft and Dupré, the balance of interest tips away from antiquity and toward more recent epochs.

However that may be, atlases of nonclassical history were still far from having an assured future. Geography was on the agenda of a committee of the Institut de France that reported to Napoleon in 1808 on "The Progress of History and Ancient Literatures since 1789."[60] The committee did not disqualify anyone from honorable endorsement on the grounds of popularization. It heaped praise on the abbé J.-J. Barthélemy's *Travels of Young Anacharsis through Greece* (1786)—a fictionalized and widely esteemed tour through the summits of Hellenic history, culture, and antiquities. Also applauded were the meticulously researched maps illustrating Barthélemy's *Travels;* they were by d'Anville's only pupil, Jean-Denis Barbié du Bocage.[61] But the committee was firm about where to draw the line: "We will not pause . . . at the *Atlas historique et géographique* of M. le Sage." Lofty sentiments justified this exclusion: "Those who work for the progress of scholarship (*science*) know that such works are virtually never more than mercenary speculations and absolutely foreign to [scholarship], since the entire life of several very learned and very hardworking men would scarcely suffice to carry out passably a plan of this scope." A century of atlases joined Lesage on the scrap heap: "When one examines in chronological order the large undertakings of the same sort [as Lesage's] that have been made for a century, one sees at first glance that the various authors of these superficial productions have done no more than copy and often disfigure the works and maps of Ortelius, Nicolas Sanson,

Guillaume de Lisle, d'Anville, and several other worthy men, to whom one must turn in order to find truly original geographers who usefully served scholarship (*la science*)."[62] Although Las Cases's atlas has shortcomings, its unequivocal disavowal in a report that praises Barthélémy suggests a misapprehension of genre rather than unswerving concern for scientific excellence.

By evoking a century of atlases—meaning general atlases—the Institut de France committee showed that its criteria continued to be those of Ortelius's day. Atlas maps were all-purpose or encyclopedic. Because history was preeminent to their clientele, specialized atlases for history were redundant. The committee was unaware that there were distinct historical collections and that the *Atlas Lesage* was one of them. Hase's atlas or Kruse's might as well not have existed. Lesage was lumped with general atlases pure and simple—the predominantly geographic works loosely called "historical," such as the Sanson-derived *Geographische en Historische Beschreyvingh* of 1683 and many more. The committee judged that, if the compilers of such works had been the creative scholarly cartographers they should have been, rather than mercenary plagiarists, they would have been unable, for lack of learning and stamina, to produce single-handedly the atlases they did. Setting up Delisle, d'Anville, and their like as models, the committee considered a historical atlas to be, basically, geographic, and to deserve respect only if it embodied estimable, original maps. The committee's dismissal seems to be extended by implication to the two (never mentioned) Robert de Vaugondys. The committee lacked the experience to judge Las Cases's atlas, or Kruse's if they had known it. Works so singular and intermittent as these were not yet recognizable, at least by this lofty body, as a distinct endeavor that deserved appraisal on a basis of its own.[63]

Even geographers might be unacquainted with works of this kind. Brion de la Tour, who was a severe critic of the *Atlas Lesage* and called himself "doyen des géographes" (he was in his eighties), complained that Las Cases's title was an abuse of language, "for the term *Atlas historique* is applied only to the voyages of discovery, whereas the one in question here contains only a few skeleton maps."[64] Brion seems to have seized upon Las Cases's lavish tracks and related these to tracks of navigators on world maps—the main cartographic home for tracks of modern adventurers until the *Atlas Lesage*. The collections of Hase, Dupré, Rizzi-Zannoni/Desnos, or Kruse were not on Brion's horizon.

The report of Napoleon's blue-ribbon committee was somewhat outdated in its appraisal of maps for history. By the time its findings were published, the atlases of Las Cases and Kruse were doing very well in the marketplace. More to the point, the journal *Allgemeine geographische Ephemeriden,* no friend of the *Atlas Lesage,* nevertheless couched its criticism (1805) in terms appropriate to the genre to which the atlas belonged.[65] Las Cases and Kruse foreshadowed the nineteenth-century outpouring of historical atlases, contrasting with the former trickle; but the wave that was to come is more visible to us than it could have been when the academicians of the Institut de France drew up their balance sheet of the humanities. Several decades went by before one or another author might, at any time, offer the public a collection of this kind.

Scholars in a Time of Ferment: C. G. Koch and C. Malte-Brun

As Napoleon's empire passed its zenith and entered its last years, two works appeared in Paris with noteworthy complements of maps for history but marginal claims to being historical atlases. The earlier one was simply a history book, an ambitious interpretive essay, accompanied by study aids that included five maps, increased to seven in the next edition. Its author was Christophe Guillaume Koch (1737–1813), professor at the University of Strasbourg and, for a time, a public figure. His mentor, J. D. Schöpflin, whose cartographic activities were sketched in chapter 4, had created a highly reputed diplomatic "academy" within the university. Koch shared in its creation and succeeded him at its head. The Strasbourg "academy" paralleled the few establishments, at Göttingen and elsewhere, that were organized to train young noblemen for a future in foreign and court service.[66]

Koch's classroom-oriented *Tableau des révolutions de l'Europe depuis le bouleversement de l'Empire d'Occident jusqu'à nos jours* was published, without maps, as early as 1771. This unsigned version, issued in Lausanne and Strasbourg, derives from lecture notes and is sometimes said to have been unauthorized.[67] Koch's name first appears on a rewritten text that covered a shorter span at greater length and that was published in 1790 as *Tableau des révolutions de l'Europe dans le moyen âge.* Maps were lacking but not forgotten; Koch regretted that he could not place one at the head of each section: "The geography of the Middle Ages is fallow ground, a real labyrinth in which it is very easy to get lost. There is no geographical work giving correct notions of the new state of affairs."[68] Full publication of the *Tableau* came in 1807 under the title of 1771. Koch's student Frederick Schoell had a hand in this edition, grown to three volumes, with chronological tables, genealogical charts, and five maps in the last volume. Schoell oversaw two more editions, one in 1814 with Koch's final additions and two more maps, and a postrevolutionary version in 1823, whose narrative Schoell continued down to the Bourbon Restoration (fig. 26).[69]

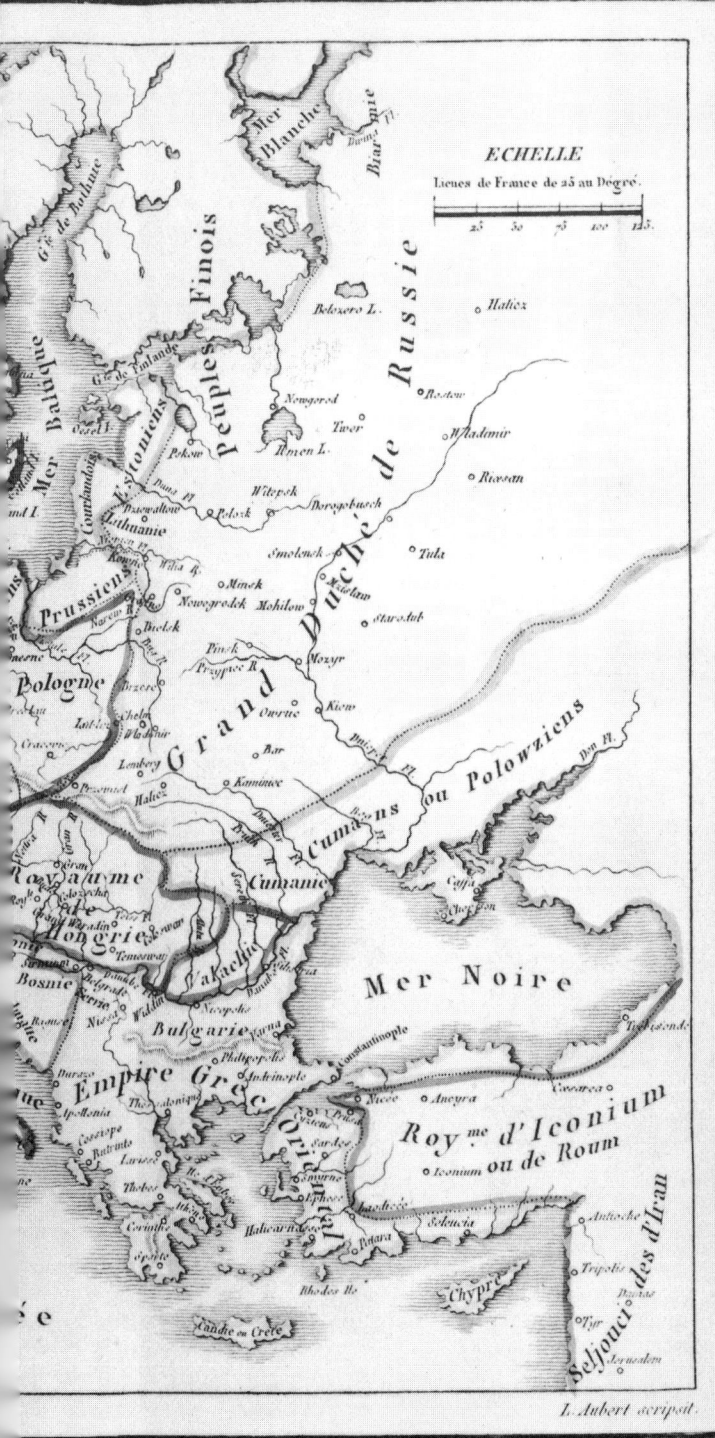

Figure 26 Christophe Guillaume Koch, Europe in A.D. 1074 (1814). Koch first published his *Portrait of the Vicissitudes of Europe from the Overthrow of the Roman Empire in the West down to Our Days* (*Tableau des révolutions de l'Europe depuis le bouleversement de l'Empire romain en Occident jusqu'à nos jours*) in 1771 and did so anonymously and without geographical aids. His book acquired five maps in 1807 and two more in 1814. They were by an otherwise unknown cartographer. The maps portrayed only medieval moments. Koch's choice of events provided the medieval scenes for many later atlases. A minor lapse of his maps is the curious "1074," a year lacking any historically noteworthy incident. Later atlases featuring that year reveal their dependence on Koch. (Courtesy of Yale University Library)

Koch's *Tableau,* despite a title referring to "revolutions" and a consecutive set of postclassical historical maps, does not properly belong among the sequential, universal collections. His history is simply complemented by maps. Kruse, in introducing his own work, evokes an image of overworked students who needed help in the little time they could devote to European history. Students of classical history and literature (and theology) had been equipped with school atlases since the 1600s; in England, at least, the triumphant publication of such works marched across the eighteenth century.[70] Thanks to Koch's work and others, postclassical history, modern as well as medieval, gained enough curricular momentum in the early 1800s to encourage the production of suitable geographic aids. Koch's maps neatly packaged the European past until 1453. Like Kruse, he featured the Continent as a whole. His choice of subjects for the seven maps proved astonishingly influential in the decades to follow. They will concern us again.[71]

Conrad Malte-Brun (1775–1826) had loftier aims than to devise aids for historical study. A poet and satirist exiled to France from Denmark and forbidden to engage in political writing, he turned polymath, but emphasized geography. Though lacking experience and even preparation, Malte-Brun made geography his specialty in France on the understanding that no other learned pursuit in the country was more backward. Master of several languages, he earned his living as a journalist and had a special gift for synthesizing the essentials of any subject from rapid reading.[72]

Malte-Brun—in Danish, Malthe Conrad Bruun—became the transmitter to France of geographic ideas and information culled from foreign writings, especially German, English, and Danish. He got his start in collaboration with the prominent geography teacher Edme Mentelle, then launched the first geographical journal in France. He was immersed in an ambitious *Précis de la géographie universelle,* not a history, when he assembled its complement of maps.[73]

Besides comprehensively surveying the current globe, Malte-Brun chose his maps so as to illustrate the history of geography and exploration and the historical context in which developments in geography took place. Ancient geography claims the lion's share; happenings after 1500 go unnoticed, except for Europe under Charles V and in 1789. Malte-Brun's allowance for maps of the Middle Ages was somewhat larger than for modern times, and he tried, by compression, to squeeze as much as he could into the space he had. The historical portion of his seventy-five-map collection had the same all-European orientation as those of his contemporaries Kruse and Koch but, faithful to the leanings of the previous century, he made room for the Muslim caliphate and the Mongols.[74]

Minor Works on the Sequential Plan

Five atlases published after Kruse's and less important than his carry forward the eighteenth-century pattern of tracing history interval by interval in a deliberate sequence. Each of them gives this format a special twist: two designers emphasized the "eastern" focus that comes naturally in a map extending across Eurasia; the three others cultivated European geographical discovery as well as the more conventional past.

In 1817, the Leipzig publisher Friedrich Gotthelf Baumgärtner issued a quarto-sized, five-map *Historical Atlas of Russia, Sweden, Poland, Austria, Turkey, etc., from 1155 to 1816.* This unassuming product displays changes on a single base map reaching from England to Kamchatka and far enough south to include the Mediterranean. The scope of the Baumgärtner base map connects it to the universal atlases, but its immediate context is more parochial. Baumgärtner, a firm with limited involvement in map publication, had recently issued atlases of Saxony and Prussia that look very much as though they were related to the circumstances of these kingdoms at and after the Congress of Vienna (1815). The atlas of 1817 fills out the rest of eastern Europe and points farther east at this fateful post-Napoleonic moment.[75]

A single frame of the Baumgärtner atlas portrays medieval times. It marks the Plantagenet possessions and French royal domain in 1155. The compiler, while emphasizing eastern Europe, does not omit salient features of the west. Attention to eastern countries was astute for cartographic reasons. Gatterer had shown the shortcomings of using a hemisphere as background for mainly western European events. Dupré's Eurasia or something comparable to it, that is, a map receding to left and right, offered a flatter, less distorted, and more central surface to eastern Europe than to western, especially when the countries in question included Russia and Turkey, with their Asian prolongations. Packing much in little may have been the main consideration. Baumgärtner, who lavished more than twenty maps of another atlas on his Saxon homeland, could manage only five for a much larger expanse.[76]

If Eurasia offered advantages to the display of Russia and associated lands, it was all the more suitable for the whole of Asia. From 1824 to 1826, Heinrich Julius Klaproth published in Paris a historical atlas of Asia from the Persian Cyrus (†529 B.C.) to the present. Its twenty-seven maps were issued in seven installments.[77]

Maps for history rarely have comparably learned auspices. Klaproth (1783–1835) was an Orientalist of the first rank, the author of a profusion of books and papers, and an eccentric. The son of a noted Berlin chemist, he faced his gymnasium examiners unable to answer any of their questions and insisting that, instead, he knew Chinese. This proved true: as a teenager he had mastered that tongue as an autodidact. Chinese was to be his greatest strength but was joined over the years by many more languages—central Asian, Caucasian, ancient.

After university studies at Halle and Dresden, Klaproth first practiced his skills for the Russian government. He was called on to join an overland embassy to China and then to travel to the Caucasus and Georgia. He also worked with the papers of the Imperial Academy, to which he won an appointment far in advance of his years. Although roaming through Asia benefited his learning (and taxed his health), Russia palled on him, the more so when he was kept from publishing the report of his mission to Georgia. In Berlin for scholarly reasons from 1810 on, he obtained an indefinite furlough from Russia in 1812 and, in exchange, was stripped of all his honors and emoluments. Late in 1814 he was at Elba paying his respects to the deposed Napoleon. The next year, he settled in Paris, permanently as it turned out. The scholar-statesman Wilhelm von Humboldt, whose brother Alexander, the noted geographer, had been publishing the reports of his scientific travels in Paris at great personal sacrifice, induced King Frederick III of Prussia in 1816 to award Klaproth a salaried professorship in oriental languages, with permission to live in Paris until his various projects were in print.

Klaproth's output in the next years was very large, including his acknowledged masterpiece, *Asia polyglotta*. Nevertheless, he was not much closer to assuming the duties of his Prussian professorship when he took sick in 1833 and died two years later. While still able to do so, he gave his personal map collection to the French royal library.[78]

Although the genesis of Klaproth's *Tableaux historiques de l'Asie* is unknown, the subject is in keeping with his interests and love of maps. He dedicated the work to the brothers von Humboldt. One would particularly like to find out whether the design was his own idea or was inspired by a model. He had equipped *Asia polyglotta* with an extensive linguistic atlas and produced a handful of sheet maps of current China and other parts of the continent, as well as a four-sheet map of Russian Central Asia. Neither Klaproth's prospectus for the *Tableaux* nor his pamphlet-length response to an unfavorable review in the German literary journal *Hermes* sheds light on his inspiration and antecedents.[79]

The *Tableaux historiques de l'Asie* has one volume of text and another of maps. Like most of the earlier universal designs, this atlas provides unchanging physical outlines as background for developments indicated by colors; it extends from Britain to Japan. (All place-names are revised on each map.)[80] Persia is Klaproth's ordinary focal point; western lands, including the extensions of Islam, are disregarded. The distribution of subjects is reminiscent of the Paris draft and Dupré: nine maps for antiquity (1–9), fourteen for the Middle Ages (10–25), and a mere two covering 1479 to his own day. Klaproth usually passes from one great domination to the other, as a few examples will suffice to show: epoch of the Umayyad caliphs (14), of the Tibetans (15), of the Abbasid caliphs (16), of the Samanids (18). He also supplies an ethnographic diagram of inner and middle Asia to A.D. 1000. The volume of text presents a "tableau général et motivé [des] révolutions [de l'Asie]." Chinese history is set out with special care, and each map is explained in a closing "aperçu général." Early readers much appreciated, and particularly praised, the commentary accompanying the ethnographic map—"Historical and Ethnographic Survey of the Peoples of Middle Asia, down to A.D. 1000; Studies of the Great Migration of Peoples."[81]

Klaproth's *Tableaux historiques de l'Asie* is, altogether, an attractive production and did not go unnoticed. Soon after its appearance, a Parisian compiler rounded out his maps for European history with insets of Asia taken from it. His explicit references to the source suggest that Klaproth was considered authoritative and essential. By Klaproth's doing, the special attention that eighteenth-century mapmakers had paid to oriental history became a learned discipline, detached from its European connections. Yet Klaproth also gave the Orient a memorable setting of its own. The originality and rarity of the subject did not lack recognition.[82]

Lithography and Other Innovations of the 1820s

Klaproth's atlas, however worthy, appealed to specialized interests. On a more general plane, Malte-Brun's *Précis de la géographie universelle* was acclaimed and resurfaced in one form or another for many years to come; but it was a bulky and comparatively austere publication whose *Atlas complet,* if it circulated in its own right, did so mainly in its geographical guise, for its maps of the current world. When Malte-Brun's atlas became available, the *Atlas Lesage,* advancing in years but blooming, still had the market for historical maps to itself, or shared it with Kruse.[83] Conditions changed markedly as the century approached its third decade: two new historical atlases turned up in Germany and France. Both ventures applied the infant techniques of lithography. As early as 1809, a work with four lithographed maps for history had appeared in Bavaria, but it had closer affiliations to pamphleteering than to scholarship or pedagogy. The lithographed maps of the 1820s were meant to teach.[84]

Friedrich Benicken's *Historischer Schul-Atlas* and its companion reference atlas will soon concern us. The lithographer in charge, Anton Falger, was a Tyrolean who had trained as a traditional engraver and learned cartographic work in the first major enterprise applying lithography to mapmaking, namely, the Bavarian Steuer-Kataster Kommission. His skill and that of the Weimar draftsman combined to produce an atlas whose claims to attention hinge on more than Benicken's historical conception and design.[85]

Its French contemporary is less commendable. Henri Selves signs himself "lithographe de l'Université" and, from 1820 to the 1840s, published a series of generally unattractive historical atlases, often with the same maps. Lithography reached France around 1817; an official inquiry seemed needed in 1825–26 to determine whether the technique was usable for maps. Some entrepreneurs had not waited for stamps of approval. The dates 1819 to 1822 appear in an atlas by Selves at the Bibliothèque nationale; its title page showing the publisher's choice of title is missing. A copy of Selves's *Atlas géographique . . . pour l'usage des collèges, Géographie ancienne, Géographie du moyen âge, Géographie moderne actuelle* at the University of Michigan contains on its inside front cover the unsigned note "must be about the earliest A[tlas] with Lithographic [plates]." [86] The annotator is not quite right, but for France, the unattractive Selves product is undeniably "pioneering." [87]

Selves's technologically precocious "geographical" atlases are of the commercial kind scorned by the Institut de France committee of 1808. They supply maps of the current world ("géographie moderne actuelle") but precede them with ancient and medieval geography, the latter also including the period we call "early modern," to 1813. "The *Géographie du moyen âge*," an early and indulgent reviewer said, "is the most interesting part of the *Atlas,* because it is at the same time [a map book and] a course in history." [88]

Ambitious History from the Weimar Geographical Institute

The two atlases by Friedrich Wilhelm Benicken, though faithful to the sequential plan of universal history, also resemble Malte-Brun's collection in their attention to the history of geography. These early examples of lithographed atlases are closely associated with the once esteemed Weimar Geographical Institute. In the late eighteenth and early nineteenth centuries, the Thuringian town of Weimar, capital of a minute principality, enjoyed a golden age, epitomized by such resident giants of German literature as Goethe and Schiller. A minor but engaging figure of this era was the locally born man of letters Friedrich Justin Bertuch (1747–1822), whose varied activities included the making and publishing of maps, for which he established a publisher (1791) and somewhat later the Weimar Geographical Institute (1804).[89] At the turn of the century, Bertuch's establishment was the focal point of map production in Germany, and its journal, the *Allgemeine geographische Ephemeriden* (in which, for example, Kruse examined Gatterer's maps of the *Völkerwanderung* and the *Atlas Lesage* was savaged), was foremost in the field. Histories of geography do not treat the Weimar Geographical Institute kindly. It came too early to share in the glory that Alexander von Humboldt and Karl Ritter brought to German geography and too late to be a pathfinder in its own right. However that may be, for several decades Weimar enjoyed great prestige in geographical publication.[90]

The institute seldom issued maps for history, least of all nonclassical ones.[91] An exception involved the F. W. Benicken just mentioned, an ex-captain in the Prussian army. To celebrate Napoleon's expulsion from the European scene, the thirty-one-year-old Benicken published in 1815 a synchronic map of Napoleonic campaigns. His *Historical School-Atlas* (*Historischer Schul-Atlas*) came five years later and was followed, very soon after, by the enriched and costlier *Historical Reference Atlas* (*Historischer Hand-Atlas*) (1820–24).[92] Both are universal in design but, unlike the works of Baumgärtner and Klaproth, take no special interest in eastern Europe or Asia. Benicken's atlases focus on western European events and propose themselves as companions to history textbooks, notably Leonhard von Dresch's *Survey of General Political History, Especially of Europe* (*Übersicht der allgemeinen politischen Geschichte, insbesondere Europens*), another Weimar publication.[93]

Benicken's oversized *Schul-Atlas* is hard to consult and seems poorly suited for classroom use; yet it was, in its time, an experimental and ambitious work. The influence of Gatterer's map collection on Benicken may be suspected from his adoption of a spherical projection, use of "methodical," (almost) scriptless maps, a march through history by intervals of events, and the treatment of much of the Middle Ages as an era of migrations (*Völkerwanderungen*). The maps of the school atlas, though lacking place-names, are profusely dotted with numbers and letters referring to a legend in the lower margin ("Explanation of the Numerals"). Tracks and lassoes (if that is the right word—see fig. 27) appear copiously on the maps; they mark incidents of conquest and exploration. Each of the fourteen maps has a detailed legend in its lower margin and is flanked by a chronological chart outlining the events of the period in question. Ample blank space is left on all sides, presumably for student jottings.

In maps 1–7, the spherical projection generates a base map reminiscent of

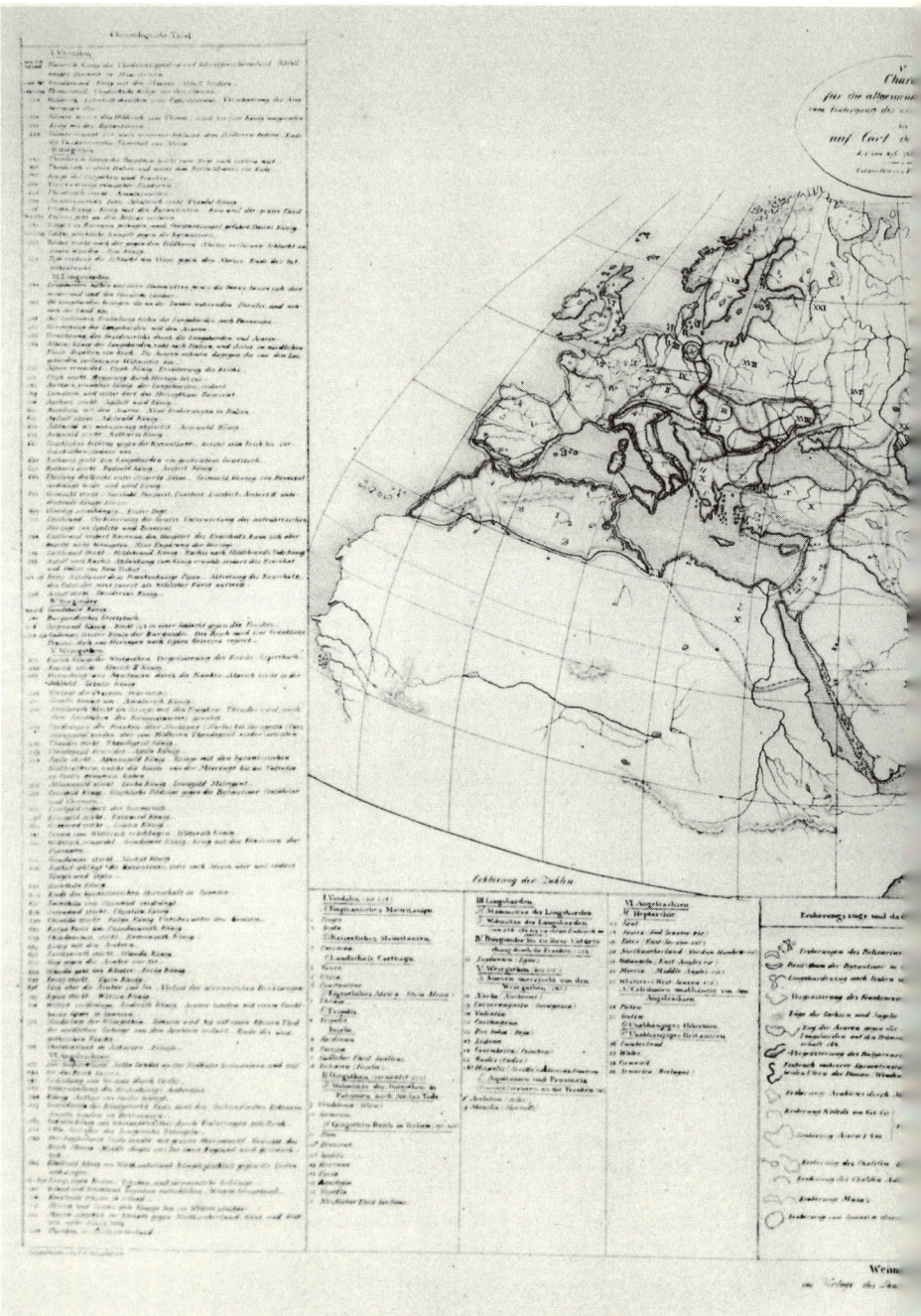

Figure 27 Friedrich Wilhelm Benicken, "Map for General History from the Downfall of the Western Roman Empire to Charlemagne (476–758)" (1820). With the support of the Weimar Geographical Institute, Benicken published two historical atlases. They are early examples of lithographed maps. Benicken modernized the sequential scheme and made generous allowance for contemporary history. The layout of the school atlas (its map no. 5 shown here) was much influenced by Lesage, in part by deploying a profusion of tracks that each terminates in a kind of lasso encircling the land acquired by conquest—a new graphic feature.

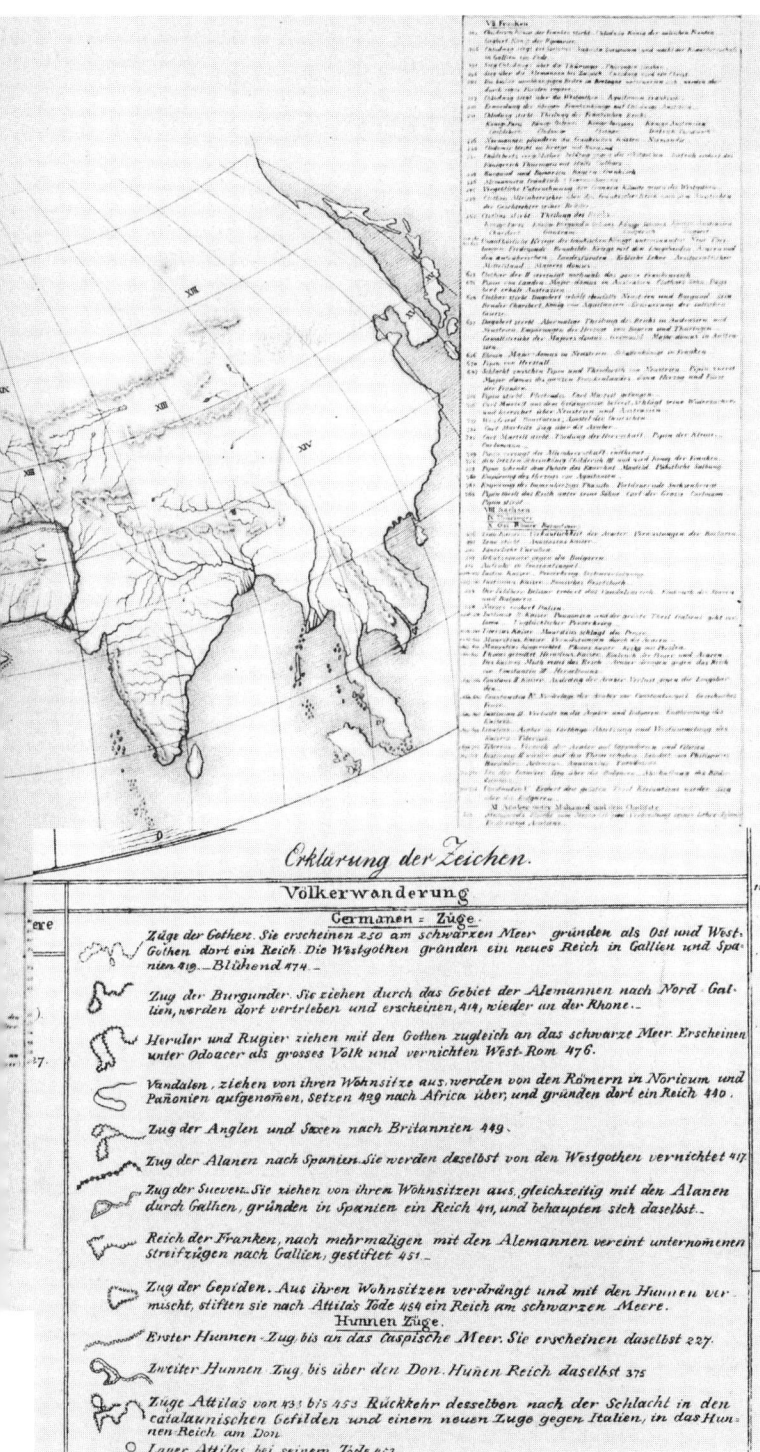

Also new is the key superimposed at bottom right, "Explanation of the Symbols" (from map 4), in which miniatures of the tracks serve to identify each of the map-sized tracks. (Courtesy of the Universitätsbibliothek, Vienna, and J. Dörflinger)

the Ptolemaic world map, often modernized under the title "The World
Known to the Ancients." The eighth map marks the passage from medieval to
modern times and coincides with the great discoveries. Its projection, and
that of all later maps, changes from spherical to Mercator; the scope widens
to include new lands; and, to compensate, the territories shown till then contract to about one-third their former size.

The title page of the *Schul-Atlas* has a table of contents composed of fourteen miniature maps, reminiscent of today's index maps, or of a cartoon strip.
The miniatures illustrate the progressive discovery of the world. On the main
maps, the presence or absence of coloring divides the historical "known"
world from that remaining to be discovered. The miniatures are more forthright. Depicting the "known" world in fourteen gradually widening stages,
they present a compact narrative-in-pictures of progressive discovery. Helped
by a better draftsman, Benicken was able to expound this theme in the reference atlas (*Hand-Atlas*) in a grander and more visually arresting way.

Both collections set out history in fourteen steps. The periods start long
and eventually narrow to as little as half a century. ("Napoleon's Campaigns"
is an unnumbered bonus for the *Hand-Atlas,* whose few surviving copies lack
the supplementary, fifteenth map, for the years 1815 to 1822.) Simple cartouches, similar to Kruse's, supply titles in the form "Map for General History
from . . . to . . . " The first interval runs from the earliest known times to the
end of the Trojan War, the twelfth from the beginning of the War of the Austrian Succession to the death of Frederick II (1740–86). Dresch's *Übersicht* is
the direct source for Benicken's divisions, though not for the correlated history of geography.[94] There are four maps each for antiquity and the Middle
Ages, then six from 1519 to the peace treaty of 1815 after Napoleon's final defeat.
Each map applies to the last date of the indicated interval, so that map 8 (1273–
1519), in which the projection changes, is more modern than medieval.[95]

Las Cases and Kruse about twenty years earlier had led the way in multiplying historical maps for the era after 1500. Benicken, too, paid generous attention to the recent past. Were the devastating events of 1789–1815 responsible? For whatever reasons, the restraint of Dupré, Gatterer, and the other
early makers of maps for history was surmounted.[96]

In another bow to innovation, Benicken resorts to tracks, intensively de-
ployed in the *Schul-Atlas.* The *Atlas Lesage* had helped to break down the in-
hibitions formerly curbing the use of this device. A Weimar reviewer censured Las Cases's fondness for *Marschrouten;* these lines on maps seemed
symptomatic of what, under Napoleon, the French were doing to Europe.
Benicken's two atlases, products of Weimar, could hardly avoid reacting to the
French best-seller, but where tracks were concerned, the Prussian ex-captain

rivaled Las Cases not by austere avoidance—as, contemporaneously, Kruse was doing and others would do in the decades to come—but by introducing more tracks than the French innovator.

Tracks and other lines give Benicken's *Schul-Atlas* a highly individual aspect and follow a distinctive and thorough program. They begin on the second map, with colors outlining the limits of ancient empires (Persians, Medes), not as precise boundaries, but as approximations of extent. Maps 4 and 6, from the age of Augustus to Pope Gregory VII (1073), combine the presumed course of "folk" migrations (Huns, Goths, Bulgarians, etc.) with the limits of their domination (such as Sueves in Spain, Vandals in North Africa, Lombards in Italy). The linking of a migration track with a territorial boundary produces the lasso effect mentioned above. The European and Mediterranean segments of these maps are very cluttered. At map 4, the unusual "Explanation of Symbols" (*Erklärung der Zeichen*) begins and continues to the end. A miniature drawing of each track is used as a symbol in the legend. First headed "Völkerwanderung," the legend titles change in pace with the events (fig. 27). At map 8 (1273–1519), as rich in tracks as the period of barbarian invasions, the legend is headed "Discoveries and Conquests"; at map 14 (1806–15), "Frontier Changes."

Benicken's program of demonstrative lines is surprisingly extended and sustained; its effectiveness is less certain. The lines are difficult to distinguish from one another, especially when crowded. Owing possibly to age or inferior materials, many colors are now faint and run into each other; even on the legends, the intended tints are often uncertain. Benicken's selection is only marginally better than Las Cases's. Every featured incident reminds one of an absentee, which may strike the modern reader as equally worthy or almost so.

Benicken's "library," or "reference," atlas improves on the school version in fullness of execution and treatment of details. Lithographed by the same Tyrolean responsible for the school version, it was beautifully drawn by Karl-Ferdinand Weiland, one of the most productive members of the Weimar enterprise.[97] The maps of the *Hand-Atlas* are supplied with appropriate placenames; its time chart is a separate publication; elaborate legends are not needed. As a result, a very large, uncontested space is available for maps. Each of the fourteen or fifteen main maps is coupled with four or more enlarged insets, and a rational scheme calls on either the large map or the insets for historical information. Printed on vast sheets (of unappealing gray paper), the *Hand-Atlas* is simpler to decipher than its precursor, but no easier to handle and consult.[98]

From start to finish, each main map is a Mercator projection of Eurasia, with the lower half of Africa sometimes cut off. The dual projection of the

Schul-Atlas is dispensed with. Eurasia is printed as a faint outline, of which, as one moves through the successive displays, more and more parts become colored and join the "known" world. Coloring as the visual equivalent to being "born to history" was foreshadowed by Dupré's *Révolutions de l'univers* and already implemented in the *Schul-Atlas*. The naively arrogant innovation of the *Hand-Atlas* involves showing the "unknown" world as a misty potential. In Weiland's firm, precise handling, the continuing and changing contrast of faint line, meaning "before," and color, meaning "after," is a visually convincing dramatization of the theme of geographic progress.

Despite the large sheets, the mass of Eurasia dwarfs the political entities— even far-flung empires—that are colored in. Larger-scale inset maps, sometimes occupying the space of the omitted parts of Africa, are used for some subjects long familiar to historical maps, such as the Exodus track and the Argonauts, and for rather new subjects, such as the campaigns of the Huns in late antiquity and the wars of Louis XIV. Tracks of trade routes and geographical explorations, such as the Phoenician circumnavigation of Africa, appear on the main maps, especially from the fifteenth century onward. The loose boundaries of the *Schul-Atlas* are dispensed with because no longer necessary. Instead, the insets focus attention on, for example, the tracks of barbarian tribes.

After having shown geographical discoveries and ordinary historical events on the same background in the *Schul-Atlas,* Benicken distinguished the two in the *Hand-Atlas* and distributed them as superimposed lines of narrative. Conventional history is in the subsidiary position. The two aspects of the past have separate roles and are rarely called upon to cooperate. The main map shows the gradual (European) discovery of the world, along with the tracks of voyagers and explorers relevant to the history of geography. The many inset maps on the lower part of each sheet form a topical historical atlas of more than fifty frames, several of them new. Benicken continues to be attentive to modern times. Fourteen insets concern antiquity and fifteen the Middle Ages, but more than twenty (as well as shorter time intervals and the main maps going with them) are granted to the five centuries from the Habsburg emperor Rudolf I (elected 1273) to 1815.

The topics of Benicken's insets extend from "Hellas before the Destruction of Troy," a subject already depicted by Nicolas Sanson, all the way to the topical "Restoration of the European State System" after Napoleon's downfall. In spite of an unprecedented breadth of scenes, there are notable omissions. The Umayyads in Spain are present, but the expansion of Islam, known to the *Schul-Atlas,* is not. This is a salient example of events edited out. Nevertheless, Benicken's *Hand-Atlas* is a visually impressive and in many respects original production.

Inventiveness and experimentation were not enough to win a favorable reception from the public. If Benicken and his sponsors thought they could divert a share of the market from Las Cases and Kruse, they were mistaken. Near the year when the *Hand-Atlas* completed publication, the *Atlas Lesage* became available in German translation and enjoyed its customary success; twenty Lesage atlases survive today for every Benicken. The two Weimar atlases had a limited future. Awkwardly sized, though not much more so than the competition, they were better suited for private contemplation than for easy and practical reference. For all their shortcomings, however, Benicken's collections may well be the most deeply thought out of the sequential atlases. They deserve to be remembered.

THE BURGEONING OF NATIONAL ATLASES

While Benicken and Selves produced works of European scope, a very different Parisian undertook a collection of a kind that had not been seen in France since Rizzi-Zannoni in the early 1760s. Between Menso Alting in the Netherlands (1697–1701) and John Andrews in London (1797), historical atlases of individual countries and regions had been rare and isolated. Maps for French history had been slighted for more than half a century when Adrien Brué repaired this neglect on an ambitious scale. Many would follow him; no country (down to 1870) approaches France in the number of its historical atlases. But others anticipated Brué in drawing up nineteenth-century maps for the history of countries.

They were inspired primarily by the vicissitudes of the Napoleonic era. In 1806 Prussia was thoroughly humbled on the battlefield, a defeat translated at the Treaty of Tilsit (1807) into territorial losses and heavy monetary indemnities. The next year, F. Handtke published a map showing "the acquisitions and alterations of the royal Prussian state from 1417 to 1807." When placed in the perspective of the territorial expansions of the previous four centuries, the disheartening present might yield to hope for the future.[99]

Napoleonic times gave unprecedented fluidity to frontiers and territories, especially in the old Holy Roman Empire. Kruse stressed how much this affected mapmaking, and Mentelle despaired of being able to draw maps again. Amid these uncertainties, Bavaria managed intermittently to win the favor of France; Napoleon even promised to make its king greater than any of his predecessors. This comment encouraged the Bavarian baron Johann Christoph Aretin to offer for examination the four "most shining moment[s] in the history of the dynasties that have ruled in this land"; so informed, Napoleon would have a better idea of how to carry out his promise. Aretin's maps were announced as "Lithographisch abgedruckt." There was reason for pride in so early an application to map printing of the new home-grown process.[100]

Saxony, which waited too long to jump from the good ship Napoleon, only just managed to weather the wrath of the victors at the Congress of Vienna. The Leipzig publisher Baumgärtner thought it timely in 1815 to issue a twenty-six map *Historischer Atlas von Sachsen [950–1815]*, subtitled "provisional (*augenblicklich*) survey" of its various possessions.[101] These were not the only circumstances making Saxony a historically interesting subject; three maps by Friedrich Zollmann (1731–32) had offered a foretaste of its cartographic attractions. The Wettin rulers of Saxony split into Ernestine and Albertine branches in the fifteenth century. The complicated territorial consequences of this partition were exacerbated by Ernestine subdivisions. In 1796 Franz Güssefeld produced a five-map survey of the vicissitudes of Ernestine territories; a nine-map treatment of the same subject was issued by M. F. Teuscher in 1825. Weimar, whose Geographical Institute published both Güssefeld and Teuscher, was in Ernestine territory.[102]

The Baumgärtner firm became cartographic under the stimulus of current affairs. A year after its survey of "provisional Saxony," it supplied both a twenty-five-map atlas of Saxony as it now was and a historical atlas of Prussia, including its enormous annexation of Saxon territory and its additional expansions after the Congress of Vienna.[103] Other compilers and publishers would turn Prussia and Saxony, Austria and Württemberg, into the subjects of historical maps or atlases—many times in the case of Prussia—before a historical collection became appropriate for Germany as a whole.

France, long united, was better suited for exercises of this kind, at least once the ice was broken. Adrien Hubert Brué (1786–1832) differed from the usually sedentary mapmakers hitherto encountered by having had firsthand acquaintance with distant lands. Paris-born, he went to sea at twelve and was on Mauritius three years later. There, a French government expedition recruited him as a midshipman for an arduous voyage of exploration along the Australian coast. The cruise shook Brué's health; he left the navy when his ship returned to Mauritius, came home, and was befriended by Louis de Freycinet, an officer on the grueling expedition, also beached for reasons of health.

Freycinet, now attached to the naval cartography bureau, nursed Brué's talents as a draftsman and taught him how to draw maps directly onto copper. From 1813 on, Brué published maps prepared by this "méthode encyprotypé" and won notice for the high quality of the results. Early in the Restoration, named "géographe de Monsieur [the king's oldest brother]," he produced a forty-map *Atlas universel de géographie physique, politique et historique, ancienne et moderne,* described by a contemporary bibliographer as the finest yet published.[104] Brué passed directly from this comprehensive collection to an effort that he regarded as "the only one of its kind," namely, an *Atlas géographique, historique, politique et administratif de la France.* In fact, Brué retraced the steps of Rizzi-Zannoni in 1764 (and knew it). The outcome was the first ambitious set of maps for history to appear in France in the twenty years since the inaugural Paris edition of the *Atlas Lesage.*[105] Brué diverged from the course set by Las Cases and also from that of the "Europeanists," Koch and Kruse.

The years in which Brué's collection made its slow way through the press saw two other works with French historical content, neither of great moment. The *Tableau chronologique, historique et géographique de la France* by J. B. L'Hermite is a large sheet on which a map of France is surrounded by lengthy text. The single sheet suggests a book illustrated by a map, rather than a map with commentary. Augustin Legrand's *Atlas géographique et géologique des quatres parties du monde et de la France en particulier* emphasizes physical relief, a cartographic preoccupation at the time. Its one map concerned with French history (or any other) is predictably overloaded and schematic.[106] Neither rivaled Brué, or sought to.

More noteworthy mapping of individual lands took place toward the end of the 1820s. Poland had been a cause célèbre since the Partition of 1772, and its grievances had not at all been remedied by the international settlements of 1814–15. In 1827 Stanislas Plater published the first of the five historical atlases of Poland that would appear by 1870. Dispassionate history was not Plater's aim; his *Atlas historique de la Pologne* concerns only the seventeenth to nineteenth centuries the better to illustrate the injuries inflicted on his homeland. Patriotism of a more placid kind is suggested by *Lothian's Historical Atlas of Scotland Consisting of Five General Maps exhibiting the Geography of the Country in the 1st, 5th, 10th, 15th and 19th Centuries* (1829). The interest in Scotland occasioned by the writings of Walter Scott (†1832) may have had something to do with this enterprise.

A Creative Burst: 1829–30

The Paris atlases of 1820–21—Selves's and Brué's—hardly equal Benicken's of 1820 in breadth and aspirations. Historical collections were beginning to be more numerous. Benicken's historical *Hand-Atlas* was completed in 1824. No less ambitious than the *Schul-Atlas,* whose Dresch-derived plan it retained, it was free from the burden of being "methodical" and, as a result, was more conventional in form. Also noteworthy, though special in scope, was Klaproth's atlas of Asian history (1826).[107] General and classical atlases show signs of widening their coverage to include a scattering of maps of medieval, but not modern, history.[108] The next burst of production came as the 1820s drew to a close. For a change, British, as well as French and German, mapmakers were involved.

Of the historical collections that were published in 1829, the least expected is the large-format *New Classical and Historical Atlas* published in Edinburgh by John Thomson (1777–184?).[109] It was an "Ancient Atlas, Classical and Sacred," but with a difference. An imposing *Atlas classica* of 1805 by the London publisher Robert Wilkinson had augmented the biblical and classical subjects with a small medieval group. Thomson was definitely inspired by Wilkinson's unusual atlas, which had been reprinted several times. He, too, extended the traditional core of subjects into the Middle Ages, added to Wilkinson's medieval maps, and, inspired by Malte-Brun, even made a small foray into modern history.[110] Thomson's important map of the Crusades will concern us later.

Within a year or two, a historical atlas comparable to Thomson's issued from the presses of the thriving Alsatian lithographer Engelmann. The compiler of this *Atlas géographique, astronomique et historique servant à l'intelligence de l'histoire ancienne, du moyen âge et moderne* was George Heck. Like many others, his general atlas had a subsidiary historical section. He paid much attention to geography and took history to mean the Bible and classical antiquity. Nevertheless, four sheets were allotted to the Middle Ages.[111] They include the same portrayal of the Crusades as in Thomson's collection, from the same or an independent source.

Three school atlases originated in Germany in 1829–30, works for which curricula evidently created a demand. The worst that may be said about them is that the compilers were sometimes more concerned with packing as much time into the least space than with ensuring adequate coverage. Arnold Möller, a clergyman, won a prize from the king of Prussia for a book with one map on the territorial expansion of Brandenburg. Soon after, he undertook at least three historical collections. His *Kleiner historischer Atlas zur allgemeinen Weltgeschichte* first appeared in 1824 and was revised in 1829. As Benicken at Weimar appealed to histories by Dresch, so Möller associated his work with Friedrich Kohlrausch's *Chronological Outline of World History.* In the spirit of abridgment, Möller fitted universal history onto ten plates, including a comparatively generous three maps for the Middle Ages. The map

"Europe, Asia, and Africa ca. 1100" is unusual in focusing thematically on religion and distinguishing lands of Christians, Muslims, and pagans. It even delineates a zone, including Wales and North Germany, of "Christendom in Combat with Paganism." Möller wrote *On Christian Edification* and similar works and devoted one of his other atlases to church history. The obscurity to which provincial origins and limited resources have condemned him may not be wholly deserved.[112]

Ferdinand von Witzleben, an army officer, had gained experience in the Prussian army's Topographical Bureau, then taught at the combined military academy. Between 1829 and 1833 he published the three installments of an atlas of world history for secondary schools—one each for antiquity, the Middle Ages, and modern times. The geographical maps are accompanied by tables of closely packed facts and figures. Folio-sized in thirteen sheets, the atlas seems ambitious in scale but impressed an early observer most of all with the poor quality of its engraving.[113]

Unlike the works encountered thus far, the *Historisch-geographischer Schul-Atlas* of Karl Kärcher divides its attention evenly between history and geography. Conventional geographical atlases had been tacitly doing the same thing for a long time—unevenly—by coupling "ancient geography" with a larger set of current maps. Now, geography as a recognizably distinct cartographic genre could be separated from historical maps; many school atlases of geography strictly speaking were coming onto the German market.[114] A predictable result in Kärcher is skimpy coverage of history: four maps for classical antiquity, three for the Middle Ages, and one each for 1500 and 1812. But Kärcher sometimes makes the most of the little space he has, such as by choosing medieval moments of maximum fragmentation and thus showing as many entities as possible.[115] Atlases spanning all history were well understood in Germany; there was no risk of condemnation by hanging juries like the French one of 1808.[116]

The notable event of 1829 in France was the publication of an atlas of European history by Maxime Auguste Denaix (1777–1844). Educated at the École polytechnique, nursery of military engineers, Denaix retired from the army in 1825 as a staff lieutenant colonel. He had headed the office of army cartography (Dépôt de la guerre) since 1821. A participant in the Paris Société de géographie from its foundation, he sat on its central committee from 1827. After retirement he applied himself to geography and was his own publisher.[117]

Denaix's *Atlas physique, politique et historique de l'Europe* is above all thematic, an eye-catching difference. Its goals were very different from a collection like Kruse's that emphasized the geographic aspects of the past. All maps for history might be called "thematic," but the ones we have examined, high-

lighting political borders at the expense of almost anything else, tend to occupy a category of their own. In the case of Denaix, many subjects are self-
evidently thematic: waters, trees, plants, mammals, ethnography, fortresses
(with sieges), battlefields, and sites of treaties, councils, concordats, con-
gresses. Nothing of the kind had been available since the publisher Desnos offered a *Tableau de la France* as complement to Rizzi-Zannoni's atlas of French
history (1764).[118] Whereas Desnos made only a suggestion—the two volumes
were sold separately—Denaix's thematic maps were between the same covers
as the historical ones. Moreover, the atlas paid intensive attention to the recent past, now clearly a subject for history. Out of twenty-one maps, including thematic ones, no fewer than seven concern the years 1795–1815. French
remembrance of the Napoleonic era had become so golden that images of its
glorious rise and fall were in demand.[119]

A cartographer as well connected as Denaix secured warm endorsements
for his *Atlas*. The most interesting of them, heading a slightly abridged "édition classique" of 1835, is from the historian François Guizot, then minister of
education and at the height of his fame (Guizot's "Avis" to Denaix's collection
is dated 1833). The great man judged the merits of historical atlases, correctly
or not, in relation to a thoughtful range of similar works: "The maps of [De-
naix's] Historical Atlas of Europe correspond to the very epochs in which notable changes took place [i.e., in contrast to Kruse's centuries]. . . . [Denaix's]
maps, all of them enhanced in the marginal columns [of text] by dates of the
formation and extinction of the states appearing in them, are thus a sequence
of contingencies following one another like the events that bring them about.
In this way the author has been able to avoid the anachronisms inseparable
from presentations [attempting to display entire] periods."[120]

The types of geographical atlases were sorting themselves into appropriate specialties. "Geography" does not appear in Denaix's title; his was a thematic complement to collections oriented strictly to geography and, in keeping with this identity, was as innocent of physical relief as Brué's atlas of
France. When translated into years, Denaix's "topical" organization, emphasized by Guizot as a special asset, in fact differs little from Kruse's double-zero
years; the two often coincide. While extolling the very recent era of French
glory, the Denaix atlas does not neglect the postmedieval ancien régime. It
could aspire to represent a new stage in historical cartography and expect to
supersede Koch and Kruse and any other atlas omitting classical and biblical
antiquity.

The historical atlases of 1829–30 are too irregular in their rate of publication and survival to be discussed in a strict order of appearance. The final author in this set, Edward Quin, far from being a pacesetter, produced the last

atlas of universal, sequential history descending from the mid–eighteenth century. He deserves a section of his own.

EDWARD QUIN: END OF AN ERA IN MAPS FOR HISTORY

England had steered clear of historical atlases on the universal plan since their inception and continued to do so for some years after Benicken's *Hand-Atlas*. Edward Quin, an Oxford graduate and London barrister, was thirty-four at his death in 1828. Two years later his *Historical Atlas in a Series of Maps of the World As Known at Different Periods, with an Historical Narrative,* somehow came onto the market. Quin's atlas was reprinted in 1836 and again, with new titles, in 1856 and 1859. In the meantime, the work obtained a face-lift. The twenty-one original maps were redrawn on a larger scale in 1846 by the well-known cartographer William Hughes for a new edition, omitting Quin's text, that was reprinted several times. In one form or the other, Quin's maps were in demand for nearly thirty years.[121] A small but striking sign of their impact, preserved in Munich, is a hand-painted Dutch copy signed and dedicated by the artist without reference to the original author.[122]

Quin's title presenting "the world as known at different periods" establishes a resemblance between his plan and that of Benicken's *Hand-Atlas.* The latter's concern with the advance of geographical knowledge is only implied, although unequivocally; Quin not only implements such a concern but announces it in his title. Since the world map in both atlases is on Mercator's projection, dependence is possible. On the other hand, the English version of universal history dispenses with inset maps and with tracks of tribes, generals, and explorers, and its total of maps more closely approximates Dupré's thirty than Benicken's fourteen.[123]

Quin lived long enough to provide his atlas with a preface, setting out his goals. His maps were meant to offset the fragmentation of history. In the successive leaves, the student would always find "the same territory in the same part of the Map, [and see] by the changes of colour, the various Empires which succeed each other." He would survey the changing historical scene "like the watchman on some beacon tower." To make this possible "it was necessary to shew [not a single 'world known to the ancients' but] at each period, only that part of the world which there is reason to believe was actually known to the geographers and statesmen of that time." Because the unknown lands nevertheless were there, clouds are used on the maps to conceal them (fig. 28).[124]

Clouds, Quin's highly personal device combining function with decoration, make him a memorable cartographer. Weiland, Benicken's fine

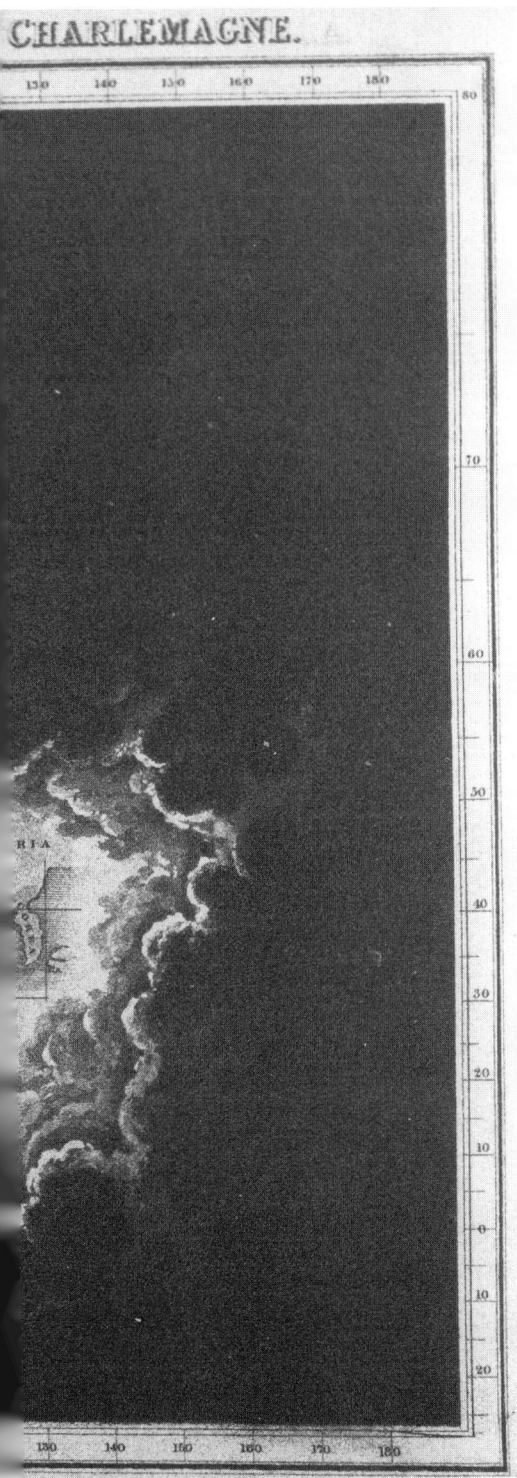

CHARLEMAGNE.

Figure 28 Edward Quin, *Historical Atlas in a Series of Maps of the World As Known at Different Periods* (1830). The maps of Quin's posthumous *Historical Atlas* are small in relation to its scope. Without adequate space to incorporate needed details, the maps were unlikely to help students of history. But Quin's use of dramatic clouds to cover and gradually disclose the known world had great success with contemporaries and posterity. The atlas had several reprintings. In 1846, William Hughes carried out a complete revision of Quin, with larger maps, projection onto hemispheres, and much-attenuated clouds. His revision sold well but did not discourage reissues of Quin's pleasing original. (Kelly Library, Saint Michael's College, University of Toronto)

draftsman, had used faint lines to half-mask the undiscovered parts of Eurasia. Quin's solution was to cover the earth with billows of thick, dramatic clouds, especially spectacular in their very dark masses. The edges, also arresting, overhang the openings through which the known tracts of the world are revealed in vivid color. The clouds break to uncover first a patch of the earth, then more and more. They continue to recede as, with the discovery of America, the map doubles in size to contain an enlarged world. The clouds part altogether at the close of the eighteenth century.[125] But the earth is not yet fully discovered. Quin invariably marks "barbarous and uncivilized countries" by flat olive shading. Much flat olive still covers the earth in the last map, signifying uncharted lands needing exploration.

Quin's theatrical staging accompanies maps so small as to limit their use for historical study. In the modification of 1846, Hughes enlarged the maps, projected and centered them onto a hemisphere, and reduced the clouds to the dimension of fluffy curtain fringes. Even though the artistic impact was much weakened in this revision, the atlas found a steady clientele. Hughes's enlargement enhanced conventional historical values, but the dramatic and aesthetic force of Quin's maps could be experienced only in his original design, for in his own images the course of history was enacted by the clouds rather than on the surface of the earth.[126]

The dark clouds, however, do not symbolize the development of civilization. We are not invited to contemplate a movement from primeval gloom to light. The clouds simply reflect the state of knowledge of the earth. When they finally dissipate in 1783, the moment almost certainly relates to the voyages of Captain Cook rather than to the outcome of the American Revolution.

An early tribute was paid to Quin by the American educator Emma Willard: she appropriated his clouds in her *Atlas to Accompany a System of Universal History,* illustrating "The Progressive Geography of the World, in a series of maps, adapted to the different epochas [*sic*] of history." Willard's history of the United States had also been accompanied by "progressive" maps, that is, a chronological sequence of the kind we have been concerned with rather than a timeless circuit of the individual states.[127] The ten maps for her *System of Universal History* are humble images, no match for those of her U.S. history; but they are made interesting by a pattern of gradually lifting clouds. Even in Willard's version, Quin's program of a world becoming known overshadows any concern with the vicissitudes of civilization. There is no return of clouds after the fall of Rome; and the clearing of the heavens in the final map, ca. 1820, probably depends on the space available to Willard rather than on any specific set of events.[128]

CONSERVATISM AND CHANGE IN THE 1830S

At the beginning of the 1830s, the burgeoning crop of maps for history tended to be of three kinds. The oldest type were historical supplements (prologues or appendixes) to general geographical collections in whose titles the words "universal atlas" often occur. More recent in conception but now central for comprehensive coverage were atlases of European history, typified by Kruse's *Atlas und Tabellen*. National or regional historical atlases formed the third category, particularly vigorous in France.

Change in literary form seems slow if traced year by year. Nothing could have been more traditional than the general atlases of geography with historical appendixes of ancient geography repeatedly published in France in the 1830s. Their cursory nod toward a century of development consisted of a few nonclassical maps. A. H. Simencourt deserted the ancients only long enough to show Charlemagne's empire. Pierre Lapie and son supplied their *Atlas universel* with three medieval items, rounded off by Europe under Charles V. They could help themselves to maps from their atlas for Malte-Brun's *Précis* (1810–12). A. H. Dufour, a commercial mapmaker who advertised that he was a pupil of Lapie, managed to publish one atlas in the 1830s wholly free of maps for history. Three, more extensive productions of his offer token coverage of the Middle Ages—nothing later than Charlemagne—plus a single map, at best, for modern times. The modern subjects in two of Dufour's atlases are ancien régime France and Europe in 1789, respectively. His postclassical set of four, ending with 1789, is found again in an "atlas universel" by his pupil, Alexandre Vuillemin. The Lyons mapmakers Monin and Frémin close their historical section with a single, overloaded, Malte-Brun-inspired map called "Europe to Aid Teaching of the Middle Ages."[129]

These thin, highly selective appendixes were token offerings rather than fully thought-out collections. They were usually called by the consecrated name of "ancient geography." Their postclassical maps were too few for useful history and too many to warrant the narrow collective title "ancient geography," which traditionally referred only to antiquity and the Bible. Large-scale, ambitious, state-of-the-art atlases of the world were then being published in Brussels and Gotha; they dispensed with historical supplements. Buyers of the more traditional collection called "atlas universel" continued to believe that ancient maps were an essential part of the package, perpetuating what, for centuries, had been ancient geography.[130]

Atlases of European history were free of these perplexities. Constantin Desjardins, assisted by J. Häufler, created a collection somewhat reminiscent of Denaix's; and the Engelmann firm in Mulhouse issued an unassuming *At-las élémentaire de géographie historique* composed of the now canonical European maps.[131]

None too soon, Kruse's pioneering collection was translated into French (1834–36). The translator, Félix Ansart, was unconcerned with the integrity of Kruse's conception and was evidently free to go his own way. He expanded the number of maps, notably into the ancient world and other continents, so as to make the atlas more comprehensive and salable. His version of Kruse's atlas may mark the first time when medieval and modern history was supplemented by classical antiquity, rather than the other way around.[132]

The surge of national atlases continued. In 1830, a revolution broke out in Brussels and detached the southern Netherlands from the north. On the eve of this revolt, two unsuspecting authors celebrated the past of the reunited Low Countries. E. Maaskamp confined himself to a one-sheet map, dedicated to the king of Holland. His design, more decorative than pedagogic, features battles, especially at sea, complete with little pictures. P. C. van der Elst, a southerner evidently attached to the regime prevailing since 1815, produced an ambitious, fifteen-sheet atlas. Its large, artistically unappealing pages thoroughly survey the history of the Low Countries. By 1836, the new, post-revolutionary Kingdom of Belgium could be attributed a slice of its own of Netherlands history. Nicolas Joseph Jusseret, a young cartographer in the Établissement géographique of Philippe Vandermaelen in Brussels, set about compiling a strictly Belgian atlas. Despite an accident that ruined his health, Jusseret brought his project to fruition in the three years left him before he died at twenty-six. Others, unnamed, must have shared his efforts: the cover page is authenticated by Jusseret's signature in ink and dated 1836, the year he died, but the final map is marked 1842. A Dutch contemporary decried Jusseret's exclusion of the northern Low Countries. Pre-1830 conditions are recalled only by a map of the Dutch East Indian colonies, perhaps as a reminder that the new kingdom had been denied a share of this empire. With confident patriotism Jusseret (or a successor) outlined the rosy prospects of an independent Belgium.[133]

Meanwhile, the French were not standing still. Brué's one-time collaborator, Joseph Guadet, compiled a new collection presenting "the political and geographic conditions of France at the main periods of its history." Guadet, a respectably qualified historian, gave the atlas a more solemn introduction than seems warranted by its fourteen rather simple sheets. He realistically warns of his weaknesses: he is without geographic pretensions; topography is completely neglected in the maps, and only such places are marked as are necessary.[134] Denaix, already encountered as the author of an atlas of Europe, published in 1836 the first form of an atlas of France that was expanded and reissued a few years later. Its chronological sequence of French historical maps was recycled as Denaix's *Petit atlas* of France. Having experienced what revolution could do to political boundaries, Denaix and his admirers realized that such boundaries were too mutable to serve as the foundations for geography. The many thematic maps of Denaix's atlas of Europe had been a step toward a different program. For France, however, Denaix managed to devise barely a handful of themes, and his physical geography was limited to the watershed theory of Philippe Buache. What fascinated Denaix, as it had Rizzi-Zannoni's historical counselor, Fontanieu, was the drama of the losses and gains of French territory, from the ostensibly vexatious inheritance partitions of the Merovingians to the colossally swollen Empire of Napoleon in 1813. This fascination with territory, somewhat muted in the major atlas, wholly dominates the *Petit atlas*.[135]

Denaix was hardly alone in being obsessed by boundaries. The *Fastes de la France, ou Tableaux chronologiques, synchroniques et géographiques de l'histoire de France* by C. Mullié was one of only a few historical atlases published outside Paris. It appeared in various editions between 1832 and 1848. Maps were first published in the edition of 1833, and their number attained stability at seven in 1836. Unblushing patriotism prevails; a statement found on the map of successive "reunions" of territory to the Crown suggests the prevailing tone: "Duchy of Tuscany, reunited in 1807." There is no hesitation in speaking about the "*re*-union" of a territory that had never been under French control in the first place. Devoted to French triumphs, Mullié's *Fastes* nevertheless records useful information and long occupied a place in the reference room of the Bibliothèque nationale.[136]

The occasional claim that Germany did not have a national atlas until the country was unified in 1871 arises from the titles of books rather than their contents. Even when fragmented, Germany had atlases in which the European past circled about the Romano-Germanic Empire—not much farther from a national interpretation of the German past than those of more cohesive countries.

An early example of this approach was by Julius Rupp (1809–84). Trained for the ministry, Rupp was made a senior teacher at the Königsberg gymnasium (1835). Soon after, he published a *Historischer Schul Atlas* to parallel a set of genealogical tables. The atlas and tables supplemented the history textbook he drew up for his classes. His seventeen maps then became separate *Charten für die Geschichte,* destined for wider sale. Rupp's focal point, the empire, was no doubt different from a centralized German state, but for purposes of historical maps it provided a rallying point comparable to the French and

English kingdoms.[137] In the 1840s, three or four German collections took the path that Rupp pioneered. Atlases of French history stayed in the numerical lead: five major ones were produced in the same decade.[138]

A New Format for Comprehensive Historical Atlases

A mutation in the form of historical atlases, especially large, ambitious ones, occurred in the 1830s, matured in the following decade, and thereafter assumed a commanding place in the genre. The emergent design involved classifying maps according to their geographical extent: to begin with, a broad, general section, perhaps continental in scope; then a continuation with particulars or details, usually resembling regional or national collections. A typical atlas of this sort might open with a section similar to Denaix's *Atlas historique de l'Europe* and continue with a sequence of compressed, regional maps. This layout had been anticipated at the opening of the century by the skeletal geographic arrangement of the *Atlas Lesage*. Its return a few decades later was designed to provide as much breadth of coverage as possible, short of going back to universality, and to blend it with the ample detail possible in maps of individual countries.

The new design was embodied in a masterpiece among comprehensive historical atlases, the work of Karl Spruner, a lieutenant in the Bavarian army. Today, more than a century and a half after its first installment, Spruner's *Historisch-geographischer Hand-Atlas* continues to be a byword for its specialty and graces the reference shelves of geography collections. It may or may not have been superseded for the ages it covers. The firm of Justus Perthes in Gotha, publishers well on their way to greatness in geography, issued the first installment of Spruner's European volume in 1837. Five more installments followed.[139]

Spruner's exceptional fame tempts one to believe that the new grouping of maps for history was inaugurated in his *Hand-Atlas*, but chronology denies him paternity. A number of collections in this format became available during the nine years in which Spruner's work was gradually published. At the very moment when his first installment appeared, a Frenchman named Antoine Houzé brought to market an atlas "classique et universel" of 101 maps organized by countries. It was sold for many years, sometimes in luxury editions embellished with historical prints by artists of varying talents. Houzé's scheme was practical or commercial rather than systematic. The maps of his collection, half the size of Spruner's (or less), could be easily broken up and sold as atlases of Greece and Italy (sixteen maps), Spain and Portugal (eight),

France (thirty), the Holy Land (fourteen), and all the other constituent parts. His maps of Spain, duly translated from French (1840), monopolized the Spanish market for a national historical atlas until challenged in 1879 by a domestic substitute. Classical antiquity and Europe did not have sections of their own.[140] Houzé's atlas was the most extensive historical collection yet published in the nineteenth century. While nowhere near Spruner's work in quality and refinement, it comes surprisingly close in organization.

Antoine Houzé is all but unknown. He steps fleetingly from the shadows in 1839 to become a member of the Paris Société de géographie and to present it with some of his works. Although his lifetime extended to the 1860s, he seems never again to have left an enduring trace of his existence; biographical dictionaries do not record him. He authored an elementary geography but was not a cartographer in his own right.[141] The maps of his atlas were merely "drawn up under [his] direction." To judge from the abundant, though concise, written matter accompanying each map, his strong suit was history, especially of an uncritical schoolmaster's kind. Place-names underscored in color are glossed in the lower margins, as, for example, "Pola, Crispus is beheaded by the order of his father [Constantine the Great]." Houzé accepted most of what he read, such as Geoffrey of Monmouth's twelfth-century account of early British history, or that Sigtunna, near Stockholm, was Odin's dwelling place.[142] He cultivated brevity in his clarifications, without coming anywhere near the professionalism of Spruner's *Vorbemerkungen*.

Houzé's maps, like those of Brué, Denaix, and many others, lack physical relief. Because sections for the continents are omitted, the new atlas design is exemplified only in its second, "particular" part. Houzé gives a historical identity to his geographical outlines by little else than the application, often rough and hasty, of one or more thin, colored boundary lines. Tracks occur for the biblical Exodus, ancient Greek wars, and barbarian migrations, but with restraint. Houzé's atlas, often unimpressive, has a total plan nearly as expansive as Spruner's and came into circulation, complete, many years earlier than the full *Hand-Atlas*.

Several additional compilers in France and Germany availed themselves of the new atlas format before Spruner's work finished its course through the press. Among them was Victor Duruy, one of the most prominent French historians of the day and a shining example of the rewards France accorded to merit. While still young, Duruy had stood in as lecturer for the celebrated historian Jules Michelet and, in time, was recruited to assist the emperor Napoleon III in writing about Julius Caesar. This post as assistant to the emperor led to Duruy's appointment as minister of education. His period in office witnessed notable reforms. He had the good fortune, besides, to return to private life in time to elude the debacle of the Second Empire in the Franco-Prussian War. Humbly born and in great need of money, Duruy earned income from writing an extraordinary number of textbooks. One of them, in 1841–42, was the *Atlas historique du moyen âge,* in fifteen maps, part of a collaborative "atlas of universal historical geography."[143]

A fuller instance of the new two-part arrangement came out of southwestern Germany in 1842. The firm of Herder in Freiburg-im-Breisgau started publishing the *Historisch-geographischer Atlas zu den allgemeinen Geschichtswerken von C. v. Rotteck, Pölitz und Becker* in the year in which its author, Julius Löwenberg, prematurely died. Johann Valerius Kutscheit, who already had one atlas to his credit, saw the original edition to completion and developed it. Löwenberg's impressive forty-map collection is explicitly divided into sections of general and particular maps (*Übersichtkarten, Specialkarten*). The *Specialkarten,* though not grouped as mini–national atlases, are predominantly of countries at various times in their history. Soon after, Kutscheit produced the *Hand-Atlas der Geographie und Geschichte der Mittelalters,* immediately followed by a second edition of Löwenberg—both in his own name. Unlike most compilers of historical atlases, he had a modestly positive reputation. In whatever edition, Kutscheit's choices of subject and presentation avoid implying a national classification. The focus moves from one part of medieval Europe and the Levant to the other as though guided only by intrinsic historical interest. Nevertheless, the dominant order for the numerous "particular maps" is not one of time or of salient events but of geography.[144]

Two French collections of these years, though less developed than Löwenberg/Kutscheit, shed light on what was going on. In around 1840 Victor Levasseur published a small, inexpensive atlas whose early maps adhere closely to the European subjects offered by Christophe Guillaume Koch. The collection issued by Levasseur four years later is reminiscent of the *Atlas Lesage:* its cursory country-by-country coverage is in keeping with what was being done in Germany and France.[145] In the next year, P. Clausolles joined with Philippe Abadie to produce a twelve-map atlas of medieval history; next on their program was a French history atlas. Although their collection includes traditional items, such as "Europe in the Tenth Century," at least half their maps are assigned to moments more narrowly defined in time and space. Their approach resembles Duruy's.[146]

As though to confirm that the grouping of historical maps by geography had abruptly come of age, the noteworthy mapmaking firm of Delamarche in Paris issued a fully developed collection ordered in this way. The Delamarches were the last link of a cartographic chain reaching back to Nicolas Sanson in the seventeenth century. The Sanson inheritance had descended to Gilles Robert and in 1786 was acquired from Didier Robert de Vaugondy by Charles François Delamarche (1740–1817). Delamarche, a lawyer by training, also bought out Jean Lattré, another well-known map publisher. He marketed the stock he had accumulated, secured a monopoly for map and globe sales to schools, produced few maps and atlases of his own, and, unlike his predecessors, achieved commercial success. Charles and his able son, Félix (1780–1835), had both died by the time the atlas in question here came on the market, but the firm continued in family hands.[147]

The Delamarche maps for medieval history were published when Spruner's *Hand-Atlas* was two years from completion; enough of it had appeared to permit imitation. By chance or intentionally, Delamarche's organization bears a close resemblance to the Gotha product. The "Cartes générales" overrepresent the early Middle Ages, as though moving on from the Roman Empire were a problem. The "Cartes particulières" are straightforwardly national: France, Great Britain, Italy (two sections), Germany, Spain. Actors not easily classed "nationally," such as Byzantium, the Islamic caliphate, the Mongols, and others, are fitted into suitable interstices. The Delamarche product does not rival Spruner in abundance or quality, but the arrangement of its contents closely resembles his.[148]

KARL SPRUNER'S *HISTORISCH-GEOGRAPHISCHER HAND-ATLAS* (1837–46)

Historical atlases organized by country, or subdivided into general and particular maps, were much more common in 1846, when Spruner's great volume of maps for European history reached its sixth and last installment, than when the opening installment had been issued nearly a decade earlier. Ostensible competitors did not detract from Spruner. His work, unanimously praised by reviewers and scholars, and sustained by a resourceful publisher, was a resounding commercial success. An atlas of ancient history and one for lands "outside Europe" (*Ausser-Europa*) rounded out the set. The one thick volume and two slim ones, encompassing the whole of history, contained 110 maps, not counting numerous insets. The total increased to 118 in the second edition. Its cartographical copiousness alone would have elevated the work above the crowd.[149]

The main volume is entitled *The History of the European States from the Beginning of the Middle Ages to the Present Day* (*Geschichte der Staaten Europa's vom Anfange des Mittelalters bis auf die neueste Zeit*). An opening section for the continent as a whole ("Thirteen General Maps of Europe") is continued by regional coverage: Germany, with the Netherlands, Austria, and Switzerland; Italy; France; the British Isles; Spain and Portugal; the northern kingdoms, including Scandinavia, Russia, and lands in between; southeast Europe and Asia Minor, including the Islamic caliphate; finally, Hungary. Regional coverage ranges from twelve maps for greater Germany, eight for the northern kingdoms, and four for Britain. The table of contents announces how many maps are assigned to the region in question. A few thematic maps are included: every region has its ecclesiastical geography, and Europe is shown once by "peoples and linguistic frontiers." The themes are neither unusual nor numerous.[150] Letterpress *Vorbemerkungen,* printed separately on small sheets, contain a succinct commentary. The whole three-part atlas, if bound as one, forms a massive oblong folio, far and away the fullest historical collection to have been published till then and perhaps unequaled to the present as authoritative cartography for history.[151]

The *Hand-Atlas* did not absorb all of Spruner's energies. A year after its opening installment, his eight-map *Atlas zur Geschichte von Bayern* was published, also by Perthes of Gotha. Among historians at large, especially in Germany, standards of documentation and critical method had been dramatically rising. Spruner won much applause for associating his work with this development: "To be as it should be, a historical atlas can and must, like a good history, be compiled only from the [primary] sources themselves." When the whole of world history was embraced, all the sources could, admittedly, not be absorbed, but mastering the best studies (*Vorarbeiten*) was obligatory. Spruner used secondary works astutely and corrected antiquated opinions. The fulfillment of his program was a milestone. As a result of his work, all historical atlases prepared on a less scholarly basis became outmoded even if they did not cease to be produced.[152]

A Württemberger by birth, Karl Spruner von Merz (1803–92) came to Bavaria as a child. He was a career lieutenant in the Bavarian army when he launched into scholarship; and a lieutenant he stayed throughout the time that the centerpiece of his *Hand-Atlas* came into print. Finally promoted to captain in 1847, he received no assignment matching his abilities for several years. Then, in 1851, the king of Prussia asked his son-in-law, Maximilian II of Bavaria, how it happened that an officer highly honored in the world of learning could be kept in obscurity in the Bavarian kingdom. This exchange nudged the reluctant Maximilian into interfering with army promotions. Spruner was seconded to the general staff, promoted to major the next year (he signed the second edition of the *Hand-Atlas* with this rank), and in 1855 named aide-de-camp to the king. He was soon a trusted adviser, especially in matters of royal support for scholarship. The Bavarian Academy of Sciences, more appreciative of Spruner than the army, elected him a member *extra ordine* as early as 1842 and a full member in 1853.[153]

When still in his twenties, Spruner became an expert in all early Bavarian documentation, notably charters, and composed *Baierns Gauen*. This study of the early cantons of Bavaria involved the contentious subject of *gau* formation that had fascinated German historical geographers for more than a century. Young Spruner butted firmly against the prevailing doctrine and proposed an alternative.[154] He was extraordinarily productive in the 1830s, during which he published not only his work on the *gaue* but also the multisheet historical atlas of Bavaria, a compendium of chronology and genealogy, a description of Charlemagne's canal in the Bavarian Nordgau, a historical map entitled "The East Frankish Duchy Divided in Its *Gaue*," and the opening installments of the *Hand-Atlas*. Retracing much of the ground covered by earlier German maps for history, he was enviably effective in outdoing them.[155]

Unusual self-assurance and energy sustained Spruner's activity down to the mid-1850s. By 1860, however, his days of scholarship were at an end. To continue development of the *Hand-Atlas,* the Perthes firm recruited Theodor Menke, a Berlin graduate, who proved an excellent choice. Menke alone was responsible for the great third edition, generally called "Spruner-Menke" in recognition of the latter's part in carrying out a total revision, complete with largely new maps. Meanwhile, Spruner aspired to succeed as a writer for the stage but fell short of finding public favor. Even with royal patronage, army promotion lagged for a soldier who may never have seen combat. Spruner persevered. He became a full general at eighty, retired three years later, and died just short of his ninetieth birthday.[156]

Spruner's *Hand-Atlas* stands out so prominently among its peers that it has been considered the model for all that has come after—the summit that has guided later compilers of historical collections.[157] This appraisal is less certain than might be imagined. An early admirer of the *Hand-Atlas* noted that this "large and valuable work is designed only for specialists (*Gelehrte von Fach*)."[158] It was an élite product, beyond the reach of students, the traditional users of historical atlases. Later atlases are more likely to bow to Spruner as a source than to reproduce its salient features. Even Spruner spin-offs, such as his *Schul-Atlas,* bear rather little resemblance to the main work. The ordering of maps by countries, which continued to be widely practiced, was not his invention. Maps with physical relief, a pronounced attraction in Spruner, have tended to be rare. It is difficult to establish in what ways the *Hand-Atlas* might be a prototype. Its mastery of historical details offered successors less incentive for imitation than encouragement to skimp on their own research.

Whether the many Spruner abridgments, translations, and adaptations proved influential is an arduous question, needing study. His own responsibility for them, as distinct from the Perthes workshop's, cannot be taken for granted.

The unabridged three-volume *Hand-Atlas* went through one revision and was marketed from 1851 to the 1870s. It then provided material for Menke's almost total overhaul. Its maps are on a noble scale and hand colored; many are a pleasure to behold and consult. The combination of scholarship, craftsmanship, and art is difficult to outclass: Spruner's *Hand-Atlas* has the look and bearing of authority and retains it to this day. Works that most closely approximate it, notably Gustav Droysen's *Allgemeiner historischer Handatlas* and especially R. L. Poole's massive *(Oxford) Historical Atlas of Modern Europe*, are those whose sponsors made determined efforts to equal Spruner's distinction in appearance as well as substance.[159]

Even in this cohort, not everyone bowed to the master. The French historical atlas directed by Franz Schrader, though carried out under auspices comparable to Droysen's and Poole's, went its own way, in a basically chronological structure. Workaday but famous classroom atlases, such as Putzger and its American counterpart, Shepherd, bear so vague a resemblance to Spruner that the idea of a filiation is more easily proposed than demonstrated.[160] Among the prestigious atlases produced since World War II—Westermann, Bayerischer Schulbuch-Verlag, Kinder and Hilgemann, *Times*—it is hard to say how Spruner specifically might fit into their ancestry.[161] The summit of a literary genre, which Spruner's *Hand-Atlas* definitely was and may still be, does not necessarily impose itself on posterity as a model.

Packing Much in Little

Historical atlases published between Spruner's European volume and 1870 had clearly marked lines of development. Some were more elaborate than others, and homogenization seems to have been avoided. Nevertheless, divergent types did not proliferate, and map subjects ran the risk of becoming monotonous. There was no trace at this time—nor would there be for many years—of atlases in which political vicissitudes are subordinated to a spectrum of social, economic, and other themes treated historically.

The great progenitors of the very early 1800s had not wholly exhausted their potency as the century approached its midpoint: Kruse atlases were reprinted, mainly in French; the *Atlas Lesage* spawned enriched editions in Italy and Belgium, as well as more or less acknowledged imitations in France and

England. Other atlases illustrating the endurance of an early prototype will concern us later.

The production of atlases of national history gained momentum. France saw them come to market at the rate of one every other year between 1841 and 1870. Germany was not far behind, with a possible twelve items in the same period. Titles like "historical atlas of Germany" were generally avoided; Spruner's *Schul-Atlas von Deutschland* was an exception. But ostensibly general German atlases focused on the Holy Roman Empire, much as French ones did on national history. Other countries also charted their past on maps. England engendered three historical atlases; Sweden and Switzerland two each; Hungary one. The Venice edition of the *Atlas Lesage* (1840) incorporated a set of maps of Italian history, and a modest second attempt at an Italian atlas followed in 1854.[162]

The combination of general and particular maps continued to find favor with compilers of impressive collections. This arrangement superseded the universal, sequential plan, from whose form it sharply diverged. Dussieux and Barberet in France, Rhode in Germany, William Hughes in England, and Huberts in Holland took this direction. Their atlases varied in size, but all aspired to combine general maps with particular (or national) ones.[163]

Historical atlases between 1846 and 1870 number almost one hundred. Among them, one small group alone, whose apparent goal was comprehensiveness in a small compass, stands out for originality.

Two German atlases in the late 1840s moved deliberately away from the expansiveness of Spruner and tried instead to dispense necessary cartography in minimal space. The *Vollständiger Atlas zur Universalgeschichte* by A. von Freyhold was composed of three large sheets, one each for antiquity, the Middle Ages, and modern times, issued in 1846, 1848, and 1850 respectively. The designer made the most of single sheets. A portrayal of Alexander's eastern conquests occupies the largest of the three large maps on the sheet for antiquity, along with a cluster of insets. The other sheets are variations of this pattern.[164]

Freyhold addresses wider concerns. The compendiousness he cultivated underscores the abandonment of the layout in successive frames that had seemed admirable in the sequential atlases of universal history. On Freyhold's single sheet with multiple maps, readers simultaneously face the equivalent of a book and are expected to focus on the frame needed for any given assignment. The author guides student-readers mainly by differentiating the maps by theme and size. A historical scenario of highs and lows, widely agreed on among historians, is reflected in the map subjects and dimensions. National highlights are welcomed even in general history so that medieval maps for

Germans emphasize German interests. As for physical relief on maps, Frey-hold's belief in its importance shows him to be a contemporary of Spruner rather than a prophet of future trends in design.

The publication of Freyhold's concentrated atlas overlapped with that of a fuller one much like it, also produced in northern Germany. Rudolph von Wedell's *Historisch-geographischer Hand-Atlas* appeared in six installments during as many years. Its foreword was by a pastor who had taught history at the royal cadet school; Wedell had been a pupil of his and was now a lieu-tenant in the Prussian army.[165] The atlas has thirty-six sheets, quite large in dimensions, two for antiquity, fourteen for the Middle Ages, and a dispro-portionate twenty for modern times. Wedell's design resembles Freyhold's in having several major maps on each sheet—as many as nine, an average of six. Comprehensive material for the period or theme in question was fitted to-gether, in a gradation of sizes, and set before the reader as an aggregate.

The end product, much leaner than Spruner's atlas, nevertheless permits an abundance of detail. Wedell's elaborate program translates into more than two hundred maps and many insets besides. The introduction underscores his very strict, perhaps penurious, use of space. Few square inches are wasted; irregular frames and careful joinery, with many variations of scale, adapt multiple major subjects to a single sheet. For the sake of historical clarity, Wedell sacrificed physical geography except when relief was essential to ex-plain historical circumstances. Color is used sparingly, mainly for frontiers and tracks; plain white backgrounds give maximum legibility to place-names and other writing, of which there is a great deal: names of dynasties and rulers, indications of trade routes and discoveries, clarifications of all kinds. Wedell considered it appropriate, for instance, that in the map of the barbar-ian invasions a line of writing should curve gracefully through the Mediter-ranean explaining, "In 384 Arianism disappeared from the Roman Empire, but began afresh among the peoples pressing forward."[166] Ostensible clarifi-cations of this kind raise more questions than they answer.

Wedell's atlas gives the impression of being less concerned with historical and geographical authority (in the manner of Spruner) than with effective and considerate pedagogy. Many of his displays anticipate the multiple maps per page found in well-regarded teaching atlases of today, such as Putzger or Westermann. He has quirks: an odd preoccupation with the possessions of the crusaders and an excessive fondness for flowing lines tracing human movements. So far as one may tell, his collection had only modest success. It suggests nevertheless that historical atlases were not ready to adopt an exclu-sive pattern. Spruner's European volume had no sooner been completed than the clientele were offered an atlas of very different design, yet comparable in scope and number of maps, ostensibly more practical to use, and surely more cost-effective in its condensed format. Spruner had no cause for concern; Perthes of Gotha did not allow its standard-setters to languish for lack of marketing. But the efforts of Freyhold and Wedell to be compact and their handling of atlas pages either influenced or foreshadowed future designs, especially those of small-size teaching collections.[167]

Maps for History Attain Maturity

Atlases to assist historical study had gained a secure, if not an honored, place by the opening of the nineteenth century. The upheavals of the revolutionary period brought rather sudden respectability to maps of recent, post-1500 history and gained them a place among specialized depictions of the past. The welcome accorded to Kruse's *Atlas und Tabellen* and the *Atlas Lesage* at the opening of the century presaged a better future for the genre.

That improvement came about haphazardly. By the 1830s, however, hardly a year went by without the publication of a new historical collection somewhere in Europe. Historical supplements to geographical atlases tended now to admit later-than-classical maps. The universal, sequential format of the eighteenth century had its last "performance" with Edward Quin. For a time, collections concentrating on Europe predominated, but they were overtaken by atlases of national (especially French) history. The most comprehensive atlases for history were soon organized as a limited general section followed by a succession of regional or national sets.

History classes in schools called for postclassical atlases as never before. The customary diet of biblical and ancient history was now supplemented by a national or European past that the turmoil of 1789–1815 had stirred to the core. School atlases initially emphasized brevity and compactness at the expense of most other virtues. In time they acquired imaginative layouts packing much in moderate space while retaining intelligibility. Quantity continued to mount as the century advanced, and more countries became involved in atlas-making; but by midcentury the forms and standing of historical atlases had attained stability.

The equilibrium of the genre gained much from the emergence—none too soon—of the specialized atlas of geography. As late as the 1850s and even beyond, atlases could still be called "historical" yet contain not a single map depicting a moment of the past. History, loosely understood, was deemed to be portrayed by all nonspecialized maps.

The meaning of the term "history," however difficult to delimit, was not the main obstacle to deciding where maps for history did or did not fit. What mattered was the intimate association of geography with history, or subordination to it, as expressed by such mottos as *historiae oculus geographia* and by the universal belief that geography was an "auxiliary" to history. This association began to unravel. The long-delayed emergence of geography as a subject with its own identity and program took place—a development attached, rightly or wrongly, to the names of Alexander von Humboldt and Karl Ritter, the giants of *Erdkunde* at the University of Berlin. The opening decades of the nineteenth century were decisive for introducing geology, meteorology, botany, zoology, and many natural sciences into the agenda of geography. In 1813 the great writer Goethe, perhaps the foremost humanist in Europe, published a picture-map comparing the main mountain peaks of the Old and New Worlds, as documented by von Humboldt (1807).[168] (Similar maps would appear in many nineteenth-century atlases.) In a geography fully opened to the natural environment, maps had loftier functions than to assist readers of history to locate the places mentioned in their books.

One set of circumstances suggests, succinctly, the magnitude of this change. During the Napoleonic age, when Weimar was a capital of German cartography, nearby Gotha, another Saxon court town, was the home of a civil servant named Adolf Stieler (1775–1836). Stieler was a fervent amateur geographer. He talked the publisher of the Gotha court calendar, Justus Perthes, into undertaking a world atlas according to the exacting specifications he laid down. The ensuing *Hand-atlas über die fünf Theile der Erde* was concerned with geographical features and political divisions, but not history. Stieler's project resembled others in scope and ambition. The Weimar Geographical Institute had produced such an atlas; Philippe Vandermaelen (1795–1869) in Brussels soon did the same. Atlases of the world were not in short supply. Stieler's *Atlas* as published by Perthes stood out because of its meticulous quality control and, soon, for its continuity of correction, revision, and improvement that made it less an atlas than a process or an institution. The "Stieler enterprise" persisted into the twentieth century, a byword for the highest standard of mapping at its scale. Prodded by Stieler's encouragement and guidance, Justus Perthes fashioned his firm into the European leader in geographic publication.[169]

Perthes steadily recruited trained experts to cultivate and enrich the works on its list. Karl Spruner's historical *Hand-Atlas* started through the presses at Gotha twenty years after Stieler's geographical atlas began publication. It shared the attention of the Perthes establishment not only with the continuous refinement of the Stieler atlas but also with other projects, notably, the *Physicalischer Atlas* (1838–48) of Heinrich Berghaus and the *Methodischer*

Hand-Atlas (1842–46) of Emil von Sydow. Offshoots of Karl Ritter's ideas via his pupils, they were as momentous for physical-thematic geography and for geographic pedagogy as Spruner's was for history.[170] The four almost simultaneous Perthes atlases of Stieler, Spruner, Berghaus, and von Sydow are symbolic of a geography defining its territory, establishing its standards, and drawing the base lines that, sooner or later, permeated humbler textbooks. However vague the idea of maps for history may still have been as late as 1800, it could hardly help becoming restricted and precise in the wake of these developments.

NOTES

1. Stendhal [Henri Beyle], *Le rouge et le noir,* ed. Jules Marsan, 2 vols. (Paris, 1923), 1:29–32, esp. 31. Émmanuel de Las Cases, *Mémorial de Sainte-Hélène,* ed. Gérard Walter, Bibliothèque de la Pléiade, 2 vols. (Paris, 1956–57).

2. Goffart, "Preliminary Report," 54 n. 11. On the length of the *Mémorial,* see Quérard, *France littéraire,* 4:589. For padding of the *Mémorial,* see A. P. P. Roseberry, *Napoleon, the Last Phase* (London, 1906), 21. Successive biographers affirm that Las Cases earned 2,000,000 francs from the *Mémorial.* (Compare the report of Eugen Weber in *Times Literary Supplement,* 25 April 1997, 10, that between 1845 and 1862 Adolphe Thiers earned 500,000 francs from his *Histoire du Consulat.*) Las Cases's work appeared when demand in France for historical memoirs was so intense that spurious ones were produced and successfully circulated: Louis Halphen, *L'histoire en France depuis cent ans* (Paris, 1914), 44–48.

3. Goffart, "Preliminary Report," 55; A. Fiero, A. Palluel-Guillard, and J. Tulard, *Histoire et dictionnaire du Consulat et de l'Empire* (Paris, 1995), 891, specify "petite noblesse." About the Quiberon raid and Volude's death as a moment of decision for Las Cases, see Emmanuel de Las Cases, *Las Cases, Le mémorialiste de Napoléon* (Paris, 1959), 78.

4. Goffart, "Preliminary Report," 56–57. Invitations for subscriptions are issued in the late 1790s by Lesage, 1799/B, 12. Las Cases, *Mémorialiste,* 81, speculates that the atlas was an expansion of the previously published Lesage, 1800/O (later bundled with, then incorporated into, the *Atlas*); this is unlikely. For a malicious and baseless rumor circulating during Las Cases's lifetime that denied his authorship of the *Atlas,* see Goffart, "Preliminary Report," 58–59 n. 23. For the wider context, see Margery Weiner, *The French Exiles, 1789–1815* (London, 1960).

5. Goffart, "Preliminary Report," 55 n. 14. See Walter in *Mémorial,* 1:xxxvii (his engagement), 1:xxxvii n.** and 2:1149 (his marriage). Las Cases's brother-in-law, Joseph de Kergariou, also a London exile, became Émmanuel's secretary while the atlas was being completed (Las Cases, *Mémorialiste,* 84). Las Cases's secret visit may explain the false allegation that he moved back to France after the coup d'état of 18 Brumaire (9 November 1799), i.e., as soon as Bonaparte was in power; Fiero et al., *Histoire et dictionnaire du Consulat,* 891: "revenu en France au début du Consulat [= 1799]." For earlier statements to the same effect, see *Arch. biog. franç.,* fiche 605, frames 172 (an item dated 1816!), 174, 179, 181–82, 184; 2d ser., fiche 390, frame 439. There is wide agreement that the atlas dates from after the Brumaire return

rather than from London. In fact, the Peace of Amiens (1802) and the ensuing amnesty made his return possible. Las Cases underwent a civil marriage in 1808, before applying to the government for preferment.

6. This South Street is now the eastern leg of Blandford Street, a little north of Manchester Square, a popular address for French émigrés at this time. Las Cases lived steps away from Manchester Street. On the organization of a team, see Las Cases, *Mémorialiste,* 83. The subscribers' list survives in two forms; see n. 8, below.

7. Las Cases, concerned to keep his "low" working life apart from his socializing among more fortunate fellow émigrés, had long used the pseudonym "Felix" but abandoned it when preparing the *Atlas.* "Lesage" was first affixed to minor publications of the 1790s. The duke of Gloucester is better remembered for having, some twenty years earlier, accepted the dedication of the first two volumes of Gibbon's *Decline and Fall of the Roman Empire.*

8. The RGS copy of Lesage, 1801, contains the subscribers' list; the CambUL copy of Lesage, 1802, also has a subscribers' list. Lesage, 1802, implies that the atlas was fuller and more mature than Las Cases admitted once he was in France. It suited French pride that the atlas should come to fruition only in Paris. The London version (so he claimed) was a mere sketch; see n. 10, below.

9. Goffart, "Preliminary Report," 53. On émigrés returning in 1802, see Jean Vidaleuc, *Les émigrés français, 1789–1825* (Caen, 1963), 135–36.

10. A guinea is £1 1s. Judgment on the London edition: Las Cases, *Mémorial,* ed. Walter, 1:279, 66. Lesage's description of the genesis of the *Atlas* is thin and says nothing we need to know: "[After abortively considering writing a novel] Je me décidai pour l'histoire, qui, dans tous les cas, m'assurait un gain moral en me procurant des connaissances positives: alors naquit l'idée mère de l'Atlas historique. Ce fut une inspiration du ciel, je lui dois le reste de ma vie. Ce ne fut d'abord qu'une simple esquisse, bien éloignée de l'ouvrage d'aujourd'hui, une pure nomenclature. Toutefois s'en fut assez pour . . . me composer même, relativement aux misères de l'émigration, une véritable fortune" (*Mémorial,* ed. Walter, 1:610). "It was a mere sketch . . . a bare outline" is unlikely to refer to the little pamphlets that preceded the London edition (Lesage, 1799/B, 1800/B) since it was the full atlas that earned the "véritable fortune."

11. The main difference between Lesage, 1801, and Lesage, 1803, is that the maps, placed in the upper left of the original version, are shifted to the center in Paris and become (permanently thereafter) a peninsula surrounded by letterpress. Though conspicuous, this was a minor improvement.

12. Las Cases returned in company with his friend and patron, Sir Thomas Clavering, Bart. (1771–1853): G. E. Cokayne, *Complete Baronetage,* vol. 3, *1649–64* (Exeter, 1903), 204–6. Las Cases, *Mémorial,* ed. Walter, 1:xxxvii, calls him "richissime," probably an exaggeration. Clavering's wife, Clare, was from Anjou. In London the Claverings lived on Manchester Square, and Las Cases, nearby, called on them daily. He reciprocated their kindness when they were stranded in France at the rupture of the Peace of Amiens and interned for the better part of ten years. Their friendship was lifelong. About Lady Clavering and the genesis of the *Atlas,* see Las Cases, *Mémorialiste,* 81–82, 84. A student of Las Cases, Miss Steers, apparently advanced half the cost.

13. Las Cases evidently did not consider his English earnings adequate for a Paris household. Lesage, 1803, has the author's rue St. Florentin address as place of publication. About his quarters there, see Jean Tulard in E. de Las Cases, *Mémorial de Sainte Hélène,* ed.

J. Schmidt (Paris, 1968), 8; see also Las Cases, *Mémorialiste,* 100. Talleyrand, the best-known resident of this street, did not arrive until 1812. A BN copy of Lesage, 1803, Ge.DD.4796 (104), has bound into it a printed "Note des cartes et ordre des livraisons [from Didot] de l'Atlas historique de A. Lesage" (presumably the "Note" is by Las Cases). A small, firm hand has written in the margin how the original installments compare with the order of Lesage, 1806a. Various dates, including 1802, are indicated on some of the maps (many are undated). The subscription list is catalogued in *BN Impr.* but was off the shelf when I called for it. Another—not a precise substitute—is in the Harvard copy of Lesage, 1806a. This list, well furnished with Napoleonic notables, includes few scholars.

14. The prices are in the review in *AGEphem.* 16 (1805): 93; the "Ordre des livraisons" (see previous note) specifies 10 livres (= francs) for each installment of the cheaper edition, 15 for the better one.

15. Review in *Allgemeine Literatur-Zeitung* (Halle) 346, no. 4 (December 1804): 521: the work has found extraordinary acceptance (*Beifall*) and support in France.

16. See ch. 3, n. 103, above.

17. *AGEphem.* 16 (1805): 93: "fürs erste besser thaten, sich an Teutsche Werke zu halten." Ibid. 27 (1808): 102–3: much as the reviewer faults the work, he recognizes that it has won great esteem in its homeland and even in Germany. Lesage, 1827, unpaged prefatory material: the atlas has had innumerable editions, "malgré son prix élevé," and enjoyed prodigious success, without ever having a detractor until recently and unjustly. *Göttingische gelehrte Anzeiger,* 1830, 1693: the work is so widely and so long known as to need no recommendation. In *AGEphem.,* 2d ser., 15 (1825): 143–44, Georg Hassel, editor of the journal, noted the success of the *Atlas Lesage* in France, England, and America but affirmed that the atlas did not satisfy the higher demands of the German public. Proving him wrong, the German translation, Lesage, 1826c, began to appear in Karlsruhe in that very year. For French detractors, led by the aged geographer Brion de la Tour, see Quérard, *France littéraire,* 4:589, and n. 32, below.

18. Las Cases, *Mémorial,* ed. Walter, 1:vii; Maurois's introduction is dated 1935. Much the same is in Frédéric Masson, *Napoléon à Sainte-Hélène, 1815–1821,* 6th ed. (Paris, 1912), 147. Neither author attempted to verify whether Las Cases's pride was justified or not.

19. Very fine copies: BN, Ge.DD.4796 (105); Harvard, Map Collection, MA 80.1F.

20. Details in Goffart, "Preliminary Report," 54 n. 10. For the 137 chamberlains appointed in 1809, see Masson, *Napoléon à Sainte-Hélène,* 147; Las Cases, *Mémorial,* ed. Walter, 1:xxxix. Napoleon was unaware of the *Atlas* until Las Cases became his companion in 1815.

21. Quérard, *France littéraire,* 4:586, notes five later editions before featuring the latest Lesage in thirty-seven "maps" (136.50 francs); Lesage, 1827a. Additional sheets to 1839, Lesage, 183-. For the Brussels version, see n. 25, below. Later variants: Lesage, 1829, 1835. A curious trace is in Paris, ArchNat., Marine, Serv. hydrog. 6JJ/86, nos. 36–48: a painstaking MS copy of a few parts of the atlas (text only); all color washes are applied, all symbols recorded. For an accounting of Las Cases's profits from the *Atlas* down to 1816, see Las Cases, *Mémorial,* ed. Walter, 1:617. Napoleon remarked that if he had known the work before his downfall: "votre ouvrage, ou certaines parties eussent inondés les lycées" (*Mémorial,* ed. Walter, 1:821).

22. Russian, Lesage, 1809; Italian, Lesage, 1813b; German, Lesage, 1826c; Spanish, Lesage, 1826d; Polish, Lesage, 1844 (see Łodyński, 3: no. 463); English, Lesage, 1813a, with six extra maps by the translator, Mme Coindé. See Goffart, "Preliminary Report," 58 n. 22. Mme

Coindé is down for fifty copies in the Paris subscription lists (n. 13, above), among the five largest orders.

23. Lavoisne, 1814; "according to the plan of Lesage, greatly improved; . . . by C. Gros of the University of Paris and J. Aspin, professor of history." Lavoisne, 1807/B, contains genealogical and other tables without maps. *LC Bibliog. of Cartog.* 3:363 suggests that Lavoisne did not live to see his maps published. Although the Lavoisne maps differ markedly from the prototype, the page layout and type of commentary are very much in the Lesage manner. On John Barfield, see Ian Maxted, *The London Book Trade, 1775–1800* (London, 1877).

Carey, 1822; Buchon, 1825. The American atlas was expressly inspired by both Lesage and Lavoisne. Carey and Lea were pioneer map publishers in the United States. Henry Charles Lea, the son of one of the compilers and named after the other, is famed as a founder of medieval studies in the United States.

24. Jarry, 1831/O ("D'après la méthode et sur le plan de l'Atlas de A. Lesage"); *NUC* lists an edition in 1826, and Quérard, *France littéraire,* 4:209, one of 1827–29. More by Jarry in the Lesage manner: *Atlas constitutionel* (Paris, 1826); *Tableau historique des révolutions nationales,* in single sheets (Portugal, Switzerland, Poland) (Paris, n.d.); and Jarry, 183-/O. A graduate of the École normale supérieure, Jarry taught history at the École des beaux-arts and in lycées. According to *NUC* 278:232, the Lesage method means "[t]ext in columns, irregularly arranged, partly in colors to indicate topical division." Weiss de la Richerie, *Atlas politique de la France . . . depuis 1789 jusqu'au règne de Charles X* (Paris, 1828). The Dutch encyclopedia—Brand, 1833—petered out after fifteen installments; neither history nor geography was privileged in its program.

25. Lesage, 1827b (see n. 21, above). Marchal states that Las Cases, preoccupied by his political career, had long left the atlas to its publisher, Sourdon, whose widow could not sustain the responsibility. This may have been said to justify piracy. The Lesage enterprise did not flag. On Marchal, eventually a member of the Belgian Academy and librarian of the Bibliothèque de Bourgogne, see A. Wauters in *Biographie nationale de Belgique,* 13:430–43. The 1820s were a low point for him.

26. Lesage, 1826b, 1843. The impresario Gianbattista Albrizzi speaks scathingly of the first Italian translation. The LC atlas is Atlas MS, 1832; the compiler avoids "Middle Ages," preferring "histoire du Bas-Empire." There is no explicit sign that it was meant for publication.

27. Philippe Gonnard, *The Exile of St. Helena: The Last Phase in Fact and Fiction* (London, 1909) (tr. of *Les origines de la légende napoléonienne,* 1906), 54; when leaving Saint Helena, Las Cases lent Napoleon 100,000 francs without security (55).

28. Quérard, *France littéraire,* 4:588; reproduced without attribution, probably from the atlas introduction. Las Cases was not concerned about there being barriers between the disciplines; they were intimately connected, and he blended them. Lenglet, 1768/B, 10:42, praises Buy, 1761, for having geography, chronology, and history uniquely advance in step.

29. Las Cases, *Mémorial,* ed. Walter, 1:65–66: "les tableaux généraux p[euvent] être difficilement surpassés par leur méthode, leur symétrie, leur clarté et la facilité de leur usage; et les tableaux généalogiques présent[ent], chacun isolément, une petite histoire entière du pays qu'ils concernent: ils en étaient tout à la fois, et sous tous les rapports, l'analyse la plus complète et les matériaux les plus élémentaires." The author was not shy in extolling his work.

A typical declaration of Las Cases's pedagogical ambitions is in the abridged student edition, Lesage, 1829, 3: "La méthode de Lesage, adoptée par le conseil royal et recommandée par le ministre de l'instruction publique, exclut toute théorie, toute abstraction, et se réduit en quelque sorte à une simple pratique: aussi est-ce ce qui la met à la portée de tous les âges, de toutes les intelligences, et l'approprie spéciallement à l'instruction primaire. C'est le mécanisme le plus simple, le mieux ordonné, souvent même élégant. Là tout arrive à l'esprit et se loge dans la mémoire par les yeux à l'aide de linéamens, de contours et de couleurs qui introduisent la clarté, écartent la confusion et assurent d'inneffaçables souvenirs." The text continues in the same vein.

30. Buy, 1761; extensive *engraved* commentaries, including full pages without maps. Courtalon, 1774; Mentelle, 1798 (not a historical atlas). Lesage, 1826c, was praised for unusually ample coverage of Germany: *Kritischer Wegweiser im Gebiete der Landkarten-Kunde* 2 (1830): 41–42.

The reviewer of a map of Persia by Brué emphasizes that it has three wide margins with commentary in letterpress (by A. Balbi) and observes that practices such as this "ursprünglich wohl in *Lesage*'s Atlas ihre Entstehung gefunden haben"; *AGEphem.*, 2d ser., 21 (1827): 412. The text-filled margins of the *Atlas Lesage* are conspicuous, but Las Cases did not invent this layout. For traces of a more complex past, see, along with Buy and Courtalon, Peter van der Krogt, "Neederlandse Editie van Crome's Produktenkarte van Europa uit 1783," *Caert Thresoor* 3 (1984): 25–26; also Harms, *Themen*, 214–15 (engraved writing).

31. Las Cases, *Mémorial,* ed. Walter, 1:65. Napoleon eventually learned the right method: as Las Cases looked on, "il s'est mis à parcourir l'Asie, faisant concorder les marges et le tableau" (2:269). The reviewer in *Bulletin de la Société de géographie* 4 (1825): 71, agreed with Las Cases's criticism of his mapmaking: "Nous sommes obligés de convenir que ces cartes ne démentent pas son assertion."

32. Review in *Allgemeine Literatur-Zeitung* (Halle) 346, no. 4 (December 1804): 526. A French critic, the eighty-four-year-old Louis Brion de la Tour (see n. 17, above), spoke harshly of the maps, considered one appropriated from one of his (with a collaborator), and regarded the price as being extortionate: *Observations curieuses et utiles avant ou après l'acquisition de l'Atlas historique de M. le Sage* (Paris, 1809), 5–8, 10, 12, 14–15 (the publisher claimed to have sold 4,000 copies in five years). Nothing substantive is added by *Réplique de L. Brion de La Tour à un libelle anonyme intitulé: Appréciation de sa diatribe (prétendue) c'est à dire de ses Observations curieuses et utiles avant etc. . . .* (Paris, 1809).

33. Mentelle, 1797/A, not a masterpiece, is a suitable example of what current cartography was capable of.

34. BN, Ge.DD.4796 (104), fasc. 5, no. 4, in margin.

35. This reordering is visible when the order of topics in, e.g., Lesage, 1801, is compared to Lesage, 1813.

36. *Allgemeine Literatur-Zeitung* (Halle) 346, no. 4 (December 1804): 522. For the likeminded commentator, see *AGEphem.*, 2d ser., 15 (1825): 143–44. He considered the *Atlas Lesage* to be *the* French atlas.

37. *Allgemeine Literatur-Zeitung* (Halle) 346, no. 4 (December 1804): 522, 523, 524 (Lesage's defective knowledge of history), 525 (the letterpress surrounding the maps is a shapeless compound of fragments of all sorts), 532 (indented quotation).

38. Las Cases should not be judged by too lofty a standard; history in France was in a trough. E.g., the then much praised fourteen-volume *Histoire de France* by Louis-Pierre Anquetil (1723–1806) was extracted from little else than the histories of Mézerai and Velly (on Anquetil, see *Biog. univ.* 3:34; *Dict. biog. franç.* 2:1372–73). See also Halphen, *L'histoire en France,* 7–9: "la plus grande attention était apporté à ne pas charger la mémoire de faits trops complexes et à présenter les anecdotes les moins sûres commes d'austères et impor-

tantes vérités." Lefebvre, *Historiographie moderne,* 154–55, 158–59, emphasizes the lack of historical schooling and of an informed reading public under the Revolution and Napoleon. François Guizot, appointed in 1811 to teach history at the Sorbonne, had never studied the subject.

39. A good example of the sort of atlas behind the *Atlas Lesage* is Mentelle, 1798. Its maps cover countries—Britain, France, Germany, Italy, Spain, and Portugal—then one map each for the continents of Asia, Africa, and America; one ancient map is assigned to each country. Philippe, 1787/A (whose historical content is only ancient), may also hover in the background.

40. Las Cases's many genealogical pages are thoroughly and excitedly criticized by Nicolas Viton de Saint-Allais, *Le Correcteur de l'Atlas généalogique de Le Sage. Brochure indispensable à ceux qui ont acheté cet ouvrage* (Paris, 1813), which reprints Viton de Saint-Allais's strictures from his *Histoire généalogique des maisons souveraines de l'Europe depuis leur origine jusqu'à présent,* 2 vols. (Paris, 1811–12).

41. Both quotes from Quérard, *France littéraire,* 4:588.

42. Dörflinger in *LGKartog.* 1:266. The Lesage maps are after the careful listing in Quérard, *France littéraire,* 4:586–88 (from the current edition, 1826): nos. 14 (England), 16 (Italy), 18 (Spain and Portugal), 21 (Germany, with campaigns of Gustavus Adolphus), 26 (Germany in 1808), 32 (Africa with tracks of recent explorers), 33 (America with early Europeans). Another map, too briefly described by Quérard, tracks the exploits of Charles XII of Sweden.

43. Recent adaptations had made Speed, 1627b, almost topical; see ch. 4, nn. 39–42, above. Las Cases need not have been aware of these offshoots. In addition to England, Spain, and Italy (above), see also Quérard, *France littéraire,* 4:586, no. 7.

44. *Historischer Atlas mit chronologischen Tabellen* (Nuremberg, 1813–14); on the fate of the Rizzi-Zannoni/Desnos atlas, see ch. 3, n. 96, above.

45. Las Cases, speaking about his atlas, is quoted in Las Cases, *Mémorial,* ed. Walter, 1:616.

46. *AGEphem.* 7 (1801): 102–15. The message is dated 13 November 1800; cited below as Kruse, Announcement, with page. For the atlas titles, see Kruse, 1802.

47. Kruse, *Allgemeine Literatur-Zeitung,* Intelligenzblatt, 1800, 190: col. 1591, promised that they would be issued before the medieval sheets were complete; this did not work out. CambUL has a fine copy of the sixth edition, complete with *Tabellen.*

48. Kruse, 1807. Kruse, Announcement, 102–15, explained that he took Homann maps as the model for his own. French tr.: Kruse, 1836. This may be a second printing, the first possibly being in 1834. Map no. 1 is new to Kruse, and no. 3 departs markedly from Kruse's form; no. 6 has a supplement for Asia. All the maps are centered farther south than Kruse's, showing more of the Mediterranean and less of Scandinavia. According to the translators, the "beau travail" of Las Cases "avait . . . plutôt signalé que rempli cette lacune," i.e., the lack of a proper historical atlas to complement the burgeoning interest in history. For recycling, see Ansart, 1840, 1847.

49. The third edition (the last in which Christian Kruse took an active part) was a turning point in the printing history. Kruse invited owners of earlier editions to upgrade their maps; for a nominal sum, they could be exchanged for the revised versions. It was generally acknowledged at the time that the new maps were better than the old. The original coloring is much to my taste (e.g., the RGS copy). MilwaukeeAGS has a fine third edition with interleaved *Tabellen.*

50. The orphanage was part of a complex of educational establishments—the Francke'-sche Stiftungen—endowed at Halle in the late seventeenth century: *Encyc. Brit.* 12:854.

51. *Biog. univ.* 22:219. This account sounds grimmer than the German equivalent (next note).

52. *ADBiog.* 17:262; *DBiogArchiv*, fiche 716, frames 268–85; K. Schwartz, in J. S. Ersch and J. G. Gruber, *Allgemeine Encyclopädie der Wissenschaften und Künste*, 2d section, pt. 40 (Leipzig, 1887), 127. The decree suppressing the duchy and creating the new department is dated 13 December 1810.

53. Quotation from Kruse, Announcement, 105. The maps after 1700, somewhat more numerous than one per century, are freed from the double-zero sequence.

54. Quotations from Kruse, Announcement, 106, 112. Against covering long periods, see Kruse, "Probe der Gattererschen Charten," 385–87. The substance of this complaint is echoed by Wolf, "Bild der europäischen Geschichte," 25, no. 7, in connection with the practice, still prevalent after World War II, of presenting maps of, say, "the tenth and eleventh centuries."

55. Quotes from Kruse, Announcement, 106–7 (including examples from 800 and 1700 of significant changes taking place right after the turn of the century), 109–10.

56. See n. 49, above.

57. Among other things, Greenland is shown too close to Iceland. Kruse may have wanted to draw it into his composition without further widening the frame. The liberty is uncharacteristic but hardly impossible. Cartouche rolled open: A.D. 600 and 1300.

58. Kruse, Announcement, 103–4 (weak reasoning about the barbarian invasions). There is an approving but critical review in *AGEphem.* 35 (1811): 95–100.

59. The quotation is from Kruse, Announcement, 108. On the paper covers for each *Lieferung,* Kruse defends himself against those who ask for divisions by historical periods. His basic defense involves asking how many periods would be needed. The Harvard copy has these covers (MA 107.802F).

60. Dacier, *Rapport historique,* 4:187–91; see ch. 1, n. 8. Christian Jacob attributes the geographical part of the report to the expert in ancient geography P. F. J. Gosselin (1751–1830). The *Biog. univ.* (17:204–8) claims that he took no part in the committee's work.

61. Dacier, *Rapport historique,* 174–76, about Barthélemy and Barbié (the period of interest to the report was "gloriously prepared and opened" by Barthélemy), 236. See Barbié, 1788. The professional antiquarian Barthélemy was an "haut vulgarisateur." The outpouring of editions recorded in *BN Impr.* is suggestive of the impact of Barthélemy's book (accompanied by Barbié's atlas). Its success is attributed to the great popularity of travel literature in France: Reinhard Frenzel, *Malthe Conrad Bruun (Malte-Brun): Ein Beitrag zur Geschichte der geographischen Wissenschaft* (Crimmitschau, 1908), 5.

62. Dacier, *Rapport historique,* 248 (dismissal of Lesage and his like).

63. There is a puzzle here. Dacier, *Rapport historique,* 249, intimated that historical atlases not only existed but were overabundant: "Malgré tous ces atlas historiques, annoncés comme donnant la géographie des différentes époques, on peut dire que la géographie du moyen âge est encore à faire." Ancient geography offering the various epochs of antiquity may be meant.

64. Brion de la Tour, *Observations curieuses,* 4.

65. *AGEphem.* 16 (1805): 81–93. The same applies to the review cited n. 17, above.

66. On Koch, see *Biog. univ.* 22:84–86 (of which *ADBiog.* 16:371–73 is mainly a translation); 84, "un des écrivains qui ont le plus contribué à éclaircir l'histoire du moyen âge"

(responsible for genuine medieval scholarship but forgotten today). Koch helped Schöpflin set up the highly selective school anticipating the elite university programs of today (Metternich was an alumnus). Appointed professor of public law at Strasbourg (1779), he entered politics when the Revolution dispersed his students (1790); sometimes a deputy and almost executed, he was a member of Napoleon's Tribunate until it was dissolved in 1808. On Schöpflin (1694–1771), see ch. 4, n. 69, above. There is much about Koch in Voss, *Universität, Geschichtswissenschaft.*

67. Koch, 1771/B (anonymous). Quérard, *France littéraire,* 4:308: "à l'insu de l'auteur, sur les simples cahiers dont il se servait dans ses leçons." Schoell, who was close to Koch, does not confirm this scenario (*Biog. univ.* 22:84–86). Voss, *Universität, Geschichtswissenschaft,* 168, 177–78, treats the work as a deliberate publication. Koch took over Schöpflin's teaching duties in 1771, the year Schöpflin died. Anonymous publication, a frequent practice, did not conceal authorship (word got around) but afforded some protection if a work was badly received. Nothing barred taking credit for one that did well.

68. Koch, 1790/B, 1:xii–xiii: "La géographie du moyen âge est une terre inculte et un vrai labyrinthe, où il est très facile de s'égarer. Il n'existe point d'ouvrage géographique qui donne des notions justes sur le nouvel état des choses que le bouleversement de l'Empire romain, la migration des peuples et leur révolutions consécutives ont introduit en Europe. Des littérateurs François et Allemands en ont bien débrouillé quelques parcelles, mais aucune des nations de l'Europe ne peut se vanter jusqu'à présent d'avoir épluché cette partie de sa géographie." Quérard, *France littéraire,* 4:308: a mass of material from this version was not carried over into later versions of the *Tableau.* Voss, *Universität, Geschichtswissenschaft,* 178: the partial publication was due to the Revolution.

69. Koch, 1807. Questions of mine about this comparatively rare edition were kindly answered by Carolyn Smith, George Peabody Librarian, Johns Hopkins University, 23 January 1996; I am very grateful for her assistance. A German translation (1807) was almost simultaneous. There were other adaptations as well. Koch's reputation stood high at least in England for several decades after his death, as witness Koch, 1831, 1828/B. About Schoell, who tried to carry forward Koch's academy but in Prussia, see the obituary in *Annales des voyages,* 2d ser., 29 (1833).

70. Ancient atlases predominate by a wide margin in *BL Maps Catal.* 15:661–72, s.v. "World, History," until the 1830s.

71. The five maps of the 1807 edition are supplemented in the 1814 set by maps for 1300 and 1453. It seems more probable that Koch's maps were conceived as a group of seven than that the additions of 1814 were afterthoughts.

72. Frenzel, *Malthe Conrad Bruun.* Obituary by G. von Ekendahl, tr. from Danish, in *AGEphem.,* 2d ser., 22 (1827): 154–60; his reputation stood high at his relatively early death (*Bulletin de la Société de géographie,* 1st ser., 9 [1828]: 133). Excellent brief appreciation by Numa Broc, "Un bicentennaire, Malte-Brun (1775–1975)," *Annales de géographie* 84 (1975): 714–20 (with bibliography).

73. His geographical journal was called *Annales des voyages, de la géographie et de l'histoire* (Paris, 1807–). Interrupted for five years after Napoleon's fall, the journal outlived Malte-Brun by twenty years. The atlas (Lapie, 1812) to Malte-Brun's *Précis* (1810) is reviewed with the parent volume in *AGEphem.* 36 (September/December 1811): 209. Pierre Lapie (1779–1851), much involved with military cartography, was followed into mapmaking by his son. Lapie's collaborator in this and other projects was Jean-Baptiste Poirson (1760–1831); see *Biog. univ.* 33:584.

74. Further, see ch. 6, n. 30, below.

75. Baumgärtner, 1817. Companion pieces, Baumgärtner, 1815, 1816. For a fuller account, see nn. 101–3, below.

76. The many maps are in Baumgärtner, 1815. *ADBiog.* 2:168: Baumgärtner (1759–1843) entered the book trade in Leipzig after a time as an attorney. He was appointed (Saxon) *Geheimer Hofrat,* an official mark of distinction, in 1820.

77. Klaproth, 1826. A prospectus appeared in *Journal asiatique,* 1st ser., 3 (1823), promising six installments at 15 francs each. The title page implies simultaneous publication in England and Germany; library catalogues do not follow suit. The atlas complements a volume of commentary (with the same title).

78. H. J. Klaproth, in *Asia polyglotta* (Paris, 1823; 2d ed. 1831). On Wilhelm and Alexander von Humboldt, see *Encyc. Brit.* 13:873–76. Wilhelm's philological interests drew him to Klaproth. On Klaproth, see C. Landresse, "Notice historique et littéraire sur M. Klaproth," *Revue asiatique* 16 (1835): 243–73; *Biog. univ.* 22:1–11; *ADBiog.* 16:51–60. His overland journeys to and from China were largely on his own, outside the delegation to which he was attached. His character had conspicuous flaws. Inclined to be very harsh toward other scholars (*Encyc. Brit.,* 9th ed., 24 vols. [New York and London, 1890], 14:107: "reckless intellectual aggressiveness"), he communicated through books and had virtually no friends. See Henri Cordier, "Un orientaliste allemand: Jules Klaproth," *Académie des inscriptions et belles-lettres, Compte-rendus,* 1917, 297–309 (venomous). He was reputed to love high living as much as arduous scholarship (the attribution of his early death to dissipation needs better confirmation than it has). Friendlier biographers maintain that Klaproth led a steady and quiet life in his last years, "wholly dedicated to scholarship." He is held responsible for two forgeries (*Encyc. Brit.,* 9th ed., as above). His views on comparative linguistics are particularly dated. Obituaries in *Journal asiatique* 16 (1835): 243–73; *Nouvelles annales des voyages,* 3d ser., 8 (1835): 5–20. His map collection: BN, Cartes et plans, *Inventaire de la Collection cédée par M. Klaproth le 9 décembre 1831: Liste par ordre de numéro.* Klaproth added to the initial gift on 7 July 1832. Many of the maps continue to bear "Klaproth" numbers.

79. H. J. Klaproth, *Antwort auf eine in Hermes abgedrukte Recension meiner Tableaux historique de l'Asie* (Paris, 1828). (This *Hermes* is not the classical journal of that name.) Klaproth's sheet maps: Quérard, *Litt. franç. contemp.,* 4:465; several are at the BL. His writings include studies of Chinese and Japanese maps, as well as reviews of maps of Asia by Arrowsmith and Brué (*LC Bibliog. of Cartog.* 3:284–85).

80. A name that stays the same (such as, "Royaume de . . . ") tends to be rearranged, and evidently engraved anew, from one map to the next.

81. "Aperçu historique et ethnographique des peuples de l'Asie moyenne, jusqu'à l'an 1000 de notre ère; Recherches sur la grande migration des peuples." See *Biog. univ.* 22:6; *ADBiog.* 16:56. Klaproth's commentary on the ethnographic map was also the subject of earnest criticism. Klaproth seems to develop ideas launched by Joseph de Guignes and still disputed (see ch. 4, nn. 14–15, above).

82. Selves, 1833, 1835, 1843. Klaproth's atlas is found in map collections much more often than Baumgärtner, 1817, and Benicken, 1820, 1824.

83. On the publication history of Malte-Brun's *Précis,* see Frenzel, *Malthe Conrad Bruun,* 67–68, 77–78. Its initial appearance was slowed by the author's need to support himself by other occupations. More than one volume remained unwritten at his untimely death. Although Kruse's atlas sold well in Germany, it did not have a full French edition until the 1830s (see n. 48, above).

84. About early lithography in atlas-making, see Walter W. Ristow, "Lithography and Maps, 1796–1850," in *Five Centuries of Map Printing,* ed. D. Woodward (Chicago, 1975), 77–112; an ancient history atlas of 1829 is given priority among historical atlases (95). According to David Woodward, in *Map Collector* 18 (1982): 2–11, "the earliest lithographed atlas" appeared in Paris in 1823. Aretin, 1809, is much earlier but perhaps does not rank as a proper atlas. Benicken's *Schul-Atlas* precedes Selves, 1821; and two of the four installments of Benicken, 1824, were issued by 1822.

85. On Falger, see Liutpold Dussler, *Die Incunabeln der deutschen Lithographie (1796–1821)* (Berlin, 1925; repr., Heidelberg, 1955), 42–43; R. Armin Winkler, *Die Frühzeit der deutschen Lithographie: Katalog der Bilddrucke von 1796–1821* (Munich, 1975), 13–14, 73–76. On the Bavarian cadastre (a large-scale mapping of the kingdom for tax assessment), see Bayerische Staatsbibliothek, *Cartographia Bavariae: Bayern im Bild der Karte,* ed. Hans Wolff (Weissenhorn [Bavaria], 1988), 217, 223–31. Benicken's maps are "in Stein graviert"; Ristow, "Lithography and Maps," 80, discusses this technique. For the wider context, see Ian Mumford, "Lithography for Maps: From Senefelder to Hauslab," *Journal of the Printing Historical Society* 27 (1998): 69–87; Falger and the Weimar atlases are not mentioned. See n. 97, below, for a contemporary who thought poorly of the graphic quality of Benicken's atlas.

86. On the beginning of lithography in France, see E. F. Jomard, in *Bulletin de la Société de géographie* 4 (1825): 317. This date coincides with the publication in France of the inventor's book. Frenchmen had studied the process earlier; F. C. Bigmore and C. W. H. Wyman, *A Bibliography of Printing* (London, 1880; repr., 1978), 346, date the start of Paris lithography to 1815–16. Selves, 1819, 1822, are lithographed historical atlases.

87. Ristow, "Lithography and Maps," 96, takes note of Selves only as responsible for producing a poor lithographed school map of the world (1821).

88. Charles Coquerel, in *Revue encyclopédique* 17 (March 1823): 635–37. He was awed by lithography, "cette nouvelle branche d'industrie."

89. On Bertuch, see *ADBiog.* 2:552–53. He founded the first fashion magazine in Germany and won special fame for a twelve-volume *Bilderbuch für Kinder.* His splendid house, in which the Weimar Geographical Institute was housed, survives as the Weimar Stadt-Museum; see Engelmann, *Heinrich Berghaus,* 25.

90. On the establishment of the Weimar Geographical Institute, see Wolkenhauer, *Leitfaden,* 58. For a brief but thorough account, see H. Arnhold, "Geographisches Institut Weimar," in *LGKartog.* 1:259–60. The parent body (Landes-Industrie Comptoir) was established in 1791; the institute proper in 1804. There were 560 titles on its stock list in 1820, and it remained significant to the 1860s for the production of globes. *AGEphem.* began in 1798, with the collaboration of an astronomer, F. X. von Zach (1754–1832). W. Wolkenhauer, "Friedrich Justin Bertuch," *Deutsche Rundschau für Geographie und Statistik* 21 (1899): 38–42 (with a portrait), stresses the centrality of the Weimar journal among contemporary German geographical periodicals. Günther, *Geschichte der Erdkunde,* 234, simply disregards the Weimar Geographical Institute. Mumford, "Lithography for Maps," 69, briefly sketches the stormy state of geography in the 1820s.

91. Vieth, 1800/A, a classical atlas, was the most prominent historical production from early Weimar. Benicken, 1829/A, is also classical. The major enterprises of the institute were multisheet maps of the world and Germany.

92. Friedrich Wilhelm Benicken, *Napoleons Heereszüge (1796–1815)* (Weimar, 1815) (BerlinPK U10885); *GV* 12:68, lists this as Weimar, 1821. The school atlas is Benicken, 1820

(maps drawn by C. C. Wendel, Jr.). The Vienna and Berlin copies have fourteen plates, in keeping with the table of contents. Łodyński, 3: no. 147, oddly specifies twenty; not recorded in *GV*. For the *Hand-Atlas,* see n. 98, below.

On Benicken, see *DBiogArchiv* 80:384: born in Schleswig in 1783, he was active as a writer (on nongeographical subjects) well into the 1830s on the fringes of the Weimar Geographical Institute or in association with persons from that circle (see also *GV*). A grudging note tells us, "Besorgte auch (1820–1826) einige Schulatlasse." He was responsible for *Die Elemente der Militärgeographie von Europa* (Weimar, 1821), with a military map of Germany; T. C. F. Enslin, ed., *Bibliotheca historico-geographica* (Berlin and Landsberg, 1825), 25.

93. The reviewer in *Allgemeine Literatur-Zeitung,* April 1822, 729–32, took Benicken's generalized reference to *Lehrbücher* as specifically denoting Dresch's *Lehrbücher der allgemeinen Geschichte insbesondere Europens,* 2d ed., 2 vols. (Weimar, 1822–24). Benicken's own phrase included Dresch's three-volume *Übersicht* (Weimar, 1814–16; 2d ed., 1822–23), the more common of his textbooks. The Weimar Geographical Institute availed itself of the press of the Landes-Industrie Comptoir. Dresch (1786–1836) was a professor of law as well as history, first at Tübingen and eventually at Munich. Benicken's maps are universal even though his focus (following Dresch) is on Europe.

94. Benicken's main maps have the names of Dresch's chapter titles. *AGEphem.,* 2d ser., 16 (1826): 191, attests that Benicken's *Hand-Atlas* was to have fifteen sheets; 1822 was the intended ending date.

95. G. H[assel], in *AGEphem.,* 2d ser., 15, fasc. 5 (1825): 143–48, ascribes fifteen maps to Benicken, 1821, one more than in the BL and NYPL copies. Map 15 filled in "newest" history from 1815 to 1822. The fourteen-map core has the same scope in both atlases.

96. See nn. 42, 53, above. Hase and Rizzi-Zannoni admitted maps of post-1500 history.

97. Karl-Ferdinand Weiland (†1847), like Benicken, was an ex-captain in the Prussian army. Yale has his *Allgemeiner Hand-Atlas der ganzen Erde* (Weimar, 1846?). Contrary to my unschooled opinion, Mees, 7, thinks very poorly of the engraving of Benicken's atlas.

98. Benicken, 1821. It is complemented by twelve date tables—a study companion that does not interpret the maps. Hassel, in *AGEphem.,* 2d ser., 15, fasc. 5 (1825): 143–48, presents the completed volume on behalf of the Weimar Institute and explains Benicken's design.

99. Handtke, 1808; cited by Engelmann, 793, and NOT SEEN. My account of Handtke's motivation is a guess.

100. Aretin, 1809: (1) Agilolfings, 772; (2) Carolingians, 900; (3) Welfs, 1136; (4) Wittelsbachs, 1343. Aretin deliberately avoids commentaries. On the context of lithographic history, see *LGKartog.* 1:451–55.

Reminiscent of Aretin's work is the small *Historische Karten,* 1816; see no. 348 bis, below on 566. The four maps, originating from Leipzig, are very neatly drawn and colored. In spite of being polar projections, they include only Eurasia. The designer appears to have been obsessed with the comparative size of empires, ancient and modern.

101. *Encyc. Brit.* 24:271, 28:54–55: the Saxon army, serving Napoleon, went over to the allies at Leipzig in 1813, and King Frederick Augustus I was taken prisoner. Saxony was administered by Russia for a year, then by Prussia. The Saxon question was pivotal at the Congress of Vienna. Prussia's demand to annex all Saxony was not granted. Its king was restored at the cost of a partition with Prussia. Relinquishing half the area of his kingdom, Frederick Augustus returned to respectability when he joined the German Confederation on 8 June 1815. For the Baumgärtner atlas, see n. 103, below.

102. For Zollmann, see ch. 3, n. 23, above. Güssefeld, 1796; Teuscher, 1825. The subject continues to occupy much space in Putzger, 1877, e.g., 1913 ed., no. 36 (several maps); Shepherd, 1929/O, no. 85.

103. Baumgärtner, 1815 (two versions), 1816. The head of the Baumgärtner firm was appointed Prussian consul general in Leipzig in 1815; *ADBiog.* 2:168. Presumably he was unsentimental about Saxon territorial losses. In Budapest I came across an unexpectedly early piece of map publication by Baumgärtner: *Receuil de plans de batailles* [by Bonaparte in Italy and Egypt, 1796–1800] (Paris and Leipzig, n.d.); call no. TA 800.

104. *Dict. biog. franç.* 7:472–73; *Biog. univ.* 5:671 (no mention of his historical atlas). Brué was eventually named *géographe du roi.* Freycinet was a career naval officer. Quérard, *France littéraire,* 1:532. After death Brué was called one of Europe's most distinguished geographers: Quérard, *Litt. franç. contemp.,* 2:455; *Annales des voyages,* 2d ser., 12 (1829): 241–45; *Annales des voyages,* 2d ser., 25 (1832): 159–60 (his obituary), corroborates Quérard. The Weimar journal, usually severe about French products, speaks with respect of Brué maps in a critical review: *AGEphem.,* 2d ser., 17 (1825): 18–24. Quérard praised Brué's *Grand atlas universel* (Paris, 1815; 1816). This publication was superseded by Brué, 1822 (2d ed., 1830), then acquired by Charles Picquet after Brué's premature death (1832).

105. Brué, 1820: 3, Brué's comment. Maps for history still fell short of being esteemed. The obituary of Brué in *Annales des voyages* omits the atlas of France from his noteworthy works.

106. L'Hermite, 1822; Legrand, 1824. Legrand's cover claims an atlas composed of twenty-four maps; the maps listed (and found) number twelve.

107. Benicken, 1824, Klaproth, 1826; see discussion earlier in this chapter. Möller, 1824, 1825 (n. 116, below) also belong to these years.

108. Kärcher, 1824 (three medieval maps). Delamarche, 1827 (two medieval maps); "moyen âge" was admitted into the atlas title in this edition. Goujon, 1828 (one medieval map). Rühle, 1827, was the inaugural, and last, installment of an intended universal history; Mees, 7, deeply regretted its suspension. On Rühle von Lilienstern (1780–1847), an accomplished and distinguished soldier, see *ADBiog.* 29 (1889): 611–15. He later produced eight wall maps on the history of the Arabs and Mohammed (1836). Vivien, 1825, confined his historical appendix to maps of the ancient world.

109. Thomson, 1829. For the compiler of this atlas and its reception, see ch. 6, nn. 53–55, below.

110. Wilkinson, 1797/A distinguishes "geographia antiqua" from "geographia historica," i.e., territories ("Africa antiqua") from events ("Alexandri Magni itinera").

111. Heck, 1830. His last map for antiquity is of the Roman Empire under Constantine (as in the *Atlas Lesage*). The main maps have insets, particularly numerous in the Crusades spread. The Mongols are tucked into the Charlemagne map. Mumford, "Lithography for Maps," 70, records Engelmann as a pioneer of lithographic printing in France.

112. Möller, 1824; ten maps, lithographed and hand colored. E.g., (map 6a) Europe ca. A.D. 511; (6b) Europe ca. 814; (7) Europe, Asia, and Africa ca. 1100; (8) Discoveries. The Chinese and East Indians are classed as "pagans" in the map of religions. Möller (1791–1864), from the Prussian Rhineland, taught history and geography until he obtained a parish in 1828. His earliest historical maps accompanied a church history: *Hierographie, oder topographisch-synchronistische Darstellung der Geschichte der christliche Kirche in Landkarten,* 2 pts. (Elbersdorf, 1822–24). See *DBiogArchiv,* fiche 850, frames 366–72. *GV* records Möller's wall maps and school atlases of Greek and Roman history, as well as nongeographic

works. I regret never having been able to see Möller, 1824, let alone his *Kleiner historischer Atlas der Lande zwischen der Maas und dem Niemen, zur Erlernung ihrer Geschichte seit der Völkerwanderung* (Münster, 1824). Friedrich Kohlrausch, *Chronologischer Abriss der Weltgeschichte*, 3d ed. (Elberfeld, 1818). The date of the first edition of Kohlrausch's book is not recorded in the main reference works. Kohlrausch's *Chronologischer Abriss* was regularly republished until the 1860s. He wrote a number of other histories. See also Muhlert, 1865.

113. Witzleben, 1829. The letterpress tables are not combined with the maps. I have never seen fascs. 2–3 and wonder whether they survive. Ferdinand August von Witzleben (1800–1859), son of a soldier/man-of-letters, was a staff lieutenant teaching at the combined military school at the time of the atlas. He was a lieutenant general at his death; see *ADBiog.* 43 (1898): 670–71 (his father, 665–66). Mees, 8, who knew the work and deplored its engraving, observed that the Netherlands always had the same shape.

114. *AGEphem.*, 2d ser., 12 (1823): 325ff., reviews recently published German school atlases of geography narrowly defined. More of the same in *AGEphem.*, 2d ser., 16 (1826): 56–59, 177–81, 191.

115. Kärcher, 1834 (1st ed., 1830). Maps 5 and 6, rather than showing an all-embracing Empire of Charlemagne, have the Frankish kingdom at a time of maximum fragmentation (early eighth century) and so indicate Aquitaine, Swabia, Bavaria, the Lombard kingdom, etc., as independent entities. Something similar is done for Europe ca. 500.

116. For an earlier atlas by Kärcher, see n. 115, above. Biographical repertories omit him.

117. *Dict. biog. franç.* 10:1013. Obituary in *Bulletin de la Société de géographie,* 3d ser., 2 (1844): 101–7. Besides maps and atlases, he published (inter alia) *Introduction à la géographie physique et politique des états de l'Europe* (Paris, 1827) and is specially credited with introducing many technical terms of geography into colloquial French. According to Mumford, "Lithography for Maps," 72, the École polytechnique gave only rudimentary cartographic training. Denaix had additional opportunities in the army.

118. Desnos, 1765c; see ch. 1, n. 62, above. Taking account of 860 languages and about 5,000 dialects, Adriano Balbi (1782–1848) won much attention with his *Atlas ethnographique du globe* (Paris, 1826); see *Dizionario biografico degli Italiani* 5:356–57. This was a mapless atlas (like some Lesage offshoots). Maps of councils and the ecclesiastical hierarchy (both synoptic in design) are good examples of early thematic maps; see ch. 1, nn. 88–93, above. Also relevant to history is a map of battle sites, Rothenburg, 1830 (distantly similar to Speed, 1627).

119. For another illustration of this mentality, see the "Précis de géographie historique—1804–1841," in Dufau, 1841, unpaged: (editorial comment about the terms of peace imposed on France after Waterloo, 1815) "Telles furent les conditions honteuses et humiliantes qu'acceptèrent les Bourbons restaurés au trône de la France. La révolution de 1830 n'a encore rien fait ni pour les changer, ni pour les modifier." Dufau, without being Bonapartist, is unapologetic about the building of the French Empire and deeply regrets its collapse. In general, see Jean Tulard, "Légende napoléonienne," in his *Dictionnaire Napoléon* (Paris, 1987), 1053–54.

120. Denaix, 1829, 1835. Guizot's "Avis" extols Denaix's atlas as a companion to Koch and affirms that his work surpasses Lesage, Kruse, and Benicken. (In the latter two cases, Guizot's sole basis of preference seems to be the written matter; Denaix's was allegedly better.) Inhibited by Las Cases's being not only alive but in politics, Guizot is convoluted about the *Atlas Lesage:* "[Lesage] est trop justement accrédité pour que, sans craindre de le déconsidérer, il soit permis de ne lui accorder que peu d'importance quant à la partie géographique, uniquement traitée comme simple canevas propre à la spécialité de l'ouvrage"

(the *spécialité* is assumed to be common knowledge). In its fourth decade on the market, the Lesage enterprise retained a public: Lesage, 1835, was augmented and lower-priced (n. 21 above), and Marchal's Brussels version (Lesage, 1827) was reissued in the 1850s.

121. On Quin (1794–1828), see *DNB* 16:548: M.A. Oxon., 1820; called to the bar, 1823; a barrister of Lincoln's Inn. (The *DNB*'s date of 1840 for the *Atlas* is probably a misprint.) There is little to be gleaned from *British Biographical Archive,* ed. Laureen Bailie and Paul Sieveking, microfiche (Munich and New York, 1984–88). Quin's atlas (1830) has twenty-one maps, six folded; some of the maps are dated 1828, suggesting that publication began in Quin's lifetime. About Hughes, see J. E. Vaughan, "William Hughes, 1818–1876," *Geog.: Biobib.* 9:49–60. I consider the maps to be basically Quin's even when redone by Hughes.

Dr. Mary Pedley (Clements Library, University of Michigan) kindly informed me that a Swann auction (New York, 16 December 1999) was offering what appeared to be Quin's original MS for this atlas. (This sale has taken place; the name of the purchaser is not public knowledge.) If the date 1818 that the MS bears is trustworthy, Quin was twenty-four when he finished the draft, and another decade passed before (posthumous) publication.

122. Palm's copy is listed with Quin, 1830. On the flyleaf, dedication in Dutch to C. Moller, dated 20 June 1839.

123. Yet Quin's medieval maps (12–15) coincide rather closely with the *Hand-Atlas* (5–8). Benicken takes one step from Charlemagne to Gregory VII (1073); Quin takes two from the same starting point to the Crusades (1100, by his reckoning). Benicken's next-to-last step is Rudolf of Habsburg (†1291); Quin's, a Mongol event of 1294. Their Middle Ages both run from the end of the western empire to the "great discoveries" (476–1492).

124. Quin, 1830: 1.

125. The reviewer for the *Literary Gazette,* 1830, 655–56, mainly quoted Quin's preface but found his own voice to comment on the artistic design: "from the Rembrandtish effect of the first map . . . to the Rubens-like diffusion of light, and of gay colours, by which the world in its present state is presented" (656). I have found no other review of Quin.

126. Black, *Maps and History,* 62, no. 14, attributes to Quin what is in fact the Hughes face-lift.

127. Willard, 1836. The accompanying *System of Universal History* appeared a year earlier (Hartford, 1835). Willard, 1828/O. Her "progressive maps" (twelve in all) have a complex printing history. Alma Lutz, *Emma Willard: Pioneer Educator of American Women* (Boston, 1964), 24, 36, 40–41, 58, 64, traces the place of geography and mapmaking in the teaching and publications of Emma Willard (1787–1870): a geography textbook first gave her financial independence; she had her students mark on maps "the paths of navigators and explorers, and the march of armies."

128. Lutz, *Emma Willard,* notes that Willard was in London in 1831, the year after Quin's book appeared (80–82), and that Willard, 1836, was well received but did less well than her history of the United States (97–98; Lutz is uninformed about the genesis of Willard's maps and even implies that Willard's chronological charts were her own invention [64]). Black, *Maps and History,* 75, devotes a page to her, without acknowledging her debt to Quin.

129. Simencourt, 1830; the Malte-Brun atlas is Lapie, 1812; Lapie, 1829 (reissued without change at regular intervals); Dufour, 1830 (no historical appendix), 1834a, 1834b, 1835; Vuillemin, 1839; Monin, 183-.

130. Examples of strictly geographical atlases—that is, atlases free of historical

content—are *Allgemeiner Hand-Atlas der ganzen Erde . . . zu A. C. Gasparis vollständigem Handbuch der neuesten Erdbeschreibung bestimmt* (Weimar, 1807–) (a collaborative Weimar work); Philippe Vandermaelen, *Atlas universel de géographie* (Brussels, 1825–27)(see L. Dankaert, in *LGKartog.* 2:850); and the standard-setting Stieler, 1824/O (see *LGKartog.* 2:782–84, and n. 169, below).

131. Desjardins, 1836; Engelmann, 1836. Häufler believed he deserved credit for most of Desjardins's atlas.

132. Kruse, 1836; maps dated 1834–35. Maps of ancient history and other continents, not by Kruse, are added. The map dates and the reference to a second edition seem to imply an edition before 1834.

133. Maaskamp, 1833; Elst, 1831. Pl. 15 in Jusseret, 1836, shows Belgium with tracks of its railroads and is dated 1842 in the table of contents. The date clashes with the autograph of someone who died in 1836—the announced date of publication. On Jusseret, see *Biographie nationale de Belgique* 10:618–19; he fractured his spine in a riding accident. Mees, 17, complains mildly that northern Netherlands is given short shrift. About responses to 1830 on both sides of the new border, see Pieter Geyl, "The National State and the Writers of Netherlands History," in his *Debates with Historians* (New York, 1958), 212–13, 215–16.

134. Guadet, 1833; prefatory comments and quotation, v–vii. Guadet, trained as a lawyer, worked as a schoolmaster, notably in the official school for the blind. He produced editions of two major medieval historians. See Quérard, *France littéraire,* 3:395, and Quérard, *Litt. franç. contemp.,* 4:183–84.

135. Denaix, 1836, 1838. The latter contains thirty-four maps of territorial changes, with commentary on the left. Denaix, 1856, is almost a reprint of the *Petit atlas. Rapports et notices sur les travaux géographiques de M. Denaix* (Paris, 1833), three pamphlets bound as one. They bear out Denaix's adherence to Buache and claim that his atlases of Europe and France were parts of a larger atlas of world dimensions in preparation. The plan of the atlas of France given there differs in arrangement from the published version.

136. Mullié, 1839 (all maps dated 1836); between 106 and 107, "Carte synthétique des fiefs et domaines réunis successivement à la couronne sous les rois de la troisième race [n.d.]," e.g., "Duché de Toscane réuni en 1807." Mullié supplies most history by showing the old provinces with indication of *réunions.* Possibly, Fr. *réunion* is equivalent to "union." If so, the *réunion* of Tuscany in 1807 would not imply the "union" of Tuscany in the distant past. Similar to this is Dussieux, 1856: 9, no. 61, the French Empire in 1815, with an inset for the "réunions de territoire à la France."

In Mullié, amid extremely varied information, only the maps are uniformly French in content. Each century has its patron: Philip Augustus and Louis XIV are predictable; less so, perhaps, are Alfred of Wessex, Tamerlane, and Peter the Great. The text, which takes up most of the space, is divided into thematic columns, including one for notable happenings outside France. According to *BN Impr.* 121:626, a copy of Mullié (1845 edition) was among reference works in the main reading room (rue de Richelieu). It was not in the current catalogue of reference works in 1997. A reference librarian advised me that entries of this kind in the printed catalogue are often outdated. Nevertheless, Mullié presumably had a place in the reference collection in 1933 and long before; it contains useful information.

137. Rupp, 1837, 1839. Königsberg is now Kaliningrad. About Rupp, see *ADBiog.* 53:635–46 and below ch. 6, n. 88. Soon after these publications, he had difficulties with his church and broke away.

138. Rodowicź, 1843; Steger, 1845; Schaarschmidt, 1846; Pompper, 1846; Bensen, 1849. French history: Dufau, 1841; Dussieux, 1843; Clausolles, 1846; Bonnechose, 1847; Maretheux, 1847; Duruy, 1849.

139. Spruner, 1837. The work was issued in installments until 1846. The first edition at NYPL is very attractively colored; the two sets of the second edition at UCB are inconsistent, one pale pastel, the other more deeply colored. On the Perthes firm, see n. 169, below.

140. Houzé, 1841, is the earliest I have seen. Quérard, *Litt. franç. contemp.*, 4:325, records a first edition of 1837–38. *NUC* 256:404, confirms this date. Luxury editions, Houzé, 184-, 185-. Houzé, 1840, binds together Spanish versions of the Spanish, French, and Holy Land sections. Artero, 1879, is Houzé's challenger in Spain. Houzé, 1840a, 1845a, are separate French and English collections. The maps in Spruner, 1837, are numbered to allow marketing separate parts, but I have never encountered a national atlas detached from this source. It would be surprising if there were none.

141. *Bulletin de la Société de géographie*, 2d ser., 11 (1839): 52: Houzé gave the society early installments of his atlas under the title *Le Monde, histoire de tous les peuples*. His geography textbook for primary schools (1839) was reprinted several times in the 1860s. A work on French place-names (1860) is his last sign of activity. See *BN Impr.* 73:1279.

142. Houzé, 1841: section C, no. 2, Geoffrey of Monmouth; section D, no. 12, execution of Crispus; section E, no. 1, Odin's dwelling place.

143. Duruy, 1841. About Duruy, see Sandra Horvath-Peterson, *Victor Duruy and French Education: Liberal Reform in the Second Empire* (Baton Rouge, 1984). His writings: *BN Impr.* 45:1002–34.

144. Löwenberg, 1839 (forty maps); 2d. ed., Kutscheit, 1844b (fifty maps). Separate from this source is Kutscheit, 1844a.

145. Levasseur, 1840, 1844. Country sections in the latter: nos. 7 (France), 9 (British Isles), 12 (Spain and Portugal), 13–14 (Germany).

146. Clausolles, 1845, 1846.

147. *Biog. univ.* 10:296; Quérard, *France littéraire*, 2:433–34; *Arch. biog. franç.*, fiche 292, frames 314–17.

148. A full table of contents of Spruner's atlas was available in advance of the finished product. Delamarche, 1844.

149. See n. 139, above.

150. The breakdown of European maps (from the 2d ed., 1854): Germany, etc., thirteen; Italy, six; France, seven; Britain, five; Spain and Portugal, seven; northern kingdoms, nine; southeast Europe, etc., eight; Hungary, five. Each map is consecutively numbered in relation to the entire *Hand-Atlas*—a tally not reproduced in the table of contents.

151. Spruner, 1871, is the most likely candidate for authority down to the present; about Theodor Menke, who was wholly responsible for this edition, see n. 156, below.

152. Spruner, 1838a; the comment about sources, *ADBiog.* 33: 326. Spruner also published *Beschreibung des Kanales von der Donau zum Maine* (Bamberg, 1836), an old subject (ch. 4, nn. 17–20, above); NOT SEEN. Other items are added to his 1830s production in n. 154, below. Spruner's practice of consulting sources had long been anticipated by Germans and non-Germans who had drawn maps based on primary research; see ch. 4, nn. 53–77.

153. *ADBiog.* 35:325–28. Spruner never married.

154. Spruner, 1831/B. On the *gaue* in historical geography, see Curschmann, "Die Entwicklung der historisch-geographischen Forschung," 129–63 (Spruner's part, 147–50). Curschmann discussed the earlier work that Spruner's study, based on charters, opposed.

A little known production of Spruner's, typical of the occupations of makers of historical atlases, is a one-sheet *Genealogisch-historische Tabelle der Regenten von Bayern* (Bamberg, 1834). In a different vein, he was responsible for an edition of the eighth-century Lombard historian Paul the Deacon.

155. As above, ch. 4, n. 74; and Spruner, 1838b.

156. *ADBiog.* 35:325–28. On Menke, see Curschmann, "Die Entwicklung der historisch-geographischen Forschung," 158–60. Menke had tried teaching and law and found content-ment as reviser of Spruner for the Perthes firm. Curschmann praises the "extraordinary erudition" shown by Menke in Spruner, 1871. Of course, he built on the Spruner founda-tion. The Perthes firm routinely recruited fresh forces to take over, carry forward, and im-prove its atlases.

157. Dean, "Sic enim est traditum," 11: "the classical prototype of most modern histori-cal atlases."

158. August Pischon in the foreword (1843) to Wedell, 1849. See also Hughes, 1869, pref-ace: "The elaborate research, and minute precision of detail [of Spruner's atlas,] . . . have deservedly made it a standard of reference. . . . These qualities, however, . . . are (not unnat-urally) carried to an extent which . . . involves needless complication of detail."

159. Droysen, 1886; Poole, 1898/O. Droysen does not seem to have closely followed the Spruner model. Quite different ideas sometimes enter his layout and design (see ch. 6, n. 132, below).

160. Putzger, 1877; Shepherd, 1929/O.

161. See Introduction, n. 5, and ch. 2, n. 155. Also Bayerischer Schulbuch-Verlag, *Grosser historischer Weltatlas,* 3 vols. (Munich, 1953; 4th ed., 1976–84).

162. Atlases of France: those in nn. 138, 143, 145–46, above, plus Guadet, 1851; Dussieux, 1854; Sanis, 1859; Babinet, 1862; Bouillet, 1865; Delamarche, 1868; Migeon, 1866. Germany: those in n. 55, plus König, 1850, 1857; Frommann, 1854; Kutscheit, 1856; Beck, 1857; Hermann, 1858; Spruner, 1866; Keppel, 1870. England: Hughes, 1849, 1863/B; Birchall, 1859. Sweden: *Kort Atlas,* 1853; Wiberg, 1856. Switzerland: Mandrot, 1855; Vögelin, 1865. Hungary: Bedeus von Scharberg, 1845. Italy: Lesage, 1826b; Maggi, 1803.

163. Dussieux, 1856; Rhode, 1861; Hughes, 1869; Huberts, 1870; Barberet, 1870; Dozy, 1870.

164. Freyhold, 1846.

165. Wedell, 1849; accompanying text. Wedell expresses special gratitude to his lithog-rapher, H. Mahlmann, head of a workshop for geographical lithography. The pastor has been cited for an opinion of Spruner's work (at n. 158, above).

166. Löwenberg, 1839, among others, used writing in this way. He is an acknowledged source, spoken of with approval by Wedell's mentor. The statement I cite seems to be influenced by a source now called the *Gallic Chronicle of 452,* Monumenta Germaniae his-torica, *Auctores antiquissimi,* ed. T. Mommsen (Berlin, 1892), 9:652 (item 51). Wedell's cap-sule account of a mysterious transfer of Arianism from Rome to the barbarians contains elements of fact but is not an accurate account of the process.

167. In view of the rich grouping of insets in such old works as Speed's *Theatre of the Empire of Great Britain* or some of the major Dutch atlases, Freyhold and Wedell's inset-rich pages cannot ultimately be called innovative. That was no obstacle to their having been in-fluential, especially in Germany.

168. Johann Wolfgang von Goethe, *Höhen der alten und neuen Welt bildlich verglichen/ Esquisse des principales hauteurs* (Paris, 1813).

169. *ADBiog.* 36:185–87; H. Haack, foreword to tenth (English-U.S.) edition of Stieler's atlas. Justus Perthes died just when Stieler, 1824, was beginning publication; Stieler and Justus's son Wilhelm were friends. Because of the continued effort at improvement, early Stieler issues are hard to come by; Harvard has one of 1824.

170. Heinrich Berghaus, *Physikalischer Atlas* (Gotha, 1838–48), 90 plates, is physical in the broadest sense of the term—not only physical relief but currents, climate, plants, wildlife, geology, etc. Von Sydow taught almost all his life at the Prussian combined military academy: *ADBiog.* 37:280–81. In his *Methodischer Hand-Atlas für das wissenschaftliche Studium der Erdkunde,* the first word is meant in our sense of "methodical," not "outline, without writing."

CHAPTER 6

Nineteenth-Century Maps of the Middle Ages

In the descriptions given in the previous chapters, maps for medieval history, from Lambarde's solitary woodcut of 1568 to the end of the eighteenth century, have been commended more for existing at all than for being accurate, comprehensive, and beautiful. As late as the 1700s mapmakers had found it more congenial to depict the birth of medieval kingdoms than their maturity and had let their attention stray to the Hungarians and Turks in preference to focusing on cathedrals, Crusades, and mendicants. These hesitations now vanished; from 1800 on, the Middle Ages ceased to be a pathless desert approached by brave explorers. The century opened with cartographers publishing medieval maps with the confidence that they were an established product for which there was a real demand. Designers inaugurated a series of maps that, with gradual improvements, continues today.

An Estrangement Not Yet Healed

The *Atlas classica* issued in 1805 by the London publisher Robert Wilkinson found space, beyond its extensive coverage of biblical and classical subjects, for "Eslam" (i.e., Islam), Saxon England, the Empire of Charlemagne, and the Crusader kingdom of Jerusalem. The map of the Kingdom of Jerusalem was rare, and "Eslam," foreshadowed by Hase, succeeded in depicting the Islamic world as a whole in an original way.[1] Wilkinson's small medieval extension was an isolated initiative; the dominant "ancient atlas, classical and sacred," continued to heed established boundaries. Change was general in at least one type of collection: atlases for history that sought to be comprehensive rarely ignored the Middle Ages any longer. Only the number of windows opened on the period, and what they contained, were matters of choice.

A compiler with a wide knowledge of eighteenth-century historical maps, and a restrained critical sense, might have complained that the accumulated labors of cartographers had yielded an embarrassment of riches. A *cartothèque* carefully assembled at the start of the nineteenth century would have contained images not only of medieval France and England but also of Holland, Ireland, and Hungary; detailed reconstructions of certain districts of Germany, France, and Italy; Charlemagne's canal project, Frankish royal palaces, and the Carolingian battle of Fontenoy-en-Puisaye; Visigothic Spain; the Islamic caliphate, several Mongol empires, the crusading principalities, western empires galore, and much else, such as three atlases, each with several dozen maps for the medieval centuries. An exhaustive hoard of this kind was never assembled, not even in someone's head or a printed bibliography. The King's Topographical Collection at the British Library, perhaps the best single mirror of conditions at the time, offers a selection of "geography of the Middle Ages" that, though generous, includes only a small fraction of the maps that had been drawn.[2]

The samples of past production that were widely known and available, such as Robert's "Charlemagne" and d'Anville's "Age intermédiaire," were rarely the best examples of the moment in question that were in existence; the better renderings were obscure and unnoticed. Early-nineteenth-century commentators were no better informed about what had been said in the past century than about what had been drawn and printed. They complained, as though they were the first to whom this thought had occurred, that medieval geography was unduly neglected. There were traces of enlightenment. When the French publisher Delamarche referred in 1844 to an "important and little known part of historical geography," he did not mean the whole Middle Ages but only its "high" centuries, characterized by "démembrement féodal" (tenth century and after).[3]

A periodization in which a "middle age" was flanked by ancient and modern times was not yet automatic as the century began. In 1809 Agricole Joseph de Fortia d'Urban set out the plan for a portable historical atlas to be issued in six duodecimo volumes.[4] Not a cartographer himself, he persuaded an experienced employee of the French army engineering office to do the maps.

Fortia's three periods—ancient, modern, current—centered on great men: Alexander, Augustus, Constantine, Charlemagne, Louis XIV, and Napoleon. Modern history was deemed to run from 401 to 1800. Fortia proposed three categories of maps: general (world and continents), particular (individual countries), and detailed. "Historical" maps belonged to the third category. Fortia meant to highlight important regions at selected moments, beginning with comprehensive maps of France for the first and second dynasties of French kings; the dates in question were 511 and 814, the deaths of Clovis and of Charlemagne respectively. These synthetic maps would be complemented by details at the four points of the compass (Duval's four-part map of 1671 comes to mind). Fortia meant to supply seven more maps for 814, depicting Greece and six details of Germany (north, south, northwest, northeast, southwest, and southeast). This scheme, without the German extension, was repeated for the third French dynasty in 1270 (the death of Saint Louis), for Francis I, and for Louis XIV.

Fortia's promise to supply maps of France in four sheets at four different historical moments may have come off the top of his head, inspired by symmetry rather than by definite plans for what the sixteen maps would contain. His atlas never got beyond a prospectus. Many of the projected maps were of medieval subjects, but the threefold periodization familiar to us eluded him. His plan was novel in classing "the present" as a distinct age, the *histoire contemporaine* now standard in France. Otherwise, it was rooted in the duo of ancient and modern times that had prevailed in the days of Ortelius and long after.

Nineteenth-century atlases were astonishingly conservative in their arrangement and choice of subjects. Some managed to be new by severely limiting their contents. In many geographic collections, the unique concession to the past kept alive early models more or less directly in the Ptolemaic tradition—such as a map of Palestine depicted as the Holy Land of biblical times.[5] Also Ptolemaic in inspiration were the atlases called "comparative," in which ancient and modern geography faced each other from page to page. They did not flourish as they once had but could still count on enough of an audience to obtain intermittent production.[6] More commonplace were atlases laid out like the *Theatrum* of Ortelius: a main section of modern maps supplemented by a prologue or epilogue of ancient geography.[7]

The continuity of these traditions in atlas-making contrasted with contemporary works hinting that a geography with its own specialized terrain might finally be coming into existence. A number of strictly geographical atlases eschewed history altogether, regardless of whether it was biblical, ancient, medieval, or recent. Collections of this austere kind might, as an exception, show the tracks of modern (circum)navigators. By documenting only the triumphs of geography, such concessions to historical events stayed within the bounds of a strictly circumscribed geography. Nevertheless, unspecialized survivals from the practices of earlier centuries held their own.[8]

The impulse of Robert Wilkinson to add medieval scenes to ancient geography is sometimes found in other collections. Token forays beyond the fall of Rome were made by "universal" atlases having modest historical supplements, as well as by a few small collections of maps for history. Most of them take a group of maps mainly concerned with antiquity and, like Wilkinson, extend it to the edge of modern times or even to Napoleon. Neither ignoring the Middle Ages altogether nor providing more than a perfunctory sample, their coverage is confined to three items or fewer. The selection of scenes in these atlases suggests which aspects of the Middle Ages then seemed to have priority; also reflected is the continuity of eighteenth-century shortcomings.[9]

These meager medieval additions were very diverse in contents. Charlemagne was everywhere, in sharp contrast to the full century that had elapsed between the first and the second Charlemagne maps (1623, 1732). Didier Robert de Vaugondy made Charlemagne the sole representative of the Middle Ages in his *Atlas universel* (1756 and later), though additional scenes were available to him.[10] The lesson was well learned in subsequent decades. Not far behind in popularity was "Europe after the Barbarian Invasions," sometimes with tracks in the manner of Lesage, more often without. The barbarians, settled down, were taken to be the start of the medieval period. After this, agreement fell off sharply. Only three themes gained as many as three to five endorsements: before the barbarians invasions (in the spirit of Delisle's "Theatrum historicum"), the disintegration of the Carolingian Empire (in intent, if not forthrightly under this title), and—surprisingly scarce—the Crusades. Additional subjects came one by one, by individual choice.[11]

Among the individualists, only the Prussian officer Rühle von Lilienstern systematically preferred events involving Asia. His Carolingian Empire was accompanied by the Islamic caliphate, while Tamerlane's realm was followed by the Mongol lands under the successors of Genghis Khan in the late Middle Ages. Rühle seemed to trail in the eighteenth-century footsteps of Gatterer. In a dependably patriotic vein, a French author illustrated the advent of Hugh Capet, whereas a German one featured Germany under its Hohenstaufen emperors. Other maps were more directly thematic. Someone tried an all-purpose medieval map, with little more success than d'Anville. A German clergyman sought, with commendable originality, to map the religions of Eurasia in the vicinity of 1100. To two others, what seemed important was "feudal France," meaning the period during which princes and principalities overshadowed the

royal power. As in the eighteenth century, the late Middle Ages were either neglected or, if noticed at all, shown without breadth or imagination.[12]

The previous century had gone far in constructing maps of medieval moments and scenes. As the nineteenth dawned, only a limited, though growing,
number of atlases gave house-room to this epoch. There were laudable innovations. Émmanuel de Las Cases was responsible for the first mapping of one
medieval scene—the barbarian invasions—that was destined to anchor a
whole sector of historical atlases. A comparably dynamic medieval event, the
Crusades, acquired its key map three decades later. Once Christian Kruse provided each century of medieval Europe with a distinct map, it became unnecessary, or at least not urgent, to devise many special designs. Atlas-makers of
the early 1800s found it less important to reconstruct individual medieval moments than to shape sets of maps into packages that the public could digest.

A Canon of Maps for Medieval Europe

The maps for history produced in the early 1800s are less impressive in quantity than in the caliber of the persons making them: Las Cases, a diminutive
French aristocrat striving in exile to avoid hunger and managing twice in his
life to produce massive best-sellers; Kruse, a north German prevailing over
poverty-stricken beginnings to win esteem and honor in his home town and
an academic chair at Leipzig; Christophe Guillaume Koch, a Franco-German
professor at Strasbourg, head of a nursery for future diplomats;[13] Conrad
Malte-Brun (to use the French form of his name), a poet banished from
Denmark, forced by exile to incessant, ill-paid writing, and acknowledged at
his early death to be France's foremost geographer.[14] That all four should be
known figures, honorably admitted into biographical dictionaries, is astonishing by comparison with the normal obscurity of historical atlas-makers.
None of the four, Kruse possibly excepted, designed maps that a geographer
would have considered estimable; Koch and Malte-Brun both relied on others to illustrate their ideas. But the programs of maps they set out proved to
be exceptionally influential.

The maps of medieval history published between 1801 and 1814 would be
unrivaled, as sets or individually, in influence and longevity. They include Las
Cases's "Transmigration of the Barbarians"; Kruse's portrayals of Europe at
the beginning of each century from A.D. 400 to 1400; Koch's five, then seven,
moments, from the Roman Empire ca. 395 to the fall of Constantinople; and
Malte-Brun's six medieval scenes.

Kruse's most influential innovation was to make the European continent the theater of his maps. That the same geographic focus was independently chosen by Koch suggests that he and Kruse expressed a preoccupation now widely shared. In Kruse's wake, maps entitled "Europe in [year]" or a similar formula—a pattern without precedent in the material we have surveyed— came close to becoming the most common single category from then on. The Asian interest notable in Hase, Gatterer, and the universal, sequential atlases from Paris did not vanish; Klaproth even compiled a twenty-six-leaf atlas for Asiatic history; but the preeminence of "Europe" became explicit and unmistakable.[15]

This focus must have been taken for granted almost at once; Kruse won scant recognition for turning historical maps in this direction. The more conspicuous role of his atlas was as a reservoir whose maps somehow suited everyone's purposes. Kruse's work was well received in all respects except its fundamental principle of displaying one map per century that recorded territorial boundaries on the last day of the double-zero year. Kruse insisted on having his maps embody an instant, nothing more. Precision in coordinating a map with historical changes, he maintained, was only possible in this way.[16] These scholarly scruples found few sympathizers. Arbitrary double-zero years were unacceptable. They were, after all, arbitrary. Historical maps had to be related to events, themes, and epochs; the moments chosen for intervals had to be charged with meaning.[17]

Though Kruse's reasoning was disdained, the maps in which he put it into effect were tacitly welcomed. "Europe near the end of . . . " or "Europe near the beginning of such-and-such a century" proved to be highly suitable phrases for captioning the named epochs and periods that Kruse, everyone agreed, should have preferred to arithmetic intervals. Kruse's maps were issued in batches of four from 1802 to 1810 and completed with five in 1818; those for A.D. 400–700 became available in 1802, 800–1100 in 1804, and 1200–1500 in 1810. No matter how emphatically his basis of choice was deplored, his maps were extraordinarily well suited to readers' needs.

Christophe Guillaume Koch's *Tableau des révolutions de l'Europe* is accompanied by seven epoch-related medieval maps. Six of them could have been (but were not) culled from Kruse's *Atlas zur Übersicht der Geschichte aller europäischen Staaten*. Even if Koch's mapmaker went his own way, others often appropriated Kruse's maps.[18] Malte-Brun graciously acknowledged the source of the three he adopted and called Kruse's atlas "excellent." The more normal practice was to distance oneself emphatically from Kruse and then plunder his work.[19]

Kruse's atlas was one source for maps of medieval history at the start of the nineteenth century; another was Koch's book, long forgotten today. Koch's *Tableau des révolutions* combines a promise of insight into current affairs with a marked emphasis on the medieval background.[20] First published in 1771, the *Tableau* still circulated in the 1830s (somewhat revised). Its reputation was enviable. The eighteenth-century taste for universal history went hand in hand with a belief that the total image should be built up country by country. Koch was among those who set about narrating the European past as a meaningful entity in itself. He believed in a continent sustained by a distinct public law and shaped by a common past.[21] The high repute of Koch's book may, but need not, have helped its accompanying illustrations to become a pattern of subjects whose traces in historical atlases can be followed far into the century.

The seven small, neat foldout maps for Koch's *Tableau* are all medieval in content, passing in seven steps from before the barbarian invasions (A.D. 395) to the fall of Constantinople in 1453. It speaks for the perceived timeliness of the medieval past that Koch's maps became available in the very years when Kruse was bringing out a work rightly recognized by contemporaries as important for the geography of the Middle Ages.[22] As early as 1790, Koch regretted that he could not secure maps for his *Tableau*. The cartographer who fulfilled Koch's wishes was A. Ire de Rosny, who is not otherwise known.[23]

Rosny's maps were supplementary to the text, tucked into a late volume together with other aids to study, such as chronological and genealogical charts. They are small and cramped, made to fold into Koch's work without being too bulky. The book is concerned with the whole of European history and extends to the 1820s in the fullest edition. The maps, however, are restricted to medieval moments. The "Koch Seven" formed an influential, compact panorama, well suited to audiences whose attention span welcomed succinct Middle Ages. Within the imposed limits, the map designer did well. Maps 1–5 were incorporated in the edition of 1807; maps 6–7 were added to the (posthumous) edition of 1814. They were considered instructive enough in their own right to be reprinted in England as late as 1831.[24]

Readers were impressed by the sequence of subjects:

1. The Roman Empire before the invasions [ca. 400]
2. Europe after the barbarian invasions [ca. 500]
3. Europe under Charlemagne [ca. 800]
4. The dismemberment of Charlemagne's empire [ca. 900]
5. Europe toward 1074
6. Europe toward 1300
7. Europe in 1453 [25]

To judge from the selection made by users, the opening cycle of unity (1, 3) and disintegration (2, 4) took pride of place. Nevertheless, the traditionally neglected "High" Middle Ages were allotted as much space as the earlier centuries (on the understanding that map 1 is ancient history). Here was a selection of medieval moments around which a teacher might hope to organize a course of historical instruction.

Koch's selection seemed to make sense in a variety of contexts. Conrad Malte-Brun was not a trained cartographer; the maps accompanying his masterpiece, a multivolume *Précis de géographie universelle*, were the work of Pierre Lapie, prominent for both civil and military maps.[26] Those done to order for the *Précis* were comparatively small and inexpensive. Malte-Brun, in response to objections that they lacked monumentality, pointed out that their low price had the advantage of widening the clientele, and their modesty was only a relative disadvantage: even the most luxurious atlas needed supplements from year to year, so that there was little point in trying to look definitive.[27] While Lapie and his associate Jean-Baptiste Poirson provided Malte-Brun's maps with geographical outlines, the author took responsibility for the historical details. "Such as it is," Malte-Brun wrote in the foreword to the full version (1812), "our Atlas is certainly the *only* existing map collection by means of which one may study the whole of ancient, medieval, and modern geography, and follow the discoveries of all the voyagers, from Herodotus to [Alexander] von Humboldt." History, not mentioned, was subordinated to geography.

Later in the century, three more cartographers retraced Lapie's footsteps and provided Malte-Brun's *Précis* with an atlas; public demand encouraged new editions. None was so receptive to the Middle Ages as the atlas of twenty-four maps whose subjects were chosen by Malte-Brun himself and published in company with the first two volumes of the *Précis* in 1810.[28] An *Atlas supplémentaire* appeared two years later, immediately joined by an *Atlas complet* combining the original twenty-four with a supplement of fifty-one. The maps concerning us were all in the installment of 1810:

17. "Europe avant l'invasion des Huns (vers l'an 370)"
18. "Europe après l'invasion des barbares"
19. "Géographie du moyen âge"
20. "Empire des Mongoles"
21. "Europe en l'an 900 et en l'an 1100"[29]

A great deal had to be fitted into the seventy-five maps allotted to Malte-Brun by his publisher, including a world atlas, ancient geography, and the history of geography. Maps of medieval subjects were allowed as historical context for the development of geographical knowledge, but even with this justification political and geographical goals were forced to coexist. Malte-Brun was sorry that the space he had for the period was restricted. It would have been better, he said, to separate the elements entering into the map of "Géographie du moyen âge," but he believed that combining Charlemagne's empire with the expanse of the caliphate and with traces of viking exploration was not without interest. Europe in 900 and 1100 shared one page as separate maps, gratefully copied from Kruse. A third Kruse map, that for 1300, was reduced to an inset as "an interesting contrast" to the Mongol Empire. Malte-Brun somehow believed that his Mongol map was completely new. The atlases of Hase and Dupré had mapped the subject before. Malte-Brun possibly thought he had outdone them, but he is more likely to have simply been unfamiliar with the earlier atlases. In his otherwise rather schematic reconstruction of the Mongol Empire under the two khans, Genghis and Kublai, the itineraries of William of Rubruck, Marco Polo, and Francesco Pegolotti stand out.[30]

The "Koch Seven" and "Malte-Brun Five (or Six)" embody a close but not identical selection of medieval moments; Malte-Brun's caliphate and Mongols are the salient differences. Malte-Brun was well versed in the works of Koch and Kruse. He spoke approvingly of both but stressed how different his own "Europe before the invasions" was from that provided by Koch's cartographer. Malte-Brun did not follow Koch in singling out 1453, and he emphasized the history of geography, a subject that, before him, consisted mainly of marking the routes of major navigators onto world maps. Nevertheless, it looks as though Malte-Brun learned from Koch what subjects to select. Because both authors were widely copied, their pattern of scenes deserve commentary.

Koch no. 1. Koch's and Malte-Bruns's maps of Europe before the invasions, or before the coming of the Huns, have basically the same subject as Delisle's "Historical Vista for A.D. 400"—a last glimpse of the ancient world, or Roman Empire, in its integrity, with attention to the "homelands" of the outlying barbarians.[31] Much the same effect might be produced by a map of "the Roman Empire at its widest extent": that, too, would be "before the invasions." Regardless of whether the moment chosen is A.D. 100, 117, 180, 370, 395, 400, or 402, a map so named belongs to ancient, not medieval, history. An author's choice of periodization determined which of these maps was chosen and in what section of the atlas it was entered.

Koch 2. The map for A.D. 500 offers a wide array of barbarian kingdoms, many of which were experiencing their final decades of existence: Britain disputed by Britons and Saxons, Vandals in North Africa, Ostrogoths in Italy,

Visigoths in southern Gaul and Spain, Burgundians in the Rhône Valley, Franks under Clovis starting their expansion, several more entities in the Danube Valley, and the surviving Roman Empire, ruled from Constantinople and controlling Egypt, much of western Asia, Greece, and more. Parts of this map had early-eighteenth-century precedents, especially in the original illustration for Daniel's *Histoire de France*. Once Las Cases's "Transmigration of the Barbarians" was in circulation, it sometimes served as an alternative for either "before" or "after the barbarian invasions," or for both at once.[32]

Koch 3. The Empire of Charlemagne, as a map scene, existed before the nineteenth century but attracted less attention than one might expect. There were five versions in two centuries. (Charlemagne's Bavarian canal was more seductive.) This obscurity ended during the era of Napoleon.[33] Helped along, perhaps, by the *Atlas universel* of Robert de Vaugondy, maps of Charlemagne's empire became obligatory fare, omitted only in unusual circumstances. Space limitations often demanded that several Carolingian themes should be combined. Triumphant portrayals of the empire as it stood from 800 to Charlemagne's death were a frequent but not an uncontested choice. In the main variants, Charlemagne's empire was combined with its dismemberment, as though telescoping Koch nos. 3 and 4. It also made a frequent pair with the Islamic caliphate, as by Malte-Brun. The form in which an atlas depicted the Islamic phenomenon (if it did so) often depended on the combination of subjects included in its Charlemagne map.

Malte-Brun's concern with the history of geographic exploration is most obtrusive when, disregarding chronology, he begins with the "Géographie du moyen âge" for the ninth century, directly follows it with the Mongol Empire of the thirteenth, and relegates the maps for A.D. 900 and 1100 to last place. Like everyone except Las Cases, he preferred to group historical maps in order of time. His preference for thematic order here may result from wishing to place his main scenes of medieval geography as close together as possible.

Koch 4. Delisle's manuscript maps for Louis XV, and Georgisch's book on the Holy Roman Empire, show how necessary it seemed to complement a Charlemagne map with another showing the dissolution of his empire through partitions, notably those of 843, 855, and 870. The lineage of France and Germany from Charlemagne's empire bifurcated in these divisions. The fragmentation of the empire coincided with attacks by Northmen on the coasts and up the rivers and, at the close of the ninth century, with the invasion of the Hungarians. Mapmakers had to decide how much or how little of these incidents to display. If Koch's thematically named map was considered unsatisfactory as a model, Kruse's "Europe in A.D. 900" offered a suitable alternative: Malte-Brun was among those who chose it.

Koch 5. Koch's selection of the year 1074 is an oddity. Because that year does not tag any major event, compilers betray Koch's influence when they attach it to a map. In Asia Minor, 1071 witnessed the Turks inflicting a calamitous defeat on the Byzantines at Manzikert. In the West, Gregory VII was elected pope in 1073; he was an aggressive prelate, whose falling out with the German emperor Henry IV, beginning in 1075–76, occasioned a crisis in the relations between the papacy and the empire in the Latin church. A looser chronology would allow one to speak of Europe "on the eve of the Crusades"; the appeals were made in 1095–96; Jerusalem was stormed in 1099. Just as Malte-Brun found the Kruse map for 900 adequate to illustrate the fate of Charlemagne's empire, so Kruse's choice of 1100 was interchangeable, for all practical purposes, with 1074. The moment might be taken to signify the passage from the earlier to the later Middle Ages, or the stirrings of religious, economic, intellectual, and martial activity in the decades leading into the prodigiously creative twelfth century.

Koch 6. From today's perspective of milestones in medieval history, a leap from 1074 (or 1100) to 1300 is too great. But by comparison with the decades-long bewilderment of cartography about what to show in Europe after Charlemagne, Koch's allotment of three maps from 1074 to 1453 merits praise. In the corresponding position (with a displacement of order), Malte-Brun placed a map of the Mongol Empire. European travelers are his focal point; the course of Mongol conquest is disregarded. The explorers' tracks are set in a vast, composite Mongol Empire (said to be Genghis's and Kublai's), whose security fostered contacts between Europe and the Far East.

In this broad perspective, the emphasis of Malte-Brun's map almost coincides with Koch's; the coincidence is total when Malte-Brun's inset map, "Europe in 1300," is preferred to the Mongols. Many noteworthy European events can be associated with the decade on either side of 1300: the fall of Acre, last crusading outpost in Palestine (1291), thus a possible "end of the Crusades"; the reign of Rudolf of Habsburg (†1291); the first Holy Year (1300), shortly followed by the papal move to Avignon (1305); even, broadly speaking, the start of the Hundred Years' War (1336). In the ongoing historiography of the Middle Ages, the year 1300 has not turned into a canonical boundary between "high" and "later" Middle Ages, but some moment to symbolize this division tends to be needed by historians—if not 1300, then another year nearby.

Koch 7. The fall of Constantinople to the Turks (1453) marks an epoch in the historical epitomes of Cellarius; typically, it is the boundary between medieval and modern. Many other historians have concurred, but by no means all. An expert comments: "In the days when historians were simple folk the

Fall of Constantinople, 1453, was held to mark the close of the Middle Ages."[34] This "simple" idea can be whimsically condensed: Greek scholars fleeing the Ottoman capture of Constantinople with whatever books they could carry took to boats, washed up in Italy, implanted Greek wisdom, and began the Renaissance. Koch's choice of 1453 for his last map had a long past, yet was fresher then than it has since become.[35] An attractive alternative, a few decades later, would be the time of the "great discoveries." Malte-Brun, still far from our vision of the end of the Middle Ages, came closest to it with a map of the empire of the Habsburg Charles V.[36] Mapmakers, like historians, could not agree on a single event in the late fifteenth century that surpassed all others in importance. Atlases reflect their diversity.

Koch's selection of maps, however imperfect, was looked upon for a time as a medieval canon. Not every segment had equal weight. The first three—barbarian invasions and Charlemagne—were unanimously welcomed. They were picked when there was no room (or desire) for more, as though what most mattered was to get the Middle Ages started; everything after 800 could be dispensed with. The full canon, however, also had takers, notably as a collection of all-European medieval moments. Hase had not been copied in this way; neither had any of the sequential atlases, including Kruse, whose cartographic abundance was not sustained by an explicit or implied narrative. Koch's book was articulated in periods, and the corresponding moments were expressed in the accompanying maps, as a helpful medieval package. In the coming decades the same cluster was often reproduced, with Koch's choice of maps or a variant, in atlases whose grouping of maps was not chronicle-like but adhered to the three established periods.

As though to document the breadth of Koch's influence, his earliest disciple appears to have been from Prussia and responsible for a small atlas of seven maps, lacking an author's name, place, and date but engraved in Berlin after the fall of Napoleon. The designer copied Koch's first five maps literally, achieving better results than Rosny's originals (Koch's 1807 edition had only these five). The last maps went their own ways: sixth, "North and Central Asia with Reference to Its Former Condition" (an intriguingly unusual subject); seventh, to represent current times, Europe in 1814. One would like to know more about this advanced and rare collection.[37]

Predictably, Koch's series affected French compilers most of all. The ugly atlases of the lithographer Selves directly espoused all of Koch's intervals. Disregarding discrepancies of date, Selves illustrated his collections with maps copied from Kruse. Equally dutiful disciples were Denaix's *Atlas de l'Europe* and Engelmann's *Atlas élémentaire de géographie historique;* the latter omitted only a map for the period before the barbarian invasions. The German

Kärcher and Englishman Quin, while not direct dependents, picked their way through the medieval period with a choice of moments whose resemblance to Koch's sequence of events is unmistakable.[38]

Koch's influence remained strong near midcentury. Victor Levasseur's *Atlas classique universel* added new subjects, such as the Byzantine Empire in 800 and the Crusades with detailed tracks, yet retained Koch signposts: Europe in 1074, 1300, and 1453.[39] The *Atlas historique du moyen âge* by the distinguished Victor Duruy was ambitious and will concern us again; its effort to provide comprehensive coverage involved six of the "Koch Seven," firm anchors amidst diversity. As late as 1865, the third edition of an 1820s atlas (Muhlert/Möller) still espoused half of Koch's repertory. More remarkably still, several of Koch's subjects surface in the masterful European survey of Karl Spruner's *Hand-Atlas* and in numerous Spruner spin-offs, such as Bretschneider's wall atlas and the school atlas translated into English (these by-products concentrated on European history as a whole). Spruner, a perceptive historian, did without 1453; he exchanged the Carolingian disintegration for the rise of the Ottonians and replaced Koch's year 1074 with a context for the Crusades. The dates that nineteenth-century historians chose for the turning points of the Middle Ages did not remain completely fixed; there was scope for individual variations, but not so much as to render the "Koch Seven" wholly obsolete.[40]

MEDIEVALIA IN THE *ATLAS LESAGE*

Émmanuel de Las Cases had little interest in the Middle Ages, let alone in a cycle of maps illustrating them. His *Atlas Lesage,* whose inaugural London edition antedates all the other nineteenth-century works discussed thus far, was found wanting, after the Bourbon Restoration (1815), in the biblical and medieval content then in demand. Las Cases eventually responded with single sheets, one for biblical history, another (by special request) for the early Caesars, and two for the Middle Ages (one general, the other specifically French).[41] The salient map of the barbarian invasions was in the original atlas of 1801. Though suitable for medieval beginnings, it was always grouped with ancient history. The Lesage collection acknowledged the Middle Ages in other ways. Its maps of European countries touch on the early German duchies and, in Spain, on the campaigns of Bertrand Duguesclin, Henry of Trastamare, and Pedro the Cruel. The designer's aim, however, was to illustrate national histories, not a collective medieval period. Las Cases could hardly be expected to contemplate the past methodically, like the professionals Koch and Kruse. He had the historical and geographical knowledge of an amateur and was appreciated by a public largely unencumbered by historical training.[42]

The sketchy medieval period embedded intermittently in the original Lesage atlas tends to be more commendable than the leaves added "by popular request" after the Restoration. Among the latter, an overloaded one-sheet supplement for medieval history and geography lumps together the empires of Ermanaric (fourth century), Charlemagne, "Mohamet" (i.e., Islam), and Byzantium, along with tracks for Hengist going to conquer England (ca. 449), Rurik going to found Novgorod (862), and Philip Augustus and Louis IX going on crusade (1190, 1270).[43] Las Cases's imagination and ability to see artistic connections do not quite redeem his tolerance for anachronism.

The Lesage sheet devoted to the first and second "races" of French kings—that is, the Frankish Merovingians and Carolingians—was designed to help schoolboys cram the indispensable details about the beginnings of the national monarchy. The original *Lesage Atlas* had dealt with the French Middle Ages in a single map illustrating the provinces of the realm, each marked with the year in which it was directly attached to the Crown. The map traced a chronology of how the present-day territory of France had taken form. A teleological design of this kind, as easily set out in a written list as graphically, calls for a resolute simplification of events. It was nevertheless a hit, less often credited to Las Cases than other inventions of his. The design was adopted inside France and out as the model for capsule portrayals of territorial development, under the title "Historical France, Spain, Europe, etc."[44]

Las Cases was most original, and enduring, in his "Map Exhibiting the Transmigration, Course, Establishment, or Distruction [*sic*] of the Barbarians, that invaded the Roman Empire." The subject had been drawn as early as 1718, but differently and in a rare, obscure work. Las Cases did not know that he had been anticipated, and neither did anyone else. The "Transmigration" map is his personal contribution to synchronic cartography and is probably the pinnacle of his mapmaking. Las Cases warmly praises his own design: "The portrait of the invasion of the Barbarians may deserve special attention by the clarity that it can effortlessly cast upon a particularly confused period of history. . . . By this new and simple method, the knowledge and study of this obscure period assumes a new face. . . . Nothing now is easier than to trace the origin of all these peoples, to follow the movement on a map, or to reflect upon their particular effects."[45] Synchronic, yet unified by its theme, the map of the invasions perfectly illustrated how, in the view of Las Cases and many others, geography could help students of history.

The "Transmigration" map has been often applauded and sometimes deplored. Either way, it has proved exceptionally powerful. The barbarians of late antiquity haunted the European historical imagination. Long believed to mark, in their settlements, the beginning of the European states, they were an aspect of the wider, also obsessive problem of the fall of Rome and documented the "northern," or Germanic, beginnings of medieval Europe.

To hear Las Cases's comments just quoted, his "Transmigration" map was compiled simply as a teaching aid, so that students might more efficiently remember the details of the invasions period. The effect of his map was more grandiose, even if not deliberately sought. He "put the barbarian invasions on the map," tightening into a visual unit what, in words, tended to be a muddle. He fashioned the self-sufficient portrait of a multicentury epoch, a synthesis that words could only ploddingly express. A map can turn centuries into a visual instant, integrate a complex past, and claim to present historical truth rather than artistic fiction. Critical observers decried Las Cases's picture of barbarian *Marschrouten*. The map was confused, to say the least, but only to a discriminating historian. To others, it brought clarity—especially if improved, as it soon was.[46] The "Transmigration" scene was an archetype, offered to the future for artistic and learned enhancement.

Maps for history generally aid in the understanding of a text; great ones escape this ancillary role. The fourth-century onset of the Huns coexists in Lesage's "Transmigration" map with the eighth-century seizure of Visigothic Spain by the Muslims, and much else.[47] Whereas a text records only sequentially, the map records at a single visual instant. At a glance, the onlooker can absorb unhesitatingly and with pleasure that which a reader, because written words are sequential, does not even have the possibility of contemplating. Las Cases achieved an image of this uncommonly powerful sort.

Koch's widely followed pattern for illustrating the Middle Ages called for one map before the barbarian invasions and one after. This format offered ample room for a track-rich "Transmigration." By the twentieth century, hardly an atlas or history book concerned with this period appeared without the invasion tracks. Changes of detail, intended as improvements, began about a decade after Las Cases's original. Some were more in keeping with historical evidence than others, and any could be readily outdone by artistic adjustments and experiments.[48] What never changed were the terms of discourse. Everyone who worked (directly or indirectly) with Las Cases's design, no matter how much they sought to improve it, espoused and propagated his premise that movements in a multicentury period should be portrayed as a simultaneous collective event.

Las Cases, who usually talked about his atlas in the language of publicity and accounting, was aware of doing something new, but he equated this contribution to a nostrum guaranteed to inculcate needed knowledge. Was he inwardly conscious of his originality? His "Transmigration" map was not the only innovation to his credit. In the 1830s, as we have seen, quite a few atlases began to be organized into one general section and several "special," that is, national, ones. Las Cases had anticipated this form. At least skeletally, he broke away from the prevailing universal, sequential type and adopted a design in which chronology was subordinated to a geographical survey of the world. Las Cases pioneered the grouping that the next generation of historical atlas-makers adopted and refined in ambitious productions.[49] Half-educated, indigent ex-naval officers rarely succeed in being so inventive.

TRACKS FOR CRUSADERS

In the late 1820s, a set of historical maps appeared in Edinburgh that sought comprehensiveness by enlarging a core of old-fashioned biblical and ancient subjects. The intense cartographic interest surrounding the Holy Land had always centered on biblical narrative. No one had paid much attention to medieval Palestine. Philippe de la Ruë, when compiling his *Terre sainte* "from the origins to the present" in 1651, filled in the period between Roman and Ottoman Syria with the Patriarchate of Jerusalem. Guillaume Delisle's project for a group of maps similar to La Ruë's replaced the patriarchate with the Latin Kingdom of Jerusalem and other crusader principalities. Delisle, on the eve of his death, compiled maps of the European enclaves in the Near East. Guillaume's map of 1764, seen into print by his brother, may have become well known. Delisle showed no boundaries, only towns; his maps were meant to accompany the study of Crusade history.[50] Portrayals of the "theater of the Crusades," with or without boundaries of principalities, began to have a place, as insets or main maps, in the attention paid by historical atlases to the crusading period. Early examples occur in Robert Wilkinson's *Atlas classica* (ca. 1805) and Benicken's *Hand-Atlas.* In 1823 and 1824 Paulus Schmidt provided Raumer's *Geschichte der Hohenstaufen* with maps for the years 1100 and 1200, embracing central and southern Europe, as well as Asia Minor. Schmidt showed the crusader principalities, along with inset plans of Antioch and Jerusalem. Delisle's posthumous, late-eighteenth-century Holy Land map for the Middle Ages, highlighting Latin beachheads, is the initial illustration of Crusade history. The crusader principalities continue to be illustrated in the atlases of today but are usually subordinated to a portrayal of Crusade expeditions, a type of map that did not exist before the 1820s.[51]

In 1829, John Thomson of Edinburgh (1777–?) published a *New Classi-*

cal and Historical Atlas with forty-nine maps on thirty-five imperial folio sheets. At eight guineas, it was expensive, but some early buyers thought it well worth the price.[52] Thomson, an established Edinburgh bookseller, had produced a large and successful geographical atlas in 1817 and had been immersed since 1822 in compiling an ambitious atlas of Scotland. The latter enterprise brought him to grief in 1830. He was forced into bankruptcy; he recovered, but not enough. A second bankruptcy followed in 1835, entailing the sale of his stock and house. His plates and atlases were acquired by the firm of W. and A. K. Johnston, soon famous in map publishing. Thomson resurfaced seven years later as author of a geographical dictionary, several times reprinted. The date of his death, perhaps in the 1840s, is not publicly recorded.[53]

The *New Classical and Historical Atlas,* issued when Thomson's Scottish project was plagued by costly delays, was too lavish to have helped ease his finances. Wilkinson's *Atlas classica* had an enlightened choice of map subjects; like it, Thomson's atlas extended a biblical and classical core into the Middle (and modern) Ages, but it went a step farther. The contrast embodied in its "classical and historical" title suggests that the old-style maps for classical studies might be distinguished from the new-style "historical" repertory of medieval and modern subjects; but a more traditional interpretation is possible.[54] The maps are handsomely produced; in keeping with the derivation from ancient geography, strict chronological order seems unattainable; and models are borrowed from wherever seemed best. Eight or nine maps concern the Middle Ages:

  36. d'Anville, "Age intermédiaire"
  37. Europe before the invasion of the Huns
  38–41. [Ancient geography: regions of the Roman Empire]
  42. Saxon England
  43. The Crusades for the Recovery and Possession of the Holy Land (Inset: Kingdom of Jerusalem and adjacent lands)
  44. "Eslem [*sic*] or the countries conquered and converted by Muhammed and His Followers"
  45. The Empire of Charlemagne
  47. Europe in 900, in 1100, under Charles V, in 1789

Thomson's medieval coverage is as full as any that had yet appeared in Britain. To judge from the ragged numbering, little effort was made to keep the period compact. Numbers 38–41 continue a series of regions of the Roman Empire earlier begun. The Crusades map, and d'Anville's for that matter, are out of place. Three sources are clear: d'Anville for the first item (a widely available classic); Wilkinson for Saxon England, "Eslem," and probably Char-lemagne; and Malte-Brun/Lapie for numbers 37 and 47.[55] The single item whose source is not apparent calls for special attention.

Thomson's "Crusades for the Recovery and Possession of the Holy Land" is a fully developed, synthetic portrayal of the Christian expeditions. At upper right, a large inset shows the Kingdom of Jerusalem and other principali-ties—the earlier Crusade map found in Wilkinson more than two decades before. Thomson's main map is a new design, namely, a portrayal of itiner-aries. Tracks are drawn in contrasting black-and-white patterns, running overland and by sea, marking each Crusade from the first to the seventh (Saint Louis's in 1270). Some lines reach as far as northern Europe, circling Spain (but partly off the map since western Spain falls outside the frame). An elaborate legend identifies each expedition and gives additional explanations.

The tracks, though more modest, graphically, than the ribbons of the *At-las Lesage,* are the eye-catching feature. Just as in Las Cases's "Transmigra-tion," space annihilates time in Thomson's Crusades map and, by the device of tracks, integrates the events of about two centuries into the privileged im-age of a homogeneous historical phenomenon. The Levant had long been shown as the "theatre" or setting of the Crusades; the new map with tracks could claim to portray the Crusades themselves.

Every other map in Thomson's medieval group has a plausible and easily identified source. Where this new, and deeply impressive, image comes from is surprisingly hard to say. The Crusades had recently gained renown in Scot-land from Walter Scott's *Tales of the Crusaders* (1825). Even so, John Thomson is unlikely to have personally designed a new map; mapping Scotland was challenge enough. Collections near in date, such as Malte-Brun, Benicken, Brué, have no trace of "Crusade itineraries."[56] Thomson's model is likely to have come from a recent publication, such as an illustrated book. His title echoes Charles Mills's four-volume *History of the Crusades for the Recovery and Possession of the Holy Land* (4th ed., 1828), but even a late, augmented edi-tion of Mills's work contains only a rudimentary map.[57] With Thomson's *New Classical and Historical Atlas* we are presented with the first appearance in an atlas of a map that soon assumed a place of honor in historical collections. A map very similar to it appeared in a French atlas in the very next year. There had to have been a common model. Identifying the prototype of the "itiner-aries," or "expeditions," design would put a face on a map of equal impor-tance to today's historical atlases as those for the Exodus, Charlemagne, and the barbarian invasions.

Contemporary events conspired to draw attention to the Crusades early in the century. Napoleon's expedition to Egypt, with his hundred-day sally into Palestine (1798–99), may have pointed in this direction.[58] The antirevolutionary surge of interest in the Middle Ages, typified by Chateaubriand's *Génie du Christianisme* (1802), is certain to have done so. Besides, a period of ideological conflict and constant European war may have brought the Crusades to mind as a "distant mirror." For whatever reasons, the theme was demonstrably current. The subject set by the Institut de France for its annual prize essay in 1806 was "Examine what was the influence of the Crusades on the civil liberty of the peoples of Europe, on their civilization, [and] on the progress of enlightenment, commerce, and industry." A German and a Frenchman shared the prize, and there was at least one runner-up. The book-length responses of all three came into circulation in 1808–9. They were joined by Jan Hendrik Regenbogen's long "commentatio" in Latin on the benefits to Europe of the "Holy War." He had set out his response to the Institut de France's subject without participating in its competition.[59]

Between 1807 and 1832, the historian and Orientalist Friedrich Wilken published a seven-volume *Geschichte der Kreuzzüge*. Outside Germany, it was more esteemed than read. Also indicative of interest in the subject was C. W. F. von Funk's four-volume *Gemälde aus dem Zeitalter der Kreuzzüge*, notable for excluding the First Crusade from its program. Despite Funk's astute choice of title, he had even fewer readers than Wilken. Yet the theme was appealing enough that the outdated Crusade history of a famous professor at the University of Göttingen was resurrected from his literary remnants and posthumously published.[60]

France identified more directly than Germany with the crusading movement; the motto "Gesta Dei per Francos" (the deeds of God by the agency of the French) was all too well remembered. Contemporaneously with Wilken's major work, the multivolume *Histoire des croisades* of Joseph Michaud was a runaway success. Often reprinted, it soon ranked as a classic, was translated into most European languages, English included, and experienced a wide and long-lasting diffusion. Extracts eventually entered anthologies for teaching French.[61] Michaud was admitted to the Académie française after volume 1. His apparent start on this subject had been an introductory "Tableau historique des trois premières Croisades" for Marie Cottin's novel *Mathilde, ou Mémoires tirés de l'histoire des Croisades* (1805), but he acknowledged a profound debt to the learned winners of the Institut de France competition.[62] Scott's *Tales of the Crusades* was part of this wave. Years earlier, his English friend William Stewart Rose had expressed his enthusiasm for medieval romance in a ballad on *The Crusade of St. Lewis*.[63] Wilken and Michaud are the

JÉRUSALEM EN 1099.

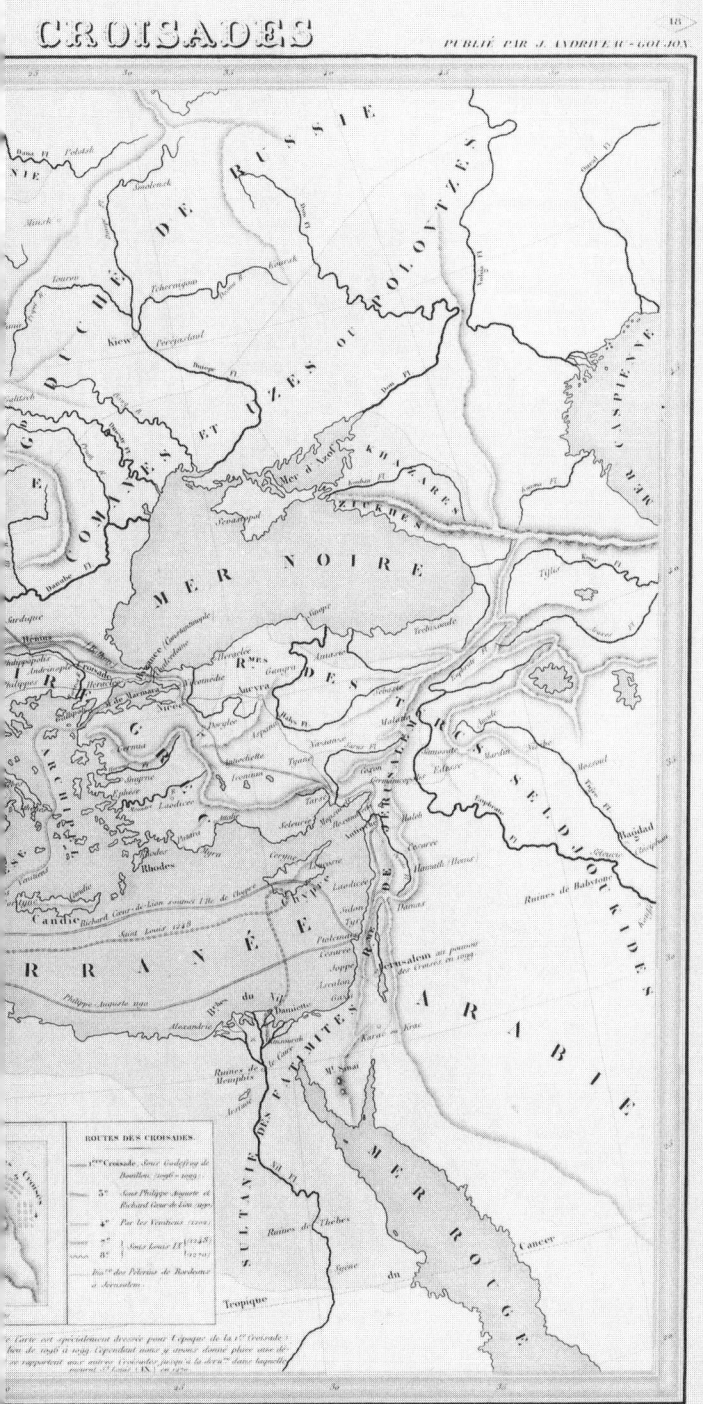

Figure 29 E. Andriveau-Goujon, "L'Europe au temps des croisades" (1840). Early maps showing the tracks of the Crusades were published within a year of each other in Scotland (1829) and France (1830). The originator of this design, however, is unknown; both atlas-makers seem to have depended on a still unrecognized source. Andriveau-Goujon's version of the Crusade expeditions map, showing eight tracks, is a good illustration of a type that by 1840 was in the public domain. It seems to have originally appeared in his *Atlas élémentaire simplifié* (1838) and was printed in many Andriveau-Goujon atlases. Andriveau-Goujon belonged to a dynasty of Paris map publishers. His Crusades map is marked as having been revised by Carl Benedict Hase, a German scholar-at-large in Paris. (Courtesy of the Map Collection, Yale University Library)

forefathers of modern Crusade histories. The impulse that moved the Institut de France to set its prize question and that elicited talented responses cannot have been alien to Wilken's and Michaud's choice of topic.[64]

The new interest in the Crusades did not inspire a surge of original research. The reviewer of one of the prize essays insisted that, owing to the abundance of earlier work, the author said nothing new and could be praised only for comprehensively retracing known ground.[65] The Institut de France's competition and other forms of attention, though comparatively barren from a scholarly standpoint, deeply affected judgments of the historical significance of the Crusades. The Crusades were thrust forward as the climactic medieval experience. Wilken's claims for the importance of his subject were unqualified: "Among all the events of the Middle Ages, none is more notable and more important in its consequences than the expeditions of the western Christians to the Promised Land. . . . The Crusades not only affected the peoples as a whole, . . . their consequences forced their way into the bosom of families. The mass of knowledge was increased; the radius of concepts was broadened. This had a positive effect on the enjoyment of life, on trade, on business, and on art. As a result of the Crusades, Europe changed form."[66] Others were no less emphatic: "[only] a small number of events sweep human society toward a total change in its customary ways of thought and action." The Crusades were among "the extraordinary causes that, like an electric shock, arouse the human spirit from its lethargy."[67] The Institut de France's question pointed unwaveringly to the conclusion that the effects of the Christian expeditions on the home front had been beneficent and radically innovative.

This interpretation has not endured. The Crusades as the all-purpose agent of high medieval developments have been shelved and replaced with less constricted accounts. For a rather long time, however, the Crusades were viewed as having been the mainspring of the High Middle Ages, confirming the direct and indirect benefits of overseas travel.[68] Perhaps the main lesson was that the *Christiana tempora* of Europe had had a revolution, too—a much more admirable one than the recent model.

The prototype for Thomson's arresting "expeditions" map probably lies still unnoticed in a mid-1820s illustrated work on the Crusades or medieval history.[69] The existence of a source seems confirmed by J. G. Heck's *Atlas géographique, astronomique et historique servant à l'intelligence de l'histoire ancienne, du moyen âge et moderne*, issued almost simultaneously with Thomson's and far from Edinburgh. Heck's few medieval sheets strive especially for dramatic effect: the Empire of Charlemagne; the Crusades, with tracks; French history, 1108–1564 (subdivided into four maps); and the great discoveries, with tracks. Heck could believe that he was presenting the most riveting me-

dieval moments, three of them French. His Crusade map, labeled as applying to 1094–1291, shows tracks for five expeditions, subdividing the Third
Crusade into three, with a line for each of the participating kings. The Fourth
Crusade (against Christian Constantinople) is disregarded. There are the
usual insets for the Christian states, the Latin kingdom, and the siege of
Jerusalem in 1099.[70] Another decade would pass until the "expeditions"
map—the Crusades illustrated by the simultaneous display of the tracks of all
or many of the crusading expeditions—made its next appearance, and al-
most two until this design found its way into a German collection (fig. 29).
Despite a slow start, a portrayal of the "expeditions" became the normative
Crusades map and assumed a central place in sequences of medieval scenes.
Where Thomson and Heck found their model is an intriguing problem.[71]

Medieval and Modern in the Last Sequential Atlases of Universal History

The two atlases produced by Friedrich Benicken at Weimar between 1818 and
1824 were advertised as complements to the historical handbooks of Leonhard von Dresch and conspicuously depend on them. Dresch's book on the
"methodical" teaching of history is echoed by the *methodische* (i.e., outline,
or scriptless) maps of Benicken's first collection, the *Historischer Schul-Atlas*.
The articulation of both collections is keyed to the chapters and periodization
of Dresch's *Übersicht der allgemeinen politischen Geschichte, insbesondere Europens*.[72] Dresch's periodization resulted in Benicken's assigning six of fourteen maps to the years after the accession of Charles V (1519). About 40 percent of the two Weimar atlases are devoted to modern history, compared to 6
percent in the archetypal Paris draft of the 1740s.

Weimar was certainly aware that Kruse's last installment, with five maps
from 1500 on, was on the way. The installment appeared in 1818, when Benicken's atlas must have already been in preparation. Kruse is a less likely source
of inspiration than the *Atlas Lesage*. There, as early as 1801, almost all maps after the classical or "Transmigration" period were modern. A shift of chronological emphasis was under way.

Benicken's use of tracks also exaggerates Lesage features. Some contemporaries looked upon Las Cases's ribbons as a breakthrough in historical
mapping. A French disciple has left an unsigned manuscript atlas documenting enthusiasm for Lesage's devices. Unreconciled to the term "Middle Ages,"
he affirmed that what Europe experienced until the fall of Constantinople in
1453 was a millennial "Late Empire" (Bas-Empire). To the Lesage maps with

tracks that he copied, he added six more, in the usual ribbon form, including Constantine the Great, Julian the Apostate, and Belisarius; the emperor Charles V and King Henry IV of France; and, not least, the Crusades, with Saint Louis given a separate spread all his own. This anonymous atlas remained an uncirculated manuscript. Nevertheless, it documents the existence of a compiler who evidently found tracks gratifying and wished to surpass Las Cases in their execution.[73] The same idea occurred to someone in Weimar during the 1810s—Dresch, Benicken, or someone else.

Tracks beyond Las Cases's dreams permeate the first of Benicken's atlases; and, in a unique, never repeated flourish, each line of travel or invasion ends in a noose circling the land finally occupied or seized. The traditional barbarian invasions, incorporated in the map running from Augustus to 476, have no exclusive claim to tracks. The medieval displays— 476 to Charlemagne, 760 to the accession of Pope Gregory VII (1073), 1073 to Rudolf of Habsburg (1273), 1273 to Charles V (1519)—are also rich in tracks of various kinds. Only the crusaders are missing (the expeditions map had not yet been conceived). From barbarians to great discoveries, Benicken's Middle Ages are threaded by a network of emphatic streaks, compelling attention and, except in the vast trackless stretches of the world, allowing few other aspects of medieval history to stand out. Here was animation outdoing the placid maps of Hase, Dupré, or Gatterer.

According to an announcement, Benicken's *Schul-Atlas* was destined for "the first course in history." The youngsters tagged by this label, whatever their age, presumably needed to be entertained as well as taught. Benicken proceeded directly after finishing the *Schul-Atlas* to executing the more sophisticated *Hand-Atlas.* Advertisements recall that the *Schul-Atlas* for beginners was still available. No one, least of all Benicken, had experienced a change of mind or considered that the second work replaced the first. The library atlas differs markedly from its "school" antecedent, but the contrast presumably results from adaptations to a more advanced public rather than from reconsiderations of the *Schul-Atlas.*

The differences between the *Hand-Atlas* and Benicken's first effort are salient but hard to explain by mere reference to their apparent audience. The fourteen divisions have one main map each, consistently on a Mercator projection, showing political boundaries at the end of the featured period. As Africa and the New World gradually enter the historical agenda, the dimensions of the main maps increase. The range of the fourteen main maps is not general history but aspects of geographical discovery, including the tracks of navigators and explorers. Historical particulars are assigned to insets, four or five per main map and about fifty in total, varying in scale and fitted into blank sections of the globe. This multiplication of insets is a first. If the smaller maps were detached and ordered as a book, they would form a spacious historical atlas. Tracks, though present, are drastically curtailed by comparison with the *Schul-Atlas* and limited to faint lines. The barbarian invasion period is cut down to two maps with tracks: "Campaigns of the Germans, Goths, and Alans" and "Hunnish Campaigns." The Hunnic track has a distinctly Lesagean appearance: the Huns advance as far as Dacia, turn back east, and ostensibly rest for years behind the Urals, before going west again for the Attilan chapter of their history.[74] Also like Las Cases, Benicken seems to think that Attila, instead of retreating from north Italy, reached Rome ("Attila vor Rom 452"). The list of faults found by an early reviewer could be expanded, but Benicken went wrong in details rather than in plan.

Most insets in the *Hand-Atlas* simply show a territory colored to reflect a period (rather than an instant). There are maps for the Merovingian Franks, the Carolingians before Charlemagne, the Heptarchy, Arab and Umayyad Spain (an unclear distinction), Lombard Italy, early and somewhat later Denmark, the Byzantine Empire at two moments, the crusader principalities, and the Mongols. High medieval Europe is dealt with in a manner similar to the practices of the eighteenth century; that is, its image brings to mind modern great powers rather than the fragmentation appropriate to medieval conditions; the western European countries are given only the large-scale political boundaries applicable toward 1000 and 1200. Benicken, eschewing the exuberant, track-rich Middle Ages of the *Schul-Atlas,* created a second medieval series whose sedateness and confinement to territory recall the regional maps of the ancient world supplied by collections of ancient geography. Benicken's diminutive insets identify medieval entities and establish their existence, little else.

The Benicken atlases are more attractive in content than in size and layout. They are too large, awkward to handle, and cumbersome to consult. The *Schul-Atlas* associates many events with animated tracks and lassoes, but it confines this drama of invasion and migration to a cramped and hard-to-examine Europe, a small protrusion from the west end of Eurasia. The universal format, for all its merits, demanded that the whole landmass be included; it stood in the way of displaying an increasingly European history. Readers of the *Schul-Atlas* might well wonder why they had to grapple with extralarge pages containing insets, instead of being given a larger number of smaller pages in which the scenes on the insets could be dealt with in isolation and greater detail. There were reasons why Benicken's productions did not win a following.

Failures or no, the Weimar atlases are endearing. They go further than

Malte-Brun in striving to combine general history with the history of geography, yet keeping the two subjects distinct. By the criteria of quantity and design, they far outdo Lesage in the game of tracks. They honor the sequential format, but temper it with Dresch's periodization, European focus, and disproportionate attention to history since 1500. Not least, the multiple insets of the *Schul-Atlas* shyly anticipate a design that came into its own a generation later. While applauding the evident sense of experimentation, one is inclined to conclude that Benicken dealt more memorably with modern than medieval history.

For all his sensitivity to current developments, Benicken produced atlases in the old universal, sequential tradition. The final atlas of this type seems almost to have been an accident. The odds against it were steep: it was compiled by a young barrister, self-taught in history and geography, who predeceased its publication. It was without precedent in the English market and in a format that was losing favor. Edward Quin's *Historical Atlas* triumphed over these handicaps, appearing in reprints and new editions for close to three decades.[75]

Quin's collection is marginally geographic; its world map is so small that political changes are hard to track. It took the intervention of William Hughes, an established cartographer also proficient in engraving, to enlarge the maps and enhance their geographic respectability.

Quin's posthumous success seems to have been artistic, not geographic or historical. His work appealed to the emotions and, if one chose to think so—Quin himself expressed no opinion—offered a vision of progress extending over the globe. His distribution of attention is closer to the Paris draft and Dupré than to Benicken; medieval history (395–1498) is allotted six maps, modern only five. Yet the maps assigned to the Middle Ages have little except historiographic interest: how the period is treated by Quin's main decorative and narrative device. He clearly informs the reader that the clouds circling and confining the maps relate to the history of geography; the parts of the globe obscured by dramatically somber billows are those regions not yet known to statesmen and scholars. Their retreat signifies annexation to the known world, not the march of civilization or enlightenment.[76] He remained faithful to this program even after the fall of Rome.

Quin's meteorology was well suited to illustrate the widespread notion of medieval times as a "Dark Age." Some regions uncovered in Roman times might, as proof, have been shown again covered over by clouds between, say, the seventh century and the twelfth or thirteenth. If this temptation ever came upon Quin, he resisted it. The wavy edges of clouds in the maps for 337, 395, 476, and 814 leave the uncovered area of the globe stable and unchanging.

Then in 912 (when the empire instituted by Charlemagne ended) the billows part toward the northwest and reveal Iceland for the first time. The gloom moves only outward, never turning back. To judge from a reading of Quin's summary of events, his knowledge of medieval history was imperfect even by the standards of the time, but he abstained from translating "Dark Ages" into an image of retrogression. His declared program was that the clouds should simply document the advance of geographical knowledge. He did not stray from this course.[77]

The Middle Ages in Maps of National History

From an early-nineteenth-century perspective, the exceptional influence of Kruse and Koch leads us to expect that future historical atlases would be European, rather than universal. This outcome yielded very quickly to an unanticipated alternative. Works mapping the history of one country, rare and isolated in the past, began to outdistance other types. Subjects from national histories even became obligatory in atlases of comprehensive history. National collections multiplied in the 1820s and never let up thereafter.

Kruse and Koch, without setting in motion a wave of atlases of Europe, nevertheless had a few imitators. One of these works on Europe was a sound, early, comprehensive historical collection; another was precociously specialized in modern history.[78] Europe was the occasion of greater originality than this. The famous geographer Karl Ritter published *Sechs Karten von Europa* in 1813. His wholly thematic work, disregarding history, set the example of using maps to chart natural features: vegetation, animals, mountain heights, and more.[79] Long after Ritter's work and within about a year of each other, the Frenchman Maxime Denaix and the Franco-Austrian Constantin Desjardins produced atlases of Europe conspicuously tilted toward thematic maps but combining them with historical ones including the Middle Ages. From a twentieth-century standpoint, their designs are very forward looking. Denaix was sufficiently appreciated that his atlas of Europe was posthumously reprinted; how Desjardins fared is less certain. Neither set an example.[80] Various thematic maps, especially of physical geography and mountain peaks, became commonplace in collections of various kinds; but historical atlases with a profuse thematic component did not take off. The chronological arrangement of national collections was better received.

France, assiduous in mapping its past, had every reason to glory in its medieval period. The ex-sailor and distinguished cartographer Adrien Hubert Brué inaugurated the nineteenth-century historical atlases of France.

Production of Brué's long atlas was curtailed by the printer's death, so that the volume contains only twenty-four maps—half as many as the title page promises—and covers history only to the very beginning of the sixteenth century. His foreword adds to the chorus deploring the condition of medieval geography: "modern geographers have involved themselves only very little with this intermediary epoch." Singling out Koch and d'Anville for praise, Brué leaves Kruse unmentioned and, in a footnote, dismisses Rizzi-Zannoni's atlas as being a negligible "trial run" (*essai*). Brué, no medievalist, or historian for that matter, was assisted by Joseph Guadet, who contributed a substantive introduction. Guadet had won his spurs in medieval research and later compiled atlases of his own.[81]

Brué's large, very clean maps disregard physical relief. The principal waterways and a limited selection of place-names accompany careful outlines and boundaries, quietly picked out in color. The main plan calls for portrayals of the frontiers of France, internal and external, in the various reigns, including Charlemagne's empire. Brué's sequence differs from Rizzi-Zannoni's in details; he is especially concerned, for example, with the Frankish partitions (as Delisle and Liébaux had been), but the level of complexity is not markedly greater. Besides showing France, Brué provides readers of Crusade histories with two maps of Europe and the Levant, delineating crusader principalities among the other countries of the time. No tracks of crusading armies are drawn. A trace of fantasy occurs in the map for Philip I (1060–1108): the Channel is filled from France to England with three parallel lines of ships, in bird's-eye view but recognizably contemporary men-of-war; they are the "Armée Navale de Guillaume le Conquérant en 1066." Brué's armada is disciplined into tight columns and suffers no casualties (fig. 30). Except for these adjustments, the vignette precisely corresponds to the casual evocation of William's crossing in Rizzi-Zannoni's scorned atlas.[82]

Fourteen extensive atlases of French history fall between Brué and 1860; production drops off oddly in the last decade of the Second Empire (1860–70).[83] They tend to focus on dynasties, individual kings, and territories,

---

Figure 30 Adrien Hubert Brué, *Atlas géographique, historique, politique et administratif de la France* (1820–28). Rizzi-Zannoni's atlas of French history (fig. 16) casually indicated the Norman invasion of England on its map concerning Philip I of France. After the revolutionary and Napoleonic periods, the conquest of England acquired a more serious significance. History would have taken a different turn if William's feat had been duplicated by Napoleon. Brué, who had little use for his precursor, Rizzi-Zannoni, and was not inclined to levity, interrupted the plainness of his maps to make room for a bird's-eye view of three columns of (William's) ships of the line on their way to subjugate Perfidious Albion. (Geography and Map Division, Library of Congress)

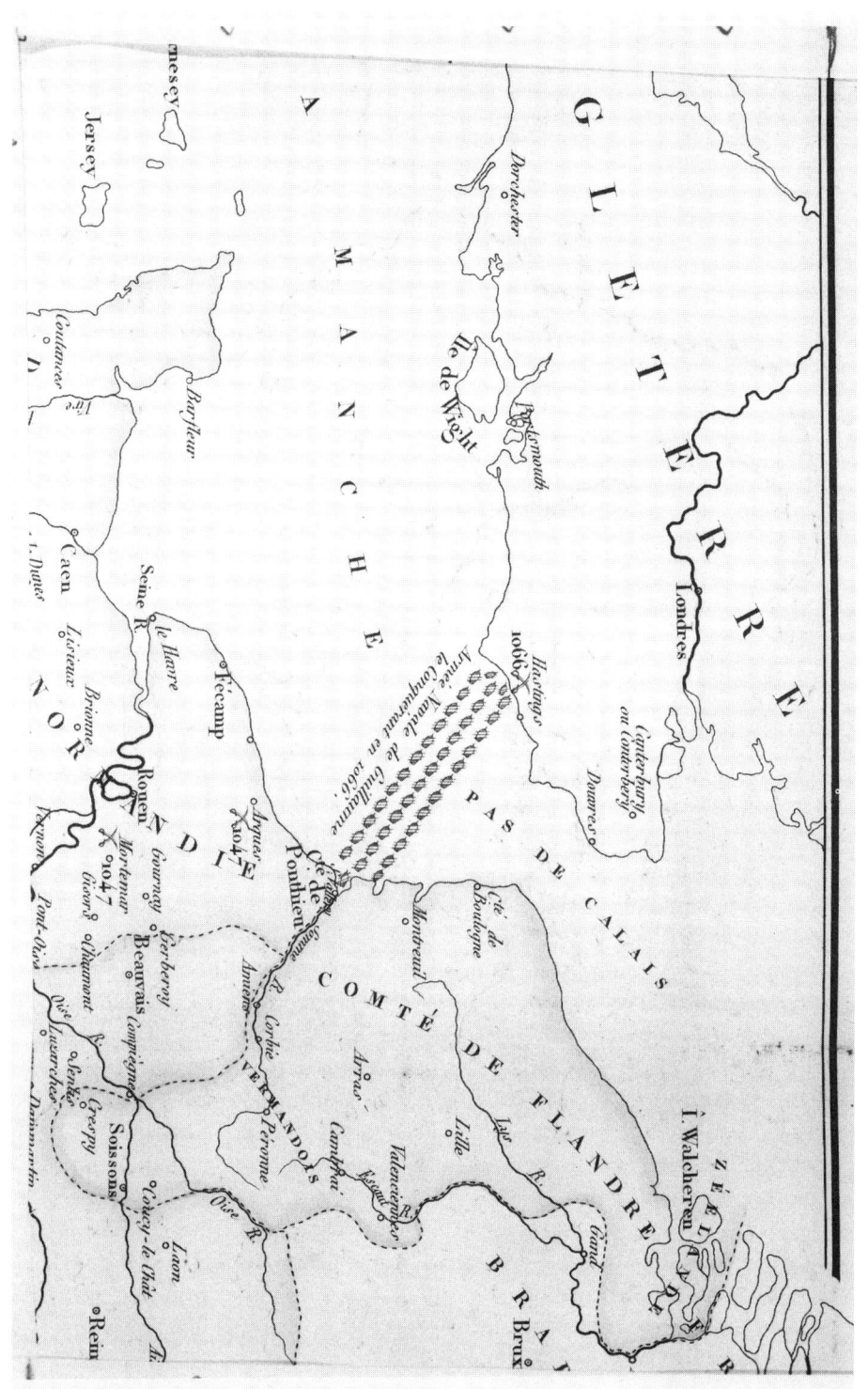

notably the royal domain and large principalities. Rizzi-Zannoni's atlas fore-shadowed most themes in nineteenth-century collections. Where the first "French," or Merovingian, kings are concerned, the preferred subject, already observed, involves the customary sharing out of the realm among the sons of the deceased king. Brué, Denaix, and Clausolles feature two major partitions (511 and 561), whereas Dussieux manages to chart as many as eight. Duruy had the good sense to skirt this overworked subject altogether; he features the 630s, widely considered to be the Merovingian zenith. In Dufau, finally, one map of the partition at Clovis's death is balanced by one of the reunification of the kingdom in 613. The theme of deplorable successoral partitions, down-played by today's historians of the Merovingians, has not gone away: it occu-pies an attractive multimap page in Kinder and Hilgemann's *Anchor Atlas*.[84]

Only one of the French atlases lacks maps of Charlemagne's empire and, separately, of its dissolution. Fragmentation is generally associated with the Treaty of Verdun in 843, but one author adds a map for 870 and a few others substitute 888, the year in which Charles the Fat was deposed and many no-blemen from outside the immediate Carolingian family were proclaimed kings on the margins of the empire. The near unanimity over these two sub-jects is hardly surprising, regardless of what date is chosen for the second: both have obligatory places in atlases of general or European history. Few French collections take notice of the Carolingians before Charlemagne; maps marking the death of Charles Martel (741) or the elevation to kingship of Pepin I (751) are rarities. Strangely, the second Frankish dynasty, with its ra-diant Charles the Great, usually rates fewer maps than the first.[85]

Rizzi-Zannoni's atlas contained twenty-four special maps thematically tracing the waning and waxing of the royal domain. Although this subject is not quite so central to the seven French history atlases (to 1870) with explicitly thematic maps, portrayals of "feudal France"—meaning the delineation of territories outside the kings' direct rule—are a favored subject. Only Denaix, in an unusual collection, offers two successive views of the medieval domain of the French kings.[86] Isolated themes include crusading times and tracks of Crusade expeditions; French ecclesiastical geography (curiously rare); and royal acquisitions of territory at the expense of the Romano-Germanic Em-pire.[87] In thematic as well as political maps, changes of boundaries attract in-ordinate attention.

Five or six compilers assign a map to the advent of Hugh Capet. The many other members of the third dynasty were gradually narrowed to a few fa-vorites. Rizzi-Zannoni had been laboriously comprehensive; Brué also wel-comed large numbers. Greater selectivity then set in. The apparent favorite of later mapmakers was Louis XI (†1483): he closed the Middle Ages and could

be credited with humbling *la féodalité*. Most attention after him went to
Philip Augustus and his grandson, Saint Louis, a tribute to the resounding
successes of their reigns. Philip VI, the first Valois, distantly follows. Most neglected are the kings of the Hundred Years' War and those epitomizing the infancy of the line (late tenth to twelfth), except for the eponymous Hugh.

The history offered by these atlases is rudimentary and traditional. If any
improvement of Rizzi-Zannoni's forgotten forerunner is discernible, it resides less in the introduction of new ideas than in conveying traditional ones
with fewer maps. France had the advantage of having a myth of national
growth and consolidation that was well adapted to cartography. The theme,
however naive, of the decline and reemergence of central power—Roman,
Frankish, royal, revolutionary, and Napoleonic—could be steadfastly elaborated. It was (so it seemed) the principle guiding more than a millennium and
a half of national striving. If the unequaled production of historical atlases of
France reflects the depth of French patriotism, some credit might be given,
too, to the peculiarly visual appeal of territorial growth.

Atlases whose contents are national did not necessarily make this focus
clear in their titles. Most German or French atlases with titles suggesting that
they are general or universal unblushingly color their selection of subjects in
indigenous ways. Rupp's *Charten für die Geschichte* starts with the Germanic
invasions and the Frankish period, then chooses intervals from German history alone: the imperial coronation of Otto I, the end of the Salian dynasty, the
accession of Rudolf of Habsburg, the election of Sigismund, the death of
Charles the Bold (at the battle of Nancy, 1477). Like Rupp, other authors before 1871 showed that Germany did not have to be politically integrated in order to be clothed with a common past.[88] Spotlighting one's country was in-
stinctive. The readers of Babinet's *Atlas universel* were treated to a display of
feudal France since Hugh Capet; those of Kutscheit's *Historisch-geographischer
Atlas* were exposed to Hohenstaufen Italy. General collections from Germany
paid attention to the creation of the Swiss Confederation and to the division
of the empire into administrative circles at the close of the Middle Ages. Neither subject was noticed by French atlases of the same kind.[89]

The medieval history of France and Germany proceeded from the same
beginnings and shared certain themes. The barbarian invasions and/or kingdoms were common ground; so were the Empire of Charlemagne (with or
without the caliphate) and its dismemberment, associated either with the
Treaty of Verdun (843) or a moment later in the century. The notable "European" subjects after the ninth century were the Crusades and the Mongol invasion. Despite these points of convergence, comprehensive collections of
this kind still reflected a distinct national character.

Because general historical atlases compiled in Germany favored subjects agreeable to their clientele, the absence of national unity detracted little from historical mapping. The greater difficulty was that the explicitly German atlases, which were few, lacked the narrative continuity made possible in France by an ancient monarchy and steady territorial acquisitions. Three works, at ten-year intervals, illustrate the problem. Pompper's four-sheet atlas gives equal space to general and German history; it shows the Romano-Germanic Empire (Italy included) from the Ottonians to the Hohenstaufen, followed by the same under the Habsburg and Luxembourg dynasties. Kutscheit features three "European" subjects—the barbarian kingdoms, Charlemagne, and the Crusades—followed by Hohenstaufen Italy and the Romano-Germanic Empire divided into circles. The name of Karl Spruner heads a *Schul-Atlas von Deutschland,* perhaps compiled at the publisher's prompting rather than carried out by Spruner himself. Here, at least, the program is explicitly German. The maps of the Merovingian and Carolingian realms that open the volume are there as German entities, as ancestors shared with France. The rest follows closely in the tracks of Pompper and Kutscheit: Germany under the Saxon and Franconian (Salian) dynasties, then the Hohenstaufen, finally the dynasties prominent in the mid–fourteenth century, and the Burgundian ascendancy in the fifteenth century. So presented, Germany suffers less from territorial fragmentation than from dynastic discontinuity. Whereas the French royal "races" flowed into each other, sustained by the majestic Capetian succession, the German epochs *seem* to exist as interrupted segments.[90]

France and Germany were hardly alone in equipping themselves with national collections. Saxony and Prussia, kingdoms if not nations, were among the earliest in the century to do so, as has already been shown. With the fifteenth century as starting point, their scenes are mainly from modern history. In the decades to come, Prussia was second only to France in the attention it received from mapmakers. Spruner's acclaimed atlas of Bavaria (1838) improved the scholarly character and medieval content of these collections.[91] In 1829, when Thomson's Crusade map was published in Edinburgh, an eight-map historical atlas of Scotland with three items for the Middle Ages also appeared. Attention in the 1830s shifted to the Netherlands (earlier provided with a notable stock of maps for history). One atlas about the reunited Low Countries appeared just when they were breaking up, and another, of an independent Belgium, followed a few years later.[92] The 1840s saw national collections in Austria, Italy, Russia, Poland, and Britain.[93] Hungary, Switzerland, and Sweden were somewhat overdue but outstripped Spain, which until 1879 had to make do with an extract from the collection of the Frenchman Houzé.[94]

Some of these atlases were negligible. William Hughes, whose work is the most disappointing, offered maps of Roman Britain and Anglo-Saxon England. These depictions had been available since at least 1600 and obtained few improvements from Hughes. The mediocre historical maps of Scotland shine by comparison.[95] Most other sets recommended themselves in various ways. The two collections from the Netherlands dealt with much the same history as Franco-German atlases but presented it from an illuminating regional perspective. For them, the Frankish kingdom of Austrasia and the Lotharingian offshoot of the Carolingian realm mattered much more than the Empire of Charlemagne; mutatis mutandis, so did Lower Lorraine and the enlarged Burgundy of the late Middle Ages. The precocious *Atlante geografico degli stati italiani* of Attilio Zuccagni-Orlandini appeared a year after his atlas of the Tuscan state. The all-Italian work vents the irredentist spirit of the age by labeling Corsica as "Italia Francese" and Malta as "Italia Inglese." Many of Zuccagni's maps claim to show regions as they were under the Romans and in the Middle Ages, without explaining what had been done to make them so. The work is more suitable as a historical document than as a reference work. A decade later, Cesare Maggi published much more conventional maps showing Italy from the Lombards to the Hohenstaufen and into the last medieval centuries.[96]

Joachin Łelewel, first famed as a historian of Poland, won a second reputation for his studies of medieval cartography, which he greatly advanced while a needy exile in Brussels. Differing from the atlas of modern Polish history by Stanislas Plater, Łelewel's atlas took all of Polish history as its program and gave ample attention to the medieval centuries. Its maps were accompanied by a two-volume narrative.[97] The *Atlas till Sveriges Historia* by C. F. Wiberg and T. von Mentzer is possibly the most visually unusual and appealing of these works. It was not first in its field. There had been a *Kort historisk atlas* a few years before, with established subjects such as Norsemen abroad in the Middle Ages. Wiberg and Mentzer saw matters more comprehensively and took pains to construct their maps from northern materials and events: Sweden on the basis of *Ynglingasaga* (a thirteenth-century account of "earliest times"), the region in pagan days (with an inset showing trade and missionary routes), the activities of the Northmen (vikings), and the Union of Kalmar. The compilers' choice of contents was expressed in outstanding artwork. Something about the atlas must have displeased its sponsors; subsequent Swedish collections offered a more conventional repertory. However that may be, Wiberg and Mentzer's first atlas makes a lasting impression.[98]

The Russian collection of I. M. Akhmatov, based on the history of N. M. Karamzin, deserves special attention for its size, old-fashioned ordering, and novel perspective. Its seventy-one very handsome, carefully colored maps start with twenty sheets devoted to Russia before Russia. The properly Russian ones inch forward from the ninth century to 1505, in a manner reminiscent of eighteenth-century sequential atlases. The first group is called "Maps for History," whereas the others are individually named. The subject of each map is spelled out briefly at the end of the atlas, together with individual scenarios for a new series of maps extending from 1505 to 1825. This second series may not have been realized.

For viewers accustomed to west European maps, Akhmatov's atlas is surprising and refreshing in its resolutely Russian focus. Akhmatov's Russia, instead of stretching out vaguely on the right side of the continent, is consistently central, visually subordinating other lands to west and east, as called for by each map subject. One longs to absorb more of its outlook, notably its carving out of medieval moments. Language is the obstacle.[99]

In the rush to record national pasts, a few atlases retained traces of the eighteenth-century fascination with Asia in the Middle Ages. Almost everyone paid attention to the Crusades, and almost as many to the Mongols. On the other hand, the mediocre atlases of Europe by Henri Selves were unusual in incorporating inset maps of Asia, borrowed from the applicable pages of Klaproth's *Tableaux historiques de l'Asie*. Even Selves seems traditional by comparison with Rühle von Lilienstern, who, as earlier noted, featured an almost wholly Asiatic Middle Ages in a five-map collection of historical maps for schools.[100] Francesco Marmocchi quite properly leaned (like Malte-Brun) toward the east in a collection emphasizing the history of geography rather than general history. Besides the Crusades, he featured the geography of al-Idrisi and, in a somewhat later atlas, paid attention to the caliphate (without Charlemagne) and the divisions of the Islamic states.[101]

The nineteenth-century collections most old-fashioned in choice of subjects appeared within months of each other in 1859 and 1860. The earlier was by Wilhelm Pütz (1806–77), professor at the Catholic gymnasium in Cologne and a prolific author of schoolbooks of many kinds; the other, by Baquol and Schnitzler, came from Strasbourg. After the opening items, both the German and the French works confined themselves to Charlemagne with the caliphate, the Crusades, and the Mongols. Such a selection should probably not be read as expressing unusual fascination with Asia. It attempted to choose subjects relevant to the whole of Europe rather than to particular countries.[102] A more focused concern seems discernible in the maps done by Dufour to illustrate the church history of Rohrbacher. A map of the conquest of the Byzantine Empire by Islam (mainly seventh century) is joined by a group of three: the Crusades, the Mongols, and the Orient in general from 636 to 1453. An even more admirably comprehensive selection occurs in a large Dutch atlas by G. J. Dozy.[103] In this case, as in Dufour/Rohrbacher's atlas, the guiding thread comes from up-to-date histories of the Middle Ages. The eastern leanings *faute de mieux* of eighteenth-century historical atlases had been overcome.

LARGE ATLASES WARMLY WELCOMING THE MIDDLE AGES

Three historical atlases from the 1840s that are not difficult to find today each contain more than 100 maps.[104] Such affluence was new. Early in the century, Koch's well-liked scheme had surveyed the Middle Ages in seven maps. When 100 maps were allocated to all epochs of the past, about one-third could be assigned to the medieval millennium. Coverage on this scale had occurred before and would later be exceeded (Shepherd's *Historical Atlas* has fifty-two medieval maps of all sizes). Nevertheless, these three spoke well for the liberal coverage major atlases provided between 1840 and 1870, presenting maps that simply did not exist at the beginning of the century.[105]

Not all historical atlases in these three decades were profuse in maps. What matters more than quantity is that compilers, in whatever country, decided to arrange the contents in geographical order and to divide them into a general and a particular section. Sometimes, the parts are "Continental and regional"; Karl Spruner arranged his major *Hand-Atlas* of Europe in this way. Compilers of shorter collections, such as Julius Löwenberg, preferred to distinguish "general and special"; they could not always afford multiple maps of one region. Antoine Houzé, an early exponent of the new scheme, skipped general coverage altogether and launched directly into sections for countries. When possible, he marketed individual regional atlases detached from his 101-map ensemble. Some collections clearly distinguished the two sections, as Löwenberg did; others adopted a different arrangement (possibly chronological) and relied on the choice of subjects to imply a division into general and particular parts. A miniature example, by Victor Levasseur, began with the barbarian invasions (after the *Atlas Lesage*) and Europe from 1074 to 1300, then added three special maps: the growth of the French royal domain, Spain before the expulsion of the Moors, and Germany from 843 to 1273. Levasseur's selection is dismal as historical atlas-making, but it compactly illustrates the two-part scheme. The new arrangement might be described schematically as starting with a Koch-style set of all-European maps, as done very precisely by

Victor Duruy, and combining it with the national collections more recently in vogue. Whatever the ingredients, almost all the midcentury atlases with ample historical components were grouped geographically and in two parts.[106]

About eighteen atlases of this kind came onto the market between the late 1830s and 1870. Their order of publication can be only approximated. Installments of Spruner's *Hand-Atlas* were in circulation for nine years before publication was completed. Houzé's 101-map atlas, though said by an early bibliographer to have been first published in 1837–38, is hard to find (even as a bibliographical entry) in a copy earlier than 1841. More than other early examples, Löwenfeld's atlas resembles Spruner's in arrangement. But Löwenfeld died in 1839, and Valerius Kutscheit completed his work (1842). These circumstances suggest that Löwenfeld, whose map style has individuality, worked independently, uninfluenced by the little of Spruner's atlas that he could have seen. Imitation of Spruner no doubt took place after the success of his atlas was assured, from the mid-1840s on. Earlier signs of copying from him are not obvious.[107] A cluster of atlas-makers adopted a geographic arrangement with two parts, general and particular. However this came about, it need not have been in Spruner's footsteps.

Owing to the repetitiveness of atlas contents, traces of inventiveness offer welcome relief. All the subjects that have been noted above as common return again and again, especially in reference to the early medieval centuries. Every now and then, Houzé succeeds in shaping geographical order into a new image. His maps are not handsome or stately, and the thirty scenes he allocates to medieval France contrast patently to the twelve granted to Germany and six to Spain. He blunders by labeling one map of the East Roman Empire as being "during the occupation of the Bulgarians": there was no such occupation except for limited districts. On the other hand, Houzé sometimes focuses on a subject not commonly depicted. The section for Greece and Italy mutes the fall of the West Roman Empire and makes astonishingly clear that Byzantine and Italian history were entwined during many medieval centuries. In this case, geographical order definitely helps historical understanding. Another remarkable map occurs in the section called "La Russie, la Suède, la Norvège et le Danemark." This setting, well suited for the exploits of Charles XII of Sweden, opens "à la fin du V siècle" with Houzé's basic barbarian invasions map, full of black-and-white arrows. The "Gothones" go from the Continent to Sweden (time unspecified); next, the Goths move from Sweden to the Continent as a block; then the arrows divide. An explanation is supplied in a big box: European Russia was full of Goths; then the Huns arrived in countless numbers and pushed them on; and the Goths attacked the Roman Empire. Houzé earns no applause for this crudely oversimplified explanation or for having the "Gothones" move from the Continent to Sweden prior to moving back. The east European perspective of his map is remarkable, however, and offers an incentive to stop thinking of the Goths as a "Germanic people" and, instead, to naturalize them as south Russians.[108]

Medium-size atlases were more common than works on the lavish scale of Houzé and Spruner. Löwenberg's, finished by Kutscheit, led the way in the bipartite format. The arrangement of this collection illustrates a defect found elsewhere, namely, maps expected to depict very long intervals of history. The record-breaker among Löwenberg's "Übersichtskarten" runs from Charlemagne (†814) to the end of the Crusades (1291). Spans of two or three centuries are normal for the four other maps. The same approach, slightly moderated, is taken by the "Specialkarten." Examples are Palestine in the twelfth and thirteenth century, France and Spain and Britain throughout the Middle Ages, Switzerland to 1789.[109] Perhaps Löwenfeld was not allowed to supply as many maps as were needed to implement an ambitious two-part scheme. Whatever the explanation, exorbitant time spans are a blemish only too often built into maps for history.[110]

Less ambitious atlases could achieve more satisfactory results. Victor Duruy did not announce two clearly delimited parts; he lined up the maps as seemed best, allowing for both chronological and regional considerations. Six were allotted to Europe, two to the Muslim world, two each to Germany and Italy, and one each to Spain and Britain. In most cases, a moment was defined; Italy "during the struggle with the Hohenstaufen" (ca. 1150–1250) is more typical than the very specific "1270" for Spain and Britain. "Germany under the Franconians" was deemed adequate even though it meant omitting the Saxon emperors and being less than comprehensive. Duruy's atlas created a coherent whole out of maps-of-one-moment, in preference to achieving a false continuity by means of multiple-century maps labeled "from A to B, from B to C," etc. It was an intelligent procedure.

Valerius Kutscheit had the misfortune to publish his main atlas of the Middle Ages just when Spruner's final installments were issued. A Bavarian reviewer shrugged him off: he just was not Spruner. Kutscheit, an atlas specialist with classical and geographical collections to his credit, championed the medieval period and disapproved of the widespread teaching that it had been a dark interlude. He first expressed such opinions in his never completed *Historisch-geographischer Atlas des deutschen Landes.* Unwittingly echoing a sentiment expressed in France in the early 1700s, he considered the Middle Ages to have much more bearing on modern conditions than antiquity. He complained that the latter had, until recently, occasioned much more study and mapping.[111]

Besides completing Löwenfeld, Kutscheit produced three atlases with medieval coverage. The contents closely overlap down to the ninth century, then vary in minor ways that need not be tabulated here. Kutscheit is not flawless. For example, a map for A.D. 700 shows a Burgundian kingdom separate from the Frankish one and Byzantium still having an enclave in the Visigothic kingdom.[112] Spain and France are neglected by comparison with the Romano-Germanic Empire. But Kutscheit is admirably concerned to play up the edges of Europe even at the expense of the core. More than half the maps of his main medieval atlas have this emphasis, among them, Scandinavia, the Slavic east, the Mediterranean and, separately, the Near East in the Crusades, Byzantium from the eleventh century, and late medieval southeast Europe. A selection like Kutscheit's conveyed a message that his peers, in their Eurocentrism, were tending to forget.[113]

Three modest collections illustrate other variables. P. Clausolles resembled Duruy in pedagogical productivity, if not in fame. He eventually undertook an atlas of French history with Philibert Abadie, whose help he also had in the medieval one that concerns us. Perhaps because the French volume was already planned, Clausolles kept his wider-ranging atlas genuinely European. Almost unique in omitting Charlemagne, he seems to have organized the opening section in such a way as to touch on each century.[114] A more detailed, country-by-country focus took over in the twelfth and thirteenth centuries: France under Philip Augustus, Spain under the Almohads, the last Crusades, Italy in the time of the Guelfs and Ghibellines, Germany under the Hohenstaufen. At the close, time regains ascendancy, with maps of Europe in the Hundred Years' War and at the end of the Middle Ages. The coverage and contents are wider and more complex if the insets are taken into account. With them or without, Clausolles assembled a creditably catholic collection.

The shoddy and unattractive volume of Friedrich Reinhold Schaarschmidt is cloaked in the vague title "small atlas as basis for historical instruction." Hardly any of its rather few maps look beyond the German Empire and its dependencies. The notable exception is a thematic map for medieval commerce. Twentieth-century atlases commonly feature this subject. Gatterer's inaugural attempt is lost; Schaarschmidt's thematic map, however indifferent in appearance, may deserve credit for being the second attempt and the first to survive. Schaarschmidt is notably parochial, but hardly unique, in depicting the Middle Ages along German lines. Much the same occurs, exchanging German for French, in the longer and commercially successful collection of Drioux and Leroix.[115]

Edward Gover, a London publisher, may have been the first mapmaker in England to recognize that the Middle Ages needed its own repertory of maps.

The medieval atlas he compiled with (licensed) Spruner maps shows much originality in coloring and remained influential in England from this standpoint. Gover offers the usual European maps of overlong periods, followed by regional maps also addressing regrettably long intervals, such as "Italy from the seventh century [to the thirteenth]." At the end, there comes a little British collection, showing Britain from the Anglo-Saxons to the sixteenth century, in four maps on two sheets, plus Scotland and Ireland. Alongside these somewhat detailed portrayals, the rest forms little more than a blurred background.[116]

The 157-map *Atlas général de géographie physique, politique et historique* by Louis Dussieux (1815–94) is almost certainly the fullest and most comprehensive historical collection produced in France up to 1870. Close to 30 percent of the contents concerns the Middle Ages. The year when the full atlas was first published is unclear; quite a few maps are dated 1848 and 1849, whereas the title pages announce 1853 and 1856. In arrangement, Dussieux's work looks very much like Spruner's *Hand-Atlas* of Europe, an impression confirmed by closer inspection. Spruner's map of Britain to 1066 has an inset showing the ethnic composition of the island in the eighth century; in a clear echo, Dussieux's inset is a "petite carte historique et ethnographique des Îles Britanniques au huitième siècle." Spain and Portugal rate just as many maps in Dussieux as in Spruner; the sets agree in subject until after no. 4, whose end point of 1453 compares with Spruner's 1479; and the latter's sixth map includes the subjects of Dussieux's fifth and sixth. Dussieux rarely strikes out on his own. Although falling far short of providing the beauty and quality of Spruner's maps, he made available the latter's coverage of time and space for a French public.[117]

Dussieux's one emphatic departure from the German model involves a lopsided thirty-eight-map section on France, within which one large thematic group stands out. Five maps of ancien régime civil, judicial, and ecclesiastical districts (nos. 56–60) are followed by eleven that plot minerals (three maps); geology; climate; forests; bovine, equine, ovine, and porcine species; viticulture; and agriculture.[118] These displays are a welcome extension of the historical atlas genre, but is Dussieux setting an example for all to emulate? He leaves us wondering whether these themes, which concern only the present day, are displayed because they are vital to an atlas of this kind—he supplies them for no other country—or merely because the land in question is France.

Spruner's plan was not always imitated quite so literally as it was by Dussieux. C. E. Rhode, an experienced atlas- and mapmaker, was responsible for a German school atlas deftly arranged as a trimmed version of Spruner. First marketed toward 1860, it had smaller dimensions and fewer maps than its model but resembled it closely enough to be rewarded by brisk sales even in North America. Within a decade, Rhode was closely imitated in the Netherlands, another compliment.[119]

History was not Rhode's strong suit. The choice of subjects is unexplained and, though a text accompanies the maps, it supplies appropriate names but sheds little light on anything else. The contents, however, are set out with a firm hand, total fidelity to Spruner's distribution by geography, and no obvious copying of Spruner's maps. The visual attractiveness of the volume adds to its appeal. Five maps carry the course of events to the division of the Carolingian Empire. Geographic distribution then takes over, with a single map for the medieval empire, followed by Britain, France, and Spain. The next move is eastward toward Slavic lands and German eastern colonization. Finally, the Near East is featured, with Byzantium, the crusader states, the Islamic caliphates, and the Mongols.[120]

Rhode did not attempt full chronological coverage; by being content to sample, he could provide more precise maps and avoid overloading. But was the arrangement ideal? After the five openers, the succession of maps is a plaything of geography, deliberately devoid of meaning. Any connections there might be between maps have to be sought by skipping from section to section. Users may profit from searching for themselves, but more deliberate linkages by the compiler might have made for better pedagogy. It is hard to tell.

Among the large, midcentury atlases we have been concerned with, Spruner's stands out for size as well as quality. If reckoned from the initial installment, it is the earliest of the bipartite type we definitely know of. By the installment of 1846, it shared shelf space with others, both French and German. His work is in a category by itself.

Spruner's *Hand-Atlas* and the Problem of Geographic Order

After 150 years, Spruner's atlas is still a force in the mapping of history. His name is still recognized in a field whose many workers have been largely forgotten. His decisive work, usually in the third edition of Theodor Menke, is on the consultation shelves of major libraries.[121] An inventory of the many Spruner abridgments, school versions, translations, and other offshoots awaits a bibliographer's zeal. Where success and readership are concerned, the only figure comparable to him is Las Cases, whose star has dimmed to near extinction while Spruner's continues to blaze.

The success of Las Cases is somewhat mysterious, that of Spruner is not, at least insofar as the dictum about better mousetraps holds true. Among the burgeoning historical atlases of the 1840s, Spruner captured the high ground. His maps were large, admirably crafted, geographically impressive, and histori-cally sounder than their competitors'. They even took serious account of phys-ical relief. Distinctive, appealing, and monumental, they were issued by the re-markable firm from whose presses, at the time, came the standard-setting atlases for geography. Spruner filled the historical slot. Perthes of Gotha, not content with an unrivaled product, also marketed it aggressively.[122]

The European volume of Spruner's *Hand-Atlas* has no "medieval" section as such; instead, more than half its maps depict segments of the Middle Ages. Spruner opens with Europe as a whole, a division that would be more widely circulated than the rest because it was singled out by school adaptations and a wall atlas. Spruner's main atlas branches out from the European continent into seven regional divisions, beginning with Germany and German-speaking lands and ending, via southeast Europe and western Islam, in Hungary. In each division chronological order prevails, modern following medieval. Spruner, unlike Las Cases, equipped each regional group with multiple maps. Europe has an exceptional thirteen to itself. The regional divisions range from a high of seven for greater Germany to a low of two for Britain. So articulated, the *Hand-Atlas* forms a succession of national or regional collections.

The geographical format contributes to the aura of authority that Spruner's atlas retains to this day. Unlike Dupré or Quin, Spruner, in his *Hand-Atlas,* did not set out to unfold history in successive maps or to display the evolving contours of frontiers. Just as its form is that of a geographical at-las of the world, so its design allows consultation by continent and land, the equivalent for reference atlases of a dictionary's alphabetical order. The atlases of geography and of history, twins in arrangement, testify to their parallel, complementary, and authoritative stocks of information. As geographical at-lases provide current places, heights, and other topographical details, so Spruner's *Hand-Atlas* was designed to be opened at a continent, region, and (befitting its genre) *date.* It could then be used to obtain needed information.

The European continent and its component lands, stocked with maps in chronological order, show historical change by a sequence of images. What Benicken and his forebears attempted for all Eurasia was parceled out here among smaller divisions. The Spruner offshoots, with smaller and fewer maps, calmly contract the magisterial order of the original; but later atlases claiming maximum authority, notably Gustav Droysen's and R. L. Poole's, im-itate Spruner's geographical arrangement.[123]

The claim to authority was not hollow. Spruner's maps would do honor to a nonhistorical atlas. Fuller and better executed than those of any (histor-ical) competitor, they convey a vivid, if not precise, sense of the ground. Physical relief is indicated by the "hairy caterpillars" that Kruse had already used. Spruner's relief does not avoid the flaw that, according to modern cartographers, usually comes with this method—excessively linear mountains coinciding too precisely with political frontiers. The blemish seems venial when balanced against the far more customary absence of physical relief. Spruner himself was responsible for the plethora of regional and place-names, some selections taking account of different languages. They are differentiated by a remarkable hierarchy of scripts; the frequent use of left-slanted lettering stands out as a special Perthes mark. Careful attention is paid to ethnography, once by a special map; tribes and other subdivisions are picked out by colored underlines or highlighting.[124]

Spruner chose historical moments with care and discernment. The Koch openers—maps for 395, 500, and Charlemagne—are found in the European section; but the resemblance fades once the various countries come into play. Multiplying maps at will cannot have been an option; Spruner made the most of space that, however generous, must have been rationed. The major western European countries are allotted an average of four maps each for the Middle Ages. Generous provision is made for peripheral regions, including lands explored by the Northmen, the Byzantine Empire (at several moments), Hungary (multiple coverage), the Islamic caliphate (east and west), and the Umayyads of Spain. Almost all the main maps have a complement of inset city plans, battles, regional details, and more.

The *Hand-Atlas* asks too much of single maps. Again and again, long time spans are compressed into a single image (fig. 31). As Kruse had rightly seen, illustrating several centuries at once on the same background is awkward. Instances of this dubious practice occur in Spruner's maps of the most fully illustrated western lands: "Germany from Rudolf of Hapsburg to Maximilian," "Italy from 1450 to 1792," "The Iberian Peninsula from 1257 to 1479"; on the European periphery, "Poland and Lithuania from 1125 to 1386." Each of these maps might be meant to do justice to the whole of these two-to-five-century periods or only to a moment; the reader is not advised.[125] Spruner admired Kruse but thought that his century-by-century intervals had grave disadvantages, defects that were avoided by dividing the *Hand-Atlas* into the most important historical periods.[126] To judge from his multiple-century presentations, both plans have their drawbacks. Whereas Kruse's single-instant maps were too limited, Spruner had no dependable method for compressing centuries into a single map.

These lapses of coherence are not unrelated to Spruner's following in the footsteps of the *Atlas Lesage* (which he despised) and adopting a geographical order. His thirty-four "medieval" maps, if projected on a wide background,

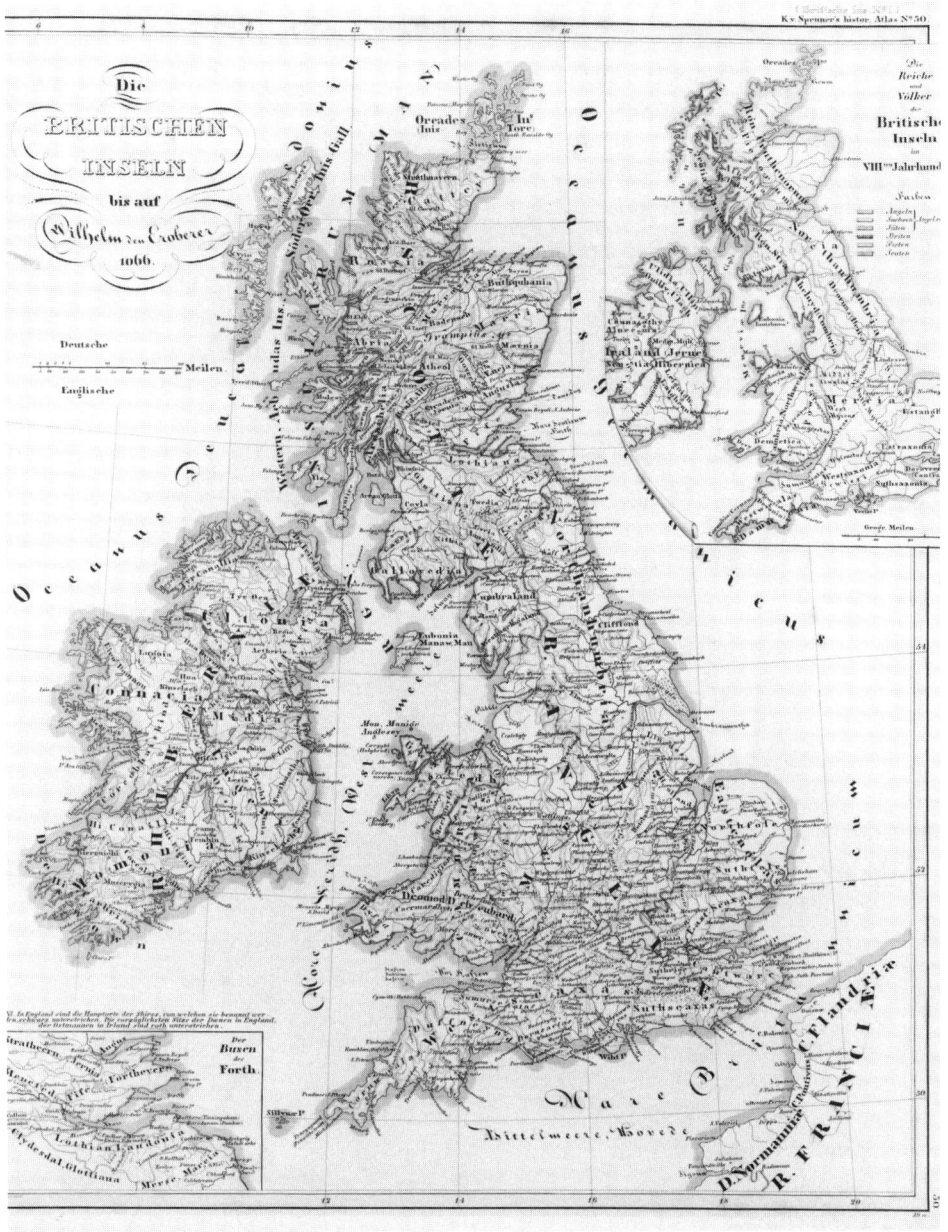

Figure 31 Karl von Spruner, the British Isles under the Anglo-Saxons, to 1066 (1837–46). Spruner's map "Die Britischen Inseln bis auf Wilhelm der Eroberer 1066" is from his *Historisch-geographischer Hand-Atlas zur Geschichte der Staaten Europa's vom Anfang des Mittelalters bis auf die neueste Zeit* (Gotha, 1837–46). After Lambarde's map of 1568 (fig. 2), the Heptarchy was often reproduced by mapmakers unconcerned that it portrayed only the first half of the Anglo-Saxon period. Although Spruner meant to depict the whole pre-Norman age, in a rare lapse he fastened on the Heptarchic epoch (in the inset as well as the main map) and lost sight of the two eventful centuries that Anglo-Saxon history still had to run until the Normans came. (Courtesy of the Map Collection, Yale University Library)

such as Kruse's Europe, might have been distributed among small time periods suited to illustrating historical change. Spruner's primary division was
spatial, however. Each region had to fit its medieval history into about four
maps, and these few had to retrace the same time period as the others. The
outcome is that, to take a simple example, "Germany under the Saxon and
Franconian Empires" in the German division was unavoidably paralleled in
the Italian by "Italy under the Saxon and Franconian Empires." Diluting time
has the effect of magnifying space. If the scale chosen had allowed Italy to be
combined with Germany, the Saxon and Franconian dynasties might have
had one map each. Similar economies could have been achieved elsewhere.

The course taken by Spruner resulted in large-scale geography and generous local details. His achievement is certain. Nevertheless, it would be
interesting to determine, if one could, whether the praise lavished on the *Historisch-geographischer Hand-Atlas* arises from its geographic merit and likeness to topographic atlases or from the contribution it made to illustrating
history.

To Condense so as to Teach

While Spruner's main atlas made its slow way to full publication, its shortcomings for pedagogy were implied and offset by a compiler whose approach
was markedly different. In a decade like the 1840s, unusually prolific in the
production of maps for history, it was too much to expect that even an ostensibly definitive collection would have the last word. Spruner's atlas, a contemporary admirer said, was for the learned only; those wanting an adequate
portrayal of the main changes in the individual realms "from the earliest to
the present times, and in the Orient as well," had nowhere to look except
Löwenberg's atlas, whose maps were worthy but too few in number. These
comments were not speculative or casual; they were written on behalf of yet
another army lieutenant—a Prussian named Rudolph von Wedell—whose
Historisch-geographischer Hand-Atlas in thirty-six sheets could rightly claim
originality in design.[127]

Much in little seemed to be Wedell's program, and not only his. A. von
Freyhold, a commercial mapmaker in Berlin, ventured in the same years to
present "universal history" in one sheet each for the now established three periods.[128] Wedell, less heroically, demarcated thirty-six significant historical
units. Their selections were not startlingly different from those of other historical atlases, but both Freyhold and Wedell, in organizing works of very dif-

ferent lengths, divided the sheets in such a way as to fit onto them all the maps they deemed relevant for the topic chosen.

Freyhold had to be drastically selective. The goal, according to his introduction, was to be compendious and to place everything before the student all at once. He explained in a personal way that, for the Middle Ages, three considerations seemed most important: to set the Crusades and the Hohenstaufen—which he believed to be, together, the medieval high point—as compellingly as possible in the foreground; to portray the Holy Roman/German Empire correctly in its various sizes and forms because it stood closest (as he said) to "us Germans"; and to present an accurate survey of height and depth in the modern sense, a feature that was just as important in historical as in purely geographic maps.

Even though "universal" was treated as synonymous with "German," Freyhold packed more information onto the medieval sheet than might have been expected. The available space was shared among five map areas, not counting insets. Because the barbarian invasions are on sheet 1 (antiquity), sheet 2 starts with the Carolingians, and though the Crusades and Hohenstaufen are preeminent (as Freyhold stipulated), the map concerning them is only slightly larger than the next in importance. Even the Mongols and Ottomans could be fitted in. Freyhold's selection of subjects was not particularly remarkable or limited; small atlases sometimes allotted no more to the Middle Ages than he did. The notable difference was that everything from Charlemagne to Tamerlane was simultaneously accessible to the eye.

Concurrent presentation also characterized Wedell's atlas, though with less drastic compression. Wedell allocated about fourteen sheets to the Middle Ages. The subjects seen all at once were the six or more fitted onto every sheet, each of which was given a definite theme, as, for example, "Map Seven. The Lands of the Mediterranean Sea until ca. 900." In this case, the one sheet contains eight maps, large and small, fitted carefully to form a thematic composite. One of Wedell's ensembles shows Spain in nine panels from the Visigoths to Ferdinand and Isabella (resembling the display of an earlier French atlas)—a mini-sequential atlas but visible as a whole (fig. 32).[129] Wedell was less enamored than Freyhold with the medieval German Empire; it is never featured alone on any of his sheets. One of his particular characteristics is a fascination with the possessions of the crusading orders, mainly those of northeast Europe (suggesting a specifically Prussian interest). Eight groups of maps depict aspects of this subject. Comparably emphasized is the distribution of peoples; six "Völkerskizze" of various kinds, especially in the invasions period, are supplied.[130]

Wedell's space-saving format gave him scope for multiplying maps, in-

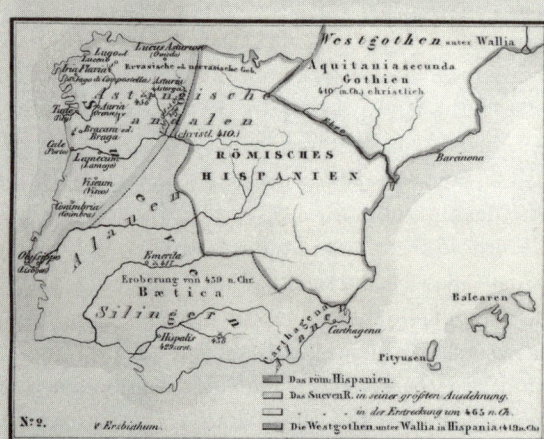

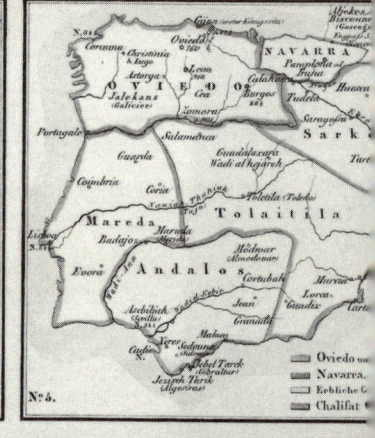

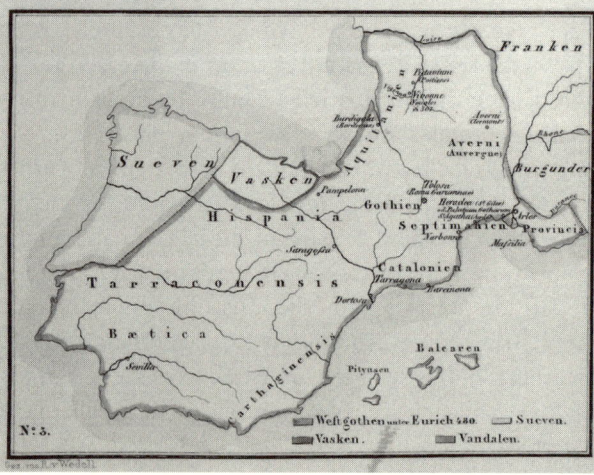

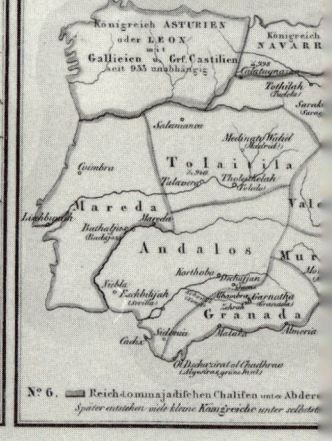

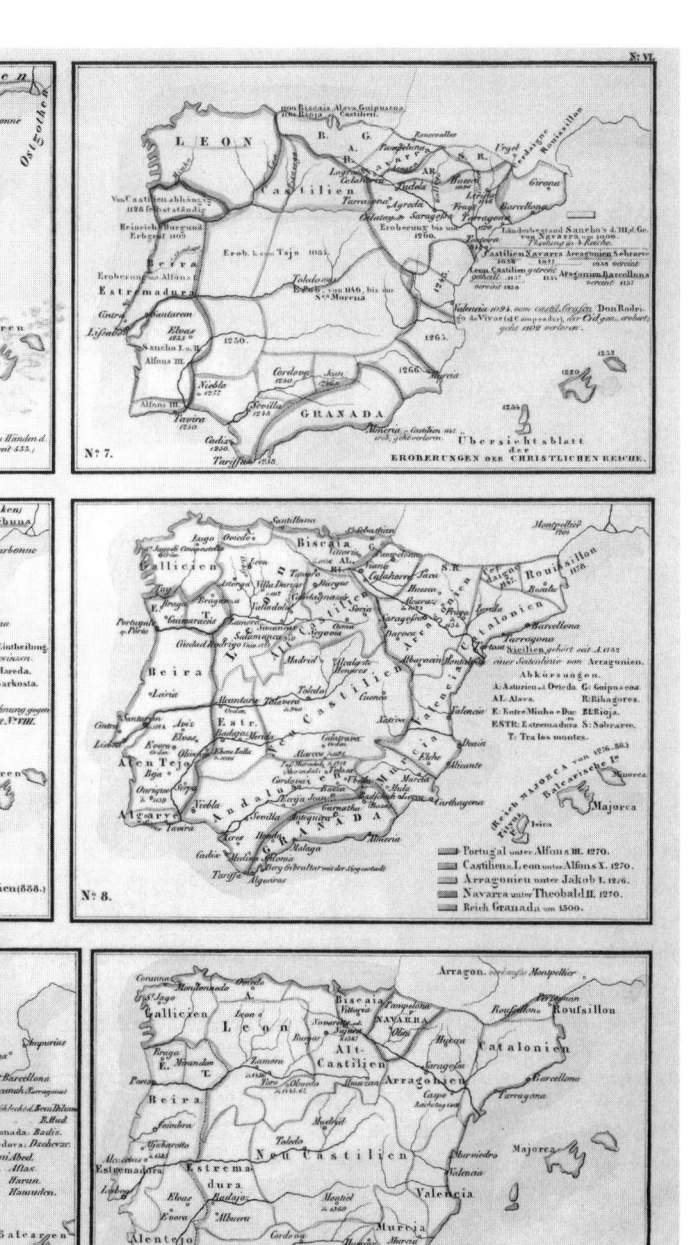

Figure 32 Medieval Spain in Rudolph von Wedell, *Historisch-geographischer Hand-Atlas* (1849). The atlas of Wedell, a captain in the Prussian army, packs a multiplicity of maps into a minimum of space. In plate 6 of his atlas, nine moments of medieval Spanish history are presented as individual units but also simultaneously to readers. Nine maps are a severe simplification by comparison with the course of historical events, but nine are better than one. This layout, which Wedell did not invent, occurs in quite a few nineteenth-century atlases, especially in France. Change is a problem in the mapping of historical information. If many years are recorded on a single background, they form either a misleading synoptic image or a palimpsest that the reader must decipher with effort. The method exemplified by Wedell, though not ideal, avoids the major disadvantages in the portrayal of change. (Harvard Map Collection, Harvard College Library)

cluding subjects, such as "The Neo-Persian [Sassanian] Empire ca. 550," rarely seen elsewhere.[131] The medieval period alone has seventy-five maps. In each, historical detail was preferred to geographical detail. Wedell was not obliged, like Kruse, Spruner, and many others, to keep displaying the same lands. Whereas Kruse, for example, repeated the whole of medieval Europe ten or more times, Wedell felt free to skip dormant areas and devote more space to scenes calling for greater attention. He divided the Middle Ages into three chronological periods—to 900, then 1300, finally 1500—but did not wholly escape the geographic inclination of the 1840s. The most glaring sign of this comes when he omits a comprehensive map of the barbarian kingdoms and prefers to scatter the kingdoms six ways. A collective map of the invasions, exuberant in tracks and other traces of migration, occurs in Karte 3; the Burgundian and Frankish kingdoms in Karte 4; the Huns in Karte 5; the Sueves and Visigoths in Spain in Karte 6; the Vandal, Ostrogothic, and Lombard kingdoms (as separate maps) in Karte 7; and the Anglo-Saxon Heptarchy in Karte 9. Multiplication allows much greater geographical detail than if the kingdoms were on one map. Some oddities, however—as when the Latin Empire and the Mongols (Karte 13) are placed before the early Crusades (Karte 14)—do not result from considerations of geography and are hard to explain.

Imperfections are outweighed by Wedell's vigorous experimentation and originality. He was doing the same thing as Freyhold but had the advantage of having many more sheets at his disposal. Refusing to be controlled by fixed blocks of space and time, he strove to create a historical atlas deliberately trimmed and shaped to suit what he regarded to be the interests of pedagogy.

Freyhold and Wedell attract attention because, in a decade when time-fragmenting geographical order was in the ascendant, they again made a stab at active narration through maps. The example did not go unheeded. Gustav Droysen's imposing atlas of 1886 (already mentioned in another context) is a high-quality folio that reveals authorial ambitions comparable to Spruner's. One of its sheets, entitled "Zur Geschichte der Völkerwanderung," has four maps of different sizes. In the center, the largest map shows the barbarian kingdoms; the one above, slightly smaller, portrays the invasions, with tracks; the lowest and smallest space is shared by two maps conveying relevant details. All four maps include exactly the same lands, in the same orientation.[132] The page reminds one of Wedell rather than Spruner. Droysen's *Völkerwanderung* is expounded in several maps presented to the reader all at once, but in a meaningful hierarchy. Someone consulting the atlas would find an account of the invasions adequately illustrated on one page. The four-map ensemble might serve merely as a companion to a written account, but a thoughtful onlooker, running his eyes over the page, might well realize that the grouped maps also formed a narrative sequence.

HONORABLY MENTIONED

The lines of development in atlas-making emphasized here were not perceptible to all contemporaries. In selecting a historical atlas or map, many buyers in the mid–nineteenth century seemed to scorn the latest innovations in preference for the past. Heneage Howe believed that there was widespread demand in 1851 for parts of the now-rare *Atlas Lesage;* he oversaw a generous selective reprint. A full edition of Lesage appeared in Brussels soon afterward under other auspices.[133] In the year of Howe's initiative, the historian who had helped Adrien Brué a generation earlier published his own second historical atlas of France. Denaix's historical atlas of Europe, more than two decades old, reappeared in a new, hardly changed edition in 1855; months later, Dufour produced yet another map-companion to Malte-Brun's universal geography; and, within the same decade, the son of Kruse's French translator, Ansart, brought out an atlas basically patterned on Kruse's fifty-year-old work.[134] Not even sequential atlases of universal history were wholly passé: the dramatic clouds of Edward Quin were reprinted in the late 1850s, occasionally under a different title.[135] The works that had begun the century retained appeal.

New maps for ecclesiastical history brought to prominence even more ancient prototypes. The French specialty of provincial and diocesan cartography had been perfected by the Sansons. Such maps and atlases had continued to be compiled; several have been mentioned. Ecclesiastical divisions were the largest cycle of thematic maps in Spruner's atlas. They were more prominent still in the main nineteenth-century atlases of church history. Johann Wiltsch's *Atlas sacer sive ecclesiasticus* went only to the Reformation. It was a Perthes publication, highly regarded and widely circulated. Selectivity is its salient characteristic. The basic maps chart the "condition (*status*) of the Christian religion" at five noteworthy moments: 311 (end of the apostolic age), 616–22 (eve of Islam), 1073 (papacy of Gregory VII), 1216 (death of Innocent III), and 1517 (Reformation). Where appropriate, colors distinguish the variety of religions. Subsidiary maps mainly show provinces, dioceses, monasteries, and councils in various parts of the Christian world.[136] The three cumbersome and heavy volumes of *L'Orbe Cattolico ossia Atlante geografico storico ecclesiastico,* by Girolamo Petri, focus even more than Wiltsch did on ecclesiastical organization and do not pretend, except in the title, to retrace its history. Petri's focal point is the Catholic *orbis* in the days of the controversial current pope, Pius IX.[137] To judge by these productions, the program of *historia ecclesiastica* continued along the track laid down in the sixteenth and seventeenth centuries.

Some atlas-makers, heedless of developments in the genre, simply made collections that seemed to be needed. After several centuries in which the Anglo-Saxons dominated historical maps of England, James Birchall compiled *A Series of Maps and Plans (Norman and Plantagenet Period)*, wide enough in scope to stray beyond the British Isles and admit maps of Europe in 1300 and 1400. Birchall supplied a long-needed addition to the available maps for medieval England. The flow of sundry atlases of French history brought to the surface an esteemed one by Duruy without stemming the tide of production, but then weakened. The spirited output of national atlases moved east. In the single decade 1858–68, Töppen, Kiepert, Leeder, and Brecher each compiled one of Prussia.[138]

The advent of two-part atlases did not prevent sets of maps for the Middle Ages along the former lines from being assembled. The works of Rodowicz, König, and Frommann offered representative selections. König, author of a modest (but well regarded) reference atlas for all history, took the greatest pains to give his Middle Ages a comprehensive scope. Alongside seven all-European subjects like "Charlemagne" and "The Crusades," only two maps were distinctly German.[139] Rodowicz and Frommann did not notably differ from him. The former had trouble breaking free of the early Middle Ages; the latter surpassed the others by paying attention to Byzantium, Islam, and the Mongols. All three managed to dispense with the guidance of an authority like Koch; they appear to have made their own selections.[140] In France Marie-Nicolas Bouillet, or his son, was responsible for an *Atlas universel d'histoire et de géographie* whose various editions found unusual favor in North America, perhaps owing to Bouillet's extensive attention to chronology, genealogy, and similar aids for historians. The book is seven centimeters thick. Maps did not fare particularly well in the Bouillet collection. By a generous count, ten are assigned to the Middle Ages. Half are concerned with subjects before Charlemagne, and the "central" Middle Ages are wholly unrepresented.[141]

Maps illustrating books, which I have drawn upon with profit to document earlier centuries, are harder to put to such use in the 1800s. Too many books became available containing too many maps; extraordinary efforts would be needed to collect a suitably representative collection of medieval subjects. Nevertheless, samples suggest that maps-in-books, when found, can widen the repertory found in atlases.

In the case (already touched on) of the Old English Orosius, the Oxford philologist Joseph Bosworth took up where Johann Reinhold Forster and Daines Barrington had left off in the later 1700s and provided a four-map attachment, done by the professional cartographer Edmund Weller, to his *A Description of Europe and the Voyages of Othere and Wulfstan written in Anglo-Saxon by King Alfred the Great*.[142] In a similar way, the maps of Gerhard Schöning were superseded by the researches into early Norse explorations by Carl Christian Rafn. *American Antiquities, or Northern Writers about Pre-Columbian Affairs in America,* with four maps, is representative of his works.[143] Even more exotic was the "Comparative Map of Modern and Eleventh-Century Sicily," produced by a collaboration between the Orientalist Michel Amari and the cartographer A. H. Dufour. Alongside modern place-names, the Arabic equivalents of medieval Sicily are printed in red and in Arabic script. Wide public sale could not have been envisaged. In a more ordinary vein, Édouard Gauttier du Lys d'Arc (1799–1843) gave the rather grand name of "Atlas of the History of the Norman Conquests in Italy, Sicily, and Greece from 1016 to 1140" to a mere two maps illustrating his history of this same subject.[144] Also collected into an atlas were the four exceedingly simple maps for Augustin Thierry's *Histoire de la conquête de l'Angleterre par les Normands,* the work at the origin of Thierry's fame.[145] The maps of Paulus Schmidt for Raumer's history of the Hohenstaufen and Heinrich Kiepert's for Giesebrecht's best-selling *History of the German Imperial Period* can be only a small fraction of the cartography applied to the illustration of histories in nineteenth-century Germany.[146] The many more books with maps in that century deserve to be further explored, without illusions, however, about the ease with which such research can be carried out.

In the 1840s a concerned atlas-maker admitted that "medieval geography" was at last being rescued from undeserved neglect. A better-informed observer would have known that maps of medieval moments and conditions were already plentiful at the start of the century but that they were so widely dispersed as to be very difficult to find. Medieval scenes had drawn much more attention since 1700 than maps of recent, postmedieval history, whose incorporation into Kruse, Lesage, and later atlases was the conspicuous innovation of the period after 1790. Recent history quickly caught up with the Middle Ages in winning attention and space. Both became sturdy tenants in the many atlases of the day.

NOTES

1. Wilkinson, 1797/A. The four maps listed are in the version of 1805. It's hard to tell whether the many issues of the *Atlas classica* between 1797 and 1842 are "editions" whose periodicity can be tracked.

2. *Catalogue of Maps, Prints, Drawings etc. Forming the Geographical and Topographical Collection . . . Presented by H.M. King George IV to the British Museum* (London, 1828), 3: index, p. ii.

3. Delamarche, 1844 ("Avertissement").

4. Fortia, 1809/B. *BN Impr.* 53:735–48, lists a large, eccentric bibliography, including *Mélanges de géographie, d'histoire et de chronologie anciennes* (Paris, n.d.), with twenty-seven maps.

5. A selection: Dufour, 1830; Thomas Milner and Augustus Petermann, *A Descriptive Atlas of Astronomy and of Physical and Political Geography* (London, n.d. [1849]), no. 30; Tallis, 1851; Völter, 1855; Gaspare Ardin, publ., *Atlante scelto di carte geografiche, politiche e fisiche, prospetti astronomici, oro-idiografici e storichi* (Rome, n.d. [1855–59]), no. 39.

6. Patteson, 1825/A; SDUK, 1871/A; Findlay, 1853/A. See also Mentelle, 1797/A, 1804a–c/A.

7. Hérisson, 1806a; Delamarche, 1809/A (all ancient geography); Playfair, 1814 (good ancient section); Bazeley, 1815 (three ancient maps); *Cyclopedia,* 1820 (ancient geography); Rossi, 1820 (four ancient maps); Bossi, 1824 (two ancient maps); Delamarche, 1824 (three ancient and biblical maps); Vivien, 1825 (appendix of ancient maps); Wyld, 1825/A (some ancient, no medieval); Meyer, 1830 (ancient before modern); Dower, 1831; Bradford, 1835 (ten ancient and biblical maps as appendix); Andriveau, 1837 (ancient before modern), 1847 (some ancient before modern); Soleil, 1851 (ancient appendix); Black, 1854 (three ancient maps before modern ones); SDUK, 1857 (ancient before modern); Murphy, 1858/A (ancient only).

8. Arrowsmith, 1804/A; Reichard, 1804/O; Gaspari, 1811/O; Johann Walch, 1811/O. Most works listed below are not in the Catalogue because they are strictly geographical. J. Dirwaldt, *Allgemeiner Hand-Atlas zum Gebrauch für die Jugend* (Vienna, 1816); Wilkinson, *An Atlas for the Use of Schools* (Stourport, 1816); Pierre Lapie and J. B. Poirson, *Nouvel atlas élémentaire à l'usage de la jeunesse* (Paris, 1816); Stieler, 1824/O (1816); Thomson, 1817/O; A. C. Gaspari, *Allgemeiner Hand-Atlas der ganzen Erde* (Weimar, 1821) (at Yale); Galletti, 1807/O; Vandermaelen, 1825/O; Sidney E. Morse, *New Universal Atlas of the World* (New Haven, 1825); Sidney E. Morse, *New Universal Atlas of Sixty Maps, Charts and Plans from the Latest Authorities* (Boston, n.d. [1835]); P. Rousset, engraver, *Atlas général des cinq parties du monde* (Paris, 1835); D. T. Ansted and C. G. Nicolay, *An Atlas of Physical and Historical Geography* (London, 1840?); Thomas Gamaliel Bradford and S. G. Goodrich, *A Universal, Illustrated Atlas* (Boston, 1842); *Crutchley's General Atlas for the Use of Schools and Private Tuition* (London, 1843); Johnston, 1846; William C. Woodbridge, *Modern Atlas, Physical, Political and Statistical* (Hartford, 1845); Chambers, 1853; George W. Colton, *Atlas of the World* (New York, 1856). School atlases of geography in Germany and nearby are listed in Astrid Badziag and Petra Mohs, *Schulatlanten in Deutschland . . . bis 1950: Ein bibliographisches Verzeichniss* (Munich, 1982).

9. Reilly, 1806; J. J. O. A. Rühle von Lilienstern, *Duodez-Schul-Atlas* (Berlin, n.d. [1821?]) (NOT SEEN; summary description in Engelmann, 87); Perrot, 1823; Möller, 1825; Delamarche, 1827; Goujon, 1828; Simencourt, 1830; Lapie, 1826; Kärcher, 1834; Dufour, 1834b,

1835; Monin, 1842; Steger, 1845; Delamarche, 1827 (1850); Vuillemin, 1839; Ohmann, 1853; George, 1852; Philip, 1855/A; Dufour, 1860; Drioux, 1867; Piré, 1868; Tardieu, 1842 (1868); Migeon, 1866. These twenty-four include thirteen atlases called "universal," and seven historical (one specifying ancient history); the remaining four pose minor problems of classification.

10. RVaugondy, 1756, introduction (a brief account of the history of cartography), 12. For Didier's own revision (1752) of his father's Charlemagne map, see ch. 4, n. 141, above. Only Piré and Migeon lack a Charlemagne map. Wolf, "Bild der europäische Geschichte," 2 (Charlemagne's empire the one subject occurring in all forty-two of the post–World War II atlases analyzed).

11. After the barbarian invasions: Möller, Delamarche, Kärcher, Dufour, Tardieu, Monin, Steger, Vuillemin, Ohmann, George, Philip, Piré. Before the barbarian invasions: Delamarche, Lapie, Dufour, Philip. Crusades: Reilly ("France at the time of . . . "), Kärcher, Piré.

12. About Rühle, see n. 9, above; known to me only indirectly. For maps of the advent of Hugh Capet: Tardieu. Germany under the Hohenstaufen: Steger. An all-purpose medieval map: Drioux. Eurasian religions ca. 1100: Möller. "Feudal France": Delamarche, Migeon. Late Middle Ages: Ohmann (in 1453), Tardieu, Piré.

13. About Las Cases, Kruse, and Koch, see ch. 5, nn. 2–4, 50–52, 66, above.

14. According to an obituary in the journal Malte-Brun had founded, his *Précis* was among the literary monuments of the age: *Annales des voyages,* 2d ser., 2 (1826): 407. It had a wide and long circulation. His fame as a geographer has inevitably dimmed; Broc, "Malte-Brun," 718.

15. See Voss, *Universität, Geschichtswissenschaft.* About Klaproth, see ch. 5, n. 78, above.

16. See above ch. 5, nn. 54, 59, above.

17. For evidence of this criticism, see ch. 5, n. 59, above. Black, *Maps and History,* 42, comments that Kruse's "maps at regular chronological intervals" were "an important innovation." Kruse's method, though possibly novel, was almost unanimously rejected; it had little or no effect in its time or later. See also Black, *Maps and History,* 93.

18. The Kruse equivalents to Koch's nos. 1–5 (i.e., to 1074) appeared by 1804; no. 6 corresponds to a Kruse map published in 1810; no. 7 (1453) has no precise Kruse equivalent.

19. Lapie, 1812: 5. Typically, Denaix, 1835, opens with a preface contrasting his sound method to Kruse's, then continues with four sections neatly coinciding with Kruse's system: 14, end of fifth century; 15, start of ninth; 16, end of ninth; 18, toward 1300.

20. One is reminded of Geoffrey Barraclough, *The Origins of Modern Germany* (London, 1946), in which four of five parts are pre-1519 (Barraclough explains in the preface to the second edition that the book "is concerned less with the history of modern Germany . . . than with historical analysis of its background").

21. Voss, *Universität, Geschichtswissenschaft,* 179: histories of Europe, such as Koch's book, were a newly blossoming category. Koch, *Tableau des révolutions de l'Europe dans le moyen âge* (Strasbourg and Paris, 1790), vii: "L'idée d'une histoire générale de l'Europe n'est pas moins juste que celle de son droit public."

22. I refer to the part of Kruse's atlas from A.D. 400 to 1500. *AGEphem.* 35 (1811): 98, underscores his contribution to the geography of the Middle Ages. The Institut de France report of 1808 is full of solicitude for the fate in France of studies of the Middle Ages (Dacier, *Rapport historique,* 182–83, 193–204).

23. Ire de Rosny's maps were judged in Weimar to be far superior to those of Köhler,

1730: *AGEphem.* 27 (1808): 145. That a comparison with Köhler's old and haphazard collection should have been made at all bears out Koch's complaint about the rarity of maps of medieval Europe. Rosny's maps for Koch are handsome and very clear (in relation to size) but not impeccable in contents, e.g., a fanciful "État des Suédois" in no. 3 (the Swedes had no state in the time of Charlemagne).

24. The reprint is Koch, 1831. Koch, 1807, is comparatively rare; see ch. 5, n. 69, above.

25. Each title begins "Tableau de l'Europe . . . ": (1) "Sous l'Empire d'Occident avant l'invasion des Barbares"; (2) "Vers la fin du 5ᵉ siècle"; (3) "Sous l'Empire de Charlemagne"; (4) "À l'époque du démembrement de l'Empire de Charlemagne vers la fin du 9ᵉ siècle"; (5) "Vers 1074"; (6) "Vers 1300"; (7) "Vers l'an 1453."

26. Lapie, 1812. Lapie (not Lapié, as Black, *Maps and History,* 40) rose to be a lieutenant colonel in the French army. For regrets that he was limited to the (lesser) modern geography, see ch. 1, n. 8, above.

27. Lapie, 1812: 11.

28. The back cover of *Annales des voyages* 13 (1811) has an emphatic advertisement for the second volume of Malte-Brun's *Précis* and its twenty-four-map atlas. Later atlas-companions to Malte-Brun's *Précis* are Huot, 1837; Tardieu, 1842; and Dufour, 1856.

29. Lapie, 1812.

30. *Annales des voyages* 20 (1813): 276. Malte-Brun said that the Mongol map was based on absolutely new information. It is hard to tell what he meant. At Lapie, 1812, no. 17, he contrasted his rendering to that of Rosny (in Koch); at no. 18, he wished he might have had three maps (Attila, Theoderic, Justinian) in place of one. Analytical comments like these are rare.

The Florentine Francesco Balducci Pegolotti (first half of the fourteenth century) wrote a famous treatise, *La Pratica della mercatura,* an account of Mediterranean commerce; ed. Allen Evans (Cambridge, Mass., 1936); about Pegolotti, see *Encyc. Brit.* 21:57–58.

31. Delisle, 1705. Malte-Brun contrasted himself to Rosny as paying special attention to barbarian homelands.

32. Daniel, 1696; Lesage, 1801.

33. Reilly, 1806, vol. 2, pt. 2, no. 698. Note the placement of Charlemagne and historical maps of Gaul as a prelude to maps of France in the Blaeu atlas (see ch. 2, nn. 67, 72, above).

34. Steven Runciman, *The Fall of Constantinople, 1453* (Cambridge, 1965), xi. He spells out the shortcomings of this concept. Referring to "simple" historians allows Runciman to bypass an account of the real past of this periodization. (The notion of 1453 as a climactic moment had been popularized by Pierre Bayle, a contemporary of Cellarius.)

35. Our Renaissance period had yet to be mapped out; see Wallace K. Ferguson, *The Renaissance in Historical Thought, Five Centuries of Interpretation* (Boston, 1948). Ferguson notes the "tendency, which lasted throughout the Enlightenment, to attribute historical developments to accidental or cataclysmic events," 72; on Cellarius, Voltaire, and 1453, 76, 91; Gibbon endorsed the date, then learned better, 104–5.

36. Lapie, 1812, no. 22.

37. *Kleiner Atlas,* 1816. I know it from only one copy in Leiden (owned originally by a Netherlander, who, in a marginal note, acknowledged the relationship to Koch). With its flimsy, dark blue cover, the atlas resembles classical atlases such as Vieth, 1800/A, and Reichard, 1824/A. The Kartensammlung of the Staatsbibliothek zu Berlin has no record of this atlas or information about it (fax to me, 23 July 1998).

³⁸ Selves, 1819–33; Denaix, 1835; Engelmann, 1836; Kärcher, 1824, 1834; Quin, 1830.

39. Levasseur, 1840. Also included are Charlemagne and the Carolingian disintegration. These are so commonplace in atlases that they cannot indicate dependence unless accompanied by other evidence, as here.

40. Duruy, 1842; Muhlert, 1865; Spruner, 1837, 1855, 1860, 1861; Bretschneider, 1856.

41. Lesage, 183-; Lesage, 1835, no. 2 bis, 8 bis, 8 ter.

42. Lesage, 1813: transmigration, no. 9; Spain, no. 12; German duchies, no. 29. The 1813 translation is cited for convenience (numbering is not consistent from edition to edition); the items cited were in Lesage, 1801. Also the French map at n. 44, below. About the public for the atlas and its lack of historical sophistication, see ch. 5, n. 38, above.

43. Lesage, 1835, no. 8 bis. This might be an expansion of Malte-Brun's map for medieval geography (Lapie, 1812, no. 19).

44. Historical France: Lesage, 1802, no. 3b. Descendants: e.g., Andriveau, 1840. Boucher, 1804a (two maps), is a more detailed development of the same design. The proximity of dates makes it hard to decide priority. I hesitate to attribute paternity of "Historical France" to Las Cases or Boucher rather than to an unidentified predecessor.

45. For the title, see Lesage, 1801, no. 16 (changed to "invasion" in 1826). The copyright date of the map is August 1800. My quotation combines extracts from Lesage, 1806b, preface, overleaf, and Lesage, 1813a. A hostile contemporary reviewer, *AGEphem.* 16 (1805): 93, believed that Las Cases particularly cherished this map. The earlier migration map: Hagelgans, 1718. See also Goffart, "Longer Look," which traces the descent of the Lesage design in later cartography. About the term "transmigration" (which occurs in Lesage, 1801) and its association with the biblical Exodus, see Goffart, "Preliminary Report," 57 n. 20.

46. Criticism: *AGEphem.* 16 (1805); *Allgemeine Literatur-Zeitung* (Halle) 304, no. 4 (December 1804): 522–32; *Göttingische gelehrte Anzeiger*, 1830, 1693. The first improvement is probably Lavoisne, 1814, with long explanations of the changes; Goffart, "What's Wrong with the Maps?" 118–19 n. 26.

Mascov, *Geschichte der Teutschen* (1726–37), is considered the first adequate history of the barbarian invasions; Las Cases was not acquainted with it at firsthand.

47. Other maps in which events of very different times are compressed into an ostensibly coherent unit by geography: Exodus, e.g., Tilleman Stella in Nebenzahl, 76–77; Crusade tracks, Heck, 1830, no. 22; Lesage, as n. 45, above.

48. For the apparently earliest revision to the "Transmigration" scene, see n. 45, above. On this subject, see Goffart, "Longer Look," 9, and Goffart, "What's Wrong with the Maps?" Among artistic changes, I particularly applaud the angular tracks of Gatti, 1851—an excellent reminder that these map symbols are themselves abstractions, not realistic routes of march.

49. See ch. 5, nn. 37, 39, above.

50. La Ruë, 1651/E; ch. 2, n. 147, above. Delisle, 1726, 1764; ch. 4, n. 89, above.

51. Wilkinson, 1797/A, no. 51 (1807). Benicken, 1821, no. 7 (inset of crusader principalities). Benicken, 1820, no. 7 (1073–1273), refers inter alia to Mongols but not to the Crusades. For the illustrations of Raumer's work, see Schmidt, 1823.

52. Thomson, 1829 (not in *NUC*). The details on price and reader satisfaction are from the front matter of John Thomson, *Atlas of Scotland* (Edinburgh, 1832), kindly supplied to me, with much else, by Margaret Wilkes, Map Librarian of the Royal Library of Scotland, for whose interest and help I am very grateful. The 1832 announcement attributes fifty maps to the historical atlas, one more than in the surviving copies.

53. The best source on Thomson is *The Early Maps of Scotland to 1856*, 3d ed., 2 vols.

(Edinburgh, 1973), 1:143, but only to the point when the Scottish maps left his hands. Omitted is Thomson's *New Universal Gazetteer and Geographical Dictionary* (Edinburgh, 1842; London, 1843, 1845, 1857). *British Biographical Archive*, fiche 1079, frame 402, knows Thomson only for the *Gazetteer*. Also see S. A. Allibone, *A Critical Dictionary of English Literature*, 3 vols. (n.p., 1859–71). Moreland and Bannister, 171, set Thomson's death in approximately 1869 (no source cited). Reliable information seems lacking.

Thomson's *General Atlas* was often reprinted; the edition of 1821 contains eight historical maps of Europe after 1815 (see LC List 731, 750, 3545). The Royal Library of Scotland has "an unpublished specimen copy of an atlas [ca. 1835] using the plates to Thomson's New Classical and Historical Atlas with some of the plates altered and bearing [W. & A. K.] Johnstons' imprint."

54. Wilkinson's *Atlas classica* has four divisions (like Padua, 1699). Thomson's adjectives "classical and historical" may reflect its distinction of "geographia antiqua" from "geographica historica," rather than the ancient/(medieval and) modern classification that we expect (see ch. 5, n. 110, above).

55. Compare Wilkinson, "Eslam, or the Countries which have professed the Faith of Mahomet," to Thomson, "Eslem [*sic*] or the countries conquered and converted by Muhammed and His Followers." It looks as though Wilkinson's title is repeated in different words. The borrowing from Malte-Brun is plain from the inclusion of the latter's only "modern" maps, for Charles V and 1789. Coupled with the two recycled by Malte-Brun from Kruse, they plainly announce their source. Anville, 1771, was available in English atlases (e.g., the world atlases printed by Laurie and Whittle in 1798 and 1807 that contain Enouy, 1797).

56. For Scott, see n. 63, below. Lesage, 183-, 1835, no. 8 bis, an ultracomprehensive supplement for the Middle Ages, includes the strictly French Crusade tracks of Philip Augustus and Saint Louis. The undated sheet is unlikely to be earlier than Thomson (or a French equivalent: Heck, 1830).

57. Mills's book had earlier editions in 1820, 1821, and 1822 (all 2 vols.), and two U.S. editions, initially from the third London edition. The map was included from the first. Mills's title partly echoes Antoine Caillot, *Tableau des croisades pour la conquête de la Terre-sainte,* 2 vols. (Paris, 1818), contemporary to Michaud (on Michaud, see nn. 60–61, below). Caillot was a polymath, specializing in abridgments; *Dict. biog. franç.* 7:867 is very uncomplimentary. His work contains illustrations but no maps.

58. A cursory foray into the literature suggests otherwise: Henry Laurens (with C. C. Gillespie, J.-C. Golin, and C. Traunecker), *L'expédition d'Égypte, 1798–1801* (Paris, 1989); Jacques Derogy and Hesi Carmel, *Bonaparte en Terre Sainte* (Paris, 1992); Pierre Constantini, *Bonaparte en Palestine* (Paris, 1967).

59. The Institut de France's question, set on 11 April 1806, is in Arnold Hermann Ludwig Heeren, *Essai sur l'influence des croisades,* tr. Charles Villers (Paris, 1808), title page; originally published as *Versuch einer Entwicklung der Folgen der Kreuzzüge für Europa* (Göttingen, 1808). Heeren was the German winner; the French one was A. U. M. de Choiseul d'Aillecourt (Daillecourt), *De l'influence des croisades sur l'état des peuples de l'Europe* (Paris, 1809). First honorable mention, Joseph Lemoine (the title of his treatise is uncertain) (Paris, 1808). Only Heeren, a professor at the University of Göttingen, was a professional historian. The Institut de France's question for 1802 had been on the consequences of Luther's reformation. Epoch-making events were on its mind (Heeren's translator, Villers, had won the Luther competition). In 1820, the question again involved the Crusades, but in a more lim-

ited context. The question for 1806 probably provoked J. H. Regenbogen (1767?–1814) to write his *Commentatio de fructibus quos humanitas, mercatura, industria, artes atque disciplinae per cunctum Europam perceperint e Bello sacro* (*Discussion of the Benefits That Culture, Commerce, Industry, the Arts, and Education throughout Europe Derived from the Holy War*) (Amsterdam, 1809). The consequences of the Crusades had been discussed before: Johann Christoph Maier, *Versuch einer Geschichte der Kreuzzüge und ihrer Folgen,* 2 pts. (Berlin, 1780–81, 1797).

60. For Wilken and Michaud, see Hans Eberhard Mayer, *Bibliographie zur Geschichte der Kreuzzüge* (Hannover, 1960), no. 1841: Friedrich Wilken, *Geschichte der Kreuzzüge nach morgenlandischen und abendlandischen berichten,* 7 vols. (Leipzig, 1807–32); it was never reprinted or translated. On Michaud, see next note. Carl Wilhelm Ferdinand von Funk, *Gemälde aus dem Zeitalter der Kreuzzüge,* 4 vols. (Leipzig, 1821–24). His subjects run from Tancred and Baldwin III to Frederick II and Saint Louis. One copy at the BN, none at the BL or in *NUC.* Ludwig Timotheus Spittler, *Geschichte der Kreuzzüge,* ed. J. G. Gurlitt and C. Müller (Hamburg, 1827); not in the standard collection of Spittler's writings and probably written in the 1780s or 1790s, when Spittler taught at Göttingen. He left academic pursuits in 1797 and died in 1810: *ADBiog.* 35:212–16.

61. What I translate as "French" is "Franks" in the Latin text. The residents of Mediterranean lands called northerners, such as the crusaders, by the name of "Franks." In spite of this well-attested usage, "French" has to be used for clarity in translation.

Mayer, *Bibliographie der Kreuzzüge,* no. 1844: J. F. Michaud (1767–1839), *Histoire des croisades,* 7 vols. (Paris, 1812[1811]–1822) (BN and NYPL list the 1st ed. as 1813–22); 4th ed. (Paris, 1825–29); 6th ed., ed. J. J. F. Poujoulat (Paris, 1840); 7th ed., ed. Poujoulat, 4 vols. (Paris, 1854), with an "expeditions" map (steel engraved); etc. Poujoulat and Michaud were close friends. A luxury edition with many etchings by Gustave Doré appeared at midcentury. Mayer lists translations into German, English, Spanish, Dutch, and Italian. Michaud was Piedmontese and never took French nationality. Each successive edition of the Crusade history was retouched, but stylistically rather than substantively: *Biog. univ.* 28:206–14.

62. Marie Cottin (1770–1807) was published by the Michaud firm (headed by Joseph's brother). Michaud's introduction was often reprinted, along with summaries of the Crusade chroniclers Vilhardouin and Joinville, in the collected works of Cottin or reprints and translations of *Mathilde.*

63. Sir Walter Scott, *Tales of the Crusaders,* 4 vols. (Edinburgh, 1825); the novels are *The Betrothed* and *The Talisman* (in which Saladin plays an extensive, heroic part). William Stewart Rose, *The Crusade of St. Lewis and King Edward the Martyr* (London, 1810); on Rose, see *DNB* 17:244–45.

64. With Michaud, a connection to the Christian revival is less likely than a Revolution-inspired impulse to take stock of the past; see Albert Thibaudet, *Histoire de la littérature française de 1789 à nos jours* (Paris, n.d. [1936]), 264–66. Michaud's consistent opposition to the Revolution had little in common with the religious inclination of Chateaubriand. The Institut's prize question hints at hopes, nourished by history, that good might come from current (Napoleonic) wars, just as it had from the Crusades.

65. G. A. Depping, review of Heeren, in *Annales des voyages* 5 (1811): 373. R. A. Peddie, *Subject Index of Books Published up to and Including 1880,* new ser., *A–Z* (London, 1948), 209, lists seventeen pre-1800 works about the Crusades; J. Bongars seems particularly important.

66. Wilken, *Geschichte der Kreuzzüge,* 1:1–2.

67. Depping, review of Heeren, in *Annales des voyages* 5 (1811): 371.

68. Its enduring effect is discernible in the "second Pirenne thesis," in whose terms the growth of medieval towns is attributed to the resumption of eastern trade and the influx of rootless merchants on the fringes of communities: Henri Pirenne, *Economic and Social History of Medieval Europe* (1933), tr. I. E. Clegg (New York, n.d.), 1– 49. A more modest perspective on the Crusades is already found in Charles Homer Haskins, *The Renaissance of the Twelfth Century* (1927; repr., New York, 1957), 14 –15.

69. Thomson is almost certainly not the creator. All other medieval maps in his atlas are from easily identified sources, and not by him. He was deeply involved at the time in an arduous map project and was beset by financial difficulties in that connection. The one— inadequate—reason for attributing the Crusade map to him is that the atlas is his.

70. Heck, 1830. The table of contents announces "Marches et combats des croisés et les royaumes chrétiens de l'Orient"; the map itself is labeled "Carte de l'Europe et du théatre de la guerre au temps des croisades (1094 –1291)."

71. Wolf, "Bild der europäische Geschichte," 6, records the arresting frequency of Crusade maps in current historical atlases (almost equal to Charlemagne maps).

72. *AGEphem.,* 2d ser., 7 (1820): 53, advertisement for Benicken's *Schul-Atlas.* Leonard Dresch, *Über den methodischen Unterricht in der allgemeinen Geschichte* (Weimar, 1818), expressed a wish for maps; Benicken has done the maps and chronological tables; announcement, April 1820, with prices. *AGEphem.,* 2d ser., 16 (1826): 191, in connection with the announcement of the *Hand-Atlas,* adds, "A smaller, similar atlas, *ohne Schrift,* for the first course of history, has already previously appeared," i.e., the *Schul-Atlas.* For Dresch, *Übersicht,* and other works, see ch. 5, n. 93, above.

73. Atlas MS, 1832. The latest date in the MS is 1840, but it looks like an addition in a second hand. I prefer 1832, the last date in the first hand. The most modern map is of Europe in 1826.

74. This error in the handling of Hunnic history is peculiar to Lesage. Instances like this one (or, e.g., that in Johnston, 1853/A, no. 20) can hardly be attributed to another source.

75. Quin, 1830; see ch. 5, nn. 121–26, above. On the formerly unknown MS of Quin's atlas, see ch. 5, n. 121.

76. For a direct quotation, see ch. 5, n. 124, above.

77. Quin's accompanying commentary conveys a rather limited knowledge of the Middle Ages. In the sample that follows, I mainly summarize but include a few direct quotes: "15th Period, From the Commencement of the Crusades, A.D. 1100, to the Division of the Mogul Empire on the Death of Kublai, A.D. 1294." Europe had been sunk in a state of barbarism for several centuries but was now emerging from it, appearances to the contrary notwithstanding. The text surveys the European kingdoms; medieval Christianity is equated to superstition; no mention of the battle of Bouvines; Crusades are referred to intermittently, but their start is mentioned only in the title; finally, very brief and summary accounts of India and China and Mongol exploits in those lands.

78. Witzleben, 1829; Imbert, 1834.

79. Ritter, 1813/O (introduction dated 1806). It complemented Ritter's *Europa: Ein geographisch-statistisches Gemählde,* 2 vols. (Frankfurt-am-Main, 1804 –7).

80. Denaix, 1829, 1835; Desjardins, 1836, 1838. Johannes Dörflinger and Helga Hühnel, *Atlantes Austriaci, Österreichische Atlanten, 1561–1918,* 2 vols. (Vienna, 1995), 1: 23 –30, cata-

logue Desjardins's production in Austria (his success in a ten-year period was unspectacular); but he also published outside Austria.

81. For Brué's comments, see Brué, 1820: 3, and ch. 3, n. 81, above. On Guadet's contribution, see Quérard, *France littéraire*, 3:495. Brué's title page credits him with supplying a "précis de la géographie historique, politique et administrative de la France" and an "analyse raisonnée des cartes." Brué, 1820, is sometimes combined with A. Lenoir, *Atlas des monumens des arts libéraux . . . depuis les Gaulois jusqu'à nos jours* (at LC), a series of line drawings.

82. The size of the map for Philip I is approximately 57.7 cm × 43.5 cm; (13, 15) Crusades; (12) William the Conqueror. The maps are often marked with broad dates: (7) 561–755; (10) 843–987; (12) "à la mort de Philippe Ier," 996–1108; (14) 1108–37; (18) 1270–1328. Equivalences (not always exact) of Brué's maps to Rizzi-Zannoni's (= z prefix): 4 = z3; 6 = z4; 7 = z6; 10 = z9; 11 = z11; 12 = z12; 16 = z15–16; 17 = z17–18; 18 = z23; 19 = z25–26; 20–21 = z27–28; 23 = z31–32; 24 = z33–34.

83. My working total includes Legrand, 1824; Guadet, 1833; Denaix, 1836, 1838; Mullié, 1839; Dufau, 1841; Dussieux, 1843; Clausolles, 1846; Bonnechose, 1847; Maretheux, 1847; Duruy, 1849; Guadet, 1851; Dussieux, 1854; Sanis, 1859. In the dry decade, reprints and new editions were probably numerous enough to mask the lack of wholly new atlases of France.

84. *Anchor Atlas*, 1:120. The traditional, negative view of Merovingian partitions is conveyed by O. M. Dalton, *The History of the Franks: By Gregory of Tours,* 2 vols. (Oxford, 1927), 1:131–46. In a more modern vein (partition ensured the survival of the dynasty), Edward James, *The Origins of France: From Clovis to the Capetians, 500–1000* (London, 1982), 134–35.

85. Denaix, 1836: no Charlemagne map. Denaix, 1838, and Clausolles, 1846: at the death of Charles Martel. Dufau: at the accession of the Carolingians to kingship.

86. Royal domain: Denaix, 1836 (at two moments); Maretheux, 1847. *France féodale:* Denaix, 1836, 1838; Dufau, 1841; Dussieux, 1843; Bonnechose, 1847; Duruy, 1849; Guadet, 1851; Sanis, 1859.

87. Crusades: Duruy, 1849; Sanis, 1859. Ecclesiastical geography: Dufau, 1841. Acquisitions from the German Empire: Guadet, 1851.

88. Rupp, 1837, 1839. Rodowicz, 1843: the special maps of medieval Germany incline toward a national atlas without quite getting there. Pompper, 1846: of its five medieval maps (there are no modern ones), one locates the seats of the major (Germanic) tribes after the barbarian invasions; another, Germany and Italy from the Saxon to the Swabian emperors; and a third, the same territories under the other imperial dynasties of the Middle Ages. National themes predominate.

89. Babinet, 1862; Kutscheit, 1856. Maps of early Switzerland: Schaarschmidt, 1846; Frommann, 1854; Dittmar, 1865. Maps of the Holy Roman Empire divided in administrative circles: Löwenberg, 1840; Kutscheit, 1856; Frommann, 1854.

90. Pompper, 1846; Kutscheit, 1856; Spruner, 1866 (Perthes is the publisher; Spruner's activity dwindled in the 1860s; see ch. 5, n. 156, above).

91. Saxony: Tutzschmann, 1852; Süssmilch-Hörnig, 1860. Prussia: Löwenberg, 1840; Voigt, 1846; Freyhold, 1853; Fix, 1855; Pawlowski, 1855; Toeppen, 1858; Kiepert, 1865; Leeder, 1866; Baur, 1868; Brecher, 1868. Bavaria: Spruner, 1838a (see ch. 4, n. 43, above).

92. Lothian, 1829, different from Thomson's atlas of Scotland (see n. 53, above), which had no historical pretensions. Low Countries: Elst, 1831; Jusseret, 1836.

93. Häufler, 1840 (cf. Spruner, 1860); Zuccagni, 1844; Łelewel, 1844; Hughes, 1849.

94. Hungary: Scharberg, 1845. Delayed in publication but completed in 1837, its design was visibly dependent on the *Atlas Lesage*. Switzerland: Mandrot, 1855—anticipated by Huber, 1830 (a one-sheet Swiss supplement for the *Atlas Lesage*). Sweden: Wiberg, 1856. Spain: Houzé, 1840; Artero, 1879: 4, comments on the long time that Spain relied on a foreigner for historical maps.

95. Hughes, 1849. Hughes, 1863/B, is a fuller English atlas but based on text rather than maps. For the atlas of Scotland, see n. 92, above.

96. Atlas of Tuscany: Zuccagni, 1832. It is not historical except in commentary; three of the margins around the maps contain writing, in the Lesage manner. Zuccagni, 1844, offers one largely undifferentiated historical map per region, e.g., "Carta del regno Veneto sotto il dominio dei Romani e nel Medio Evo." Maggi, 1803: miscellaneous maps bearing dates down to 1854. Ancient and Italian history are included; nos. 30–33 are medieval.

97. Łelewel, 1844. For his work on medieval geography, see Łelewel, *Géographie du moyen âge*, and Łelewel, 1849/O. Plater, 1827/O (history of the seventeenth century only); Jarry, 1832, is a shorter, similar work, seen by me only in an Italian version.

98. *Kort atlas*, 1853, nos. 5–7. Wiberg, 1856, in collaboration with Mentzer. Mentzer, 1871, is very different.

99. Akhmatov, 1845. The maps may be as much as 45 cm on each side. The designer has close-ups of relevant districts (such as the Sea of Azov, in the ancient history section). He also avails himself of tracks, sometimes in zigzag form (found notably in the *Atlas Lesage*). No. 54, concerned with the Mongols, shifts its focus to Asia. The work is remarkably lavish.

100. Selves, 1833, 1835, 1843; Klaproth, 1826; Rühle von Lilienstern, *Duodez-Schul-Atlas*.

101. Marmocchi, 1840. Editions of various dates to the 1870s.

102. Pütz, 1859; Baquol, 1859. On Pütz, see *DBiogArchiv*, fiche 986, frames 60–67. Much influenced by Karl Ritter, he traveled extensively in Europe. The fortune he made by his writing was bequeathed to his alma mater, the University of Bonn.

103. Dufour, 1860. Dozy, 1870; largely derived from Rhode, 1863. Dozy's maps involving Muslims and the East illustrate the Arab conquest; the eastern caliphate, eighth to ninth centuries; the western caliphate, from 950; the crusader kingdoms; the Latin Empire of Constantinople; the Mongols. The selection is unusually rich amid an equally ample set of maps of Europe. The twenty-nine-volume *Histoire de l'Église catholique* (Paris, 1842–49) of René François Rohrbacher is deplored for being ultramontane and excessively apologetic; it reached nine editions (1899–1903): *Lexikon für Theologie und Kirche*, 1st ed., 8 : 621; *New Catholic Encyclopedia*, 12 : 557. Rohrbacher (1789–1856) has the merit of sometimes drawing attention to neglected facets of the Middle Ages.

104. Houzé, 1841; Spruner, 1837; Wedell, 1849.

105. Koch, 1814: seven maps. As usual, I cite Shepherd, 1929/O.

106. In order of appearance: Spruner, 1837; Löwenberg, 1839; Houzé, 1841 (extracts, Houzé, 1845a–b); Levasseur, 1844; Duruy, 1841. Straight chronological schemes occur in, e.g., Rupp, 1839; Rodowicź, 1843; König, 1850; Drioux, 1852; Beck, 1857. These are often national atlases masquerading as general ones.

107. A full table of Spruner's atlas was probably available (e.g., as a prospectus) long before 1846.

108. Houzé, 1841, no. D 5–13, H 1. Houzé errs, not by endorsing the traditional Gothic migration, but by deviating from it with an "original" migration from the Continent to Scandinavia.

109. Löwenberg, 1839, nos. A5 (Charlemagne), B11 (Palestine), B22–24 (France, etc.), B26a (Switzerland).

110. For recent examples of multiple centuries on one map, see John Haywood et al., *The Cassel Atlas of the Medieval World, A.D. 600–1482* (London, 1998), 3.25, "The Kingdoms of Southeast Asia, 600–1500"; 3.27, "Toltecs and Aztecs, 800–1520."

111. Kutscheit, 1844a, foreword; the unfinished German atlas, Kutscheit, 1842. For his productivity, see *GV* 83:112–13. Anonymous review in Bayerische Akademie der Wissenschaften, *Gelehrte Anzeigen* 85 (27 April 1844): 686–87 (Kutscheit is reviewed together with fascs. 5–6 of Spruner, 1837). For the French opinion (expressed in relation to Delisle's medieval maps), see ch. 4, n. 3, above.

112. The independent Burgundian kingdom was conquered by the Franks in ca. 532. The Byzantine enclave was eliminated in ca. 624.

113. Seventh-century west European kingdoms (15), Scandinavia (21), Slavic east (22), the Mediterranean (24) and, separately, the Near East during the Crusades (19), and Byzantium from the eleventh century (23).

114. Clausolles, 1845: barbarian invasions (sixth century), Arab invasions (eighth), dismemberment of the Carolingian Empire (ninth), Europe in the tenth century, the First Crusade (eleventh century).

115. Schaarschmidt, 1846. The commerce map is no. 26; for Gatterer's lost map of commerce, see ch. 3, nn. 104, 108, above. Comparably limited coverage, Drioux, 1851.

116. Nos. 16–20 in Gover, 1853, are the British set. The coloring is mentioned in the preface as a special concern. The colors are, in practice, somewhat muddled as a system of signs but very attractive. Poole, 1898/O, seems influenced by Gover. Gover, 1854a, preface, "For the Geography of the Middle Ages, it may be observed, that the *matériel,* although abundant, was of a very doubtful and conflicting character."

117. Dussieux, 1856. Britain: nos. 95–96; Spain and Portugal, nos. 108–14 (the last map in Dussieux is modern and does not include the ecclesiastical divisions in Spruner, 1837).

118. Dussieux, 1856: 9, nos. 36–73. See also Dussieux, 1845.

119. The Dutch copy is Dozy, 1870.

120. Rhode, 1861 (the following letters indicate sections of Rhode's atlas, containing separate maps): (*a*) 500; Merovingians; Pepin; Italy in 700; Charlemagne and 843. (*b*) The next is an explicitly German section, but not richly furnished: Germany under the Saxons and Franconians to 1138, Switzerland. (*c*) National series begins: Anglo-Saxon Heptarchy; Great Britain; France to 1180; France in 1461; Spain in 711–1028, 1157, 1252. (*d*) Moves eastward and stays there: eastern Europe in 1250; Germans in the East; Byzantium in 700; crusader states and Latin Empire; Islamic conquest; caliphate; Seljuks to Mongols; Mongols in ca. 1284. It is a Spruner in miniature.

121. Impressive examples are at the Radcliffe Camera at Oxford (special history collection), UCB, and UWis-Madison. But even with Spruner, there is no unanimity. *Cartographia Bavariae* (1988) has a little room for Seutter, 1745b, but slights both Spruner, 1838a, and Spruner himself.

122. Osvaldo Baldacci, "Atlanti geografi e atlanti storici in bibliotheche della Puglia e della Basilicata," *Atti del XIX Congresso geografico italiano* (Como, 1965), 3:429, notes the success of the firm of Loescher (Turin) in marketing Spruner as late as the twentieth century. It presumably acted for Perthes.

123. Droysen, 1886; Poole, 1898/O. See Dean, "Sic enim est traditum," 12.

124. Spruner, 1837, "Europe," no. 11; see also Gover, 1853.

125. Spruner, 1837, 2d ed.: Germany, no. 7; Italy, no. 5; Spain and Portugal, no. 4; northern kingdoms, etc., no. 5. British Isles, no. 1, has a large inset to the right showing England in the eighth century, i.e., the Heptarchy (see Goffart, "First Venture," 60). The main map, rather than being significantly different, is a modestly modified version of the inset.

126. Spruner, 1838a: 2.

127. August Pischon in the foreword (1843, unpaged) to Wedell, 1849.

128. For other publications by Freyhold, see *GV* 41:251. Freyhold, 1846. A collaborator, A. von Schmidt, is sometimes indicated.

129. Andriveau, 1840: nos. 19 (Europe, four panels), 20 (France, nine panels). For another instance, see Shepherd, 1929/O, nos. 82–83: four panels from 910 to 1492.

130. Wedell, 1849: (listed by sh[eet] and m[ap]) ethnography, sh. 4, m. 3–4; sh. 5, m. 4, 6–7; sh. 7, m. 7. German orders, sh. 10, m. 2; sh. 11, m. 4; sh. 12, m. 3–6; sh. 14, m. 3–4.

131. Precedents: *Atlas complet*, 1747, no. 37; Klaproth, 1826, no. 9. Other instances exist but are rare and diluted by other matter.

132. Droysen, 1886, no. 19. The lower maps are of Europe at the death of Theoderic and at the death of Justinian. Fuller discussion in Goffart, "What's Wrong with the Maps?" 170–73. Dean, "Sic enim est traditum," 11, acknowledges Droysen as a "standard" atlas but considers it "not at all well organized in content."

133. Howe, 1851; Lesage, 1827b, an improved and enlarged version, ed. Joseph Marchal (Brussels, 1827), and reissued in 1853.

134. Guadet, 1851 (the first is Guadet, 1833); Denaix, 1855, new ed. by Richard Wahl, repr., 1860 (the first ed. is Denaix, 1829); Dufour, 1856; Ansart, 1861.

135. See Quin, 1830. My catalogue entry lists a number of these, too few of which have been seen by me: *Atlas of Universal History* (London, 1857), Newberry; BL, Maps 48.c.39 (1859). Another variant title: *Ancient and Medieval History* (London and Glasgow, n.d.); LC List 5562 (1856). Quin's original design is sometimes confused with Hughes's revision, but Quin's design is represented among these late printings.

136. Wiltsch, 1843/E. The basis of selection is sometimes hard to understand. No sign is given in no. 2 that a war with Persia was raging at the indicated time or that three patriarchates were occupied by a non-Christian power. Nevertheless, this is a thoughtful work.

137. Petri, 1858/E. These volumes are exceptionally heavy, physical mass largely outweighing contents. A part of Petri's dedication is quoted in ch. 1, n. 97, above. Pius IX, whom I was taught to consider the archreactionary of the nineteenth century, was beatified in 2000.

138. Birchall, 1859; Duruy, 1846. Prussian atlases: Töppen, 1858; Kiepert, 1865; Leeder, 1866; Brecher, 1868. It needs to be determined whether Prussian school curricula called for historical atlases of the kingdom.

139. König, 1850 (a 5th ed. appeared in 1857).

140. Rodowicź, 1843; Frommann, 1854.

141. Bouillet, 1865. As an example of popularity, editions 1–3 are at PhiladUL, piously collected, though identical in contents.

142. See ch. 4, n. 25, above. Bosworth (1789–1876) was a clergyman by calling, whose studies of Anglo-Saxon were carried out as a sideline to his clerical duties. His appointment to Oxford took place several years after he published *A Description of Europe; see Encyc. Brit.* 4:299.

143. Rafn, 1837 (*Antiquitates Americanae, sive Scriptores Septentriones rerum Ante-Columbianarum in America*).

144. Dufour, 1859; Gauttier, 1830 (*Atlas de l'histoire des conquêtes des Normands en Italie, en Sicile, et en Grèce de l'année 1016 à 1140*). The Dufour-Amari map of Arab Sicily is "medieval geography" of a very scholarly kind. It presupposes an Arabic-reading audience.

145. Thierry, 1826. Thierry, who went blind early in life, stands high among French historians of the first half of the century.

146. Schmidt, 1823; Kiepert, 1863. Nothing would be known of Schmidt in this capacity if his name were not catalogued in BerlinDSB. The illustrated works: Raumer, *Geschichte der Hohenstaufen;* Wilhelm von Giesebrecht, *Geschichte der deutschen Kaizerzeit,* 5 vols. (Braunschweig and Leipzig, 1855–88).

CONCLUSION

Map Types and Written Glosses

Nineteenth-century maps for medieval history added nothing new to the ways in which the past is portrayed. Examples of these methods, or types, can be detected in the earliest period and continue to appear in the historical atlases of today. Four types of maps seem to be enough for ordering the possibilities: *passive*, such as Ortelius's "Expedition of Alexander the Great"; *poster* (or catalogue), such as John Speed's "Invasions of England and Ireland"; *comparative*, such as atlases in which Ptolemaic and modern maps of the ancient world face each other; and *synchronic* (or dynamic), such as portrayals of the biblical Exodus. "Passive" maps (which outnumber active maps by about 20 to 1) adapt geography to history; "active," or "dynamic," ones, more intimately historical, try to convey a semblance of time and change.

The most commonplace and routine maps for history are passive—prototypical aids to reading or study: "Italy under the Hohenstaufen," "France under Louis XI," "The Anglo-Saxon Heptarchy," "Europe after the Treaty of Westphalia," and many more. They are the bread-and-butter of historical atlases. Ptolemy's provincial maps of "ancient geography" may serve as examples, with the difference that no date was affixed to them; rather, they were only implicitly qualified as "ancient." A major improvement came

443

when someone took a portrayal of the ancient eastern Mediterranean, fastened a suitable inscription to it, and, by this device, turned geography into a historical guide to the journeys of Saint Paul.

Maps tend to be static; this shortcoming can be most simply overcome by an explanation in a text box. Ortelius depicted the Roman Empire and added a box summarizing Roman expansion; Camillo Pellegrino showed the Duchy of Benevento along with a box relating the changes it had experienced. Passive maps may have come closest to being scientific when Kruse offered his atlas as the historico-geographic equivalent to a surveyor's baseline: Europe century by century on the last day of the double-zero year. From that base, users might work out changes with the help of the provided written matter. The passive map is like a dictionary; as the latter offers ranges of meaning, so maps provide geographical context—a humble but pressing necessity for readers, students, and scholars.

"Poster," or "catalogue," maps are simply a subdivision of passive maps. Catalogues of battles, church councils, events of imperial history, and so forth, such as Speed's battles map, address multiple years without pretense of conveying sequential, let alone flowing, history. Some maps acquire chronological coherence only if we peel away layers of anachronism. Unlike them, the displays of Speed and his cohort are undemanding and effective in their simplicity. Their maps offer a pleasing way to learn a little about English battles and Christian councils or, in the case of Jaillot, to call to mind the pioneers of monastic asceticism. Decently historical, they give value without ostentation. Several centuries' worth of battles projected on a map needs the support of a written list but is informative all the same. Maps of this kind instruct by itemizing facts rather than by telling stories; they achieve helpful results without toxic side-effects.[1] The historical guidance they provide may be more in keeping with the possibilities of the medium than the attempts of "dynamic" maps to convey movement in time.

Pictures are expected to pack more meaning into their lines, shapes, and colors than cumbersome words manage to express. The surprising thing is how rarely nonverbal eloquence has been attempted, let alone achieved. One truly articulate map, already mentioned several times (see fig. 1), marks our transition from passive to comparative maps. Ortelius's "Geographical Figure of the Ancient Epoch" ("Aevi veteris typus geographicus," 1590) shows the Ptolemaic world in detail centered on a grid demarcating the four-times-larger world that was now known: "Here, O onlooker, is a grid of the bounds of the entire earth and a geography of only as much [of it] as was known to the ancients until the year of grace 1492." An unambiguous visual device sets off the world of the past from that of Ortelius's present day and vividly intimates the extent of change since Columbus. Other maps for history may equal this one in conveying an important message by visual means; none that outdoes it comes to mind.

Maps nourish the historian's search for differences through time by allowing the eye to compare the lay of the land at discrete moments. Plans of developing cities are very well suited to this exercise. Lamare's eight distinct displays of Paris, a worthy beginning, are less instructive than the portrayals of Amsterdam and Strasbourg in which city growth is charted on a single sheet in a series of insets that viewers see both separately and together. The unexpected but long-established equivalence of "comparative" and "historical" seems to have proved its force by long practice. Typically, the universal, sequential atlases invited exercises in comparison by their chronological succession of frames. Their shortcoming was a cumbersome format, better suited to viewing a single display than to examining two or more maps side by side.

Comparison is the most practical and cartographically safest way to convey a message without words. Several sheets of Mejer's *Atlas of Schleswig-Holstein* (1652) illustrate a district as it then was and set beside it an earlier form, such as that of 1240. Mejer's images, though unscientific, dramatically document Frisian fears and apprehensions vis-à-vis angry seas. Placing maps of the same scene at different dates side by side invites the reader's engagement with the material and conveys a message of change. Each map may be impeccably tranquil, but proximity produces an "active" display. Whatever activity takes place, however, depends on the viewer, who alone can do the comparing. The maps, passive in themselves, are simply positioned in such ways as to stimulate the viewer's activity.

History involves duration, motion, change. Passive maps are useful for reference and rarely exact a price for the information they convey; but makers of maps for history can hardly help wishing to illustrate process and activity. Their aspirations have resulted in repeated efforts to exceed the limits of the medium. Would it not be splendid if the visual impact of cartography were transposed from the display of lands at rest to "dynamic" maps, showing vibrant happenings moving in time and space? Multiple events spread over long periods have been condensed into single images, with or without integration by arrows, tracks, and other devices.

The main outcome of these longings is a fourth kind of map for history, namely, the "dynamic" type. No early example is more conspicuous than the Exodus. The track of the Israelites crossing the desert involves at least forty years. Usually the twelve tribal territories are shown, so that the map, in its instantaneousness, encompasses everything from the persecution of the Jews in

Egypt to the conquest and distribution of the Holy Land. The Exodus was not history like any other; everyone was familiar with it. Large and small Exodus illustrations had long been a part of Christian art. Like them, Exodus maps were narratives; they exemplified the age-old, respectable artistic practice of depicting a succession of separate moments in the same space.[2] The background of an artist's image is timeless and malleable; with a subject as steeped in tradition as the Exodus, the same pliancy might be offhandedly attributed to a geographic ground. But the privilege allowed to biblical images does not extend in all directions. When scenes from profane and modern history are in question, it is harder to forget that geography is time specific; it is "obviously synchronic" and has "synoptic force."[3] If activities are shown taking place on a geographic background, that background does not necessarily mean that the activities were chronologically simultaneous, but geography is not chronologically neutral or indifferent. A map showing both Naples and Smyrna intimates that the two places exist at the same time, not that (for example) Naples is shown as though in 137 B.C. and Smyrna as though in A.D. 203. To this extent, a geographic background suggests, not that the lands in question are chronologically malleable, but that the foreground scenes are as simultaneous as the lands onto which they are projected.

Later than the Exodus, the compression of time and illusion of change have outstanding examples in depictions of the barbarian invasions and the expeditions of the crusaders. For both subjects, lines and arrows insistently call attention to the activities they suggest: "Look, I'm moving! See me go from the Baltic to North Africa!" Anyone with historical training realizes that an image concurrently showing, in the case of the Crusades, tracks from at least 150 years is a fiction blighted by anachronism. The excitement such maps convey is, at best, an unintended distortion, the by-product of a worthy goal. Las Cases, limited to a single map for the barbarians, assumed that the viewer would take pains to unravel the simultaneous tracks and sort them into the appropriate order of time. He and others overlooked the side effects. Even if viewers dutifully disentangle the scene, and a fortiori if they do not, the map in its synoptic exuberance makes a lasting, perhaps indelible impression. Its (misleading) total image risks eclipsing any learning imparted by the details. The most stubborn, often unintended, and profoundly false teaching of the barbarian invasions map is that movement or migration was the only memorable feature of early Germanic life.

The shortcomings of synchronic maps are best experienced in less controversial examples. Mapmakers need to economize; a single map may have to do the work of five or more. The consequences are well illustrated by Karl Spruner's great atlas. Spruner's map of Anglo-Saxon England embraces more than six centuries, from the invasions, starting ca. 449, to the Norman conquest of 1066. His huge European section has many similar displays, such as medieval France in three maps: first, the Merovingians alone, from 482 to 751; next, France from 750 to 1180; finally, from 1180 to 1461. As often happens, the first French map embodies the shortest time span; yet the Merovingians surely do not provide more mappable information than the thirteenth, fourteenth, and fifteenth centuries. Spruner's English map portrays the Anglo-Saxon Heptarchy, which existed only for 350 years, more or less, of the early Anglo-Saxon period. His written commentary may qualify and rectify, but to anyone glancing at his map, this fraction of the Anglo-Saxon period stands for the whole.

Much is omitted and distorted when many centuries are melted into one map. Depictions of less than a century can be just as misleading as a long span, as Kruse observed. Trace onto a single map the vicissitudes of the four years of World War I or the six years of World War II: condensed images of this sort cannot avoid having a "total" appearance, a look that, owing to the proximity of heterogeneous details to each other, is bound to confuse the novice, at least at first glance. Experts are not automatically immune. The drawbacks of compression are unintended, just the regrettable outcome of limited space. Desired or not, such a cramming in of details means that users have to apply multiple filtration; so as to learn from the palimpsest before them, they must mentally blot out all lines, arrows, and notations irrelevant to their purposes, focus on the years or theater that concern them, and painstakingly sort out what they wish to find out. That is much to ask.

Synchronic maps are worrisome. They include some of the most striking images that have come our way, but visual and dramatic qualities are not guaranteed to complement, enhance, and respect historical coherence. Judged by comparison with written history, most attempts to show time and motion on maps have been quixotic or imprudent—memorable failures.

Works prior to 1870 rarely anticipate the thematic maps that are the pride of twentieth-century historical atlases. Gatterer, after his twenty-four maps for the Middle Ages, supplied three such maps for the "current world." This was generous and went beyond what most historical atlases would do for a long time to come. The European atlases of Denaix and Desjardins, and the few other collections venturing noticeably in this direction, did not inspire waves of imitation.[4] Precisely what makes a map thematic? Some commentators justifiably take any map for history to fall under this rubric, the theme being "history." Others might think that such inclusiveness is unhelpful. A book rarely concerned about thematic maps is not the place to polish a definition. Widening the thematic category inevitably blurs the line between older and more recent styles of mapmaking. Statistics and social sciences are the pillars of today's thematic designs; by 1870 they had not yet moved from the margins to center stage. The *Putzger historischer Schul-Atlas* become notably thematic only in the 1920s.[5] The role of thematic designs in early maps for history will be best studied from the perspective that twentieth-century thematic mapping affords.

"Dynamic," or "active," maps have not fared well in my typology, but it is possible to take leave of them on a positive note. Historical change seems to be most effectively suggested on maps by a series of chronologically successive panels fitted together and integrated on a single page. A good, but not ideal, example is the display "Zur Geschichte der Völkerwanderung" in Droysen, 1886—multiple, related maps set out side by side as an ensemble. Readers, wishing to compare, do not have to memorize details as they turn from page to page; the relevant maps are before them. Wedell's medieval Spain—a "bande dessinée" (or cartoon strip) of nine panels (see fig. 32)—projects the *Reconquista* to the eye. Similar displays can be of as few as four panels. Panels in sequence respect the rigidity of maps and allow their effective use for depicting a development. Projecting time on a single background reduces decades or centuries to a moment and creates synthetic "events" that never were. Panels in sequence obviate compression. This way, at least, one may hope to map events that transpired over many years and yet shape them into an aggregate that does not obliterate intervals. Despite the resistance of maps to the portrayal of time, this may be one way in which, without excessive distortion, something more arduous may be shown than, for example, "The Eastern Mediterranean in the Time of Saint Paul."

My emphasis has been on the maps of historical atlases, and I have assumed that the (almost unattainable) ideal map for history is one that speaks for itself. The same assumption has guided my intermittent ranking of maps in order of excellence. Second thoughts about this primacy of maps are probably called for. The written commentaries that usually accompany maps are not minor embellishments. It may not be wholly accidental that almost the only large early collection that has not survived is the atlas of Christoph Gatterer, consisting of twenty-four maps completely free of text.

> In the last analysis [Catherine Hofmann comments], what power does the map have to reflect historical phenomena? A synoptic vision able to show the "theater of history," the map—obviously synchronic—displays objects that are supposed to be permanent and contemporaneous. Thus, even today, in spite of indisputable technological and semiological improvements, the map has difficulty expressing chronological changes, at least those, such as essen-

tially unforeseeable events, that cannot be reduced to homogeneous series. Under these circumstances can the map do without a text in its various forms . . . ?[6]

The most careful atlases of today tend to bear out these reflections. The deservedly admired Bavarian atlas of world history features scores of excellent maps, colored in multiple shades and assisted by carefully traced lines, arrows, and other symbols. In spite of these multiple means of visual eloquence, the Bavarian atlas cannot forgo volumes of written commentary, such that words outweigh maps.[7] Hofmann draws the appropriate moral:

> [F]rom A. Ortelius to T.-F. de Grace (1713–1798), [authors] underscore the synoptic force of the map, which thanks to "the eyes of the body" directly impresses "the image of the site" on the memory[; nevertheless, this same force] proves powerless by itself to express historical time in all its complexity. As a result the authors of historical atlases for the most part had recourse to various sorts of textual methods to make up for the inadequacies [of maps]."[8]

Far from being self-explanatory, maps for history have involved a dialogue between words and images. They are most likely to be useful and instructive in the future if this partnership is acknowledged and deliberately developed.

The alliance of map and letterpress is found as early as the sixteenth century. It probably resulted from the mainly verbal nature of geography itself, rather than from a realization, rapid or gradual, of the inadequacy of maps for historical portrayal. William Lambarde provided his map of the Anglo-Saxon Heptarchy with a substantial commentary (1568). Ortelius initiated the practice in the Dutch atlas tradition of printing written notes on the reverse of each map and adding supplementary text pages when necessary (1570). Four pages of text are associated with Bertius's "Empire of Charlemagne" (1623); more writing, addressed specifically to the map, would not have been superfluous. The atlas of Schleswig-Holstein (1652), with its compelling maps of battered coastlines, suffered from not coordinating the comments of Danckwerth with Mejer's maps. Something similar occurred in connection with the Amsterdam *Atlas historique* (1705–13). The anonymous designer, who presumably had reasons for his choice of maps and other engravings, entrusted the extensive columns of text to a professional writer who was less concerned with the maps before him than with pursuing a personal feud against Louis XIV and the Catholic church.

No clear doctrine or usage emerged from early practices. Individual cartographers furnished explanations, did not do so, or offered too few to make a difference. Guillaume Delisle's pioneering *Theatrum historicum ad* A.D. *400* traveled with a booklet of clarifications, whose pages were sometimes glued to the sides of the two sheets of the map. His contemporaries Cellarius and Köhler compiled maps of medieval territories but omitted commentaries. Hagelgans, a practiced writer, illustrated his *Atlas historicus* with four maps, leaving their rationale unstated. Single maps were likely to travel alone. Zollmann's admirable pair about Saxony badly want explanation; they have none. Falcke's maps of Old Saxony (1752) are baffling and, though accompanied by much writing, never clarified. The universal, sequential atlases encouraged comparisons of map to map and so had built-in eloquence. Almost all of them, from the Paris draft (ca. 1747) to Quin (1830), supplemented the maps with commentaries of varying length. Las Cases, with a different plan, pushed textual assistance to an unparalleled extent, to the evident liking of his readers. Many pages of his atlas were without maps, devoted to chronology, genealogy, and other information. Especially noteworthy was his page layout, in which each map formed a peninsula in a lake of print. This design, pioneered by others, came to be attributed to him and much imitated. The proximity of words and maps was a genuine advantage. Today's atlases often take the same course. Unlike Las Cases, however, they generally allot considerably less space to commentary than to maps.

The need for written explanations has been no more widely agreed upon in the twentieth century than it was in the past. Shepherd's atlas (1911) is composed only of maps, with an extensive index. Its progenitor, Putzger's school atlas (1877), also lacks explanations. Writing in school atlases risks interfering with the textbooks that teachers using the atlas might assign. Atlases meant for general consultation are not burdened by this restriction, yet take much the same course. One example, *Hammond's Historical Atlas,* was widely distributed for half a century either in freestanding form or as a supplement to Hammond's world atlases.[9] It contains only maps. Almost the same holds for the artistically lavish and graphically innovative *Times Atlas of World History.* Few words accompany its many maps.

Much greater allowance for text is made in other very recent works. MacKay's *Atlas of Medieval Europe* (1997) offers numerous plain, black-and-white maps, each one alongside an extensive narrative. The success of this commendable scheme depends on how closely the text is coordinated with the map it is supposed to explain.[10] The sumptuous, much subsidized *Historical Atlas of Canada* takes full advantage of the means now available for printing lines of type directly onto colored maps. The result is a superb combination of image and text. None of the maps and atlases surveyed here implements as full and comprehensive a combination of text and illustration as does the

Canadian atlas. It gives hope that maps, well and extensively glossed, may yet be able to master the problem of depicting historical change.

RECAPITULATION

Legend has it that scholars reading a learned work in their field need only run their eyes over the footnotes in order to gain an ample grasp of the text overhead. The anecdote points up the inverted layout of scholarly books rather than the proficiency of scholars. The weighty stuff of learned monographs is at the bottom of pages or at the back of the volume, while the early and central pages contain the lighter, plain words that guide lay readers and sustain (or lose) their attention.

My book has this typical inverted layout. As the text draws to a close, grave matter still lies ahead: the Catalogue of Maps and Atlases records my efforts to assemble and inspect as many maps and atlases for history as possible, to distinguish relevant from inapplicable items, to unchain ancient geography from history, and to detect links between collections. Even though my discussion of historical atlases cannot address all existing evidence, it takes into account a comprehensive body of material, larger than has been considered before. Many maps and atlases with connections to history have been examined, and a reasoned effort has been made to judge which ones warrant a place in the development of historical atlases and which do not. The maps and atlases are presented, century by century, along with associated material, in the final pages. I hope some readers will decide that my center of gravity is in this Catalogue.

Cartobibliography is full of surprises. Some maps by Bellin, Gatterer, and Chanlaire are known in considerable detail from descriptions but remain unfindable, whereas some commonplace items, safeguarded in secure collections, seem to have been excluded hitherto from discussions of maps for history. Three inset maps in Ortelius's *Theatrum,* published in 1570, are conspicuously historical in content; they show aspects of the Netherlands in the times of Augustus, the flood of 1277, and Count Gui de Dampierre. Scholars who consider the *Parergon,* which appeared nine years later, to be the first historical atlas leave them out of account. If they were noticed and studied, they would undermine the preeminence of the *Parergon* as a supplement of maps illustrating history.[11] Another anomalous case involves Jacques Chiquet's *Nouveaux et curieux atlas géographique et historique* (1719). In a brief sketch of historical atlases, Chiquet is praised for producing an atlas improving somewhat on the standard of his day. Chiquet, a Paris bookseller, has few claims to distinction; he simply reissued the atlas compiled about six years earlier by his associate Antoine Ménard. Ménard and Chiquet's atlas was as remote from innovation as from history, except in the diffuse sense common to other world atlases.[12] A third case should be enough to demonstrate my point. The dramatic clouds of Edward Quin's historical atlas earn a color reproduction in a recent book. The chosen specimen, attributed to Quin in the caption, was taken from the redrawn edition by William Hughes, in which the dramatic qualities that won Quin's atlas its fame are largely effaced.[13] Continual attention and verifications are needed when gathering together the evidence for these three centuries of specialized cartography.

My Catalogue should encourage readers familiar with map collections to locate items that I mention but could not find. It may also draw attention to additional items whose existence passed me by. A list that falls short of its aspiration to be complete can at least serve as the springboard for the eventual attainment of completeness.

The even-numbered chapters of this book have examined maps for medieval history, from William Lambarde to Rudolph Wedell. Their value for my subject resides not specifically in being about the epoch we call the Middle Ages but in incorporating all history outside the agenda of ancient geography. Such maps were long attributed to an undifferentiated "modern" history, beginning with the fall of Rome. The epoch we call "modern"—from ca. A.D. 1500 onward—used to be a short appendix easily compressed into "the present."

These particular maps have been subordinated to the story they make possible, namely, the origin of historical atlases. They are better understood as unique undertakings sharing in the wider enterprise of making maps for history. Instigated by individual initiatives, and showing miscellaneous parts of the world, maps for medieval history appeal to those willing to become alternately fascinated with Frisian floods; Austrian imitators of Dom Mabillon; celebrations of "Austrasia," Flanders, or the Palatinate; a barely begun Carolingian canal; Orientalists at the French royal court; a refugee naval officer turned Bonapartist; the dilatory promotion of a brilliant Bavarian soldier; the formation of a canon of medieval scenes; and much else. My accounts of these particular stories, often concerned with the minutiae of editions and problems of identification, resemble a narrative catalogue moving from subject to subject. Though documenting an essential aspect of the development of maps for history, they mainly complement the tale of atlas origins, with an argumentative start and a qualified finish, that structures the entire book.

Besides drawing attention to individual examples, the discussion of maps depicting medieval history is also a study in historiography—how history is written. More particularly it is concerned with "medievalism," the changing view held in modern times about our medieval past.

Without special concern for "middle" or any other ages, chronology was the dividing line between an ancient geography that took no account of time—until La Ruë broke ranks—and proper historical atlases, led off by Alting's. Ancient geography did not treat chronological order as though it were an obligatory frame. One or another compiler adopted it without changing the ways of his peers. Geography did not, in general, have to be controlled by time. Ancient geography, when showing scenes (such as Alexander, early Rome, and the Sicilian expedition), was not compelled to sort them into a pictorial progression. Chronology, on the other hand, mattered much to atlases of comprehensive history, especially in the period of experimentation after Hase. In the century of sequential atlases, when mapmakers believed that a permanent structure had been attained, historical atlases embraced chronology as their primary framework. In the 1830s, geographical order, as found in atlases of the world, took the lead in historical collections, especially in expensive productions relegating chronology to a subordinate level of organization. Both chronology and geography had disadvantages as plans for historical atlases; but they offered atlas-makers a choice. Designers might even be eclectic.

The initial appearance of the Middle Ages in maps was almost always attached to the past of particular countries or provinces; early English kingdoms, threatened Frisia, the Beneventan duchy, and "Austrasia" are typical. Although this regional quality of cartography continued, a widening of horizons took place in the eighteenth century, when d'Anville and others decided to fill the gap separating ancient and modern geography. Ortelius had casually mentioned the historical period called *media aetas,* and Lubin, long after, had a developed sense of its place and importance in the past. Nevertheless, fewer than ten maps depicting medieval scenes had accumulated by 1700, and the designers of most of them would have been surprised to hear them called "medieval." Conditions changed remarkably in the eighteenth century. Maps of medieval scenes multiplied to such an extent, in a scattered, unmethodical way, that the items existing by 1800 might have filled several specialized medieval collections. Such a product was not yet in demand, however, except on a limited scale.

Until Kruse's time, mapmakers comprehensively depicting the Middle Ages did so out of the duty to fill in the relevant years, rather than out of a desire to communicate a traditional or original account of medieval history. In the sequential atlases they produced, the Middle Ages lay within a history beginning with the sons of Noah and ending in the present. Without being any less committed to the medieval part of their atlases than to the other ages, these authors ordered time by "intervals" rather than in relation to notable historical events. Kruse thought it more scientific to periodize by neutral centuries rather than by events; an excellent written handbook was an integral part of his plan. It took the influential maps of Koch, in 1807 and 1814, to articulate the Middle Ages in seven thematically determined steps. Guided by this point of departure, nineteenth-century mapmakers learned to subdivide the Middle Ages into characterized periods instead of more or less arbitrary clumps of years.

Maps illustrate other problems experienced in the early efforts to comprehend the Middle Ages. An interval like the twelfth and thirteenth centuries—the High Middle Ages to us—proved much more daunting to cartographers than did the early barbarian kingdoms. The problem of depicting the "high" centuries, if not wholly skirted, was sometimes attenuated by shifting the focus of interest to Genghis Khan, Tamerlane, and western travelers to Asia. A medieval history focusing on Asia was preferred in the eighteenth century. Depicting the more favored early medieval period was not without its problems, however. The lands in question ran the risk of being portrayed as barely different from late Roman provinces. An associated tendency was that the history of some country was begun only to be stopped almost at once. The Anglo-Saxon Heptarchy, early in being mapped and often repeated and reprinted, was for some time the sole representative of all the English Middle Ages. France was long given the same asymmetric treatment. After collections of maps about the Middle Ages ceased being rarities, they seem in general to have done without the guidance of written histories and nourished each other instead. Even Gatterer, a prolific academic historian, did not derive his twenty-four-map atlas of what he called "the geography of the Migration Age" from any one of his works.[14] This detachment of maps from established narratives is perhaps an enduring feature of the genre, observable also in the twentieth century.

Loose ends can be more appealing than conclusions. The maps examined in this book have raised many particular problems, not all of which have been fully resolved. What story is embedded in Zollmann's map of high medieval Saxony? Was there a particular occasion for his mapmaking, or for the commemoration of the Swiss battle of Saint Jacob? Who was the English mapmaker who decided to compile a map of medieval France in 1756, and what induced him to do so? The ethnic map of the Slavs and the map of the medieval Duchy of Milan reproduced in the *Algemeine Welt-Historie* probably derive from books, but which ones? Franz Ludwig Güssefeld is a much praised professional cartographer; his maps of dynastic partitions in Saxony have been called particularly fine. But whoever decides to study this part of his output will have a hard time locating his work. Also hard to document in full, but potentially rewarding for the study of the relationship between historical maps and politics, are the mapmaking activities of the Baumgärtner firm in Leipzig before and soon after the Congress of Vienna. In these and other problems, there is no lack of loose ends. The designer of the Paris draft of 1747 is an outstanding figure in the early development of history atlases: can he be identified?

Where the origin of historical atlases is concerned, what matters most in the three-century development I have traced is not that Ortelius designed erudite maps of ancient geography but that it took until, perhaps, Wilkinson's atlas of 1797 for the tradition of Ortelius's *Parergon* to produce a descendant in which a substantial selection of medieval and modern history was included. Wilkinson was not a turning point. Most collections of ancient geography continued thereafter to be chastely ancient. Only rare individuals turned such works into comprehensive historical collections. Their departures from the pure *Parergon* design were insignificant by comparison with historical atlases of nonclassical origin, in which our kind of maps for history came into being.

The absence of activity, the eloquence of silence, speaks more loudly than momentary traces of initiative. Half a century passed between Lambarde's map of the Heptarchy and the next portrayal of a medieval entity, namely, Bertius's "Empire of Charlemagne." After Bertius more than a century went by until Charlemagne was mapped for a second time. In 1651, La Ruë arranged a thin atlas in chronological order; it took five decades for a map collection disposed in the same way to be produced by Alting. Three atlases of medieval geography or history were planned between 1700 and 1720. These serious projects, addressing the need for maps of subjects that had been bypassed until then, never came to fruition. When the initiators died or lost energy, no one stepped into their shoes and carried on. Fifty empty years separate Alting's atlas of ancient and medieval Frisia from Hase's *Imperia maxima*. After Hase, a certain continuity was sustained. Even amid these beginnings, significant gaps did not cease to occur. The atlas of French history by Rizzi-Zannoni (1764) is not much worse in contents and design than the several dozen French history atlases marketed in the nineteenth century. Rizzi-Zannoni's trailblazing work was reprinted a few times but was never revised, expanded, or copied. No competitor saw profit in contesting Rizzi-Zannoni's monopoly. It took almost six decades for the next collection of French history to be produced. The belief that historical atlases were known and appreciated from the late 1500s onward is inconsistent with the halting, limping, intermittent advance of a type of map and map book that became a widely produced and continuously published product only between 1800 and 1830.

Early maps—some types, such as marine charts, excluded—were valued adjuncts to history regardless of how they were packaged. To a student reading about some land or region, even a simple modern map might prove helpful and enlightening. These undemanding conditions entitled many maps to be called "historical." Ortelius's motto, "Geography is the eye of history," implied that geography "saw for" history by means of the maps provided for everyone. Modern history, it was thought, did not need distinctive maps.

"Ancient" geography formed a diptych with "modern"; maps of both kinds had historical dimensions. Ortelius's ancient geography supplement was no more historico-geographic than the *Theatrum* proper. It simply served a narrower, more learned audience of humanists in need of portrayals of the world before the end of Rome. Because mastery of the classical languages and philological training were the condition for practicing ancient geography, an aura of scholarly preeminence clung to the subject. Nevertheless, modern geography was as historical as ancient. The ancient branch, calling for erudition, differed mainly in esteem. In the words of Augustin Lubin, "Ancient geography, *vetus,* [is] the finest and most learned."

Within geography, the preferred historical technique was comparison, notably the comparison of ancient and modern territories. Atlases engaging in this exercise marched triumphantly across the early modern era and even into the nineteenth century. The insistence on comparing ancient and modern might pall, but the method of comparison was as appealing to standard geography with its single maps-applicable-to-all-history as it was to the creators of special maps for history. Mejer in Schleswig placed a plan of Oldenburg in its medieval prosperity alongside one of the shrunken town of his time. The Paris draft universal atlas, in its sixty-six maps of Eurasia from a uniform angle, invited comparison from map to map as the measure of historical change. Hughes believed that universal, sequential atlases, such as Quin's, were the foundation of a geography called "Comparative." That branch held the promise of making maps for history an academic discipline in their own right. Although this potential was not realized, the method of comparison is not outgrown today. A permanent issue in mapping the past is how to facilitate comparisons of moments of time.

Specialized historical atlases were anticipated by single maps for history. These could be drawn without any intention that they should be collected. It took a rare cartographer, portraying an unusual district, to need more than a single map to show the geographical past of his homeland. Mejer, faced with ferocious seas that, within living memory, had battered an inhabited island into mudflats, gave historical perspective to the mapping of the Schleswig coast. The special and often tragic circumstances that turned the shores of Zeeland and Frisia into material for historical chronicles furnished an unusual incentive for maps comparing one age with the next.

Even without the encouragement of unstable ground, individual scholars and mapmakers before 1700 supplemented general maps with separate portrayals of moments of the past. The maps are few, and their subjects isolated.[15] However admirable some of these initiatives were, none points toward the historical atlases we know. Ancient geography itself only slowly expanded its repertory. The main addition of the seventeenth-century descendants of the *Parergon* to the received themes was an ecclesiastical geography demarcating patriarchates and bishoprics.

The early work pointing most directly toward future historical atlases is La Ruë's six-map *La terre sainte* (1651). Half a century later, individual initiatives resulted in the same innovative chronological ordering of maps as La Ruë, with refinements. The crucial figures were Alting, Lamare, and Delisle. Before them, Lubin had announced the existence of a "medieval geography," and some decades afterward, a German author published a book on this subject, though only in a German context.[16] Equipped with a compartment of its own, medieval geography became respectable. But it was hardly possible yet to take the term for granted. The great d'Anville spoke in 1771 as though this "intermediary period," whose established name seemed to stick in his throat, had been recently discovered by him.[17]

The works of Alting, Lamare, and Delisle were turning points in the creation of a body of specialized maps for history, not as finished models but as indications of things to come. The ideas operative in them were chronological order, narration, and historicity—not necessarily all at once. Where Alting and Lamare were concerned, La Ruë had definitely been the forerunner, even if unnoticed. All three offered sets of maps illustrating an ambitious narrative: the Holy Land from the earliest times to the present; the ancient and medieval development of the (future) United Provinces down to the great floods; the growth of Paris from the Gauls to Louis XIV. Stories and maps had kept company before, but as pictures on maps or in their margins. La Ruë, Alting, and Lamare arranged maps in sequences. They told stories by imitating the course of history.

Delisle's special contribution to these developments was historicity. Chronological precision had not been prized in cartography. A very old practice of world atlases, continuing past 1800, was the pious anachronism of showing the biblical Holy Land in lieu of the current Levant. Later times were similarly treated. Bertius's "Empire of Charlemagne" belonged to a loose Carolingian period; Pellegrino's "Duchy of Benevento" concerned multiple centuries; Alting gave his maps flexible dates, such as "the thirteenth-century floods," rather than exact ones. History even in specialized maps was not firmly anchored. Delisle changed this practice in one of his most famous productions. The time of his "Historical Vista for A.D. 400," though flexible, was chosen for a precise historical reason. The two-sheet map represented the Roman Empire and its neighbors in the vicinity of 400, at a moment, Delisle explained, before the fateful Rhine crossing of 405. Delisle did not try to show movement or any kind of change in his map but took pains to fix the situation applicable to the chosen date. He took a notable step away from contentment with the diffuse history of older maps and toward maps embodying earnest likenesses of historical moments.

The passage from all-purpose atlases to specifically historical ones was foreshadowed in the opening decades of the eighteenth century by the authors just mentioned; it was fully realized in the 1740s by Hase in the *Imperia maxima* and associated maps—works meant for no other purpose than to illustrate the successive configurations of the Eurasian past at definite moments from antiquity to 1740. Soon afterward, historical maps in chronological sequence with narrative implications were realized in quantity by the Paris draft, Dupré, Rizzi-Zannoni, Gatterer, and a few others.

The existence and merit of the works just mentioned did not outweigh their small number and limited impact on the book trade. Specialized historical atlases had not taken root by the 1770s. What a historical atlas should contain was still known only to a few. The maps collected to accompany the bestselling *Histoire de France* of Velly and Villaret (1787) were in the all-purpose mold of world atlases, and Andrews's *Historical Atlas of England* (1797) took a halting route, past fisheries and rural deaneries, to what it considered history. The one-size-fits-all-epochs approach exemplified by d'Anville in his single map for medieval western Europe is a long step backward from Delisle's sense of precise time. In spite of the advances made near 1700, new-style atlases were few and unappreciated a century later, and their design was far from universally accepted. Not only did historical atlases not begin in the sixteenth century, but they were still tender shoots in 1800.

The practice of special geographico-historical works set in motion by Alting, Hase, Dupré, and Gatterer was borne into the nineteenth century initially by the atlases of Las Cases and Kruse. The warm welcome given to their works makes Las Cases and Kruse look epoch-making in the history of the genre. Their popularity, still based on anecdote and hearsay rather than on hard numbers of sold copies, was a novel and encouraging occurrence. Their influence, however, was not so direct and recognizable that it may be unequivocally assumed.

Kruse's atlas wholly excluded the ancient world and focused on Europe rather than Eurasia. Except for these weighty changes, his program comes from the sequential atlases of Dupré and Gatterer. The chronological rigor he sought to impose won fewer friends than the excellent execution of his sequential maps and accompanying letterpress tables.

Las Cases was more innovative. The eccentricity of his work is apparent at first glance: he produced a decidedly nongeographic historical atlas, replete with genealogies, time lines, and other forms of printed matter—less a geographical aid to history than a historical reference work with a few maps. Las Cases stressed that his maps were subordinate. Critics pointed out that they had rudimentary outlines, lacked grids, and were outclassed by those of almost any other atlas. Las Cases knew what he was doing. His simplified maps were overlaid with conjectured tracks, in the form of wide ribbons, and other narrative devices. The *Atlas Lesage* was so nongeographic that imitators such as Jarry de Mancy and Huber-Salantin dispensed with geography altogether. Las Cases took the momentous step of paring down the geographical content of the historical atlas and expanding the associated text. Gone was "good" cartography, such as Kruse's; the *Atlas Lesage* offered practical, uncritical, informative history, accompanied by simple, schematic, and talkative maps.

Also discarded in full by Las Cases was the convention according to which current geographic maps were adequate for historical purposes—the belief that long made atlases of specialized history superfluous. All modern history, that is, after the fall of Rome, had once been in this category. Ever since the subject matter of "medieval geography" was recognized, the line between times that needed mapping and times that did not had been drawn at about 1500. Current maps continued to be adequate for times later than the Middle Ages. Las Cases, by introducing maps of the Thirty Years' War, Spain in the War of the Spanish Succession, and Napoleon and Suvorov in Italy, eliminated the historical privilege of ordinary maps after 1500. All the past, including just yesterday, belonged to specially drawn maps and the atlases containing them. Maps for recent times rapidly proliferated after 1801. Kruse, Malte-Brun, Benicken, and others may not have needed the guidance of Las Cases in order to design them, but the *Atlas Lesage* led the way in exemplifying a sense of history that other atlas-makers rapidly shared.

When do gestation and labor cease, and when does the actual birth—of historical atlases—occur? Hase's *Imperia maxima* deserves the name of a historical atlas; so does Alting's. On the other hand, world atlases containing only all-purpose maps were entitled "historical" in Edinburgh as late as 1854 and in Paris in 1866. However that may be, by the 1830s, several specialized historical atlases were appearing every year in various European countries, influenced by one or the other of the available models. The volume of production that was conspicuously absent in the days of Hase, Dupré, and Rizzi-Zannoni now became manifest. A wave of historical atlases began to sweep over Europe and has never receded.

This book has traced the origin of historical atlases and, as the means to that end, has examined as many as possible of the early maps that attempt to reconstruct moments of the Middle Ages. Few of these publications include explanations detailing how the authors went about designing maps for history. Maps were drawn and atlases were compiled, it appears, without profound attention to ways of portraying the past, let alone to questions of their character or purpose. Particular circumstances inspired invention. Spruner spurned Las Cases's ribbons, disagreed with Kruse's arbitrary moments, and took sources seriously. In turn, his views, though noticed and praised, launched no debate, or even conversations, among interested and skilled parties. Every now and then a map for history has been an aesthetic triumph. Quin's atlas, almost useless for study of the past, sold well by being visually striking. Scholarly success is harder to gauge. How many flaws or errors may be condoned if a work is deemed (flaws aside) to have scholarly or pedagogic merit? How much compacting of time can a cartographer engage in before he subverts the usefulness of his product? These and better questions might be posed and have gone unanswered in the absence of established yardsticks and academic discussions. The making of maps for history lacked norms in the past; these norms are still lacking, with perhaps even more serious consequences. No corpus of maps for history exists on whose basis one might select and foster types likely to harmonize geography and history to their mutual advantage. I hope that my account of the first three centuries of historical atlases provides a certain conceptual unity to this genre of mapmaking and history writing and encourages reflection on how such maps and atlases may be better designed and more effectively created.

NOTES

1. There can be too much of a good thing, however: Rothenburg, 1830, plots (on a map far exceeding normal dimensions) every battle, siege, and other military action in Germany between 113 B.C. and A.D. 1830. The sense to be made of this ensemble is obscure.

2. See ch. 2, nn. 154–55, above.

3. Quoted from the comments of Catherine Hofmann, below, nn. 6 and 8.

4. Denaix, 1829, 1835, took an early lead but did not persevere (ch. 5, nn. 120, 135, above). See also ch. 5, n. 131, above (Desjardins) and ch. 6, n. 118 (Dussieux).

5. Wolf, "100 Jahre Putzger," 707–8, about the Putzger edition for 1923.

6. Hofmann, "Genèse de l'atlas historique," 100. The subject of this part of her article is "De la nécéssaire complémentarité de la carte et du texte."

7. Bayerischer Schulbuch-Verlag, *Grosser historischer Weltatlas.* The 4th ed. has three volumes of maps and four volumes of *Erläuterungen.*

8. Hofmann, "Genèse de l'atlas historique," 108 (my translation is a little free, but not, I think, unfaithful to Hofmann's sense). Her comments on the Westermann atlas prompt mine on the Bavarian one. She also invokes a "dialogue between image and text." I owe much to her discussion.

9. The atlases cited are not an exhaustive list of Hammond atlases with historical maps: *Hammond's Modern Atlas of the World* (New York, 1904–5); C. S. Hammond and Company, *Hammond's Modern Atlas of the World* (New York, 1911); C. S. Hammond and Company, *The New Reference Atlas of the World* (New York, 1926); C. S. Hammond and Company, *New World Atlas* (Garden City, N.Y., 1947); *Hammond's New International World Atlas* (New York and Maplewood, N.J., 1952); Hammond Incorporated, *Hammond's Ambassador World Atlas* (Maplewood, N.J., [1954]); Hammond Incorporated, *Hammond Atlas of World History* (Maplewood, N.J., ca. 1976).

10. Angus MacKay, *Atlas of Medieval Europe* (London, 1997).

11. See ch. 1, n. 34, above.

12. Ménard, 1713; Chiquet, 1719. Dean, "Sic enim est traditum," 9. One copy of the Chiquet atlas was turned into an *atlas composé* with the addition of quite a few sheets of mainly ancient history (BN, Ge.FF.8344).

13. Black, *Maps and History,* 62.

14. There are exceptions, of course. Benicken explicitly cites his guide, the books of Dresch. Koch wrote a book only and engaged a specialist for its maps.

15. Duval's "La carte de l'empire des Sarrazins" might be an exception; but the second Arab map in his collection has no clear relation to the first. The maps are linked only by being Islamic in subject.

16. Lubin, *Mercure géographie* (see ch. 2, n. 149, above). For the treatise on medieval geography, see ch. 3, n. 42.

17. See ch. 4, nn. 118, 120, above.

The list is classified by seven letters: **H, M, G, A, E, O,** and **B.** The letters **H, M,** and **G** apply to maps and atlases that are the central concern of this book. **H** stands for **H**istorical atlases; **M** for **m**aps for history; and **G** for **g**eographical atlases with historical additions. The other letters apply to shorter, subordinate lists. **A** stands for atlases of the type "Ancient Atlas, Classical and Sacred," and the maps they contain, **E** for ecclesiastical and sacred maps and atlases; **O** for other noteworthy items; and **B** for books, usually without maps.

The maps and atlases of central concern (**H, M, G**) are catalogued as a combined list subdivided by century. The distinguishing letters are provided for purposes of classification, so that, for example, a single map is not mistaken for an atlas. Alphabetical order and date apply in each chronological section. In assigning titles or authors to anonymous works for purposes of identification, the main consideration has been the convenience of readers consulting this catalogue and the notes to the book; bibliographic precision is not guaranteed.

The order of each section of the Catalogue depends on the brief citation—author's name (or, rarely, title) and date—heading each entry. When the same compiler is responsible for several titles, the entries are ordered by date (as best as may be determined), not by title. These brief citations by name and date have been used in the text and notes to refer to works fully listed in this Catalogue (these citations in the notes are collected in Index 1). An explanation of the four types of abbreviations used in this book is given on p. xix.

In brief citations to works in the subordinate lists (**A, E, O,** and **B**), each identifying name and date is followed by a slash and the letter signifying the list in question; for example, Smith, 1860/E. When no slash and letter follow a citation, the reference is to a work in the main section—**H, M, G,** classified by century. Maps and atlases later than 1870 are flagged by a diamond (♦).

H, M, G	Sixteenth and Seventeenth Centuries	464
H, M, G	Eighteenth Century	472
H, M, G	Nineteenth Century	500
A	Ancient, Sacred, and Comparative Atlases (Sixteenth to Nineteenth Centuries)	540
E	Ecclesiastical and Sacred Collections (Exclusively)	550
O	Other Relevant Atlases and Maps	554
B	Books, Usually without Maps	562

Historical Atlases and Maps; Atlases with Historical Additions

H, M, G   SIXTEENTH AND SEVENTEENTH CENTURIES

1. *Atlas factice,* 1675.   *Atlas factice* (n.p. [London?], 1673–75).

    M—RGS, 264.h.15. Mapmakers include Ortelius, Hondius, Janssonius, Joseph Moxon, Blaeu, Morden, Danckertsz, W. Berry, Thornton, Westenbert. Specially noteworthy are nos. 67–68, profane adaptations by Philip Lea of Haraeus, 1617/E: "Lumen historiarum per occidentem" (the principal places of modern Europe); "A New Map of Antient Europ to all historiographers" (classical antiquity).

2. Bergeron, 1634.   Pierre Bergeron, *Voyages faits principalement en Asie dans les XII, XIII, XIV, et XV siècles par . . .* (Paris, 1634).

    M—Toronto-Fisher, E-10 1572. Repr., Leiden, 1729; The Hague: Neaulme, 1735. A possible source for Aa, 1729b.

3. Bertius, 1623.   Petrus Bertius (Pierre Bert), "Imperium Caroli Magni cum vicinis regionibus" (Paris, 1623).

    M—BN, Ge.D.12164, Jean Boisseau, 1 sheet; Ge.AF Pf207 (1036 a–d), Jean Picart, 4 sheets. BL, Maps 37.c.1., nos. 34–35 (Paris: Boisseau), 4 sheets, cartouche dedication to Louis XIII; K.3.94, 4 sheets. Repr. in Blaeu and Janssonius atlases; Koeman, 1:100, e.g., *Novus Atlas, Das ist Abbildung und Beschreibung von alle Ländern des Erdreichs,* pt.2 (Amsterdam: Wm. Blaeu, 1635), no. 155; Henr. Hondius and Joannis Ianssonii [Mercator], *Atlantis novi pars secunda exhibens Germaniam inferiorem, Galliam, Helvetiam atque Hispaniam* (Amsterdam, 1638), no. 149 (cf. Koeman, 2:400); *Le Theatre du Monde ou Nouvel Atlas,* ed. Guillaume and Jean Blaeu, pt.2 (Amsterdam, 1640). Newberry.

    Second after Lambarde's Heptarchy as a medieval subject. Two more Charlemagne items deserve mention in this connection:

An item entitled "Caroli Magni imperium" (BL, Maps 10003.d.2) is said to be associated with *Cosmographia, das ist Wahrhafte eigentliche und kurze Beschreibung dess gantzen Erdbodens* (Frankfurt, 1576). There is an impressive effigy of Charles on the cover, but no map. The very different second item, RomeSGI, Cart. Z 2/30, is entitled "Imperium Caroli Magni" and dated 1629. It is a twentieth-century product. The SGI acquired it by gift in 1934. The map records lands of "Tcheco sclavi" and "Yougosclavi," as well as the post-Trianon Treaty boundaries between Hungary and Rumania. The hoax deserves investigation.

4. Boisseau, 1641.   Jean Boisseau, "Europe françoise ou description générale des Empires, royaumes, estats, et grandes seigneuries, qui ont esté possédées et régies, en divers temps par les descendans de la royalle, et tres illustre famille de France" (Paris, 1641).

M—BN, Ge.D.12165. Shows every land ever ruled by French kings, with genealogies superimposed on the geographical background.

Briet, Philippe. *See* subdivision A.

5. Camden, 1600.   William Camden, *Britannia, magna accessione adaucta,* 5th ed. (London, 1600, 1607).

M—BL, "Englalond, Anglia, Anglosaxonum Heptarchia" (fol. 107), by Wm. Rogers. Enlarged ed., 1607, Toronto-Robarts, STC/f, pp. 98–99, by Wm. Hole. Anglo-Saxon characters (with a key in upper left). *Britannia,* 1st ed., 1586. See Shirley, *Maps of Britain to 1650,* no. 232, p. 95 with pl. 86 (Rogers, 1600), model for Hole; no. 280, p. 113 with pl. 99, Hole map.

6. Camden, 1689.   William Camden, *Britannia,* ed. Edmund Gibson (London, 1689).

M—UCB-Bancroft, ffDA610.C25.B75 1695 (2 copies). "Britannia Saxonica," copy 1 between cxx and cxxi; copy 2 opposite (tipped in between) i–ii.

7. Clüver, 1611.   Philip Clüver, *Commentarius de tribus Rheni alvei et ostiis item de quinque populis quondam accolis* (Leiden, 1611).

M—Göttingen, 8vo H. Holl. I, 40. 3 maps. Supplement to Clüver's *Germania antiqua* (1616). See esp. "Trium Rhenis alveorum et populorum quondam accolentium accurata descriptio." *Catalogus mapparum* places this in P. Schrijver (Scriverius), *Inferioris Germaniae provinciarum unitarum antiquitates* (Leiden, 1611), BL, 154.1.8.(1.); less perfect copy, 1054.h.7.

8. Danckwerth, 1652.   Caspar Danckwerth, and Johann Mejer, "Orbis vetus cum origine in eo gentium a filiis et nepotibus Noe" (Husum, 1652).

M—Main entry at Mejer, 1652. Fragment, with plate, in Rodney W. Shirley, *The Mapping of the World: Early Printed World Maps, 1472–1700* (London, 1983), 409, no. 388. Also relevant is Kircher, 1675.

9. Dangeau, 1695.   Louis de Courcillon, abbé de Dangeau, "Histoire du Royaume de France depuis le traité de Verdun jusqu'à Robert, roi de France" (ca. 1695 [BN cat.]).

M—MS. BN, Ge.D.15787. Unsigned, untitled, 1 sheet, colored. In ink, upper-left margin, "De M. l'abbé de Dangeau." Hugh Capet is the focus of interest.

10. Dangeau, 1697.   Louis de Courcillon, abbé de Dangeau, *Nouvelle metode de geografie historique pour apprendre facilement, & retenir longtemps / La geografie moderne et l'ancienne / l'histoire moderne et l'ancienne / Le Gouvernement des Etats / Les interets des Princes, leur Généalogie &c.* (Paris, 1697).

H—BN, G.1859. Rare complete copy; no author's name; maps of the world and Europe precede many maps of France. A pedagogical atlas containing outline maps (*cartes muettes*) of many kinds with gaily colored boundaries. Described in the BN Cartes et plans catalogue as "Cartes de l'administration civile et ecclésiastique au XVIIᵉ siècle." BN, another ed., 8vo G.9499; also Ge.DD.1212, incomplete copy with maps only of France; Ge.DD.1799, also incomplete; Ge.DD.5571, 8 leaves only. UMich-Clements has an anonymous copy, described as "School atlas of France." It resembles the incomplete BN copies. BL, L.2.c.8. (Courcillon Dangeau, *Nouvelle metode*). The alphabetical list of provincial governors in BN, G.1859, is for 1699, that in the UMich-Clements copy is for 1693. The title without accents points to Dangeau as a pedagogical experimenter.

11. Daniel, 1696.   Gabriel Daniel, S.J., "Description de la France par rapport au règne de Clovis et de ses Enfans" (Paris, 1696).

M—Le Long, 1:34, no. 392. BN, Ge.F.carte. 5623; BL (under Berey, fils, engrav.). From Daniel's *Histoire de France depuis l'établissement de la monarchie française dans les Gaules* (Paris, 1696). For other maps in Daniel's history, see Daniel, 1729, and Liébaux, 1728.

12. Duval, 1665.   Pierre Duval, "Cartes pour les itinéraires et voyages modernes" (Paris, 1665).

M—Pastoureau, *Atlas français*, 154. LC; BL. Part 3 of Duval, 1665/A. Begins with maps of the early Islamic Empire.

13. Duval, 1679.   Duval, Pierre, *Cartes de géographie les plus nouvelles et les plus fidèles* [5th ed.] (Paris, 1679 [1671?])

M—Pastoureau, *Atlas français*, 147, nos. 79–82; Le Long, 1:35, nos. 398 and 399, 1:36, nos. 415 and 419. Newberry (the atlas date is earlier than the dates on the four maps listed here). Prior publication in broadsheet form is likely. Hauber, *Versuch einer Historie*, 139, states "[Duval] a. [*anno*] 1671 published a map of medieval France." The Newberry copy of Duval's map includes "La France avec ses anciennes et ses nouvelles bornes. Le Royaume d'Aquitaine" (42); "Royaume de Bourgogne et d'Arles" (43); "Royaume de la France Orientale, dit autrement Austrasie, avec partie de celui de Neustrie" (44); "Royaume de la France Occidentale dit autrement Neustrie" (45). The same maps are in the BL: Maps C.39.e.4 (general atlas of Duval maps; maps numbered in pencil), nos. 27, 28, 29, 30.

14. Ewich, 1652.   Hermann Ewich, "Patriae antiquae inter Julii et Caroli Magni Caesarum" (Amsterdam, 1652).

M—BL, 410.(5.); many more copies. Occurs in Jan Janssonius, *Accuratissima orbis antiqui delineatio* (Amsterdam, 1652), no. 28; Janssonius, *Novus Atlas,* vol. 6 (1658), no. 37, and descendants (not H. de Hondt, 1741). See Koeman, 2:185, 500– 502. An ostensibly historical map of "The Ancient Homelands between the Emperors Julius and Charles the Great," with few historical traits. Shows the Lower Rhine Valley, chiefly the Netherlands.

15. Fabius, 1641.   Nicasius Fabius, "Nova antiquae Flandriae geographica tabula qualis sub Baldwino Ferro et Judith . . . fuit" (Amsterdam, 1641).

M—Stevens, *Bibliographia geographica,* pt. 1, no. 1058. BL, K.3.91, detached map. NYPL; Williams College, Chapin Library, from Antoine Sanderus, *Flandria illustrata sive descriptio comitatus istius,* 2 vols. (Cologne [Amsterdam: Blaeu], 1641–44). Engraving probably by the Hondius firm. Repr., 3 vols. (The Hague, 1732) (Harvard); also (The Hague, 1735) (NYPL?). A reprint in Brussels (1735) is not confirmed. Dutch tr., *Verheerlykt Vlaandre* (Leiden, 1735) (not in NUC).

16. Gibson, 1692.   Edmund Gibson, "Map Accurately Showing Places Mentioned in the Saxon Chronicle" (Oxford, 1692).

M—UCB-Bancroft, Toronto-Fisher, in E. Gibson, ed., *Chronicon Saxonicum ex MSS Codicibus* . . . (Oxford, 1692). Gibson affirms that the map is limited to the places established in the included "Explicatio nominum locorum."

17. Jaillot, 1689.   Hubert Jaillot, *Atlas nouveau contenant toutes les parties du monde* (Paris, 1689).

G—BSB. Sanson maps, all modern except "Terra sancta" divided among twelve tribes.

18. Janssonius, 1644.   J. Janssonius, *Des Nieuwen atlantis* (Amsterdam, 1644).

M—LC List 4258. No. 51 is Bertius, *Imperium Caroli magni.*

19. Keere, 1617.   Pieter van den Keere, *Germania inferior* (Amsterdam, 1617).

G—Facsimile, ed. Cornelis Koeman; Cornell-maps, G1025 T37++ Ser. 3 v. 3; MilwaukeePL. An atlas of the whole Netherlands published during a truce in the war with Spain; similar to English county atlases but with an "ancient" prelude.

20. Kircher, 1675.   Athanasius Kircher, S.J., *Arca Noë magna rerum varietate explicata* (Amsterdam, 1675).

M—BL, 460.c.9. Rich in illustrations and maps. At p. 222, "Tabula geographica divisionis gentium et populorum per tres filios Noë"; p. 192, "Geographia conjecturalis de orbis terrestris post diluvium transformatione ex variorum geographorum sententia" (world map with shaded areas, some labeled as once land but now ocean, others as once ocean and now land); p. 196, "Topographia Paradisi terrestris juxta mentem et conjecturas auctoris."
La Ruë, 1651. *See* Subdivision E.

21. La Ruë, 1653.   Philippe de La Ruë, "Armenia vetus in quattuor partes distincta ad tempora Justiniani" (Paris, 1653).

M—BN, Ge.DD.2686 (*atlas composé,* Église de Paris), no. 23. Another copy, BN Ge.AF.Pf36 (100).

22. Lambarde, 1568.   William Lambarde, "Angliae Heptarchia" (London, 1568).

M—Shirley, *Maps of Britain to 1650*, nos. 83b, 106, 115, 134, 145, 190. BL, 508.c.7; also 8vo (1576) 578.f.1., (1596) 578.f.2. (single sheets for binding into a book). BL, *Archaionomia* [Greek capitals] *sive de priscis Anglorum legibus* (London, 1568), opp. fol. Ej (= E1); "names and supposed areas of the kingdom superimposed [on a modern background]." Copies of the *Archaionomia* normally contain a one-page explication of the map. BL-North, G.3995, Lambarde, *A Perambulation of Kent* (London, 1576; enl. ed., 1596), opp. p. 1, "Angliae Heptarchiae."

23. Lazius, 1561.   Wolfgang Lazius, *Typi chorographici provinciarum Austriae* (Vienna, 1561).

H—Facsimiles: Eugen Oberhummer and Franz R. von Wieser, eds., *Wolfgang Lazius Karten der österreichischen Lande und des Königreichs Ungarn aus den Jahren 1545–1563* (Innsbruck, 1906) (ViennaÖNB). Ernst Bernleiter, ed., *Wolfgang Lazius Austria Vienna 1561* (Amsterdam, 1972). 11 maps, ostensibly historical, to celebrate Habsburg dominion over its lands. The facsimiles have major introductions and commentaries.

24. Malbrancq, 1639.   Iacobus Malbrancq, S. J., *De Morinis et Morinorum rebus*, 3 vols. (Tournai, 1639–54).

M—Oxford. 1 map each of ancient St. Omer in vols. 1 and 2. Only pt. 1 of vol. 3 was published; a fourth volume existed in MS at Malbrancq's death.
Manesson-Mallet, 1683. *See* Subdivision A.

25. Mejer, 1652.   Johannes Mejer, "Ditmarsiae tabula ann. 1559"; "Ditmarsiae tabula ann. 1651"; "Frisia Borealis in ducatu Sleswicensi sive Frisia cimbrica Anno 1651"; "Frisia Borealis in ducatus Sleswicensi Anno 1240"; "Frisia Cimbrica antiqua"; "Landtcarte von dem Nortfrieslande in dem Hertzogthumbe Slesswieg. Anno 1651"; "Landtcarte von dem alten Nortfrieslande Anno 1240"; "Frisia Cimbrica antiqua; Helgelandia A° 1649"; "Helgelandt in annis Christi 800 1300 et 1649"; "Nordertheil vom Alt Nordt Frießlande biß an das Jahr 1240"; "Sondertheil Vom Alt Nord Frießland biß an das Jahr 1240" (Husum, 1652).

M—In Johannes Mejer and Caspar Danckwerth, *Newe Landesbeschreibung der zweij Hertzogthümer Schleswich und Holstein* (Husum, 1652). BL, Maps C.24.g.20, p. 28; separately, 798.k.14; Maps 13.e.21, 98.k.14. StockholmRL. Facsimile: LC, G.1923.S4 D3 1963, *Die Landkarten von Johannes Mejer, Husum, aus der neuen Landesbeschreibung etc.*, ed. K. Domeier and M. Haack, Introduction by Christian Degn (Hamburg and Bergedorf, 1963). Also (because the plates were sold to Blaeu) in a fascimile of the Blaeu major atlas (Toronto-Fisher): *Géographie Blaviane*, vol. 3, *L'Allemagne* (= Europe, bk. 8) (Amsterdam, 1663), 68a; more Mejer historical material, *Géographie Blaviane*, 3:61a, 64, 68b. Bibliographical details of the Blaeu version, Koeman, 1:248. A partial fascsimile (BerlinDSB): *J. Meyer's Karte des alten Nord-Frieslande bis an das Jahr 1240 mit Angabe der jetzigen Lage der Inseln* (Tondem and Westerland-Sylt, n.d. [1880]). See also Danckwerth, 1652, and Mejer, 1942/O.

26. Mercator, 1585.   Gerard Mercator, "Le royaume d'Arles" (Antwerp, 1585).

M—In the 1st ed. of the original Mercator atlas. Koeman, 2:287, no. 3, and 300, no. 3, "Aquitania australior, Arelatense regnum. Aquitania australis. Regnum Arelatense cum confiniis." LC List 5918; BL C.3.c.3. A facsimile exists.

27. Metellus, 1594.   Johannes Metellus (Jean Matal), "Franciae, Austrasiae et Helvetiae descriptio" (Cologne, 1594).

M—Reproduced in P. Meurer, ed., *Atlantes Colonienses,* 168 (Toronto-Fisher, fGA312.Z9M48 1988). Part of a small atlas of France. "Austrasia" is a medieval name, an archaism by the sixteenth century.

28. Ortelius, 1570.   Abraham Ortelius, *Theatrum orbis terrarum* (Antwerp, 1570).

H—Koeman, 3:29–68. Facsimiles (Toronto-Fisher): R. A. Skelton, ed., *"A. Ortelius, Theatrum orbis terrarum," Antwerp, 1570* (Amsterdam, 1964); R. A. Skelton, ed., *Abraham Ortelius, "Theater of the Whole World," London, 1606* (Amsterdam, 1968). Large bibliography. Koeman attributes part of its success to the combination of text with maps.

29. Ortelius, 1579.   Abraham Ortelius, *Parergon in quo veteris geographiae aliquot tabulae* (Supplement containing several maps of ancient geography) (Antwerp, 1584).

H—Koeman, 3:69–70. Facsimile, R. A. Skelton, ed., *Abraham Ortelius, "Theater of the Whole World," London, 1606* (Amsterdam, 1968). "[A] series of maps illustrating ancient history, sacred and secular" "supplementary" to the modern geography of Ortelius, 1570, "quo tantum hodiernum regionum situm exhibere proposueram." Earliest ed., 1579, 3 maps. 1584, 12 maps. 1592, 26. 1595, 32. 1601, 35. 1624, 41. Ortelius was personally responsible for 38. *Parergon* maps were still being reprinted in the eighteenth century.

30. Ortelius, 1590.   Abraham Ortelius, "Aevi veteris, typus geographicus" (Antwerp, 1590).

M—Koeman, 2:70. YaleB (1592); in Ortelius's *Parergon.* I list two ephemeral publications because they were very valuable in my documentation: the Huntington Library calendar for 1991 (Petaluma, Calif., 1990); also *Maps of the Ancient World,* The Smith and Osher Cartographic Collections, Catal. no. 95071 (1995), Pomegranate Calendars (Rohnert Park, Calif., 1994). Ortelius projects the Ptolemaic world map onto a grid marking the bounds of the world then known; the reader is invited to marvel at expansion since 1492. An exceptionally eloquent and effective map for history.

31. Pellegrino, 1644.   Camillo Pellegrino, *Descriptio antiqui ducatus Beneventani* (Naples, 1644).

M—*NUC.* CambUL, Pellegrino, *Historia ducum Langobardorum,* 2 parts (Naples, 1643–44), 2:1. Van der Aa repr of *Historia ducum,* BL, 585.1.17, map at col. 215. Map repr Aa, 1729a, BL, 214.f.5. vol. 17.

32. Pontanus, 1599.   Jacobus Pontanus, S. J., "Aeneae Navigatio."

M—Sommervogel, 6:1013, no. 8. Oxford, in *Symbolarum libri XVII quibus P. Virgilii Maronis Bucolica Georgica Aeneis . . . declarantur* (Augsburg, 1599), cols. 726–27. Brandmair, *Bibliographische Untersuchungen,* 154, followed by Dainville, *Géogra-*

phie des humanistes, 65, believed Ortelius took his maps of "literary geography" from existing editions, such as Pontanus. Meurer, *Fontes Cartographici Orteliani,* 21–25, demurs.

33. Rossi, 16 –.   Giovanni Giacomo de Rossi, publ., *Mercurio geografico overo guida geografica in tutte le parti del Mondo conforme le tavole geografiche del Sansone Baudrand e Cantelli* (Rome, 16 –).

G—RomeMar, 5166. The main Italian atlas of the seventeenth century. Many editions with varying contents, predominantly modern.

34. Sanson, 1651.   Nicolas Sanson (d'Abbeville), *La France, l'Espagne, l'Italie, l'Allemagne et les isles Britanniques . . . Descrites en plusieurs cartes et differens Traittés de Géographie, & d'Histoire, suivant les plus belles & les principales distinctions, qui se peuvent remarquer dans tous les Autheurs anciens, & nouveaux,* 2d ed. (Paris, 1651).

H—Pastoureau, *Atlas français,* 430. BN, Ge.DD.1857; BerlinPK; when dated, the maps at BN and Berlin PK are marked 1641 (1651 at Yale). First ed. title: *Les Isles Britanniques, l'Espagne, la France, l'Italie et l'Allemagne, Descrites en plusieurs cartes* (1644).

35. Sanson, 1654.   Nicolas Sanson (d'Abbeville), "Anciens royaumes de Mercie et Est-Anglie, Anciens royaumes de Kent, d'Essex et de Sussex; Heptarchie" (Paris, 1654).

M—Shirley, *Maps of Britain to 1650,* no. 506, "rather scrappy work." BL, K.5.: 50-1, "Anciens royaumes de Kent, d'Essex et de Sussex où sont aujourdhuy les comtés de . . . [old kingdoms marked over modern counties with colored borders, 1654]"; 50-2, "Provinces d'West autrefois Royaumes d'Westsex où sont aujourdhuis etc" (n.d.); 50-3, "Anciens royaumes de Mercie et Est Angles etc." (1654); 50-5, "Ancien royaume de Northumberland etc." (date obliterated in alteration); 51, "Le royaume d'Angleterre divisé dans les sept royaumes ou Heptarchie des Anglois-Saxons et la principauté de Galles." Harvard, Sanson, *Atlas de 100 Cartes* (Paris, 1646–86), nos. 9–13.

36. Sanson, 1658.   Nicolas Sanson (d'Abbeville), *Cartes générales de toutes les parties du Monde* (Paris, 1658).

G—LC List 4260; Pastoureau, *Atlas français,* 407, nos. 98–113, are the appendix of ancient geography.

37. Sanson, 1665.   Nicolas Sanson (d'Abbeville), *Cartes générales* (Paris, 1665).

G—Pastoureau, *Atlas français,* 155–71, lists a notable number of maps for ancient and biblical geography, including de La Ruë's set for the Holy Land. The title of the atlas was changed in 1675 to *Cartes générales de la géographie ancienne et nouvelle.*

38. Sanson, 1675.   Nicolas Sanson (d'Abbeville), Guillaume Sanson, and Nicolas Sanson Jr., *Cartes générales de la géographie ancienne et nouvelle,* 2 vols. in 1 (Paris, 1675).

G—LC. Modern and ancient geography only.

39. Sanson, 1683.   *Geographische en Historische Beschryvingh der vier bekende werelds-deelen* (Utrecht, 1683).

G—Pastoureau, *Atlas français*, 429 (Sanson VII F *bis* 1683). BL, Maps 48.b.39. YaleB: catalogued under Sanson. 62 Sanson maps, strictly modern.

40. Schotanus, 1664.   Bernhard Schotanus à [van] Sterringa, "Typus Frisiae veteris inter Rheni medium ostium et Amisiam, Itemque Insulae Batavorum, ut fuerunt habitu terrarum et cultu, cum hic Romani, et postea Franci, rerum potirentur" (n.p., 1664).

M—Koeman, 3:122–23; Bodel Nyenhuis and Eekhoff, *Algemeene Kaarten van Friesland*, 5, no. 8. Map in Christian Schotanus à Sterringa, *Beschrijvinge van de heerlykheed von Friesland Tusschen 't Flie en de Lauwerz* (n.p., 1664), with 35 maps and 11 city plans (NOT SEEN). Wholly revised version: Bernhard Schotanus, *Friesche Atlas* (Leeuwarden, 1698), repr., F. Halma, publ., *Uitbeelding der Heerlijkheit Friesland* (Leeuwarden, 1718) (BSB); facsimile (Schotanus listed as author) (Amsterdam and Leeuwarden, 1979). Schotanus's first map: Halma, 1725; its maps only, BL Maps 25.d.20. See also Alting, 1701.

41. Speed, 1611.   John Speed, *The Theatre of the Empire of Great Britain* (1st ed., 1611).

M—Shirley, *Maps of Britain to 1650*, nos. 316–18, 343, 345, 352, 379, etc. Thomas Chubb, *The Printed Maps in the Atlases of Great Britain and Ireland: A Bibliography, 1579–1870* (repr., London, 1966), 23–45. Colored reproduction in Nigel Nicholson and Alasdair Hawkyard, *The Counties of Britain: A Tudor Atlas by John Speed* (London, 1988). Speed's county atlas complements his *Historie of Great Britaine* (London, 1610).

42. Speed, 1627a.   John Speed, *A Prospect of the Most Famous Parts of the World* (London, 1627).

H—3d ed of Speed, 1611. All dated maps marked 1626. LC List 5928; BL, Maps c.7.c.b.(2); YaleBAC, Atlas (in folio) 32, beautifully colored specimen of *Prospect*, with the battles map but without the *Theatre*. Facsimile: R. A. Skelton, ed., *John Speed, A Prospect of the Most Famous Parts of the World, London, 1627* (Amsterdam, 1966).

43. Speed, 1627b.   John Speed, "The invasions of England, with al their Ciuill wars since the Conquest" (London, 1601, 1627).

M—Shirley, *Maps of Britain to 1650*, nos. 255, 316, 317 (pl. 102), 261. The map was initially published in 1601 (one copy survives). It was reengraved in Holland by Cornelis Danckertsz and included in Speed, 1627a. The map immediately precedes Speed, 1611, in the BL copy noted at Speed, 1627a, and the third item of Speed, 1611, in the edition of 1650–62.

44. Tassin, 1633.   Christophe Tassin, *Cartes générales des royaumes et provinces de la haute et basse Allemagne* (Paris, 1633).

M—Pastoureau, *Atlas français*, 437, 445–47. Newberry; AmstUL. My concern is no. 34, "Austrasie"; see also Metellus, 1594, above.

45. Tavernier, 1634.   Melchior Tavernier, *Théatre géographique du royaume de France* (Paris, 1634).

   M—BL; Newberry. No. 64, all the bishoprics of France (map date 1622).

46. Tavernier, 1645.   Melchior Tavernier, "Carte de l'ancien royaume d'Austrasie, Le vray & primitif heritage de la Couronne de France" (Paris, 1645).

   M—Pastoureau, *Atlas français*, 470–71. BN. (Modern) Lorraine is the main subject of the map.

47. Wendelinus, 1649.   Gottefridus Wendelinus, "Leges Salicae illustratae illarum natale solum demonstratum" (Antwerp, 1649).

   M—Princeton, Annex IV 7619.964q. Map at p. 1, but marked for p. 108 (an unsuitable location). Wendelinus's subject is where the Frankish laws originated.

48. Wree, 1647.   Olivario Wree (Vrée, Vredius), "Francorum primae sedes" (Bruges, 1647).

   M—BL, 153.g.8–9. Two-page map from Wree's posthumous and incomplete *Historia comitum Flandriae pars prima: Flandria ethnica a primo consulatu Caj. Jul. Caesaris usque ad Clodovaeum primum Francorum regem Christianum per DLIV annos*, 2 pts. (Bruges, 1650), 1:236. Repr., Aa, 1729a, pt. 4, no. 1 (1729).

H, M, G   EIGHTEENTH CENTURY

49. Aa, 1707.   Pieter van der Aa, *Cartes des itinéraires et voyages modernes . . . dans toutes les parties du monde* (Leiden, 1707).

   M—LC List 5482. LC, G1036.A25 1707 Vault. Medieval items are nos. 86, 87. Repr. from van der Aa, *Naaukeurige Versameling der gedenkwaardigste zee en landreysen na Oost en Westindien* (Leiden, 1707) [NOT SEEN]; also in Solomon Bor, ed., *(Seer) Aanmerkelyke Reys-Beschryvingen van Johan der Plan Carpin en Br. Ascelin* (Leiden, 1706). NYPL, BAC/+ p.v.113, no. 22. New repr. from the Dutch, with added French titles in van der Aa, *Atlas nouveau et curieux des plus célèbres itinéraires . . . depuis 1246 jusqu'à l'an 1696*, 2 vols. (Leiden, 1714; repr., 1728 [RLScot]), 1:43–44. See also Bergeron, 1634.

50. Aa, 1729a.   Pieter van der Aa, *Galerie agréable du monde*, 66 pts. in 27 vols. (Leiden, 1729).

   M—Koeman, 1:15–29; LC List 3485. Ecclesiastical provinces and bishoprics; repr. of Pellegrino, 1644, and Wree, 1647.

51. Aa, 1729b.   Pieter van der Aa, *Recueil de divers voyages curieux*, 2 vols. in 1 (Leiden, 1729).

   M—NYPL-rare, *KB/1729. Repr., *Voyages faits principalement en Asie dans les XII, XIII, XIV, et XV siècles par . . .* (The Hague, 1735). Toronto-Fisher, E-10 1572. See Bergeron, 1634.

52. *Aloude Holland*, 1745.   *Atlas van het aloude Holland, en deszelfs waare gelegenheid, zoo als die was, onder de Regeering der Keyseren, Koningen, Hertogen en Graven, naaukeurig verbeeld in IX Landkaarten* (The Hague, 1745).

H—French title, *Atlas de la Hollande et de sa veritable situation, telle qu'elle étoit sous la domination des empereurs, rois, ducs, et comtes representé en IX cartes.* Koeman, 2:148, "copied from Menso Alting's work . . . [the maps] are newly engraved in a poorish way." This view needs checking. Göttingen, fol. H. Holl I, 141; BL, Maps C.9.d.3. (6 of the 9 maps), Maps 13.e.z. The medieval items are no. 7, "Landkaart van het t'hans genaamde Holland, zoo als dat, in't begin der vijfde eeuwe door' de Vriezen, Saxers, Warners, Britten, Engelschen, Sclaaven ent'overschot der Batavierns, Salers, Quaden en Franken bewoond wierdt"; no. 8, "Landkaart von Gallië, zoo in den tijd als Klovis zich van t'zelve meester maakte" (to go by the title, the source is Liébaux, 1728, rather than Alting); no. 9, "Landkaart verbeeldende het Vriesche Hertogdom van het thaans genaamde Holland, en des zelfs twaalf onderhoorige graafschappen zoo als dat aan Keyzer Lotharis, in't jaar 839 by zijns vaders gemaakte rijksverdeeling was te beurt gevallen." The "ancient" items reshuffle Alting.

53. Alting, 1701.   Menso Alting, *Notitia Germaniae inferioris antiquae qua hodie in dicione VII Foederatorum,* 2 vols. (Amsterdam, 1697–1701).

H—BL, 156.R.4. The earliest atlas of "national" history; vol. 2 contains the first sequence of maps of medieval history. Alternative titles: *Descriptio agri Batavi et Frisii omnisque regionis quae hodie est in dicione VII. Foederatorum cis et ultra Rhenum,* or *Descriptio agri Batavi et Frisii seu notitia Germaniae inferioris; Germaniae inferioris antiquae quae hodie est in dicione VII Foederatorum [descriptio].* Repr., F. Halma, 1725: BSB; BL, Maps 25.d.20; UCB-Bancroft (incomplete).

54. Andrews, 1797.   John Andrews, *Historical Atlas of England* (ancient and modern . . . from the Deluge to the Present Time) (London, 1797).

H—BL, Maps 9.c.13 (incomplete); Oxford. 13 plates. The number of plates may increase if additional copies come to light.

55. Anville, 1719.   [J. J.] Bourguignon d'Anville, *La France ancienne* (Paris, 1719).

M—L. C. J. Manne and J. D. Barbié du Bocage, *Notices des ouvrages de M. d'Anville* (Paris, 1802), 53, 57 (NYPL). Illustrations for Louis Dufour de Longuerue, *Description historique et géographique de la France ancienne et moderne,* 2 vols. in 1 (Paris, 1719), 9 maps, LC-rare. For individual maps, see also Beaurain, 1749. MS maps on this theme, BN, Ge.D.10412–14.

56. Anville, 1757.   Bourguignon d'Anville, *Cartes pour l'Histoire de St. Louis par Joinville* (Paris, 1757).

M—Manne and Barbié, *Notices des ouvrages de M. d'Anville,* 57–58; NUC. Oxford. Jean de Joinville, *Histoire de Saint Louis* [and Guillaume de Nangis, *Annales de son règne . . .*], ed. Anicet Melot et al. (Paris, 1761), p. 1, d'Anville, "Carte pour la croisade de Saint-Louis en Égypte et en Palestine"; p. 41, "Carte particulière pour l'expédition de Saint-Louis en Égypte."

57. Anville, 1758.   Bourguignon d'Anville, *Cartes pour l'Histoire de Saladin par Marin* (Paris, 1758).

M—Manne and Barbié, *Notices des ouvrages de M. d'Anville,* 57. DumbOaks; Princeton. Three illustrations for François Louis Claude Marin, *Histoire de Saladin,*

Sulthan d'Egypte et de Syrie, 2 vols. (Paris, 1758): "Environs de Ptolémaïs"; "Ptole-maïde ou Acre (plan)"; "Jérusalem (plan)."

58. Anville, 1762.   Bourguignon d'Anville, "Carte pour le mémoire sur l'expédition d'Héraclius en Perse" (Paris, 1762).

M—Manne and Barbié, *Notices des ouvrages de M. d'Anville*, 63. LC, AS162.P3, d'Anville, "Recherches géographiques concernant l'expédition de l'empereur Héra-clius en Perse," Paris, Académie des inscriptions et belles-lettres, *Histoire et mémoires* 32 (1768): 559–72, map at 559. The map was ready years before d'Anville's study was published. He would probably have considered its subject to be ancient geography.

59. Anville, 1771.   Bourguignon d'Anville, "Germanie, France, Italie, Espagne, Isles brittaniques dans un âge intermédiaire de l'Ancienne géographie et de la Moderne" (Paris, 1771).

M—Manne and Barbié, *Notices des ouvrages de M. d'Anville*, 56–57. The map complements d'Anville's *États formés en Europe après la chute de l'Empire romain en occident* (Paris, 1771), BN, G.4094. D'Anville's main contribution to the geography of the Middle Ages, this map is very widely circulated. BL, K.3.97., K.3.98. (an English edition); LC, in d'Anville's *Atlas général* (Paris, 1743–80), no. 11; repr., *Atlas antiquus Danvillianus* (Nuremberg, 1784), no. 12. Many other locations.

60. Arles, 17–?   "Table généalogique des royaumes d'Arles et de Bourgogne et de leurs parties démembrées."

M—MS, undated. ArchNat, Marine, Serv. hydrog., 6JJ/86, no. 21. Very large, linen-backed sheet, with 2 small maps and much writing along their sides.

61. *Atlas complet,* 1747.   *Atlas complet des révolutions que le globe de la terre a éprou-vées depuis le commencement du monde jusqu'à présent* (en 66 cartes gravées toutes pareilles) (n.p., n.d. [catal. 1747]).

H—Unpublished. BN, Ge.CC.1370. Acquisition record at BN, Cartes et plans, Journal des Entrées des Cartes, Registre C 14094 à 14970, no. 14526 "du 23 Mai 1854 Reçu à titre gratuit du Départ. des imprimés." An unfinished draft; each map is composed of 2 folio printed sheets pasted together. Handwritten explanations be-tween maps. Nos. 32–62 are about the Middle Ages. Pedley, *Vaugondy*, 233, strongly doubts the attribution to Gilles Robert. For later atlases of this kind, see Luneau, 176–, and Dupré, 1763.

62. *Atlas novus,* 1730.   *Atlas novus ad usum serenissimi Burgundiae ducis—Atlas françois à l'usage de M. le Duc de Bourgogne. Atlas nouveau . . . par Guillaume de l'Isle* (Amsterdam, 1730).

G—AmstUL. Amsterdam University Library, *Catalogue of Atlases*, vol. 1, ed. Al-bert H. Sijmons (Amsterdam, 1987), no. 087; Koeman, 2:54–55. Includes Delisle's medieval canon.

63. *Atlas of England,* 1720.   Atlas of England and Wales in 50 maps (n.p., ca. 1720 [catal.]).

M—BN, Ge.D. 2654. No.2, "Britannia Saxonica," from Gibson, 1692. "Roman Britain" is the only other historical map.

64. Bachiene, 1785.   W. A. Bachiene, ed., *Atlas tot opheldering der hedendaagsche Historie* (Amsterdam, 1785).

M—AmstUL, VIII-B-26. Adapted from Bowen, 1766/O. The Holy Land in twelve tribes (also for the travels of Jesus Christ).

65. Banduri, 1711.   Anselmo Banduri, O. S. B., "Urbis Constantinopolitanae in xiv regiones divisae qualis fuerit sub Honorio et Arcadio necnon eiusdem suburbiorum delineatio"; "Urbis Constantinopolitanae in tres partes divisae, qualis ab Auctore Anonymo, qui Alexii Comneni tempore vivebat, describitur, delineatio" (Paris, 1711).

M—DumbOaks. In Banduri, *Imperium Orientale sive antiquitates Constantinopolitanae,* 2 vols., Corpus Byzantinae historiae, vol. 24 (Paris 1711), 2:448. The full contents are listed in *Repertorium fontium historiae Medii Aevi, Series collectionum* (Rome, 1962), 184. See also Delisle, 1711.

66. *Bardengow,* 17–.   "Bardengow pagus antiquae Saxoniae cum finitimis nonnullis" (eighteenth century).

M—HABW, 30,21. From an unidentified book: lower-left margin, "Ad Part. I Cap. IV."

67. Barrington, 1773.   "A Map of Europe for the Illustration of King Ælfred's Anglo-Saxon translation of Orosius" (London, 1773).

M—CambUL, Z.21.7. Done by Reinhold Forster for Daines Barrington, ed., *King Ælfred's Orosius: The Anglo-Saxon Version from the Historian Orosius by Ælfred the Great, together with an English Translation* (London, 1773). See also Forster, 1788.

68. Basel, 1748.   "Eigentliche Vorstellung der Schlacht und Gegend by St. Jacob vor Basel" (Nuremberg, 1748).

M—Julien, *Nouveau catalogue,* 65. Göttingen, Mappe 235¹,C. The map shares equal space with an *Erklärung* coded by letters.

69. Beaurain, 1749.   Jean de Beaurain, publ., *Atlas géographique contenant les cartes pour servir à l'intelligence de la géographie sacrée et prophane* (n.p. [Paris], 1749).

H—BN, Ge.BB.565 I, *atlas de choix* in 16 vols. Vol. 15, profane antiquity; vol. 16, 63 historical maps by noted mapmakers. J. A. Dezauches (†1829) signed the table of contents.

70. Bellin, 1759.   Nicolas Bellin, "Carte d'une partie du globe pour l'intelligence des migrations des peuples" (n.p. [Paris], 1759).

M—NOT SEEN. The title continues, "suivant l'histoire des Huns, des Turc, des Mogols &c de M. de Guignes 1758." Two sheets 4to. *Journal des sçavans,* August 1759, 129–30, says that the map is absolutely necessary for understanding Joseph de Guignes, *Histoire générale des Huns, des Turcs, des Mogols et des autres Tartares occidentaux,* 4 vols. (Paris, 1756–58). A similar (incomplete?) map is in ArchNat, Marine, Serv. hydrog. 6JJ/43, nos. 5A and 5B. The surviving map and the one reviewed in the *Journal* are unlikely to be identical; many features described in the review are absent.

71. Bessel, 1732.   Johann Georg Bessel, with F. J. Hahn, *Chronicon Gotwicense seu Annales liberi et exempti monasterii Gottwicensis* (Tegernsee, 1732).

M—2 vols. in 1, continuously paged, 3 maps. *NUC* 51:141; Le Long, 1:35. BL, 477.h.9, 10, and 206.i.5, 6; Princeton-rare; UCB-Bancroft. For the map titles, see 285nn. 21–22, 289nn. 55–56, above. The time span of the maps reaches the Interregnum (1254). About the author, see P. Emmeram Ritter, O.S.B., "Gottfried Bessel—der 'deutsche Mabillon,'" in *Gottfried Bessel (1642–1749): Diplomat in Kurmainz—Abt von Göttweig—Wissenschaftler und Kunstmäzen*, ed. F. R. Reichert (Mainz, 1972), 203–14; Peter G. Tropper, "Abt Gottfried Bessel (1714–49)," *Stift Göttweig Jubiläumausstellung: 900 Jahre Stift Göttweig, 1083–1983, ein Donaustift als Repräsentant benediktinischer Kultur* (Stift Göttweig, 1983), 644–86.

72. Blair, 1768.   John Blair, *Fourteen Maps of Ancient and Modern Geography for the Illustration of the Tables of Chronology and History* (London, 1768).

M—RGS, 1.A.16. For Blair, *Chronology and History of the World* (London, 1768). No. 8, "Imperium Caroli Magni occidentis imperatoris ad finem saeculi post Christum VIII." Entries for Blair's book: LC List 3305, records 1st ed., 1754; LC, 1768, 15 maps, 56 tables, with lists of rulers, etc.; BN, 1779; BPL.

73. Bonne, 1762.   Rigobert Bonne, Jean Denis Janvier, and G. A. Rizzi-Zannoni, *Atlas moderne* (Paris, 1762; augm. ed., 1771).

G—*NUC;* LC List 5986. BL, K.118.e.25.; UMich-Clements; Princeton; LC; Newberry; Yale. More appropriately listed under the publisher's name. *Journal des sçavans,* in April 1762 announces it as forthcoming and, in November 1762, as published; very favorable notice, ibid., July 1771. Delamarche, 1806b, is a nineteenth-century reissue. The 1771 ed., again by Lattré (LC List 5984), is somehow associated with the publisher Desnos.

74. Bonne, 1787.   Rigobert Bonne, and Nicolas Desmarest, *Atlas encyclopédique contenant la géographie ancienne, et quelques cartes sur la géographie du moyen âge, la géographie moderne et les cartes relative à la géographie physique (= Encyclopédie méthodique,* vols. 33–34) (Paris, 1787–88).

G—LC; Newberry. Anonymous title page. Ancient geography followed by modern and physical; nothing medieval. A recycled Bonne, 1762.

75. Bourgogne, 1712.   "Nouvelle Carte de differents etats de la monarchie françoise sous Jules Cesar sous les rois de la premiere, de la seconde et la troisieme race avec leur généalogies jusqu'à Hugues Capet. Plusieurs observations pour l'intelligence de l'histoire" (n.p., n.d. [Amsterdam? 1712?]).

M—LeidenUL, Portf. 82, no. 2. Two oblong folios pasted together into a very wide field, engraved so as to fit together. Discreetly colored. Left sheet marked in pen, "Tom 1 Nr 27a"; right, "Tom 1 Nr 27b." Three genealogical trees with the House of Burgundy in the central place. Four small maps of France, of the pre-1700 variety. The list of battles of the third dynasty ends with 1712 Denin (i.e., Denain). The subject suits the *Supplement* to Chatelain, 1705 (vol. 7), but is not from there. The duke of Burgundy, heir to Louis XIV, died in 1712.

76. Bray, 1711.   Thomas Bray, ed., *Papal Usurpation and Persecution . . . Designed as Supplemental to the Book of Martyrs* (London, 1711–12).

M—RLScot. Pt. 2 is by Jean Paul Perrin and entitled *History of the Old Waldenses and Albigenses,* tr. S. Lennard. Two maps: before p. 1, "The Valleys of Piedmont and France which were the seat of the Waldenses or Vaudois both Ancient and Modern" (John Senex); between pp. 72 and 73 (end of Waldensian history, beginning of Albigensian), "The Seat of the Albigenses So Famous for Their Sufferings by the Papal Croisados and Massacres in the 13th Centurie." Districts shown in their current guise.

77. Brion, 1766a.   Louis Brion de la Tour, *Atlas général, civil et ecclésiastique* (Paris, 1766).

G—*NUC.* LC List 640, 3509. LC; Harvard has a so-called 3d ed. dated 1772 (maps dated 1766; *BN Impn.* records a 3d ed. in 1768); UIll. The promised coverage of France boils down to one historical map by Rizzi-Zannoni and various maps for current features (e.g., arsenals).

78. Brion, 1766b.   Louis Brion de la Tour, *Coup d'œuil général sur la France . . . faisant partie de l'Atlas historique de la France dirigé par le Sieur Desnos* (Paris, 1766).

G—*NUC.* LC. Its relationship to Desnos, 1765c, is unclear.

79. Brion, 1777.   Louis Brion de la Tour, *Atlas, et tables élémentaires de géographie, ancienne et moderne (adopté par plusieurs Écoles royales militaires),* new ed. (Paris, 1777).

G—LC, 20 maps.

80. Brion, 1786.   Louis Brion de la Tour, *Atlas général et élémentaire pour l'étude de la géographie et de l'histoire moderne,* nouvelle édition avec description en marge de chacune des cartes (Paris, 1786).

G—BL, Maps 143.c.15; BN, Ge.DD.583. The side panels are neatly pasted on; the lower margin is not used. Brion is named as reviser on the maps but not on the title page. Not historical.

81. Buy, 1761.   Claude Buy de Mornas, *Atlas historique et géographique dédié à M. le président Hénault; Atlas méthodique et élémentaire de géographie et d'histoire* (Paris, 1761–70).

G—*NUC.* LC (incomplete); BL (3 copies); RLScot; BSB; PhiladUL-rare (Lea). Luxurious 4-vol. production with the usual Desnos border. Much marginal text (engraved in wide margins, sometimes occupying a full page), genealogical tables, etc. Repr., 1783; without Buy's name, 1793. Predominantly ancient and so slow-moving that the last volume reaches only to the emperor Augustus. On this atlas, see Hofmann, "La genèse de l'atlas historique en France," 120–21.

82. Cellarius, 1776.   Christoph Cellarius (Keller), *Appendix triplex notitiae orbis antiqui Christophori Cellarii* (Leipzig, 1776).

H—BerlinPK; YaleB. 18 plates. The 3d appendix publishes the plates, long languishing at the Gleditsch firm, of Keller's unfinished atlas of medieval geography. J. B. Homann had engraved them. The original *Notitia* is Cellarius, 1701b/A.

83. Châtelain, 1705.   [Zacharie] Châtelain, *Atlas historique, ou nouvelle introduction à l'histoire, à la chronologie et à la géographie ancienne et moderne, représentée dans de nouvelles cartes où l'on remarque l'établissement des états et empires du monde, leur durée, leur chute etc. par M. C******, 7 vols. (Amsterdam, 1705–20; repr., 1738).

G—Koeman, 2:33–38 (full printing history); *NUC*. NYPL-rare; Newberry; Harvard; Toronto-Royal Ontario Museum; Yale (3 copies). First direct pairing of the words "atlas" and "historique," but predominantly concerned with the world of its own day. A best-seller, often attributed to "Henri Abraham" Châtelain. For the attribution to Zacharias, see *Acta Eruditorum* (Leipzig), July 1709, 294–95. Few and incidental historical maps; in particular, 1: nos. 7, 10; 2: nos. 1, 3, 4, 5; 3: nos. 42, 45, 46; 7: nos. 28, 34–36. Many maps are surrounded by text. The ample and often controversial explanatory matter was by Nicolas Guedeville, a fugitive from a French monastery (he was replaced for vol. 7).

84. Chiquet, 1719.   Jacques Chiquet, *Le nouveau et curieux atlas géographique et historique ou le Divertissement des Empereurs, Roys, et Princes tant dans la guerre que dans la paix* (Paris, n.d. [1719]).

G—LC List 4279. BerlinPK, 24 maps. A modern world atlas; at the end, table of ecclesiastical districts. A recycling of Ménard, 1713, which has the same title.

85. Coquart, 1705.   A. Coquart, *Lutèce ou Premier-[Huitième] Plan de la ville de Paris par M. L. C. D. L. M.* (n.p., n.d. [Paris, 1705]).

M—*NUC*. Harvard; Milwaukee AGS; Columbia-rare (incomplete). "M. L. C. D. L. M." stands for "Monsieur le Commissaire de La Mare." It is almost certain that Coquart produced plans 1–7 of Paris for Lamare, 1705 (q.v.). See also Fer, 1705. Cartouches in 4 corners (in most of the maps): upper left, title with indication of sources; lower left, "renvoys," i.e., numbers on the map identified with particular places; upper right, "Description," lower right, "Suite de la description." Léon Vallée, *Bibliothèque nationale, Catalogue des plans de Paris* (at BN, Cartes et plans) (Paris, 1908): 3d plan, no. 2463; 4th, no. 2460; 5th, no. 2461; 6th, no. 2458. Vallée indexes these plans under de Fer (as publisher?). Entries indicate sources, e.g., no. 2460, "Tiré de Rigord, de Knobelsdorf, de Rodolphe Boteree, de Raoul, de Praeles, de Paul Merula, de Guaguin, the Pithou, de Papire Masson, de Corrozet de Dubreuil de Duchesne, des Memoriaux, et autres Anciens Registres de la Chambre des Comptes, et des Archives de l'Archevêché, du Chapitre de Notre Dame, et des Anciennes Abbayes."

86. Courtalon, 1774.   Courtalon, *Atlas élémentaire où l'on voit sur des cartes et des tableaux relatifs à l'objet l'Etat actuel de la constitution politique de l'Empire d'Allemagne* (Paris, 1774).

H—ViennaÖNB; BN, G.13662; LC; Harvard. Mees, 3, lists it as a type of historical atlas. Courtalon, an abbé with an unrecorded first name, is unusually careful to explain his procedures. He makes clear that history is found in his (carefully color-coded) genealogical tables rather than in the maps. His atlas probably influenced Lesage, 1801. For the 2d ed., see Mentelle, 1789. Courtalon was tutor to the pages of two lofty ladies at court, one of them Saxon: *Dict. biog. franç.* 5:1094.

87. Daniel, 1721.   Gabriel Daniel, S. J., "Ionction de l'Océan et de la Mer Noire," in *Histoire de la milice française et des changements qui s'y sont faits depuis l'établissement de la monarchie franque en Gaule,* 2 vols. (Paris, 1721), 2:621.

M—LC-rare. Reproduced in Harms, *Themen,* no. 62. Charlemagne's attempted canal to link the Rhine and Danube.

88. Daniel, 1729.   Gabriel Daniel, S. J., *Histoire de France depuis l'établissement de la monarchie française dans les Gaules,* nouv. ed. "enrichie de cartes géographiques," 10 vols. (Paris, 1729).

M—For editions, *NUC;* Sommervogel, 2:1806–10. LC, DC37.D178; Princeton. For its earliest map, see Daniel, 1696. For the widely circulated later ones, see Liébaux, 1728. A final set of maps for this work is Robert, 1755. Daniel's full *Histoire* was first published in 3 vols. (Paris, 1713).

89. Delisle, as listed below.   Guillaume Delisle, atlases containing all or most of his major maps of medieval subjects.

   a. Delisle, 1700. [Atlas de géographie] (1700–1712). LC List 535, 636, 3456, ad hoc gatherings of Delisle maps; including "Theatrum historicum ad A.D. 400" (1705). StockholmRL, LF 78, fine copy of an atlas, 1700–1713; full canon of medieval maps.

   b. Delisle, 1733. *Atlas nouveau,* 2 vols. (1733). NYPL; Yale, 2: nos. 131–32, medieval Toul and Dauphiné; no. 133, the eastern empire under Constantine Porphyrogenitus; no. 135, Byzantine Empire divided into themes after Heraclius. UCB-Bancroft, same title but in brackets; UCB catalogue gives the dates "1700–1726," but 1731 is on the map of Alexander's empire.

   c. Delisle, 1730. See *Atlas novus,* 1730, above.

   d. Delisle, 1754. Guillaume Delisle and Philippe Buache, *Atlas général de géographie ancienne et moderne* (n.p., n.d. [Paris, 1754]). UMich-Clements. Most Delisle maps are 1745 reprints by Buache, "gendre de l'auteur." No. 126, Ph. Buache, "Carte générale pour . . . l'Histoire Sainte" (1754).

   e. Delisle, 1769. Guillaume Delisle and Philippe Buache, *Atlas géographique et universel* (Paris, 1769–99). LC List 3512. Same title, Paris 1789–90, LC List 3525. Nos. 36–37, "Theatrum historicum." Modern geography followed by ancient. Another printing, untitled, [*Atlas géographique et universel*] (Paris, 1700–1762). LC List as *a* above, nos. 87–88, "Theatrum historicum," East and West (1705); no. 100, the eastern empire under Constantine Porphyrogenitus.

90. Delisle, 1705.   Guillaume Delisle, "Theatrum historicum ad annum Christi quadringentesimum, pars occidentalis, pars orientalis," 2 sheets (Paris, 1705).

M—Author's commentary in a separate book: G. Delisle, *Remarques sur le Théatre historique pour l'an 400* (Paris, 1705), whose pages are sometimes pasted to the sides of the map sheets. The maps can be found at LC, NYPL, and widely elsewhere; *BL Maps Catal.* 8:1104. Frequent in Delisle atlases. Noteworthy parallel transmission: John Senex, *Modern Geography* (London, 1708–25), nos. 30–31; Moll, 1709, no. 30; Faber, 1711, no. 46, "Scena historiarum orientalis quinti seculi P.N. Chr."; no. 47, " . . . occidentalis" (LC List 5963); Köhler, 1720/A, nos. 43, 44; the "Scena" printed by

Weigel (Köhler's printer), without attribution to Delisle, has a Caspian Sea that differs markedly from Delisle's outline.

91. Delisle, 1707.   Guillaume Delisle, "Civitas Leucorum, sive pagus Tullensis" (Paris, 1707).

M—Le Long, 1:35, no. 408. Harvard. For Benoît Picart, *Histoire ecclésiastique et politique de la ville et du diocèse de Toul* (Toul, 1707), with a 13-page "Avertissement." The map is in many atlases (e.g., Beaurain, 1749).

92. Delisle, 1710a.   Guillaume Delisle, "Carte de l'Asie Septentrionale pour servir à l'histoire de Ghengis Khan composée par M. Petis de la Croix" (Paris, 1710).

M—BN, Ge.AF.Pf37(94). For *Histoire du grand Genghizcan, premier empereur des anciens Mogols et Tartares . . . contenant la vie de ce grand can . . . traduite et compilée de plusieurs auteurs orientaux et de voyageurs européens,* ed. F. Pétis de la Croix (Paris, 1710); see also *NUC.* Map reproduced in F. Petis de la Croix, *The History of Genghizcan the Great, First Emperor of the Antient Moguls and Tartars,* anon. Eng. tr. [by Penelope Aubin] (London, 1722; facsimile repr., Beijing: Wendiange Bookstore, 1938), Toronto-Victoria, DS J41PMS. Beijing was under Japanese occupation in 1938; my former colleague Professor Tim Brook advised me that the celebration of Mongol heroes was in keeping with Japanese wishes to win over elements of the population. BN, Ge.D.16739, pen and ink MS with boundary washes, "Carte des Conquestes de Ginghiz Khan par Guillaume de Lisle corrigée en plusieurs endroits et rendu conforme à l'histoire de Ginghiz."

93. Delisle, 1710b.   Guillaume Delisle, "Tabula Delphinatus et vicinarum Regionum, distributa in Principatus, Comitatus, Baronias, &c cum iisdem nominibus quae in antiquis chartis, sub Principibus Delphinis" (Paris, 1710).

M—Le Long, 1:36, no. 421. Oxford [J.-P. Moret de Bourchenne de Valbonnais], *Mémoires pour servir à l'histoire de Dauphiné sous les Dauphins de la Maison de la Tour du Pin* (Paris, 1711); the work was published anonymously. The map is in many atlases (e.g., Beaurain, 1749).

94. Delisle, 1711.   Guillaume Delisle, *Orbis Romani descriptio seu divisio per themata . . . post Heraclii tempora; Imperii orientalis sub Constantino Porphyrogenito . . . descriptio* (Paris, 1711).

M—DumbOaks. For Banduri, *Imperium orientale* (see Banduri, 1711). MS sketch connected to these maps: ArchNat, Marine, Serv. hydrog. 6JJ/70, no. 18.

95. Delisle, 1717.   Guillaume Delisle, *De Clovis à Charlemagne,* 5 maps (1717).

M—Unpublished maps. To locate Delisle's MSS at the ArchNat, see Étienne Taillemite, "Les cartes anciennes du Service hydrographique de la Marine conservées aux Archives nationales," in *La carte manuscrite et imprimée du XVIᵉ au XIXᵉ siècle: Journée d'étude sur l'histoire du livre, Valenciennes 1981,* ed. Frédéric Barbier (Munich, 1983), 29–31. Le Long, 1:34–36, announces each map as having been "dressé[e] en 1717 pour l'usage du roi [Louis XV]." At the time he drew the maps, Delisle was teaching, or was about to teach, geography to the young king. See *Mercure de France,* March 1726, 485. ArchNat, Marine, Serv. hydrog. 6JJ/70 and /71/1.

Chronologically, the set should begin with Le Long, no. 393, "Carte des premiers établissemens des François dans les Gaules dressée en 1717 pour l'usage du Roi"; not found by me at ArchNat or BN. The following maps survive:

"Carte de l'Empire de Charlemagne" (n.p., 1717). MS. ArchNat, Marine, Serv. hydrog. 6JJ/70, no. 17. The title is penciled in at left, unsigned. See also no. 12c: pen draft, colored MS without legend, very faulty history. No. 7: corrected, improved; gaily colored boundaries; the attribution to Delisle is not certain. No. 10B: "France sous Charlemagne," old-fashioned enough to have the unreformed, pre-1700 outline of France (typically, a drooping Brittany). No. 10C: small, very unfinished pen sketch, possibly for a "Theatrum historicum" in the time of Charlemagne; marked "Europe ancienne." Note also Le Long, no. 413, "Carte de l'Empire de Charlemagne, avec le partage de ses petits-fils et arrière-petits-fils" [not found].

"Carte de la France partagée à la mort de Dagobert en Neustrie et Austrasie" (n.p., 1717). MS. ArchNat, Marine, Serv. hydrog. 6JJ/71/1, no. 26. One of two important pieces of this set; the map runs far into the margins.

"Carte de la France suivant le partage des enfans de Clotaire" (n.p., 1717). MS. ArchNat, Marine, Serv. hydrog. 6JJ/71/1, no. 25. Another copy, black-and-white, more carefully done than the Archives draft: BN, Res. Ge.D.7830. Faint pencil note in the cartouche, "Executée par Phil. Buache." Very carefully finished. The map extends far into the margins.

"Division de l'Empire" (n.p., n.d. [1717]). MS. ArchNat, Marine, Serv. hydrog. 6JJ/70, no. 11 (marked "Portefeuille 53"). Gaily colored frontiers. The maps of times after the death of Chlotar I and Dagobert seem to have been destined for expansion: lines and colors reach across the borders.

"France partagée aux enfants de Clovis" (n.p., n.d. [1717]). MS. ArchNat, Marine, Serv. hydrog. 6JJ/71/1, no. 23. Same style and colors as the map of Charlemagne's empire, but smaller, less finished than nos. 25, 26.

96. Delisle, 1721.   [Guillaume Delisle? Draft maps about later medieval France (n.d. [ca. 1721]).

M—None is positively attributed to Delisle (i.e., no association with his papers). BN, Ge.C.6189: Delisle, "Carte de France" (sous Philippe le Bel, 1285–1314), barely begun; Delisle, "Atlas historique sur papier" (Klaproth gift, 1832), nothing usable. ArchNat, Marine, Serv. hydrog. 6JJ/71/1, no. 19: on rear, "Carte pour le règne de Louis XI," undated; France mainly in the fifteenth century; historical notes.

97. Delisle, 1726.   Guillaume Delisle, *Cartes pour l'intelligence de l'*Histoire de Malte (Paris, 1726).

M—For René d'Aubert de Vertot, *Histoire des chevaliers hospitaliers de S. Jean de Jérusalem appelez depuis . . . les chevaliers de Malte,* 4 vols. (Paris, 1726); CathU-rare. 1:1, "Carte des pays où les chevaliers de Malte ont portés leurs armes"; 1:662, "Carte particulière de la Syrie et de l'Isle de Chypre"; 1:662, "Carte des Isles Rhodiennes"; all 3 are signed. Loose copies, BN, Ge.AF.PF 41 (63, 64), Ge.AF.Pf 30 (70). A MS of the first: ArchNat, Marine, Serv. hydrog. 6JJ/70, no. 22.

98. Delisle, 1764.   Guillaume Delisle, "Carte générale de la Syrie, de la Palestine et de l'Isle de Chypre . . . pour servir à l'histoire des croisades et . . . des royaumes de Jérusalem du Temps des croisés," ed. Joseph-Nicolas Delisle (Paris, 1764).

M—BL, K.3.99. Completed 1726, presumably in connection with the Vertot commission (Delisle, 1726). Le Long, 1:36, no. 423, gives the title, "Carte pour servir à la lecture de l'Histoire des Croisades, et à la connoissance particulière du Royaume de Jérusalem du temps des Croisés, et des Comtés et Commanderies que ce Royaume contenoit."

99. Delisle, 1766.   Guillaume Delisle, "Carte de Babylonie nommée aujourd'hui Hierac-Arab avec les noms tant anciens que modernes et les routes des expéditions de Cyrus et de Julien l'Apostat aussi bien que celles de Texeira, Benjamin et autres voyageurs modernes" (Paris, 1766).

M—BN. Published by J.-N. Delisle. A few historical notes running from Cyrus via Julian the Apostate to the modern traveler Texeira.

100. Desing, 1731.   Anselm Desing, O.S.B., *Kürziste Universal-Historie nach der Geographia auf der Land-Karte zu erlernen; Geschichtskarte* (Freising, 1731).

M—*GV* 28:240–41. BSB. Engraved at the Freising lyceum. One synchronic map with many numbers keyed to 215 events of universal history. Often reprinted. Last issue of the enlarged ed. (NOT SEEN), *Anleitung zu die Universalhistorie nach der Geographie auf der Landkarte zu erlernen, mit einer Universalkarte,* ed. Franz Xaver Jann (Augsburg, 1808). For a full cartobibliography, see Johannes Dörflinger, "Die Karten und Globen von Anselm Desing," in *Anselm Desing (1699–1772): Ein benediktinischer Universalgelehrter im Zeitalter der Aufklärung,* ed. Manfred Knedlik and Georg Schrott (Kallmünz, Germany, 1999), 322–27.

101. Desnos, 1765a.   L. C. Desnos, *Atlas général* (Paris, 1765–68).

G—StockholmRL, Qv. 18 1–10. A collection of 10 small atlases (1765–68), including Brion, 1766/O, and Rizzi-Zannoni, 1764. RGS, 1-vol. Desnos, *Atlas général* (Paris, 1790), with maps from the 1760s.

102. Desnos, 1765b.   L. C. Desnos, *Atlas historique, géographique et chronologique de la France ancienne et moderne contenant les évênements de notre histoire . . . pour l'intelligence de l'Abrégé chronologique de Hénault (NUC adds, Dédié et présenté à M. le président Hénault)* (Paris, 1765).

H—*NUC.* UIll, 912.44 / 766dA (reported as mislaid, 2002): 32 maps. Listed in *Catalogue des ouvrages tant anciens que modernes du Fond du Sr. Desnos* (Paris, 1765), 5–6 (Harvard, MA 336.765). See also Rizzi-Zannoni, 1764. The book referred to is Jean François Hénault, *[Nouvel] Abrégé chronologique de l'histoire de France* (Paris, 1744; 6th ed., 1761). Hénault acknowledged authorship of the *Abrégé* in 1756; its reputation stood high into the nineteenth century.

103. Desnos, 1765c.   L. C. Desnos, publ., *Tableau analytique de la France* (Paris, 1765).

H—LC G1838.B73 1765. A 4-part guidebook to current France. Advertised as the complement to Rizzi-Zannoni's historical atlas. In 2 of the 4 parts, maps signed by

Rizzi-Zannoni are dominant: the "Tableau" proper—all French administrative, military, ecclesiastical, and other internal boundaries—and the *Petit Neptune français.* The work was designed for popular sale and survives widely. Parts were incorporated in Brion, 1766a. *Journal des sçavans,* August 1765, 570, records a single-sheet "Tableau historique et géographique de la France," issued by Lattré (1764?), incorporating a spectrum of French circumscriptions and other details, including the "principales eaux minérales" [NOT SEEN].

104. Desnos, 1766.   L. C. Desnos, *Atlas chronologique de la France contenant les diverses accroissemens qu'a éprouvés le Domaine de la Couronne dans son étendue depuis la formation des Fiefs vers la fin de la seconde Race jusqu'à leur entière réunion sous Louis XV, servant de supplément à l'Atlas historique pour servir à la Lecture de l'abrégé chronologique de la réunion des grands Fiefs [de] M. Brunet* (Paris, n.d. [1766?]).

   H—Harvard, MA 336.765. 24 maps. These are the "cartes de répétition" from Rizzi-Zannoni, 1764, issued separately by Desnos as a companion to Pierre Nicolas Brunet, *Abrégé chronologique des grands fiefs de la couronne de France* (Paris, 1759); BL, 1319.d.15.; *NUC,* many copies.

Desnos, 1771. *See* Bonne, 1762.

105. Dheulland, 1756.   Guillaume Dheulland, "Plan en perspective de la ville de Paris telle qu'elle étoit sous le règne de Charles IX" (Paris, 1756).

   M—Julien, *Nouveau catalogue,* 44, specifies, "Paris in the 14th century." NYPL. The map was alleged to be either of fourteenth-century Paris or, more narrowly, of the time of Charles V and VI. These dates were rejected in *Journal des sçavans,* February 1757, 225–27, and in *Journal de Trévoux,* April 1757, 1131–34; the date was corrected to the reign of Charles IX. The map was based on an early (postmedieval) wall hanging at the Abbey of Saint Victor.

106. Dubos, 1742.   Jean-Baptiste Dubos, *Histoire critique de l'établissement des Francs dans les Gaules,* new ed., 2 vols. (Paris, 1742; 1st ed., 1734).

   M—Toronto-Fisher. Foldout map: "Theatrum Galliarum vicinarumque regionum ad annum Christi quadringentesimum septimum" (Vista of the Gauls and neighboring regions in A.D. 407). Admirable handling of Netherlands inland waters.

107. Dupré, 1763.   Dupré (pseudonym), *Les Révolutions de l'univers en trente cartes* (Paris, 1763; repr., 1775).

   H—BN; RGS, 14.B.117 (probably the copy offered for £2 10s. by Francis Edwards, Ltd., in *Ancient Geography: A Catalogue of Atlases and Maps of All Parts of the World* (London, [1929]), 55, no. 126); CambUL, Atlas 2.77.1 (intact title page); Göttingen, folio, Hist. univ. I, 191 (excellent condition, title page); BL, 118.f.13 (repr., 1775). Thirty identical two-part maps of Eurasia in chronological sequence with shifting lines of political division. The repeated map is by Michel Picaud of Nantes, under whose name the atlas often travels. The attribution of the atlas to Philippe de Prétot (Barbier, *Dictionnaire des ouvrages anonymes*) is improbable. Associated commentary, *Les révolutions de l'univers ou remarques et observations sur une carte géographique destinée à l'étude de l'histoire générale* (Paris, 1763), in-12mo.

108. Enouy, 1797.   J. Enouy, "The Invasions of England and Ireland" (London, 1797; 2d ed., 1801).

M— Oxford. Single sheet. Other copies: Thomas Kitchin, *A New Universal Atlas,* 2d ed. (London, 1798), no. 9 (YaleB; LC, G1015.K56 1798 Vault), also 1799 (BL), 1802; *New and Elegant Imperial Sheet Atlas comprehending General and Particular Maps of Every Part of the World,* new ed. (London, 1807). An updating of Speed, 1627. Boxes on the map face briefly identify the battles shown. Enouy, poorly documented, is credited with 4 maps (1805–7): *BL Maps Catal.* 5:613.

109. Faber, 1711.   Samuel Faber, *Atlas scholastichoeporicus* (Nuremberg, 1711–16).

G—LC List 5963, 52 maps; UMich-Clements, 1740, 71 maps. The long Latin word means *Schul-und Reisen-*. Brief historical section in closing (news maps rather than history). Without mentioning Delisle, it includes his "Theatrum historicum" and the Byzantine Empire divided into themes (medieval provinces). J. D. Köhler took charge of this atlas after his friend Faber's death (1716).

110. Faden, 1783.   William Faden, publ., "A Chronological Map of Europe Describing the Revolutions of Its Principal States" (London, n.d. [1783]).

H—BL, (microfiche) K.IV.95. The outlines of each country as frame for a brief historical synopsis.

111. Falcke, 1752.   Johann Friedrich Falcke, "Pars Saxoniae veteris necnon Angariae in orientali regione in pagos distributa" (vertical); "Pars Saxoniae antiquae in orientali regione in pagos distributa" (horizontal); "Pars Saxoniae antiquae sive Westfaliae nec non Angariae in occidentali regione in pagos distributa" (horizontal); "Pars Saxoniae veteris sive Angariae in occidentali regione in pagos distributa" (vertical); "Frisiae antiquae et veteris Brabantiae pars in pagos distributa" (vertical) (Leipzig and Wolfenbüttel, 1752).

M—HABW, folio Germ 94. In Falcke, *Codex traditionum Corbeiensium notis, diplomatibus ac tabb. geog. et genealog. illustratus.* HABW has some loose copies (30, 33; 30, 39; 30,56; 30,39). Falcke disputes with Lauenstein, 1745.

112. Falckenstein, 1733.   Johannes Heinrich von Falckenstein, "Delineatio Nordgoviæ veteris prout eius facies sæculo XI et XII fuit" (Frankfurt, 1733).

M—BSB. Frontispiece for Falckenstein, *Antiquitates Nordgavienses,* 2 vols. (Frankfurt, 1733) (Harvard); also *Codex diplomaticus antiquitatum Nordgaviensium* (Frankfurt and Leipzig, 1733). Loose copy in Homann, 1753 (LC). Inset map of Charlemagne's canal.

113. Fer, 1705.   Nicolas de Fer, *Lutèce, ou premier [-huitième] plan de la ville de Paris* (Paris, 1705–14).

M—*NUC;* Pastoureau, *Atlas français,* 179–80 (about *Atlas curieux,* nos. 188–92; see also *Map Collector* 9 [1979]: 52–58). In de Fer's *Atlas curieux,* BL, Maps c.39.c.2, p. 1, nos. 15–22; Newberry. Repr., in de Fer, *Les beautés de la France* (1724); see Pastoureau, *Atlas français,* 195–96. Separately, NYPL. On these maps of Paris, see also Coquart, 1705, and Lamare, 1705. De Fer's contribution in 1705 was no. 8 (contemporary Paris). Nos. 1–7, dated 1714 and of inferior quality, were probably copied from Coquart's.

114. Forster, 1788.   Johann Reinhold Forster, "Carte de l'Europe pour servir d'éclaircissement à la géographie du moyen-Age et à la Traduction Anglo-Saxonne du roi Alfred d'Orose" (n.p., n.d. [Paris, 1788]).

     M—BN, Ge.D.12166. Probably from Forster, *Histoire des découvertes et des voyages faits dans le nord,* tr. M. Broussonet, 2 vols. (Paris, 1788), 1:124. Repr., J. R. Forster, *Geschichte der Entdeckungen und Schiffahrten im Norden* (Frankfurt-an-der-Oder, 1784) (Toronto-Fisher); English tr., *History of the Voyages and Discoveries Made in the North* (London, 1786), 74–75 (map) (Toronto-Fisher). See Barrington, 1773. A map of Europe with Latin and Anglo-Saxon names for the parts named.

115. *Gallia,* 1720.   "Gallia medii aevi" (n.p., n.d. [London, 1720?]).

     M—BL, 870.(1.). Single sheet, without identification.

116. Gatterer, 1776.   Johann Christoph Gatterer, Charten zur Geschichte der Völkerwanderung (Göttingen, ca. 1776).

     H—NOT SEEN. Descriptions: *Historisches Journal von Mitgliedern des königlichen historischen Instituts zu Göttingen* 8 (1776): IV. Stück: 17–19; Gatterer, 1775/B, xiii–xvii; Christian Kruse, "Probe der Gattererschen Charten zur Geschichte der Völkerwanderung," *AGEphem.* 16, no. 4 (1805): 377–99 (color reproduction of Gatterer's eighteenth map); and Woltersdorf, 144–45.

117. Gatterer, 1789.   Johann Christoph Gatterer, *Gatterers Planiglobien* (n.p., 3d ed., 1789; 1st ed., 1773?).

     M—Göttingen, folio Geog. 189: 1 Rara. Nine items related to Gatterer's lost historical atlas. For additional survivors, see Gatterer, 17–/O.

118. Georgisch, 1732.   Peter Georgisch, "Regnum et imperium Merovingo-Carolingicum"; "Divisio Regni Lotharii anni DCCCLXX iuxta tabulas Procraspidianas cura cum divisionum anteriorum annis 843. 855. et 858. rationibus"; "Gallia et Germania ultimorum Romani Imperii temporum aequalis" (Halle, 1732).

     M—In Peter Georgisch, *Versuch einer Einleitung zur Römisch-Teutschen Historie und Geographie in chronologischer Ordnung nebst zugehörigen Land-Charten der alten u. mittleren Zeiten* (Halle, 1732), 92, 120, 262; Göttingen, 8vo Hist. Germ. un. IV, 2660. Four other maps fall outside the Middle Ages. Modern provinces are faintly marked to orient readers. The second-earliest map of Charlemagne's empire. Budapest, TA 353 (an *atlas factice*) has loose copies of 3 Georgisch maps, 2 of them Carolingian.

119. Gibson, 1758.   John Gibson, *Atlas minimus* (n.p. [London], 1758; 2d ed., 1792).

     G—BL, Maps c.7.a.20 (1758). Pocket maps of modern states with historical notes. Attractive book.

120. Giustiniani, 1739.   Francisco Giustiniani, *El Atlas abreviado o el nuevo compendio de la geografía universal, politica, historica, i curiosa, segun el estado presente del mundo* (Lyons [Leon de Francia], 1739).

G—LC List 3483, 4275, 5965. Newberry. 3 vols. 12mo, 43 maps, with a brief section of ancient geography.

121. **Gratiolus, 1735.**   Petrus Gratiolus, "Mediolani, ut ante Aenobarbi cladem extitit, iconographia" (Milan, 1735).

M—*NUC.* RomeBNVEmm, 1.40.K.13.; BL, 660.i.16., 179.i.6. In his *De Praeclaris Mediolani aedificiis quae Aenobarbi cladem antecesserunt dissertatio* (Milan, 1735), 17, Gratiolus provided a rather scanty picture-map of the major Roman monuments believed destroyed when Frederick I Barbarossa razed Milan in 1162.

122. **Güssefeld, 1796.**   Franz L. Güssefeld, "Geographische Übersicht der in dem Herzoglich-Sächsischen Hause Ernestinischer Linie vorgegangenen Landes-Theilungen und Darstellung derselben durch eine . . . neuentworfene fünffache genaue Special-Charte" (Weimar, 1796).

M—NOT SEEN. The Berlin-BSB-Göttingen map data base provides the reference BerlinDSB, Kart. U 16 440. Very rare. *Catalogus mapparum*, 181, characterizes Güssefeld's *Geographische Übersicht* as the first, and for its time highly successful, attempt to draw up a historical special map on a scientific basis, with reference to unpublished charters—an outstanding piece of historical geography (F. Curschmann). On the same theme, Engelmann, 880, adds M. F. Teuscher, *Geographisch-historische Uebersicht aller Ländertheile in dem Hause Sachsen, Ernestinische Linie:* 9 small maps and text on 1 sheet (Weimar, 1825). See also Güssefeld, 1796/O.

123. **Hagelgans, 1718.**   Johan Georg Hagelgans, *Atlas historicus* (Frankfurt, 1718).

H—Princeton-rare; Jerusalem, SX folio 93 C 5427. *NUC* 225:619. Repr., Frankfurt, 1737, BPL; enlarged, Frankfurt and Leipzig, 1751–52, LC (LC List 3308). Primarily a chronicle in pictures. Four maps of "the development of Europe," including the earliest reconstruction of the "migrations of nations."

124. **Halma, 1700.**   François Halma, *Description de tout l'univers* (Amsterdam, 1700 [1709]).

G—Koeman, 2:127. No. 3, Heptarchy.

125. **Halma, 1725.**   François Halma, *Toneel der Vereenigde Nederlanden en onderhorige Landschappen* (Leeuwarden, 1725).

H—BSB; BL, Maps C.24.f.25, the maps only; text, 10270.h.8. UCB-Bancroft, maps incomplete. See also Schotanus, 1695. Repr., of Alting, 1701, incomplete, somewhat rearranged, and combined with maps by others. Halma did not acquire all the Alting plates. This "theater" is basically a dictionary, with images or portraits of major figures in Dutch history, as well as maps.

126. **Hase, 1743.**   Johann Matthias Hase (Haas, Hasius), *Historiae universalis politicae quantum ad eius partem I ac II. Idea plane nova et legitima tractationem summorum imperiorum exhibens . . . in lectionum academicarum usum,* ed. A. G. Böhme (Nuremberg, 1743).

H—BN; Harvard, MA 80.743s (fine copy). 28 maps of "the greatest empires" (*summa imperia*), from antiquity to 1740. Woltersdorf, 46–47, wrongly lists as 1734. For 1743 as the received date, see Sandler, "Die homännischen Erben," 403, the BN and Harvard copies, and the BL copy of Hase, 1750. Hase, 1750, at BSB has a 1746 reprint of Hase, 1743. For an authoritative review of Hase's bibliography and appraisal of his work, see Johannes Dörflinger, "Das geschichtskartographische Werk des Johann Matthias Hase," in *Geschichtsdeutung (Archäologie und Geschichte) auf alten Karten* (proceedings of the 46th Wolfenbütteler Symposion, 26–29 October 1999) (forthcoming).

127. Hase, 1750a.   Johann Matthias Hase, *Atlas historicus, comprehendens imperia maxima seu monarchias Orbis Antique historice, chronologice et geographice repraesentatas* (Nuremberg, 1750).

H—*NUC* (at "Hasius"). BL, Maps 3.d.35 (imperfect); BSB (excellent). An omnibus collection including Hase, 1739/E, 1743, 1745/O, and 1750b.

128. Hase, 1750b.   Johann Matthias Hase, *Mappae VII. geographicae pro illustrandis totidem periodis historiae Germaniae sub Carolo M. Ottone I Conrado II Friderico II Friderico III et Carolo VI* (Nuremberg, 1750, 1752).

H—BN, Ge.DD.2481; Harvard, MA 1237.750F, 7 maps. Has been called the first atlas of German history. The year 1736 appears in Homann Heirs, *Atlas Germaniae specialis,* 6 vols. (Nuremberg, 1732), 1: nos. 13–19 (BL). Sandler, "Die homännischen Erben," 403, states (to the contrary) that the maps of the empire were a fragment in Hase's *Nachlass.* Publication in 1749 according to Julien, *Nouveau catalogue;* later dates in Hase, 1750a, and elsewhere.

129. Hell, 1771.   Miksa (Maximilian) Hell, S.J., "[Primordia Regni] Ungariae ab Anno Xti 887 ad Annum 907/sub Arpado primo Ungariae DUCE / ex / Historia anonymi Belae Regis Notarii (Tabula geographica Hungariae veteris ex Historia anonymi)" (Vienna, 1771). Engraved by Ant. Schlechter.

M—Budapest, T351 (*atlas factice,* maps of Hungary). Opens with two versions of Hell's map; notable differences in the cartouche, the shape of mountains. The second version is reproduced in Árpad Papp-Váry and Pál Hrenkó, *Magyarország régi térképeken* [Hungary in Old Maps] (n.p., 1989), 120–21. Little circulated when first published, Hell's map became better known in the posthumous *Tabula geographica Ungariae veteris ex historia Anonymi Belae regis notarii* (Pest, 1801).

130. Henry, 1799.   Robert Henry, *The History of Great Britain from the First Invasion of It by the Romans,* 3d ed., 10 vols. (London, 1799–1800).

M—LC-rare. (Table of contents, "A Map of Britain according to the Saxon Chronicle"). 4:405, Appendix to book 2, map, "Saxon England according to the Saxon Chronicle," antiqued with Old English letters. Index of places with their meaning. (Vol. 2, Appendix to book 2, maps of Roman Britain.)

131. *Hist. abrégée,* 1701.   *Histoire abrégée des Provinces unies des Pays-Bas* (Amsterdam, 1701).

M—BL, Cup. 404.b.14.2 maps. Author, Zacharie Châtelain (see also Châtelain, 1705). Probable source of Harms, *Themen,* no. 49 (from the Châtelain *Atlas historique*).

132. *Hist. universelle*, 1779.   *Histoire universelle depuis le commencement du monde jusqu'à présent* . . ., nouv. trad., par une société de gens de lettres, 126 vols. (Paris, 1779–89).

M—Tr. of the cooperative *Universal History from the Earliest Times,* ed. George Sale and Thomas Salmon (London, 1730–36); see also *Welt-Historie,* 1744. Bibliography in Barbier, *Dictionnaire des ouvrages anonymes,* 2:836–37. *NUC.* LC, D20.U595; BN, under Ussieux, Louis d', *Hist. univ.* (1779–91), 125 vols., G.14300–14424. Relevant maps: 46:1, 69:79, 74:1 (see Liébaux, 1728), 75:1.

133. Homann, 1753.   J. B. Homann and Heirs, *Atlas Germaniae specialis* (Nuremberg, 1753).

H—Harvard (1753); PhiladAmPhil (also PhiladLibCo). This name was given to Homann Heirs, *Atlas maior,* vol. 2. Historical items include nos. 3, "Germania ecclesiastica"; 4, "Germania secundum religiones"; 5, Carl, 1732/E; 46–48, Zollmann, 1732; 74, Falckenstein, 1733. A 6-vol. collection at BL also bears this name (see *GV* 6:168).

134. Homann, 1762.   Homann Heirs, *Atlas Homannianus Mathematico-historice delineatus* (Nuremberg, 1762?).

G—BL, Maps 32.e.30. The Holy Land in twelve tribes is the only historical item.

135. Hugier, 1756.   Hugier, "Allemagne et Franks sous les fils de Louis le Debonnaire" (n.p., 1756).

M—Manuscript. BN, Ge.F.2228 carte. Forms a pair with "Allemagne du Temps de Charlemagne" (Ge.F.2227 carte). Incomplete drafts. The author's name, otherwise unknown, is supplied by the catalogue.

136. Kitchin, 1756.   Thomas Kitchin, engrav., "A Map of the Antient Dominions of the Kings of England in France with some adjacent countries" (n.p., 1756).

M—BL, 870.(2.). The boundaries of French principalities, e.g., Limousin, Bordelais, to the Rhône only.

137. Kitchin, 1798.   Thomas Kitchin, *A New Universal Atlas,* 2d ed. (London, 1798).

M—YaleB. No. 1, "A New Chart of the World on Wright's or Mercator's Projection in which are exhibited all the parts hitherto explored or discovered, with the tracks of the British circumnavigators Byron, Wallis, Carteret and Cook etc." (London, 1794). A very large, beautiful map. Other editions of the Kitchin atlas: 1770, 1780, 1788, 1795, 1796, 1802.

138. Köhler, 1718.   Johann David Köhler, *Schul- und Reisen-Atlas zur Erlernung der alten, mittlern und neuen Geographie* (Nuremberg, 1718).

G—LC (3 copies); UCB-Bancroft; BL, Maps 38.e.4.; BAV, RG Geog I.31; HABW, Cb folio 17. Cf. Faber, 1711. "Bequemer" is sometimes the first word of the title; alternative title, *Atlas (manualis) scholarius et itinerarius.* The reference to "medieval" in the title corresponds to nothing in the atlas. Ends, like Faber's, with news maps ("historischer Zeitungs-Atlas").

139. Köhler, 1730.   Johann David Köhler, *Kurtze und gründliche Anleitung zu der alten und mittleren Geographie nebst XII. Land Chärtgen (Compendium geographiae antiquae et mediae)*, 3 pts. (Nuremberg, 1730–65).

   H—*NUC*. BerlinPK (pts. 1–2 only); Göttingen, 8vo Geog. 317 (same 2 pts.); UIll 911/K82k/1745, with 2. verm. u. verb. aufl. (of vol. 1) (Nuremberg, 1745–65), 3 vols. in 1. Vol. 2, ed. Georg Martin Raidel; vol. 3, ed. Georg Andreas Will; vol. 3, publ. J. D. Tyroff, heir to Weigel. Woltersdorf, 52, records a Weigel reprint in 1790. Arthur Kühn, *Die Neugestaltung der deutschen Geographie im 18. Jahrhundert* (Leipzig, 1939), 18: printing history. Includes a 9-piece anthology of maps for medieval history. Loose detached sheet, BerlinPK, U14240. "Germania in seculo V p. Ch. n."

140. Kremer, 1778.   Christoph Jakob Kremer, "Herzogtum der Rheinfranken in seine Gauen abgetheilt mit den angränzenden Provinzen" (Mannheim, 1778).

   M—BL, 1311.g.3. From Kremer's *Geschichte des Rheinischen Franziens unter den Merovingischen und Karolingischen Königen bis in das Jahr 843 (als eine Grundlage zur Pfälzischen Staats-Geschichte)*, ed. Andreas Lamey (Mannheim, 1778).

141. Lamare, 1705.   Nicolas de Lamare (La Mare, Delamare), *Traité de la police, où l'on trouve l'histoire de son établissement etc. On y a joint une description historique et topographique de Paris et huis plans gravés*, 4 vols. (Paris, 1705).

   M—BL. An anonymous book with many later editions. Vol. 1 provides the factual underpinnings for the 8 maps of Paris; Coquart, 1705, and Fer, 1705. It is most likely that the drawing and engraving of maps 1–7 are by Coquart and map 8 by de Fer: *Catalogue of Maps and Charts in the Library of Harvard University* (Cambridge, Mass., 1831), 114–15; Vallée, *BN, Catalogue des plans de Paris*, confirms that nos. 3–6 are by Coquart. Loose copies of maps 3–8, so attributed, are in MilwaukeeAGS. The relationship of Lamare, Coquart, and de Fer is not straightened out by Pastoureau, *Atlas français*, 179–80. The "Collection du commissaire Nicolas de Lamare," BN, 261 MS vols., concerns the history of municipal government (alias "police"). *Catalogus mapparum*, 212–93, bears out the special advantages of early city maps in charting development.

142. Lamey, 1766.   Andrew Lamey, 8 maps of medieval *gaue* in the Rhineland (Mannheim, 1766–94).

   M—"Pagi Lobodunensis qualis sub Carolingis maxime regibus fuit descriptio," *Acta Academiae Theodoro Palatinae* 1 (1766): 215–42 (map, 217); "Pagi Wormatiensis qualis . . . [as above]," ibid., 243–300 (map, 243); "Pagi Rhenensis . . . [as above]," ibid. 2 (1770): 153–86 (map, 253); "Pagi Spirensis qualis antiquis temporibus fuit, descriptio," ibid. 3 (1773): 228–80 (map, 228); "Pagi Craichgoviae qualis antiquis temporibus fuit descriptio," ibid. 4 (1778): 104–246 (map, 104); "Pagi Navensis qualis . . . [as above]," ibid. 5 (1783): 127–86 (map, 127); "Elzengoviae Franciae Rhenensis pagi, qualis medio aevo maxime fuerit, descriptio," ibid. 6 (1789): 91–111 (map, 91); "Wingartheibae veteris pagi Franciae novae, ex monumentis medii aevii, descriptio," ibid. 7 (1794): 41–67 (map, 41) (BL; PhiladAmPhil, 1–4 only). The maps differ considerably from one another in size and layout. They are supplements to A. Lamey, ed.,

Codex Laureshamensis diplomaticus, 3 vols. (Mannheim, 1768–70), LC-rare, DD901.L8C6. Official name of the journal: *Historia et commentationes Academiae Electoralis scientiarum et elegantiarum litterarum Theodoro-Palatinae.*
Lattré, publ., *Atlas moderne.* See Bonne, 1762.

143. Lauenstein, 1745.   Joachim Berward Lauenstein, *Specimen geographiae medii aevi diplomaticae, hoc est descriptio dioecesis Hildesheimensis per antiquos suos pagos* (Hildesheim, 1745).

H—HABW, 4to Gn 6884. Long title set in an elaborate cartouche: "Dioecesis Hildesheimensis medii aevi tabula, in qua ex variis medii aevi diplomatibus nomina villarum, castrorum, civitatum, etc. collegit, et ad ductum diplomaticum in pagos tam majores quam minores, curate divisit, nomina quoque fluvium, sylvarum, etc. ad dialectum mediae aetatis expressit, Joach. Berward Lauenstein ad S. Michael in Hildesh. Pastor" (*Dioecesis Hildesheimensis medii aevi tabula*). Cf. Falcke, 1752.

144. Le Masson, 1717.   Le Masson du Parc, collection of maps (n.d., after 1717).

M—LC List 560, 5933. Includes Delisle, 1705.

145. Liébaux, 1728.   Henri Liébaux, Maps of the beginnings of France (Paris, 1728).

M—Le Long, 1:35, nos. 394–95. BL, K.3.85–87. From Daniel, 1729:(1) "Carte de la France où les limites du Royaume sont marqués suivant les Traités d'Utrecht de Rastat et de Bade conclus en 1713 et 1714 par Henri Liébaux Géographe 1728."
(2) "Carte des Gaules où l'on voit les Dominations auxquelles elles étoient soumises, lorsque CLOVIS vint y jetter les Fondements de la Monarchie Françoise par Henri Liébaux 1728"; under this title, BL, K.3.86.; *Welt-Historie,* 1744, 35:3; redrawn, *Hist. universelle,* 1779, 74:1 (loose, PhiladFree); *Aloude Holland,* 1745, no. 8; Prétot, 1768, no. 3; Velly, 1787, 1:6. (3) The original Daniel, 1696, map; "Estat des Gaules contenant les trois monarchies qui les partagoient quand Clovis en fit la conqueste" (loose, BL, K.3.85); *Welt-Historie,* 1744, 35:584, "Die Staaten in Gallien der drey zertheilten Monarchien bey Clodwigs Eroberung" (very close to Daniel/Berey); modernized but retaining design and title, "Carte de la Gaule Représentant l'étendue des trois Dominations qui la partageoient à l'époque où Clovis en fit la conquête," *Hist. universelle,* 1779, vol. 74 (loose, PhiladFree). Nos. 2 and 3 have a very similar subject. The most notable feature of Liébaux's version (no. 2) is a wide road, almost a causeway (*agger*), marked "Route de Clovis." (4) "Carte de la France pour la fin du règne de Clovis et pour le Partage des ses Etats entre ses Enfans"; BL, K.3.87.; BN, Ge.F.carte.13022; *Welt-Historie,* 35:32, "Charte von Franckreich zu Ende der Regierung Clodwigs, und zu Anfangs der Theilung unter dessen Söhne."

146. Luneau, 1760.   Pierre Joseph François Luneau de Boisjermain, *Atlas historique, ou cartes des parties principales du globe terrestre, assujetties aux révolutions séculaires qu'il a éprouvées,* pour servir à l'histoire des temps qui ont suivi la création etc. (Paris, 1760–61).

H—*NUC.* PhiladAmPhil, L965ter Large, "Carte des parties principales du globe terrestre pour servir à l'histoire des deux premiers siècles depuis la création du monde" (Paris, 1765). Murphy D. Smith, *"Realms of Gold": Catalogue of Maps (and*

Atlases) in the Library of the American Philosophical Society (Philadelphia, 1991), 81, no. 65. No sign of provenance. Single sheet. Upper right: Première feuille. Inset lower left: "Supplément à la carte du Paradis terrestre." There is no trace of the 20-map collection advertised in *Journal de Trévoux,* January 1761, 175–76. *Nouvelle biographie générale* 32 (1860): 254–55, credits Luneau with an *Atlas historique* (Paris, 1761; expanded title for new ed., 1767), *ou Trois Cartes élémentaires pour trouver en peu de mots ce qu'il est contenu dans le cours d'histoire et de géographie.* The 3 maps from the Creation are said to be attached to Luneau's *Cours d'histoire Universelle* (Paris, 1765–68, 2 vols. 8vo; 1779, 3 vols. 8vo), but their presence is not yet confirmed. None appears in the copies of BN, G.14506–7, or BL. The Philadelphia map has a date suitable for attachment to Luneau's *Cours d'histoire universelle;* upper-right corner: "Première feuille"; lower left, "Dans cette carte et dans les suivantes, on a adopté la méthode dont on se sert pour dresser les Cartes réduites. . . ."

147. Macpherson, 1796.   David Macpherson, *Geographical Illustrations of Scottish History* (London, 1796).

   M—LC, DA869.M22 (lost?); NYPL. One historical map.

148. Ménard, 1713.   Antoine Ménard, *Le Nouveau et curieux atlas géographique et historique, ou le divertissement des empereurs, roys et princes tant dans la guerre et dans la paix, dédié à François Blouet de Camilly* (Paris, n.d. [1713]).

   G—BN, Res.G.1051 (the BN Réserve contains 2 more copies of the atlas); Jerusalem, 92 C 1402. Title and contents anticipate Chiquet, 1719. Almost all Ménard's maps are dated 1711. Catherine Hofmann (BN, Cartes et plans) kindly supplied me with details of the three editions of the Ménard/Chiquet atlas.

149. Mentelle, 1782.   Edme Mentelle, *Atlas nouveau* (Paris, 1782).

   G—BN; RomeMar (imperfect). Luxurious. Some maps of ancient geography.

150. Mentelle, 1785.   Edme Mentelle, "France au tems de Clovis" (Paris, ca. 1785).

   M—BN, Ge.F.carte.5683.

151. Mentelle, 1798.   Edme Mentelle and P. G. Chanlaire, *Atlas élémentaire de l'Empire d'Allemagne,* 2d ed. (Paris, 1798).

   H—BN; BL, Maps 38.e.5. 1st ed., Courtalon, 1774. The revision was prompted by the Congress of Rastatt (December 1797–April 1799). All maps are dated 1774, i.e., taken from Courtalon.

152. Mille, 1771.   Antoine Étienne Mille, *Abrégé chronologique de l'histoire de Bourgogne,* 3 vols. (Dijon and Paris, 1771–73).

   M—BL. At 1:1, "Carte de l'ancien royaume de Bourgogne." The markings are so subdued that one cannot tell which kingdom of Burgundy is meant.

153. Moll, 1709.   Hermann Moll, *The World Describ'd* (London, 1709–36).

   M—LC List 3469. No. 30, "An historical map of the roman empire and the neighbouring barbarous nations in the year of our Lord four hundred when the empire began to be rent with foreign invasions. By monsieur William Del Isle . . . 1709." Enlarged copy of Delisle, 1705.

154. Muratori, 1727.   Ludovico Muratori, "Tabula Italiae medii aevi Graeco Langobardico Francici," accurante Societate Palatina (Milan, 1727).

   M—Le Long, 1:36, no. 422. Newberry. In Ludovico Muratori, *Rerum Italicarum
scriptores ab anno aerae christianae quingentesima*, 25 vols. in 28 (Milan, 1723–51),
1: col. i–ii. Map by Gasparo Beretti, O.S.B. Accompanying text (cols. i–cccxxxvi),
"De Italia medii aevi dissertatio chorographica pro usu tabulae Italiae Graeco-
Langobardico-Francicae, ut a Graecis et Langobardis ad Carolum M. translatae . . .
auctore Anonymo Mediolanense."

155. Netherlands, 17–.   Untitled map (Viking Devastations in the Netherlands,
ninth to eleventh centuries).

   M—AmstUL, 61-28-23. Catalogue: Netherland in tenth/eleventh century. From
a book. Fine delicate engraving. Dates (839, 889, 923, 969, 985, 1018) keyed to specific
districts or towns. A line northwest of Ghent is labeled "Gracht v. Otto" (Otto's
ditch).

156. Netherlands, 1792.   "Geographische Tafel der Midden-Eeuwe van Holland,
Zeeland en Friesland" (n.d., [1792]).

   M—AmstUL, 61-28-08. Date from the Catalogue. From a book. Apparently
based on a map of Roman Netherlands. A part of "Otto's ditch" is shown. The cartographer is preoccupied with the frontiers of Holland proper and with water and
river mouths.

157. Nolin, 1722.   Jean-Baptiste Nolin, Five maps for a history of Tamerlane (Paris,
1722).

   M—Toronto-Fisher. Sharaf al-Dīn ʿAlī Yazdī, *Histoire de Timur-bec connu sous
le nom du grand Tamerlan*, tr. François Pétis de la Croix, ed. Alexandre Pétis de la
Croix, 4 vols. (Paris, 1722); English tr., J. Darby (London, 1723). Cf. Delisle, 1710.

158. Nolin, 1733.   Jean-Baptiste Nolin, "Roiaume et duché de Septimanie" (Paris,
1733?).

   M—Le Long, 1:36, no. 416. Single sheet, BN, Ge.D.17517; plus another (both
loose copies are dated ca. 1700 in the BN catalogue). In book context, Claude de Vic
and Joseph Vaissete, *Histoire générale du Languedoc* (Paris, 1733), 2:1 (original edition
only); BN, folio L²k 832 (colored and clean). Vol. 4 contains an unsigned folded
map, "Le Languedoc, divisée suivant ses anciennes sénéchaussées."

159. Nolin, 1746.   Jean-Baptiste Nolin, *Le Théatre du monde dédié au roi* (Paris,
1746).

   G—Ed. of 1709, HABW, Cb Folio 93. Ed. of 1746, RomeBNVEmm, 9.Banc.2.12;
maps from 1688 to 1745. Many war maps, including internal ones in France; Greece
old and new. No. 76 (later than 1690), dedicated to M. de Catinat, lieutenant-general
of the king's armies in Italy: "Les vallées du Piemont Habitées par les Vaudois ou
Barbets dressees sur les Memoires de Valerius Crassus et de Jean Leger Ministre des
Vaudois" (Paris, n.d.). Also see Bray, 1712.

160. Nolin, 1783.   Jean-Baptiste Nolin, *Atlas général à l'usage des collèges et maisons
d'éducation, pour l'intelligence de l'histoire ancienne et moderne* (Paris, 1783).

G—LC List 4297. AmstUL, Atlas catalogue, no. 238, 48 maps; LC, 44 maps. A small portable atlas, ending with ancient geography.

161. Palairet, 1755.   Jean Palairet, *Atlas méthodique* (Paris, 1755).

G—BPL. BN adds to the title "composé pour l'usage [du Prince d'Orange et de Nassau]" and notes London, Berlin, Amsterdam, The Hague as places of publication. No history except for the Holy Land divided among twelve tribes.

162. Philippe, 1768.   Étienne André Philippe de Prétot, *Cosmographie universelle* (n.p. [Paris], 1768).

G—LC. The sole approach to the Middle Ages is from Liébaux, 1728.

Philippe de Prétot. *See also* **A,** below.

Picaud, Michel. *See* Dupré, 1763.

163. Plancher, 1739.   Urbain Plancher, O.S.B., "Carte des pays de la Germanie occupez par les Vandales et les Allemans et où les Anciens Bourguignons ont fait leurs demeures avant que de passer le Rhin pour se venir Etablir dans les Gaules"; "Carte de l'ancien Royaume de Bourgogne" (Dijon, 1739).

M—*NUC;* Le Long, 1:36, no. 418. Oxford. Two maps in Urbain Plancher, *Histoire générale et particulière de Bourgogne,* 4 vols. (Dijon, 1739–81, repr., Farnborough, Hants, 1968), 1:1, 27. Plancher shows separate sources for the Upper and Lower Rhine (!). Earlier in the book: "Grande première demeure des Bourguignons" from the Vistula deep into Russia; first Vandal dwelling, from the base of Jutland to the Vistula; second Burgundian dwelling has its center from the Sudetenland through Bohemia.

164. Prévost, 1749.   Abbé (Antoine François) Prévost and others, *Histoire générale des voyages . . . pour former un système complet d'histoire et de géographie moderne,* 80 vols. (Paris, 1748–89).

M—Toronto-St. Michael's-rare. Maps by Nicolas Bellin. One map relevant to the Middle Ages, 22:338, "Carte pour les voyages de Rubruquis, Marco Polo Jenkin-son &c Insérés dans le VIIe volume de l'histoire générale des voyages."

165. Rapin, 1724.   Paul de Rapin-Thoyras, "Carte d'Angleterre sous les Saxons" (The Hague, 1724).

M—After Gibson, 1792. YaleB, *Histoire d'Angleterre,* 10 vols. (The Hague, 1724–27), 1:91. Oxford: Rapin's history tr. N. Tindal, 2d ed., vol. 1 (London, 1732), "Britannia Saxonica."

166. Remondini, 1784.   Remondini (firm), *Atlas universel dressé sur les meilleures cartes modernes* (Venice, 1784).

G—ViennaÖNB. (Variant title given at Table of contents: *Nouvel atlas géographique en grande feuille atlantique.*) 6 maps by Rizzi-Zannoni (1774) of the Ottoman Empire in Europe. Appendix of ancient geography, nos. 54–63.

167. Rizzi-Zannoni, 1762.   Giovanni Antonio, Rizzi-Zannoni, *Atlas géographique contenant la mappemonde et les quatres parties,* 12mo (Paris, n.d. [1762]).

G—UIll-rare; UMich-Clements. Small maps, well colored and clear. No history.

168. Rizzi-Zannoni, 1764.   Giovanni Antonio Rizzi-Zannoni, *Atlas historique et géographique de la France ancienne et moderne; Atlas historique et géographique de la France; Atlas historique de la France ancienne et moderne* (Paris, 1764, 1765, 1766).

H—Le Long, 1:3, nos. 3, 4. "Tableau varié des Revolutions qu'a subies la Monarchie Française." 60 maps (also available in a 36-map version). Brief comments to each map. BL, Maps 24.d.6. (1764), Maps c.24.d.22, also 118.c.5. (1765); Göttingen (1764, 1766); Harvard, MA 336.765 (1766); NYPL (1766; another in main collection); Princeton-rare (1766); RomeCasan (old catal.), P.V.8 in CCC (1766); StockholmRL (1766; maps, 1764); UIll-rare (1766); Yale (1765). The 1764 copy of this atlas at the BL has the same map no. (6), later changed, as Desnos, 1765b.

169. Robert, 1740.   Gilles Robert, "État de la France sous les rois de la première race tiré des observations de Dom Bouquet et des dissertations de M. Lebeuf" (Paris, 1740).

M—Pedley, *Vaugondy,* Catalogue, no. 88. UIll, in *Receuil des historiens des Gaules et de la France,* ed. Martin Bouguet et al., 24 vols. (Paris, 1738–1904), 3:1. Incorporates information for most of the sixth century.

170. Robert, 1742.   Gilles Robert, "Carte générale de la monarchie des Gots tant dans les Gaules qu'en Espagne" (Paris, 1742).

M—Pedley, *Vaugondy,* Catalogue, no. 89. BL, 178.c.7-16; BN, 4to Oa.80. For Juan de Ferreras, *Histoire générale d'Espagne,* tr. d'Hermily, 10 vols. (Amsterdam, 1742–51), 1:1; NYPL. Spanish ed., Madrid, 1700–1727 (BL); German tr., 1754–72.

171. Robert, 1743.   Gilles Robert, "Imperium Caroli Magni occidentis imperatoris complectens universam Galliam, in Hispania quiquid a Pyrenais jugis occurrit ad Iberum et Rubricatum usque fluvio etc." (Paris, 1743).

M—Pedley, *Vaugondy,* Catalogue, no. 91. UIll, in Martin Bouquet, ed., *Receuil des historiens des Gaules et de la France,* 5:1. Better base map than Bertius, 1623; little improvement otherwise.

172. Robert, 1745.   Gilles Robert, "Carte de la Germanie sous les empereurs Carlovingiens et Saxons" (Paris, 1745).

M—Pedley, *Vaugondy,* Catalogue, no. 246. BN, Ge.D.11033, Ge.D.11032. For Joseph Barre, *Histoire générale de l'Allemagne,* vols. 2–3 (Paris, 1745) (NYPL). Accompanies Robert's "Carte de la Germanie ancienne pour servir à la lecture de l'histoire d'Allemagne du R.P. Barre."

173. Robert, 1753.   Gilles Robert, "Le royaume d'Angleterre divisé selon les sept Royaumes ou Heptarchie des Saxons avec la Principauté de Galles et subdivisé en Shires ou comtés" (Paris, 1753).

M—Pedley, *Vaugondy,* Catalogue, no. 48. BL, K.5.64. 2 sheets.

174. Robert, 1755.   Gilles Robert, [cartes pour l'*Histoire de France* de Père Daniel].

M—LC-rare, DC37.D179: Gabriel Daniel, *Histoire de France,* 17 vols. (Paris, 1755–57), vol. 1, unpaged, letter of Robert to the publishers about maps accompanying histories. The maps, all modern and modest in size, are in vol. 17:(1) "Carte générale de l'Europe"; (2) "Partie occidentale de l'Allemagne"; (3) "Italie"; (4) "France." All foldout; France is much the largest. Cf. Daniel, 1696, and Liébaux, 1728.

175. **Robert, 1757.**   Gilles Robert, *Cartes de l'ancienne Irlande* (Paris, 1757).

  M—Pedley, *Vaugondy*, Catalogue, nos. 58–63. 6 maps for Ma-Geoghegan (James MacGeoghegan), *Histoire de l'Irlande ancienne et moderne*, 3 vols. (Paris, 1758), LC, DA910.M143. Pedley nos. 58, "Carte de l'ancienne Irlande dressée pour la lecture de l'Histoire de ce Pays"; 60, "Province de Connacie [Connaught] divisée en Dynasties pour les premiers siècles du Christianisme"; 61, "Provinces de Midie et de Lagenie [Leinster] divisées en Dynasties"; 62, "Province de Momonie [Munster] divisée en Dynasties pour les premiers siècles du Christianisme"; 63, "Province d'Ultonie [Ulster] divisée en Dynasties pour les premiers siècles du Christianisme"; 59, "Royaume d'Irlande . . . XII–XVII siècle" (tipped into vol. 2 in CambUL, Hib.4.758.1–3).

176. **RVaugondy, 1752.**   Didier Robert de Vaugondy, "Imperium Caroli magni occidentis imperator" (Paris, 1752).

  M—Pedley, *Vaugondy*, Catalogue, no. 93. BL, K.3.96. Enlarged version of Gilles's map and the sole sample of medieval geography in RVaugondy, 1757. 2 sheets. Gallia/Germania as basic divisions. Little attempt at subdivisions. Coloring suggests great unity. The empire takes in all Bohemia and reaches far to the east.

177. **RVaugondy, 1756.**   Didier Robert de Vaugondy, "Germania antiqua in quatuor magnos populos, in minores et minimos distincta et regiones Danubium inter et mare Adriaticum contentae" (n.p., 1756).

  M—Pedley, *Vaugondy*, Catalogue, no. 247. BerlinPK. Single sheet, black and white. Colored in *Atlas universel*, no. 7 (BerlinPK). Resembles the Sanson map of the same Ptolemy-derived subject.

178. **RVaugondy, 1757.**   Didier Robert de Vaugondy, *Atlas universel* (Paris, 1757 and later).

  G—LC List 4292, 5996; Pedley, *Vaugondy*, Catalogue, nos. 227–29. Toronto-Fisher; BerlinPK; PhiladFree (1757). 2d ed, enlarged, by C. F. Delamarche (1796). Begun by 12 maps of ancient geography; about medieval geography, see RVaugondy, 1752.

179. **RVaugondy, 1797.**   Didier Robert de Vaugondy, *Atlas d'étude, pour l'instruction de la jeunesse* (Paris, 1797–98).

  G—LC List 3530. Assembled by Delamarche from Vaugondy stock, the atlas closes with 11 maps for history.

180. **Santini, 1776.**   P. Santini, publisher, *Atlas universel dressée sur les meilleures cartes modernes* (Venice, 1776).

  G—RomeSGI, Z.2.III.1. Good quality; anthology of mainly French maps. BAV, RC Geog. S.91 (1–2). Vol. 2, 53 maps of modern geography followed by 9 (10) "cartes de la géographie ancienne"; territories only, no "literary geography." The standard Italian atlas of its time.

181. Schannat, 1724.   Joannes Fridericus Schannat, "Nova veteris Buchoniae tabula" (Leipzig, 1724).

M—CalStLibSutro. In Schannat, *Corpus traditionum Fuldensium*, complementary part (same page) *Patrimonium s. Bonifatii sive Buchonia vetus una cum suis confinii ex traditionibus Fuldensibus eruta . . . a temporibus Pippini regis ad initium saeculi XIV*, 318 and 319.

182. Schatz, 1749.   Johann Jacob Schatz (Schazius), *Schul-Atlas von zwantzig General- and Special-Landkarten* (Atlas scholasticus) (Nuremberg, 1749).

G—AmstUL. Composed of a selection of Homann maps. Only the Terra Sancta map with 12 tribal territories is explicitly historical.

183. Schöning, 1763.   Gerhard Schöning, "Facies Europae potissimum Borealis, ad mentem veterum Graecorum, eorum precipue quibus Melae Pliniique nituntur testimonia. Expressa a G. Schöning"; "Facies Orbis Septentrionalis ad mentem Ptolemæi expressa a G. Schöning" (Trondheim, 1763).

M—Toronto-Fisher. 2 maps in August Ludwig Schlözer, *Fortsetzung der Algemeine Welt-Historie* (see *Welt-Historie*, 1744, below), 31:101A, 176B.

184. Schöning, 1771.   Gerhard Schöning, *Det gamle Norge med dets Graendser og Lande ferstillet* (n.p., 1771).

M—Cornell, Ic DL46 S37N8, vol. 1: a map in G. Schöning, *Norges Riiges Historie*, første Deel, indeholdende Riigets aeldste historie etc. (Sorøe, 1771), 68. This may be the preliminary version of Schöning, 1777.

185. Schöning, 1777.   Gerhard Schöning, "Facies trium regnorum borealium Europae, ad normam eorum Scriptorum expressa"; "Norwegia antiqua a fluvio Gotelf ad Halogalandiam delineata a G. Schönning [*sic*]."

M—Cornell, Ic PT7276 A1++ 1777, from Snorri Sturluson, *Heimskringla edr Noregs Konunga-sögor*, ed. G. Schöning, 6 vols. (Copenhagen, 1777–78), vols. 1 and 2: 1 map each.

186. Schöning, 1779.   Gerhard Schöning, "Facies Europae et finitimarum regionum ad mentem et nomina veterum Norvegicorum s[eu] Islandicorum scriptorum" (Copenhagen, 1779).

M—BerlinDSB. Single sheet, black and white, full of writing. BerlinPK, Kart. U3303; border colors: "Geographie des Mittel Alters vorzüglich des 9ten und 10ten Jahrhundert." From Iceland to east of the Caspian, including the Mediterranean.

187. Schöpflin, 1751.   Johann Daniel Schöpflin, "Alsatiae Francicae ducatus in pagos et comitatus suos divisus, cum oppidis, castris, palatiis, monasteriis, vicis tabula geographica" (Colmar, 1751).

M—Le Long, 1:35, no. 407. From Schöpflin, *Alsatia illustrata*, 2 vols. (Colmar, 1751–62), 1:619 (Oxford, Meerman 86, 87; LC-rare, DD801 A34S3, vol. 1 lacks the map). Vol. 2 contains a map of Strasbourg with 8 insets showing the growth of the city (also see Lamare, 1705). The map of Alsace is oriented west, to parallel the Rhine. Schöpflin's fellow academician d'Anville supplied the finishing touches.

188. Senex, 1708.   J. Senex, *Modern Geography* (London, 1708–25).

G—LC List 550. Closes with a supplement of ancient history including the 2-sheet Delisle, 1705.

189. Senex, 1721.   J. Senex, *New General Atlas Containing a Geographical and Historical Account of All the Empires, Kingdoms, and Other Dominions* (London, 1721).

G—LC; UChi. Extensive text. Suggestive of a stripped-down Châtelain, 1705. No maps for history.

190. Seutter, 1720.   Matthaeus Seutter, *Atlas geographus* (Augsburg, 1720).

G—Harvard. Modern maps followed by no. 32, map of the Holy Land with twelve tribes (small Exodus track); nos. 37–39, chronological lists of popes (with symbols of their characters), kings, electors, dukes, as well as genealogical trees.

191. Seutter, 1745a.   Matthaeus Seutter, *Atlas novus* (Augsburg, 1745).

M—LC List 5976.   Vol. 1, no. 24 (Heptarchy), "Brittaniae sive Angliae Regnum tam secundum prisca Anglo-Saxonicum Imperia quam recentiorum provinciarum divisionem." LC List 5973. Rogg, 1740/E, is in an atlas with the same title.

192. Seutter, 1745b.   Matthias Seutter, "Historia Circuli Bavariae necnon et finitimarum regionum" (Augsburg, ca. 1745).

M—BSB; YaleB, 1983, fol. 50. Little figures on a map, like Seutter, 1745c. Numbers refer to a legend on a second sheet, not provided.

193. Seutter, 1745c.   Matthaeus Seutter, "Historia imperii Romano-Germanici . . . in mappa exhibita" (Augsburg, 1745).

M—BL, 26905.(26.); YaleB, 1983, fol. 50 (an atlas); HABW, 30,18 (2 sheets). Records the death of Emperor Charles VII, January 1745 (no. 286.Ff). 2d sheet, "Erklaerung deren Figure, so sich auf der historischen Land-Charten von Teutschland, und derent Angraenzenden Laendern befinden"; a legend of 306 events in chronological order. Small color reproduction in BSB, *Cartographia Bavariae: Bayern im Bild der Karte,* ed. Hans Woff (Weissenhorn [Bavaria], 1988), 379. See also Desing, 1731.

194. Seutter, 1770.   Matthaeus Seutter and F. Lotter, atlas factice (n.p., 1720–70).

M—UMich-Clements. No maps of ancient geography; many decorated lists of popes, emperors, European kings, etc. Tracks of navigators: nos. 224, 225.

195. Spener, 1717.   Jakob Karl Spener, "Rhetia et Noricum ad veteris mediique aevi rationes" (lib. 6, cap. 2:245); "Gallia Belgica ad veteris, mediique aevi rationes" (lib. 6, cap. 4:271); "Germaniae medii aevi priorum seculorum rationibus attemperata" (vol. 2:369).

M—Le Long, 1:35, no. 403. Princeton. Maps from Spener, *Notitia Germaniae antiquae ab ortu rei publicae ad regnorum Germanicorum in Romanis provinciis stabilimenta Germaniae et Germanicarum civitatum statum et conditionem plene declarans* [the author] *ex fide dignis documentis argumentum perfecit et novis tabulis geographicis instruxit accessit conspectus Germaniae mediae qualis seculo VI et post paulo sequentibus seculis fuit,* 2 vols. in 1 (Halle, 1717).

196. Strebel, 1761.   Johann Sigmund Strebel, "Pagus Rangowe ex medio aevo restitutus cum locis, quorum in vetustis chartis mentio fit, secundum pronuntiationem et orthographiam veterum Francorum addita brevi expositione historica" (n.p. [Schwabach?], 1757).

M—*GV* 141:142. HABW, 30,30. This map is a separate publication, anticipating Strebel's book, *Franconia illustrata oder Versuch zur Erläuterung der Historie von Franken aus zuverläßigen archivalischen Documente,* vol. 1 with a map (Schwabach, 1761). HABW supplies the map separately from the book.

197. Tardieu, 1798.   P. F. Tardieu, "Carte des descentes faites en Angleterre et en Irlande depuis Guillaume le Conquérant" (n.p. [Paris], an VI [1798]).

M—Oxford. A descendant of Speed, 1627; cf. Enouy, 1797. Foldout, handsome map of Ireland and England. Little groups of eighteenth-century ships, with numbers keyed to an accompanying text. The Tardieus were a clan of engravers, reaching back at least to the early eighteenth century. The individual members have not been precisely identified.

198. Tomka, 1751.   János Tomka Szászky, *Parvus atlas Hungariae* (Bratislava [Pozsony, Pressburg], 1750–51).

H—Budapest, T35 (18 maps). Dörflinger, *Österreichische Kartographie,* 1:60 (19 maps). Purgina, *Tvorcovia Kartografie Slovenska* (19 maps listed by title, with 3 black-and-white reproductions). Kuchař, *Mapová Sbírka Molla,* 316 (13 maps listed by title). Affirmed to be "The First Historic [*sic*] Atlas of Hungary" by Árpad Papp-Váry and Pál Hrenkó, *Magyarország régi térképeken* [Hungary in Old Maps] (n.p., n.d. [date stamp 1991]), 98–99 (1 color reproduction).

199. Tomka, 1781.   János Tomka Szászky, *Introductio in geographiam Hungariae antiqui et medii aevi . . . e veteribus monumentis eruta et VI tabulis illustrata* (Bratislava [Pozsony, Pressburg], 1781).

H—Göttingen, 8vo H. Hung, I, 78. Small 8vo, 45 pp. of text. Maps dated 1750 or 1751. Assembled from Tomka, 1751, by István Jeszenák: Lajos Széntai, *Atlas Hungaricus 1528–1850: Magyarország nuomtatott térképei 1528–1850* [Hungary in Printed Maps], 2 vols. (Budapest, [1996]), 2:620.

200. Vaissete, 1749.   Joseph Vaissete, "Carte du Languedoc avec les provinces voisinnes où l'on à [*sic*] marqué la division du Roy^m et Duché de Septimanie et celles des Trois Senechaussés anciennement comprises sous le nom de Languedoc" (Paris, 1749).

M—BN. From Joseph Vaissete, *Abrégé de l'histoire générale du Languedoc* (Paris, 1749), 1:1. The basic map seems modern. Cf. the unabridged version, Nolin, 1733.

201. Velly, 1787.   Paul François Velly, Claude Villaret, and Jean-Jacques Garnier, *Recueil de cartes pour l'étude de l'histoire de France, destiné principalement à celle commencée par MM. Velly et Villaret [et complétée par] Garnier,* 2 vols. (Paris, 1787).

H—*NUC, BN Impr.* 205:136. BN, Ge.DD.3012; Princeton. 85 maps, "extrait[es] de l'Atlas universel dirigé par Philippe de Prétot." A world atlas with few concessions to history. The Middle Ages are represented by Gaul at the advent of Clovis (Lié-baux, 1728); the Empire of Charlemagne (Robert, 1743); Netherlands in the Middle Ages (Wagenaar, 1749).

202. Wagenaar, 1749.   J. Wagenaar, *Vaderlandsche Histoire*, p. 2 (Amsterdam, 1749).

M—BN. A major narrative. P. 2, frontispiece: "Oude Kaart der nu Vereenigde Nederlande tot opheldering der Vaderlandsche Historie in de Middeleeuwe" (Amsterdam, 1749). Cf. *Aloude Holland*, 1745. This map circulated widely apart from Wagenaar's book, e.g., PhiladFree; AmstUL, 28-29-02.

203. Wastelain, 1761.   Charles Wastelain, S.J., "Gallia Belgica ad historiam medii aevi concinnata" (Lille, 1761).

M—Le Long, 1:35, no. 404: NYPL. From Charles Wastelain, *Description de la Gaule-Belgique selon les trois âges de l'histoire, l'ancien, le moyen et le moderne, avec des cartes de géographie et de généalogie* (Lille, 1761). Wastelain situates the "middle age" between Clovis and ca. 1050. He regarded the strengthening of post-Carolingian monarchies as inaugurating the modern period (preface, p.i).

204. Wells, 1700.   Edward Wells, *New Sett of Maps both of Antient and Present Geography* (London, 1700).

G—NYPL. Beautiful maps on superb paper. There are later editions. Nothing medieval.

205. *Welt-Historie*, 1744.   *Algemeine Welt-Historie von Anbeginn der Welt bis auf gegenwärtige Zeit*, 57 vols. (Halle, 1744–91).

M—*NUC* 8:505 (sets in United States). Toronto-Fisher, J-10 235 (vols. 1–47, 50–51, 53–55). For the English original, see *Hist. universelle*, 1779. The German publisher obtained the illustrations from the Dutch edition. Detached maps UCB-Bancroft, PhiladFree. Medieval subjects (Toronto): 17; 21:475; 22:294; 34; 35 (same as Liébaux, below); 36:15; 36:128; (Fontenai, 1772), "Carte Géographique des Lieux voisins d'Auxerre nommés dans Nithard à l'occasion de la Bataille de Fontenai, et autres Lieux;" 40:preface (same as Muratori, 1727); 41: preface, "Ager Mediolenensis Medii Ævi" (1779); 51.

206. Zatta, 1779.   A. Zatta, publ., *Atlante novissimo* (Venice, 1779–85).

G—LC List 650. The clump of ancient geography at the close includes no. 62, "Empire of Charlemagne" (1795).

207. Zollmann, 1732.   Friedrich Zollmann, "Ducatus Saxoniae superioris prout ipsius conditio fuit ab Anno 1000 usque ad 1400, sive intra saeculum X et XV, ex historia maxime mediae aetatis erutus;" "Ducatus Saxoniae superioris ut status ipsius antiquissimus fuit, per saecula X priora, scilicet post Christum natum ad A. 1000 usque, ex historiae Saxonicae monumentis compilatus" (Nuremberg, 1732).

M—Abundant copies, e.g., BL, BN, BSB; see the Homann atlases, especially Homann, 1753. Zollmann placed a map of modern Electoral Saxony at the head of the two historical ones, to form a Saxon trio. The reversed chronological order was deliberate.

H, M, G   NINETEENTH CENTURY

208. Akhmatov, 1845.   I. M. Akhmatov, *Atlas Geografickeskii, Istoricheskii i Chrono-logicheskii Rossiiskago* [Geographical, historical, and chronological atlas of the Russian state], after the history of N. M. Karamzin. New ed. by I. Einerling (St. Petersburg, 1845).

H—BL, 71 maps. 20 "Maps for (Ancient) History," followed by 51 individually titled maps for Russian history ending in 1505. Descriptions are given for the contents of additional maps from 1505 to 1825 (apparently not carried out).

209. Andriveau, 1837.   J. Andriveau-Goujon, publ., *Atlas classique et universel de géographie ancienne et moderne* (Paris, 1837).

G—UIll, 37 maps. The date 1835 also occurs in scattered catalogue listings. Ancient and modern only. YaleB, new ed. (1847), 45 maps, dated 1845.

210. Andriveau, 1840.   E. Andriveau-Goujon and E. Soulier, *Atlas élémentaire simplifié de géographie ancienne et moderne* (Paris, 1840).

M, G—BL, Maps 38.e.10, Maps 41.e.10. (1868). The coverage of ancient and medieval history includes maps of the barbarian invasions and Crusades (with tracks), then France and Europe each in a page of little maps from the early Middle Ages to the Revolution.

211. Andriveau, 1841.   E. Andriveau-Goujon, *Atlas de choix, ou recueil des meilleures cartes de géographie ancienne et moderne* (Paris, 1841–62).

H—Łodyn´ski, 3; no. 507 (1836–47); LC List 6093. LC, G1019.A64 1862; UMich 186-? very large dimensions; BN, Ge.CC.1414. Ancient history prolonged by medieval and early modern; overlap with Andriveau, 1840. See also Goujon, 1828.

212. Ansart, 1840.   Félix Ansart, *Atlas historique, ancien et moderne, Cour de troisième, Histoire du moyen âge* (Paris, 1840).

H—Łodyński, 3: no. 459. BN. Ansart was co-translator of Kruse, 1802; Ansart's first names are Charles Boniface Félix. Part 2 of *Atlas historique ancien et moderne dressé pour l'usage des collèges,* 12 maps dated 1834–35.

213. Ansart, 1844.   Félix Ansart, *Petit atlas historique et géographique ancien et moderne* (Paris, 1844).

H—Columbia, 911An81. 38 maps, ancient, Roman, medieval, modern, France; plus 14 of contemporary geography. Emphatic border coloring on outline background.

214. Ansart, 1847.   Félix Ansart, *Atlas historique ancien et moderne, renfermant toutes les cartes anciennes, du moyen âge, et moderne. . .* (Paris, 1847).

H—Harvard. Ansart, 1844, augmented by the 12 maps done for Kruse, 1836.

215. Ansart, 1859.   Félix Ansart, *Atlas historique et géographique,* new ed. by Edmond Ansart (Paris, n.d. [1859]).

H—BN. To be put together into distinct atlases of different subjects, differently priced.

216. Ansart, 1861.   Félix Ansart, *Atlas historique universel dressé d'après l'Atlas historique des états européens de Kruse* (Paris, n.d. [ca. 1861]).

H—Columbia, 911.4K94 F, 19 maps. The last map, Europe in 1861, is an Andriveau-Goujon production, dated 1861, very different from the Ansart maps. The latter, dated 1833, have Ansart's name and are for Kruse, 1836.

217. Aretin, 1809.   Baron Johann Christoph (J. C. A. M.) von Aretin, *Baierns grösster Umfang unter den Agilolfingern, Carolingern, Welfen und Wittelsbachern, in 4 geographische Karten dargestellt zur Erläuterung einer merkwürdigen Aeusserung Napoleons des Grossen* (Munich, 1809).

M—BL, Maps 8.b.12. The maps are announced on the title page as lithographed; a very precocious instance of this technique applied to maps for history. Aretin, probably the younger of two brothers with the same name, was head of the Bavarian Academy.

218. Artero, 1879.   ◆ Juan de la Gloria Artero y Gonzalez, *Atlas histórico-geográfico de España desde los tiempos primitivos hasta nuestros dias* (Granada, 1879).

H—UCB, DP27.A75. 6 ancient maps, 10 medieval, 7 modern. Artero held the chair of historical geography at the University of Granada.

219. Atlas MS, 1832.   *Atlas contenant des cartes géographiques et chronologiques propres à faciliter l'Étude de l'Histoire ancienne et moderne* (France, 1832? 1840?).

H—Manuscript. LC List 97 (acquired 1903), LC, G1030.A32 1842 Vault. Contents: maps, time charts, genealogical tables. Latest date: 7 June 1840 (in a later hand?): Latest date on the time chart: (Portugal) 1832 "Expédition de Dom Pedro" (an episode of the Miguelite Wars, 1828–34: Dom Pedro, after abdicating the Brazilian throne, captured Oporto, 8 July 1832). The atlas is incomplete; many blank genealogical charts. Influenced by the *Atlas Lesage,* several of whose maps are copied more or less fully. Other historical episodes are depicted in the Lesage manner. "Middle Ages" is not mentioned; classical antiquity is followed by the "Bas-Empire."

220. Babinet, 1862.   Jacques Babinet, *Atlas universel de géographie physique, politique et historique* (Paris, 1862).

H—LC, G1019.B125 1861; also BN. Current geography: 1–25. Historical: ancient, 1–14; medieval, 15–24; modern, 25–35.

221. Baquol, 1859.   J. Baquol and J. H. Schnitzler, *Atlas historique et pittoresque ou histoire universelle disposée en tableaux synoptiques,* 3 vols. (Strasbourg, 1859–60; new ed., 1861–64).

H—LC List 99–100. BN (vol. 1 only). About 12 maps; 4 for antiquity. Two-part periodization; the Middle Ages (3 maps) also including the beginning of modern times.

222. Barberet, 1870.   Ch. Barberet and Ch. Périgot, *Atlas général de géographie physique et politique, ancienne, du moyen âge et moderne* (Paris, [ca. 1870]).

H—CambUL, UMich. Contents suggest later publication than the advertised ca. 1865. 95 maps. Medieval general, 6; medieval France and other lands, 12.

223. Baumgärtner, 1815.   Baumgärtner Buchhandlung, publ., *Historischer Atlas von Sachsen, oder augenblickliche Uebers[icht] der verschiedenen Besitze dieses Landes* [950–1815] (Leipzig, 1815).

H—NOT SEEN. *GV* 6:171 (s.v. "Atlas"). 26 maps. *GV* records a copy at Berlin. New ed., 1816: *Historischer Atlas von Sachsen in 25 colorierten geographischen Karten, mit Erläuterungen über die Vergrösserungen und Verkleinerungen dieses Landes von 950– 1815* (also recorded at Berlin). (Many *GV* entries are prewar; the postwar division of the Berlin library is not taken into account.)

224. Baumgärtner, 1816.   Baumgärtner Buchhandlung, publ., *Historischer Atlas von Preussen* [1273–1816] (Leipzig, 1816).

H—NOT SEEN. *GV* 6:171. 12 maps. Recorded at Berlin.

225. Baumgärtner, 1817.   Baumgärtner Buchhandlung, publ., *Historischer Atlas von Russland, Schweden, Polen, Oesterreich, der Türkei u.s.w. vom Jahre 1155 bis zum Jahre 1816* (Leipzig, 1817).

H—Łodyński, 3: no. 125. BerlinDSB. 5 maps, all on an identical background extending from England to Kamchatka.

226. Baur, 1868.   C. F. Baur, *Historisch-geographische Karte von Württemberg* (Stuttgart, n.d [1860?–68]).

M—BL, 30284 (1.2.3.). One large historical wall map with two variations.

227. Bazeley, 1815.   C. W. Bazeley, *New Juvenile Atlas* (Philadelphia, 1815).

G—LC List 4310. 28 maps. Nos. 2, Roman Empire; 3, Ancient Greece; 28, Palestine and Holy Land.

228. Bazin, 1854.   François Bazin and Félix Cadet, *Atlas spécial de la géographie physique, politique et historique de la France,* 2d ed. (Paris, 1855).

H—BN, Ge.DD.5504 (unbound), Ge.DD.5503, Ge.DD.2828 (1864). A detailed atlas of current France with historical background in no. 18, "France historique. Agrandissements et réunions des provinces à la Couronne." Some history in no. 19 and later. BN, L^{14}.60, 2-page prospectus (Soissons, 1854).

229. Beck, 1856.   Joseph Beck, *Historisch-geographischer Atlas für Schule und Haus* (Freiburg im Breisgau, 1856–57).

H—ViennaUL, 10 ancient, 7 medieval, 8 modern maps. BN, 2d ed. (1870); 3d ed., 26 maps (1878). Beck, a secondary school teacher, published only this (*GV* 10:94).

Bedeus, 1845. *See* Scharberg.

230. Benicken, 1820.   Friedrich Wilhelm Benicken, *Historischer Schul-Atlas oder Uebersicht der allgemeinen Weltgeschichte* (Weimar, 1820).

H—Łodyński, 3: no. 147 (records a copy at Wrocław; its count of 20 plates is anomalous). BerlinDSB, BSB, ViennaUL. 14 maps; lithographer, Anton Falger; drawn by C. C. Wendel. Announced, *AGEphem.,* 2d ser., 7 (April 1820): 53 (14 maps). An early lithographed atlas. "Methodical" (i.e., scriptless), maps for teaching purposes, but colored, all identifications supplied in the margins. Unusual index map on the cover sheet.

231. Benicken, 1821.   Friedrich Wilhelm Benicken, *Historischer Hand-Atlas zur Versinnlichung der allgemeinen Geschichte aller Voelker und Staaten nebst Zeitrechnungstafeln über alte, mittlere, neuere, und neueste Geschichte* (Weimar, 1821–24).

H—Engelmann, 75, lists as *Handatlas der allgemeinen Weltgeschichte,* 4 fascs.; Łodyński, 3: nos. 161, 176; Mees, 7. BL, Maps 46.f.9.; NYPL-main; StockholmRL, AB 45 (3d installment only); BerlinDSB, gr 2° U166 (catal. listing only). Lithographer, Anton Falger; drawn by Karl-Ferdinand Weiland. For the companion vol., see Benicken, 1822/B. Some copies of the atlas (e.g., BL) have 15 sheets by the inclusion of a bonus, "Napoleons Heereszüge (1796–1815)" (Weimar, 1815), Benicken's first map.

232. Bensen, 1849.   Heinrich Wilhelm Bensen, *Historisch-geographischer Atlas von Europa* (Stuttgart, 1849).

H—*GV* 12:101. BerlinPK. Only the 1st of 5 installments is historical (5 maps, 2 of which are medieval). Fascs. 2–4, modern regional maps; fasc. 5, historical tables. Bensen's method resembles the contemporary compilations of Freyhold and Wedell.

233. Bergh, 1852.   ♦ L. Ph. C. van den Bergh, *Handboek der Middel-Nederlandsche Geographie naar de brounen bewerkt* (Leiden, 1852; 2d ed., The Hague, 1872; 3d ed., 1949).

M—BL, 10270.e.10. Map, "Nederland in het Frankische Tijdvak." 2d ed., The Hague, 1872.

234. Birchall, 1859.   James Birchall, *The Student's Atlas of English History: A Series of Maps and Plans (Norman and Plantagenet Period)* (London, n.d. [1859]).

H—BL, Maps 15.b.8. 12 maps; attention is paid to France and Europe.

235. Black, 1854.   A. Black and C. Black, *General Atlas of the World,* new ed. (Edinburgh, 1854).

G—LC List 4334. 3 historical maps, ancient and sacred.

236. Blumenthal, 1840.   J. Blumenthal, *Atlas abrégé de géographie et d'histoire universelle, comprenant les principes de la géographie astronomique, physique, l'histoire de la géographie, la géographie physique et historique de la France et des autres états européens, etc., etc.* (Paris, 1840).

H—BN, G.1192, 1st and 2d installments. 10 maps.

237. Boiste, 1806.   P. C. V. Boiste, *Atlas du dictionnaire de géographie universelle ancienne, du moyen âge et moderne, comparées* (Paris, 1806).

G—LC, G1019.B655 1806. Ancient, biblical, and modern only.

238. Bonnechose, 1847.   Émile de Bonnechose, *Géographie physique, historique et politique de la France* (Paris, 1847; 2d ed., 1877).

H—BN. 18 maps, of which 10 medieval (4 on one plate).

239. Bossi, 1824.   Luigi Bossi, *Nuovo atlante universale della antica e moderna geografia* (Milan, 1824).

G—LC List 743. LC, G1019.B69 1824. 2 maps of ancient history.

240. Bostwick, 1826a.   Henry Bostwick, [Atlas of Historical Maps and Charts] (New York, 1826, 1827).

M—NYPL, 3 copies without title page. Maps out of Lavoisne, 1814. Bostwick also authored *Historical and Chronological Atlas* (New York, 1838) without maps (NYPL).

241. Bostwick, 1826b.   Henry Bostwick, *A Historical and Classical Atlas illustrating, by a series of maps and charts, ancient history and geography, both sacred and profane; and also, Grecian and Roman mythology* (New York, 1826–30).

H—NUC. Princeton. Many charts, few maps.

242. Bosworth, 1855.   Joseph Bosworth, *A Description of Europe and the Voyages of Othere and Wulfstan Written in Anglo-Saxon by King Alfred the Great* (London, 1855).

M—CambUL, LE.32.44. See also Barrington, 1773. 4 foldout maps by Edward Weller, attractively tinted.

243. Boucher, 1804a.   N. Boucher and P. Picquet, "Carte synthétique des accroisse-ments succéssifs de la puissance des Francs en Gaule sous les rois des 1ère et 2ème races"; "Carte synthétique des principaux fiefs et domaines qui, par réunion succés-sive à la couronne, ont relevé la monarchie Française, sous les Rois de la 3ème Race"; "Ensemble les nouvelles possessions cédées à la République par le Traité de Lunéville, sous le consulat de Bonaparte" (Paris, 1804).

M—BerlinPK. Two sheets. Narrative in margins. Fixed districts. Letters so placed as to illustrate "Accroissements" of first kings, then "Réunions," from "A" by Hugh Capet to "O" by Louis XVI, "R" by Convention nationale, "S" by Directoire exécutif, and "T" by gouvernement consulaire (early Napoleon).

244. Bouillet, 1865.   Marie-Nicolas Bouillet, *Atlas universel d'histoire et de géogra-phie* (Paris, 1865).

H—*NUC.* LC, Harvard, Yale, Princeton, PhiladUL; NYPL. 88 maps, 6 medieval. Geography by Ernest Desjardin; the work was completed by Bouillet's son. Explica-tory comments. Widely sold; see 1877 ed. Careful review by Ernest Desjardin, *Bul-letin de la Société de géographie*, August 1866, 147–56.

245. Bouillet, 1877.   ♦Marie-Nicolas Bouillet, *Atlas universel d'histoire et de géogra-phie*, 3d ed. (Paris, 1877).

H—BN; LC PhiladFree; UMich-Clements, 3 in. thick, encyclopedic extra mat-ter. 2d ed. by Ph. Bouillet, maps by Vuillemin (1872). 87 maps. PhiladUL has eds. 1, 2, and 3.

246. Bradford, 1835.   Thomas Gamaliel Bradford, *A Comprehensive Atlas, Geo-graphical, Historical and Commercial* (Boston, New York, and Philadelphia, 1835).

G—NYPL, 69 maps; at the end, 10 ancient and biblical maps.

247. Brand, 1833.   J. P. Brand Eschauzier, ed., "Geschiedkundige Kaart der Groote Volksverhuizing overgenomen uit den Atlas van Lesage" (Amsterdam, 1833–39).

M—AmstUL, 28-34-40; BL, Maps 49.f.16. Inaugural installment, left sheet, of Brand, ed., *Encyclopédische Atlas* (Amsterdam, 1838–41); 28 installments published. Brand's encyclopedia, with no definite thematic limits, includes very few maps.

248. Brecher, 1868.   A. Brecher, "Darstellung der territorialen Entwicklung des Brandenburgisch-Preussischen Staates vom Jahre 1415" (Berlin, 1868).

M—BL, 26907.(30.); 33129.(5.). Prussian development is shown on a single map with a series of insets. The 2d call number is for a wall map.

249. Bretchneider, 1856.   C. A. Bretschneider, *Historisch-geographischer Wand-Atlas, nach Carl Spruner* (Gotha, 1856).

H—*NUC* "Cartographic Materials"; Engelmann, 1082. Cornell, Map G 5700 B7 1–10; LC, G1796.S1B76 1856. 10 wall maps of European history.

250. Brewer, 1853.   John S. Brewer, *An Elementary Atlas of History and Geography* (London, 1853–54).

H—BPL. Maps by Edward Weller (q.v.). Explanatory text. Columbia-rare, new ed. (London, 1871), said to be up to date to 1865. Weller thought that no one before him had represented "to the eye the political state of the world *at successive eras.*" Despite making historical atlases, he was no better informed than anyone else about their development.

251. Brué, 1820.   Adrien Hubert Brué, *Atlas géographique, historique, politique et administratif de la France* (Paris, 1820–28).

H—LC List 8405; LC, G1841.S1B7 1820. Dated 1820, 1821. Down to Louis XI only, BL, Maps 20.e.13. Planned for 12 installments of 4 maps each, it started publication in 1820 and ended after 6 installments (Louis XI) owing to the death of the publisher. Quérard, *Litt. franç. contemp.,* 2:455, attributes 7 fasc. to the atlas. Cf. Quérard, *France littéraire,* 1:532.

252. Brué, 1822.   Adrien Hubert Brué, *Atlas universel de géographie physique, politique et historique, ancienne et moderne* (Paris, 1822, 1838–39).

G—LC List 4321. LC, 64 maps; Yale; RGS; BN Ge.DD.4796 (253) (1822), 2d ed, Ge. CC.1158, another, Ge.CC.1176. 3-map ancient prelude to modern geography. *BN Impr.,* not Cartes et plans, has the best holdings of Brué atlases.

253. Cacciatore, 1831.   Leonardo Cacciatore, *Nuovo atlante storico,* 3d ed, 3 vols. (Florence, 1831–33; 4th ed., 1832–33, 1835–36).

G—LC List 103; RomeMar, *Catalogo analitico . . .* (1913); 146 plates, 32 maps. NYPL (1836), BAV. Mainly a picture book.

254. Cassini, 1801.   Giovanni Maria Cassini, *Nuovo atlante geografico universale,* 3 vols. (Rome, 1792–1801).

G—LC; BN, Ge.DD.309. Some ancient geography in vol. 3.

255. Chaix, 1850.   Paul Chaix, *Atlas élémentaire géographique et historique* (Paris, ca. 1850).

G—*NUC;* Łodyński, 3: no. 572. BPL, 28 maps; RGS. "Historique" means "political."

256. Chambers, 1853.   W. and R. Chambers publ., *Atlas of Modern and Ancient Geography* (London and Edinburgh, 1853).

G—UIll. All maps modern.

257. Chanlaire, 1803.   P. G. Chanlaire, "Carte des Isles Brittanique et des côtes, qui les avoisinent, servant á l'intelligence de l'Histoire des descentes faittes [*sic*] dans

ces Isles depuis les Romains jusqu'á présent" (Paris, 1803–4, *revue et augmentée en l'an XII*).

M—NOT SEEN. Reviewed in *AGEphem*. 18 (1805): 214–16. "Depuis les Romains," in the title, means "since but not including."

258. Chantreau, 1809.   Pierre Nicolas Chantreau, "Plan de Paris qui en indique les accroissements successifs depuis Jules César jusqu'á nos jours" (n.p., n.d. [Paris, 1809?]).

M— Oxford, 2374.4.160/1. Bound into the author's *Histoire de France abrégée et chronologique . . . jusques en septembre 1808*. Superimposes many centuries of change on a single sheet; confusing.

259. Charle, 184-.   Charle, *Atlas classique universel et complet de géographie ancienne, du moyen âge et moderne* (Paris and Lyons, n.d. [184-]).

G—NYPL. Steel engraving; border coloring. Historical complement to 25 modern maps. Charle's first name is not in *BN Impr*.

260. Chevalier, 1867.   Henry Chevalier, *Atlas de géographie ancienne, moderne et contemporaine* (Paris, n.d. [1867]).

G—BL, Maps 38.d.16. Historical maps start at no. 21.

261. Chevalier, 1881.   ◆ Henry Chevalier, *Atlas complet de géographie contemporaine, ancienne, du moyen âge, moderne*, new ed. (Paris, ca. 1881).

G—CambUL. History complements geography but is highlighted. 4-map medieval coverage.

262. Clausolles, 1845.   P. Clausolles and Philibert Abadie, *Atlas historique et géographique du moyen âge* (Paris, 1845).

H—BN, G.5656; MilwaukeeAGS, At.050./B-1846. 12 maps, steel engraving.

263. Clausolles, 1846.   P. Clausolles, and Philibert Abadie, *Atlas historique et géographique de la France: Les changements succéssifs de la monarchie aux principales époques* (Paris, 1846).

H—BL, Maps 4.aa.13. Middle Ages in 6 maps.

264. Colart, 1841.   L. S. Colart, *Histoire de France et d'Angleterre comparée*, 2d ed. (Paris and London, 1841).

H—LC List 4010. LC, G1841.31Cb 1841; BN. 2 maps only. Many "tableaux," predominantly textual.

265. Colbeck, 1885.   ◆ C. Colbeck, *The Public School Historical Atlas*, 2d ed. (London, 1885).

H—Łodyński, 3: no. 1119. BL, 101 small maps, 20 medieval.

266. Collier, 1875.   ◆ William F. Collier, [and Leonhard Schmitz], *International Atlas* (London, n.d. [ca. 1875]).

H—Toronto-Fisher, F-10 4965. Modern geography followed by "historical" and "ancient"; 8 maps of medieval history in the "historical" section.

267. *Comprehensive Atlas*, 187-.   ◆ *The Comprehensive Atlas and Geography, Modern, Historical, Classical, and Physical* (New York, 187-?).

G—NYPL. Letterpress by James Bryce, William F. Collier, and Leonhard Schmits. Maps by Edward Weller (q.v.) from his publications.

268. Cortambert, 1870.   Eugène Cortambert, *Petit atlas de géographie du moyen âge* (Paris, 1870).

H—First names, Pierre François Eugène. Harvard. 15 maps, identical to those in Cortambert, 1878. No. 3 (barbarian invasions) is better than Cortambert, 1878, no. 18 (probably of 1865 or earlier).

269. Cortambert, 1872.   ◆Eugène Cortambert, *Petit atlas de géographie ancienne, du moyen âge et moderne* (Paris, n.d. [1872]).

H—Łodyński, 3: no. 918. BL. Nos. 16–30 are medieval and well designed.

270. Cortambert, 1878.   ◆Eugène Cortambert, *Nouvel atlas de géographie contenant, en 80 cartes la géographie ancienne, la géographie du moyen âge, la cosmographie etc.*

H—Łodyński, 3: no. 889. LC (1865); Yale, 37 maps (Paris, 1867, 1878); BL, 98 maps (1878); BPL, 100 maps (Paris, 1886). In BL copy, nos. 17–31 are medieval.

271. *Cyclopedia, 1820.*   *The Cyclopedia, vol. 6, Ancient and Modern Atlas,* ed. Ephraim Chambers. (London, 1820).

G—Toronto-Robarts. Black-and-white, ancient followed by modern.

272. Delamarche, 1806a.   Charles François Delamarche, *Atlas élémentaire,* 3d ed. (Paris, 1806).

G—BN, G.8065. 4th ed., 1816, 33 maps, basically of France. BN, G.3227.

273. Delamarche, 1806b.   Charles François Delamarche, *Petit atlas moderne ou collection de cartes élémentaires dédié à la jeunesse* (Paris, n.d. [1806]).

G—BAV, Chigi IV, 4292. Adaptation of Bonne, 1762, published by Lattré, whose stock Delamarche acquired. Dating depends on no. 7, "L'empire français divisé en 111 départements." The only history is biblical.

274. Delamarche, 1824.   Charles François Delamarche, *Atlas de la géographie ancienne et moderne* (Paris, 1824, 1826).

G—Łodyński, 3: no. 205, 36 maps; LC List 4317, C. F. Delamarche, *Atlas de la géographie ancienne et moderne* (n.p., n.d. [1829?]); maps dated 1811–20. CalStLib (1824): ancient history maps, nos. 21–33, dated 1783–1822.

275. Delamarche, 1827.   Charles François Delamarche, *Atlas de la géographie ancienne, du moyen âge, et moderne* (Paris, 1827, 1828, 1832, 1833–39, 1850). For schools.

G—LC List 6058. Continues the foregoing, a long-running atlas. LC, G1019.D38 1827: the 1827 title first includes "Middle Ages"; it and that of 1834 begin to have arrows indicating directions of barbarian movements. Historical maps run from nos. 37 to 55; nos. 53–55 are medieval. Harvard; BSB, 1832, 36 maps; LC, 1850. Delamarche labels himself (correctly) "successeur de Robert de Vaugondy."

276. Delamarche, 1844.   Charles François Delamarche, *Atlas de géographie historique du moyen âge* (Paris, 1844).

H—BN (also Z.Renan.233). The Middle Ages are a "partie importante et peu connue de la Géographie historique." 6 "cartes générales"; 6 "particulières," with many subdivisions.

277. Delamarche, 1868.   Félix Delamarche, and Augustin Grosselin, *Atlas de géographie physique, politique et historique* (Paris, 1868).

H—Harvard, 105 maps, nos. 31–47 medieval; MilwaukeeAGS, new ed., 69 maps. See Quérard, *Litt. franç. contemp.*, 3:209–10.

278. Delgeur, 1848.   Louis Delgeur, "Carte historique de la Belgique au Moyen Âge" (Brussels, 1848; 2d ed., 1863).

M—NOT SEEN. Recorded in Institut néerlandais, 2: no. 124. Delgeur's map, for T. David, *Manuel de l'histoire de Belgique,* shows the southern Netherlands under the dukes of Burgundy, and Burgundy under Philip the Good.

279. Denaix, 1829.   Maxime Auguste Denaix, *Atlas historique de l'Europe* (Paris, 1829).

H—BL, Maps 144.c.1; LC; NYPL, BAI/++.

280. Denaix, 1835.   Maxime Auguste Denaix, *Atlas historique de l'Europe: Édition classique* (Paris, 1835).

H—UCB-Bancroft, fG1796.S1.D4 1835, 21 maps. Half the price of Denaix, 1829. Introduction by François Guizot. 9 medieval maps; early in including a group of thematic maps.

281. Denaix, 1836.   Maxime Auguste Denaix, *Atlas physique, politique et historique de la France* (Paris, 1836–37).

H—BL, Maps 150.e.14, 10 pl.; BL (1836–42), Maps 16.e.19, 16 maps. LC List 4019, new ed. (by Richard Wahl), 1855, 17 maps; Yale (1855), 14 maps. *BN Impr.:* M. A. Denaix, *Atlas . . . France, Prospectus* (1836).

282. Denaix, 1838.   Maxime Auguste Denaix, *Petit atlas historique de la France* (Paris, 1838).

H—BN, 34 maps, of which 15 are medieval. Written supplement, BN 8vo L^{14}.49.

283. Denaix, 1855.   Maxime Auguste Denaix, *Atlas physique, politique et historique de l'Europe,* new ed. by Richard Wahl (Paris, 1855; repr., 1860).

H—*NUC.* LC List 3965, 3966; BL, Maps 2.e.1., 30 maps. The thematic maps include a "Tableau de l'élévation et de l'abaissement des cent principaux états de l'Europe."

284. Denaix, 1856.   Maxime Auguste Denaix, *Atlas historique de la France* (Paris, 1856).

H—UMich. Somewhat longer than Denaix, 1838, but very similar.

285. Desjardins, 1836.   Constantin Desjardins, with J. Häufler, *Geographisch-historischer Atlas von Europa* (Vienna, 1838), p.2 of Desjardins, *Physisch-statistischer u. politischer Atlas von Europa* (Vienna, 1836–38).

H—Łodyński, 3: no. 364; *GV* 28:238–39; Engelmann, 262. BerlinPK, Kart. F1130. In p.2, 6 historical maps from Augustus to Napoleon, of which 3 are medieval. Many insets; written matter surrounds the maps on 3 sides. The maps of pt. 1 are thematic.

286. Dittmar, 1864.   Heinrich Dittmar and Daniel Völter, *Historischer Atlas,* 2d pt., *Atlas der mittleren und neueren Geschichte,* 4th ed. (Heidelberg, n.d. [1864]).

H—Łodynski, 3: no. 700. LC List 4133. 1st ed. possibly 1856 (map date). BL, Maps 48.b.53.: 6 medieval maps.

287. Dower, 1831.   J. Dower, *New General Atlas of the World* (London, 1831, 1838).

G—LC List 6083, 6094. Called Teesdale after the publisher. Few historical maps.

288. Dozy, 1870.   G. J. Dozy, *Historische Atlas der Algemeene Geschiedenis: Afbeeldingen en Kaarten* (Zutphen, n.d. [1870]).

H—Koeman, 6:9. AmstUL, 951 D 20. Nos. 22–45 medieval. Largely derived from Rhode, 1861.

289. Dressler, 1868.   O. Dressler, publ., [maps of medieval Italy] (Milan, 1868).

M—NOT SEEN. BerlinDSB. Four maps from Pasquale Villari, ed., *Storia generale d'Italia,* 8 vols. (1877–82), in *L'Italia sotto l'aspetto fisico, militare, storico, letterario, artistico et statistico,* 2d pt., no. 6 (Milan, 18–). Related to Vallardi, 1867.

290. Drioux, 1851.   Claude Joseph Drioux and Charles Leroy, *Atlas universel et classique d'histoire et de géographie ancienne, romaine, du moyen âge, moderne et contemporaine* (Paris, 1851 [1852]).

H—Łodyński, 3: nos. 614, 629, 830, 997. 1st ed., 56 maps: LC; BN, Ge.FF.4590, 4623 (4 reprints). New ed. (1864), 76 maps: YaleB. New ed. 1878, 76 maps: BL; Harvard, *G1020.D75.1884.

291. Drioux, 1867.   Claude Joseph Drioux and Charles Leroy, *Atlas d'histoire et de géographie (année préparatoire)* (Paris, 1867).

H—*NUC.* BL, Maps 3.b.11; BN, Ge.FF.4328, 4327.

292. Drioux, 186-.   Claude Joseph Drioux and Charles Leroy, *Atlas de géographie et d'histoire: Classe de 3ᵉ* (Paris, 186-).

H—YaleB. A selection of the maps in Drioux and Leroy, *Atlas universel.*

293. Droysen, 1886.   ♦Gustav Droysen, *Allgemeiner historischer Handatlas* (Bielefeld and Leipzig, 1886).

H—*NUC.* MilwaukeePL; etc. Exceedingly well represented in U.S. collections. A major production that took up the slack from the aging Spruner, 1871.

294. Dufau, 1841.   Louis Dufau, *Atlas historique et géographique de la France depuis les temps les plus reculés jusqu'à nos jours* (Paris, 1841).

H—UMich; UCB-Bancroft, 14 plates, 22 maps, 16 medieval.

295. Dufour, 1830.   Auguste Hippolyte Dufour, *Atlas de la géographie* (Paris, 1830?).

G—YaleB. No history except for the Holy Land in 12 tribes.

296. Dufour, 1831.   Auguste Hippolyte Dufour and Léonard Chodžko, *Atlas historique, politique et statistique de la Pologne ancienne et moderne indiquant ses divers démembrements et partages* (Paris, 1831).

H—BN, Ge.DD.1771. Dedicated to Joachim Łelewel. Handsome large folio; on binding, "Atlas des partages de la Pologne." Maps span 1768 (provinces) to 1815 (new Polish kingdom). No medieval content.

297. Dufour, 1834a.   Auguste Hippolyte Dufour [and Ch. Picquet], *Atlas classique et universel* (1839: *de géographie*) (Paris, 1834, 1839).

H—BN, G.1252; BL, Maps 38.d.30. (1839). 60 maps, 4 medieval. Picquet's name appears on title page after the 1st ed.

298. Dufour, 1834b.   Auguste Hippolyte Dufour, *Atlas élémentaire de géographie ancienne, du moyen âge et moderne* (Paris, n.d. [1834]).

G—BN, 40 maps, modern, then historical from no. 26, with 3 medieval.

299. Dufour, 1835.   Auguste Hippolyte Dufour, with E. Jomard and A. Balbi, *Le Globe, Atlas classique universel de géographie ancienne et moderne pour servir à l'étude de la géographie et de l'histoire* (Paris, 1835).

G—NYPL; LC; Yale. Ambitious, handsome production; Jomard and Balbi were prominent names. 9 historical maps at the start; 2 medieval. Spanish ed.: *El globo* (Madrid, 1852).

300. Dufour, 1840.   Auguste Hippolyte Dufour and Th. Duvotenay, *La Terre, Atlas historique et universel de géographie ancienne, du moyen âge et moderne* (Paris, 1840).

H—BL, Maps 48.e.17., 44 maps, 3 medieval; UCB, fG63.D8; MilwaukeeAGS. Printing in red as well as black is advertised as a special asset.

301. Dufour, 1856.   Auguste Hippolyte Dufour, *Géographie universelle de Malte-Brun: Atlas dressé et divisé par nationalités contenant tous les chemins de fer* (Paris, 1856).

G—YaleB. Few historical maps, 1 medieval. See also Huot, 1837; Tardieu, 1802, 1842. Huot, 1837;

302. Dufour, 1859.   Auguste Hippolyte Dufour and Michel Amari, "Carte comparée de la Sicile moderne avec la Sicile au XIe siècle d'après Idrisi et d'autres géographes arabes" (Paris, 1859).

M—Oxford, Bodleian 3.DELTA.1075. Large tipped-in map of Sicily with modern place-names alongside the Arabic equivalents *in Arabic characters.* Extensive topographic index.

303. Dufour, 1860.   Auguste Hippolyte Dufour, *Atlas universel physique, historique et politique de géographie ancienne et moderne* (Paris, 1860).

G—NYPL, LC, BL, BN, G.1541bis (1856–58), repr., 1875. Oversize; 40 maps, 9 historical.

304. Dufour, 1870.   Auguste Hippolyte Dufour, *Atlas géographique dressé pour l'Histoire de l'Église catholique de l'abbé Rohrbacher* (Paris, 1870).

H—*NUC.* BL-main. Good 9-item medieval coverage.

305. Dufour, 1878.   ♦ Auguste Hippolyte Dufour, *Atlas historique de la France* (Paris, 1878).

H—BL, Maps 4.b.10., 14 maps, 6 medieval.

306. Duruy, 1841.   Victor Duruy, *Atlas de géographie historique universelle,* pt. 3, section 1, *Atlas historique du moyen âge* (Paris, 1841).

H—Łodyński, 3: no. 403. *BN Impr.* G.23177, *Histoire universelle,* 47 vols. (1846–1918), *Atlas,* 3 pts. in 1 vol. (1841–43). Fifteen sheets. France is downplayed.

307. Duruy, 1846.   Victor Duruy, *Atlas historique de la France* (Paris, 1846).

H—Mees, 5: 1st ed., 1846; next, 1849. BL, Maps 11.c.23, 14001.d.8. (1849); Harvard and Harvard Law (explanatory text); BN, Ge.FF.4595 and companion vol. A limited but highly regarded selection of maps, 6 for the Middle Ages.

308. Duruy, 1877.   ♦Victor Duruy, *Histoire de France,* new ed., 2 vols. Histoire universelle; *Histoire du moyen âge depuis la chute de l'empire d'Occident jusqu'au milieu du XVe siècle,* 13th ed. Histoire universelle (Paris, 1877, 1890).

M—Newberry. Maps by Vuillemin; 5 medieval in each vol.

309. Dussieux, 1843.   Louis Dussieux, *Géographie historique de la France, ou Histoire de la formation du territoire français* (Paris, 1843).

H—NYPL; BN, 8vo L⁷.12. 14 maps for the Middle Ages.

310. Dussieux, 1854.   Louis Étienne Dussieux, *Atlas pour servir à l'étude de l'histoire et de la géographie de la France* (Paris, 1854).

H—BL, Maps 25.d.4. French part of the *Atlas général.* Nos. 38–51 are concerned with the Middle Ages.

311. Dussieux, 1856.   Louis Étienne Dussieux, *Atlas général de géographie physique, politique et historique* (Paris, n.d. [1856]).

H—UMich; BN, G.1265 (1854). The publication date is not entirely clear. A major historical collection vying with Spruner in fullness of coverage.

312. *Educational Maps,* 1847.   *Educational Maps for the Use of Schoolmasters* (London: SPCK, 1847).

H—RLScot-maps. The third of three parts is historical. Although the atlas is not of ecclesiastical history, its outlook is unusually religious. No. 5, "The Christian Empire, at the Rise of the Mohametan Religion"; 6, "The Mohametan Empire. The Hegira, A.D. 622."

313. Elst, 1831.   P. C. van der Elst, *Atlas des Pays-Bas représentant l'État géographique et politique des XVII provinces aux différentes époques de l'histoire* (Brussels, 1831).

H—AmstUL (unbound portfolio), 15 maps, 6 medieval.

314. Elton, 1825.   Charles Abraham Elton, "The Migration of the Barbarians who invaded the Roman Empire, showing the place of their departure, and that of their establishment, or of their destruction" (London, 1825).

M—RLScot; *NUC.* Adapted from Lesage, 1801, to illustrate the author's *History of the Roman Emperors from the Accession of Augustus to the Fall of the Last Constantine;* Elton was brother-in-law of the noted historian Henry Hallam. See also Whittaker, 1825.

315. Engelmann, 1836.   Godefroi Engelmann and Jean Engelmann, *Atlas élémentaire de géographie historique* (Mulhouse, n.d. [1836]).

H—BN, Ge.FF.4587 pièce. 20 maps, nos. 13–18 medieval. The compilers were father and son, owners of a lithographic establishment. Their atlas of modern geography (1835) is BN, G.3348.

316. Fayard, 1875.   ♦J. A. Fayard, and Alphonse Baralle, *Atlas universel Fayard* (Paris, n.d. [1875]).

G—Łodyński, 3: no. 1003 (1877). BN, Ge.F.5295, a sheaf of 10–15 wrinkling sheets; LC, 1877 ed., 79 maps, 9 historical, of which 3 are medieval. Vuillemin was the main cartographer.

317. Fix, 1855.   W. Fix, "Wand-Karte zur Geschichte des preussischen Staats insb. seit 1415" (Berlin, 1855?)

M—BL, 33129 (1.).

318. Foncin, 1888.   Pierre Foncin, *Géographie historique (Leçons en regard des cartes): Antiquité–Moyen Age–temps modernes–période contemporaine* (Paris, 1888).

H—Łodyński, 3: no. 1163. Harvard; LC. 48 maps, 13 medieval. An outline atlas by Foncin is also at Harvard.

319. Fortunatus, 18–.   "Reis Route des Venantius Fortunatus von Augsburg durch Rhaetien nach Agunt in Noricum in Iahr Christi 564" (n.p., n.d.).

M—BerlinPK, Kart. U5000. Unsigned, undated. Must have accompanied a book. Shown are parts of the Duchy of Alamannia, much of that of Bavaria, some of that of Trent. Fortunatus was a Latin poet of the late-sixth century.

320. H. Freudenfeldt and F. Pfeffer, *Die Erwerbungen des preussischen Staates,* eine Karte zum Gebrauche für die Schule beim Unterricht in der vaterländischen Geschichte, zunächst für die beim Hugo Bieler . . . erschienene Tabelle "Preussen unter den Regenten aus dem Hause Hohenzollern" (Berlin, n.d. [after 1849]).

H—BN. Ge.D.11143bis. Very attractive coloring in discreetly contrasting tints; the colors are mainly keyed to reigns.

321. Freyhold, 1846.   A. von Freyhold and A. von Schmidt, *Vollständiger Atlas zur Universalgeschichte* (Berlin, 1846–50).

H—*GV* 41:251; Engelmann, 75. BL, Maps 48.f.14. 3 parts—ancient, medieval, and modern, with several subdivisions and insets—on 3 sheets. 58 pages of text.

322. Freyhold, 1848.   A. von Freyhold and A. von Schmidt, "Karte zur Geschichte des Mittelalters" (Berlin, 1848).

M—*GV* 41:251. BerlinPK. Single sheet, 5 maps and 3 insets (one large). P.2 of Freyhold and Schmidt's *Vollständiger Atlas.*

323. Freyhold, 1853.   A. von Freyhold and A. von Schmidt, *Neue Karte von Deutschland, zugleich historisch-geographische Karte von Preussen* (Berlin, 1853).

M—NOT SEEN. *GV* 41:251. Only the map of Prussia is called "historical."

324. Frommann, 1854.   M. Frommann, *Historischer Atlas nach Angaben von Heinrich Dittmar,* 2d enl. ed. (Heidelberg, 1854).

H—Engelmann, 75. CambUL. 16 maps (1st ed., 1852), 6 medieval. Accompanies Dittmar, *Geschichte der Welt,* 4th ed., ed. D. Völter, 18 maps (Heidelberg, 1862–63); 7th ed. (Heidelberg, 1894).

325. Gage, 1869.   William L. Gage, *Modern Historical Atlas; A Modern Historical Atlas for the Use of Colleges, Schools . . .* (New York, 1869).

H—Toronto-Victoria; Newberry; Harvard. Based on Fromman, 1854. 5 maps for the Middle Ages. Announces that the minutiae of Spruner, 1837, are omitted.

326. Gall, 1850.   Gall and Inglis, publ., *The Edinburgh Imperial Atlas, ancient and modern* (Edinburgh, n.d. [map dates, 1850]).

G—Yale. 47 maps; the first 10 are ancient and austerely regional.

327. Garollo, 1890.   ♦Gottardo Garollo, *Atlante geografico storico dell'Italia dell'Istituto geografico italiano* (Milan, 1890).

H—Łodyński, 3: no. 1188. Harvard, Hist. of Science Library, Earth Sci., Garo 110. 24 maps, nos. 20–22 medieval; small pocket size. Few but high-quality historical maps.

328. Gatti, 1851.   Ferdinand Gatti, "Die Hauptzüge der in der Völkerwanderung erscheinenden Völker" (Graz, 1851).

M—BN, Ge.C.8358. Single sheet. Unusually angular tracks. No context. *GV* 44:10 records a book by Gatti (Vienna, 1857), a geographic aid to the study of medieval history.

329. Gauttier, 1830.   Louis Édouard Gauttier du Lys d'Arc, *Atlas de l'histoire des conquêtes des Normands en Italie, en Sicile, et en Grèce de l'année 1016 à 1140* (Paris, 1830).

M— Oxford. Accompanies the never completed book of this name. Maps, no. 1, "Carte politique de l'Italie méridionale au XIe siècle pour servir à l'histoire etc." (1829); no. 2, "Partie de la Grèce septentrionale." Modern authorities listed in lower margin.

330. George, 1852.   F. E. George, ed., *Atlas illustré destiné à l'enseignement élémentaire* (Paris and Lyons, n.d. [1852?]).

G—UCB, 48 maps, including 13 historical, 3 medieval. An anthology of noted French cartographers.

331. Goldsmith, 1813.   J. Goldsmith, *Atlas for Schools* (London, 1813).

G—BL, Maps 38.b.2. A few ancient maps.

332. Görringer, 1840.   Mich. Görringer, *Statistisch-Historischer Schul-Atlas für den allgemeinen geographischen Unterricht* (Zweibrücken, 1840).

H—NOT SEEN. Engelmann, 75; *GV* 48:353. 12 maps.

333. Goujon, 1828.   J. Goujon and J. Andriveau, *Atlas de choix ou receuil de cartes de géographie ancienne et moderne* (Paris, n.d [ca. 1828]).

G—BN. 42 maps, by various cartographers, including 10 ancient and Brué's "Charlemagne." A handsome production.

334. Gover, 1853.   Edward Gover, *Historic Geographical Atlas of the Middle and Modern Ages . . . Based on Spruner* (London, 1853).

H—LC; BL, 2059.f. 20 maps. Gover's name is on the maps, not the title page. "To render the Work easy of reference, special attention has been given to Tinting and Colouring the Maps," especially to tag the various invading tribes. The color scheme, though unusually attractive, is not wholly consistent.

335. Gover, 1854a.   Edward Gover, publ., *Atlas of Universal Historical Geography embracing Sacred and Classical—Medieval—and Modern Geography* (London, 1854).

H—LC, 28 maps. The introduction complains of the difficulties of medieval geography. Nos. 18–21 concern the Middle Ages.

336. Gover, 1854b.   Edward Gover, *The University Atlas, or Historical Maps of the Middle Ages* (n.p. [London], n.d. [1854? 1871?]).

M—BL, Maps 4.d.24. Anonymous. 1, "Engla-Land (A.D. 450–1066)"; 2, "England and Wales (A.D. 1066–1485)."

337. Gross, 1853.   Rudolf Gross, *Historischer Schul-Atlas* (Stuttgart, 1853).

H—NOT SEEN. Łodyński, 4: no. 158; Engelmann, 75; *GV* 51: 56.9 folio sheets, colored. Gross was a major map publisher.

338. Guadet, 1833.   Joseph Guadet, *Atlas de l'histoire de France* (Paris, 1833).

H—Quérard, *Litt. franç. contemp.*, 4:183–84. BN *Impr.*, 8vo L¹⁴.39; LC. 14 maps, 10 medieval.

339. Guadet, 1851.   Joseph Guadet, *Atlas élémentaire de l'histoire de France*, new ed. (Paris, 1851).

H—*BN Impr.*, 8vo L¹⁴.57. The 15 maps are lacking, but there is a table of subjects, 7 medieval. A revision of the atlas of 1833.

Halluvin. *See* Wautier.

340. Handtke, 1808.   F. Handtke, "Historische Karte von der Erwerbungen und Veränderungen des königl. Preussischen Staats vom J. 1417 bis zum Jahre 1807" (n.p., 1808).

M—NOT SEEN. Engelmann, 793. Prussia suffered grave defeats in 1806.

341. Häufler, 1840.   Joseph Häufler, *Historisch-geographisch Tableau des oesterreichischen Kaiserstaates: eine übersichtliche Darstellung der Geschichte dieses Staates, seine Gebietsveränderungen, Wappen und Orden* (Vienna, 1840).

M—BN. Mees, 10. 2 very large sheets; the 1st has historical information; the 2d, a map (with insets), "Historischer Uebersichts-Karte der Entstehung der oesterreichischen Monarchie" (allowance is made for the Babenbergers, for Spain and the Low Countries; the colors reflect these complications).

342. Heck, 1830.   J. George Heck, *Atlas géographique, astronomique et historique servant à l'intelligence de l'histoire ancienne, du moyen âge et moderne* (Paris, London, and Mulhouse, 1830–31).

H—Łodyński, 3: no. 430; Engelmann, 75; *NUC*. ViennaÖNB; RomeBNVEmm 202.Banc.5.A.21 (1st ed., 62 maps); NYPL; BN. 4 medieval maps, interestingly chosen. Other eds., 1835; 1842, 66 maps.

343. Heck, 1838.   George Heck, and L. Plée, *Atlas des familles: La France géographique, industrielle et historique, avec des cartes physiques, politiques et historiques, le plan de Paris, etc.* (Paris, 1838).

H—BN, L14–47; 2d ed., 1843? Mees, 4 (very approving). Said to be done on a new plan; text by Plée. "La France monarchique avant 1789" is similar to the widespread synoptic map "France historique" that shows provinces and acquisition by the Crown. Many maps relevant to present-day France. Much reading matter.

344. Hérisson, 1806a.   Eustache Hérisson, *Atlas du dictionnaire de géographie universelle, ancienne, du moyen âge et moderne, comparées* (Paris, 1806).

G—BN, Ge.FF.18936 (new ed., 1809); Harvard, MA 19.06s (1806). 45 maps, including 4 ancient and biblical.

345. Hérisson, 1806b.   Eustache Hérisson, *Atlas portatif contenant la géographie universelle ancienne et moderne* (Paris, 1806).

G—BN, Ge.FF.4420, Ge.FF.12449 (1807); *BN Impr.,* G.3228, G.8068 (1809). 49 maps, nos. 1–5 ancient.

346. Hérisson, 1816.   Eustache Hérisson, *Nouvel atlas portatif [et élémentaire], contenant [comprenant] la géographie universelle, ancienne et moderne,* 3rd [and 4th] ed. (Paris, 1811 [1816]).

G—Łodyński, 3: no. 120. Harvard, "Augmentée par H. Brué." The same ancient maps as the foregoing.

347. Herrmann, 1828.   August Leberecht Herrmann, *Ein historischer, chronologischer, geographischer Hand-Atlas zur Versinnlichung der Deutschen Geschichte von den ältesten bis zu den neuesten Zeiten, nebst Text und genealogischen Tafeln . . .,* ed. von Ehrenkreuz (Münster, 1828).

H—NOT SEEN. Engelmann, 438. (Not in *GV* and *NUC.*) 1 installment only with 4 lithographed leaves, roy. folio. The title resembles both Lesage, 1801, and Benicken, 1821. If completed, the work would have been the earliest explicit atlas of German history. Herrmann also has a *Lehrbuch der allgemeinen Weltgeschichte,* 2d ed. (Meissen, 1840), 8 maps (1st ed., 1832). *GV* 60:57–55, *NUC* 213:2. NYPL.

348. Herrmann, 1845.   August Leberecht Herrmann, *Geschichte des Königreichs Sachsen in gedrängter Uebersicht* (Meissen, 1845).

M—Newberry, F/4798.39. Folded at the end: "Charte vom Königreiche Sachsen dessen Theilung seit 1815 betreffend." Colored boundaries.

349. Hofdijk, n.d.   W. J. Hofdijk, [Historical maps] (Amsterdam, n.d.).

H—AmstUL, 23-10-05. 1. "De Nederlanden tijdens de Romainsche overheersching." Very large, 6 insets (lithographer, P. W. M. Trap, Leiden). 2. "Nederland in zijne go-verdeeling omstreeks de 10e eeuw." 7 insets. The 2 sheets form a small atlas. Hofdijk (1816–88) has a history of the Dutch people, 3 vols. with maps.

350. Houzé, 1840.   Antoine Houzé, *Atlas historico de España* (Barcelona, 1840).

H—LC. Selection from Houzé, 1841.   These maps of Spain are bound with a Spanish tr. of the maps of France and of the Holy Land. Only the latter has its full complement of 14 maps.

351. Houzé, 1841.   Antoine Houzé, *Atlas universel historique et géographique* (Paris, 1841).

H— Quérard, *Litt. franç. contemp.* 4:325, has a 1st ed., Paris: Dumesnil, 1837–38; cf. *NUC* 256:404, "Librairie universelle, 1837–38" (LC), a pen-and-ink addition in square brackets. LC List 110, 101 maps. LC, 1848; also UMich; Yale; UIll, 1849; LC, 1850; BPL, 1857. BL, Maps 48.d.11, 100 maps. *BN Impr.* has better holdings than Cartes et plans, but not comprehensive.

352. Houzé, 184-.   Antoine Houzé, and Victor Adam, *Atlas universel, historique et géographique; Galerie historique* (Paris, n.d.).

H—CalStLib. Stamped on binding: "Atlas Houzé / Galerie V. Adam." Original binding, mint condition. Each of the 101 maps faces a historical image by Adam. On Adam, see Emmanuel Benezit, *Dictionnaire critique et documentaire des peintres, sculpteurs,* new ed. (Paris, 1976), 1:32.

353. Houzé, 1845a.   Antoine Houzé, *Atlas historique de l'Angleterre composé de dix cartes géographiques* (Paris, 1845).

H—*BN Impr.,* folio N.294. Extracted from Houzé, 1841. Only 5 of the 10 maps.

354. Houzé, 1845b.   Antoine Houzé, *Atlas historique et universel de la France* (Paris, 1845).

H—Harvard (*NUC*). 30 maps from Houzé, 1841. *BN Impr.,* folio L^{14}.67, 8 maps only.

355. Houzé, 185-.   Antoine Houzé, *Le Musée—Atlas universel historique et géographique* (Paris, n.d.).

H—Harvard. The 101 maps of Houzé, 1841, complemented by 500 steel engravings.

356. Howe, 1851.   Heneage Howe, *A Concise Historical, Biographical and Genealogical Atlas of the Principal Events in the Histories of England, France, Spain, and Portugal, Germany and Italy (incl. the valuable Historical summary and observations of Lesage)* (London, n.d. [1851]).

H—CambUL. Responds to continuing popular demand for the *Atlas Lesage.*

357. Huber, 1830.   Jean Huber-Salantin, "Tableau chronologique de la Confédération suisse" (Paris, 1830).

M—NOT SEEN. Quérard, *France littéraire,* 4:588; *BN Impr. GV* 60:110 records J. Huber, *Chronologische Gemälde der Geschichte der Schweizer Eidgenossenschaft: Als Ergänzungsblatt zu A. Lesage's Atlas* (Bern, 1832), to complement Lesage, 1826b. No indication of a map.

358. Huberts, 1870.   W. J. A. Huberts, *Historisch-geographische Atlas der algemeen en vaderlandsche Geschiedenis,* 4th ed. (Zwolle, 1870).

H—Koeman, 6:182. BL. 1st ed., 1855, see Thierry, 1858. 33 maps. Nos. 13–22 medieval.

359. Hughes, 1849.   William Hughes, "Roman Britain," "Saxon England in the Seventh Century," National Society for Promoting the Education of the Poor, Geographical Illustrations of British History no. 2 (London, 1849).

M—Oxford.

360. Hughes, 1869.   William Hughes, *A Popular Atlas of Comparative Geography comprehending a chronological series of maps (based on Spruner's Hand-Atlas); A Popular Atlas of Comparative History. A Chronological Series of Maps of Europe and Other Lands at Successive Periods (based on Spruner)* (London, n.d. [1869, 1870]).

H—*NUC.* Newberry; CambUL; BL, Maps 3.c.5. On binding: Philip's Historical Atlas. 28 maps, nos. 2–20 medieval.

361. Hunt, 1860.   F. W. Hunt, *J. H. Colton's Historical Atlas: A Practical Class-Book of the History of the World* (New York, 1860).

H—BL. A time chart with maps as subsidiary insets. Much attention to barbarian invasions.

362. Huot, 1837.   Jean Jacques Nicolas Huot, *Atlas complet du précis de la géographie universelle de Malte-Brun* (Paris, 1837).

H— Quérard, *France littéraire*, 4:163, and Quérard, *Litt. franç. contemp.*, 4:347–48. RGS, 7H11; Oxford; NYPL (2 copies, 1839; one marked Brussels). 72 maps. Same program as the original companion atlas (Lapie, 1812), but differences of detail.

363. Issleib, 1874.   ♦ Wilhelm Issleib, publ., *Historisch-geographischer Schul-Atlas enthaltend in 36 Karten die alte, mittlere und neuere Geschichte* (Gera, 1874).

H— *GV* 70:216. BL. The only edition. Designed by Theophil König (q.v.), who worked on Issleib's biblical atlas the previous year. Issleib published a wide range of atlases and geographies.

Jarry de Mancy. *See also* B listings.

364. Jarry, 1832.   Adrien Jarry de Mancy, *Atlante istorico politico e statistico della Polonia antica e moderna con la indicazione dei suoi diversi smembramenti e divisioni* (Capolago [Switzerland], 1832).

H—RomeBNVEmm, 9.Banc.5.3. Foldout 1st sheet: "Quadro istorico . . . (Metodo di Lesage . . .), per A. J. de Mancy." Maps of Poland in the eighteenth and early nineteenth centuries.

365. Jausz, 1874.   ♦ György Jausz, *Historisch-Geographischer Schul-Atlas für Gymnasien, Realschulen und verwandte Lehr-Anstalten* (Vienna, 1872, 1873–74).

H— *GV* 68:142–43. BL, Maps 48.b.63, 32 maps, 10 medieval. Hungarian version, 1873; 2d ed., 1876; repr., 1903.

366. Johnston, 1846.   Alexander Keith Johnston, *The National Atlas of Historical, Commercial and Political Geography*, augm. ed. (Edinburgh, 1846).

G—LC List 4325; *DNB* 10:958–59. BL, Maps 41.f.10. 1st published 1843 (LC List 4323). Nothing overtly historical except an ethnographic map of current Europe.

367. Johnston, 1880.   ♦ Alexander Keith Johnston, *Johnston's Historical Atlas*, 2 vols. (Edinburgh and London, 1880).

H—LC List 114. Toronto-Robarts. Nos. 4–11 are medieval. Vol. 2 contains the explanatory text.

368. Johnston, 1889.   ♦ Alexander Keith Johnston, *Cosmographic Atlas of Political, Historical, Classical, Physical and Scriptural Geography and Astronomy*, 4th ed. (Edinburgh and London, 1889).

G—Toronto-Robarts, G1046 F2J65 1889 folio. 71 maps, including 21 historical, of which 3 are medieval.

369. Jusseret, 1836.   Nicolas Joseph Jusseret, *Atlas historique de la Belgique ancienne et moderne, depuis Jules-César Jusqu'à nos jours* (Brussels, 1836).

H—Mees, 17. Göttingen, folio H. Belg. I, 15. 14 maps. The date of publication is questionable; information dated 1842 is featured on the last map.

370. Kan, 1881.   ◆ J. B. Kan, *Historisch-Geographischer Atlas* (Amsterdam, 1881).
H—Koeman, 6:18. BL. 1st ed., 1867. 24 maps concern the Middle Ages.

371. Kärcher, 1824.   Karl Kärcher, *Orbis terrarum antiquus et Europa aevi medii in usum scholarum* (Karlsruhe, 1824).
H—*GV* 72:336; Łodyński, 3: nos. 187, 224; Engelmann, 78, lists a large-size version, 24 maps (Karlsruhe, 1827); other printings, 1831, 1839. BSB (NOT SEEN); BerlinPK; BAV, RG Geog Strags. 155: 23 maps, mainly ancient; 4 medieval.

372. Kärcher, 1834.   Karl Kärcher, *Historisch-geographischer Schul-Atlas* (Karlsruhe, 1830; 2d ed., 1834).
H—*GV* 72:336; Engelmann, 75. ViennaUL. 9 historical maps from antiquity to 1812, complemented by 9 geographic maps. Possibly connected with Gross, 1853 (see above). Kärcher published much on many subjects for secondary schools.

373. Kausler, 1831.   Fritz von Kausler, *Atlas des plus mémorables batailles, combats et sièges des temps anciens, du moyen âge et de l'âge moderne* (Karlsruhe and Freiburg im Breisgau, 1831–37).
H—BL, Maps 144.d.38, Maps 144.d.22 (1839); 200 sheets, French and German. Also a text vol. An officer in the Württemberg army, Kausler published *Versuch einer Kriegsgeschichte aller Völker und Zeiten,* 4 vols. (1825–33), with accompanying charts, tables, lexicon, and other aids.

374. Keppel, 1870.   Karl Keppel, *Geschichts-Atlas* (Nuremberg, n.d. [ca. 1870]).
H—*GV* 74:225–26, 7th ed., 1885; 18th ed., Munich, 1904; Łodynski, 3: no. 917, 27 maps. BL (1876?), imperfect; MilwaukeeAGS, At 600 B 1882, maps marked, "Keppel's Geschichts-Atlas für Mittelschulen." Basically German history.

375. Keppel, 1874.   ◆ Karl Keppel, *Geschichts-Atlas für Mittelschulen* (Hof, 1874; 2d ed., 1878).
H—*GV* 74:225–26. BL. 24 maps; one part each for ancient and German history.

376. Keppel, 1876.   ◆ Karl Keppel, *Atlas zur Geschichte des deutschen Volkes für Mittelschulen,* 2d ed. (Hof, 1876).
H—BL. 13 maps, of which 8 are medieval.

377. Keppel, 1877.   ◆ Karl Keppel, *Kleiner Geschichts-Atlas von Europa,* 2d ed (1877).
H—NOT SEEN. BL, Maps 10.b.40.

378. Kiepert, 1863.   Heinrich Kiepert, *Das Römische Reich Deutscher Nation um das Jahre 1000: Übersichtkarte zu W. Giesebrecht's Geschichte der deutschen Kaiserzeit,* vols. 1–2 (Braunschweig, 1863).
M—BerlinDSB, U S4030 (as a single-sheet map, detached from Giesebrecht's book); NYPL (in Giesebrecht).

379. Kiepert, 1865.   Heinrich Kiepert, *Historische Karte des brandenburgisch-preußischen Staates: Nach seiner Territorial-Entwickelung unter den Hohenzollern* (Berlin, 1865).

M—NOT SEEN. *GV* 75:40. Repr., e.g., 1866, 1873, 1889. Kiepert, a specialist in classical antiquity, rivals Spruner as the century's leading German maker of maps for history.

380. Kiepert, 1879.   ♦Heinrich Kiepert and Carl Wolff, *Historischer Schulatlas zur alten, mittleren und neueren Geschichte* (Berlin, 1879); 5th ed. (1890).

H—NYPL. Inconsiderately small. 36 maps, 12 medieval by Wolff (q.v.).

381. Klaproth, 1826.   Julius Heinrich Klaproth, *Tableaux historiques de l'Asie depuis la monarchie de Cyrus jusqu'à nos jours. Atlas* (Paris, 1826).

H—Łodyński, 3: no. 210. BL; Princeton (text only); Oxford. 27 maps. A uniform map extending from Britain to Japan or Kamchatka. Names and colored boundaries change. A handsome production. Persia is at the geographical center. BL, Klaproth, *Prospectus: Tableaux historiques de l'Asie* (no. 23 in a volume of offprints largely by him). Catalogues disagree on the order of his first names.

382. *Kleiner Atlas,* 1816.   *Kleiner historischer Atlas von Europa vor dem Einbruch der Barbaren bis zum Jahre 1814* (n.p., n.d. [Berlin, ca. 1816?]).

H—LeidenUL, Atlas 632. 7 maps, the first 5 corresponding to Koch, 1807; no. 6 is of central Asia. A very precocious small atlas of European history. The engraver, Karl Jättnig, belonged to a family of Berlin engravers; a map specialist, his years of known activity were 1805–16.

383. Knorr, 1860.   Julius Knorr, [Polemical map of France vs. Germany] (n.p., 1860).

M—BSB, Mapp. II, 129. 4 maps in 1 sheet, of which 1 medieval. They portray Germany as it might be and as Napoleon III wanted it to be. See also BL, "Europe As It Will Be" (1856).

384. Koch, 1807.   Christophe Guillaume Koch, *Tableau des révolutions de l'Europe depuis le bouleversement de l'Empire romain en Occident jusqu'à nos jours,* new ed., 4 vols. (Paris, 1814; 1st ed., 1807).

M—BL, 211.b.2–5; LC, D105.K76 (1814; 1807 ed. missing). 7 maps by Ire de Rosny. 1807 ed.: George Peabody Library, Johns Hopkins University, Baltimore, 5 maps. See also Koch, 1771/B, 1790/B.

385. Koch, 1831.   Christophe Guillaume Koch, *Maps and Tables of Chronology and Genealogy Selected and Translated from Mons. Koch's* "Tableau des Révolutions de l'Europe," for the Use of Harrow School (London, 1831).

H—BL.

386. Koeppen, 1854.   Adolphus Louis Koeppen, *The World of the Middle Ages: An Historical Geography* (New York, 1854).

H—BPL; MilwaukeePL; Northwestern (1856). Mainly text. Six of the maps are freely adapted from G.W. Green's tr. of Spruner's medieval maps (the maps are only in the BPL copy). See also under Spruner.

387. König, 1850.   Theophil König, *Historisch-geographischer Hand-Atlas zur ältern, mittlern und neuern Geschichte: Zu den Geschichtswerken von Schlosser, Becker, Nösselt, Pölitz, Rotteck, Volger u. Andern* (Wolfenbüttel, 1850).

H—Łodyński, 3: no. 591; Engelmann, 75; *GV* 78:354–55. BerlinDSB, oblong 4to U371, 27 maps; ViennaUL. Further issues: 2d ed., 1851; 3d ed. 1853; 4th ed., 1855. YaleB, 5th ed. (1857), 2. Abt. *Zur mittlern u. neuern Geschichte,* 30 maps in both parts. About 7 maps of medieval subjects (plus insets).

388. *Kort Atlas,* 1853.   *Kort historisk atlas: Ett supplemént till Helds och Covins illustrirede Verldhistoria* (Stockholm, n.d. [1853]).

H—StockholmRL. 12 maps, 4 for each period.

389. Kruse, 1802.   Christian (Karsten) Kruse [and Friedrich Kruse], *Atlas zur Übersicht der Geschichte aller europäischen Staaten; Atlas und Tabellen zur Übersicht der Geschichte aller europäischen Länder und Staaten von ihrer ersten Bevölkerung an bis zu den neuesten Zeiten,* 2 vols. (Oldenburg and Halle, 1802–10); 3d ed. (Halle, 1822); 6th ed. (Leipzig, 1841).

H—Łodyński, 3: no. 12; Engelmann, 76. Harvard, MA107.802F; RGS, 3.H.15; StockholmRL, AF 207 (3d ed., 1822, complete, 17 maps); MilwaukeeAGS, At 050 B 1822 (3d ed., excellent copy, text interleaved); LC, G1796.S1K78 (4th ed., 1827); BL, Maps 24.e.9 (5th ed., 1834), 1853.e.15 (6th ed., 1841); CambUL, Atlas 3.84.1 (6th ed.; fine copy with *Tabellen* continued to 1840). French adaptation: BL, Maps 12.f.3.; Maps 25.c.2.

390. Kruse, 1807.   Christian (Karsten) Kruse, *Atlas historique de tous les états de l'Europe depuis leur origine jusqu'à l'an 1800* (*Tables historiques,* tr. F. d'Apples) (Oldenburg, 1807–8).

H—*BN Impr.* 13 tables and maps. This is the original French version, meant to appear in parallel with the German. It lapsed after the first installment.

391. Kruse, 1836.   Christian (Karsten) Kruse and Friedrich Kruse, *Atlas historique des états européens,* tr. Ph. Lebas and F. X. Ansart (Paris, 1836).

H—UMich. Maps dated 1834, 1835. See also Ansart.

392. Kutscheit, 1842.   Johann Valerius Kutscheit, *Vollständiger historisch-geographischer Atlas der deutschen Landes und Volkes* (Berlin, 1842).

H—BL, Maps 20.3.3. 5 leaves only. The same at BN. Mees, 10–11, was very hopeful about this production and regretted that the first installment was the last.

393. Kutscheit, 1844a.   Johann Valerius Kutscheit, *Hand-Atlas der Geographie und Geschichte des Mittelalters für den Schul- und Privatgebrauch,* 3 fasc. (Berlin, 1844–47).

H—*GV* 83:112–13; Stevens, *Bibliographia geographica,* pt. 1, no. 1589; Engelmann, 76. BerlinDSB; BL, Maps 143.d.3. Some parts of this atlas began to appear in 1840. The BL copy has more than the 14 maps listed by Engelmann. Were there separate school and reference editions? Kutscheit was a very productive, full-time map- and atlas-maker.

394. Kutscheit, 1844b.   Johann Valerius Kutscheit, *Historisch-geographischer Atlas zu den allgemeinen Geschichtswerken von C. von Rotteck, Pölitz u. Becker,* 2d ed. (Freiburg im Breisgau, 1844; 3d ed., 1854).

H—*GV* 83:112. ViennaUL. 50 maps; nos. 13–27 concern the Middle Ages. For the 1st ed., see Löwenberg, 1839.

395. Kutscheit, 1856.   Johann Valerius Kutscheit, *Historisch-geographischer Atlas zu den Lehrbüchern der Weltgeschichte von Johannes Bumüller* (Freiburg im Breisgau, 1856).

H—Engelmann, 1083. BerlinPK. 25 maps, 5 medieval.

396. L'Hermite, 1822.   J. B. L'Hermite, "Tableau chronologique, historique et géographique de la France," 5th ed. (Paris, 1822).

M—BerlinPK. Single sheet with 1 map and letterpress in 4 margins; the new and old boundaries of France. Minimal interest.

397. Labberton, 1872.   ♦Robert H. Labberton, *An Historical Atlas Containing a Chronological Series of One Hundred Maps* (Philadelphia, 1872).

H—NYPL; MilwaukeeAGS; UWis-Madison, F.6L11 (2 copies). The colored maps (22 for the Middle Ages) were lithographed in Philadelphia and very effectively convey the information they were meant to impart.

398. Labberton, 1886.   ♦Robert H. Labberton, *New Historical Atlas and General History* (New York, 1886).

H—NYPL. A variant: UWis-Madison, F.6L11.3, *Historical Atlas 3800 B.C. to 1886 A.D.*, 14th ed. (New York, 1889). The 1872 atlas was destroyed in a fire and had to be replaced. Considerable changes, including integration of maps into the text. Much less eye-catching maps than in Labberton, 1872.

399. Lapie, 1812.   Pierre Lapie and Jean-Baptiste Poirson, *Atlas complet du Précis de la Géographie universelle de M. Malte-Brun* (Paris, 1812).

H—BerlinDSB; RGS, 7F13; BAV, RG Geog S.44. 75 maps. Prepared under Malte-Brun's direct supervision. RomeSGI, Z.3.I.3: 1st installment (1810), 24 maps, 5 medieval. BN, G.626:2d installment, *Atlas supplémentaire du Précis . . .* (1812), 51 maps. Most buyers and collections obtained this atlas as a single volume with 75 maps (1812).

400. Lapie, 1829.   Pierre Lapie and A. E. Lapie, *Atlas universel de géographie ancienne et moderne* (Paris, 1829).

G—Łodyński, 3: nos. 360, 409. BL; ViennaÖNB; BAV, RG Geog S.89 (1829), S.45 (1841), also 1836 and 1838. 50 maps. Earlier ed., *Atlas classique et universel de géographie ancienne et moderne* (1812). Łodyński, 3: nos. 89, 260. LC; RGS. From one edition to the next, hardly any change can be detected. 20 historical maps, of which 1 is medieval.

Las Cases, É. de. *See* Lesage.

401. Lavoisne, 1814.   C. V. Lavoisne, *Complete Genealogical, Historical, Chronological, and Geographical Atlas . . . according to the plan of Lesage Greatly Improved: A New Edition enlarged with 11 new Historical and twenty-five geographical maps* (London: J. Barfield, 1814).

H—BL, 1851.d.5. Issued by the original Lesage publisher and improved by Lavoisne (†1807?), Ch. Gros, and J. Aspin. 2d and 3d U.S. eds., Philadelphia: Carey and Lea, 1820–21. UIll (London, 1822). Enlarged by John Satchell, 1840 (*NUC*, 1830). U.S. eds. are found widely in U.S. collections; e.g., LC (multiple copies); Milwau-keeAGS; MilwaukeePL; NYPL; PhiladFree; PhiladUL-Rare; PhiladLibCo; Princeton; UCB-Bancroft; etc. Lavoisne inspired the first historical atlas of the United States; see Carey, 1822/O. The Philadelphia editors often mention an otherwise unrecorded London, 1817, ed. of Lavoisne. This may have been the actual completion year of the ed. officially dated 1814.

402. Leeder, 1866.   E. Leeder, *Atlas zur Geschichte des preussischen Staates* (Weimar, 1866).

H—BL, Maps 12.d.23. 10 plates, for schools; 2d ed. (1869), Maps 12.c.16. Leeder is also responsible for a *Wandkarte zur etc.*, 12 sheets (Glogau, 1863), BL 33129.(2.). The first sheet, divided in 3, is medieval.

403. Legrand, 1824.   Augustin Legrand, *Atlas géographique et géologique des quatres parties du monde et de la France en particulier* (Paris, 1824).

G—RomeBNVEmm, 1.45.G.14; BN, G.1394. The final map of the atlas is a 3-in-1 history of France from Gaul to the *départements*. An unusual attempt to compress all French history in 1 map.

404. Łewelel, 1844.   Joachim Łewelel, *Histoire de Pologne: Atlas* (Paris and Lille, 1844).

H—LC List 9145. LC, DK414L54, 16 maps, 7 medieval.

405. Lesage, 1801.   A. Lesage, *Genealogical, Chronological, Historical, and Geographical Atlas* (London, 1801).

H—BL, K.208.b.8, Maps 48.e.14 (1802); BPL (imperfect); RGS. The 1st ed. of Las Cases's atlas. Despite the author's claim that it was a mere draft, it clearly foreshad-ows the last ed. Assertions (e.g., by Gianbattista Albrizzi and the *Catalogus mapparum*) that the atlas was launched in Paris are mistaken. The "A" abbreviation in the pseudonym "A. Lesage" stands for only itself.

406. Lesage, 1802.   A. Lesage, *Select Maps out of Lesage's Complete Historical Atlas, particularly calculated for the use of Schools, Teachers and Pupils, Presenting the most complete and the easiest Method of improvement in History, Geography, Genealogy, and Chronology* (London, 1802).

H—CambUL. 18 maps, £3 10s. (full atlas, 25 maps, £4 14s.). Includes an updated subscription list, possibly later than the list in Lesage, 1801, at RGS.

407. Lesage, 1803.   A. Lesage, *Atlas historique, généalogique, chronologique et géo-graphique de A. Lesage* (Paris, 1803–04).

H—Łodyński, 3: no. 22 (Polish public libraries report 32 copies of the 1st French ed.). BL, K.210.i.1, 20 sheets; LC (called 1st Paris ed., but seems to be a composite); BN, Ge.DD.4796 (104) (includes "Note des cartes et ordre des livraisons de l'Atlas historique de A. Lesage"); PhiladAmPhil; StockholmRL 1807, 33 sheets (stamp of Christian XV, 1807); RLScot (1808); BPL (1809, 1823, 1829); MilwaukeeAGS (1809); BrusselsBR (1813); BL, Maps 48.f.12 (1814); BAV (1825, 1835); ViennaÖNB (unbound).

See also Lesage, 1827a. Careful MS copy of parts of the atlas, with color washes, in ArchNat, Marine, Serv. hydrog. 6JJ/86, nos. 36–48 (no maps).

408. Lesage, 1806a.   A. Lesage, *Atlas historique, généalogique, chronologique et géographique de A. Lesage* (Paris, 1806).

H—BN; Harvard. Both copies consulted are exemplary, systematically ordered, very neatly colored. Dates on sheets are 1802–6, but a reference to 1807 occurs. The Harvard copy includes a list of ca. 1000 subscribers.

409. Lesage, 1806b.   A. Lesage, *Atlas historique chronologique géographique et généalogique par M. Lesage avec corrections et additions* (Florence, 1806–7).

H—Prospectus dated 3 April 1805, 14 pp. (BAV, Ferraioli IV 9732 int. 11). Subscription price, 135 silver *paoli* of Florence, barely half the French price of 127 francs (in Paris). The atlas: LC; AmstUL; BAV (3 copies). 1st pirated ed., usually cited "1806." Widely available, especially in Italy. The publisher does not seem to know that Lesage is a pseudonym.

410. Lesage, 1809.   [A. Lesage], *Istoricheskii, genealogicheskii, khronologicheskii, geograficheskii atlas* [of Lesage], tr. E. Shavrov (St. Petersburg, 1809–12).

H—NOT SEEN. LC, G1030.L356 1809.

411. Lesage, 1813a.   A. Lesage, *Historical, Genealogical, Chronological and Geographical Atlas,* tr. from the last and much improved French ed. (London, 1813).

H—Tr. by Mme Coindé, who personally supplies 6 extra maps. NYPL; BPL; PhiladLibCo; UCB fD12L3 1813 (without title page and miscatalogued as Lavoisne). A reviewer deplored the translator's English. Repr., 1818 (RLScot). P. 2, footnote: "N.B. The Editor will give an Explanation of the Work and the manner of using it to any of the Purchasers. She likewise will recommend Teachers of History according to the method of M. Lesage."

412. Lesage, 1813b.   A. Lesage, *Atlante storico geografico genealogico e cronologico del A. Lesage* (Florence, 1813–14).

H—LC List 3540. LC. Italian translation of the pirated Lesage, 1806b (a "barbara traduzione," according to Albrizzi, in Lesage, 1826b).

413. Lesage, 1826a.   A. Lesage, *Atlante storico geografico genealogico e cronologico del A. Lesage,* con notabili correzioni ed aggiunte (Naples, 1826).

H—BAV, Rossiana 2666; RomeAngel, Bancone Stampa 227. Derived from Lesage, 1813b.

414. Lesage, 1826b.   A. Lesage, *Atlante storico geografico genealogico cronologico e letterario di M. A. Lesage,* in ogni sua parte corretto, ampliato e proseguito sine all'anno corrente (Venice, 1826–40).

H—BAV, RG Geogr I 16; LC, G1030.L355 1826; RomeMar, no. 7055 Emeroteca (pt. 3 only). 1st Venice ed., by Gianbattista Albrizzi. Many maps, with alterations, take up full pages; text columns moved elsewhere. Added maps, especially American ones from Buchon, 1825/O, and 4 for Italy in 1400, 1700, 1789, and after 1814. Pt. 3 (1840) is Albrizzi's expanded adaptation of Jarry, 1831/B, *Atlante storico, letterario, biografico, archeologico . . . in più di due terzi compilazione originale.*

415. Lesage, 1826c.   A. Lesage, *Historisch-genealogisch-geographischer Atlas von Lesage, Graf Las Cases,* ed. and tr. Alexander von Dusch (Karlsruhe, 1826–27).

H—Łodyński, 3: nos. 233–34, 289, 328, 354; Engelmann, 76; LC List 5526; *GV* 87:159. LC, G1030.L353 1835; BSB. 33 sheets, unnumbered; new ed., with Joseph Eiselin, 1829, 35 sheets; 4th ed., 1835, 42 sheets; repr., 1843. Reviewed, *Kritischer Wegweiser im Gebiete der Landkarten-Kunde* 2 (1830): 41–46. Dusch worked at improving the atlas.

416. Lesage, 1826d.   *Atlas histórico, genealógico, cronolózico, geográfico, y estadistico di Lesage . . . traducido, corregido y augmentado, por un Español americano* (Paris, 1826).

H—*NUC.* 22 maps. Small differences in *BN Impr.* (Didot as publisher), lacking a map count. The translator is Antonio de Arcos, a Chilean.

417. Lesage, 1827a.   A. Lesage, *Atlas historique, généalogique, chronologique et géographique de A. Lesage* (Paris, 1827).

H—*Bibliographie de la France, ou journal général de l'imprimerie et de la librarie* (repr., New York, 1966), no. 6179: "l'atlas de Lesage se compose de deux feuilles pour les faux titres, titre et introduction, et de trente-sept tableaux imprimés successivement depuis deux ans, soit chez M. Pochard, soit chez M. Didot ainé. [Price, 120 francs]." Noted as 1826 (Leclerc) by Quérard, *France littéraire,* 4:586–88, who lists the titles of the plates in full (letterpress as well as maps) and a price of 136.50 francs. J. C. Brunet, *Manuel de libraire,* 6 vols. (Paris, 1860–65), 3:859, gives 1826 as the final edition of the *Atlas.*

418. Lesage, 1827b.   A. Lesage, *Atlas historique, généalogique, chronologique et géographique de A. Lesage,* ed. Joseph Marchal (Brussels, 1827).

H—Émile van der Vekene, *Cosmographies, Théatres du Monde et atlas: Catalogue* (Luxembourg, 1984), 157–58, no. 127. Newberry. Reissued 1837, 1853 (BrusselsBR IV/20.006/D, II/3692). Was it authorized? Material about Netherlands history and other improvements are added.

419. Lesage, 1829.   A. Lesage, *Atlas élémentaire géographique, historique, chronologique et généalogique* (Paris, 1829).

H—BPL. 10 sheets. Cf. *Extrait de l'Atlas historique . . . ou Cartes les plus classiques,* ed. H. Duval (Paris, 1823), 9 maps; cited by Quérard, *France littéraire,* 4:589.

420. Lesage, 183-.   A. Lesage, *Feuilles supplémentaires et complémentaires* (Paris, 1827–39).

M—*Bibliographie de la France = Bib.Fr.;* Quérard, *Litt. franç. contemp.,* 4:626–27; *BN Impr.* 89:561. These sheets contain the last 15 years' work on the Lesage atlas other than the edition of 1835 and are based mainly on reference works. Many items listed consist only of letterpress. *1827,* extra sheet for current Germany (*Bib.Fr.,* no. 4609); *1828,* new sheet for current Russia; complementary sheet for 1828 (*Bib.Fr.,* no. 2116); Asia, Africa (=Lesage sheets 31, 32) (*Bib.Fr.,* no. 5852); *1829,* Lesage sheets 1, 2, 5, 6, 7, 8, 12, 29, sacred history (*Bib.Fr.,* no. 5769); "Tableau spécial de l'histoire sainte" (*Bib.Fr.,* no. 3839); *1830,* "Tableau de l'histoire universelle" (*Bib.Fr.,* no. 3563);

1833, "L'Europe actuelle" (Lesage 30bis) (*Bib.Fr.*, no. 3963); "Tableau général de l'histoire universelle moderne" (*Bib.Fr.*, no. 4837); ancient Greece with its divisions (Lesage 6) (*Bib.Fr.*, no. 4031); "Transmigrations des barbares" (Lesage 8) (Quérard, *Litt. franç. contemp.*); *1836*, "Tableau spécial de l'histoire sainte" (*Bib.Fr.*, no. 5845); *1839*, "Tableau d'histoire universelle ancienne," "Tableau d'histoire universelle moderne."

Single additions, without publ. date, from *BN Impr.* 89:561: "Carte pour l'intelligence de la géographie et de l'histoire du moyen âge, à l'usage des collèges"; "Tableau des premiers Césars, à l'usage spécial des classes de latinité"; "Tableau de l'histoire sainte à l'usage des collèges"; "Tableau spécial des Mérovingiens et des Carlovingiens pour l'intelligence du moyen âge, à l'usage des collèges."

421. Lesage, 1830.   A. Lesage, *Le Manuel ou l'Indispensable de l'histoire et de la géographie de France, pour l'éducation primaire: Tableaux détachés du grand Atlas* (Paris: Leclère, 1830).

H— Quérard, *Litt. franç. contemp.*, 4:627 (1831 instead of 1830); *Catalogue de l'histoire de France* (Paris, 1855), 1:56. One of the many spin-offs.

422. Lesage, 1835.   A. Lesage, *Atlas historique, généalogique, chronologique et géographique de A. Lesage (comte de Las Cases)*. Édition populaire et d'étude (Paris, 1835).

H—BAV; LC. Quérard, *Litt. franç. contemp.*, 4:626, 33 *tableaux.* Possibly the last full ed., not basically different from Lesage, 1827a.

423. Lesage, 1843.   A. Lesage, *Atlante storico, geografico, genealogico, cronologico e letterario*, 2d Venice ed. (Venice, 1843–45).

H—BAV, RG Geogr S.1. No significant changes from Albrizzi's ambitious 1st ed.

424. Lesage, 1844.   A. Lesage, *Atlas historyozny, genealogiczny, chronologizny, i geograficzny* (Wilno, 1844).

H—Łodyński, 3: no. 463, records 5 maps. BAV, RG Geogr S.129, lacks the maps.

425. Levasseur, 1840.   Victor Levasseur, *Atlas classique universel de géographie ancienne et moderne* (Paris, n.d. [1840]).

H—BSB, 86 maps, black-and-white, pocket size; Columbia-rare, (Plimpton) B912 L57, 87 maps, border coloring. The Crusades expedition map (no. 22) stands out in quality; the scale elsewhere is too small to permit showing anything but "the great powers."

426. Levasseur, 1844.   Victor Levasseur, *Atlas de cartes historique pour servir de complément au Dictionnaire de dates* (Paris, 1844).

H—CambUL. (Also called *Atlas du Dictionnaire des dates.*) A slightly modified derivative of the *Atlas Lesage.*

427. Lévi-Alvarès, 1840.   D. Lévi-Alvarès, *Nouvel atlas complet de géographie ancienne et moderne* (Paris, 1840).

H—Łodyński, 3: no. 399 (specifies the date and 9 maps). BN, 7 maps, loose in portfolio, black-and-white.

428. Liskenne, 1853.   Charles Liskenne, et Jean Baptiste Balthazard Sauvan, *Biblio-thèque historique . . . dédiée à l'armée et à la garde nationale de France,* vol. 8, *Atlas* (Paris, 1853).

M—LC, U15B5. Many maps of ancient and modern military history, 4 medieval: 4:64, 109, 122, 135. The maps and plans by Th. Duvotenay are very angular and explicit. Commentaries to these 4 are in Liskenne's vol. 4.

429. Lizars, 1831.   Daniel Lizars, *Edinburgh Geographic and Historical Atlas* (Edinburgh, 1831).

G—LC; Newberry.

430. Longnon, 1885.   ♦ Auguste Longnon, *Atlas historique de la France* (Paris, 1885–89).

H—UWis-Madison; BL, Maps 20.e.8, also 1400.1.32 (with text vol.); Toronto-Robarts. Lavishly praised and still considered the only scholarly historical atlas of France. The prospectus lists more maps than were completed; only no. 15 (death of Charles V) was reached, well short of A.D. 1500. Émile van der Vekene (*Cosmographies, Théatres du Monde et atlas: Catalogue* [Luxembourg, 1984], 167) exaggerates when he calls Longnon (a prolific and esteemed medievalist) the creator of "géographie historique."

431. Lothian, 1829.   J. Lothian, *Lothian's Historical Atlas of Scotland Consisting of Five General Maps exhibiting the Geography of the Country in the 1st, 5th, 10th, 15th and 19th Centuries* (Edinburgh, 1829).

H—LC list 5213. LC, G1826.S1L65 1829. 8 maps.

432. Löwenberg, 1839.   Julius Löwenberg, *Historisch-geographischer Atlas zu den allgemeinen Geschichtswerken von C. v. Rotteck, Pölitz u. Becker* (Freiburg im Breisgau, 1839–42).

H—Łodyński, 3: no. 378; LC List 134; Engelmann, 76; GV 90:355. LC, G1030.L7 1839; BL, Maps 48.e.15; BAV; HABW, Cb folio 18; MilwaukeeAGS. 40 maps; imposing dimensions. Completed by J. V. Kutscheit. 2d ed. by Kutscheit, 50 maps (Freiburg im Breisgau, 1844). Löwenberg also produced a 24-map picture and map atlas of history for youth (Berlin, 1844; repr., 1847).

433. Löwenberg, 1840.   Julius Löwenberg, *Kleiner historischer Atlas des preußischen Staates* (Berlin, 1840–42).

H—GV 90:355. BN, Ge.FF.5187; 16 maps (to 1840), with a lengthy text. Map no. 1: "*Die Völkersitze* between the Rhine and the Vistula in the Fifth Century." Engelmann, 794, records a possibly identical *Historischer Taschen-Atlas des Preußischen Staates* (Berlin, 1842), with the same number of maps.

434. Maaskamp, 1833.   E. Maaskamp, *Geschiedkundige kaart der Nederland* (Amsterdam, 1833).

M—AmstRijksprent. Black-and-white microform. Mees, 23 (dates the map to 1830). Sites and dates of battles, little pictures; modern background map.

435. Maggi, 1803.   Cesare Maggi, [Atlas of miscellaneous maps] (Turin, 1803–54).

G—UCB, fG.1025.A9. Maps of different sizes and provenance. The subjects include ancient and Italian history; nos. 30–33 are medieval. Maggi may have been the compiler and was probably the publisher,

436. Mandrot, 1855.   A. de Mandrot, *Atlas historique de la Suisse de l'an 1300 jusqu'en 1789* (Geneva, 1855).

H—BerlinDSB, 7 very large sheets; nos. 1–4 medieval.

437. Maretheux, 1847.   L. Maretheux, *Géographie de la France ancienne et moderne* (Paris, 1847).

H—BN. *BN Impr.*, 8vo L⁷.14. 5 black-and-white sketch maps, 3 medieval.

438. Marmocchi, 1840.   Francesco Constantino Marmocchi, *Atlante di geografia-storica universale* (Florence, 1840, 1845).

H—*NUC*. Princeton, 1018.616f (lost?); BL; RomeBNVEmm, 203.4.k.6. (1845). Editions of various dates to the 1870s. Concern for the history of geography. Comprehensive medieval coverage. Intended to complement his *Corso di geografia-storica antica, del medio-evo e moderna, con atlante*, 3 vols. (Florence, 1845); HABW, GK24/0025.

439. Meissas, 1842.   Achille Meissas and Auguste Michelot, *Petit atlas universel de géographie ancienne, du moyen âge et moderne et de géographie sacrée* (Paris, 1842; 1855; n.d. [1875?]).

H—1842? *BN Impr.*, G.6381, 10 maps. NOT SEEN. 1855, *BN Impr.*, G.31549, lacks the 54 maps. 1875, nos. 45–49, medieval Europe.

440. Mentelle, 1801.   Edme Mentelle, *Atlas de tableaux et de cartes gravé par F. Tardieu pour le cours complet de cosmographie, de géographie, de chronologie et d'histoire ancienne et moderne*, 2d ed (Paris, 1804–5).

G—Łodynski, 3: no. 24; LC List 6011. BL; RLScot (maps only). Comparative. Only medieval item, "Héraclius ou commencement du moyen âge." Also *Cartes pour le cours de géographie et d'histoire de Mentelle* (n.p., n.d. [Paris, 1801?]) (RLScot). Cf. *BN Impr.*, 4 to Atlas G.309 and G.9760–62 (Paris, 1800–1801); G.3589, 15 maps. RomeBNVEmm, 14.23.0.M.21: 15? maps; no title page; comparative. Mentelle's bibliography is voluminous and chaotic.

441. Mentelle, 1816.   Edme Mentelle and C. Malte-Brun, *Atlas de la géographie universelle ancienne et moderne, mathématique, physique, statistique, politique et historique des cinq parties du monde* (Paris, 1816).

G—BL, Maps 31.c.21. Some editions before 1816. Nothing medieval.

442. Mentzer, 1871.   ◆Thure Alexander von Mentzer, *Svensk historisk atlas* (Stockholm, n.d. [1871]).

H—StockholmRL. 18 maps, nos. 1–5 medieval. See also Wiberg, 1856, 1859.

443. Mertens, 1864.   Franz Mertens, *Das Abendland während der Kreuzzüge, Denkmal-Carte mit der Angabe der Landgebiete der Kunststyle [sic] oder der entsprechenden Bauschulen in dieser Zeit* (Berlin, 1864).

M—*GV* 95:87. BerlinDSB. A guide to medieval antiquities. 2 very large sheets. Later editions, 1868, 1873, and a supplement, 1875. Complements Mertens's work of 1850–51 on German architecture.

444. Meyer, 1880.   ◆ Clemens Friedrich Meyer, *Atlas zur deutschen Geschichte* (Essen, 1880).

H—Łodyński, 3: no. 1047. NYPL. 16 maps, 5 medieval.

445. Meyer, 1830.   Joseph Meyer, ed., *Neuester grosser Schul-Atlas über alle Theile der Erde* (Hildburghausen and New York, n.d. [1830–38]).

G—LC, G1019.M45 1838. 7 ancient maps, 28 modern, no medieval.

446. Meyer, 1849.   Joseph Meyer, ed., *Neuester Zeitungs-Atlas für alte und neue Erdkunde* (Hildburghausen, Amsterdam, and New York, n.d. [1849]).

G—NYPL. 123 maps, none medieval.

447. Migeon, 1866.   Migeon (firm), *Géographie universelle: Atlas-Migeon (revu par Vuillemin) historique, scientifique, industriel et commercial* (n.p. [Paris], 1866–70).

G—Łodyński, 3: nos. 949, 1099; *NUC*. UIll, 38 maps. Vuillemin (q.v.) was an established cartographer. 4 maps are historical, of which 1 is of medieval France.

448. Möller, 1824.   Arno Wilhelm Möller, *Kleiner historischer Atlas der Lande zwischen der Maas und dem Niemen, zur Erlernung ihrer Geschichte seit der Völkerwanderung* (Münster, 1824).

H—NOT SEEN. Engelmann, 76: 16 maps, 16mo.

449. Möller, 1825.   Arno Wilhelm Möller, *Kleiner historischer Atlas zu der allgemeinen Weltgeschichte für den Schulgebrauch* (Elberfeld, 1825; rev. ed., 1829).

H—Łodyński, 3: nos. 249, 468; Engelmann, 76. BerlinDSB. 10 maps; 2 medieval. BSB 4 Mapp. 73k. For the ed. by Karl Franz Muhlert (1844); see Muhlert, 1865. Especially to accompany Fr. Kohlrausch, *Chronologischer Abriss der Weltgeschichte für den Jugend-Unterricht,* 3d ed. (Elberfeld, 1818 and later).

450. Moeller, 1850.   Jean Moeller, *Atlas de géographie historique* (Brussels, n.d. [ca. 1850]).

H—BN, nos. 16–20 medieval. Also *BN Impr.*, G.6434 (Brussels, 1868). Related to the author's history textbook; *GV* 98:150–51, *Manuel d'histoire du moyen âge* (Louvain, 1837); also *Geschichte des Mittelalters* (Mainz, 1844), in French (Louvain, 1846).

451. Monin, 183-.   C. V. Monin and A. R. Frémin, *Atlas universel de géographie* (Paris, n.d. [183-?]).

G—NYPL. 1 map for the Middle Ages.

452. Monin, 1842.   C. V. Monin, *Atlas classique de la géographie ancienne, du moyen âge, et moderne,* nouv. ed. (Paris and Lyons, n.d. [1842–43?]).

H—Toronto-Pontifical Institute of Mediaeval Studies. 40 maps, nos. 37–39 medieval.

453. Monin, 1871.   ◆ C. V. Monin and A. Vuillemin, *Atlas de géographie ancienne et moderne à l'usage des collèges et de toutes les maisons d'éducation* (Paris, n.d. [1871?]).

G—"Vers 1850" according to the BN Cartes et plans catalogue, but traces of the Franco-Prussian War are unmistakable. Cf. *BN Impr.*, DD.1038 (1842; same title but fewer maps). A few medieval maps at the end.

454. Muhlert, 1865.   Karl Friedrich Muhlert, *Historiskt geographiska atlas med 22 chartor*, tr. and ed. N. v. Hess (Stockholm, 1865).

H—StockholmRL. See Möller, 1825. I have not seen Muhlert's German-language original: *Kleiner historischer Atlas zu der allgemeinen Weltgeschichte für den Schulgebrauch* (Leipzig, 1844). 25 maps on 10 leaves (1st fasc., 1842). ViennaUL has a copy. It is regarded by the publisher as prolonging but also improving and enlarging Möller's work, to the extent of being entirely new.

455. Mullié, 1839.   C. Mullié, *Fastes de la France ou tableaux chronologiques, synchroniques et géographiques de l'histoire de France depuis l'établissment des Francs jusqu'à nos jours*, 4th ed. (Lille, 1839).

H—*BN Impr.* folio, L^{33}.21 and .21A to E, 6 eds. from 1832 to 1848. LC, G1841.S1M7 1839. 7 maps, of which 4 medieval, disposed one at a time in an ambitious time chart or chronicle of French history. A new 7th ed., 4 vols. (1856–58), changes title at "depuis": *précédées de l'histoire de la Gaule depuis l'arrivée de la race celtique en Europe jusqu'à l'établissement des Francs* (NOT SEEN).

456. Murphy, 1863.   William Murphy, *The Historical and Statistical School Atlas* (Edinburgh, 1863).

G— Oxford. No historical content except 2 Holy Land maps.

457. Naymiller, 1867.   Filippo Naymiller, Pietro Allodi, and Vincenzo de Castro, *Grande atlante di geografia universale cronologico, storico, statistico e letterario* (Milan, n.d. [1867]).

H—LC List 851. LC, G1019.C27 1867; RomeBNVEmm, 214.Banc.8.D.1. Historical maps, ancient and medieval, distributed in no fixed order.

458. Netherlands, 18–.   "Nederland ten Tijde der Franken, Saksers en Friezen."

M—AmstUL, call no. 64-33-51. Marked no. 2 in upper-right margin. From a book. Good account taken of historical hydrography. Preoccupied with Migration Age peoples: along the coast, in echelon, "Warnen, Scleven, Sueven"; inland, Bructeri, Sicambri. The main peoples from the north down are Frisians, Saxons, Salians, Franks.

459. Nichols, 1811.   Francis Nichols, *A New Atlas Adapted to the Use of Students of Geography and History* (Philadelphia, 1811).

H—NOT SEEN. LC List 3375.

460. Oger, 1868.   Félix Oger, *Géographie de la France et géographie générale physique, militaire, historique, politique, administrative et statistique* (Paris, 1860; 2d ed., 1868).

H—*BN Impr.*, L^8.44. 2d ed., Harvard, MA 310.868F. Few historical maps; none medieval.

461. Oger, 1878.   ♦ Félix Oger, *Atlas géographique et historique*, 2d ed (Paris, 1878).

G—BL. No. 8 compares the old provinces to the departments of France.

462. Ohmann, 1853.   C. F. Ohmann and F. W. Kliewer, *Historisch-geographischer Schul- und Hand-Atlas der alten Welt* (Berlin, 1853).

H—Jerusalem. Not homogeneous. Most maps dated 1846. A classical atlas with a small extension into the Middle Ages: 7 maps from the barbarian kingdoms to 1453.

463. Pawlowski, 1855.   J. N. Pawlowski, *Historisch-geographische Karte vom alten Preussen während der Herrschaft des deutschen Ritterordens . . .*, 2d ed. (Gdańsk [Danzig], n.d. [1855?]).

M—BL, 895 (1.); another copy, 895 (2.). The frame of reference extends well beyond what the title suggests in its reference to the rule of the Teutonic Order (ca. 1190–1410).

464. Pearson, 1869.   Charles H. Pearson, *Historical Maps of England during the First Thirteen Centuries with explanatory notes and indices* (London, 1869).

H—BL Maps, 4.d.35.; Newberry, 3d ed., 1883. On Pearson (1830–94), *DNB* 15:607–9.

465. Perrot, 1823.   A. M. Perrot, *Atlas de géographie ancienne et moderne* (Paris, 1823).

G—RomeBNVEmm, 1.7.A.27. 1 medieval map. Perrot is also responsible for *Atlas de toutes les parties du monde pour servir à l'étude de la géographie et de l'histoire* (1824) (BAV, RG Geog I.78), with no historical content. In view of the date, the title is interesting, old-fashioned.

466. Piré, 1868.   L. Piré and C. Callewaert, *Atlas classique de géographie physique et politique ancienne et moderne*, 9th ed. (Brussels, 1868).

G—YaleB. *BN Impr.* 40 maps, 10 historical, 3 medieval.

467. Pompper, 1846.   Hermann Pompper, *Geographisch-historischer Handatlas als Leitfaden zum Geschichtsunterricht* (Leipzig, 1846–49).

H—*GV* 110:188; Engelmann, 77. BL, Maps 48.c.34. Fasc. 1, ancient; fasc. 2 (1849), 5 medieval maps.

468. Potel, 1850.   Felice Potel, *Atlante universale di geografia antica e moderna* (Naples, 1850).

G—LC List 802. LC, G1019.P683 1850. 50 maps, 2 medieval.

469. Pütz, 1859.   Wilhelm Pütz, *Historisch-geographischer Schul-Atlas,* pt. 2, *Die mittlere und neuere Zeit* (Regensburg, 1859).

H—*GV* 112:148–54; Łodyński, 3: nos. 737, 873; *NUC; DBiogArchiv,* N.F., fiche 1032, frame 26. BL. *Atlante geografico-storico ad uso delle scuole,* pts. 1–2, tr. Felice de Angell (Regensburg, 1858–59); Dutch ed., 1858; Hungarian ed., 1861. Extensive explanations. 16 main maps, 4 medieval, with multiple insets. 2d ed. 1861, 1866–69.

470. Putzger, 1877.   ♦ F. W. Putzger, *Historischer Schul-Atlas* (Bielefeld and Leipzig, 1878).

H—Multiple editions to the present. Titles have varied. The first was *Historischer Schulatlas zur alten, mittleren und neuen Geschichte.* An unusual and short-lived title is *F. W. Putzger's historischer Schul-Atlas: Neue Ausgable mit besonderer Berücksichtigung der Geopolitik, Wirtschafts- und Kulturgeschichte* (1930–31); see also ch. 2, n. 104. The atlas was published in the German Democratic Republic after World War II. Putzger's work, designed for secondary, rather than university, education, was a byword for the genre and very widely disseminated since its first appearance. For its American offshoot, see Shepherd, 1929/O.

471. Quin, 1830.   Edward Quin, *An Historical Atlas in a Series of Maps of the World As Known at Different Times, with an Historical Narrative* (London, 1830).

    H—*NUC;* LC List 4139 (1836), 141 (1846), 5562 (1856). 21 maps, 6 medieval. Engraved by Sidney Hall (maps dated 1828). Repr., 1836. New ed. by William Hughes, 1846. 4th ed., 1850, repr., 1856, 1857. NYPL (1857); BL, 581.1.17. (1830), Maps 8.bb.35. (1836), Maps 143.c.14. (1846); RLScot (1830, 1846); Oxford; Jerusalem (1830); MilwaukeeAGS, At/050/B/1836m (1836); Harvard (1830); Toronto-St. Michael's College (1830); Yale, AC/G1030/+Q56/1836 (1836). New title, *Atlas of Universal History* (London: Griffin, 1857), Newberry; BL, Maps 48.c.39 (1859). Another variant title, *Ancient and Medieval History* (London and Glasgow, n.d.).

    Partial hand-painted copy (without attribution): MS with dedication (in Dutch) 20 June 1839: John D. Palm, *Historical Maps*, BSB, folio Mapp. 287b. See also 374n. 121.

472. Rafn, 1837.   Carl Christian Rafn, "General Chart Exhibiting the Discoveries of the Northmen in the Arctic Region of America during the 10th, 11th, 12th, 13th, and 14th centuries; A Map of Vinland from Accounts Contained in Old-Northern MSS" (Copenhagen, 1837).

    M—HABW, 30,9 and 30,12. Detached, probably from *Antiquitates Americanae, sive scriptores septentriones rerum ante-Columbianarum in America,* ed. C. C. Rafn (Copenhagen, 1837) (Cornell).

473. Rafn, 1845.   Carl Christian Rafn, *Americas arctiske Landes Gamle geographie efter de Nordiske Oldskrifter* (Copenhagen, 1845).

    M—Cornell, Ic/g/742/R13/ 1845. Reduced versions of the general maps of 1837 and 3 detailed maps.

474. Rasche, 1874.   ♦ Emil Rasche and Reinhold Zimmerman, *Historischer Atlas zum Gebrauch beim Geschichtsunterricht in höheren Volksschulen* (Annaberg, 1874).

    H—BL. Rudimentary sketch maps, 2 for the Middle Ages.

475. Reilly, 1806.   Franz Johann Joseph von Reilly, *Schauplatz der fünf Theile der Welt,* 2 vols. (Vienna, 1789–1806).

    M—BSB. Vol. 2, pt. 2, sec. 4 (Vienna, 1806), no. 698 (vol. 2, pt. 2, is in 4 pts.), "Das Reich der Franken unter Carl dem Grossen"; no. 699, "Frankreich zur Zeit der Kreuzzüge oder ums Jahr MCCC [*sic*]." These maps are a succinct historical prelude to the part of the atlas concerning France. Copies of the atlas at LC, BL, and Toronto-Fisher do not reach this far; the copy at UMich-Clements seems complete.

476. Rhode, 1861.   C. E. Rhode, *Historischer Schul-Atlas zur alten, mittleren und neueren Geschichte* (Glogau, 1861).

H—*GV* 117:132–33. LC List 4140, 4141: 8th and 11th eds. PhiladUL, 2d ed., 1862; BL; Yale, 9th ed.; UWis-Madison, 12th ed. 89 maps on 30 sheets, and explanatory text; nos. 27–80 medieval. Region-by-region coverage, a miniature Spruner. Good, clear maps. Wide circulation in America.

477. Rodowicź, 1843.   Theodor Rodowicź, *Historischer Schul-Atlas mit Bezug auf Dielitz geographisch-synchronistische Uebersicht der Weltgeschichte* (Berlin, 1843).

H—*GV* 118:292. BL. 3 fascs.; 22 maps, of which 8 are medieval.

478. Rossi, 1820.   Luigi Rossi, *Nuovo atlante di geografia universale in 52 carte* (Milan, 1820–21).

G—LC List 739. LC, G1019.R885 1821; RomeMar, 5385 Emeroteca. Closes with a few maps of ancient history.

479. Rothenburg, 1830.   Friedrich Rudolf von Rothenburg, "Übersichts-Karte aller in Deutschland und den angränzenden Ländern seit dem Jahre 113 v. Chr. bis 1830 vorgefallenen Schlachten, Belagerungen und Gefechten" (Berlin, 1830).

M—*GV* 120:77–78. LC. Two millennia of battles on one map, a dot for each. The dots badly need the explanations in Rothenburg's *Wörterbuch aller in Deutschland vorgefallenen Schlachten,* 6 vols. (Vienna, 1835–38).

480. Rothenburg, 1840.   Friedrich Rudolf von Rothenburg, *Schlachten-Atlas,* 2 vols. (Vienna, 1840; 5th ed., Berlin, 1853).

H—*GV* 120:77–78: 1st ed., 1830–31. Yale. Vol. 2, maps, includes 2 medieval battles; the design and coloring of "The battle of Lechfeld (955)" are very successful, an exemplary map.

481. Rupp, 1837.   Julius Rupp, *Historischer Schul Atlas* (Kaliningrad [Königsberg], 1837).

H—NOT SEEN. BerlinDSB Catal; BerlinPK, U S290 (only sheet 12 survives). 17 maps, "Zunächst zur Gebrauche bei den Lehrbücher der Geschichte des Herausgebers," i.e., his *Uebersicht der allgemeinen Geschichte für die oberen Klassen der Gymnasium* (Kaliningrad [Königsberg], 1837).

482. Rupp, 1839.   Julius Rupp, *Charte für die Geschichte* (Kaliningrad [Königsberg], 1839).

H—*GV* 121:74; Engelmann, 75, lists as 1837. Northwestern, 17 maps, of which 7 are medieval. Unsophisticated. All Rupp entries may refer to the same 17 maps.

483. Sanis, 1859.   J. L. Sanis and Delaleau de Bailliencourt, *Géographie historique de la France: Atlas spécial* (Paris, n.d. [1859?]).

H—BL, Maps 198.f.15. *BN Impr.,* 4to L^{14}.68 (1858), 4to L^{14}.68A (1859). 7 maps for the Middle Ages. Unattractively produced.

484. Schaarschmidt, 1846.   Friedrich Reinhold Schaarschmidt, *Kleiner historisch-geographischer Atlas als Grundlage für den Geschichtsunterricht* (Meissen, 1846; 2d ed., 1852).

H—*GV* 123:305; Engelmann, 77. NYPL. Small dimensions. 26 maps, 7 medieval. An early "History of Commerce, 10th to the end of the 15th Century" (no. 26).

485. Schade, 187-. ◆ Th. Schade, *Atlas zur Geschichte des preussischen Staats* (Glogau, n.d. [after 1871]).

H—BN, Ge.DD.1949. 12 sheets. Nos. 1–2 concern the Middle Ages. Schade stresses that charting all historical change on a single sheet is doomed to fail, because the multiplicity of colors necessarily confuses; multiple sheets are important, he says, for clarity of exposition.

486. Scharberg, 1845.   Joseph Bedeus von Scharberg, *Historisch-genealogisch-geographischer Atlas zur Uebersicht der Geschichte des ungrischen Reichs, seiner Nebenländer und der angrenzenden Staaten und Provinzen* (Sibiu [Hermannstadt, Nagy Szeben], 1845–51).

H—Engelmann, 1008; *GV* 10:276; *NUC* 42:649; BN, Ge.DD.1686; BL, 1852.c.25. Very large format. The MS was given to the publisher in 1837 and unchanged thereafter. 8 maps surrounded by text are in pts. 1 and 3. The atlas is compared to those of Lesage, Kruse, and Löwenberg. The resemblance to Lesage is obvious.

487. Schmidt, 1823.   Paulus Schmidt, "Mittel- und Süd-Europa nebst Klein Asien für das Jahre MC; Mittel- und Süd-Europa nebst Klein Asien für das Jahr MCC" (Leipzig, 1823, 1824).

M—Cartographer identified in BerlinDSB catal. BL, 173.i.12. Single sheets from vols. 1 and 3 (1823, 1824) of F. L. G. von Raumer, *Geschichte der Hohenstaufen,* 6 vols. (Leipzig, 1823–25).

488. Schmidt, 1800.   Wilhelm Theodor Schmidt, *Karte von Hessen und der Wetterau samt den angrentzenden Laendern nach der Geographie des Mittelalters vom VIII–XIIten Jahrhundert* (n.p. [Darmstadt], 1800).

M—BerlinPK; Kart. U16290. Divisions in *pagi* (*gaue*), names in Latin; many important places shown. To illustrate Helfrich Bernhard Wenck, *Hessische Landesgeschichte,* 3 vols. (Darmstadt, 1783–1803).

489. SDUK, 1857.   Society for the Diffusion of Useful Knowledge, *School Atlas of Classical and Modern Geography* (London, 1857).

G—RLScot-maps.

490. Selves, 1819.   Henri Selves, *Atlas écrit en couleurs* (Paris, 1819–22).

H—BN, Ge.DD.5239. Lithographed, title penciled in. Many maps dated 1821. A general historical atlas starting in antiquity; 6 maps for the Middle Ages.

491. Selves, 1822.   Henri Selves, *Atlas géographique dressé pour l'usage des collèges comprenant les trois parties suivantes: géographie ancienne, géographie du moyen âge, géographie moderne actuelle* (Paris, 1822–29).

H—*NUC;* LC List 6049. LC, G1033.S34 1822 (maps dated 1823). UMich. BN, 2 copies (repr., 1830). Lithographed. 6 maps for the Middle Ages.

492. Selves, 1833.   Henri Selves, *Atlas géographique dressé . . . pour l'usage des collèges, Géographie ancienne, Géographie du moyen âge, Géographie moderne actuelle* (Paris, 1833).

H—BN, Ge.DD.1030–1033. Maps of Europe on the Kruse pattern. Many maps have Asian insets from Klaproth, 1826.

493. Selves, 1835.   Henri Selves, *Atlas géographique,* pt. 2, *Géographie du moyen âge* (Paris, 1835).

    H—BN. Small-scale reproduction of the atlas of 1833, 16 maps. Insets from Klaproth, 1826.

494. Selves, 1843.   Henri Selves, *Atlas géographique pour l'usage des collèges* (Paris, ca. 1843).

    H—CambUL. Probable repr. of the atlas of 1833, with a modern supplement. Insets from Klaproth; the maps with them are dated 1832–34.

495. Sheahan, 1873.   ♦ James Washington Sheahan, *Universal Historical Atlas of the World: Genealogical, Chronological, and Geographical* (New York and Chicago, 1873).

    H—*NUC.* LC; NYPL; MilwaukeeAGS. The maps, including 7 for the Middle Ages, are preceded by a lengthy historical narrative.

496. Simencourt, 1830.   A.-H. de Simencourt, *Atlas universel complet de géographie ancienne et moderne* (Paris, 1830).

    G—BN, Ge.DD.1027. 48 maps, including 11 ancient maps in appendix; the last of the 11 is medieval. Alternative titles, *BN Impr.,* G.3347: *Atlas classique de géographie ancienne, moderne et du moyen âge; à l'usage des collèges royaux . . . ; Atlas classique élémentaire de géographie ancienne et moderne.*

Society for the Diffusion of Useful Knowledge. *See* SDUK and SDUK/A.

497. Soleil, 1851.   Carlo Soleil, *Atlante geografico antico e moderno* (Turin, n.d. [catal. 1851]).

    G—RomeMar, 9127 Emeroteca. 36 sheets (5 lacking in this copy). Ancient geography follows modern. Nothing medieval.

498. Somerhausen, 1828.   H. Somerhausen and N. Lamiraux, *Germanie ancienne et moderne: Tableau historique, géographique et statistique de l'Allemagne depuis les temps les plus reculés jusqu'à nos jours, dressé à l'instar du cours du temps du prof. Strass* (Brussels, 1828).

    M—BerlinDSB; BSB, 2Germ.g.118m and Germ.g.331, *Notice explicatif du Tableau* (1829). 2-sheet wall hanging; 3 little maps, much printing. Details about German noble houses flank the "River of Time" by Friedrich Strass (1766–1845). "Der Strom der Zeiten, oder bildlich Darstellung der Weltgeschichte" (1804)—contemporary to the *Lesage Atlas*—was a celebrated, widely circulated chart of the "flow" of world history (Strass was a gymnasium teacher in Prussia). *See ADBiog.* 36:496–99, and esp. Harms, *Themen,* 111 (with partial illustration).

499. Spruner, 1837.   Karl Spruner, *Historisch-geographischer Hand-Atlas zur Geschichte der Staaten Europa's vom Anfang des Mittelalters bis auf die neueste Zeit* (Gotha, 1837–46; 2d ed., 1854–55).

    H—NYPL; Yale, AC/Fol/G1030/S66/1846; Oxford, Radcliffe Camera; UWis-Madison, Map Reference. 73 maps, of which 34 are medieval. Companion volume: Spruner, *Vorbemerkungen zu dem historischen-geographischer Hand-Atlas* (Gotha, 1837–46) (Toronto-Robarts; StockholmRL, Qv. 14). Issued in 6 installments. 2d ed. (1854–55), 118 maps: UCB, PhiladFree. (Pt. 1) *Atlas antiquus* (= Spruner, 1850/A),

(Pt. 3) *Ausser-Europa* (1851, 10 maps; 2d ed., 1854, 18 maps). I count principal (main) maps only; some leaves have 8 or more detail insets (which are not included in my totals). For the 3d ed., see Spruner, 1871.

500. Spruner, 1838a.   Karl Spruner, *Atlas zur Geschichte von Bayern* (Gotha, 1838).

H—BSB; LC; *BN Impr.* 7 plates with 8 maps; *Vorbemerkungen* (7 pp.). Also see Spruner, 1831/B.

501. Spruner, 1838b.   Karl Spruner, "Das Herzogthum Ostfranken in seine Gauen eingetheilt" (Bamberg, 1838).

M—HABW, 30,30. Single sheet. Brief comment on methodology. Many place-names but few distinctions among them (Mainz and Frankfurt-am-Main are not easily located). Lithographed by Rösser.

Spruner, 1853. See Gover, 1853.

H—UMich, LC. Main entry, Gover, 1853.

502. Spruner, 1855.   Karl Spruner, *Historico-Geographical Atlas of the Middle Ages,* ed. Adolphus Louis Koeppen (New York, 1855).

H—NYPL. The first 6 maps of Spruner, 1837. The same maps are Koeppen, 1854.

503. Spruner, 1860.   Karl Spruner, *Historisch-geographischer Schul-Atlas* (Gotha, 1860).

H—BL. Earlier ed., same title, 22 maps (Gotha, 1856: BN); Italian tr. (Gotha, 1865); 12 maps (Gotha, 1858); many editions to the 1890s. A version for Austria: *Spruner's historisch-geographischer Schul-Atlas des Gesammtstaates Österreich* (Gotha, 1860). BL, Maps 4.aa. 17., 13 maps.

504. Spruner, 1861.   Karl Spruner, *Historico-Geographical Hand Atlas,* anon. English tr. (New York, London, and Gotha, 1861).

H—BPL, Newberry. 26 maps from 476 on. Tr. of Spruner, 1860, but given the title of Spruner, 1837. Slightly different version, Yale: ends with a special section of English history. Large-type letterpress faces each map; minimal clarifications.

505. Spruner, 1866.   Karl Spruner, *Historisch-geographischer Schul-Atlas von Deutschland,* 2d ed. (Gotha, 1866).

H—BerlinDSB. 12 maps, 6 medieval. There is an ed. of 1858, NOT SEEN.

506. Spruner, 1871.   ♦ Karl Spruner and Theodor Menke, *Hand-Atlas für die Geschichte des Mittelalters und der neueren Zeit,* 3d ed. of Spruner's *Hand-Atlas,* newly revised by Th. Menke (Gotha, 1871–80).

H— Oxford, Radcliffe Camera. 90 main maps. Menke alone was responsible for this definitive and still famous and consulted edition. Further editions appeared under the aegis of Wilhelm Sieglin.

507. Spruner, 1872.   ♦ Karl Spruner, *Historico-Geographical Hand Atlas* (Gotha, 1872).

H—CambUL, Atlas 5.87.21. 72 maps in English tr.

508. Starke, 1850.   Otto Starke, *Kleiner Geschichts-Atlas von Europa* (Leipzig?, 1850).

H—NOT SEEN. Łodyński, 3: no. 576. 18 maps. No trace found outside the Łodyński repertory.

509. Steger, 1845.   E. Steger, *Kleiner historischer Schul-Atlas* (Leipzig, 1845).

H—BL, Maps 48.b.40; Harvard. To accompany Steger's *Weltgeschichte für das deutsche Volk.* 12 sheets, 3 for the Middle Ages.

510. Streit, 1843.   F. W. Streit, *Geographisch-historischer Schulatlas* (Berlin, 1843–44).

H—NOT SEEN. Engelmann, 77. 4 fasc.; 28 colored leaves; explanatory text in the margins. Streit collaborated with W. Fischer in a mapless, strictly modern *Historischer u. geographischer Atlas von Europa,* 2 vols. in 3 (Berlin, 1836–37): BL-main, 10108.bb.9.; rev. ed. (1841), 1851.a.5.

511. Stüwe, 1835.   Friedrich Stüwe, *Die Handelszüge der Araber* (Berlin, 1835).

M—BerlinPK. *D BiogArchiv,* fiche 1245, frame 219. Winner of a prize competition of the Gesellschaft der Wissenschaften in Göttingen.

512. Süssmilch-Hörnig, 1860.   M. von Süssmilch-Hörnig, *Historisch-geographischer Atlas von Sachsen und Thüringen,* 3 pts. (Dresden, 1860).

H—BL, Maps 10.d.24. 2 medieval maps in p. 2. Accompanying text (Dresden, 1860–62).

513. Tallis, 1851.   John Tallis and Co., publ., *Illustrated Atlas and Modern History of the World,* ed. R. Montgomery Martin (London and New York, n.d. [1851]).

G—LC. The only historical map is of the Holy Land.

514. Tardieu, 1802.   P. F. Tardieu, engrav., *Nouvel atlas universel de géographie ancienne et moderne pour la nouvelle édition de la géographie de Guthrie* (Paris, 1802 [an X]).

G—BN, Ge.DD.2348. 60 maps. Starts with a few maps of ancient geography, mainly by Mentelle and Chanlaire.

515. Tardieu, 1842.   Ambroise Tardieu, *Atlas universel de géographie ancienne et moderne pour l'intelligence de la Géographie universelle par Malte-Brun* (Paris, 1842).

H—NUC. LC, G1019.T25 1842. 27 maps, 1 sheet (in 4 frames) for medieval France. Later editions, with a few more maps: *BN Impr.,* Yale, UChi. Enlarged by A. Vuillemin without reference to Malte-Brun (Paris, 1868): RLScot.

516. Thalheimer, 1874.   ♦ Marie Elsie Thalheimer, *Eclectic Historical Atlas* (Cincinnati and New York, 1874).

H—LC List 153. LC, G1030.T4 1874. 25 maps, 6 medieval (by A. von Steinwehr). Maps from Thalheimer's history textbooks. She longs for an American equivalent to Spruner.

517. Thierry, 1826.   Augustin Thierry, *Histoire de la conquête de l'Angleterre par les Normands, des ses causes et des suites jusqu'à nos jours, en Angleterre, en Écosse, en Irlande et sur le continent,* 4 vols., 2d ed, with atlas (Paris, 1826).

M— Quérard, *France littéraire,* 9:427 *(Quérand, Litt. franç. contemp.,* 6:465, records a 6th ed. [1843], in 4to atlas, 14 maps, 34 pls. NOT SEEN). Toronto-Robarts, 4 black-and-white lithographed outline maps of northern Europe, England, and France.

518. Thierry, 1858.   Heirs of Thierry and Mensing, publ., *Historisch-Geographische Atlas der Algemeene en Vaderlandsche Geschiedenis* (The Hague, 1858).

H—Koeman, 6:28. BL. Probably the earliest Dutch comprehensive historical atlas. 40 maps, 8 medieval.

519. Thomson, 1829.   John Thomson, publ., *New Classical and Historical Atlas* (Edinburgh, 1829).

H—BL, Maps 36.f.9; RLScot, WD.12 (3 copies). 47 maps, 7 medieval, from various sources; scales vary with the sources. Ancient history augmented by medieval. The map of the crusading expeditions (no. 43) seems unprecedented but is fully developed.

520. Töppen, 1858.   Max Pollux Töppen, *Atlas zur historisch-comparativen Geographie von Preussen, nach den Quellen dargestellt* (Gotha, 1858).

H—GV 146:248. YaleB (imperfect). 5 maps, 2 medieval. Töppen was a professional historian of Prussia.

521. Tutzschmann, 1852.   Max. Moritz Tutzschmann, *Atlas zur Geschichte der sächsischen Länder, Atlas* (Grimma, 1852).

H—GV 147:318. Princeton, 1586.807.92q. 5 maps, 4 medieval. Separate *Erläuterungen* (Grimma, 1853). *GV* and Engelmann, 868, specify 22 maps (1853).

522. Vallardi, 1867.   Francesco Vallardi, publ., *Atlante corografico, iconografico, storico e geologico (L'Italia sotto l'aspetto fisico, militare, storico, letterario, artistico e statistico*, pt. 3) (Milan, n.d. [1867–74?]).

H—LC List 3063–64. Cf. Naymiller, 1867, LC, G1985.U5 1867 (maps, 1868). RomeBNVEmm, 214.Banc.18.D.1., *Atlante corografico, orografico, idrografico e storico dell'Italia* (Milan, n.d. [1869?]). On verso, *L'Italia sotto l'aspetto fisico, storico, letterario, artistico e statistico* (no author given). Parte terza: *l'Atlante corografico iconografico storico e geologico*. This atlas is related to Dressler, 1868. 15-sheet map of Italy, then a historical atlas of Italy, including 10 maps for the Middle Ages.

523. Vat, 1863.   L. Vat, *Nouvel atlas classique, politique, historique et commercial divisé en trois parties conformes au programme du baccalauréat ès lettres et ès sciences* (Paris, 1863).

H—BN. Maps coupled with outline maps (*cartes muettes*) of the same subject; 4 for the Middle Ages.

524. Vivien, 1825.   L. Vivien de Saint-Martin, *Atlas universel, pour servir à l'étude de la géographie et de l'histoire anciennes et modernes* (Paris, 1825; repr., 1827).

G—LC List 4315. The final maps, nos. 40–48, are historical and limited to classical antiquity.

525. Vögelin, 1864.   J. K. Vögelin and Gerd Meyer von Knonau, *Historisch-geographischer Atlas der Schweiz* (Zurich, 1846–64?)

H—BerlinDSB (incomplete). Several installments with 14 maps, 6 medieval.

526. Völter, 1855.   Daniel Völter, *Hand-Atlas der Erd-, Völker- und Staatenkunde*, 4th ed. (Esslingen, 1855).

G—*GV* 152:329–33. BSB. 38 maps, none historical, but (modern) Palestine is adapted to the Bible. 1st ed., 1841. See also Fromman, 1854, and Dittmar, 1864.

527. Voigt, 1846.   Friedrich Voigt, *Historischer Atlas der Mark Brandenburg* (Berlin, 1846).

H—*GV* 151:599. YaleB. 7 maps, predominantly medieval.

528. Voigt, 1877.   ♦ Friedrich Voigt, *Historisch-geographischer Schul-Atlas der mittleren und neueren Zeit,* 2d ed. (Berlin, 1877).

H—Engelmann, 1083 (1st ed., 1856); *GV* 151:600 (2d ed, 1862). BL, 3d ed., 17 maps (in all editions), 7 medieval.

529. Vuillemin, 1839.   Alexandre Vuillemin, *Atlas universel de géographie ancienne et moderne à l'usage des pensionnats* (Paris, 1839).

G—Łodyński, 3: no. 393. BN (1839); LC, G1019.V8 1847; UIll (1849); BL, Maps 38.e.12. (1858). Revision of Tardieu, 1842. 50 maps; 15 for "géographie ancienne," including 2 medieval.

530. Wautier, 1857.   Édouard Wautier d'Halluvin, *Les deux yeux de l'histoire, ou guide chronologique et géographique de l'histoire universelle: Moyen âge* (Paris and Lyons, 1857).

H—BL. Wautier devised a pedagogic method based on his 12mo atlas, which he recommended as much better for study than large atlases. The front cover advertises 28 maps, not at BL or BN or in *NUC.* Many other works by him are listed in *BN Impr.* NYPL identifies him as Halluvin.

531. Wedell, 1849.   Rudolph Wedell, *Historisch-geographischer Hand-Atlas von 36 Karten* (Berlin, 1843–49).

H—*GV* 154:365; Engelmann, 1083. BL, Maps 144.d.18 (with *Erläuternder Texte,* 4to); HABW, Cb folio 41 (incl. text volume); 2d rev. and enl. ed. (Glogau, 1856); LC, G1030.W4 1857; Harvard. Executed at the Geographisch-lithographische Anstalt of H. Mahlmann. The 3d ed. reached only fasc. 1, 1862. Each sheet contains multiple maps and insets. Nos. 3–16 concern the Middle Ages.

532. Weller, 1871.   ♦ Edward Weller, ed., *Pocket Atlas of Historical Geography* (New York, 1871).

H—NYPL; RLScot (maps). 16 uncomfortably small maps, 6 medieval. See also Weller, *Crown Atlas of Historical Geography* (London, 1871). Yale. A pair with the New York *Pocket Atlas;* identical maps.

533. Wesenfeld, 1840.   F. Wesenfeld, "Historisch-geographische Charte der Völkerwanderung" (Magdeburg, n.d. [ca. 1840]).

M—NOT SEEN. BerlinDSB, Catal., U S3780 (not now in Berlin). Possibly no. 2 of Wesenfeld's *Allgemeine Weltgeschichte in vier Tableaux* (Magdeburg, 1838–39), recorded in *GV* 156:10.

534. Whittaker, 1825.   G. B. W. Whittaker, publ., *Historical Map of Europe Showing Routes of the Ancient Invaders* (London, 1825).

M—BL, 1035.(161.). Designer unknown. Title from BL catalogue. Based on Lesage, 1801, with efforts at improvement. The tracks, lightly colored, are sometimes indistinct. See also Elton, 1825.

535. Wiberg, 1856.   Carl Fredrik Wiberg and Thure Alexander von Mentzer, *Atlas till Sveriges Historia* (Stockholm, n.d. [1856]).

H—Yale; StockholmRL, S.Fol. 24. 19 maps, 5 medieval (1, unnumbered, "ancient"); *Förklaringar,* 24 pp. Very appealing and effective in design.

536. Wiberg, 1859.   Carl Fredrik Wiberg and Thure Alexander von Mentzer, *Atlas öfver allmänna historien i sammendrag,* 2 vols. (Stockholm, 1859–62).

H—StockholmRL, S.AF.12–13. UMinn, Quarto G1019.W53X 1862. Clear, attractive design. Vol. 1 (1862), ancient history, 9 maps; vol. 2 (1859), medieval and modern, 13 maps (5 medieval). Many maps subdivided. An ed. of 1872 (StockholmRL S.Fol.7) introduces "northern" history and changes the medieval and modern maps (e.g., new insets). The coloring also changes.

537. Willard, 1836.   Emma Willard, *Atlas to Accompany a System of Universal History* (Hartford, Conn., 1836).

H—*NUC.* NYPL; BL. 6 maps based on the receding clouds of Quin, 1830. (*NUC* lists copies with 8 maps.)

538. Witzleben, 1829.   Ferdinand August von Witzleben, *Geschichtlich-geographischer Atlas von Europa: Von der Errichtung der ersten Staaten bis zu den neuesten Zeiten, zum Gebrauch für höh. Schulen,* fascs. 1–3 (Berlin and Poznań, 1829–33).

H—*GV* 142:442–43 (full publication history); Mees, 8, no. 19. BerlinPK, Kart. 19855: only fasc. 1 (1829), tables 1–5, maps 1–4 (out of 16 tables, 13 maps); table 5 reaches 768. I have NOT SEEN the historical atlases of Poland and Prussia that were among Witzleben's other productions.

539. Wolff, 1872.   ◆Carl Wolff, "Die mitteleuropäischen Staaten nach ihren geschichtlichen Bestandtheilen des ehemaligen römisch-deutschen Kaiserreiches" (Berlin, 1872).

M—*GV* 158:42; BL, 1035. (45.). A rather incomplete comparison on 1 sheet of the new German Empire with the old. Most of Italy and southeastern France are left out.

540. Wolff, 1877.   ◆Carl Wolff, *Historischer Atlas: Neunzehn Karten zur mittleren und neuern Geschichte* (Berlin, 1877).

H—*GV* 158:40–42 (under Wolf). NYPL; BL, Maps 49.b.5 (1889). 19 maps, 8 medieval. Artistically impressive. Minor differences between this atlas and Kiepert, 1879. Wolff took over later editions of Kiepert's classical atlas.

541. Zinkernagel, 1802.   Karl Friedrich Bernard Zinkernagel, "Pagus Retiensis ex chartis medii aevi restitutus" (Augsburg, 1802).

M—*GV* 160:309. BerlinPK. From Zinkernagel's *Historische Untersuchung der Grenzen des Riesgaues und seiner Grafen, in den Zeiten des Mittelalters* (Nördlingen, 1802).

542. Zuccagni, 1832.   Attilio Zuccagni-Orlandini, *Atlante geografico, fisico et storico del Granducato di Toscana* (Florence, 1832).

H—RomeBNVEmm, 215.Banc.37.B.2. 20 very large sheets with maps showing modern conditions; history in writing only. Letterpress in three margins. Osvaldo Baldacci calls the author "the Magini of the nineteenth century."

543. Zuccagni, 1844.   Attilio Zuccagni-Orlandini, *Atlante geografico degli stati italiani . . . per servire di corredo alla corografia fisica storica e statistica dell'Italia,* 2 vols. (Florence, 1844).

H—RomeSGI, Z.3.III.9–10. Very large format. 1 generally undifferentiated historical map per region, e.g., "Carta del regno Veneto sotto il dominio dei Romani e nel Medio Evo" or "Carta del Piemonte e della Liguria Marittima avanti il dominio dei Romani sotto il governo dei medesimi e nel medio evo." Picture supplement, *Atlante illustrave,* 3 vols. (Florence, 1845).

Ancient and Ecclesiastical History

A   ANCIENT, SACRED, AND COMPARATIVE ATLASES
           (SIXTEENTH TO NINETEENTH CENTURIES)

544. Adam, 1794.   Alexander Adam, *Summary of Geography and History, both Ancient and Modern . . . with an abridgement of the fabulous history of mythology of the Greeks . . . to which is prefixed, an historical account of the progress and improvement of astronomy and geography . . . designed chiefly to connect the study of classical learning with that of general knowledge* (Edinburgh, 1794).

A—*NUC.* LC, D9A19; BAV, 2d ed (London, 1797). Maps either ancient or modern. On Adam (1741–1809), see *DNB* 1:84.
Anon. *See* ClassAtl, below.

545. Anville, 1769.   Bourguignon d'Anville, *Géographie ancienne abrégée* (Paris, 1769).

A—Toronto-Fisher. Celebrated; widely circulated.

546. Anville, 1806.   Bourguignon d'Anville, *Complete Body of Ancient Geography . . . improved by inserting the modern names of places under the ancient* (London, 1806).

A—*NUC.* PhiladUL-Rare (Lea); editions listed from 1771 to 1818.

547. Arnz, 1829.   J. Arnz, *Atlas der alten Welt . . . zum Gebrauch in Gymnasien* (Düsseldorf, 1829?).

A—LC List 5590; BN. 16 maps.

548. Arrowsmith, 1828.   Aaron Arrowsmith, *Orbis terrarum veteribus noti descriptio: A Comparative Atlas of Ancient and Modern Geography . . . for the Use of Eton School* (London, 1828).

A—BL, Maps 38.d.3. Ancient opposite modern. Medieval content treated as ancient: no. 6, "Engla Land sive ea Britanniae pars quae citra Tuedam fl. Saxonibus subjecta est."

549. *Atlas minor, 1806.*   *Atlas orbis antiqui minor in usum scholarum Austriacarum (Kleiner Atlas der alten Welt für Schulen)* (Vienna, 1806).

   A—NOT SEEN. Łodyński, 3: no. 47; Engelmann, 78. 13 leaves.

550. Barbié, 1788.   Jean-Denis Barbié du Bocage, *Recueil des cartes géographiques, plans, vues et médailles de l'ancienne Grèce, relatifs au voyage du jeune Anacharsis, précédé d'une analyze critique des cartes* (Paris, 1788).

   A—*BN Impr.* 7:387–88. PhiladLibCo; MilwaukeeAGS, 1799; Harvard, MA 2117.819, 1819, also 1861. To accompany Jean-Jacques Barthélémy's fictional account of Anacharsis's tour through ancient Greece. Thrace, and Asia Minor (editions in *BN Impr.*). After Barbié, Ambroise Tardieu and C. V. Monin compiled new atlases for Anacharsis's travels.

551. Barentin, 1807.   Charles Paul Nicolas Barentin de Montchal, *Atlas de la géographie ancienne et historique* (Paris, 1807).

   A—Łodyński, 3: no. 51. BL: d'Anville maps for the histories of Rollin and Cluvier. *NUC* lists *Géographie ancienne et historique . . . d'après les cartes de d'Anville* (Paris, 1807), 25 maps (LC); 2d ed., Paris, 1823 (Peabody Institute, Johns Hopkins University).

552. Benicken, 1829.   Friedrich Wilhelm Benicken, *Orbis terrarum antiquior . . . in usum scholarum* (Weimar, 1829).

   A—BL, S.T.W.(2); BN. 18 very large black-and-white wall maps.

553. Bertius, 1618.   Petrus Bertius, *Theatrum geographiae veteris,* 2 vols. (Amsterdam, 1618–19).

   A—Koeman, 1:63. BL; NYPL; YaleB. A source book of ancient geography. Mainly Ptolemy in vol. 1, mainly Ortelius in vol. 2, with many shorter items.

554. Bertius, 1628.   Petrus Bertius, *Geographia vetus ex antiquis et melioris notae scriptoribus nuper collecta* (Paris, 1628, 1645).

   A—Pastoureau, *Atlas français,* 65. LC, YaleB, BL. 20 maps. Each map is ascribed to a named source (Strabo, etc.); 1 page of text with each map. Some authorities are medieval or modern (unnamed). Influential and often reprinted.

555. Blair, 1853.   D. Blair, *A Classical Atlas on an Entirely New Plan: The Ancient and Modern Names of Places Being Given on the Same Map . . . in Different Coloured Inks* (London, 1853).

   A—RGS. A lithographed version of Jones, 1830. Comparative. The allegedly new plan had been implemented as for back as the sixteenth century.

556. Boiste, 1806.   P. C. V. Boiste, *Atlas du dictionnaire de géographie universelle ancienne, du moyen âge et moderne, comparées* (Paris, 1806).

   A—LC, G1019.B655 1806. Only ancient, biblical, and modern.

557. Bonne, 1783.   Rigobert Bonne, *Receuil de cartes sur la géographie ancienne* (Paris, 1783).

   A—*NUC* 66:21. LC. 20 maps; repr. in *Atlas encyclopédique; Encyclopédie méthodique.* Appended: Th. François de Grace, *Tableaux historique et chronologique des principales révolutions . . . jusqu'au moyen âge.*

558. Briet, 1647.   Philippe Briet, *Parallela geographiae veteris et novae,* 3 vols. (Paris, 1647–49; later ed., 2 vols., 1748–49).

A—Pastoureau, *Atlas français,* 89–95. BL, K.304.k.1, 2; Princeton-rare; New-berry; PhiladLibCo. A great success in the seventeenth century. On unpublished parts, see Sommervogel, 2:156. The maps, separately, 1653: *Théatre géographique de l'Europe,* 79 maps, and *Theatrum geographicum Europae veteris,* 60 maps.

559. Butler, 1822.   Samuel Butler, *Atlas of Ancient Geography* (London, 1822).

A—BL, Maps 37.b.9 and 37.b.10; PhiladFree, 910 An 86s Rare Books. 21 maps (many eds.). Repr., Philadelphia, 1851 (PhiladUL).

Buy de Mornas, Claude. *See* Buy, 1761.

560. Cellarius, 1701a.   Christoph Cellarius, *Geographia antiqua in compendium redacta* (Leipzig, 1701).

A—Various titles and dates. LC List 24, 25; Toronto-Fisher. Starting point: *Nucleus geographiae antiquae et novae* (1676), superseded by *Geographia antiqua iuxta et nova* (Jena, 1692); YaleB. A small book of comparative geography without maps. Cellarius, *Geographia antiqua . . . tot chartas ex maiori auctoris geographiae antique quot ad minorem hanc illustrandum requirebantur,* 6th ed., ed. Samuel Patrick (London, 1731); enl. ed. with 27 maps, none historical. Further printings (all at BL), 1745, 1764; Rome, 1774, 1782, 1786, 1790, 1802, 1808, 1812.

561. Cellarius, 1701b.   Christoph Cellarius, *Notitia orbis antiqui sive geographia plenior,* 2 vols. (Leipzig, Cambridge, and Amsterdam, 1701–6; ed. L. Io. Conrad Schwartz, Leipzig, 1731).

A—Toronto-Fisher, D-10 5314 (1703–6). 34 maps. Newberry (Schwartz ed.) "[T]he first complete and systematic treatise on [ancient geography]" (*Grand Larousse du XIXe siècle* 3:682).

562. ClassAtl, 1630.   97 French maps illustrating ancient geography (Paris, 1630–90).

A—BL, Maps 37.e.1.: *atlas factice,* sacred and ancient history. On spine, "Cartes anciennes," many by Sanson. Nos. 20–31, patriarchates and bishoprics; 34–35, Bertius, 1623 (Charlemagne).

563. ClassAtl, 168-.   *Atlas factice,* Église de Paris.

A—BN, Ge.DD.2686. From Notre-Dame. Ancient history, sacred and profane, in French maps. Cartographers include La Ruë and Duval.

564. ClassAtl, 17–.   Scrapbook of Maps and Plates of the World [ancient geography only] (n.p. [U.K.], n.d. [eighteenth century]).

A—UKans, D 1676. Systematically ordered atlas of ancient geography composed of black-and-white maps of various sizes cut out of English books (mainly multivolume). Plans of ancient cities and archaeological sites (also cut out) are added. Maps tabbed for easy reference.

565. ClassAtl, 174-.   *Atlas factice de géographie ancienne.*

A—UMich-Clements. 114 maps. Latest entries by Gilles Robert, including some of his maps for medieval history. Discussed by Pedley, *Vaugondy,* 233.

566. Clüver, 1616   Philip Clüver, *Germania antiqua* (Leiden, 1616; 2d ed., 1631).

A—Toronto-Fisher; Newberry, 2d ed. 11 maps of Germany in antiquity. This book, researched at Oxford, made Clüver's reputation and won him the equivalent of a research professorship at the University of Leiden.

567. Delamarche, 1809.   C. F. Delamarche, *Description géographique et historique des peuples les plus renommés de l'Europe ancienne, et des lieux les plus remarquables* (Paris, 1809).

A—BN, G.3225. 19-map ancient geography.

568. Duval, 1665.   Pierre Duval, *Diverses cartes et tables pour la géographie ancienne* (Paris, 1665).

A—Pastoureau, *Atlas français,* 156. LC. 3-part small-sized atlas; notably, "Cartes géographiques dressées pour bien entendre les historiens." 28 chronologically ordered maps plus (sketchy) chronological tables. For pt. 3, see Duval, 1665.

569. Duval, 1667.   Pierre Duval, *Cartes de géographie les plus nouvelles* [3d ed.] (Paris, 1667).

A—Contents in Pastoureaux, *Atlas français,* 141–44. Collections with this title, at first wholly modern, incorporate many maps of ancient and sacred history by 1667. Maps by Clüver, La Rüe, and Sanson, as well as by Duval.

570. Findlay, 1853.   Alexander G. Findlay, *Comparative Atlas of Ancient and Modern Geography* (London, 1853).

A—BL, Maps 38.d.6; Northwestern. 54 maps, ancient facing modern.

571. Findlay, 1854.   Alexander G. Findlay, *Classical Atlas to Illustrate Ancient Geography* (London, 1854).

A—LC. 25 maps. Various editions.

Funke, Karl Philip. *See* Vieth, 1800/A.

572. Gail, 1815.   Jean-Baptiste Gail, *Atlas pour servir à l'étude de l'histoire ancienne et à l'intelligence des auteurs grecs et latins* (Paris, 1815).

A—LC List 5606; YaleB. Diverse maps focusing on the works of Xenophon.

573. Götz, 1729.   Andrea Götz, *Brevis introductio ad geographiam antiquam in iuventutis usum* (Nuremberg, 1729).

A—LC. 10 maps, attractively colored.

574. Grenet, 1779.   Abbé Grenet, *Atlas portatif à l'usage des collèges* (Paris, n.d. [1779–82]).

A— 44 maps by R. Bonne, as a comparative atlas. YaleB (London, 1790), 86 maps "pour servir à l'intelligence des auteurs classiques," comparative. Newberry, 4 vols., 1779–91? (many maps dated 1790); vol. 1 is ancient geography.

575. Holle, 1852.   L. Holle, publ., *Historisch-geographischer Schulwandatlas zur alten, mittleren und neuen Geschichte* (Wolfenbüttel, 1852–54).

A—NOT SEEN. Engelmann, 75. The maps listed are all ancient.

576. Hondius, 1607.   Jodocus Hondius [Joost Hondt], *Atlas minor Gerardi Mercatoris* (Amsterdam, 1607).

A—Koeman, 2:508–10, 513; BSB; Yale; Harvard (German ed., 1609); NYPL-rare (Latin, 1610). "I have added the following maps of antiquity (*vetus aevum*), drawn by me, for the sake of initiates in both ancient sacred and ancient profane history." The first extension of ancient geography beyond Ortelius, 1579. Plates sold and reused in England, 1621.

577. Janssonius, 1652.   Joan Janssonius, *Accuratissima orbis antiqui delineatio sive geographia vetus, sacra et profana* (Amsterdam, 1652).

A—LC List 5613–14; Koeman 2:185–88. LC, G1033.H6 1654 (black-and-white); AmstUL; also Janssonius, 1658. A major atlas of ancient geography. Contains Bertius's "Charlemagne" (no. 52) and Speed's "Anglo-Saxon Heptarchy" (modified) (no. 43). 1 year after the 1st ed., Georg Horn supplied the atlas with written commentaries. Eds. (*BL Maps Catal.*, s.v. "Horn"): 50 maps (1653), 49 maps (1660), 47 maps (1684); *Full and Exact Description* (1700); 50 maps (The Hague: de Hondt, 1740); *Description exacte de l'Univers* (1741); *Compleat Body of Ancient Geography* (1741).

578. Janssonius, 1658.   Joan Janssonius, *Novus atlas,* vol. 6 (Amsterdam, 1658).

A—Koeman 2:499–502. A packaging of Janssonius, 1652. For another (1662), see Koeman, 2:503–7 (vol. 10).

579. Jenner, 1645.   Theodore Jenner, publ., *A New Booke of Mapps exactly describing Europe Both the present as now it standeth and ancient state thereof . . . together with the mutation of them by sundry accidents of time* (n.p. [London?], n.d. [1645]).

A—YaleB. Comparative Ptolemaic and modern maps.

580. Johnston, 1853.   Alexander Keith Johnston, *School Atlas of Classical Geography* (Edinburgh and London, 1853).

A—LC List 21. Toronto-Fisher, E-10 4964; PhiladUL, 911.3 J64. The last map (no. 20), the barbarian invasions, faithfully follows Lesage into his most glaring error but is admirable in draftsmanship. Toronto-Robarts, G1033 J65 1881: 2d ed., 23 maps.

581. Jones, 1830.   Jones and Co., *Jones's Classical Atlas, on an Entirely New Plan.*

A—NOT SEEN. Recorded by Black, *Maps and History,* 50. Prototype for Blair, 1853.

582. Kausler, 1826.   Fritz von Kausler, *Atlas der alten Welt* (Ulm, 1826).

A—BL, Maps 37.f.11. Maps for wars in the ancient world.

Keller, C. *See* Cellarius.

583. Kiepert, 1848.   Heinrich Kiepert, *Historisch-geographischer Atlas der alten Welt zum Schulgebrauch,* 8th ed. (Weimar, 1848).

A—NYPL. 16 plates.

584. Kiepert, 1859.   Heinrich Kiepert, *Atlas antiquus.* [*Acht Karten zur alten Geschichte*] (Berlin, 1859).

A—GV 75:36–44; NYPL; UCB, 1859, 8 maps; 3d ed., 1863, 10 maps; 5th ed., 1869, 12 maps. French, Italian, U.S. (4th, 1867, 6th, 1878) eds. all fed by sheets from the Berlin publisher, Dietrich Reimer. Outstanding in its time (preferred to Spruner's classical atlas).

585. Knapton, 1721.   J. Knapton and P. Knapton, *Geographia classica. The Geography of the Ancients so far described as it is contained in the Greek and Latin classics,* 3rd ed. (London, 1721).

A—LC (8th ed., 1747); PhiladLibCo (1723); BL, Maps 37.b.2. (1st ed.). Maps by John Senex. Small, for classroom use. A very full collection of "literary geography."

586. Knapton, 1742.   *Geographia antiqua et nova, or a System of Antient and Modern Geography* (London, 1742).

A—BL, 793.i.2. 33 maps by Cellarius, text by Lenglet Dufresnoy, tr. P. Morant. 2d ed. and later eds. to 1812 by J. Rivington and F. Rivington.

587. Köhler, 1720.   Johann David Köhler, *Descriptio orbis antiqui in xliv tabulis exhibita* (Nuremberg, 1720).

A—LC (imperfect). BSB. Alternative name: *Atlas minor geographiae antiquae.* 44 plates. Very attractive anthology of ancient geography.

588. Köhler, 1759.   Johann David Köhler, *Atlas minor antiqui et medii aevi in usum scholarum editus* (Nuremberg, n.d. [1759?]).

A—BSB, 8vo Mapp. 22t. Budapest, TA 148 (and others); Köhler is not mentioned. No medieval contents. Said to be designed for youth unable to afford the 12 d'Anville maps.

589. Le Clerc, 1705.   J. Le Clerc, *Atlas antiquus, sacer, ecclesiasticus et profanus* (Amsterdam, n.d. [1705]).

A—LC List 5628; Koeman, 3:17–18 (under Mortier). LC; NYPL. The compiler is often called "Clericus." Very comprehensive as a reprint collection; includes much older cartography.

590. Lenormant, 1868.   François Lenormant, *Atlas d'histoire ancienne de l'Orient antérieurement au guerres médiques* (Paris, 1868).

A—LC List 31a; BN. 24 plates.

591. Letronne, 1827.   A. J. Letronne, *Atlas de géographie ancienne pour servir à l'intelligence des œuvres de Rollin* (Paris, 1827).

A—LC List 5629. 17 maps, by A. H. Dufour.

592. Macpherson, 179-.   David Macpherson, *Geographia antiqua* (n.p., 179-?).

A—YaleB (poor condition); NYPL. 61 maps. Comparative ancient and modern.

593. Macpherson, 1806.   David Macpherson and A. Macpherson, [Atlas of Ancient Geography] (Philadelphia, 1806).

A—LC List 711, 4312. LC, G1019.R575 1806 vault. An anthology. Source for *Cyclopedia,* 1820.

Magini. *See* Ptolemy, 1596A.

594. Manesson-Mallet, 1683.   Allain Manesson-Mallet, *Description de l'univers, contenant les systèmes du monde . . . ,* 5 vols. (Paris, 1683).

A—LC List 3447; Pastoureau, *Atlas français,* 319. Newberry; BL. Mainly comparative ancient and modern; extensive text. Its maps of the ancient Burgundian kingdom concern the Middle Ages. German tr. (1719), LC List 4280.

595. Mayo, 1814.   R. Mayo, *Atlas of Ten Select Maps of Ancient Geography, both Sacred and Profane* (Philadelphia, 1814).

A—LC List 4101. LC.

596. Mentelle, 1778.   Edme Mentelle, *Géographie comparée, ou Analyze de la géographie ancienne et moderne des peuples de tous les pays et de tous les âges* (Paris, 1778–84).

A—UCB, UMich (incomplete); BN, G.9740–46, plus atlas, G.3588. The title clearly indicates the contents.

597. Mentelle, 1797.   Edme Mentelle and P. G. Chanlaire, *Atlas universel de géographie physique et politique, ancienne et moderne* (Paris, 1797–1801).

A—LC List 712, 3539. LC; UMich-Clements; Harvard, MA 18.98.2. Basically comparative, but modern and physical maps are stressed.

598. Mentelle, 1804a.   Edme Mentelle, *Abrégé élémentaire de géographie et d'histoire* (Paris, 1804).

A—BN, G.23913, G.6387. RomeBNVEmm, 14.23.0.M.21. 15 maps. Comparative.

599. Mentelle, 1804b.   Edme Mentelle and C. Malte-Brun, *Atlas de la géographie universelle ancienne et moderne* (Paris, 1804).

A—LC. Comparative. 45 maps, bibliography.

600. Mentelle, 1804c.   Edme Mentelle, *Atlas de tableaux et de cartes gravé par F. Tardieu pour le cour complet de cosmographie, de géographie, de chronologie et d'histoire ancienne et moderne*, 2d ed (Paris, 1804–5).

A—Łodyński, 3: no. 24. BL; RLScot (Maps). 20 maps. Comparative.

601. Mitchell, 1845.   S. Augustus Mitchell, *Mitchell's Ancient Atlas, Classical and Sacred* (Philadelphia, 1845).

A—YaleB; Northwestern (1863); PhiladLibCo. 12 maps; ends with a map of the barbarian invasions.

602. Moll, 1721.   Hermann Moll, *Geographia antiqua* (London, 1721).

A—Also called *Thirty-two Maps of the Geography of the Ancients*. LC List 5639–40, LC. Another ed., 1739, with no. 30, William Stukeley, "Dispersion of the sons of Noah."

603. Möller, 1851.   Johann Heinrich Möller, *Orbis terrarum antiquus: Schul-Atlas der alten Welt*, 23d ed. (Gotha, n.d. [1851]).

A—LC List 5638, LC; BL, Maps 37.b.29. 10 maps.

604. Münster, 1540.   *Geographia universalis, vetus et nova* (Basle, 1540).

A—Oxford. Introduction by Sebastian Münster. A core of Ptolemaic maps complemented by 20 *novae tabulae*. NYPL-rare, *LB/+/1540. Facsimile available.

605. Murphy, 1858.   William Murphy, *The Classical-Historical School Atlas* (Edinburgh, 1858).

A—BL, Maps 37.b.22. Classical only.

606. Murphy, 1832.   William Murphy, *Comprehensive Class Atlas* (Edinburgh, n.d. [1832?]).

A—BL, Maps 37.a.5. Comparative ancient and modern.

607. Murphy, 1850.   Murphy, William, *School and College Atlas of Ancient and Modern Geography* (Edinburgh, 1850).

    A—BL, Maps 38.b.3. Comparative, ancient and modern.

608. Ohmann, 1846.   C. Ohmann, *Historische-geographische Karte der alten Welt* (Berlin, 1846).

    A—BL, S.T.W.(2). 4 wall maps, probably derived from an atlas. See also Ohmann, 1853. (Spelled Olimann in *BL Maps Catal.* 15:672).

609. Ottens, 1720.   "Carte des quatres grandes monarchies des Assiriens, des Perses, des Grecs, et des Romains: Dressé pour bien entendre l'Histoire Sainte, l'Histoire Prophane, et particulièrement Celle de Flavius Josephe" (Amsterdam, n.d. [ca. 1720]).

    A—BerlinPK, S1970. All-purpose map of ancient history; extends from Morocco to Pakistan with few names of lands and cities.

610. Padua, 1699.   Padua, Seminario vescovile, *Tabulae geographicae* (Padua, 1699).

    A—LC List 5641, LC. Well-organized, representative collection of ancient and ecclesiastical maps, mainly from the Sanson circle. Not rare.

611. Park, 1823.   John Park, *Outlines of Ancient History and Chronology, Selected from Le Sage's Atlas, in which his errours are corrected, and the Proper Names accented* (Boston, 1823).

    A—Columbia-rare, A902/1823 (lacks the one colored plate). Very short. The title tells it all. For the ladies of the Boston Lyceum.

612. Patteson, 1825.   Edward Patteson, *A General and Classical Atlas, Exhibiting in the Same Maps the Principal Features Both of Ancient and Modern [sic] Geography,* 2d ed. (London, 1825).

    A—BL, Maps 38.d.1. (1801) and 38.d.2.; LC, G1019.P25, "Richmond and London, 1804 i.e. 1806." Comparative.

613. Perthes, 1834.   Justus Perthes, publ., *Schul-Atlas der alten Welt,* 9th ed. (Gotha, 1834).

    A—LC List 5645. NYPL. 10 plates. LC, 4th ed., 1827.

614. Philip, 1855.   George Philip and son, *Philip's Atlas of Classical, Historical, and Scriptural History* (Liverpool, 1855).

    A—NYPL. Ends with conditions before and after the barbarian invasions.

615. Philippe, 1787.   Étienne André Philippe de Prétot, *Atlas universel: Pour l'étude de la géographie et de l'histoire ancienne et moderne* (Paris, 1787).

    A—LC. Black-and-white. Basically comparative. Some maps by other authors.

616. Playfair, 1814.   J. Playfair, *New, General Atlas, Ancient and Modern* (London, 1814).

    A—LC List 137, 4309.

617. Potocki, 1805.   Jean Potocki, *Atlas archéologique de la Russie européenne* (St. Petersburg, 1805).

A—NOT SEEN. *AGEphem.* 19 (1806): 264; 20 (1806): 313–16. An ambitious, idiosyncratic venture by a major scholar. It may never have been marketed.

618. Ptolemy, 1596.   Claudius Ptolemaeus, *Geographiae universae tum veteris tum novae absolutissimum opus,* ed. G. A. Magini, 2 vols. (Venice, 1596).

H—LC List 403, cf. 404–5, "In secundo volumine insunt Cl. Ptolemæi antiquæ orbis tabulæ XXVII. ad priscas historias intelligendas summè necessarie. Et tabule, XXXVII. recentiores . . . . " NYPL-rare, *KB/1596; *BL Maps Catal.* 15:341; RGS (Cologne, 1597). The Ptolemy 27 are described as "ancient geography" par excellance, essential for understanding history. The atlas, according to its editor, is a companion to reading ancient history, like Ortelius's *Parergon.*

619. Reichard, 1823.   Christian Gottlieb Reichard, *Atlas antiquus* (Nuremberg, 1818–20).

A—*GV* 115:21–22. Incomplete.

620. Reichard, 1824.   Christian Gottlieb Reichard, *Orbis terrarum antiquus cum thesaurus topographicus* (Nuremberg, 1824).

A—LC List 3286. LC; HABW, Cb 26 folio, 11 maps; StockholmRL (1818–31), 19 maps.

621. Reichard, 1830.   Christian Gottlieb Reichard, *Orbis terrarum antiquus veteribus cognitus, in usum iuventutis* (Nuremberg, 1830; 2d ed., 1833).

A—LC List 5646. LC, 21 plates. Austerely limited to the geography of ancient lands.

622. Rhode, 1772.   J. C. Rhode, *Orbis veteribus notus* (Berlin, 1772).

A—BL. 35 maps. Done at the behest of the Berlin Academy; trilingual.
Robert de Vaugondy. *See* RVaugondy.

623. Rühle, 1827.   Johann Jacob Otto August Rühle von Lilienstern, *Geographische Darstellungen zur ältesten Geschichte und Geographie von Aethiopien und Aegypten,* fasc. 1 (no more published) (Berlin, 1827).

A—*NUC;* Engelmann, 76; Mees, 7–8, no. 15: first installment of *Universal Historischer Atlas oder anschauliche Darstellung der gesammten Weltgeschichte nach wissenschaftlicher Entwicklung von den frühesten Sagen bis auf die gegenwärtige Zeit in Charten, Tabellen und anderen graphischen Constructionen.* Princeton, Yale, NYPL. The 9 plates include very few maps.

624. RVaugondy, 1753   Didier Robert de Vaugondy, "Antiquorum imperiorum tabula in qua prae caeteris Macedonicum, seu Alexandri Magni, imperium et expeditiones exarantur" (Paris, 1753).

A—Pedley, *Vaugondy,* 196, Catalogue, no. 382. BL, K.3.40. A traditional composite. Referred to by Dupré, 1763, as a map of "empires" by Boudet. In RVaugondy, 1757. Cf. Ottens, 1729/A.

625. SDUK, 1851.   Society for the Diffusion of Useful Knowledge, *Maps of the Society for the Diffusion of Useful Knowledge,* 3 vols. (London, 1846–51).

A—LC List 4326. Various "ancient maps." UIll. 2 vols., 1844: maps of the ancient world, Britain, and Italy (several) compared to modern maps; the rest of the world gets no comparison; "residual Briet," but a handsome product.

626. SDUK, 1871. ♦ Society for the Diffusion of Useful Knowledge, *Complete Atlas of Modern Classical and Celestial Maps* (London, 1871).

A—Columbia, Phoenix P912 A F. Maps comparing the ancient and modern world, Britain, France, Italy (4 exposures), Spain, Greece, Asia Minor, Syria, Persia.

627. Sickler, 1846.   F. C. L. Sickler, *Schulatlas zur alten Geographie* (Kassel, [1846]).

A—BL, Maps 143.c.22.

628. Smith, 1809.   C. Smith, *Smith's Classical Atlas Containing Distinct Maps of Countries Described in Ancient History Both Sacred and Profane* (London, 1809).

A—BL, Maps 37.d.6. (1809); LC List 5650 (1835), 14 plates.

629. Spruner, 1850.   Karl Spruner, *Atlas antiquus* (Gotha, 1850; 2d ed., 1855).

A—LC List 150–51, 5651. 3d ed. by Theodor Menke (1865). First part of the full Spruner atlas.

630. Stackhouse, 1790.   Thomas Stackhouse, *Universal Atlas . . . Maps for Ancient and Modern Geography* (London, 1790).

A—LC. Comparative. No. 12, "A synopsis of the repeopling of the world by the descendants of Noah."

631. Tanner, 1834.   S. H. Tanner, *Atlas Classica* (Philadelphia, 1834–38).

A—PhiladAmPhil. Exact copy of Wilkinson, 1797/A. LC List 3289 (1840), 46 maps; for the title, cf. LC List 3290.

632. Tardieu, 1819.   Pierre Tardieu, *Atlas de géographie ancienne pour servir à l'intelligence de l'Histoire des empereurs par Crevier* (Paris, 1818–19).

A—NOT SEEN. LC List 5599; Łodyński, 3: no. 139. 5 maps.

633. Tardieu, 1821.   Pierre Tardieu, *Atlas pour servir à l'histoire ancienne, romaine et du Bas-Empire par M. le comte de Ségur* (Paris, 1821).

A—LC List 7, 146 (1822), 4143 (1840); Łodyński, 3: no. 184 (1824), 10 maps. BN, Ge.F.5297. Title varies; e.g., *Atlas pour l'histoire universelle par le Comte de Ségur, Partie ancienne, romaine et du Bas-Empire.*

634. Vieth, 1800.   Gerhard Ulrich Anton Vieth and Karl Philip Funke, *Atlas der alten Welt; Atlas du monde ancien* (Weimar, 1800).

A—LC List 5654. RGS (title in French); HABW, Cb folio 2; Cb 26, 2d improved ed. (Weimar, 1806); 4th ed., 1819; 5th ed., 1823. Always 12 maps and Funke's dictionary of ancient geography. Much explanatory text.

635. Walckenaer, 1839.   Charles Athanase Walckenaer, *Atlas de la géographie ancienne, historique et comparée des Gaules cisalpine et transalpine* (Paris, 1839).

A—LC List 2958. LC, G1841.S1W3 1839. 9 maps concerning the migrations of Gauls into Italy.

636. Wilkinson, 1797.   Robert Wilkinson, publ., *Atlas classica, being a collection of Maps of countries mentioned by the ancient authors both sacred and profane with their various subdivisions at different periods* (London, 1797).

A—*NUC;* LC List 54, 3290. PhiladLibCo; BL, Maps 49.d.23.; NYPL; Jerusalem, folio 72C10195. Ca. 50 maps; many editions between 1797 (LC), 1808 (PhiladLibCo), and 1842 (NYPL). Cf. Tanner, 1834/A. 4 memorable maps for the Middle Ages are included (1801–7).

637. Willard, 1822.   Emma Willard, *Ancient Atlas* (Hartford, Conn., 1822).

A—Yale. 6 maps. Published for many years and very widely found in U.S. libraries. Many titles, all beginning *Ancient Atlas.* William C. Woodbridge is sometimes listed as author. The maps are said to be adapted from d'Anville, Adam, Lavoisne, and Malte-Brun.

638. Worcester, 1839.   Joseph E. Worcester, *Ancient Classical and Scripture Atlas* (Boston, 1839).

A—RGS. Very limited contents.

639. Wyld, 1825.   J. Wyld, *General Atlas* (Edinburgh, n.d. [ca. 1825]).

A—BL, Maps 26.c.41. Important periods of ancient history only.

E   ECCLESIASTICAL AND SACRED COLLECTIONS
           (EXCLUSIVELY)

640. Anville, 1732.   Bourguignon d'Anville, "Géographie sacrée" (Paris, 1731–32).

E—DumbOaks (Rare Books), Michel Le Quien, *Oriens Christianus in quattuor patriarchatus digestus,* 3 vols. (Paris, 1731–32), 1:1 (map of the Patriarchate of Constantinople), 2:329–30 (Alexandria), 2:369–70 (Antioch), 3:1 (Jerusalem). Repr. reduced to 8vo (Graz, 1958).

641. AtlFact, 17–.   *Atlas factice* (Italian?), mainly of ecclesiastical and sacred subjects, to eighteenth centuries.

E—RomeBNVEmm, R.D. 82. Several items by La Ruëe, including "Égypte moderne." Miscellaneous items, such as Switzerland.

642. Bochart, 1707.   Samuel Bochart, *Geographia sacra, seu Phaleg et Canaan,* 4th ed. by Petrus de Villemandy (Leiden and Utrecht, 1707).

E—PhiladUL-rare, 13 maps.

643. Brion, 1766.   Louis Brion de la Tour, *Atlas ecclésiastique comprenant tous les évêchés des quatres parties du monde* (Paris, 1766).

E—Alternative title: *Atlas ecclésiastique . . . ouvrage nouveau adapté à la géographie de l'abbé Nicolas de la Croix &ca.; et servant de supplément à l'atlas général* (1760). *NUC;* LC List 80. LC, 41 maps; BN, Ge.FF.3033. A world atlas in which ecclesiastical provinces and bishoprics have been marked in red ink. Part of Brion, 1766a.

644. Buache, 1754.   Philippe Buache, *Carte générale pour . . . l'Histoire sainte* (Paris, 1754).

E—No. 126 in Guillaume Delisle and Philippe Buache, *Atlas général de géographie ancienne et moderne* (n.p. [Paris], n.d. [1754]). UMich-Clements. Draws tracks for Abraham, the Exodus, the Magi, the apostle Paul.

645. Calmet, 1724.   Augustin Calmet, *Commentaire littéral sur tous les livres de l'ancien et du nouveau testament*, 8 vols. (Paris, 1724–26).

  E—HABW, Jc folio 3; Jerusalem. (First appeared without maps in 20 vols. in 8vo.) Many maps, including Moulart-Sanson's "Sons of Noah" (1:1, 1704), "The Exodus" by Liébaux according to Calmet's design (1:181), "The Promised Land" (2:1), and many more.

646. Calmet, 1822.   Augustin Calmet, *Atlante per la storia dell'Antico, nuovo Testamento et degli Ebrei*, new ed. by Guiseppe Battaglia (Venice, 1822).

  E—ViennaÖNB. Jerusalem, folio 22V3432, Venice 1823. 9 maps. A poor, derivative ed.

647. Carl, 1732.   Rupert Carl, *Gallia, Germania, Italia Benedictina* (Nuremberg, 1738, 1732, [1750?]).

  E—3 maps of Benedictine houses. Harvard, Map coll. 290.1738.2; 1215.1732; 645.1750. Current conditions are recorded; no attention to dates of foundation.

648. Delisle, 1764.   Guillaume Delisle, "Paradisi terrestris et circumiacentium regionum situs," opus posthumus (Paris, 1764).

  E—BL. Drawn in 1719 to illustrate the revisionist theory of the Jesuit Jean Hardouin that situated Paradise near Palestine. Published by J. N. Delisle, *Mémoires sur trois cartes nouvellement publiées en juillet 1764* (Paris, 1764). On the three theories about the location of Paradise, see J. N. Delisle, as above; Delano-Smith and Ingram, *Maps in Bibles, 1500–1600*, 2–24 (Eden).

649. Duval, 1672.   Pierre Duval, *Le monde chrestien ou sont les cartes des archeveschez et des eveschez de l'univers* (Paris, n.d. [before 1672]).

  E—Pastoureau, "Atlas en France avant 1700," 166, and Pastoureau, *Atlas français*, 143–44. BN, Ge.DD.1221–1231 (specifically 1222bis). Maps of ecclesiastical districts.

650. Fer, 1700.   Nicolas de Fer, *Ancienne et nouvelle Thebaïde* (Paris, 1700).

  E—BN, Ge.BB.565–XVI (45,1 and 2). Desert fathers and the Trappist mother house. Repr. in Le Clerc, *Atlas antiquus* (Le Clerc, 1705). A 1696 printing has eluded me but may well exist. See also Jaillot, 1693/E.

651. Guppenburg, 1657.   Wilhelm Guppenburg, *Atlas Marianus, sive de imaginibus Deiparae per orbem Christianum miraculosis*, 2d ed. (Ingolstadt, 1657).

  E—Harvard-Houghton. No maps.

652. H . . . . , 1730.   M. A. H., *Germania Augustiniana* (Augsburg, 1730).

  E—Harvard. Divided by provinces. See also Lubin, 1659/E.

653. Haraeus, 1617.   Franciscus Haraeus [Verhaer], "Lumen historiarum per Occidentem; Lumen historiarum per Orientem" (Antwerp, 1617, 1624).

  E—BL K.2.(21.), K.2.21(2). See the account of Nebenzahl, 114–15 (illustration). The maps illustrate the Christian churches in the West and the East. The eastern half was published in the 1624 ed. of Ortelius's *Parergon*. Nebenzahl points out admiringly that the presence of 3 scenes on 1 map makes this unique among maps for history. Haraeus is definitely a precursor of such groupings, which became frequent in the nineteenth century (Freyhold, Wedell).

654. Hartzheim, 1758.   Josef Hartzheim, S.J., "Charta chorographica episcopatuum Germaniae ab anno 300 ad 1500" (Cologne, 1758).

E—BSB. From Johann Friedrich Schannat, *Concilia Germaniae*, 11 vols (Cologne, 1759–90), 1:24. Indicates provinces, not dioceses. Hartzheim saw the *Concilia* to completion after Schannat's death.

655. Hartzheim, 1762.   Josef Hartzheim, S.J., "Mappa chorographica omnium epi-copatuum Germaniae, 1500–1760" (Nuremberg, 1762).

E—Harvard. Single sheet, border coloring. Special attention to suppressed or transferred bishoprics. ViennaÖNB, in Homann Heirs, *Atlas geographicus maior* (Nuremberg, 1784), no. 18.

656. Hase, 1739.   Johann Matthias Hase, *Regni Davidici et Salomonaei descriptio geographica et historica* (Nuremberg, 1739).

E—*GV* 56:308. Harvard. Repr. in Hase, 1750a. For an authoritative account, see Dörflinger, "Geschichtskartographische Werk von Hase" (forthcoming).

657. Jaillot, 1693.   Hubert Jaillot, *Les Deserts d'Égypte de Thebaïde d'Arabie de Sirie etc.: Les lieux habités par les saincts pères des deserts* (Paris, 1693).

E—Pastoureau, *Atlas français*, 140–42, 250 (Jaillot I D [107]); LC List 5953. Nebenzahl, 136–37 (illustration). See also Fer, 1700.

658. Joly, 1784.   Joseph Romain Joly, *Atlas sur la géographie sacrée et sur l'histoire sainte* (Paris, 1784).

E—NOT SEEN. LC List 22. Jerusalem.

659. Lafréri, 15–. *Atlas factice* of Lafréri type.

E—HABW, Cb folio 9. sixteenth-century Italian. Map of Palestine with twelve tribes called "modern." No. 46, "Tabula Moderna Terrae Sanctae"; oriented east, divided into twelve tribes. Many very brief notes on Old Testament and New Testament events. Under Tribus Dan: "Hic fuit reducta archa dei de terra philistinorum." At Jordan, "Hic orabat Johannes baptista." No. 47, Exodus map with the Israelites marching in regiments (as Nebenzahl, 74–75, no. 25).

660. La Ruë, 1651.   Philippe de La Ruë, *La Terre sainte en six cartes géographiques* (Paris, 1651).

E—Pastoureau, *Atlas français*, 293; Laor, nos. 415–21. Newberry, UMich, Yale, Harvard, MA 3225.651F, RGS, 5.H.6. Name spelled Rué in Le Clerc, 1705/A. Explanatory text, 11 pp.

661. Lubin, 1659.   Augustin Lubin, *Orbis Augustinianus, sive conventum ordinis eremitarum S. Augustini chorographica et topographica descriptio* (Paris, 1659).

E—Pastoureau, "Atlas en France avant 1700," 64. Harvard-Houghton. Geography of the Augustinian Order. 54 maps. Lubin was a member of the order.

662. Lubin, 1660.   Augustin Lubin, *Martyrologium Romanum illustratum sive tabulae aecclesiasticae geographie tabulis et notis historicis explicatae quibus sanctorum sive mortis sive dispositionis tempus et locus exactissime exprimuntur* (Paris, 1660).

E—BN, Ge.FF.1141; H.3397. Places of martyrdom marked region by region.

663. Lubin, 1691.   Augustin Lubin, "Italia ecclesiastica in suas viginti distincta provincias" (Rome, 1691).

E—RomeSGI, Z.2.I.10$^{3-4}$. 2 sheets dedicated to the reigning pope.

664. Maximinus, 1649.   Maximinus a Guchen, "Chorographica descriptio provinciarum et conventuum fratrum minorum S. Francisci Capucinorum" (Turin, 1649).

E—LC, G1793.M3 1649 vault. Geography of the Capuchin order (a sixteenth-century branch of the Franciscans).

665. Neander, 1847.   August Neander, "Karte zur Geschichte des apostolischen Zeitalters" (Berlin, 1847).

E—Stevens, *Bibliographia geographica,* pt. 1, no. 1926. Toronto-Trinity College, 260N26: Heinrich Kiepert, "Karte des römischen Reiches für die ersten Jahrhundert der Kirchengeschichte zu Prof. Dr. Neander's Geschichte des apostolischen Zeitalters" (Hamburg, 1847), added to 4th ed. of Neander's *Geschichte der Pflanzung und Leitung* (1st ed., 1832–33). Basically, a map of the travels of the apostle Paul.

666. Nolin, 1715.   Jean-Baptiste Nolin, Jr., [Dioceses and Ecclesiastical Provinces of France] (Paris, 1715, 1720, 1728, etc.).

E—For 2d ed. of *Gallia Christiana in provincias ecclesiasticas distributa,* 13 vols. (Paris, 1715–85). *BL Maps Catal.* (long entry); BN, Ge.DD.2987 (192–94, 196–99, etc.). Nolin's maps are in the recent repr. of Gallia Christiana, 16 vols. (Farnborough, Hants., 1968). Cf. Sainte-Marthe, 1656/E.

667. Oort, 1884.   H. Oort, *Atlas voor Bijbelsche en Kerkelijke Geschiedenis* (Groningen, 1884).

E—Koeman, 6:214. BL, 40 plates.

668. Petri, 1858.   Girolamo Petri, *L'Orbe Cattolico ossia Atlante geografico storico ecclesiastico,* 3 vols. (Rome, 1858–59).

E—RomeSGI, Z.3.IV.1–3. Enormous and very heavy. Presented to Pope Pius IX in a political spirit: "queste carte, che presentano a colpo d'occhio le conqueste mirabili dell'Evangelo per le cure incessanti del Pontificato, etc." Statistics rather than history.

669. Renner, 1823.   Renner (publ.), "Mappa Calcographica conventuum provinciae s. Ioannis a Capistrano anno 1823."

E—In Gergely Csevapovics, *Recensio observantis minorum provinciae S. Joannis A Capistrano per Hungariam* (Buda, 1830), reproduced in Lajos Széntai, *Atlas Hungaricus: Magyarország nuomtatott térképei, 1528–1850* [Hungary in Printed Maps], 2 vols. (Budapest, [1996]), 2:506; small pictures of the monasteries are on the map, accompanied by "list of the monasteries and their dates of foundation."

670. Rogg, 1740.   Gottfried Rogg, "Deserta Aegypti, Thebaidis, Arabiae, Syriae, etc. ubi accurate notata sunt loca inhabitata per sanctos patres anachoretas" (Augsburg, ca. 1740–50).

E—BerlinPK. This map often occurs in Seutter, 1745a. Copied from Jaillot, 1693/E; cf. Fer, 1700/E.

671. Sanson, 1676.   Nicolas Sanson, *Cartes particulières de la France* (Paris, late 1650s, 1660, 1676).

E—Pastoureau, *Atlas français,* 417–19. BPL, at least 10 as loose maps; BN, Ge.DD.1181. Dainville, *Cartes anciennes des l'églises de France,* 38–39. Bishoprics are divided (by dots and wash) into their large subdivisions: archpresbyterates, "mandements," deaconries, whatever they might be. Within the district, ecclesiastical establishments are identified. The boundaries to be colored—ancient, ecclesiastical, modern, or profane modern—were at the buyer's choice; the corresponding line(s) of the cartouche would be highlighted.

672. Sainte-Marthe, 1656.   Scevole Sainte-Marthe and Louis Sainte-Marthe, *Gallia Christiana, qua series omnium archiepiscoporum . . . ad nostra tempora,* 4 vols., ed. Pierre Abel and Nicolas Sainte-Marthe (Paris, 1656).

E—UMich-rare. Map, vol. 4, frontispiece, "La France divisée en archeveschés eveschés abbaies Par les Srs de S.M. pour servir au livre de Gallia Christiana à Paris DCLVI." The map extends from Antwerp to St. Gall. Ecclesiastical provinces as main divisions. A small inset (upper left) shows the location of great Paris abbeys.

673. Spanheim, 1701.   Fridericus Spanheim, *Geographia sacra et ecclesiastica* (Leiden, 1701).

E—MilwaukeeAGS. Part of *Opera quatenus complectuntur Geographiam, chronologiam, et historiam sacram atque ecclesiasticum utriusque temporis,* 3 vols. (Leiden, 1701–3), 1:5–61. Maps mostly of dioceses but also of biblical history. Verhaer. *See* Haraeus.

674. Vialart, 1641.   Charles Vialart (de Saint-Paul), *Geographia sacra sive notitia antiqua episcopatuum Ecclesiae universae* (Paris, 1641).

E—Dainville, *Cartes anciennes des églises de France,* 114–15. BN. Repr., Amsterdam, 1704. The author, bishop of Avranches, was called Saint-Paul in his religious order (the Theatines).

675. Wiltsch, 1843.   Johann Elieser Theodor Wiltsch, *Atlas sacer sive ecclesiasticus* (Gotha, 1843).

E—Newberry; BL Maps 49.f.3. (title, *Kirchenhistorischer Atlas*); MilwaukeeAGS. Up to the Reformation. Basically an atlas of provinces and dioceses. Authoritative and widely circulated.

Other Noteworthy Works with and without Maps.

O   OTHER RELEVANT ATLASES AND MAPS

676. Arrowsmith, 1804.   Anon., *A New and Elegant Atlas comprising all new discoveries (for Pinkerton's Geography)* (Philadelphia, 1804).

O—YaleB. PhiladFree, 910.9 An86s (p. iv): "This latter division of the subject [medieval geography], however, has only been so far attended to as was rendered absolutely necessary for the clearer understanding of particular cases. To have entered more seriously into mediaeval details would have made the work too voluminous, without adding much to its utility." Maps by Arrowsmith and Lewis. No historical content. John Pinkerton, *Modern geography: a description of the empires, kingdoms, states and colonies, with the oceans, seas* . . . (Philadelphia, 1803, and later eds.).

677. AtlFact, 16−.   Casanatense Atlas of Europe.

O—Rome Casan, Rari 1131 (già BB II 43; già K.II.43), handwritten title "Orbis terrarum variis variorum . . . delineata . . .," i.e., atlas of Europe. Pike symbol for regiments, nos. 26 and 27 (1566−67), 43, 55.

678. AtlFact, 165-.   Augmented Magini atlas of Italy (sixteenth to seventeenth centuries).

O—RomeBNVEmm, 71.5.G.25. Mid−seventeenth century. 53 items.

679. Aubert, 1771.   Jean Louis Aubert, *Les traits de l'histoire universelle sacrée et profane d'après les plus grands peintres et les meilleurs écrivains etc., Histoire poétique,* new ed. (Paris, 1771).

O—LC, BL720.A8. (The book is anonymous). Picture history. Each recto page has a scenic engraving, figures, etc. (verso is blank), with brief characteristics of the scene in Latin and French. E.g., no. 21, "Apothéose d'Hercule," with passages of Ovid and their translation (more writing here than in most other scenes).

680. Blake, 1826.   John L. Blake, *A Geographical chronological and historical atlas on a New and Improved Plan* (New York, 1826).

O—LC List 747. LC, G1019.B63 1826; PhiladUL, 912.B58. 18 charts "done on the plan of Lesage," 1 world map.

681. Bodenehr, 1735.   Gabriel Bodenehr, engrav., *Curioses Staats und Kriegs-Theatrum durch unterschiedlich Geographische, Topographische und Historische Carten Abriss und Tabellen erläutert* (Augsburg, n.d. [1735?]).

O—RomeBNVEmm, 71.3.H.19[1]. Companion to the campaigns of the War of the Spanish Succession.

682. Boutruche, 1837.   A. Boutruche, *Atlas chronologique et synchronique d'histoire universelle* (Paris, 1837, 1844).

O—Łodyński, 4: no. 178. NYPL (1837); UMich. Time chart with 2 maps only (by P. Bineteau).

683. Bowen, 1766.   E. Bowen, and T. Kitchin, *An Universal History: Maps and Charts to the Modern Part* (London, 1766).

O—BL, Maps 48.e.13; LC, G1015.U5 1766 Cage. No medieval content.

684. Buache, 1767.   Philippe Buache, *Cartes et tables de la géographie physique ou naturelle presentées au Roi le 25 mai 1767* (Paris, 1767).

O—YaleB. Buache's theories on physical geography blazed a trail and were very influential; endorsed by Gatterer, Denaix, and other makers of maps for history.

685. Buache, 1768.   Philippe Buache, *Géographie physique politique et mathématique des états et royaumes de l'Europe: Nouvelle manière de consider la Terre par la disposition naturelle de ses parties* (Paris, 1762, 1768).

O—RGS. More in the same vein, BN, Ge.DD.5400, [Ph. Buache], *Cartes et tables de la géographie physique ou naturelle présentées au Roi le 15 mai 1757.*

686. Buchon, 1825.   J. A. Buchon, *Atlas géographique, statistique, historique et chronologique des deux Amériques* (Paris, 1825).

O—UCB-Bancroft, ffE14.B9x. Tr. of Carey, 1822/O. *NUC* (many copies in the United States). The tr. and Carey and Lea design was related to Lesage, 1801. Buchon acknowledges Lesage, not Carey.
Buret de Longchamps. *See* Vaudoncourt.

687. Carey, 1822.   Henry Charles Carey and Isaac Lea, *A Complete, Historical, Chronological, and Geographical American Atlas . . . to the year 1822* (Philadelphia, 1822).

O—*NUC;* LC List 3660a. NYPL; MilwaukeeAGS, 75-B. The first historical atlas of the United States; very widely represented in U.S. collections. Modeled on Lesage's "celebrated atlas" and Lavoisne, 1814, but more geographical than historical.

688. Colart, 1841.   L. S. Colart, *Histoire de France et d'Angleterre comparée,* 2d ed (Paris and London, 1841).

O—LC List 4010. LC, G1841.31Cb 1841 (not found); 2 sheets? BN: "tableaux," no maps.

689. Coronelli, 1686.   Vincenzo Coronelli, O.F.M., *Memorie istoriografiche dei Regni della Morea Negroponte e littorali fin a Salonichi,* 2d ed. (Venice, n.d. [1686]).

O—RomeBNVEmm, 7.1.G.45. Done "nel laboratorio" of Coronelli. Many foldout maps. Not obviously a war atlas. 37 maps, none of battles. Coronelli discusses the recent history of the shores of Greece.

690. Coronelli, 1687.   Vincenzo Coronelli, O.F.M., *Memorie istoriografiche della Morea riacquisitata dall'armi veneti* (Venice, 1687).

O—RomeBNVEmm, 10.5.N.12. Ordered by geography, not chronology. A victory book with battle and especially fortress maps. Text overshadows maps. See also BL 45125.(10.).

691. Coronelli, 1691.   Vincenzo Coronelli, O.F.M., *Atlante veneto,* 2 vols. in 3 (Venice, 1691, 1695–97).

O—LC. Nothing apparently historical in vol. 1. The others are an *Isolario.* For Coronelli's Heptarchy map, see Shirley, *Maps of Britain, 1650–1750,* 47 (Coronelli 2).

692. Dartois, 1750.   Jean Dartois, *Mappe-Monde historique, ou carte chronologique, géographique et généalogique des états et empires du monde,* rédigée par le Sr. Barbau de la Bruyère [the actual author] (Paris, 1750).

O—BL, K.4.46.2. Time chart meant to be striking. No maps, though provision is made for them.

693. Dodaly, 1765.   Anon., *Geography and History of England* (London, 1765).

O—Princeton. History in the text. The maps are only geographical.

694. Elwe, 1792.   Jan Barent Elwe, *Atlas* (Amsterdam, 1792).
O—Koeman, 2:104. BL, Maps 40.f.7. Tables but no historical maps, except for the Holy Land.

695. Foresti, 1711.   Antonio Foresti, S.J., *Mappamondo istorico cioè Ordinata narrazione dei Quattro Sommi imperii del Mondo*, 10 vols. (Parma, 1711).
O—RomeMar, 6967. Sala C, Sc. 61. No illustrations. For information, see Sommervogel.

696. Gaspari, 1811.   Adam Christian Gaspari, *Neuer methodischer Schul-Atlas*, Erster Kursus, 11th printing (Weimar, 1811; overwritten, 1816).
O—ViennaÖNB, 2 vols. in 1; StockholmRL. Outline maps of the whole world to complement A. C. Gaspari's complete *Handbuch* of contemporary geography (Weimar, 1804–6).

697. Gatterer, 1775.   J. C. Gatterer, *31 Landkarten von Gatterer zum Gebrauch seiner geographischen Vorlesungen* (Göttingen, 1775).
O—BerlinPK. No printed title page. 4 maps are of ancient Germany.

698. Gatterer, 178-.   Johann Christoph Gatterer, *Gatterers Kleine Special-Karten* (n.p., n.d.).
O—Göttingen, 4to Geog. 189: 2 Rara. 27 items. A small, neat atlas, opening with physical geography; often comparative.

699. Gatterer, 1780.   Johann Christoph Gatterer, *[Gatterers] Methodische Landkarten* (n.p., n.d. [Göttingen? 1780?]).
O—Harvard, MA 100.780s. Colored, no lettering. Several maps in 2 copies with different coloring.

700. Güssefeld, 1796.   Franz L. Güssefeld, *Charte über die Länder des herzoglich sachsen-ernestinischen Hauses nach astronomischen Beobachtungen, geometrischen Messungen und andern bewährten Hülfsmitteln neu entworfen* (Weimar, 1796).
O—NOT SEEN. BerlinPK, Mapp 7029; reference from Berlin-BSB-Göttingen map data base, Göttingen. See also Güssefeld, 1796.
Halluvin. *See* Wautier.

701. Hase, 1745.   Johann Matthias Hase, *Vorstellung der Grundrisse von denjenigen weltberühmten Städten* (Nuremberg, 1739).
O—LC List 5388. BSB. Repr. in Hase's *Atlas historicus*. City plans on a uniform scale.

702. Henselius, 1741.   Godefrid Henselius, "Europa polyglotta, linguarum genealogiam exhibens" [alternative title, scribendi modos gentium exhibens] (Nuremberg, 1741).
O—AmstUL. In *Atlas Homannianus Mathematico-historice delineatus* (1762), 4: no. 21. For clarification, see Harms, *Themen*, 76, with illustration.

703. Hocquart, 1826.   E. Hocquart, *[Tableaux historiques]* (Brussels, 1826).

O—NOT SEEN. ViennaÖNB, K III 112.952. 5 leaves without collective title, incl. (4) "Tableau de l'histoire de France depuis les Gaulois jusqu'à nos jours"; (5) "Tableau historique et chronologique du Royaume des Pays Bas." Letterpress. The same in NaplesBN, Bᵃ 8/50, under the name Hecquart (Paris, 1822).

704. Imbert, 1834.   Charles Imbert de Mottelettes, *Atlas synchronistique, géographique et généalogique pour servir à l'étude de l'histoire moderne de l'Europe depuis l'avènement de François Iᵉʳ, jusqu'à la Restauration, 1515–1815* (Paris, 1834).

H—*BN Impr.*, G. 172; BL, Maps 12.f.3 (to 1713 only). Possibly the first atlas limited to modern history. Map outlines by A. H. Dufour.

705. Johnston, 1850.   Alexander Keith Johnston, *Atlas to Alison's History of Europe* (Edinburgh and London, 1850).

O—NOT SEEN. Łodyński, 3: no. 575. 108 maps. BL, 1310.e.1,2. Alison's *History* starts from the French Revolution.

706. Jones, 1875.   ♦ Charles H. Jones, *Historical Atlas of the World* (Chicago, 1875).

O—LC List 4136. Northwestern; LC. No historical maps. Also listed with a joint author, T. F. Hamilton, and the title *General Atlas of the World Illustrated.*

707. Julien, 1765.   Roch-Joseph Julien, publ., *Grand Atlas* (Paris, 1765).

O—*Journal des Sçavans,* July 1765: announcement by Julien of subscriptions for a 10-vol. atlas of the best maps from everywhere. Twenty sets to be readied at 1,200 livres each. I have found no record that this project was carried out.

708. Keller, 1814.   Heinrich Keller, "Erdkarte nach der Bonne'schen Projection alle für die Erdkunde ergiebigen Entdeckungsreisen zu Wasser und zu Lande von der Mitten des 9ten Jahrhunderts bis jetzt" (Weimar, 1814).

O—BerlinPK. Single-sheet with many color-coded lines.

709. Kilian, 1758.   G. C. Kilian, *Kriegsatlas* (Augsburg, 1758).

O—UMinn-rare, 912.43 K553. Maps of European lands, followed by a longer set of battle maps. Not very attractive but popular. Fine coloring of the war maps.

710. Lafréri, 1980.   Anon., *Carte, Piante e stampe storiche delle raccolte Lafrerriane della Biblioteca nazionale di Firenze* (n.p., 1980).

O—RomeBNVEmm, geography reference mentioned here to illustrate early battle maps. For the same purpose: Istituto Veneto di scienze, lettere ed arti, *Carte geografiche cinquecentesche a stampa della Biblioteca Marciana e della Biblioteca del Museo Correr di Venezia* (Venice, 1954).

711. Lapie, 1805.   Pierre Lapie, *Histoire des guerres en Italie—Atlas: Atlas de l'histoire des guerres des Gaulois et des Français en Italie* (Paris, 1805).

O—UIll (imperfect). Maps by Tardieu. This title evoking early Roman history heads a celebration of Napoleon's Marengo campaign (1805).

712. Łelewel, 1849. Joachim Łelewel, *Géographie du moyen âge, Atlas* (Brussels, 1849).

O—Toronto-Robarts (2 copies). 35 pls. Pioneering major book on the history of geography and maps in the Middle Ages. The atlas is largely of Arabic and Jewish maps, which are emphasized in Łelewel's argument.

713. Martignoni, 1721.   Girolamo Andrea Martignoni, *Spegiazione della carta istor-ica dell'Italia e di un parte della Germania dalla Nascita di Gesu' Cristo fino all'anno MCCC . . . Con un triplicato metodo di rinvenire la Storia: cioè da Successo in Successo, da Secolo in Secolo, e da Signoria in Signoria* (Rome, 1721).

O—RomeMar, 7930, Sala C, Sott. 62. An effort to combine geography with a time chart. The chart anticipates Sass's "River of Time." Chart 3 has the map of Italy that is announced in the title. Also reminiscent of Hagelgans, 1718.

714. Mees, 1865.   G. Mees, *Historische Atlas van Noord-Nederland, van de XVIe eeuw tot op heden* (Rotterdam, 1865).

O—Koeman, 2:274. Koeman records the one Rotterdam edition, plus a some-what shorter edition (NOT SEEN), Leiden, 1881. UCB, fDH16M4. Especially valuable for its critical review of historical atlases (the category is widely interpreted).

715. Mejer, 1942.   Johann Mejer, [Map of the Danish Kingdom] (Copenhagen, 1942).

O—Columbia-rare, B912.489 M479. N. E. Nølund, ed., *Johannes Mejers Kort over det Danske Riget, udgivet med Støtte af Carlsbergfondet,* 3 vols. facsimile, Geo-dåetrek Instituter Publikationer, 1–3, 1 roll with 3 maps. An extraordinarily lavish publication, with an important historical introduction in Danish.

716. Mentelle, 1788.   Edme Mentelle, *Atlas de la monarchie prussienne* (London, 1788).

O—RomeCasan. Mentelle is responsible for 10 maps out of the 93 plates; no his-torical content. The atlas supplements Gabriel Honoré Mirabeau, *De la monarchie prussienne, sous Frédéric le Grand,* 7 vols. in 8 (London, 1788).

717. Novacco, 19–.   Anon., *Antiche carte geografiche topografiche e storiche della collezione Franco Novacco,* ed. Valeria and Piero Bella (Milan, n.d.).

O—RomeBNVEmm. Geography reference. No. 13, precedent for Speed's massed pikes, 1558; also no. 30 (1557).

718. Plater, 1827.   Stanislas Plater, *Atlas historique de la Pologne accompagné d'un tableau comparatif des expéditions militaires dans ce pays pendant le XVIIe, XVIIIe et XIXe siècle* (Poznań, 1827).

H—Łodyński, 3: no. 213. BL; BN, Ge.DD.3374. 11 maps of modern history. The underlying map seems printed from one plate.

719. Poole, 1898.   ◆R.L. Poole, *Historical Atlas of Modern Europe* (Oxford, 1898).

O—Very large, long, and authoritative. Widely circulated. Dean, "Sic enim est traditum," 12 (citing the 1902 ed.), calls it "the highest plateau of the von Spruner prototype."

720. Reilly, 1791.   F. I. I. Reilly, *Schulatlas* (Vienna, 1791–92).

O—YaleB. All maps modern.

721. Remondini, 1801.   G. Remondini, publ., *Atlas géographique . . . à l'usage des écoles* (Venice, 1801).

O—LC List 3377. LC, G1019.R595 1801 vault. 60 maps. No historical content.

722. Renard, 1739.   Louis Renard, *Atlas de la navigation et du commerce qui se fait dans toutes les parties du monde* (Amsterdam, 1739).

O—YaleB. No attempt to "show" commerce. Tracks of circumnavigators on a world map.

723. Ritter, 1813.   Karl Ritter, *Sechs Karten von Europa* (Schnepfenthal, 1813; new ed., 1820).

O—NOT SEEN. NYPL, YaleB. The preface is dated 1806. The atlas consists of thematic maps emphasizing physical aspects of Europe. An expanded title appears on the original wrappers of the new edition (YaleB): *über Producte, physicalische Geographie und Bewohner dieses Erdtheils, neue, in Hinsicht der politischen Grenzen berichtigte Ausgabe.* Considered epoch-making in geography.

724. Rizzi-Zannoni, 1761.   Giovanni Antonio Rizzi-Zannoni, *Atlas géographique et militaire, ou théâtre de la guerre présente en Allemagne* (Paris, 1761).

O—NOT SEEN. Reviewed in *Journal de Trévoux,* April 1761, 1123–24; announced with much praise in *Journal des sçavans,* May 1761, 211–12. 20 maps, for years 1756– 61. Engraving specially commended. Partly documents Rizzi-Zannoni's earliest years in Paris.

725. Rizzi-Zannoni, 1762.   Giovanni Antonio Rizzi-Zannoni, "Mapa dos Reynos de Portugal et Algarve" (Paris, n.d. [1762]).

O—UCB-Bancroft, G 6540 1762 Z3 Case XB. 1 sheet of at least 2 sheets. The map shows the march routes of opposing armies in the Iberian Peninsula during the Seven Years' War. The UCB-Bancroft copy is unfinished. *Journal des sçavans,* August 1762, 505–6, notes that this map will open a 12-sheet atlas of Spain.

726. Roost, 1830.   J. B. Roost, *Allgemeiner Hand- und Schulatlas* (Kempten, n.d. [1830]).

O—BSB. 30 maps. Preoccupied with altitude; almost every map has an inset relief cross section. One world map (Mercator projection) "als moralisch-politisch Erd Karte." Shading distinguishes 5 levels of "Civilizations-Zustand." Symbols for religion and mode of government. No other historical content.

727. Rousset, 1742.   Jean Rousset de Missy, ed., *Nouvel atlas géographique et historique* (Amsterdam, 1742).

O—*NUC;* LC List 143. LC, 26 maps, only modern. Also under title *Nieuwe astronomische geographische en historische atlas.* Quérard, *France littéraire,* 8:242 ter, "publiciste et compilateur infatigable." His works listed in Quérard and *BN Impr.* include no atlas.

728. Rühle, 1825.   Johann Jacob Otto August Rühle von Lilienstern, *Allgemeiner Schulatlas* (Berlin, 1825).

O—BL. Handsome, expensive product; much attention to heights (*hachures*). Not historical. *GV* lists it with 26 sheets (Berlin, 1826); also n.d., 29 sheets. *AGEphem.* 26 (1826): 56–59, notes flaws in lithography; reproduction leaves much to be desired.

729. Rühle, 1834.   Johann Jacob Otto August Rühle von Lilienstern, *Historiogramm des Preussischen Staats von 1280 bis 1830 n. Chr. in synchonistischem Verhältniss zu den Nachbarstaaten* (Berlin, n.d. [1834]).

O—BN, Ge.CC. 2199. 6 very large sheets individually entitled "Zeitstrom der Geschichte des preussischen Staats." A time line packed with details in complicated coloring.

730. Seutter, 174-.   Matthäus Seutter, *Mappa geographica quibus in locis per totam Germaniam bello tricennali Rex Sueciae Gustavus Adolphus [proelia commiserit]* (Augsburg, 174-).

O—YaleB, 1983 fol. 50 (an unindexed, unpaged atlas). The battles of Gustavus Adolphus of Sweden during the Thirty Years' War were a rare subject for cartography at the time. No march routes. Explanation in German, including minor clashes.

731. Shepherd, 1929.   ♦ William R. Shepherd, *Historical Atlas,* 7th ed. (New York, 1929).

O—Marked "Printed in Germany." 1st ed., 1911; latest (9th ed.), 1973. The core of Shepherd's atlas is stable. The copy on my shelves is the 7th ed.

732. Stieler, 1824.   Adolf Stieler, *Hand-Atlas über alle Theile der Erde* (Gotha, 1824).

O—*NUC;* LC List 3381 (1841), 25 maps. UIll; Harvard; no ed. before 1824 seems available in America. *Prospectus* (Gotha, 1816) at BrusselsBR (catal.), III/83.078/B/50.

733. Thomson, 1817.   John Thomson and Co., *New General Atlas* (Edinburgh, 1817).

O—Yale, 16. 1st map, the world, shows "tracks of the last circumnavigators" (dated 1814). Otherwise "pure" geography. Excellent workmanship.

734. Vaissete, 1755.   Joseph Vaissete, *Géographie historique, ecclésiastique et civile, ou Description de toutes les Parties du Globe Terrestre, enrichie de Cartes Géographiques,* 12 vols. (Paris, 1755).

O—Northwestern (Special Collections). 72 foldout maps from the Robert de Vaugondy *Atlas portatif* (Paris, 1748 – 49). "Historical" geography understood in the same way as Châtelain, 1705.

735. Vaudermaelen, 1825.   Philippe Vaudermaelen, *Atlas universel de géographie,* 6 vols. (Brussels, 1825–27).

O—Prominent in its time. An early application of lithography. Vandermaelen founded the Établissement géographique de Bruxelles in 1830; see *LGKartog.* 2:850.

736. Vaudoncourt, 1825.   Guillaume de Vaudoncourt, *Atlas universel pour l'étude de l'histoire et de la géographie ancienne et moderne* (Buret de Longchamps, *Fastes,* 2e partie) (Brussels, 1825).

O—BN. The three maps offer glimpses of the twentieth and fifth centuries B.C. and the sixteenth century A.D.

737. Walch, 1811.   Johann Walch, publ., *Allgemeiner Atlas* (Augsburg, 1811).

O—BN. No historical maps. A geographical gazette advertising astronomical place determination, new discoveries, secularizations of ecclesiastical lands, and compensations. An unavowed tribute to Napoleonic reshufflings of Europe.

738. Wautier, 1806.   Wautier, [unidentifiable], *Geochronology of Europe* (London, 1806).

O—BL, 1078.(15.). Outline map of Europe with summary boundaries. Lists of rulers written on each country so that dynastic tables lie "on" or "within" the land to which they apply. See also Faden, 1783.

739. Weiland, 1824.   Karl Ferdinand Weiland and Georg Hassel, *Geographisch-statistisch-historischer Atlas der Staaten des deutschen Bundes* (Weimar, 1824–28).

O—BSB. Excellent maps, flanked by writing in wide margins. Skimpy historical content. The project may not have been finished.

740. Willard, 1828.   Emma Willard, *A Series of Maps to Willard's History of the United States* (New York, 1828).

O—LC, G1201.S1W5 1828 vault. 12 maps. No. 1 appears to adapt the barbarian tracks of Lesage (or its Lavoisne derivative) to the movements of Indian tribes.

741. Worcester, 1827.   Joseph E. Worcester, *Historical Atlas* (Boston, 1827).

O—*NUC* (1863). Many historical charts, but no maps. Originally destined to accompany Worcester's *Elements of History, Ancient and Modern.* Worcester published many similar works.

B   BOOKS, USUALLY WITHOUT MAPS

742. Avity, 1644   Pierre d'Avity, *Nouveau théatre du monde contenant les estats, empires, royaumes et principautez* (Paris, 1644).

B—YaleB. Title in the privilege: "Les états et empires, ou nouveau théatre du monde." Dedication to Mazarin by Claude Malingre, official historiographer of France. Descriptive only; no maps. Ancient, as well as modern, states. See also Châtelain, 1705.

743. Benicken, 1822.   Friedrich Wilhelm Benicken, *Zeitrechnungs-Tafeln für den historischen Hand-Atlas* (Weimar, 1822).

B—NYPL; StockholmRL. 12 tables of dates to accompany Benicken, 1821.

744. Delisle, 1718.   Claude Delisle, *Tables généalogiques et historiques des patriarches, des rois, des empereurs . . . depuis la création du monde* (Paris, 1718).

B—BN. By the father of Guillaume and Joseph-Nicolas. The tables, uniform in format, are engraved rather than typeset. No pictures of any kind.

745. Dufrène, 17–.   Maximilian Dufrène, S.J., *Geographischer Anfang [Rudimenta geographica] oder Kurtze und leichte Weise, die Catholische Jugend in der Historie zu unterrichten, für die Schulen der Gesellschaft Jesu . . . verfasset* (Fünfte Werklein)(Augsburg, 1729).

B—Jerusalem. See Sommervogel, 3:263–70, and Desing, 1731. The 1st *Werklein* (1727) was called *Rudimenta historica* in its Latin edition.

746. Duruy, 1839.   Victor Duruy, *Géographie politique du moyen âge* (Paris, 1839).

B—BN, G.20602. No maps. Brief appendix on medieval commerce.

747. Fischer, 1836.   W. Fischer and F. W. Streit, *Historischer und geographischer Atlas von Europa*, 2 vols. in 3 (Berlin, 1836–37).

B—LC List 2789, "no maps." BL, 8/88, 10108.bb.9. (1836–37); 1851.a.5 (rev ed., 1841). No maps and strictly modern. But see Streit, 1843.

748. Fortia, 1809.   Agricole Joseph de Fortia d'Urban, *Plan d'un atlas historique portatif* (Paris, 1809).

B—BAV, RG Storia V, 2316. The plan (never carried out) calls for 6 vols., 12mo. Fortia's title (*Portatif*) seems to imply that historical atlases normally were not.

749. Galletti, 1807.   J. G. A. Galletti, *Allgemeine Weltkunde oder geographish-statistisch-historische Übersicht aller Länder* (Leipzig, 1807; Pest and Vienna, 1823).

B—Łodyński, 3: no. 48. 20 maps (1807). BL: a few maps, none historical. Brief historical sketch of each country, brisk and statistical, not always accurate (e.g., U.S. presidents are attributed six-year terms).

750. Gatterer, 1775.   Johann Christoph Gatterer, *Abriss der Geographie* (Göttingen, 1775–78).

B—BL, 304.g.10. Includes Gatterer's own description of his (now lost) historical atlas.

751. Gatterer, 1789.   Johann Christoph Gatterer, *Kurzer Begriff der Geographie*, 2 vols. (Göttingen, 1789).

B—BL. According to Gatterer, wholly different from the *Abriss* of 1775; a country-by-country survey.

752. Hase, 1739.   Johann Matthias Hase, *Phosphorus historiarum, vel prodromus theatri summorum imperiorum* (Leipzig, 1742).

B—AmstUL (catalogue date, 1740); my date is later than Dörflinger's. This is Hase's account of the background to Hase, 1743, his main historical atlas.

753. Horn, 1658.   Georg Horn, *Historia totius orbis antiquus* (Amsterdam, 1658).

B—NYPL. For Horn's main fame, see Janssonius, 1652/A.

754. Hughes, 1863.   William Hughes, *The Geography of British History. A Geographical Description of the British Islands at Successive Periods from the Earliest Times to the Present Day: with a Sketch of the Commencement of Colonisation on the Part of the English Nation* (London, 1863).

B—*NUC*. Cornell, arV2992; Harvard, new ed., 1866. A famous book in its time. A few maps as illustrations. Battlefields stressed in the medieval section.

755. Jarry, 1831.   Adrien Jarry de Mancy, *Atlas historique et chronologique des littératures anciennes et modernes, des sciences et des beaux-arts* (Paris, 1831).

B—UCB-Bancroft, ffCB71.J37 1835; UMich; PhiladUL-Lea; PhiladAmPhil. No geographic maps. Format derived from the *Atlas Lesage,* whose layout is adopted. Jarry assiduously applied the "Lesage method" to large-sheet presentations of many subjects: listed (with praise) by Quérard, *France littéraire,* 4:209–10, and Quérard, *Litt. franç. contemp.,* 4:390. See also Lesage, 1826b (Italian tr.).

756. Jarry, 183-.   Adrien Jarry de Mancy, *Travail, industrie et commerce . . . Atlas historique des Français des divers états selon la méthode de A. Lesage, Cte de Las Cases, et du professeur A. de Mancy* (Paris, n.d.).

B—NOT SEEN. BN, Z.282 (38), single sheet. No maps.

757. Josseran, 1830.   F. A. Josseran, "Carte chronologique de l'histoire universelle"; "Carte des origines, inventions et découvertes" (Amsterdam, 1830).

B—Stevens, *Bibliographia geographica,* pt. 1, no. 1562. BL, 999 (58). A chart, meant for university students.

758. Koch, 1771.   [Christophe Guillaume Koch], *Tableau des révolutions de l'Europe depuis le bouleversement de l'Empire d'Occident jusqu'à nos jours* (Lausanne and Strasbourg, 1771).

B—NYPL. Reviewed in *Journal des sçavans,* January 1772, 31–42. Koch kept this first version anonymous. Histories of Europe were newly blossoming. Scope: from 406 to the Peace of Belgrade, 1739.

759. Koch, 1790.   Christophe Guillaume Koch, *Tableau des révolutions de l'Europe dans le moyen âge* (Strasbourg and Paris, 1790).

B—BL. No maps. Fuller version of the foregoing but covering only the Middle Ages. Later times omitted as politically dangerous in the Revolution.

760. Köhler, 1762.   Johann David Köhler, *Anweisung für reisende Gelehrte* (Frankfurt-am-Main, 1762).

B—NYPL; Newberry. Posthumous. Köhler gave a famous course on this subject at Göttingen; the lecture notes were probably published by his son. Foreign places are not mentioned. Typically, the traveler is equipped with a rundown of schools of painting. *NUC* records a version for young travelers.

761. Lavoisne, 1807.   C. V. Lavoisne and C. Gros, *A New Genealogical, Historical, and Chronological Atlas . . . complete in thirty-six maps* (London, 1807).

B—LC List 3315. BL, 1852.c.17. No geographical maps. See Lavoisne, 1814. Lavoisne took over the Lesage atlas wholesale and meant to produce an improved edition. He died before achieving this goal.

762. Lawrence, 1685.   J. Lawrence, *Orbis imperantis tabellas geographico historico genealogico chronologicae* (London, 1685).

B—Newberry. Few maps; many tables of emperors and kings. A special feature is a table of symbols, or codes, identifying salient character traits of the rulers in the tables (e.g., a circle with a cross pointing down signifies "lustful" [*voluptuosum*]). The same symbolic alphabet occurs in Hagelgans, 1718, and widely elsewhere.

763. Lenglet, 1740.   Nicolas Lenglet du Fresnoy, *Méthode pour étudier l'histoire avec un catalogue des principaux historiens,* 4 vols. (Paris, 1740?).

B—Many editions. 4to ed. (Toronto-Fisher) reproduces some maps, such as Augustin Calmet, "The Dispersion of Peoples (Genesis 10)."

764. Lenglet, 1768.   Nicolas Lenglet du Fresnoy, *Méthode pour étudier l'histoire,* ed. Étienne-François Drouet, 15 vols. 12mo (Paris, 1768).

   B—BN. Lenglet's scope includes maps.

765. Lenglet, 1965.   Nicolas Lenglet du Fresnoy, *Catalogue des meilleures cartes géographiques générales et particulières* (Amsterdam, 1965).

   B—Toronto-Robarts, Z6028 L4 1965. Repr. from Lenglet, *Méthode pour étudier la géographie,* vol. 1, pt. 2, 3d ed. (Paris, 1742).

766. Lesage, 1799.   A. Lesage, *A Short Guide to Mr. Le Sage's Historical Maps, which will enable every one to learn history by himself, as well as to teach it to any one in a very short time* (London, n.d. [1799]).

   B—BL, 568.e.16 (7.). 12 pp., 2 plates (no maps). The date is inferred from internal evidence.

767. Lesage, 1800.   A. Lesage, *The Geography of History . . . from the Christian Era to the 11th Century* (London, 1800?).

   B—BL, 1078.(17.). Soon incorporated into the atlas.

768. Marie, 1767.   Guillaume Marie, *Abrégé de l'histoire universelle sacrée et profane* (London, 1797).

   B—BL, 9005.aaa.29. Also *Tableau chronologique de l'histoire universelle, à l'usage de la jeunesse* (London, 1798), single sheet (BL, Tab 539.c.). Cited as sources by the *Atlas Lesage.*

769. Marmocchi, 1845.   Francesco Constantino Marmocchi, *Corso di geografia-storica antica, del medio-evo e moderna, con atlante,* 3 vols. (Florence, 1845).

   B—HABW, GK24/0025 (geography reference collection).

770. Mentelle, 1801.   Edme Mentelle, *Précis de l'histoire universelle pendant les premiers siècles de l'ère vulgaire, ou Introduction à l'histoire moderne des différents États d'Europe* (Paris, an IX [1801]).

   B—BN.

771. Merian, 1662.   Matthaeus Merian, ed., *Theatrum Europae oder . . . denkwurdigen Geschichten (1617–17–),* 21 pts. (Frankfurt-am-Main, 1662–17–).

   B—ViennaÖNB. Decorated with portraits (especially), city panoramas, battle scenes, city fortifications (sieges), some maps.

772. Mézeray, 17–.   François Eudes de Mézeray, "La géographie historique ou sont marquéz, selon la chronologie les [*sic*] plus exacte, les divers changements arrivez dans tous les états du monde jusqu'à présent," MS.

   B—*NUC.* UChi, G114.M6, Manuscript Room. Undated, 207 pp. No maps. The meaning of *géographie historique* in the title is unclear.

773. Muller, 1863.   Frederick Muller, *De Nederlandsche Geschiedenis in Platen, Beredeneerde beschrijving van Nederlandsche Historieplaten, Zinneprenten en Historische Kaarten,* 4 vols. (Amsterdam, 1863–81).

B—Harvard. Guide to early Dutch "historical atlases," i.e., individual collections of prints and maps concerning Dutch history.

774. Murillo, 1752.   Pedro Murillo Velarde, S.J., *Geographica historica, donde se describen los reynos, provincias, ciudades . . . con la mayor individualidad, y exactitud, y se refieren las guerras . . . las paces . . . los frutos, las requezas . . .*, 10 vols. (Madrid, 1752).

B—NYPL. Illustrates the loose sense of "historical geography." No maps. Similar to Avity, 1644/B.

775. Reischl, 1758   Marcellinus Reischl, *Atlas historicus utramque ab orbe condito historiam ecclesiasticam et profanam et omnes scientias ad historiam necessarias complectens* (Augsburg, 1758)

B—YaleB, call no. 1971 13. Reischl was a Benedictine at the monastery of Ettal in Bavaria. The book, a small octavo with no maps, contains a rundown of disciplines auxiliary to history plus a chronicle.

776. Sax, 1768.   Peter Sax, "Beschreibung der Insul Helgoland" (Copenhagen and Leipzig, 1768).

B—BL North, 269.c.15,16,17, *Altes und neues von gelehrten Sachen aus Dännemark*, vol. 2, *14, 505–64; introd., 506–10. (Anon.), "Nachricht von die Insel Helgelandt im Jahre 1699," 538–64. Related to Mejer, 1652.

777. Schaevius, 1659.   Heinricus Schaevius, *Sceleton geographicum in usus poëticos et historicos,* 2d ed. (Hanover, 1659).

B—BAV, RG Geog II, 11. No maps. Ancient geography set out diagrammatically.

778. Spruner, 1831.   Karl Spruner, *Bayerns Gauen nach den drey Volksstaemmen der Alemannen, Franken und Bajoaren aus den Urkunden nachgewiesen, Gegen Herrn Ritter v. Lang's Bayern's Gauen etc. aus den alten Bisthums Sprengeln nachgewiesen* (Bamberg, 1831).

B—UCB, DD801 B35L38. Spruner's major contribution to *gaue* research; 3, "I believe that the division into *gaue* has existed since the earliest times, and did not first become general in the 10th century" (nur glaube ich daß die Eintheilung in Gauen eine seit den Urzeiten her bestehende, und nicht erst im 10ten Jahrh. allgemeiner gewordene Eintheilung sey.).

779. Tyson, 1845.   J. Washington Tyson, *Atlas of Ancient and Modern History, presenting in a chronological series the rise, progress, revolutions, decline and fall of the principal states and empires of the world* (Philadelphia, 1845).

B—Yale, Harvard, NYPL. No maps, 5 double tables.

348bis. *Historische Karten,* 1816.   *Historische Karten zur geographischer Kenntniss der vornehmsten älteren und neueren Reiche;* um solche sowohl unter einander als mit den verschiedenem gegenwärtigen Staaten Europes, Asiens, Nord-Afrika's u.a.w. zu vergleichen (Leipzig, Schreibers Erben, 1816).

H—Budapest, TA 1579 (no other copy encountered). The title is descriptive: "historical maps for geographical knowledge of the greatest ancient and modern empires, so as to compare them with each other and with the various states of to-day's Europe, Asia, North Africa, etc." Four maps only, small, attractive, precise, and nicely colored. The projection is polar, but the map includes only Eurasia.

INDEX OF MAPS AND ATLASES

This index tracks the references in the notes to the maps and atlases listed in the Catalogue of Maps and Atlases. See page xix for an explanation. The references indexed here, normally limited to "Name, Year," lead to the full entries in the Catalogue.

Aa, 1707, 293n.101
Aa, 1729a, 122n.85, 124n.106, 293n.99
Aa, 1729b, 293nn. 100, 103
Akhmatov, 1845, 438n.99
Aloude Holland, 1745, 128n.136, 296n.142
Alting, 1701, 128n.136, 129n.152, 180n.60,
  298n.167
Andrews, 1797, 115n.24, 301nn. 204, 206,
  302n.208
Andriveau, 1837, 1847, 430n.7
Andriveau, 1840, 433n.44, 440n.129
Ansart, 1840, 1847, 366n.48
Anville, 1719, 176n.17, 294n.110, 301n.197
Anville, 1732/E, 48n.92
Anville, 1757, 291n.88, 294n.111
Anville, 1758, 291n.88, 294n.111
Anville, 1762, 291n.82, 294n.111
Anville, 1771, 116n.35, 178n.35, 294nn. 116, 120,
  121, 367n.55
Anville, 1795, 43n.28
Aretin, 1809, 370n.84, 371n.100
Arrowsmith, 1804/O, 430n.8
Arrowsmith, 1828/A, 43n.28
Artero, 1879, 376n.140, 438n.94

Atlas complet, 1747, 180nn. 61, 62, 294n.179,
  440n.131
Atlas MS LC, 1832, 364n.26, 436n.73
Atlas of England, 1720, 123n.23
AtlFact, 16–/O, 123n.94

Babinet, 1862, 377n.162, 437n.89
Bachiene, 1785, 286n.29
Banduri, 1711, 118n.46, 291n.83
Baquol, 1859, 438n.102
Barberet, 1870, 377n.163
Barbié, 1788, 367n.61
Barringon, 1773, 285n.24
Basel, 1748, 296n.144
Baumgärtner, 1815–17, 369nn. 75, 76, 82,
  372n.103
Baur, 1868, 437n.91
Bazeley, 1815, 430n.7
Beaurain, 1749, 46n.69, 180nn. 58, 59, 294n.111
Beck, 1857, 377n.162, 438n.106
Bedeus von Scharberg, 1845, 377n.162, 438n.94
Bellin, 1759, 284n.14, 293n.98
Benicken, 1820, 369n.82, 370n.92, 433n.51,
  461n.12

Benicken, 1821, 371nn. 95, 98, 433n.51, 461n.12

Benicken, 1824, 369nn. 82, 84, 372n.107

Benicken, 1829/A, 370n.91

Bensen, 1849, 376n.138

Bergeron, 1634, 293n.100

Bertius, 1618/A, 42n.19, 46n.74, 120n.69

Bertius, 1623, 114n.19, 116n.35, 120nn. 67, 69, 70

Bertius, 1628/A, 42n.19, 120n.69

Bessel, 1732, 119n.55, 177n.24, 285nn. 20, 21, 22, 288n.53, 289nn. 55, 56

Birchall, 1859, 377n.162, 440n.135

Black, 1854, 430n.7

Blaeu, 1635, 120nn. 67, 72

Blair, 1768, 296n.141

Blair, 1853/A, 43n.28

Boisseau, 1641, 116n.33, 124nn. 102, 103

Bonne, 1762, 286n.32

Bonnechose, 1847, 376n.138, 437nn. 83, 86

Bossi, 1824, 430n.7

Boucher, 1804, 433n.44

Bouillet, 1865, 377n.162, 437n.138

Bowen, 1766/O, 286n.29

Bradford, 1835, 430n.7

Brand, 1833, 364n.24

Bray, 1711, 285n.23

Bretschneider, 1856, 433n.40

Briet, 1647/A, 42n.25, 47n.78, 115n.29, 119n.59

Brion, 1766/E, 1766a, 48n.91

Browne, 1700, 46n.60, 114n.18

Brué, 1820, 183n.81, 186n.110, 285n.22, 372n.105, 437n.81

Brué, 1822, 44n.38, 372n.104

Buache, 1767/O, 1768/O, 176n.14, 186n.107

Buchon, 1825, 364n.23

Buy, 1761, 364n.28, 365n.30

Carey, 1822, 364n.23

Carl, 1732/E, 49n.95

Cassini, 1792, 46n.70

Cellarius, 1701a/A, 293n.105

Cellarius, 1701b/A, 177nn. 27, 28, 178n.30, 293n.105

Cellarius, 1776, 114n.19, 177n.30, 180n.51, 292nn. 104, 105, 106

Chambers, 1853, 430n.8

Châtelain, 1705, 44n.39, 174nn. 1, 2, 175n.4, 177n.19, 180n.116, 283nn. 6–9

Chiquet, 1719, 461n.10

ClassAtl, 1675/A, 47n.77

ClassAtl, 168-/A, 46n.72

Clausolles, 1845, 1846, 276nn. 138, 146, 437nn. 83, 85, 439n.114

Clüver, 1611, 128n.133

Clüver, 1616/A, 123n.91, 126n.123, 128n.133, 177n.20

Coquart, 1705, 287n.44, 289n.46

Coronelli, 1689, 1691, 113n.14

Courtalon, 1774, 365n.30

Cyclopedia, 1820, 430n.7

Danckwerth, 1652, 126n.122

Dangeau, 1697, 45n.61, 185n.99, 295n.128

Daniel, 1696, 294n.123, 296nn. 130, 132, 432n.32

Daniel, 1721, 284n.18

Daniel, 1729, 295n.132

Delamarche, 1809/A, 1824, 430n.7

Delamarche, 1827, 372n.108, 430n.9, 431nn. 11, 12

Delamarche, 1844, 376n.148, 430n.3

Delamarche, 1868, 377n.162

Delisle, 1705, 177n.32, 182n.71, 282n.1, 377nn. 156, 161, 432n.31

Delisle, 1707, 176n.16, 238n.48, 288n.162, 301n.197

Delisle, 1710a, 176n.16, 178n.32, 292n.93, 297n.156

Delisle, 1710b, 176n.16, 288n.48, 301n.197

Delisle, 1711, 176n.16, 291n.85

Delisle, 1717, 119n.56, 177n.19, 178n.36, 294n.80, 295nn. 126, 128, 130, 297nn. 156–57

Delisle, 1718/B, 179n.39, 299n.154

Delisle, 1721, 295n.127

Delisle, 1726, 177n.18, 291n.87, 297n.156, 433n.50

Delisle, 1730, 126n.16

Delisle, 1733, 180n.59

Delisle, 1754, 176n.14

Delisle, 1764, 292nn. 89, 91, 297n.156, 433n.50

Delisle, 1764/E, 48n.84, 292n.90

Delisle, 1766, 292n.90

Denaix, 1829, 373n.120, 437n.80, 461n.3

Denaix, 1835, 373n.120, 431n.19, 432n.38, 436n.80

Denaix, 1836, 1838, 375n.135, 437nn. 83, 85, 86

Denaix, 1838, 437n.86

Denaix, 1856, 375n.135

Desing, 1731, 286nn. 35, 36

Desjardins, 1836, 1838, 375n.131, 437n.80, 461n.3

Desnos, 1765b, 182n.79, 184n.89

Desnos, 1765c, 46n.62, 183nn. 84, 88, 186n.108, 302n.206, 373n.118

Desnos, 1766, 184n.92

Dheulland, 1756, 288n.47

Dittmar, 1865, 437n.89

Dower, 1831, 430n.7

Dozy, 1870, 377n.163, 438n.103, 439n.119

Drioux, 431n.12

Drioux, 1851, 439n.115

Drioux, 1852, 438n.106

Drioux, 1867, 431n.9

Droysen, 1886, 377n.159, 439n.123

Dubos, 1742, 128n.132, 295n.136

Dufau, 1841, 373n.119, 376n.138, 437nn. 83, 86, 87

Dufour, 1830, 430n.5

Dufour, 1830, 1834a, 1834b, 1835, 374n.129, 431n.9

Dufour, 1835, 42n.22, 431nn. 9, 11

Dufour, 1856, 432n.28

Dufour, 1859, 440n.140

Dufour, 1860, 431n.9

Dufour, 1861, 439n.103

Dupré, 1763, 131nn. 67, 75, 283n.5, 299nn. 178, 181

Duruy, 1841, 376n.143, 438n.106

Duruy, 1842, 433n.40

Duruy, 1846, 440n.135

Duruy, 1849, 376n.138, 438nn. 83, 86, 87

Dussieux, 1843, 376n.138, 437nn. 83, 86

Dussieux, 1845, 439n.118, 461n.3

Dussieux, 1854, 379n.162, 437n.83

Dussieux, 1856, 375n.136, 377n.163, 439nn. 1 17, 118

Duval, 1665, 47n.78, 125nn. 112, 113, 129n.147, 291n.87, 293nn. 101, 102, 461n.13

Duval, 1667/A, 47n.78, 124n.109

Duval, 1672/E, 48n.91

Duval, 1679, 119n.61

Elst, 1831, 375n.133, 437n.92

Engelmann, 1836, 375n.131, 437n.38

Enouy, 1797, 287n.39

Enouy, 1798, 287n.41, 434n.55

Ewich, 1652, 123n.90

Faber, 1711 [1740], 282n.1, 291n.82

Fabius, 1641, 122nn. 87, 89

Faden, 1783, 291n.84, 301n.203

Falcke, 1752, 289n.66

Fer, 1700/E, 42n.24, 125n.116, 126n.120

Fer, 1705, 287n.45, 288nn. 46, 47

Findlay, 1853/A, 43n.28, 430n.6

Fix, 1855, 437n.91

Forster, 1788, 285n.25

Fortia, 1809/B, 430n.3

Freyhold, 1846, 437n.164, 440n.128

Freyhold, 1853, 437n.91

Frommann, 1854, 377n.162, 437n.89, 440n.137

Galleti, 1807/O, 430n.8

Gallia, 1720, 294n.112

Gaspari, 1811/O, 185n.99, 430n.8

Gatterer, 1775, 186n.106

Gatterer, 1775/B, 185n.102

Gatterer, 1775/E, 186n.107

Gatterer, 1775/O, 1780/O, 185n.100, 186n.107

Gatterer, 1789, 186n.106, 300n.190

Gatterer, 1789/B, 300n.189

Gatterer, 1789/O, 185n.102

Gatti, 1851, 433n.48

Gauttier, 1830, 440n.140

George, 1852, 431nn. 9, 11

Georgisch, 1732, 291n.80, 296n.140

Goujon, 1828, 372n.108, 430n.9

Gover, 1853, 439nn. 116, 124

Gover, 1854a, 439n.116

Guadet, 1833, 375n.134, 437n.83

Guadet, 1851, 377n.162, 437nn. 83, 86, 87

Guppenburg, 1657/E, 49n.95

Güssefeld, 1796, 290n.78, 372n.102

Hagelgans, 1718, 175n.5, 178nn. 29, 34, 182n.71, 283nn. 9, 12, 433n.45

Handtke, 1808, 293n.99

Hartzheim, 1758, 1762, 49n.94

Hase, 1739/B, 180n.51, 288n.5

Hase, 1743, 180n.53, 182n.71, 298nn. 165, 167, 169

Hase, 1750a, 180n.49

Hase, 1750b, 180n.56, 299n.171

Häufler, 1840, 438n.93

Heck, 1830, 372n.111, 433n.47, 434n.56, 436n.70

Hell, 1771, 301n.202

Hérisson, 1806a, 430n.7

Hermann, 1858, 377n.162

Historische Karten, 1816, 371n.100

Hist. universelle, 1779, 11n.15

Homann, 1753, 289n.62

Hondius, 1607, 10n.11, 47n.77, 48n.83, 186n.108

Hondius, 1607/A, 47n.77

Houzé, 184-, 185-b, 376n.140

Houzé, 1840, 376n.141, 438n.94

Houzé, 1841, 376nn. 141, 143, 438nn. 104, 106, 108

Houzé, 1845a–b, 376n.140, 438n.106

Houzé, 1850a, 376n.140

Huber, 1830, 438n.94

Huberts, 1870, 377n.163

Hughes, 1849, 379n.162, 438nn. 93, 95

Hughes, 1863/B, 379n.162, 438n.95

Hughes, 1869, 114n.22, 115n.29, 379nn. 158, 163

Hugier, 1756, 296n.141

Huot, 1837, 432n.28

Imbert, 1834, 436n.78

Jaillot, 1693/E, 49n.96, 125n.116, 126n.120

Jaillot, 1695, 45n.59

Janssonius, 1638, 120nn. 71, 72. *See also* Bertius, 1623

Janssonius, 1652/A, 46n.69, 120nn. 69, 72, 298n.161

Jarry, 183-/O, 364n.24

Jarry, 1831/O, 364n.24

Jarry, 1832, 438n.97

Johnston, 1846, 44n.38, 430n.8

Johnston, 1853/A, 436n.74

Jones, 1830, 43n.28

Jusseret, 1836, 375n.133, 437n.91

Kärcher, 1834, 372n.108, 373nn. 115, 116, 430nn. 9, 11, 432n.38

Keppel, 1870, 377n.162

Kiepert, 1859/A, 10n.6

Kiepert, 1863, 440n.140

Kiepert, 1865, 437n.91, 440n.135

Kircher, 1675, 130n.156

Kitchin, 1756, 297n.149

Klaproth, 1826, 234n.15, 369n.77, 372n.107, 438n.100, 440n.131

Kleiner Atlas, 1816, 432n.37

Koch, 1771/B, 290n.67

Koch, 1790/B, 290n.68, 431n.21

Koch, 1807, 368nn. 69, 71, 431nn. 18, 23, 432nn. 24, 25, 461n.12

Koch, 1814, 368n.71, 432n.24, 438n.105

Koch, 1828/B, 368n.69

Koch, 1831, 368n.69, 482n.24

Koeppen, 1854, 10n.6

Köhler, 1718, 179n.42

Köhler, 1720, 282n.1

Köhler, 1720/A, 179n.41

Köhler, 1730, 114n.19, 122n.88, 178n.35, 179nn. 40, 43, 285n.20, 288n.52, 291nn. 82, 84, 294nn. 113, 115, 431n.23

König, 1850, 377n.162, 438n.106, 440n.136

König, 1857, 377n.162

Kort Atlas, 1853, 377n.162, 438n.98

Kremer, 1778, 290n.77

Kruse, 1802, 366n.47, 438n.18

Kruse, 1802 (3d ed.), 366n.49

Kruse, 1807, 366n.48

Kruse, 1836, 366n.48, 375n.132

Kutscheit, 1842, 439n.111

Kutscheit, 1844a, 376n.144, 439n.111

Kutscheit, 1844b, 376n.144

Kutscheit, 1856, 377n.162, 437nn. 89, 90

Lafréri, 1980/B, 123n.94

Lambarde, 1568, 112nn. 1, 3

Lamey, 1766, 11n.15, 240n.75

Lapie, 1812, 368n.73, 374n.129, 431nn. 11, 19, 432nn. 26, 27, 29, 30, 36, 433n.43, 434n.55

Lapie, 1826, 430n.9

Lapie, 1829, 374n.129

La Rüe, 1651/E, 124n.110, 129n.147, 178n.33, 433n.50

La Rüe, 1653, 124n.110, 180n.54

Lauenstein, 1745, 289n.65

Lavoisne, 1807/B, 364n.23

Lavoisne, 1814, 364n.23, 366nn. 46, 48

Lazius, 1561, 126n.104

Le Clerc, 1705/A, 125n.116

Leeder, 1866, 437n.91, 440n.135

Legrand, 1824, 372n.106, 437n.83

Łelewel, 1844, 438nn. 93, 97

Łelewel, 1849/O, 438n.97

Lenglet, 1740, 48n.88
Lenglet, 1752/O, 283n.3
Lenglet, 1965/B, 121n.73
Lenglet, 1768/B, 48n.88, 49n.47, 115n.25, 120n.66, 182n.77, 183n.79, 364n.28
Lesage, 1799/B, 361n.4, 362n.10
Lesage, 1800/B, 362n.10
Lesage, 1800/O, 361n.4
Lesage, 1801, 123n.95, 182n.32, 362nn. 8, 11, 365n.35, 433nn. 42, 45, 47, 436n.74
Lesage, 1802, 362n.8, 433n.44
Lesage, 1803, 362nn. 11, 13, 373n.120
Lesage, 1806a, 362n.13, 433n.47
Lesage, 1806b, 433n.45
Lesage, 1809, 363n.22
Lesage, 1813, 365n.35, 433n.42
Lesage, 1813a, 363n.22, 433nn. 45, 47
Lesage, 1813b, 363n.22
Lesage, 1826b, 364n.26, 377n.162
Lesage, 1826c, 363nn. 17, 22, 365n.30
Lesage, 1826d, 363n.22
Lesage, 1827, 363n.17, 374n.120
Lesage, 1827a, 363n.21
Lesage, 1827b, 364n.25
Lesage, 1829, 363n.21, 364n.29
Lesage, 183-, 363n.21, 433n.41, 434n.56
Lesage, 1835, 363n.21, 373n.120, 433nn. 41, 43, 56
Lesage, 1843, 364n.26
Lesage, 1844, 363n.22
Levasseur, 1840, 376n.145, 433n.39
Levasseur, 1844, 10n.8, 376n.145, 437n.106
Lévi-Alvarès, 1840, 123n.95
L'Hermite, 1822, 372n.106
Liébaux, 1728, 177n.19, 179n.44, 295nn. 132, 133, 134
Lizars, 1831, 42n.36
Lothian, 1829, 437n.92
Löwenberg, 1839, 276n.144, 377n.166, 438n.106, 439n.109
Löwenberg, 1840, 437nn. 89, 91
Lubin, 1659/E, 49n.95, 125n.111
Lubin, 1660/E, 49n.95, 125n.111
Luneau, 1760, 181nn. 65, 66

Maaskamp, 1833, 375n.133
Maggi, 1803, 438n.96
Maggi, 1854, 377n.162
Malbrancq, 1639, 122n.82

Malte-Brun. *See* Lapie, 1812
Mandrot, 1855, 377n.162, 438n.94
Manesson-Mallet, 1683/A, 114n.19, 119nn. 63, 65
Maretheux, 1847, 376n.138, 437nn. 83, 86
Marmocchi, 1840, 438n.101
Maximinus, 1649/E, 49n.95
Mejer, 1652, 126nn. 122, 123, 127n.124
Mejer, 1942/O, 126n.122
Ménard, 1713, 461n.10
Mentelle, 1778/A, 43n.27
Mentelle, 1782, 294n.122
Mentelle, 1797/A, 365n.33, 430n.6
Mentelle, 1798, 365n.30, 366n.39
Mentelle, 1801/A, 43n.28
Mentelle, 1804a–c/A, 430n.6
Mentzer, 1871, 438n.98
Metellus, 1594, 115n.26
Meyer, 1830, 430n.7
Migeon, 1866, 44n.38, 377n.162, 431nn. 9, 12
Mitchell, 1845, 46n.68
Moll, 1709, 282n.1
Möller, 1824, 372nn. 107, 112
Möller, 1825, 372n.107, 430n.9, 431nn. 11, 12
Monin, 183-, 374n.129
Monin, 1842, 431nn. 9, 11
Muhlert, 1865, 372n.112, 433n.40
Muller, 1863/B, 128n.136, 175n.9, 176n.11, 296n.148
Mullié, 1839, 375n.136, 437n.83
Muratori, 1727, 178n.44, 288nn. 50, 51
Murphy, 1858/A, 42n.22, 430n.7

Netherlands, 17–, 284n.16
Netherlands, 1792, 297n.153
Nolin, 1715/E, 48n. 92, 49n.93, 289n.63
Nolin, 1722, 292n.95
Nolin, 1733, 289n.63
Nolin, 1746, 285n.23
Novacco, 19–/O, 123n.94, 124n.107

Ohmann, 1853, 430–31nn. 9, 11, 12
Ortelius, 1570, 39n.1, 40n.11
Ortelius, 1579, 39n.1, 44n.45, 46n.74, 47n. 77
Ortelius, 1590, 40n.11

Padua, 1699, 46n.71, 47n.79, 180n.58, 434n.54
Patteson, 1825/A, 43n.28, 430n.6
Pawlowski, 1855, 437n.91

Pellegrino, 1644, 124nn. 106, 107, 108
Perrot, 1823, 431n.9
Petri, 1858/E, 49n.97, 440n.134
Philip, 1855/A, 431nn. 9, 11
Philippe, 1767, 1768, 181nn. 70, 71
Philippe, 1768, 1787, 181n.70
Philippe, 1787, 43n.27, 47n.77, 366n.39
Piré, 1868, 431nn. 9, 11, 12
Plancher, 1739, 113n.13
Plater, 1827, 438n.97
Playfair, 1814, 430n.7
Pompper, 1846, 376n.138, 437nn. 88, 90
Poole, 1898/O, 377n.159, 439nn. 116, 123
Pütz, 1859, 438n.102
Putzger, 1877, 9n.6, 124n.104, 128n.132,
    292n.94, 372n.102, 377n.160

Quin, 1830, 374nn. 121, 122, 123, 124, 432n.38,
    436nn. 75, 77

Rafn, 1837, 440n.143
Reichard, 1804/O, 430n.8
Reichard, 1824/A, 432n.37
Reilly, 1806, 430n.9, 431n.11, 432n.33
Reischl, 1758/B, 181n.64
Renner, 1823/E, 49n.95
Rhode, 1861, 377n.163, 438n.103, 439n.120
Ritter, 1813/O, 436n.79
Rizzi-Zannoni, 1761, 1762, 300n.185
Rizzi-Zannoni, 1764, 114n.24, 184n.96,
    295n.131, 299n.175, 366n.44
Robert, 1740, 1742, 1743, 296n. 138, 301n.197
Robert, 1742, 284n.116
Robert, 1745, 116n.33, 296n.143
Robert, 1753, 114n.17
Robert, 1757, 297n.151
Rodowicź, 1843, 301n.137, 376n.138, 438n.106,
    487n.88
Rogg, 1740/E, 49n.96, 125n.116
Rossi, 1820, 430n.7
Rothenburg, 1830, 373n.118, 460n.1
Rühle, 1827, 372n.108
Rupp, 1837, 1839, 370n.88, 372n.106, 375n.137
RVaugondy, 1752, 296n.141
RVaugondy, 1756, 296nn. 141, 143, 431n.10
RVaugondy, 1757, 114n.17

Sainte-Marthe, 1656/E, 48n.92
Sanis, 1859, 377n.162, 437nn. 83, 86, 87

Sanson, 1651, 45n.56, 113n.13
Sanson, 1654, 1665, 113n.13
Sanson, 1683, 43n.36
Schaarschmidt, 1846, 376n.138, 437n.89,
    439n.115
Schannat, 1724, 288n.49, 298n.162
Schmidt, 1800, 291n.79
Schmidt, 1823, 433n.51, 440n.140
Schöning, 1763, 285n.26
Schöning, 1771, 1779, 286n.27
Schöpflin, 1751, 290n.69
Schotanus, 1664, 1698, 128n.131
SDUK, 1851/A, 43n.28
SDUK, 1857, 430n.7
SDUK, 1871/A, 430n.6
Selves, 1819, 1822, 370n.86
Selves, 1819–33, 432n.38
Selves, 1821, 370n.84
Selves, 1833, 1835, 1843, 369n.82, 438n.100
Seutter, 1720, 1770, 287n.38
Seutter, 1745b, 286n.37, 439n.121
Seutter, 1745c, 286nn. 37, 38
Shepherd, 1929, 10n. 6, 292n.94, 372n.102,
    377n.160, 438n.105, 440n.129
Simencourt, 1830, 376n.139, 430n.9
Soleil, 1851, 430n.7
Speed, 1611, 112nn. 8, 9
Speed, 1627, 112n.8, 130n.154, 373n.118
Speed, 1627b, 123n.94, 366n.43
Spener, 1717, 293nn. 107, 108
Spruner, 1831/B, 376n.154
Spruner, 1837, 10n.6, 291n.81, 376nn. 139, 140,
    150, 433n.40, 438nn. 104, 106, 439nn. 111,
    117, 124, 440n.125
Spruner, 1838a, 287n.43, 376n.152, 437n.91,
    439n.121, 440n.126
Spruner, 1838b, 377n.155
Spruner, 1850/A, 46n. 74, 47n. 80, 291n.81
Spruner, 1855, 1860, 1861, 433n.40
Spruner, 1860, 438n.93
Spruner, 1861, 10n.6
Spruner, 1866, 377n.162, 437n.90
Spruner, 1871, 376n.151, 377n.156
Spruner, 1872, 10n.6
Stackhouse, 1790/A, 43n.28
Steger, 1845, 376n.138, 430n.9, 431nn. 11, 12
Stieler, 1824/O, 430n.8
Strebel, 1761, 283n.6
Süsmilch-Hörnig, 1860, 437n.91

Tallis, 1851, 430n.5
Tardieu, 1798, 287n.40
Tardieu, 1842, 431nn. 9, 11, 12, 438n.28
Tassin, 1633, 115n.26
Tavernier, 1645, 44n.47, 120n.66
Thierry, 1826, 440n.140
Thomson, 1817/O, 430n.8
Thomson, J., 1829, 372n.109, 433n.52, 434n.55, 436n.69
Toeppen, 1858, 437n.91
Tomka, 1751, 46n.62, 180n.60, 301n.198
Tomka, 1781, 180n.60, 301nn. 198, 201
Töppen, 1858, 440n.135
Tutzschmann, 1852, 437n.91

Vaissete, 1749, 435n.64
Vandermaelen, 1825/O, 430n.8
Velly, 1787, 181n.70, 182n.78, 296n.148, 461n.16
Vialart, 1641/E, 48n.90
Vieth, 1800/A, 370n.91, 432n.37
Vivien, 1825, 46n.70, 372n.108, 430n.7
Vögelin, 1865, 377n.162

Voigt, 1846, 437n.91
Völter, 1855, 430n.5
Vuillemin, 1839, 374n.129, 431nn. 9, 11

Wagenaar, 1749, 296n.148
Walch, Johann, 1811/O, 430n.8
Wastelain, 1761, 294n.117, 300n.185
Welt-Historie, 1744, 11n.15, 291n.80, 297n.149
Wendelinus, 1649, 122n.86
Wiberg, 1856, 377n.162, 438nn. 94, 98
Wilkinson, 1797/A, 129n.143, 372n.110, 430n.1, 433n.51, 434nn. 54, 55
Willard, 1828/O, 10n.7, 374n.127
Willard, 1836, 374n.127
Wiltsch, 1843/E, 440n.133
Witzleben, 1829, 392n.113, 436n.78
Wree, 1647, 122n.85
Wyld, 1825/A, 430n.7

Zollmann, 1732, 175n.23, 289n.59, 372n.103
Zuccagni, 1832, 1844, 438nn. 93, 96

INDEX OF SECONDARY LITERATURE
CITED IN THE NOTES

Full bibliographical details are supplied for each work in the first citation.

Acta eruditorum, 174n.1, 283n.6

ADBiog., 45n.51, 126n.122, 128nn. 133, 135, 170n.10, 177nn. 26, 27, 179n.40, 184n.97, 185n.101, 285nn. 24, 25, 289nn. 58, 59, 61, 66, 290nn. 67, 69, 72, 297n.160, 363n.17, 367n.52, 367n.66, 369n.76, 370n.89, 372nn. 104, 108, 373n.113, 377n. 156, 378nn. 169, 170, 435n.60

AGEphem., 43n.24, 287n.42, 288n.48, 302n.205, 363nn. 14, 17, 365n.30, 366n.47, 367nn. 58, 65, 368nn. 72, 73, 372n.104, 373n.114, 431nn. 22, 23, 433nn. 45, 46

Aharoni, Y., and M. Avi-Yonah, *Macmillan Bible Atlas,* 130n.157

Aimoin, *Historia Francorum,* 115n.29

Akerman, J. R., "From Books with Maps to Books as Maps," 10n.9

Algemeine Welt-Historie, 285n.26, 286nn. 28, 29, 288nn. 32, 52, 291n.80, 293n.149

Alinhac, G., *Historique de la cartographie,* 176nn. 12, 13

Alisio, G., and V. Valerio, *Cartografia napoli-tana,* 183n.86, 299n.185

Allgemeine Literatur-Zeitung (Halle), 186n.110, 363n.15, 365nn. 32, 36, 37, 371n.98, 433n.46

Allgemeiner Hand-Atlas der ganzen Erde, 374n.130

Allibone, S. A., *Critical Dictionary of English Literature,* 437n.53

Almagià, R., "Historical Map," 127n.127

Almagià, R., *Le pitture geographiche murali,* 11n.14

Alpers, S., *Art of Describing,* 10n.8, 129n.142

Anchor Atlas. See Kinder, H., and W. Hilgeman

Annales des voyages (Paris), 368nn. 69, 73, 372nn. 105, 106, 433nn. 28, 30, 481n.14

Annals of St-Bertin, 122n.89

Ansted, D. T., and C. G. Nicolay, *Atlas of Physical and Historical Geography,* 430n.8

Anton, H. A., "Austrasia/Austria," 115n.27

Ardin, G., publ., *Atlante scelto di carte ge-ografiche,* 430n.5

Arnhold, H., "Geographisches Institut Weimar," 370n.90

Badziag, A., and P. Mohs, *Schulatlanten in Deutschland,* 430n.8

Bagrow, L., *Geschichte der Kartographie,*

174n.1; *History of Cartography,* 39n.4, 41n.13; *Ortelii Catalogus,* 117nn. 37, 38, 127n.126

Bailey, M., *"Per impetum maris,"* 126n.121

Bainbridge, J., *Canicularia,* 125n.114

Baker, A. R. H., "On Historical Geography," 11n.16

Balbi, A., *Atlas ethnographique,* 378n.118

Baldacci, O., "Atlanti geografi e atlanti storici," 439n.122

Banfi, F., "Maps of Wolfgang Lazius," 45n.50, 117nn. 37, 39, 42, 118nn. 46, 47, 51, 119n.53

Banse, E., *Entwicklung der Geographie,* 48n.86

Barber, P., in *Map Collector* 48 (1989), 47n.81

Barber, R., and J. Barker, *Tournaments,* 289n.62

Barbier, A. A., *Dictionnaire des ouvrages anonymes,* 181n.70, 286n.29, 288n.48

Barraclough, G., *Origins of Modern Germany,* 431n.20

Bartlett, T., in *Times Literary Supplement* (1997), 297n.150

Baumgärtner, F. G., *Receuil de plans de batailles,* 372n.103

Baumgartner, F. J., *Henry II, King of France,* 116n.32

Baustaedt, B., *Richelieu und Deutschland,* 120n.72

Bayerische Akademie der Wissenschaften, *Gelehrte Anzeigen* (1844), 439n.111

Bayerischer Schulbuch-Verlag, *Grosser historischer Weltatlas,* 377n.161, 461n.6

Beck, F., *Karlsgraben,* 284nn. 18, 19, 285n.20

Beck, H. *Geographie,* 10n.8

Benicken, F., *Elemente der Militärgeographie,* 371n.99; *Napoleons Heereszüge,* 370n.92

Berghaus, H., *Physicalischer Atlas,* 378n.170

Bernard, J. F., publ., *Recueil des voyages au Nord,* 39n.5, 292n.96

Bernleithner, E., *W. Lazius Austria,* 45n.50, 117nn. 37, 39, 41, 118n.46

Bertius, P., *Breviarium totius orbis,* 120n.69; *Commentarii rerum Germanicarum,* 120n.69

Bibliotheca sanctorum, 125n.118

Bigmore, F. C., and C. W. H. Wyman, *Bibliography of Printing,* 370n.86

Biographie nationale de Belgique, 120nn. 67,

69, 122nn. 82–84, 87–89, 364n.25, 375n.133

Black, J., "Historical Atlases," 11n.13; *Maps and History,* 11n.21, 39n.2, 43n.28, 121nn. 75, 79, 374nn. 126, 128, 431n.17, 432n.26, 461n.12

Blaeu, W. J., *Theatrum orbis terrarum,* 46n.69

Blaeu, W. J., and J. Blaeu, *Theatrum orbis terrarum* (1645), 113nn. 10, 11

Blessich, A., "Un geografo italiano," 183n.83

BN, Cartes et plans, *Inventaire de la Collection Klaproth,* 369n.78

BN, Département des imprimés, *Catalogue de l'histoire de France,* 287n.44

BN, *Inventaire de la Collection Anisson,* 184n.91

Bodel Nyenhuis, J. T., and W. Eekhoff, *Kaarten van Friesland,* 48n.89, 128nn. 131, 132, 134, 130n.156, 296nn. 142, 148

Bombi, G., et al., *Atlas of World History,* 284n.15

Bonacker, W., "Johann Matthias Haas," 118n.52, 297nn. 159, 160, 298n.166, 299n.170

Boxhorn, M. Z. van, *Theatrum sive Hollandiae description,* 122n.88

Bradford, T. G., and S. G. Goodrich, *Universal, Illustrated Atlas,* 430n.8

Brandmair, E., *Bibliographische Untersuchungen,* 39n.11, 46n.67, 47n.76, 48n.82

Braun and Hogenberg. *See* Skelton, R. A.

Bréhier, L., in *Dictionnaire d'histoire et de géographie ecclésiastique,* 291n.83

Breisach, E., *Historiography,* 298nn. 163, 164, 299n.175

Brion de la Tour, L., *Observations curieuses,* 365n.32, 367n.64; *Réplique de L. Brion de La Tour,* 365n.32

British Biographical Archive, 374n.121, 433n.53

Broc, N., *Géographie de la Renaissance,* 40nn. 6, 7, 41n.13, 47n.81; "Malte-Brun," 368n.72, 431n.14

Bromberg, S. A., "Philipp Clüver and the 'Incomparable' *Italia antiqua,"* 128n.133

Brué, A. H., *Grand atlas universel,* 372n.104

Brühl, C., *Palatium und civitas,* 285n.22

Buache, P., *Idée générale de la carte de la Terre Sainte,* 292n.90

Buczek, K., *History of Polish Cartography*, 300n.186

Bulletin de la Société de géographie, 365n.31, 368n.72, 373n.118, 376n.142

Butterfield, H., *Man on His Past*, 10n.8, 177n.27, 184n.97, 185n.98, 186n.109, 286nn. 28, 30, 299n.175

Caeneghem, R. L. van, *Kurze Quellenkunde des Mittelalters*, 122n.83

Caert thresoor, 120n.68

Caillot, A., *Tableau des croisades*, 434n.57

Cambridge Medieval History, 299n.175

Cambridge Modern History, 299n.175

Camden, W., *Britannia*, 112n.6, 114n.22, 127n.124

Campbell, E. M. J., "Early Development of the Atlas," 39nn. 2, 3

Campbell, T., *Earliest Printed Maps*, 40n.6, 42n.23, 47n.81

Cappon, L. J., "The Historical Map in American Atlases," 10nn. 6, 7

Cartographia Bavariae, 370n.85, 439n.121

Catalogue of Maps and Charts of Harvard, 288n.46, 295n.132

Catalogue of Maps, Prints, Presented by King George IV, 430n.2

Catalogus mapparum, 10nn. 12, 13, 41n.13, 124n.122, 127n.127, 285nn. 26, 27, 287n.43, 290n.78

Catrou, F., and P. J. Rouillé, *Histoire romaine*, 295n.132

Centennia (computer program), 182n.74

Chanlaire, P. G., *Carte des Isles*, 287n.42

Châtelain, Z., *Histoire des Pays-Bas*, 174n.1

Chevallier, P., *Louis XIII*, 120n.72

Childers, E., *Riddle of the Sands*, 126n.121

Choiseul d'Aillecourt, A. U. M. de, *De l'influence des croisades*, 434n.59

Chubb, T., *Printed Maps in Great Britain*, 112n.9

Clüver, P., *Germania antiqua*, 123n.91, 177n.20

Codex Iustinianus, 124n.110

Cogswell, T., *The Blessed Revolution*, 124n.101

Cokayne, G. E., *Complete Baronetage*, 362n.11

Collins Atlas of World History, 9n.5

Colton, G. W., *Atlas of the World*, 430n.8

Constantine VII Porphyrogenitus, *De administrando imperio*, 291n.85

Constantini, P., *Bonaparte en Palestine*, 434n.58

Coquerel, C., review of Selves, 370n.88

Corbett, J. S., *England in the Seven Years' War*, 297n.149

Cordier, H., "Un orientaliste allemand," 369n.78

Corpus Byzantinae historiae, 291n.83

Cottin, M., *Mathilde*, 435n.62

Cox, E. L., "Dauphiné," 119n.58

Croke, B., "Origin of the Christian World Chronicle," 175n.6

Crutchley's General Atlas, 430n.8

Curschmann, F., "Die Entwicklung der historisch-geographischen Forschung," 114n. 24, 115n. 25, 289n. 54, 290nn. 74, 75, 376n.154, 377n.156

Dacier, H. J., *Rapport historique*, 40n.8, 44nn.37, 194n.121, 367nn. 60–63, 431n.22

Dahl, E. H., "The Original Beaver Map," 130n.158

Dainville, F. de, "L'Alsace comme la voyaient les cartes anciennes," 181n.71; *Cartes anciennes des églises de France*, 48n.90, 49nn. 93–95; *Géographie des humanistes*, 39n.4, 47nn. 76, 78, 116n.31

Dalton, O. M., *The History of the Franks: By Gregory of Tours*, 437n.84

Dansk Biografisk Leksikon, 126n.122

Dean, W. G., "Sic enim est traditum," 11n.13, 377n.157, 439n.123, 440n.132, 461n.11

Degn, C., 127n.127; *Schleswig-Holstein*, 136n.122, 137n.125

Delanglez, J., "Sources of the Delisle Map," 179n.14

Delano-Smith, C., "Maps as Art and Science," 49n.45, 130n.154; "Maps in Bibles," 44n.45, 48nn. 82, 83

Delano-Smith, C., and E. M. Ingram, *Maps in Bibles, 1500–1600*, 44n.45, 48n.83

Delisle, G., *Carte d'Artois*, 289n.48; *Remarques sur le Théatre*, 292n.4

Delisle, J., *Mémoire sur la carte de l'ancienne Palestine*, 292n.89; *Mémoires sur trois cartes*, 292n.89

Denaix, M. A., *Introduction à la géographie*, 373n.117

Depping, G. A., review of Heeren, 435n. 65, 436n. 67

Derogy, J., and H. Carmel, *Bonaparte en Terre Sainte*, 434n.58

Derolez, A., *Lambertus qui librum fecit*, 122n.82

Desnos, L. C., *Atlas moderne*, 292n.90; *Catalogue des atlas historiques*, 184n.96; *Catalogue des ouvrages* (1765), 46n.62, 182n. 78, 79, 184nn. 90, 96; *Catalogue des ouvrages* (1775), 183n.80; *Catalogue raisonné*, 184n.96

De Vrij, M., *World on Paper*, 10n.10, 41n.15, 176n.12

Dictionary of American Biography, 46n.68

Dictionary of the Middle Ages, 119nn. 53, 58

Dictionnaire des lettres françaises: Le XVIIIe siècle, 43n.26, 183n.82

Dierauer, J., *Geschichte der Schweizerische Eidgenossenschaft*, 296nn. 144, 145, 147

Digot, A., *Histoire de Lorraine*, 116n.32

Dirwaldt, J., *Allgemeiner Hand-Atlas*, 430n.8

Dizionario biografico degli Italiani, 291n.83, 373n.118

Doederlein, J. O., "Conspectus Fossae Carolinae," 285n.20

Dollinger, P., *Histoire de l'Alsace*, 118n.52

Domeier, K., and M. Haack, *Landkarten von Johannes Mejer*, 127n.127

Dopsch, A., "Naissance et formation de l'État autrichien," 115n.28

Dörflinger, J., "Geschichtsatlanten," 11n.13; "Geschichtskarte," 11n.13, 366n.42; "Karten von Anselm Desing," 286n.35; *Österreichische Kartographie*, 301n.198; "Werk des Johann Matthias Hase," 179n. 48, 180n.50, 297n.159, 298n.170

Dörflinger, J., and H. Hühnel, *Atlantes Austriaci*, 436n.80

Douglas, D., *English Scholars*, 39n.1, 40n.6, 42nn. 20, 21

Dresch, L. *Übersicht*, 371n.93; *Lehrbücher der allgemeinen Geschichte*, 371n.93; *Über den methodischen Unterricht*, 430n.7

Duby, G., *Atlas historique Larousse*, 284n.15

Du Cange, C. D. de, *Glossarium mediae Latinitatis*, 115n.29

Dürrenmatt, P., *Schweizer Geschichte*, 296n.145

Dussler, L., *Incunabeln der deutschen Lithographie*, 370n.85

Duval, P., "Tabula geographica concilii," 49n.94

Early Maps of Scotland, 433n.53

Eckert, M., *Die Kartenwissenschaft*, 39n.5

Eckhardt, J. G. von, *Commentarii de rebus Franciae Orientalis*, 285n.20

Editing Early Atlases, ed. Winearls, 10n.9, 11n.21

Eeghen, I. H. van, *De Amsterdamse Boekhandel*, 174n.1

Elkhadem, H., "La naissance d'un concept," 41n.12

Enciclopedia Italiana, 122n.87

Encyclopaedia Judaica, 179n.41, 292n.90

Engelmann, G., *Heinrich Berghaus*, 186n.107, 370n.89

Engelmann, W., *Bibliotheca Geographica*, 371n.99, 430n.9, 431n.12

Enslin, T. C. F., *Bibliotheca historico-geographica*, 371n.99

Ersch, J. S., and J. G. Gruber, *Allgemeine Encyclopädie*, 124n.101

Evans, I. M., and H. Lawrence, *Christopher Saxton*, 112n.5

Falckenstein, J. H. von, *Antiquitates Nordgavienses*, 177n.26, 285n.20, 289n.62; *Codex diplomaticus antiquitatum Nordgaviensium*, 177n.26

Fawtier, R., *Les Capétiens et la France*, 340n.188

Fer, N. de, *Histoire des rois de France*, 175n.11

Ferarius, P., and M. A. Baudrand, *Lexicon geographicus*, 115n.30

Ferguson, W. K., *Renaissance in Historical Thought*, 432n.35

Fick, K. E., "Die kartographische Darstellung," 43n.32

Fiero, A., et al., *Histoire et dictionnaire du Consulat*, 361nn. 3, 5

Flower, R., "Laurence Nowell," 112n.1

Fockema Andreae, S. J., and B. van't Hoff, *Geschiedenis der Kartografie*, 121nn. 75, 77, 79

François, É., *Protestants et catholiques*, 297n.159

Fredegar, *Chronicon*, ed. A. Kusternig, 116n.32

Freed, J. B., "Austria," 119n.53

Frenzel, *Malthe Conrad Bruun*, 367n.61, 368n.72, 369n.83

Fréret, N., "Lettre," 176n.13, 177nn. 18, 19, 178nn. 31, 33, 38

Fueter, E., *Geschichte der neuere Historiographie*, 182nn. 78, 79

Funk, C. W. F. von, *Gemälde aus dem Kreuzzüge*, 435n.60

Gallic Chronicle of 452, 377n.166

Gallois, L., *Géographes allemands*, 40nn. 6, 7

Gambi, L., and A. Pinelli, *Galleria*, 11n.14, 130n.154

Gargon, M., *Walchersche Arkadia*, 121n.77

Gaspari, A. C., *Allgemeiner Hand Atlas*, 430n.8

Gatterer, J. C., *Universalhistorie*, 185n.98, 300n.193

Gautier Dalché, P., 39n.3

Gebhardi, L. A., in *Algemeine Welt-Historie*, 286n.29

Gelzer, H., *Genesis der Themenverfassung*, 124n.110

Geographia universalis, vetus et nova (1545), 41n.13

Geographica historica (Bonn), 11n.16

Géographie Blaviane, 127n.124

Geyl, P., "Writers of Netherlands History," 375n.133

Gibb, H. A. R., "Abu'l-Fida," 125n.113

Gibbon, E., *Decline and Fall*, 284n.15

Gibson, E., *Chronicon Saxonicum*, 114n.20

Gimbutas, M., "Slavic Religion," 289n.60

Goethe, W., *Höhen des alten und neuen Welt*, 377n.168

Goffart, W., *Barbarians and Romans*, 300n.192; "Breaking the Ortelian Pattern," 11n.21, 287n.45; "Christian Pessimism," 11n.14, 125n.117; "First Venture," 112-3nn. 1, 9, 440n.125; "Longer Look," 175n.5, 283n.9, 433nn. 45, 48; "The Plot of Gatterer's 'Charten,'" 184n.97; "Preliminary Report," 175n.8, 283n.10, 361nn. 2–5, 362n.9, 363nn. 20, 22, 433n.45; "Theme of 'Barbarian Invarions,'" 117n.42; "What's Wrong with the Maps?" 284n.15, 433nn. 46, 48, 440n.132

Gonnard, P., *The Exile of St. Helena*, 314n.27

Gooch, G. P., *History and Historians*, 274n.123

Gordon, G., "*Medium aevum* and the Middle Ages," 177n.27

Goss, J., *Blaeu's "The Grand Atlas,"* 113n.10

Göttingische gelehrte Anzeiger, 42n.20, 363n.17, 433n.46

Grafton, Anthony, *Defenders of the Text*, 130n.156

Grand dictionnaire universel (Larousse), 177nn. 27, 28, 178n.29

Grand livre de l'histoire mondiale, 9n.5

Grand théâtre généalogique, 174n.1

Greaves, J., *Binae tabulae geographicae*, 125n.114; *Epochae celebriores*, 125n.114

Gregory of Tours, *Historiae*, 295n.134

Grell, C., "J. D. Schoepflin et l'historiographie française," 290n.69

Günther, S., *Geschichte der Erdkunde*, 130n.158, 179n.48, 183n.86, 370n.90

GV, 185n.98, 296n.139, 370n.92, 439n.111, 440n.128

Györffy, G., *Magyarország Történeti*, 301n.202

Haas, G. Z., *De Danubii et Rheni conjunctione*, 284n.19

Haddon, A. C., *Wanderings of Peoples*, 283n.10

Hager, J. G., *Geographischer Büchersaal*, 173n.30

Hallam, H., *Supplemental Notes*, 114n.23

Halma, F., *Toneel der Vereenighde Nederlanden*, 128n.131; repr., *Uitbeelding* (1979), 128n.131

Halphen, L., *L'histoire en France*, 361n.2, 365n.38

Hamberger, G. C., and J. G. Meusel, *Das gelehrte Teutschland*, 185n.100

Hamlyn Historical Atlas, 9n.5

Hammond historical atlases, 461n.8

Harley, J. B., and D. Woodward, *History of Cartography*, 47n.81

Harms, H., *Künstler*, 176n.12, 177nn. 27, 28, 179nn. 40–42; *Themen*, 49n.94, 126n.121, 177n.19, 183n.86, 184n.88, 186n.108, 284n.18, 285n.23, 365n.30

Harper Atlas of the Bible, 130n.157
Harper Atlas of World History, 9n.5, 128n.132
Harris, R. L., *Chorus of Grammars*, 114n.20
Harvey, P. D. A., *History of Topographical Maps*, 47n.81; *Mappa Mundi*, 47n.81
Hase, J. M., "Anmerkungen über seine Landkarten," 287n.161; *Historische Atlas* (1813–14), 366n.44; *Phosphorus historiarum*, 180n.51, 298n.169
Haskins, C. H., *Renaissance of the Twelfth Century*, 436n.68
Hassel, G., in *AGEphem.*, 363n.17, 371nn. 95, 98
Hauber, E. D., *Versuch einer Historie*, 44n.47, 45n.49, 115n.25, 116n.34, 119n.61, 120n.66, 121n.73, 130n.156, 282n.1, 294n.109
Haywood, J., et al., *Cassel Atlas of the Medieval World*, 439n.110
Heather, P., "The Huns and the End of the Roman Empire," 300n.192
Heeren, A. H. L., *Essai sur l'influence des croisades*, 434n.59
Heinemejer, W. F, et al., *Amsterdam in Kaarten*, 129n.153
Hénault, J. F., *Abrégé de l'histoire de France*, 182n.79
Hill, D., *Atlas of Anglo-Saxon England*, 113n.12, 114n.24
Historical Geography, The 11n.16
Historische Atlas, ed. G. L. Wieberdink, 11n.19
Historisches Journal Göttingen, 185nn. 99, 105
Hodgkiss, A. G., *Understanding Maps*, 47n.31
Hof, J., *De abdij van Egmond*, 122n.80
Hofe, U. im, in *Handbuch der Schweizer Geschichte*, 296n.147
Hoffmann, H. H., *Kaiser Karls Kanalbau*, 235n.20
Hofmann, C., "Genèse de l'atlas historique," 9n.1, 11n.17, 39n.2, 43n.27, 44n.36, 46n.71, 129n.149, 180n.61, 181nn. 63, 65, 294nn. 120, 122, 297n.154, 299nn. 174, 180, 461nn. 5, 7
Homann, J. B., *Atlas methodicus*, 185n.99
Horn, G., *Introductio ad geographiam antiquam*, 120n.69
Horrabin, J. F., *Atlas of European History*, 284n.15
Horvath-Peterson, S., *Victor Duruy*, 376n.143

Hübner, J., *Museum geographicum*, 136n.120, 282n.1
Huguenin, A., *Histoire d'Austrasie*, 115n.27
Hümmörder, C., "Lambert von St-Omer," 122n.82
Hydatius Lemici, *Chronicon*, 294n.114

Inventaris . . . Rijks-Archiev, 121nn. 76, 77

James, E., *Origins of France*, 437n.84
Janssonius (1638), 120nn. 71, 72; (1646), 113n.11
Jarry de Mancy, A., *Atlas constitutional*, 364n.24; *Tableau des révolutions nationals*, 364n.24
Jervis, W. W., *World in Maps*, 183n.88, 299n.185
Jomard, E. F., in *Bulletin de la Société de géographie*, 291n.86
Journal Asiatique (Klaproth), 369nn. 77, 78
Journal des sçavans, 10n.9, 174n.2, 181n.64, 182n.75, 186n.107, 288n.47
Journal de Trévoux, 44n.39, 174nn. 1, 2, 178n.29, 181nn. 65, 71, 182nn. 79, 183nn. 80, 87, 288n.47, 299nn. 178, 181, 182
Journal of Historical Geography, 11n.16
Julien, *Nouveau catalogue*, 181n.67, 183n.82, 288nn. 47, 52, 296n.144, 299n.170
Juncker, C., *Geographie der mittleren Zeiten*, 179n.42, 461n.15

Kalma, J. J., in *Uitbeeldig* facsimile (1979), 128n.134
Karten alter Meister, 47n.79
Keere, P. van der, "Germania inferior," 283n.6
Keuning, J., "Bernardus Schotanus," 128n.131; "Cornelis Danckerts," 124n.100; "The History of an Atlas," 10n.8, 41n.17
Keynes, G., *Library of Edward Gibbon*, 43n.12, 294n.119
Keynes, S., "Rædwald the Bretwalda," 114n.23
Kinder, H., and W. Hilgemann, *Anchor Atlas*, 9n.3, 292n.94, 437n.84; *DTV Atlas zur Weltgeschichte*, 9n.5; *Penguin Atlas*, 9n.5
Klaproth, J., *Antwort auf eine Rezension*, 369n.79; "Aperçu historique des peoples," 369n.81; *Asia polyglotta*, 369n.78

Knowles, D., *Great Historical Enterprises,* 295n.137

Koch, A. C. F., "Egmond," 122n.80

Koch, C. J., *Tableau des révolutions de l'Europe dans le moyen âge,* 431n.21

Koeman, 1: 120nn. 67–69, 72, 293n.99; 2: 112n.8, 120nn. 68, 72, 124n.100, 174n.1, 396n.142; 3: 41n.16, 45n.48, 113n.15, 117n.38, 128n.131

Koeman, C., *Collections of Maps and Atlases,* 10n.10, 42n.18, 43n.36,175nn. 9, 10; *History of Ortelius,* 40n.6, 42n.21, 46nn. 66, 67

Köhler, J. D., *Teutsche Reichs-Historie,* 174n.40; *(Wochentliche) Historische Munz-Belustigung,* 179n.40

Kohlrausche, F., *Chronologischer Abriss,* 372n.112

Kratochwill, M., in *LGKartog.,* 117n.43

Kretschmer, P., "Austria und Neustria," 115n.27

Kritischer Wegweiser im Gebiete der Landkarten-Kunde, 365n.30

Krumbacher, K., *Geschichte der byzantinischen Litteratur,* 291n.84

Kruse, C., 366nn. 46–48, 367nn. 53–55, 58, 59; in *Allgemeine Literatur-Zeitung,* 366n.47; "Probe der Gattererschen Charten," 185nn. 101, 103, 300nn. 191, 193, 195, 367n.54

Kuchař, K., *Mapová Sbírka Molla,* 301nn. 198, 200

Kühn, A., *Neugestaltung der deutschen Geographie,* 117n. 40, 123n.99

Lamare, N., *Traité de la police,* 287n.44

Lamb, H. H., *Climate: Past, Present and Future,* 126n.121, 127n.125, 128n.138

Landmarks in Learning, 112n.3, 6

Landresse, C., "Notice sur M. Klaproth," 369n.78

Las Cases, É. de, *Mémorial,* ed. Walter, 361n.1, 362nn. 10, 11, 363nn. 18, 20, 21, 364nn. 29, 365n.31, 366n.45

Las Cases, E. de, *Mémorialiste,* 361nn. 3–5, 362nn. 6, 12, 13

Laurens, H., et al, *Expédition d'Égypte,* 434n.58

Lazius, W., *Gentium aliquot migrationibus,* 117n.42; *Rerum Viennensium commentarii,* 118n.52; *Typi,* 45n.50, 117n.39

LC List, 41n.17, 43n.30, 45n.53

Lefebvre, G., *Historiographie moderne,* 291n.87, 365-6n.38; *Napoléon,* 287nn. 41, 42; *Révolution française,* 287n.39

Łelewel, J., *Géographie du moyen âge,* 125n.113, 178n.34, 431n.9

Le Long, 44n.47, 115n.25, 119nn. 56, 61, 127n.129, 135n.11, 117nn. 19, 25, 179n.44, 182n.79, 183nn. 82, 83, 87, 184n.94, 283n.13, 288n.51, 289n.63, 291n.80, 298n.89, 294n.124, 295nn. 126, 134, 135, 296nn. 138, 141

Lenglet Dufresnoy, *Méthode pour étudier la géographie:* 1st ed., 42n.25; 3d ed., 43nn. 25, 26

Lenoir, A., *Atlas des monuments,* 437n.81

Levy, F. J., "Making of Camden's *Britannia,*" 45n.51, 112nn. 1, 5

LGKartog., 42n.23, 45n.55, 125n.115, 126n.122, 128n.133, 176nn. 12, 14, 185n.99, 285n.24, 371n.100, 374n.130, 375n.131

Literary Gazette (review of Quin), 374n.125

Lockyer, R., *Buckingham,* 124n.101; *Early Stuarts,* 124n.101

Łodyński, 364n.22, 370n.92

Longnon, A., *Formation de l'unité française,* 116n.32; *Géographie de la Gaule,* 116n.32, 295n.129

Lot, F., *Les invasions germaniques,* 284n.15

Louis XV, king, *Cours des fleuves,* 295n.128

Lubin, A., *Mercure géographique,* 129nn. 149–51, 178n.29, 298n.162, 461n.15

Luck, J. M., *History of Switzerland,* 296nn. 144, 145

Lugge, M., *"Gallia" und "Francia" im Mittelalter,* 115n.27

Lutz, A., *Emma Willard,* 374nn. 127, 128

Ma-Geoghegan (MacGeoghegan), J., *Histoire de l'Irlande,* 297nn. 150, 152

MacGeoghegan, J., *History of Ireland.* See Ma-Geoghegan, J.

MacKay, A., *Atlas of Medieval Europe,* 461n.9

Maier, J. C., *Versuch einer Geschichte der Kreuzzüge,* 434n.59

Mallart, A., *Coutûme générale d'Artois,* 288n.48

Malte-Brun, C., *Précis de la géographie universelle*, 368n.73, 431n.14, 432n.28

Martonne, E. de, *Traité de géographie physique*, 46n.65

Mascov, J. J., *Geschichte der Teutschen*, 283n.11, 433n.46

Masson, F., *Napoléon à Sainte-Hélène*, 363nn. 18, 20

Maxted, I., *The London Book Trade*, 364n.23

Mayer, H. E., *Bibliographie der Kreuzzüge*, 435nn. 60, 61

McCormick, M., *Eternal Victory*, 123n.98

McKisack, M., *Medieval History in the Tudor Age*, 112n.1

McKitterick, R., "The Study of Frankish History," 115n.33

McKittrick, R., *The Frankish Kingdom under the Carolingians*, 122n.89

Mees, 130n.156, 371n.97, 372n.109, 372n.114, 375n.134

Merguet, H., *Handlexikon zu Cicero*, 129n.141

Mertens, V., "Markgraf Heinrich III von Meissen," 289n.61

Meurer, P. H., *Atlantes Colonienses*, 10n.9, 115n.26; *Fontes cartographici Orteliani*, 40n.9, 41n.16, 43nn. 63, 64, 44n.76, 127n.130; "Ortelius," 11n.20, 39nn. 1, 2, 40nn. 9, 11, 41n.16, 45n.54, 46nn. 64, 73, 47n.75, 112n.7, 117n.43

Michaud, J. F., *Histoire des croisades*, 435n.61

Michaux, G., in *Dictionnaire du Grand Siècle*, 435n.118

Milanesi, M., *Tolomeo sostituito*, 40n.7, 43n.28

Mills, C., *History of the Crusades*, 434n.57

Milner, T., and A. Petermann, *Descriptive Atlas of Astronomy*, 430n.5

Mirot, L., and A. Mirot, *Manuel de géographie historique*, 184n.96

Moir, A. L., *World Map in Hereford Cathedral*, 47n.81

Möller, A., *Kleiner historischer Atlas*, 372n.112

Monkhouse, F. J., *Regional Geography*, 126n.121, 128n.132

Monnier, F., *Dictionnaire du Grand Siècle*, 287n.44

Monumenta Germaniae historica. *See* Hydatius Lemici; *Gallic Chronicle of 452*

Moreau, J., *Dictionnaire de géographie*, 115n.27

Moreland and Bannister, 10n.11, 40n.6, 113nn. 10, 11, 433n.53

Moréri, L., *Grand dictionnaire historique*, 115n.28, 116n.33

Morse, S. E., *New Universal Atlas*, 430n.8

Moulart-Sanson, P., *Introduction à la géographie*, 44n.44

Mourre, M., *Dictionnaire encyclopédique d'histoire*, 115n.27

Mumford, I., "Lithography for Maps," 370nn. 85, 90, 372n.111, 373n.117

Muratori, L. A., *Rerum Italicarum scriptores*, 288n.50

Musset, L., *Les invasions*, 284n.15

National Maritime Museum [Greenwich]: *Catalogue of the Library*, 44n.39

Nebenzahl, 10n.11, 42n.24, 47n.81, 48n.82, 85, 119n.63, 125nn. 116, 119, 129n.144, 130n.154, 179n.41, 301n.197, 433n.47

Nelson, J. L., *Charles the Bald*, 122n.89; "The Work of Nithard," 291n.80

Neue deutsche Biographie, 297n.160

New Catholic Encyclopedia, 120n.80, 125nn. 117, 118, 438n.103

New Universal Atlas, 430n.8

New York Public Library, *Dictionary Catalog of the Map Division*, 288n.46

Niceron, *Mémoires pour servir à l'histoire*, 176n.13

Nicholson, N., and A. Hawkyard, *Tudor Atlas by John Speed*, 123n.99

Nicolay, C. G., and D. T. Ansted, *Atlas of Geography*, 42n.22

Nieuw Nederlandsch Biografisch Woordenboek, 120n.67, 128nn. 131, 135

Notices sur les travaux de M. Denaix, 375n.135

Nouvelles annales des voyages, 369n.78

Novellae Justiniani, 124n.110

NUC, 185n.98, 285n.23, 288n.47, 364n.24, 376n.140

Oberhummer, E., and F. R. von Wieser, *Wolfgang Lazius Karten*, 45nn. 50, 52, 117nn. 36, 38–41, 43, 108nn. 45–48, 51, 52

Ohler, N., "Historische Atlanten," 9n.3

Ortelius, A., *Aurei saeculi imago*, 177n.20; *Epitome theatri*, 10n.11

Ostrogorsky, G., *History of the Byzantine State*, 291n.83

Oxford Classical Dictionary, 40n.7

Oxford English Dictionary, 10n.9

Pariset, J. D., *Relations entre la France et l'Allemagne*, 178n.32

Parisse, M., ed., *Histoire de Lorraine*, 115nn. 29, 30

Partsch, J., *Philipp Clüver*, 42n.18

Pastoureau, M., "Atlas en France avant 1700," 42n.25, 47n.78, 49n.95, 119n.62; *Atlas français*, 42nn. 19, 20, 45n.58, 47n.78, 49nn. 93, 94, 113n.13, 119nn. 60, 62, 63, 120n.67, 124nn. 102, 109, 125n.116, 287n.45, 288n.46, 289n.63, 295n.132; "La France divulguée," 48n.89, 184n.91, 294n.125; in *LGKartog* 2:699–701 (Sanson), 45n.55; in *Nicolas Sanson, Atlas du monde*, 129n.148

Patrides, C. A., *Grand Design of God*, 298n.163

Patze, H., and W. Schlesinger, *Geschichte Thüringens*, 289n.61

Peddie, R. A., *Subject Index of Books*, 435n.65

Pedley, *Vaugondy*, 180n.62, 181n.63, 182n.72, 286n.31, 288n.47

Pegolotti, F. C., *La Pratica della mercatura*, 432n.30

Pelletier, M., "Les géographes et l'histoire," 106n.141

Peschel, O. F., *Geschichte der Erdkunde*, 106n.107

Philips' Atlases of Comparative Geography, 43n.30

Piggott, S., "Camden and the *Britannia*," 112n.1, 114n.21

Pirenne, H., *Economic and Social History*, 436n.68; *Histoire de Belgique*, 121n.77

Placide, *Cartes de géographie*, 119n.62

Plaut, F., "Where Is Paradise?" 48n.83

Plummer, C., *Two of the Saxon Chronicles*, 114n.20

Pontanus, J., *Symbolarum libri XVII*, 47n.76

Ptolemy, Claudius: (Florence, 1480) 40n.6; (Strasbourg) 40n.10; (ed. Bruscelli, Venice) 42n.23

Purgina, J., *Tvorcovia Kartografie Slovenska*, 301nn. 198, 199, 201

Quérard, *France littéraire*, 119n.63, 181n.65, 287n.44, 361n.2, 363nn. 17, 21, 364nn. 24, 28, 366nn. 41–43, 368nn. 67, 68, 372n.105, 375n.135, 376n.148, 437n.81; *Litt. franç. contemp.*, 369n.79, 372n.105, 375n.135, 376n.141

Rand-McNally *Atlas of World History*, 9n.5

Rapin-Thoyras, P., *Histoire d'Angleterre*, 114n.23

Reader's Digest World Atlas, 9n.5

Recueil des historiens des Gaules, ed. M. Bouquet, 295n.137, 296n.138

Regenbogen, J. H., *Commentatio de fructibus e Bello sacro*, 435n.59

Reischl, *Atlas historicus*, 181n.64

Reland, A., *Palaestina illustrata*, 179n.41

Renkhoff, O., "Johann Georg Hagelgans," 283n.9

Révolutions de l'univers, 181n.68

Reyzenbach (Reigersbergh), J., *Chronyk van Zeeland*, 121nn. 76, 77

Richter, P., "Ueber Johann Georg Hagelgans," 175n.7

Rijn, G. van, and C. van Ommeren, *Atlas von Stolk*, 113n.10

Ristow, W. W., "Lithography and Maps," 183nn. 84, 85, 87

Ritter, K., *Europa*, 436n.79

Ritter, P. E., "Gottfried Bessel," 177n.25

Robert de Vaugondy, D., *Essai sur la géographie*, 97n.78

Robinson, A. H., *Elements of Cartography*, 43n.31; "Mapmaking and Map Printing," 44n.40

Rogers, E., "Maps Connected with Buache," 116n.107

Rohrbacher, R. F., *Histoire de l'Église catholique*, 438n.103

Romm, J., "Mythe, cartes et histoire," 40n.11

Rose, W. S., *Crusade of St. Lewis*, 435n.63

Roseberry, A. P. P., *Napoleon, the Last Phase*, 361n.2

Rosenberg, A., *Gueudeville*, 174nn. 1, 2

Rousset, R., *Atlas général*, 430n.8

Rühle von Lilienstern, J., *Duodez-Schul-Atlas*, 431nn. 9, 12, 438n.100

Runciman, S., *Fall of Constantinople*, 432n.34

Ruska, J., "al-Tusi," 125n.113

Sale, G., and T. Salmon, *An Universal History*, 286n.28

Sanders, A., *Flandria illustrata*, 122nn. 87, 88

Sandler, C., "Die homännischen Erben," 179n.48, 180n.49, 298n.159; *Reformation*, 42nn. 20, 24

Sanson, G., *Geographia synodalis*, 49n.94; *Introduction à la géographie*, 44nn. 42, 43, 48n.87, 186n.108, 286n.31

Sanson, N., *Cartes générales*, 113n.13; "Cartes particulières de la France," 49n.93, 129n.145; *Les cinqs royaumes*, 113n.13; *Europe divisée*, 48n.87; *Royaume de France*, 45n.59

Sarton, G., *Introduction to the History of Science*, 125n.113

Sax, P., *De gestis Frisiorum*, 127n.127

Schmithüsen, J., *Geographischen Wissenschaft*, 39n.3, 113n.14

Schöning, G., *Kiøbenhavnske Selskab Skrifter*, 285n.26

Schöpflin, D., "Jonction du Danube avec le Rhin," 284n.19

Schotanus, B., *Uitbeelding der Heerlijkheit Friesland*, 128n.131

Schwartz, K., in Ersch and Gruber, *Allgemeine Encyclopädie*, 367n.52

Scott, W., *Tales of the Crusaders*, 435n.63

Senex, J., *Modern Geography*, 282n.1

Seybold, C. F., "Al-Idrisi," 178n.34

Shirley, R. W., *Mapping of the World*, 41n.17; *Maps of Britain to 1650*, 112nn. 2, 3, 6, 113nn. 10, 11, 13, 114n.18, 123n.94; *Maps of Britain to 1750*, 45n.60, 113nn. 13–16, 114nn. 18, 19, 22

Skelton, R. A., "Early Atlases," 11n.14; *G. Braun and F. Hogenberg,* "*Civitates orbis terrarum*," 43n.33, 295n.134; in *John Speed, A Prospect 1627*, 112n. 8, 123n.94; in Ortelius, "*Theatrum 1570*," 10n.9; in Ptolemy, "*Geographia*," Strassburg, 1513, 40n.6, 41n.12

Smallegange, *Nieuwe cronijk van Zeeland*, 121n.77

Smith, C. *See* Delano-Smith, C.

Spittler, L. T., *Geschichte der Kreuzzüge*, 435n.60

Spruner, K., *Beschreibung des Kanales*,

376n.152; *Genealogisch-historische Tabelle*, 376n.154

Stanley, E. G., "Continental Contribution to the Study of Anglo-Saxon Writings," 285nn. 24, 25

Stegmann, I., *Anselm Desing*, 129n.141, 286n.35

Stendhal (Henri Beyle), *Le rouge et le noir*, 361n.1

Stevens, H., *Bibliographia geographica*, 122n.89

Stoddart, D., *On Geography*, 48n.85

Storia universale, 286n.28

Struve, B. G., *Bibliotheca historica*, 115n.30

Suter, H., *Die Mathematiker der Araber*, 125n.113

Taillemite, É., "Les cartes anciennes de la Marine," 292n.89

Talbert, R., ed., *Barrington Atlas*, 46n.73

Tapié, V.-L., *France in the Age of Louis XIII*, 120n.72

Teixeira, P., *Relaciones*, 292n.91

Thibaudet, A., *Histoire de la littérature*, 439n.64

Thieme, V., and E. Becker, *Allgemeines Lexikon der bildenden Künstler*, 123n.90

Thomson, J., *General Atlas*, 433n.53; *New Universal Gazetteer*, 433n.53

Times Atlas of World History, 9n.5, 128n.132, 130n.155

Tooley, R. V., "The De l'Isle, Buache, Dezauche Succession," 179n.39; *Mapmakers*, 124n.100, 295n.132; *Maps and Map-Makers*, 176n.12

Tooley, R. V., and C. Bricker, *Landmarks of Mapmaking*, 47n.81

Tulard, J., "Légende napoléonienne," 373n.119; in É. de Las Cases, *Mémorial de Sainte Hélène* (1968), 362n.13

Unger, W. S., *Catalogus van den historisch-topografischen Atlas*, 121nn. 75, 76

Vajda, L., "Zur Frage der Völkerwanderungen," 284n.15

Valerio, V., *Istituzioni cartografiche*, 183nn. 84–86, 299n.85

Vallée, L., *BN, Catalogue des plans de Paris*, 288n.46

van den Bergh, L. P. C., *Handboek der Middel-Nederlandsche Geographie*, 128n.134

Vanden Broecke, M. P. R., *Ortelius Atlas Maps*, 43nn. 34, 35, 44nn. 45, 46, 45n.48, 47n.76, 127n.130

Vander Aa, A. J., *Biographisch Woordenboek der Nederlanden*, 120n.67, 128nn. 131, 135

Vander Heijden, H. A. M., "Ortelius and the Netherlands," 43n.34

Vander Krogt, P., "Neederlandse Editie van Crome's Produktenkarte," 365n.30

Vandermaelen, P., *Atlas universel de géographie*, 394n.130

Vann, J., "Mapping under the Habsburgs," 117n.44

Varga, L., *Das Schlagwort vom Finsteren Mittelalter*, 177n.27

Vaughan, R., *Charles the Bold*, 118n.52

Velly, P. F., and C. Villaret, *Histoire de France*, 182n.78, 184n.96

Vertot, R., *Histoire des chevaliers hospitaliers*, 175n.11, 177n.18

Vidaleuc, J., *Les émigrés français*, 40n.9

Viton de Saint-Allais, N., *Correcteur de Le Sage*, 366n.40

Vivien de Saint-Martin, L., "État actuel de la cartographie," 176n.12

Voss, J., *Universität, Geschichtswissenschaft*, 119n.58, 290nn. 69, 70, 72, 73, 74, 76, 367n. 66, 368nn. 67, 68, 431nn. 15, 21

Waard, C. de, *Rijksarchief in Zeeland*, 121nn. 76, 77, 182n.81

Wagner, H., *Lehrbuch der Geographie*, 41n.12

Wain, J., "Alternative Poetry," 127n.128

Walter, G., in Las Cases, É., *Mémorial de Ste Hélène*, 361n.5

Weber, E., in *Times Literary Supplement*, 112n.2

Wegele, F. X. van, *Geschichte der deutschen Historiographie*, 180n.57, 299n.170

Weigel, C., *Historiae veteris et novi Testamenti*, 175n.6; *Sculptura historiarum*, 175n.6

Weigel, C., and J. D. Köhler, *Welt in Ein Nuss*, 175n.6

Weiland, K.-F., *Allgemeine Hand-Atlas*, 371n.97

Weiner, M., *The French Exiles*, 361n.4

Weiss de la Richerie, *Atlas politique de la France*, 364n.24

Wellens-De Donder, L., "Un atlas historique: Le *Parergon*," 39n.2

Werner, J. W. H., *Leo Belgicus*, 123n.92

Werner, K. F., *Les origines*, 119n.64

Westermann's Atlas zur Weltgeschichte, 130n.155

Wheeler, H. F. B., and A. M. Broadley, *Napoleon and the Invasion of England*, 287nn. 39, 41, 42

Wieder, F. C., *Nederlandsche Documenten in Spania*, 121n.78

Wilken, F., *Geschichte der Kreuzzüge*, 435nn. 60, 66

Wilkinson, *Atlas for Schools*, 430n.8

Willard, E., *System of Universal History*, 374n.127

Winkler, R. A., *Frühzeit der deutschen Lithographie*, 370n.85

Witt, W., in *Lexikon der Kartographie*, 11n.18, 119n.54

Wolf, A., "100 Jahre Putzger," 9n.6, 124n.104; "Bild der europäische Geschichte," 9n.2, 367n.54, 431n.10, 436n.71

Wolfram, H., "Gothic History and Historical Ethnography," 283n.11

Wolkenhauer, W., "Friedrich Justin Bertuch," 370n.90; *Leitfaden*, 183n.86, 370n.90

Woltersdorf, 175n.3, 185nn. 100, 101, 104, 105

Woodbridge, W.C., *Modern Atlas*, 430n.8

Woodward, D., in *Map Collector* 18 (1982), 370n.84; "Reality, Symbolism, Time," 130n.155

Wurzbach, C. von, *Biographisches Lexikon Oesterreich*, 301n.199

Zotz, T., "Austrien," 115n.27

The notes and the Catalogue of Maps and Atlases are not indexed as subjects.

Aa, van der (Amsterdam publisher), 30, 231

Abadie, P., 352, 408, 416

Abbassids. *See* Caliphate, Islamic (map subject)

Aeneas, 251. *See also* maps, literary

Africa: continent 3, 16, 112, 231, 257, 317, 320, 402; North, 91, 116, 200, 247, 259, 267; Vandal invasion of North, 191, 351, 387, 446

Akhmatov, I. M., 411

Alaric I and II, kings, 110, 236, 246

Alexander the Great: map subject, 33, 458; in the *Parergon*, 25, 104, 133, 148; treatment by mapmakers, 35, 133, 148, 257, 335, 381

Ælfred, 197, 430

Algemeine Welt-Historie, maps in, 209, 219, 229, 454

Alps, 65, 67, 70, 71, 250, 267

Alsace, 58, 65, 66, 216–19, 222, 261

Altdorf. *See* Köhler, J. D.

Alting, M., 75, 99–101, 110, 152, 187, 250, 259, 262, 266, 275–77, 337, 455, 457, 458

Amari, M., 429

America, 12, 16, 111, 196, 202, 267, 346, 419, 429, 430

Amiens, breakdown of Peace of, 204

Amsterdam, 84; map of, 445; as publication center, 23, 54, 78, 80, 100, 132, 133, 194, 230, 449

Andalusia, 238

Andrews, J., 274, 278–80, 337

Anglo-Saxons: chronicle, 56–57; Heptarchy, 52–57, 80, 146, 279, 411, 417, 426, 428, 443, 446, 447, 449, 454; as Saxons, 29, 105, 189, 191–92, 197, 200, 232, 278; tracks of invaders, 279, 280, 192; translation of Orosius, 197, 200, 429. *See also* Camden, W.; Forster, J. R.; Gibson, E.; Lambarde, W.; Ohthere; Speed, J.; Wulfstan

Anthony, Saint, 91

Apulia. See Pellegrino, C.

Aquitaine, 25, 66, 67, 246, 271

Archaionomia. See Lambarde, W.

Aretin, J. C., 338

Argonauts, 33, 133, 336

Arles, kingdom of, 66–68

Armada, the: Spanish, 110, 203, 313; of William the Conqueror, 406

Armenia, 266; as map subject, 86–87, 106

Arminius (Hermanns), J., 69

Arrowsmith, A., 21

Asia Minor, 85, 257, 333, 389, 394

Atlas complet des révolutions. See Paris draft
atlases and maps: "Ancient Atlas, Classical
   and Sacred," 31, 37, 133, 140, 147, 148, 155;
   comparative, 16, 20–22, 33, 37, 38, 95,
   168, 232, 382, 430, 443–45, 456; de choix,
   274; Dutch 10, 55, 149; and film animation, 159; geographical, 337–38, 381–82;
   historical (*see* historical atlases); Ptolemaic, 14–17, 21, 25, 32, 34, 37, 104, 105,
   197, 250, 258, 334, 382, 443, 444; national,
   23, 35, 51, 100, 110, 134, 167, 172, 209, 246,
   254, 271, 275, 337, 347, 349–53, 357, 359,
   392, 393, 395, 406, 410, 411, 413, 414, 420,
   429; universal, general, or world (with
   "ancient" preludes or appendices), 1, 22–
   24, 31, 32, 70, 111, 132, 134, 314, 324, 328,
   382, 388, 409, 452, 457, 458, 459–60. *See
   also* cartography; geography; historical
   atlases; *specific maps; individual atlas-
   makers*
Atlas Lesage. See Las Cases
Augustinian Order, 87
Ausser Europa (vol. 3 of Spruner's *Hand-
   Atlas*), 222
Austrasia, 60, 65, 77
Austria, 26, 57, 59–62, 64–66, 204, 234; as
   Noricum 60, 233; as Ostmark 27, 65
auxiliary sciences, 146, 167, 311; geography, as
   aid to history, 3, 15, 131
Avars, (nomadic people), 277
Azov, Sea of, 85

Babenberg, counts of, 27, 64, 65
Baldwin I and II, counts of Flanders, 79
Baltic Sea, 191, 201, 233, 238, 250, 446
Bantry Bay (Ireland), invasion via, 203
Baquol, J., 412
Barbarian invasions: d'Anville, 238; and *Atlas
   historique,* 134, 190–91, 243; Bellin, 193–
   94, 229; Benicken, 320–21, 402-3; G.
   Delisle, 188–89; Freyhold, 423; Gatterer,
   273; Hagelgans, 191–93, 243; as historical
   period, 382, 385, 387–88, 390, 393, 409;
   Houzé, 331; Köhler, 236; Las Cases, 306–
   9, 383, 388, 391–93; Lazius, 64; map subject, 282, 393, 410, 413, 414, 486; Plancher,
   193; *Révolutions de l'univers,* 266; Wedell,
   336, 426; "Sea of Barbarians," 238. See
   also *Völkerwanderung*

Barbié du Bocage, J. D., 20, 321
Barfield, J., 307
Barrington, Daines, 197, 430. *See also* Anglo-
   Saxons
*Barrington Atlas of the Greek and Roman
   World,* 32
Barthélémy, J. J., 322. *See also* Barbié du
   Bocage, J. D.
Basel, 251
battles, 107, 432, 444; near Basel, 251; Dutch,
   328; Speed's map of, 38, 80, 81, 84–86,
   104, 110, 201, 203, 281, 308, 443, 444. *See
   also* atlases and maps
Bavaria, 139, 322; canal in, 147, 154–55, 333–
   34; historical atlas of, 205, 333, 410; in
   Lazius, 26, 59, 65; and lithography, 318;
   Nordgau, 139, 212; poster maps from,
   202–3; and Spruner, 330, 333, 452. *See
   also* Bayerischer-Schulbuch-Verlag;
   Maximilian II
Bayerischer-Schulbuch-Verlag, historical at-
   las of, 356, 440
Bayle, P., 68
Beaurain, Jean, 31, 149, 152, 234, 274
Bel, M., 275
Belgium, 78; as atlas subject 348, 411; medie-
   val, map of, 237. *See also* Flanders; Neth-
   erlands, southern
Bellin, N., 193, 194, 451
Benedictine Order: as problematic map sub-
   ject, 37, 281. *See also* Bouquet, M.;
   Desing, A.; Mabillon, J.; Plancher, U.;
   Vaissette, J.; Vic, C.
Benevento, 86, 87, 92, 104, 247, 444, 457
Benicken, F. W., 318–22, 324, 396, 401–4,
   419, 459
Benjamin of Tudela, 226, 230–31
Beretti, G. *See* Muratori, L.
Berey, N., *fils,* 240, 242–43
Berlin, 168, 219, 327, 328, 355, 391, 423; Univer-
   sity of, 353, 359
Berlinghieri, F., 20
Bertius, Petrus (Pieter Berts, Pierre Bert), 14,
   17, 32, 69–71, 74, 106, 246, 247, 249, 261,
   455. *See also* Empire, Byzantine; Empire
   of, Charlemagne
Bertuch, F. J., 313. *See also* Weimar
Bessel, G. (abbot), and the *Chronicon Got-
   wicense,* 139, 196, 209, 210, 213, 216, 239

Bibliothèque historique de la France. See Le Long, J.

Bibliothèque Nationale, 152, 242, 259, 267, 315, 324; as Bibliothèque Royale, 134

Biondo, F., 26, 78, 79

Birchall, J., 428

Blaeu, W. J., 54, 62, 71, 93, 104

Blommaerts, J., 98

Bodel Nyenhuis, J. T., 100, 252

Böhme, A.G., and Hase's atlas, 148, 259

Boisseau, J., 85, 86

Bonapartist cause. *See* Las Cases, É. de

Bor, S., 230. *See also* Carpini, G. de Plano

Bosworth, J., 429

Bouches-du-Weser. *See* Oldenburg

Boudet, A., 153, 253

Bouillet, M. N., 428

Boullemier. *See* Le Long, J.

Bouquet, M., 246, 249

Bourbon dynasty, Restoration of (1814), 312, 318, 319, 391, 392

Bourgogne, Burgundy, 240; early kingdoms, 25, 56, 66–68, 88, 240, 246, 264, 416, 426 (*see also* Arles, kingdom of); French province, 270; Valois duchy, 410

Brandenburg. *See* Prussia

Bratislava. *See* Pozsony

Bray, T., map collection of, 196

Brecher, A., 428

Bredow, G. G, 168, 271

Briet, P., 20, 21, 59, 66, 67, 104, 265

Brion de la Tour, L., 36, 311, 314

Britain: historical maps of, 55, 327, 410, 415, 417, 418, 419; Roman, 56, 57, 62, 276-77, 327, 387, 411 . *See also* Anglo-Saxons; Speed, J.

Britannia, geographical term, 60; *Saxonica,* map of, 56. *See* Camden, W.

British Library, King's Topographical Collection at, 380

Brittany: drooping nose outline of, 240, 241; and Las Cases, 303, 304

Brno. *See* Tomka Szásky, J.

Browne, C., 29, 55

Brué, A. H., 396, 405, 406, 408, 427; career, 319; and scorn for Rizzi-Zannoni's atlas, 160, 161, 271, 318, 406

Brunet, P. N., 166

Bruun, M. C. *See* Malte-Brun, C.

Buache, P., 135, 138, 171, 241, 275, 324. *See also* Delisle, G.

Buchon, J. A., 307

Budapest. *See* Tomka Szászky, J.

Bulgaria, 267, 317; in Byzantine period, 201, 250, 263, 413

Burgundians, origins of, 88, 193, 293, 388. *See* Plancher, U.

Burgundy. *See* Bourgogne, Burgundy

Büsching, Anton Friedrich, 172

Byzantine Empire: Angelus dynasty, 264; and Bulgaria, war with, 201; defeat at Manzikert, 389; as map subject, 71, 222, 224, 250, 254, 256, 260, 274, 332, 391, 392, 403, 412, 414, 416, 419, 420, 428; in southern Italy, 87, 209; studies of, 222. *See also* Constantine VII; Constantinople; Procopius

Caert Thresoor (Tabulae contractae). See Bertius, P.

Caesar. *See* Ewich, H.

Caliphate: Islamic, 313, 327, 380, 418, 420; Abbasid, 259, 314, 382, 387–88, 409, 412; of Cordova, 238, 250, 420; Ummayad, 88, 260, 314

Calvin, John, and maps of Paradise, 35

Calvinists: and the abbey of Egmond, 76; and Bertius, 69, 74

Camden, William, maps in his *Britannia,* 26, 52, 54–57, 62, 64

canal, of Charlemagne, 139, 147, 194–96, 212, 240, 328, 380, 388, 452

Capet, Hugh (king of France), 382, 408, 409

Capetian (French dynasty), 86, 238, 239, 249, 271, 410

Carey, H. C., 304. *See also* Buchon, J. A.

Carolingian dynasty: 211, 218, 235, 241, 242, 250, 255, 271, 392, 403, 408, 423; and Austrasia, 58, 59, 60, 66; division of empire, 59, 67, 219, 246, 249, 252, 255, 380, 382, 388, 391, 408, 418; empire as map subject, 61, 70, 71, 74, 246–47, 382, 388, 410, 411, 457; and Frisia, 101; palaces mapped, 155; Spanish March, 71, 162, 202, 239. *See also* atlases and maps; canal, of Charlemagne; Franks

Carpini, G. di Plano, and travels to Asia, 230, 231

cartography: Danish, 32; Dutch, 74, 75, 253; French, 193; German, 360; "historical," 7–8, 30, 34, 92, 98, 280, 342; history of, 14, 30, 61, 74, 75; medieval, 411; sacred and ecclesiastical, 36, 91, 92, 427; specialized for history, 8, 104, 354

Cassini, G. M., 31

Cassini, J. D., 135

Catalogue: of dealers, 160–62, 165, 166, 251; of libraries, 4, 8, 132, 152, 194, 235, 240, 273; map type, 81, 86, 201–4, 251, 277, 443, 444. *See also* maps, catalogue or poster

Catherine, Saint, monastery of (Mt. Sinai), 68

Cellarius (Keller), C., 21, 56, 106, 139–41, 146, 147, 231, 232–33, 235–37, 239, 254, 257, 265, 309, 390, 450; his periodization, 209, 236, 389

Chamber of Deputies. *See* Las Cases, E. de

Chanlaire, P. G., 204, 451

Charlemagne (Charles the Great), 27, 59, 77, 85, 135, 149, 190, 219, 233, 239, 277, 400, 405, 412; and Austria, 64, 65; as common European ground, 409, 410, 411, 415, 420, 428; division of his empire, 138; and *Histoire de France* (Daniel), 242; and persecution of paganism, 211; as a term in periodisation, 133, 139, 148, 209, 236, 388–90, 402–3; ubiquity of, 382, 396, 408, 416. *Maps of empire, by cartographer:* Beaurain, 274; Bertius, 5, 69–74, 77, 85, 106, 194, 449, 455, 457; Brué, 406; Delisle, 255–56; Freyhold, 423; Georgisch, 247–49; Hase, 59–61; Koch, 385; Las Cases, 392; Malte-Brun, 387; Rizzi-Zannoni, 162, 165, 270; de Vaugondy, 246, 380. *Other maps:* 2, 141–42, 328, 380, 381, 395–96. *See also* canal, of Charlemagne

Charles I, king of England, 84

Charles V, Holy Roman Emperor, 149, 260, 267, 395, 401, 402; map subject for Malte-Brun, 313, 323

Charles V, king of France, 205, 270

Charles VI, Holy Roman Emperor, 148, 149, 260, 261, 275

Charles VI, king of France, 205, 270

Charles VII, Holy Roman Emperor, 202

Charles VII, king of France, 251, 270

Charles XII, king of Sweden, 306, 414

Charles Martel (grandfather of Charlemagne), 408

Charles the Bald, king of West Francia, 61, 79

Charles the Bold, duke of Burgundy, 409

Chiquet, J., 451, 452. *See also* Ménard, A.

Chlothar I, king of the Franks (Merovingian), 241, 246

chorography (celebration of a country), 26, 62, 64, 104

Christ: as dating element, 106, 211; possible depiction of, 92

Chronicon Gotwicense. See Bessel, G.

chronology, 2, 104, 105, 132, 148, 155, 205, 256, 277, 330, 389; absent from maps for history, 25, 27, 104, 105, 388; handbooks of, 3, 333, 428, 450; present in maps and atlases, 105, 106, 107, 148, 196, 212, 254, 282, 308, 392; subordinate to geography, 394, 453. *See also* genealogy, as aid to history

Cicero, quoted by Ortelius, 16

civil wars: Carolingian, 219; English, 80, 84; Swiss, 251

Clausolles, P., 352, 408, 416

Clovis, king of the Franks, 65, 79, 381; as map subject, 138, 235, 239–41, 243, 250, 270, 277, 388, 408. *See also* Daniel, P.; Delisle, G.; Liébaux, H.; Mentelle, E.

Coindé, Madame, 307

Collège royal (renamed Collège de France), 69, 227. *See* Bertius, P.; Pétis, F.

Cologne, 412; atlas publications from, 58, 64, 78

Congress of Vienna (1814–15), 313, 318, 319, 455

Constantine I, the Great 139, 325, 381, 402

Constantine VII, Porphyrogenitus, treatises by, 135, 222, 224

Constantinople (Byzantium), 85, 105, 141, 147, 222–24, 255, 265, 266, 383, 385, 388–90, 401

Cook, Captain J., 198, 322

Copenhagen. *See* Mejer, J.

Coquart, A., 208

Coronelli, V., 55

Corsica, 70; called "Italia francese," 411

Corvey, Saxon abbey, 214, 254

Cosmographia. See Ptolemy, C.

Councils, church, 32, 36, 37, 202, 277, 322, 427, 444

Covens and Mortier. *See* Delisle, G.

Croix (Crois), François Pétis de la. *See* Pétis, F.

crusades, 2, 35, 36, 85, 222, 224, 225, 256, 259, 272, 379, 382, 389, 391, 406, 409, 410, 412, 413, 416, 423, 426, 428; within Christendom, 196; end of, 389, 415, 416; expeditions, 7, 320, 382, 391, 394–401, 402, 408, 410, 446; Fourth, 264, 265, 272, 401; principalities established by, 225, 255, 388, 394, 396; projected, 141, 148; of St. Louis, 225, 403; Third, 191, 277

Cyrus, king of Persia, 226, 313

Dacia, (ancient name for Transsylvania), 276, 404; learned name for Denmark, 58

Dagobert, king of the Franks, 241

Danckerts, C., 84

Danckwerth, C. *See* Mejer, J.

Dangeau, Louis Courcillon de, 29, 241

Daniel, Pierre: and Charlemagne's canal, 195; *Histoire de France,* 160, 162, 271; illustrations for, 240, 242, 243, 250, 271, 388. *See also* Berey, N.; Liébaux, H.; Vaugondy, R.

Danube river, 28, 65, 191, 212, 232, 273; and Charlemagne's canal, 139, 194, 240; mouths of, 71; valley of, 152, 389

Dauphiné, 69, 239, 267, 271; G. Delisle's map of, 208, 212, 216, 223, 254, 273, 277

Delamarche, 326–27, 380

Delisle, C., (father of Guillaume), 135, 140

Delisle, G., 5, 20, 66, 74, 135, 146–48, 149, 156, 187, 205, 223, 240, 246, 258, 265, 274, 280, 322, 457–58; early French kingdoms, 240–42, 249, 406; the Holy Land, 225; maps of the Byzantine Empire, 224; *Theatrum historicum,* 188–90, 243, 265, 272, 382, 387, 450; Toul, 208–9, 212, 222; unrealized projects, 140–41, 173, 232, 235, 254–56, 259, 260, 276, 394. *See also* Buache, P.; historical atlases, *theatrum historicum;* Louis XV

Delisle, J. N., (brother of Guillaume), 140, 225–26

deluge. *See* flood

Denaix, M. A., 325, 326, 329, 331, 405, 408, 447

Denmark, 58, 154, 311, 354, 403; Frisian, 92; history of, 265, 267; mapping of, 93, 98. *See also* Malte-Brun, C.; Mejer, J.

Dépôt de la guerre (French army map bureau). *See* Denaix, M. A.

Dépôt de la Marine (French naval map bureau), 225, 241, 243, 338

Desing, A., 202–3

Desjardins, C., 323; thematic maps by, 405, 447

Desnos, L. C., 311; first historical atlas of France, 160–62, 165, 172, 269–71; indifferent reception, 166, 276, 307; and thematic maps, 29–30, 321. *See also* Rizzi-Zannoni, G. B.

Deventer, J. van, 75, 98

Didot, J., 304

discoveries, 15, 200, 230, 319, 390, 400; Captain Cook's, 197; displayed on maps, 230, 320, 321, 336, 386, 402

dispersion. *See* Noah, dispersion of sons

Doederlein, J. A., 195. *See also* canal, of Charlemagne

domain (French royal), 162, 165, 168–71, 313, 408, 413

Dozy, G. J., 413

Dresch, G. L. von, 316, 320, 401–2, 404

Drioux, C. J., 416

Droysen, G., 419, 426, 448

Dufau, L., 408

Dufour, A. H., 323, 412–13, 427, 429

Dufour de Longuerue, L., (mentor of d'Anville), 135, 152, 234–35, 238

Dupré, (pseudonym), *Révolutions de l'univers:* 155, 159–62, 165–67, 171, 172, 262, 265–69, 271, 272, 306, 308–17, 322, 336, 387, 402, 404, 419, 457, 459, 460; identity, 156. *See also* Picaud, M.

Duruy, V., 326, 391, 408, 413, 415, 416, 428

Dussieux, L., 329, 408, 417; and thematic maps, 30, 417

Duval, P., 33, 67; 3-part atlas, 33, 88, 90, 105; 4-sheet map of France, 25, 26, 58, 66, 67, 68, 104, 234

East Indies, 202, 329. *See also* India

ecclesiastical geography: 25, 29, 32, 36–37, 105, 327, 427, 457; English, 56, French, 405, 417. *See also* Brion de la Tour, L.; Nolin, J. B.; Petri, G.; Rizzi-Zannoni, G. B.; Sanson, N.; Wiltsch, J.

ecclesiastical history, 140, 254–55. *See also* Delisle, G.; Rohrbacher, R. F.

Edinburgh. *See* Thomson, J.

Eekhoff , W. *See* Bodel Nyenhuis, J. T.

Egmond, abbey of (Holland), 75, 76

Egypt, 110, 263, 264, 267, 388; crusade to, 225; deserts of, 20, 90, 91; Exodus from, 35, 446; Napoleon in, 204, 397

Eidgenossen (Conferederates). *See* Switzerland

Elbe river, 80, 233, 276

Electoral Saxony. *See* Georgisch, P.

Elector Palatine. *See* Karl Theodor, Elector Palatine

Elizabeth I. *See* Speed, J.

Elst, P. C. van der, 347

Empire, 24, 71, 132, 133, 148 152, 154, 167, 190, 229, 256, 257-9, 262, 317, 335, 343, 380, 455; Byzantine, 71, 87, 133, 135, 222, 224, 250, 254, 263, 264, 388, 391, 403, 412, 414, 420; Carolingian, 59, 61, 67, 71, 219, 282, 408, 418; of Charlemagne, 2, 5, 31, 69, 71, 79, 85, 106, 138, 141, 219, 242, 246, 247, 249, 255, 256, 259, 261, 270, 274, 347, 379, 385, 387, 388, 389, 392, 395, 400, 405, 406, 408, 409, 411, 449, 455, 457; Dutch East Indian, 348; *of Great Britain* (Speed atlas), 54, 81, 84; Islamic, 88, 141, 148, 255, 259, 382, 387; Latin of Constantinople, 105, 141, 255, 426; of Napoleon, 313, 314, 323, 349; Ottoman, 148, 227, 257, 259, 2 77; Persian, 87, 149, 426; Roman, 34, 79, 87, 104, 105, 133, 148, 153, 159, 188, 189, 190–91, 193, 194, 209, 232, 238, 240, 258, 266, 270, 273, 313, 323, 353, 358, 383, 385, 387, 392, 395, 401, 414, 444, 451; Romano-Germanic, Holy Roman, medieval German, 60, 67, 71, 74, 138, 139, 141, 148, 149, 190, 202, 249, 250, 257, 259, 260, 261, 267, 271, 273, 275, 308, 337, 349, 357, 388, 389, 390, 408, 409, 410, 416, 418, 423; Second French, 352, 406; Seleucid, 266; of Tamerlane, 207, 229, 230, 255, 258

Ems river, 92, 100, 101

Engelmann, G., 320, 327, 390

English Channel, 71, 76, 110, 193, 203, 253, 271, 406

Enlightenment, the, 21, 35, 191, 222, 381, 398, 405

Enns river, 62, 65

Enouy, J., 203–4

Ernst August, duke of Saxe-Weimar, 211. *See also* Zollmann, F.

Établissement géographique. *See* Vandermaelen, P.

Eurasia, 312, 403; in Dupré, 159, 171, 272, 316; and the Middle Ages, 222, 231, 382; in the Paris draft, 152, 154, 262, 456; in universal, sequential atlases, 155, 316, 321, 327, 419, 458

Europe 14, 71, 149, 191, 192, 194, 216, 222, 229, 238, 249, 256, 260, 267, 273, 275, 304, 315, 319, 401, 419, 460; Europeans, 14, 15, 55, 230–31, 280; as map subject, 2, 3, 5, 20, 51, 52, 230–31, 239, 250, 255, 259, 273, 277–78, 279, 281, 307, 309–13, 315–16, 317, 319–23, 328–30, 331–38, 382–97, 400, 403–6, 408–10, 412–17, 419–20, 422, 423, 426, 428, 429, 444, 447, 458–60; publication in, 3, 52; thematic maps of, 35, 36, 86, 329, 447

Eusebius of Caesarea: chronicle of, 133, 264; map of Palestine by, 34

Ewich, H., 79–80

Exodus, map of, 25, 30, 35, 51, 106, 155, 396; synchronism in, 110, 111, 443, 445, 446; tracks in, 34, 317, 325

exploration, geographical, 104, 230, 319, 322; history of, 313, 316, 317, 387, 388, 429. *See also* Anglo-Saxons; discoveries

eye of history. *See* geography

Fabius, N., 79, 104

Faden, W., 277

Falcke, J. F., 214, 215. *See also* Corvey, Saxon abbey

Falckenstein, J. H., 139, 195, 212

Falger, A. *See* Benicken, F. W.

Fembo, C. (successor to Homann Heirs), 307

Fer, N. de, 20, 91, 208

Ferdinand I, Holy Roman Emperor, 26, 61; of Aragon and Castile, 423

Findlay, A. G., 21

Flanders, as map subject, 74–77, 79, 80, 146, 452

Flevus (Vlie) river, 101, 238

flood: after the biblical, 101, 275; classical, of Deucalion, 155, 451; in Frisia, 79, 95, 98, 108, 111, 450. *See also* Noah, dispersion of sons

Fontanieu, G. M. de, and Rizzi-Zannoni, 161, 268, 271, 324

Fontenoy, battle of, as map subject, 219, 380

Forster, J. R., and King Alfred's Orosius, 197, 200, 429. *See also* Anglo-Saxons

Fortia d'Urban, A. J. de, 380–81

Fourth Crusade. *See* crusades

Francia (pre-Capetian Frankish kingdom), 59, 60, 70, 78, 139, 195, 209, 215, 218, 232, 235, 238, 239

Francis I, king of France, 381

Franciscan Order, 37

Franconia, 70, 209, 215, 239

Frankfurt, 218, 235. *See also* Hagelgans, J. G.

Franks: by-gone kingdoms of the, 26, 58, 68 411; Carolingian, 69, 139, 196, 218, 235, 246, 247, 249, 259, 408; early, 76–79, 86, 138, 189, 205, 215, 238, 240, 271; east, 355; equivalent to French, 71, 85, 235, 242–43; in Frisia, 99, 101; in Italy, 209–10; Merovingian, 58, 64, 68, 242, 256, 388, 392, 403, 406, 409, 416, 426; and names for the winds, 70, 249; palaces, 380; Salian 101, 233; Salic law of the, 77, 78; west, 61, 213. *See also* Francia; Neustria; *individual Frankish dynasties and kings*

Frederick, duke of Gottorp, 93

Frederick II, Holy Roman Emperor, 149, 261

Frederick II, king of Prussia, 316

Frederick III, 251, 314

Freiburg im Breisgau. *See* Herder, J. G.

Frémin, A. R. *See* Monin, C. V.

French Revolution, 3, 35, 58; and frontiers, 98; wars of the, 203, 310

Fréret, N. (friend of G. Delisle), 254

Freycinet, L. de. *See* Brué, A. H.

Freyhold, A. von, 357, 359, 422, 423, 426

Frisia, Friesland, 75, 77, 92–94, 95, 98–101, 194, 208, 214, 231, 238, 246, 250, 252, 253, 453, 455, 457. *See also,* Mejer, J.; Zeeland

Frommann, M., 428

Fulda, abbey, charters of, 208, 214

Galiani, F., and Rizzi-Zannoni, 269

Gallia Christiana. See ecclesiastical history

Gatterer, J. C., 131, 147, 167, 168, 171, 172, 272–74, 309, 321, 327, 334, 383, 385, 403, 447, 448, 451, 454, 458, 459

gau, gaue (district), 139, 196, 208, 209, 211–16, 218, 219, 280, 355

Gaul, 77, 246; inhabitants of, 77, 192, 266, 457; as map subject, 28, 31, 80, 138, 146, 147, 165, 193, 234–38, 240, 242, 250, 260, 271, 388; meaning France, 58, 66, 236, 242, 243, 282

Gauttier du Lys d'Arc, É. , 429

genealogy, as aid to history, 29, 64, 167, 305, 333, 428, 450. *See also* chronology

Genghis Khan, 135, 148, 227, 228, 264, 265, 267, 273, 281, 382, 454

geography: from 1700, 139, 232–33, 234, 237, 238, 249, 328, 381, 399; and ancient history, 5, 13–18, 451, 452, 455, 456; historical, 6, 8, 24, 25, 29, 87, 160, 219, 238, 352, 380; modern, 5, 15–17, 20–22, 38, 61, 111, 132, 146, 171, 234, 239, 280, 381, 386, 453, 456; nature of, 105, 106, 110, 111, 453, 456; physical, 23, 135, 171, 278, 349, 358, 405; as preoccupation of early modern mapmakers, 4, 5, 15–20, 30–38; Ptolemy and, 13–17, 443. *See also* atlases and maps; cartography; history

Georgisch, P., 74, 246, 247, 249, 261

Germany, 60, 80, 201, 216, 219, 380; cartography in, 69, 262, 318, 319, 331–33; "Great" (*Magna Germania*) (*see* Ptolemy, C.); history of, 388, 409; location, 92, 243, 251, 267, 273, 305, 310, 311, 323, 325, 336, 397, 410; national atlas of, 329–30, 335, 409, 410; and school atlases, 324. *Maps of, by mapmaker:* Alting, 100; Bertius, 70; Bessel, 209–10; Blaeu (*Theatrum*), 31; Clausolled, 416; Delisle, G., 141, 255; Duruy, 415; Forster, 197; Fortia, 381; Hase, 147, 148, 257; Homann atlas, 139, 212; Köhler, 147, 286; Las Cases, 308; Levasseur, 413; Mejer, 93; Ortelius, 22; Pommper and Kutscheit, 410; Sanson, 27–28; Schannat, 37; Seutter, 202; Spener, 138, 233–34; Spruner, 414, 419–20, 422; Vaugondy, 60, 249, 250

Ghent, 54, 70, 75, 76, 78. *See also* Flanders

Gibson, E., 56, 57

Gleditsch, J. F., 148

Goethe, J. W., 330, 360

Gotha, map publication in, 347, 350, 353, 354, 359, 360, 419. *See also* Perthes, J.

Goths, 232, 403; in Gaul and Spain, 236–37, 240, 246, 274, 380, 388, 393, 416, 423, 426; of Italy, 277, 387, 426; migrations of, 191, 192, 335, 414–15

Göttingen, 160, 275, 315, 397; University of, 131, 141, 167, 168, 172, 272. *See also* Büsching, A. F.; Gatterer, J. C.; Köhler, J. D.

Göttweig, abbey, 139, 210, 213. *See also* Bessel, G.

Gover, E., 416, 417

Grande Compagnie. *See* Quebec (as French colony)

great discoveries. *See* discoveries

Greaves, J., 88, 90, *See also* Duval, P.

Greece, 31, 64; as map subject, 20, 71, 250, 279, 313, 330, 381, 388, 414, 429

Greenland, 230, 320

Gregory (historian) of Tours, 243

Gregory V, pope, 214

Gregory VII, pope, 335, 389, 402, 427

Groningen, 98, 100, 152

Gros, C., 307, 308. *See also* Lavoisne, 307

Guadet, J., 348, 406

Gueudeville, N., 132

Gui de Dampierre, count of Flanders, 75, 451

Guignes, J. de, 193, 194

Guizot, F., 342

Güssefeld, F., 219, 338, 454

Gustavus Adolphus, king of Sweden, 308, 311

gymnasium, 314, 315, 327, 349, 412

Haas, G. Z., 195. *See also* Hase, J. M.

Habsburg, 139, 336, 390, 391, 403, 410, 411; and Lazius 26, 61, 62, 64–66

Hagelgans, J. G., 133–35, 191–94, 209, 236, 450

Hahn, F. J. *See* Bessel, G.

Halle, University of, 138, 139, 306, 314, 328. *See also* Cellarius (Keller), C.; Spener, P. J.; Wittenberg, University of

Halma, F., 56, 99, 100

Handtke, F., 337

Hannibal, campaigns of, as map subject, 33, 133, 267, 312

Haraeus (Verhaer), 34

Hartzheim, J., 37

Hase, J. M., 74, 147–49, 156, 161, 172, 190, 236, 239, 254, 265, 271, 280, 314, 340, 402, 453, 458, 460; "Greatest Empires," 154, 155,

166, 167, 173, 229, 256–62, 274, 275, 276, 280, 317, 322, 380, 384, 387, 455, 458, 459; posthumous atlas of, 134, 152, 313. See also *Seven Maps* (historical atlas of the German empire)

Hauber, E. D., 25, 26, 58, 60, 61, 67, 74, 104

Häufler, J., 347

Heck, G., 340, 400, 401

Henry III, Landgrave of Thuringia, 212

Henry IV, king of France, 67, 277, 389, 402

Heraclius, East Roman emperor, 222, 224, 234

Herder, J. G., 321

Hereford cathedral, world map at, 34

Hildesheim. *See* Lauenstein, J. B.

Hilfswissenschaft. See auxiliary sciences

Hilgemann. *See* Kinder, H.

historical atlases: and chronology, 105–10; definition of, 7–8, 13, 132–34, 147, 155, 190, 313–14, 326; design of, 167, 173, 307–8, 312, 330–32, 335–36, 381, 422, 426, 453; *Geschichtsatlanten, Geschichtskarte*, 7, 16, 17, 39; after Spruner, 334; *theatrum historicum*, 135, 140, 141, 190, 224, 243, 272, 383, 388, 450; of today, 1–6; universal, sequential, 155, 172, 274, 326, 335, 337, 384, 394, 404, 445, 450, 456; and writing, 448–51. *See also* atlases and maps; cartography; geography

history, contemporary, 381; modern, 2, 24, 51, 131, 148, 155, 311, 340, 347, 381, 401, 405, 410, 446, 455, 456, 459

Hole, W., 52

Holy Land, 14, 20, 33–35, 88, 91, 92, 104, 110, 140, 191, 224–26, 273, 329, 381, 394–96, 446, 457

Homann, J. B., 55, 62, 139, 147, 161, 195, 212, 231, 251 256, 257, 262, 309, 310, 312

Homann Heirs (*Heredes Homanniani, Homman Erben*), 139, 147, 161, 251, 256, 257, 309, 310

Hondius, J., 33, 54, 62, 70, 78, 79

Hopper, J., 98

Houzé, A., 330, 331, 410, 413–15

Howe, H., 427

Huber-Salantin, J., and the *Atlas Lesage*, 459

Huberts, W. J. A., 357

Hughes, W., 21, 343, 346, 357, 405, 412, 452, 456

Humber river, as Saxon landing place, 191, 280

Humboldt, A. von, 331

Humboldt, W. von, 328, 331, 360, 386

Hundred Years' War, 390, 410, 417

Hungary, 59, 65, 229, 233, 267, 273; historical atlas of, 152, 274–77, 300, 410; as map subject, 62, 354, 357, 379, 388, 419, 420; as Pannonia, 58, 60, 276, 277

Huns, 233; history of the, 193, 266, 272, 273, 277; as map subject, 386, 387, 393, 395, 414, 426; tracks of the, 110, 335, 336, 403

Husum. *See* Mejer, J.

Huxley, T., 35

Iceland: colonization, 230, 279; literature, 200; as map subject, 200, 277, 279, 317, 320, 405. *See also* Northmen

Idrisi, al (the "Nubian geographer," Arab geographer), 88, 141, 255, 256, 413

India, 2, 148, 225, 227, 263, 267. *See also* East Indies

Insselin, C., 67

Islam, 105, 192, 229, 256, 427; sources from, 228; world of, as map subject, 90, 141, 154, 238, 255–56, 260, 317, 322, 332–33, 380, 382, 388, 392, 412, 418, 419, 420, 428. *See also* Caliphate, Islamic; Muhammad; Muslims

Isle, de l'. *See* Delisle, G.

Jabłonowsky, J. A., and Rizzi-Zannoni, 269

Jaillot, H., 28, 29, 37, 90–92, 240, 444

Janssonius, J., 31, 32, 70, 71, 80

Japan, 159, 193, 231, 314, 328

Jarry de Mancy, A., 307, 459

Jesuits, 37. *See also individual Jesuits:* Briet, P.; Daniel, P.; Hartzheim, J.; Jouvancy, J.; Kircher, A.; Malbranq, J.; Wastelain

Jeszenák, 275. *See* Tomka Szászky, J.

Johnston, A. K., 396

Jørden, M., 95

Journal des sçavans, 141, 161, 166, 193, 263

Journal de Trévoux, 155, 160–62

Jouvancy, J. de, saying by, 104, 105

Julian the Apostate, Roman Emperor, 226, 402

Julien, R. J., catalogue of, 157, 161, 209, 251

Julius Caesar. *See* Ewich, H.

Jusseret, N. J., 348

Justinian, empire of, 87, 148, 260, 261

Kadesiah, battle of, 226

Karamzin, N. M., 411

Kärcher, K., 341, 391

Karl Theodor, Elector Palatine, 217

Keere (Kaerius), P. van den, 70

Keller. *See* Cellarius (Keller) C.

Kiepert, H., 428

Kinder, H., 2, 4, 11, 13, 262, 356, 408

Kircher, A., 111

Klaproth, J., 172, 327–29, 331, 384; and the Humboldt brothers, 328

Koch, C. G., 216, 323, 326, 339, 342, 352, 383–85, 387, 388, 391, 392, 405, 406, 413, 420, 428, 454. *See also* Rosny, A.

Köhler, J. D., 56, 139, 141, 146, 147, 152, 167, 195, 209, 222, 235–37, 254, 255, 260, 265, 274, 315, 450

Kohlrausch, F., 340

König, T., 428

Kremer, C. J., 218

Kruse, Christian (Karsten): atlas of, 313, 314–23, 329, 337, 341, 348, 356, 385, 388–89, 390, 401, 406, 420, 426, 427, 429, 444, 453, 458; biography, 315–16, 383; century intervals, 342, 382, 384, 387, 390, 420, 454, 460; European scope, 339, 347, 359, 383–84, 405, 422, 459; and Gatterer's atlas, 168, 171, 172, 272, 273, 331; and Las Cases's atlas, 306, 310, 312;

Kutscheit, J. V., 352, 410, 413, 415, 416

Lafréri (atlas type), 16

Lamare (Delamare), N. de, 75, 110, 205, 217, 262, 276, 457

Lambarde, W., 52, 54, 57, 104, 146, 449, 452

Lamey, A., 217, 218

Languedoc, maps of, 212, 213, 228, 240

Lapie, P., 15, 347, 387, 397

La Ruë, P. de, 87, 88, 100, 105–7, 141, 205, 256, 395, 453, 455, 457

Las Cases, É. de, and the *Atlas Lesage* , 315, 317, 320, 321, 334, 335, 339, 401, 402; and barbarian "Transmigration," 382, 383, 388, 392–93, 396, 446; biography, 313–18, 383; dismissed, 274, 313, 321–23, 325, 331, 334; historical training, 312, 313; layout,

Las Cases, É. de (*continued*)
303, 310, 312, 314, 350, 356, 388, 401, 419, 427, 429, 450, 454, 459; Lesage trademark, 304; and medieval history, 391–94; success, 315, 322, 337, 356, 359, 382, 391–93; and synchronic maps, 312

Latin Empire of Constantinope, 105, 141, 256, 427

Lattré, J., 262, 265, 353

Lauenstein, J. B., 213–15

Laurie and Whittle, 203

Lavoisne, C. V., and the *Atlas Lesage,* 307

Lazius, W., 27, 58, 61, 62, 64–66, 68, 104, 105

Lea, Isaac. *See* Carey, H. C.

Le Clerc (Clericus), 91

Leeder, E., 428

Legrand, A., 339

Leiden, 78, 194; University of, 17, 69, 70, 99

Leipzig: publishers at, 139, 148, 316, 322, 455; University of, 256, 311, 383

Łelewel, J., 411

Le Long, J., 25–27, 67, 161, 241, 242

Lenglet du Fresnoy, N., 21, 25, 26, 36, 58, 74

Leroy, C. *See* Drioux, C. J.

Lesage. *See* Las Cases, É. de

Levant, the, 141, 226, 256, 257, 352, 397, 407, 457

Levasseur, V., 352, 413

L'Hermite, J. B., 339

Library of Congress, manuscript atlas at, 307

Libya. *See* Africa, North

Liébaux, H., 138, 146, 235, 240, 242, 243, 246, 407

Lipsius, J., 61, 70

Lisle, de. *See* Delisle, G.

lithography, 318, 324; and maps, 321, 322, 390

Lithuania, 153, 268, 421

Liutpold, Saint. *See* Babenberg, counts of

Lizars, D., 23

Lombards: before Italy, 192, 233; in Italy, 58, 87, 146, 209, 320, 426; as map subject, 222, 280

London, 54, 279, 304, 306, 307, 326; bishopric of, 56; maps in (British Library), 242, 278; as publication center, 4, 105, 188, 235, 252, 277, 304–6, 322, 324, 379, 391, 416

Longnon, A., 205

Longuerue. *See* Dufour de Longuerue, L.

Lorraine, 26, 58–61, 65, 66, 69, 141, 153, 156, 159, 208, 210, 239, 256, 268, 412

Lorsch, abbey of, cartulary 217. *See also* Lamey, A.

Lothar I and II, Carolingian kings, 59, 61, 149, 233, 242, 247, 250, 262

Louis I, the Pious, Carolingian emperor, 61, 219, 250

Louis IX, Saint, king of France, 85, 86, 225, 271, 382, 403, 410. *See also* crusades

Louis XI, king of France, 241, 272, 409, 443

Louis XIII, king of France. *See* Bertius, P.

Louis XIV, king of France, 68, 69, 91, 132, 135, 205, 222, 235, 336, 382, 449, 457

Louis XV, king of France, 165, 255, 271; tutored by G. Delisle, 66, 138, 188, 240, 241, 276, 388

Low Countries. *See* Netherlands

Löwenberg, J., 352, 413

Maaskamp, E., 347

Mabillon, J., 139, 216, 452

MacGeoghegan, J., 253

Maggi, C., 411

Malbrancq, J., 76, 77

Malta, 411; knights of, 138, 225

Malte-Brun, C., 173, 323, 326, 340, 347 384, 385, 387–91, 397, 405, 413, 459

Mancy, A. Ire de. *See* Koch, C. G.

Manesson-Mallet, A., 56, 68, 193

maps: as cartoon strips ("bande dessinée"), 316, 448; cartouche (inscription in a decorated frame), 26, 52, 54, 67, 162, 165, 192, 211, 247, 271, 310, 316; catalogue or poster, 80, 81, 86, 91, 201–4, 233, 251, 277, 281, 311, 443, 444; dynamic or synchronic, 7, 81, 85, 93, 101, 192, 201, 264, 308, 319, 382, 392, 443, 444–48; engravers, 26, 54, 62, 84, 90, 196, 202, 205, 240, 252, 257, 330; grid on, 14, 16, 104, 308, 444; for history, defined, 7–8; literary, 33, 34, 51, 149; medieval, 14, 34, 35, 75, 411; military 2, 29, 64, 68, 81, 161, 251, 341, 386; "new" (*tabulae novae*), 15, 17; outline (methodical), 168, 272, 331, 340, 401; passive, 7, 280, 443–45; projections, 12, 14, 24, 257, 279; synchronism in, 4, 110, 311, 331; thematic, 11, 16, 23, 29, 36, 81, 88, 133, 165, 172, 255, 271, 278, 341, 342, 349, 354, 356, 361, 383,

389, 406, 409, 417, 418, 424, 428, 447, 448; tracks on, 34, 110, 111, 133, 191, 192, 225, 226, 229, 231, 243, 261, 266, 279–81, 311–13, 322, 331, 334–36, 343, 351, 358, 383, 390, 392–95, 397, 401–5, 407, 409, 411, 427, 445, 446, 459. *See also* atlases and maps; cartography; geography; historical atlases
Marchal, J., 307, 308. *See also* Las Cases, É. de
Marco Polo, as map subject, 229–31, 388
Marmocchi, F. C., 412
Maximilian, Holy Roman Emperor, 420
Maximilian II, king of Bavaria, 353
Mejer, J., 92–98, 99, 100, 104, 111, 208, 231, 445, 449, 456
Mémorial de Sainte Hélène. See Las Cases, É. de
Ménard, A., 452
Menke, T., 355, 418. *See also* Spruner von Merz, K.
Mentelle, E., 21, 239, 308, 326, 337
Mentzer, T. A. von, 411
Mercator, G., 12, 17, 20, 26, 33, 58, 66, 334, 335, 403
Mercure de France, 140, 255
Metz. *See* Austrasia
Meurer, P., 30, 33
Mézerai (Mézeray), F. E. de, 160, 162, 240
Michaud, J., 397, 400
Michelet, J., 351
Middle Ages, medieval (selective), abbreviated (to 9th century only): 106, 133, 139, 209, 236; canon of scenes, 208–19, 385–91, 454; emphasis on Asia, 148, 154, 173, 229, 255, 259, 260, 329, 382, 384, 412–13, 454; geography, 36, 56, 69, 100, 111, 139, 146, 187, 215, 219, 231, 238, 266, 280, 330, 406, 429, 455, 457, 459; included in "modern," 2, 6, 24, 51, 52, 138, 316, 326, 381; mapmakers' aversion to "high," 141, 147, 236, 239, 241, 249, 252, 255, 260, 261, 263, 280, 380, 386, 389, 403, 428, 454; medievalism, 6, 453
migration. See *Völkerwanderung*
Mikoviny, S., 275
Mitchell, S. A., 31
Mohamet. *See* Muhammad, prophet
Möller, A. W., 340, 341, 391. *See also* Muhlert, K. F.
Mongols, 148, 228–30, 256, 259, 260, 263, 264,

265, 267, 273, 326, 353, 380, 382, 387–89, 403, 409, 412, 413, 418, 423, 426, 428
Monin, C. V., 347
Monsieur (royal honorific), 228, 339. *See also* Brué, A. H.
Morden, R., 56
Moulart-Sanson, P. (third Sanson generation), 24. *See* Sanson, cartographic dynasty
Muhammad, prophet, 392, 395
Muhlert, K. F., 391. *See also* Möller, A. W.
Muller, Frederick, book dealer, 134
Mullié, C., 349
Muratori, L., 146, 209, 222
Muslims, 70, 231, 236, 261, 341 394. *See also* Islam

Nancy, battle of, 409. *See also* Austrasia
Napoleon, 313–14, 317, 329, 381, 382, 390; age of, 303, 311, 314, 320, 322, 325, 337, 388, 409; and Las Cases, 303–5, 307–9, 320; as map subject, 308, 309, 319, 320, 321, 459; wars of, 203, 204, 312, 397
Napoleon III (Louis Napoleon), 331
Nassir-Eddin, 88, 90. *See also* Duval, P.
Navarre, kingdom of, 85, 250, 264, 268, 269, 271
Near East, 192, 227, 395, 417, 419
Netherlands (Low countries), 70, 71, 74, 100, 149, 187, 200, 347, 348, 418; as map subject, 7, 75, 80, 110, 194, 250, 252, 275, 337, 354, 410, 411, 451; northern, 98, 99; Roman, 253; southern (Belgium), 76, 78, 348; threatened by the sea, 74, 80, 92, 98, 99; united, 99, 252, 254, 267. *See also* Frisia, Friesland; United Provinces; Zeeland
Neustria, Frankish kingdom, as map subject, 25, 26, 39, 58, 59, 66, 67, 238, 241
new world, 16, 153, 403
Nicaea I, church council, 37, 202
Nijmwegen, Peace of, 69
Noah, dispersion of sons, as map subject, 35, 93, 111, 141, 156, 159, 254
Nolin, J. B., (father and son mapmakers), 212, 216, 228, 229, 261, 280
Norden, J., 81
Nordgau. *See* Bavaria
Noricum. *See* Austria

Normans, and Normandy, 192, 278, 429, 430, 447

North Sea, 92, 101, 191, 201, 238

Northmen (Norse, Norsemen, viking): aggression 70, 194, 280, 388, 411; colonisation, 280, 387; exploration and, 411, 420, 429, place names, 200, 201. *See also* Iceland

"Nouvelle Holland," 228

Nubia. *See* Idrisi, al

Nuremberg, map publication at, 133, 134, 139, 141, 146, 161, 167, 188, 195, 211, 217, 251, 257. *See also* Homann, J. B.; Weigel, C.

O'Flaherty, R., 253

Ohthere, 197. *See* Anglo-Saxons

Oldenburg, duchy of, 94, 95, 306, 313–15, 456. *See also* Kruse, Christian; Mejer, J.

Old English. *See* Anglo-Saxons

Orient, 141, 148, 159, 172, 222, 247, 256, 266, 273, 329, 414, 423

Ortelius, A., 26, 36, 37, 138, 146, 189, 190, 265, 282, 313, 443, 449, 453; and ancient geography, 7, 13–17, 20, 30, 31, 32, 33, 35, 274, 281, 455; and combining writing with maps, 22, 51, 64, 87, 444, 449; and history, 1, 22, 24, 25, 30, 38, 104, 105, 111, 131, 132, 173, 314, 381, 456; *Orbis vetus* map (1590), 16, 34, 104, 444; various maps of the *Theatrum*, 63, 74, 75, 95, 98, 451. See also *Parergon*

Ottoman Empire, 227, 277, 390; as map subject, 148, 257, 259, 260, 273, 423; and Syria, 106, 384

Oxford, University of, 56, 81, 88, 278, 343, 356, 430

Pagi (districts), 79, 209. See also *gau, gaue*

Palestine, 34, 35, 91, 146, 224, 225, 382, 390, 395, 398, 416. *See also* Holy Land

Pannonia. *See* Hungary

Paradise, 35, 155, 226

Parergon (supplement of Ortelius's *Theatrum*), 17, 25, 26, 30–35, 37, 38, 104, 263

Paris, 135, 152, 156, 161, 209, 222, 225, 240, 241, 242 (BN), 251, 253, 262, 328, 337; as place of publication, 149, 155, 160, 172, 195, 204, 212, 265, 306, 308, 323, 327, 329, 339, 349,

362, 384, 451, 460; plans of, 100, 110, 205–8, 216, 217, 415, 457. *See also* Lamare (Delamare) N.; Paris draft; *Société de géographie*

Paris, M., 34

Paris draft, unpublished universal, sequential *Atlas complet*, ca. 1747, 154–56, 159, 160, 166, 167, 171–73, 190, 263, 265–67, 269, 270, 272, 281, 314, 320, 321, 329, 387, 402, 405, 450, 455, 456, 458. *See also* atlases and maps

Patrick, J. *See* Cellarius (Keller), C.

Paul, Saint, travels of, as map subject, 25, 26, 33, 36, 91 133, 444. *See also* Vialart, C.

Pellegrino, C., 86, 87, 444

Pepin I, Carolingian king, 208, 409

Persia, 90, 190, 222, 264, 265, 268, 328

Perthes, J., 350, 354–56, 359–61, 419, 420, 427

Pétis, A., 227, 256

Pétis de la Croix, F., 226–27, 229, 256

Petri, G., 37, 428

Peutinger Table, 28, 34, 275, 276

Pharamond, proto-Frankish king, 270, 271

Philadelphia, as publication center, 31, 307

Philip II, Augustus, king of France, 205, 270, 392, 409, 416

Philip IV, king of France, 241

Philip VI, king of France, 409

Philippe de Prétot, F. A., 21, 156

Picaud, M., of Nantes, 156, 159, 265 266

Piccolomini, Aeneas Silvius (Pope Pius II), 251

Pius IX, pope. *See* Petri

Placide de Sainte Hélène, Father, 67

Plancher, U., 193, 194

Plater, S., 339, 411

Poirson, J. B. *See* Lapie, P.

Poland: as map subject, 267, 316, 327, 410, 411, 420; partitions of, 323–24; succession, 251. *See also* Jabłonowsky, J. A.; Lelewel, J.; Plater, S.

Poole, R. L., 356, 419

Porphyrogenitus. *See* Constantine VII

Portugal, 68, 85, 237, 268, 350, 354, 418

Pozsony (Pressburg, Bratislava), 152, 275

Provence, 68, 212, 271

Prussia, 153, 331, 390, 422; army, 319, 320, 325, 336, 382; king of, 317, 333; as map subject, 323, 324, 410, 423, 428

Ptolemy, Claudius ("Stoleme"), 35, 111, 279; atlas of, 14, 21, 32, 34, 37, 105, 197; core of ancient geography, 13–17, 20–22, 24–25, 28, 30; "Magna Germania," 22, 138, 149, 250; maps of territories, 34, 443; source of map prototypes, 57, 100, 104, 146, 149, 210, 319, 381

Putzger, F., 86, 230, 356, 358, 448

Quebec (French colony), 85

Quin, E., 313, 326, 327, 337, 391, 404, 419, 427, 450, 452

Rafn, C. C., 429

Rancé, A. de, 91, 92.

Rapin-Thoyras, P., 57

Reconquista (Christian conquest of Spain), 448

Recueil des historiens des Gaules. *See* Bouquet, M.

Reformation: Catholic 91, Protestant, 171, 427

Rerum Italicarum scriptores. *See* Muratori, L.

Restoration. *See* Bourbon dynasty

Révolutions de l'univers. *See* Dupré

Reyzenbach. *See* Zeeland

Rhaetia (Rhaëtia). *See* Switzerland

Rhine, 79, 80, 99, 135, 139, 194, 209, 216, 240, 252, 315; Austria on the, 58, 62, 64–66; boundary, 58, 59, 60, 70, 71, 78, 100, 193, 197, 238, 243, 457; Palatinate of the, 217; Rhineland, 70; valley, 77, 189. *See also* Franks

Rhode, C. E., 335, 417–18

Rhône river, 67, 68, 225, 246, 252, 389

Ritter, K., 331, 360, 405

Rizzi-Zannoni, G. B., 74, 172, 237, 280, 329; atlas of France, 160–67, 239, 241, 242, 269, 271, 281, 309, 312, 314, 322, 325, 408, 409, 455, 458; atlas of Poland, 269; atlas of the Two Sicilies, 162, 269; and the sequential plan, 161, 162, 172, 302; life of, 161–62, 209. *See also* Brué, A. H.

Robert, G., *See* Vaugondy, Robert de

Rodowicź, T., 428

Rogers, W., 52

Rogg, G., 90

Rohrbacher, R. F., 412

Romano-German Empire, *See* Empire, Romano-Germanic

Rome, 17; fall of, 2, 6, 14, 24, 32, 38, 131, 188, 192, 194, 209, 313, 328, 382, 393, 404, 452, 456, 459; historical incidents, 31, 192, 200, 236, 403, 453; patriarchate of, 36; as the Roman Empire, 141. *See also* Empire, Holy Roman

Rosny, A. Ire de, 385. *See also* Koch, C. J.

Royal domain. *See* domain (French royal)

Rubruck, William of, travels of, as map subject, 229–31, 388

Rudolf I of Habsburg, Holy Roman Emperor, 66, 336, 390, 403, 410, 421

Rühle von Lilienstern, J. J. O. A., 382, 412

Rupp, J., 349, 350, 409

Russia, 85, 229, 305; historical atlas of, 172, 316, 410–12; J. Delisle in, 135, 225; Klaproth in, 316–17; as map subject, 22, 85, 148, 193, 201, 224, 257, 259, 267, 316, 333, 414, 415

Saint Jacob, battle of, 251, 252, 454

Salians. *See* Franks

Sanders, A., 78, 79

Sanson, cartographic dynasty, 25, 56, 87, 427; Guillaume, 24, 25, 36, 37, 90, 202; map hoard belonging to, 25, 29, 55, 90, 236, 250, 332; Nicolas, the patriarch, 20, 23–25, 27–29, 33, 36, 37, 55, 88, 90, 105, 146, 189, 277, 314, 321, 332. *See also* Duval, P.; La Ruë, P.; Moulart-Sanson, P.; Vaugondy, Robert de

Saracens, 8, 277

Savery (Savry), S., 80

Saxony, 213, 246, 337; Carolingian, 80; duchy of, 139, 210–12; electors of, 110, 219, 247; historical interest, 322–23; imperial dynasty (Ottonians) from, 27, 209, 214, 250, 410, 415, 422; king of, 257; as map subject, 222, 316, 410, 450, 454, 455. *See also* Anglo-Saxons; Gotha, map publication in; Güssefeld, F.; Weimar; Zollmann, F.

Scandinavia, 2, 95, 279, 354, 416

Schaarschmidt, F. R., 416

Schannat, F., 208

Scheldt river, 74, 100, 252

Schenck, P., 55

Schleswig, 92–94, 99, 445, 449, 456, 457. *See also* Mejer, J.

Schmidt, P., 395, 430

Schnitzler, J. H. *See* Baquol, J.
Schoell, F. (editor of Koch), 323
Schöning, G., 200, 430
Schöpflin, J. D., 195, 216, 217, 323
Schrader, F., 334
Scotland, 81, 153, 268, 277, 339, 396, 397, 411, 412, 418
Scott, Walter, 339
Second Empire. *See* Empire, Second French
Seleucid Empire, 266
Seljuk Turks, 263, 273
Selves, H., 330, 337, 390, 412
Senex, J., 196
Septimania (province), 212, 216, 228, 237
Seutter, M., 55, 62, 90, 202, 203, 281
Seven Maps (historical atlas of the German Empire), 14, 149, 205, 247, 260–62, 265, 271, 275 326, 327, 390, 413. *See also* Hase, J. M.
Sharaf al-Din 'Alī Yazdi, 228
Simencourt, A. H. de, 347
Société de géographie (Paris), 341, 351
Soissons, 77, 203, 243, 282
Spain: 68, 76, 105, 134, 190, 191, 268, 330, 396, 410; Armada, 110, 202, 308; army, 70; as map subject, 27, 71, 88, 152, 153, 159, 162, 225, 237, 238, 249, 250, 264, 273, 274, 320, 322, 330, 332, 333, 380, 388, 391, 392, 393, 404, 418–26, 448, 459; Roman, 71, 147, 236; War of the Succession, 459
Speed, J., 54, 55, 57, 80, 81, 84, 105, 165, 203, 204, 444; *Prospect* (Speed atlas), 81, 84, 195. *See also* Empire, Great Britain
Spener, P. J., 138, 233, 235
Spruner von Merz, K., 219, 336, 417, 420, 460; *Atlas antiquus,* 32, 34; atlas of Bavaria, 255, 333; atlas of Germany, 335, 410; biography, 333–34; *Hand-Atlas,* 21, 29, 222, 390, 331, 332–34, 336, 338, 391, 413–14, 415, 417–22; and thematic maps, 422
Stiedbeck, J., 216. *See also* Schöpflin, J. D.
Stieler, A., 360, 361
Strabo, 15, 30, 31
Strasbourg, 315, 412; edition of Ptolemy, 15, 16; as map subject, 216–17, 445; University of, 195, 216, 314, 315, 383. *See also* Koch, C. G.; Schöpflin, J. D.
Strebel, J. S., 215, 216
Stuart Pretender, 196, 279, 313

Sturt, J., 56
Switzerland, 251, 354, 357, 408, 410, 415; as Rhaetia, 27, 233
Sydow, E. von, 361
Syria, 85, 90, 91, 106, 141, 255, 266, 394
Szászky. *See* Tomka Szászky, J.

Talbert, R. J. A., 32
Tamerlane, conquests of, as map subject, 141, 227–29, 255, 259, 260, 267, 423, 454
Tardieu, P. F., 204
Tassin, C., 58
Tavernier, M., 26, 36, 58, 60, 105
Tegernsee, abbey of. *See* Bessel, G.
Teixeira, P., travels of, as map subject, 226
Ten Thousand, retreat of the, as map subject, 34, 105, 133
Teuscher, M. F. *See* Güssefeld, F.
Thomson, J., 340, 394–96, 401
"Thousand and One Days." *See* Pétis de la Croix, F.
Tilleman Stella, 110
Tomka Szászky, J., 152, 274, 275, 277
Topographical Bureau, of Prussian army, 341
Töppen, M. P., 428
Toul, 60, 208, 209, 212, 216, 222, 223, 255, 274
Travels of Young Anacharsis through Greece (Barthélémy, J. J.), 20, 321
Two Sicilies, kingdom of. *See* Rizzi-Zannoni, G. B.
Tycho Brahe. *See* Mejer, J.

Ulug Bey, 88. *See* Duval, P.
Ulysses, 33, 34, 104, 111. *See also* maps, literary
Ummayad Caliphate. *See* Caliphate, Ummayad
United Provinces. 71, 75, 98–100, 104, 457. *See also* Netherlands (Low countries)

Vaissette, J., 212, 213, 216
Vandalitia (learned name for Andalusia), 238
Vandals, 191–93, 233, 236, 335, 388
Vandermaelen, P., 348, 360
Vatican Map Gallery, 91, 104, 110
Vaugondy, Robert de, G. and D., 156, 265, 280; *Atlas universel,* 249, 382; authorship of Paris draft denied, 152–53; and Charlemagne map, 246, 249, 380, 382; historical maps by, 60, 236, 250, 251, 274;

and Sanson hoard, 55, 250, 332, 353; thematic map by, 201
Velly, P. F., 160, 162, 166, 458
Verdun, bishopric, 60;
Verdun, Treaty of (843), 242, 409, 410
Vertot, R., 138, 225, 255, 256
Vesconti, P., 34
Vialart (de Saint-Paul), C., 36
Vic, C. de, 212, 216
Vienna, 27, 61, 216, 275, 327, 338, 455
Villaret, C., *See* Velly, P. F.
Visigoths, 110, 236, 240, 246, 274, 388, 423, 426
Visscher, N., 55
Vlie. *See* Flevus (Vlie) river
Völkerwanderung, 171, 192, 272, 273, 316, 331, 335, 427, 448
Volude, J. H. de Lage de (friend of Las Cases), 304
Vredius. *See* Wrée (Vredius) O.
Vuillemin, A., 23, 347

Wagenaar, J., 252
Walid, Ummayad Caliph, 88. *See also* Caliphate, Ummayad
Wastelain, 237
Wedell, R., 358, 359, 422, 423, 426, 452
Weigel, C., 133–35, 146, 188. *See also* Köhler, J. D.
Weiland, K. F., 313
Weimar, 139, 211, 325; Geographical Institute at, 219, 318, 319, 320, 321, 322, 323, 337, 401, 402; journal of, 168, 228, 310, 320

Weiss de la Richerie (copier of Lesage design), 307
Weller, E., 429
Wendelin, G., 77, 78
Weser river, 92, 243, 252, 315
Westermann, 356, 358
Westphalia, 214; Treaty of (1648), 7, 39, 171, 443
Wettin, Saxon dynasty, 212, 338
Wiberg, C. F., 411
Wilken, F., 397
Wilkinson, R., 340, 379, 382, 396, 455
Will, G. A., 147. *See* Köhler, J. D.
Willard, A., 327
William the Conqueror, duke of Normandy, 57, 110, 203, 271, 313, 406
Wiltsch, J., 427
Wittenberg, University of, 93, 147, 148, 257, 258
Witzleben, F. A. von, 341
Wolkenhauer, W., 162, 214
Woltersdorf, E. G., 168, 171, 272
Woutneel, H., 81, 84
Wrée (Vredius), O., 77, 78
Wulfstan, 197, 430

Xenophon. *See* Ten Thousand, retreat of the

Zannoni. *See* Rizzi-Zannoni, G. B.
Zeeland, 74–76, 92, 98, 101, 246, 253, 457
Zollmann, F., 139, 211, 212, 338
Zuccagni-Orlandini, A., 411
Zuyder Zee, 80, 92, 99, 101, 252